Handbook of Historical Animal Studies

Handbook of Historical Animal Studies

Edited by
Mieke Roscher, André Krebber and Brett Mizelle

DE GRUYTER
OLDENBOURG

ISBN 978-3-11-108704-7
e-ISBN (PDF) 978-3-11-053655-3
e-ISBN (EPUB) 978-3-11-053436-8

Library of Congress Control Number: 2021931937

Bibliographic information published by the Deutsche Nationalbibliothek
The Deutsche Nationalbibliothek lists this publication in the Deutsche Nationalbibliografie;
detailed bibliographic data are available on the Internet at http://dnb.dnb.de.

© 2022 Walter de Gruyter GmbH, Berlin/Boston
This volume is text- and page-identical with the hardback published in 2021.
Cover Image: Double entendre, boy with pig attached to rope, by Briton Riviere (1840–1920),
illustration from magazine The Graphic, volume XIII, no 329, March 18, 1876. DEA/
BIBLIOTECA AMBROSIANA/Kontributor/Getty Images.
Printing and binding: CPI books GmbH, Leck

www.degruyter.com

Preface and Acknowledgements

A handbook, more than any other collaborative piece of publishing, refers to a community which it consolidates, circumscribes and projects further into the future, without being the last word on it, of course. Mieke Roscher and Brett Mizelle first met in 2005 at the "Animals in History: Studying the Not So Human Past" conference in Cologne, a meeting that eventually led to the initial collaboration of the editors in the context of a special issue on animals of the journal *WerkstattGeschichte* in 2011. The conversations continued when Brett joined Mieke and André Krebber at the University of Kassel as a visiting professor in 2016, where the ideas on how to compose a handbook of this format were initially discussed, attesting to the long and generative relationships that the human-animal studies (HAS) community fosters. Indeed, we have experienced the HAS and animal history communities as exceptionally welcoming and supportive, which is far from a given in academia, as we all know. We believe that such openness is not a coincidence and is related to the field's primary concern: understanding how humans treat and have treated nonhuman animals in the interest of transforming those relationships and promoting multispecies justice. Despite the wide range of attitudes and practices of HAS scholars when it comes to animal activism and social transformation, we believe that history provides the bedrock for deciphering these relations and figuring out how to promote change. Concomitantly, we insist that the study of history remains wanting without consideration of its animals.

We want to express our deep gratitude to the HAS community for providing us so generously with a collegial and productive intellectual home, one which has made this volume possible. The relationships we have built over the years have been indispensable to the development of this handbook – it has been shaped by our many conversations, collaborations and interactions with other scholars and research groups over the years. A special place is reserved in all our hearts for the *New Zealand Centre for Human-Animal Studies*, who have impacted on our thinking on animals so deeply and with which Kassel in particular has a close relationship through a scholar exchange program. Likewise, the meetings of the *British Animal Studies Network* have become a reliable fixture in our academic calendars. In Germany, we have drawn immensely on the conversations during the regular meetings of *FITT* – Research Initiative on Animal Theory – *Forum Tiere und Geschichte* – Forum Animals and History – as well as with our colleagues in Kassel from the now concluded LOEWE research project "*Tier-Mensch-Gesellschaft*" – Animal-Human-Society. We have equally been fortunate to have worked with a number of European HAS initiatives over the years, among them the *HumAnimal Group* at Uppsala University in Sweden; the *Human-Animal Studies Group* at Innsbruck University and the *Messerli Institute* in Vienna, Austria; the *Centre for Human-Animal Studies* at Edge Hill University and the *Sheffield Animal Studies Research Centre* in the UK; as well as outside of Europe the *Human Animal Research Network* at Sydney University in Australia. In 2019, we were able to bring members from all of these groups together in Kassel for the week-long *European Summer School Interspecies Relationality*, which we co-hosted with the US-American *Animals & Society Institute*. The biennial meetings of *Living with Animals* in Kentucky and triennial meetings of *Minding Animals* at locations across the globe have also proven to be generative sites of scholarly exchange. Finally, the *H-Animal Network*, founded and edited by one of our editors as part of the H-Net Humanities and Social Sciences Online scholarly association, has played a vital role in promoting intellectual exchanges, collaborative productions, and the dissemination of new knowledge. These groups and institutions describe a constellation of our own work in historical animal studies that might help readers to better locate this handbook as well as our approaches in the wider human-animal studies landscape. We are grateful to them for the impact they have had on our thinking about animals and for enabling us to get to the point of being able to return to the community this handbook as a meaningful resource.

It is the authors of the chapters in this volume who we are most thankful to, however, for not only do they present its heart and soul, but they have also been enormously patient and lenient with our occasionally stubborn editorial demands. Likewise, we are thankful to our publisher, De Gruyter Oldenbourg, and especially to our editor-in-chief – Bettina Neuhoff – who proclaimed and offered the project to us in the first place and has been patiently waiting while we negotiated the demands of such a taxing project with our other duties. Last but not least, we would like to thank our research assistant Martina Freitag for spending tireless hours unifying the footnotes and copy editor Ian Copestake for his careful attention to polishing the prose and grammar of the text.

Mieke Roscher, André Krebber, Brett Mizelle
November 2020, Kassel and Long Beach

Contents

Mieke Roscher, André Krebber and Brett Mizelle
Writing History after the Animal Turn? An Introduction to Historical Animal Studies —— 1

Part I: Timelines

Erica Hill
Pre-Domestication: Zooarchaeology —— 21

Abel Alves
Domestication: Coevolution —— 37

Etienne S. Benson
Post-Domestication: The Posthuman —— 53

Part II: Regional Approaches

Andreas Hübner
American Studies —— 69

Sandra Swart
African Studies —— 85

Anna Boswell
Australasian and Pacific Studies —— 101

Barbara R. Ambros and Ian Jared Miller
(East) Asian Studies —— 117

Part III: Historical Fields

Mieke Roscher
Social History —— 133

Helen Cowie
Cultural History —— 147

Aaron Skabelund
Political History —— 165

Heinrich Lang
Economic History —— 181

Nadir Weber
Diplomatic History —— 197

Julia Hauser
Global History —— 213

Joanna Dean
Public History —— 227

Dorothee Brantz
Urban (and Rural) History —— 243

Mitchell G. Ash
History of Science —— 259

André Krebber
History of Ideas —— 275

Part IV: Historical Approaches

Anna-Katharina Wöbse
Environmental History —— 293

Philip Howell
Historical Animal Geographies —— 309

Aritri Chakrabarti
(Post)Colonial History —— 325

Dominik Ohrem
Feminist Intersectionality Studies —— 341

Kit Heintzman
Material Culture Studies —— 357

Silke Förschler
Visual Culture Studies and Art History —— 375

Laura McLauchlan
Multispecies Ethnography —— 393

Sarah D. P. Cockram
History of Emotions —— 409

Part V: **History of Human-Animal Interactions**

Amir Zelinger
History of Pets —— 425

Takashi Ito
History of the Zoo —— 439

Peta Tait
History of Circus Animals —— 457

Linda Kalof
History of Animal Iconography —— 471

Andrew Gardiner
History of Veterinary Medicine —— 493

Axel C. Hüntelmann
History of Experimental Animals and the History of Animal Experiments —— 509

Veronika Settele
History of Agriculture —— 525

Annette Leiderer
History of Animal Slaughter —— 539

Gesine Krüger
History of Hunting —— 555

Ryan Hediger
History of War —— 571

Janet M. Davis
History of Animal Fights and Blood Sports —— 587

Andrew Wells
History of Animal Collections/Animal Taxonomy —— 603

List of Contributors —— 619

Index —— 625

Mieke Roscher, André Krebber and Brett Mizelle
Writing History after the Animal Turn?
An Introduction to Historical Animal Studies

1 Introduction and Overview

Time and again, animals have crossed over into the realm of human history. As we write this introduction, our world is in the grip of the SARS-CoV-2 virus. Pandemics of such proportion have often left lasting effects on societies, making them prominent and powerful reference points for historiography.[1] Just like the plague has become connected inseparably (and somewhat falsely) with rats as the root of the disease,[2] the pangolin has emerged as the index species that might sadly become associated with the COVID-19 pandemic of 2020 for future historians. Yet, whereas rats were made into pests and exterminated, the pangolin today, under changed historical conditions, does not draw the same wrath and hate for transmitting the virus. Highly endangered and trafficked illegally for their meat and perceived medicinal properties, the focus turned instead to the wet market where the poor pangolin was sold, located conveniently for the West in China.[3] The pangolin, easily susceptible to capture by humans and to coronaviruses, indexes both the ongoing and intensifying environmental crisis that increasingly highlights the fateful interdependence of human and animal histories and the ongoing exoticization and exploitation of nonhuman animals by humans.[4] Animals and the relationships humans have with them surface not only as powerful lenses for unpacking history, but as powerful forces in shaping history in the first place. Indeed, in the case of these pandemics, the animals and our interactions with them seem to provide a scapegoat and explanation, an anchoring point and junction for the uncontrollable and untraceable paths along which diseases and pandemics travel, but also a point zero, an origin, a source from where to

---

[1] Cohn, Jr., Samuel K. *Epidemics: Hate and Compassion from the Plague of Athens to AIDS*. Oxford: Oxford University Press, 2018; Deb Roy, Rohan. *Malarial Subjects: Empire, Medicine and Nonhumans in British India, 1820–1909*. Cambridge: Cambridge University Press, 2017.
[2] Burt, Jonathan. *Rat*. London: Reaktion Books, 2006.
[3] Quammen, David. The Sobbing Pangolin: The Brutal History of an Elusive Creature. *The New Yorker*, August 31, 2020: 26–31.
[4] Cf. Sivasundaram, Sujit. The Human, the Animal and the Prehistory of COVID-19. *Past & Present* 249, no. 1 (November 2020): 295–316; Blanchette, Alex and Gabriel N. Rosenberg. Working Pigs and Humans in the Age of Covid-19: A Conversation between Alex Blanchette and Gabriel N. Rosenberg. *Radical History Review*, The Abusable Past, September 4, 2020. URL: www.radicalhistoryreview.org/abusablepast/working-pigs-and-humans-in-the-age-of-covid-19/ (December 7, 2020).

begin our histories.⁵ That the new coronavirus might have mutated originally in bats, from where it moved on to pangolins and then on to humans (although research suggests a different, more cryptic spread and places the zoonotic event in human-human transmission),⁶ reminds us that "animal infections can become human infections, because humans are animals."⁷ It also highlights how whatever happens in nature often remains irrelevant in the face of the stories humans create about events.⁸

However, in both the case of the plague and rats as well as the case of COVID-19 and pangolins, approaching the history of pandemics by centering these animals complicates the projections and narratives, making animals powerful allies in brushing history against the grain while also challenging the separation of animal from human history.⁹ Thus rats, rather than being the origin of the plague, were its first victims, equally beset by the fleas that transmit the virus as humans.¹⁰ And whereas the pangolins were pitied, at least in the West, for their fate as commodities, exploited for their leather, meat, and scales used in traditional Chinese medicine, such sympathy can shift quickly when the species (and space) of concern changes. In a further grim twist on the animal dimension of the human pandemic, the most recent development in the unfolding of COVID-19 are fears for further mutation of the virus occurring in farmed minks in Denmark and elsewhere. Whereas the endangered pangolin draws our sympathy, farmed minks in Europe do not receive the same courtesy. They are being preemptively killed by the millions in a bid to suppress further mutation and transmission.¹¹ What separates the socio-ecological situation of plague

---

5 Cf. Boseley, Sarah. Origin Story: What Do We Know Now About Where Coronavirus Came From? *The Guardian*, December 12, 2020. URL: www.theguardian.com/world/2020/dec/12/where-did-coronavirus-come-from-covid (December 21, 2020).
6 Andersen, Kristin G., Andrew Rambaut, W. Ian Limpkin, Edward C. Holmes and Robert F. Garry. The Proximal Origin of SARS-CoV-2. *Nature Medicine* 26 (March 2020): 450–452. doi.org/10.1038/s41591-020-0820-9; Boni, Maciej F. Philippe Lemey, Xiaowei Jiang, Thommy Tsan-Yuk Lam, Blair W. Perry, Todd A. Castoe, Andrew Rambaut and David L. Robertson. Evolutionary Origins of the SARS-CoV-2 Sarbecovirus Lineage Responsible for the COVID-19 Pandemic. *Nature Microbiology* 5 (July 2020): 1408–1417. doi.org/10.1038/s41564-020-0771-4.
7 Quammen, The Sobbing Pangolin, 31.
8 Lynteris, Christos (ed.). *Framing Animals as Epidemic Villains. Histories of Non-Human Disease Vectors*. Cham: Palgrave Macmillan, 2019.
9 Cf. Kirksey, Eben. The Emergence of COVID-19: A Multispecies Story. *Anthropology Now* 12, no. 1 (April 2020): 11–16. doi.org/10.1080/19428200.2020.1760631.
10 Dean, Katharine R., Fabienne Krauer, Lars Walløe, Ole Christian Lingjærde, Barbara Bramanti, Nils C. Stenseth, and Boris V. Schmid. Human Ectoparasites and the Spread of Plague in Europe during the Second Pandemic. *PNAS* 115, no. 6 (February 2018): 1304–1309.
11 Gorman, James. Denmark will Kill all Farmed Mink, Citing Coronavirus Infections. *The New York Times*, November 4, 2020. URL: www.nytimes.com/2020/11/04/health/covid-mink-mutation.html (November 26, 2020). There is a long history of mass killings of animals amidst panics about zoonotic diseases, killings that often take place when the spread of disease and its actual effects on humans and animals are not yet known. For the case of pigs, see Mizelle, Brett. *Pig*. London: Reaktion Books, 2011, 47–54. As Nicole Shukin (*Animal Capital: Rendering Life in Biopolitical Times*. Minneapolis, MN: University of Minnesota Press, 2009, 46) notes, "A fixation in pandemic discourse on zoonotic diseas-

and rat from COVID-19, pangolin and mink, of course, is history that weighs differently on animals in different places at different times owing to the different relationships that humans have with them. Thus, the lives, experiences, and deaths of animals become a powerful lens to understand and explain human histories, ideas and practices that affect human relationships to each other and to animals themselves. Animals hence not only cross over into human history, they are inseparably intertwined with human history and present exceptionally strong prisms for the study of history and its complexities. History matters to animals as animals matter to history.

This realization has developed its own scholarly history by now. It travels from Harriet Ritvo and Nigel Rothfels to Erica Fudge, Hilda Kean, Linda Kalof, Susan Nance and Dorothee Brantz, to name just a few. These scholars' research aligns closely with the burgeoning field of human-animal studies that seeks to trace and make visible the active engagement and entanglement of animals on and within human societies, alongside the recognition of the impact and dependency of human cultural practices and institutions on animals and their ability to live their lives in accordance with their desires and needs. Accordingly, when we invited the contributions to this handbook, we expected a number of prominent theoretical references – to Jacques Derrida, Donna Haraway, Bruno Latour, Gilles Deleuze and Félix Guattari, John Berger and Thomas Nagel, among others. These names are definitely present in the chapters that follow, yet two scholars stood out most consistently and indeed overwhelmingly: Donna Haraway and Harriet Ritvo. As we asked our authors to reflect on the impact of the animal turn, this might seem unsurprising. Ritvo discussed the movement of taking animals more seriously as social actors and as a research focus in the humanities in her important 2007 paper "On the Animal Turn". According to her,

> during the last several decades, animals have emerged as a more frequent focus of scholarship in the humanities and social sciences, as quantified in published books and articles, conference presentations, new societies, and new journals. With this change in degree has come a potential change in kind.[12]

The chapters that follow pay testimony to and unfurl such change. Yet what the overwhelming references to Ritvo especially mean, to us at least, is that her work appears not just to have been enormously generative to historical animal studies today, including to a new generation of scholars, but also that the discourses of human-animal studies and historical animal studies, just like all the other disciplines involved in the reevaluation of the lives of animals and our relationship with them, past and present, are not identical. Rather, they inform one another. What we aim at with this

---

es – diseases capable of leaping from animal to human bodies via microbial agents such as the H5N1 avian flu virus – is symptomatic of how formerly distinct barriers separating humans and other species are imaginatively, and physically, disintegrating under current conditions of globalization."
[12] Ritvo, Harriet. On the Animal Turn. *Daedalus* 136, no. 4 (Fall 2007): 118–122, 119.

handbook, then, is to gather and make accessible the contribution of historical research to the field of human-animal studies as well as the contribution of human-animal studies to the study of history. History as a discipline and scholarly venture, in other words, can no less ignore animals than animal studies can history. Our titular reference to historical animal studies, rather than to historical human-animal studies, reflects this as well as making it clear that it is also the animals themselves whose histories we are concerned with. It also acknowledges, however, that in the compiling, recording, displaying and narrating of history, humans are an omnipresent factor that cannot be left out.

Thus, our handbook hopes to open a new chapter in the development and institutionalization of human-animal studies and historical animal studies as research lenses. The scholarly establishment of such an "animal lens," as Joshua Specht has called it,[13] is manifest within a number of general introductions and handbooks to the field that already exist.[14] Yet none of these have taken a structured approach to the introduction of the field from the perspective of its research challenges and practices. *The Routledge Companion to Animal–Human History* as the closest historical forerunner to the present handbook, for example, focuses on introducing the history of animal-human relations, striving in particular to make visible the perspective and experience of animals within these relations. Likewise, *The Routledge Handbook of Human-Animal Studies* and *The Oxford Handbook of Animal Studies* do not emphasize practical and methodical approaches but where and in what capacities animals feature in scholarship; they draw on the wealth of interdisciplinary research on animals and provide overviews of the research that is pursued in human-animal studies rather than a methodological account of the field itself. An exception might still be Margo DeMello's book *Teaching the Animal: Human-Animal Studies Across the Disciplines*, which, however, operates on the macro-level of disciplines, just like *The Edinburgh Companion to Animal Studies*. The present handbook, by contrast, aims to take the next step by approaching animal-human history from the perspective of history as a field of research at the intersection of history and human-animal studies. Our aim is to thereby highlight the role of history as a discipline in human-animal studies while ordering the existing research in an attempt to provide a structured introduction to the historical study of animals and their relationships with humans. As

---

13 Specht, Joshua. Animal History after Its Triumph: Unexpected Animals, Evolutionary Approaches, and the Animal Lens. *History Compass* 14, no. 7 (September 2016): 326–336.

14 Cf. Calarco, Matthew. *Animal Studies: The Key Concepts*. Abingdon and New York, NY: Routledge, 2021; DeMello, Margo (ed.). *Teaching the Animal: Human-Animal Studies Across the Disciplines*. New York, NY: Lantern Books, 2010; Gruen, Lori. *Critical Terms for Animal Studies*. Chicago, IL and London: University of Chicago Press, 2018; Howell, Philip, and Hilda Kean (ed.). *The Routledge Companion to Animal–Human History*. London and New York, NY: Routledge, 2018; Kalof, Linda (ed.). *The Oxford Handbook of Animal Studies*. Oxford: Oxford University Press, 2017; Marvin, Garry and Susan McHugh (ed.). *Routledge Handbook of Human-Animal Studies*. New York, NY: Routledge, 2014; Turner, Lynn, Ron Broglio and Undine Sellbach (ed.). *The Edinburgh Companion to Animal Studies*. Edinburgh: Edinburgh University Press, 2018.

a consequence, the book is structured more stringently, perhaps, than previous human-animal studies handbooks along the lines of historical research discourses. This can be seen in both the structure of the book's content as well as in each of the chapters, which reflect on both research topics and themes and on historical methods and approaches.

We believe such a step to be appropriate not only in relation to the historical development of the field, but also consider it necessary today if the field does not want to make itself obsolete. The explosion of animal studies literature in recent years has made it difficult to discern a specifically human-animal studies research concern other than that we are dealing with animals. The attraction of animals as research topics seems little surprising. People are interested in animals precisely because they have a central place in human cultures and animal-centered projects make for attractive research monographs in the struggle for academic visibility and the push to justify the common worth of our work: growing public recognition of this work turns into a self-fulfilling prophecy. Yet the downside of this growth can be a blurring of the questions, approaches and challenges that human-animal studies as a research concern tackles. The change in degree that Harriet Ritvo noted in 2007 threatens to overshadow, hide and outdo the change in kind. However, if human-animal studies wants to retain its specificity as a research perspective rather than dissolve back into an unspecific consideration of animals that we can see throughout the history of scholarship, a position for which we believe there is not only social but scientific merit and in fact necessity, it is time to be more decisive about what human-animal studies and, for us in particular, historical animal studies as research concerns stand for and to be more affirmative about how to do (historical) animal studies. While this volume hardly claims to be the last word on this, we hope it provides an impulse for moving human-animal studies as a field of research forward.

2 Topics and Themes

This handbook spans a wide spectrum of animal related topics and themes, some of which are already spelled out in the titles, some more hidden under broader concepts used by historians at large. The articles also cover a variety of different animal-human relationships that are in themselves subject to chronological, cultural and spatial particularities. These relationships are also profoundly influenced and regulated by the different species of nonhuman animals covered by the following entries. While most articles derive their clues on how to write animal centered history from relationships with what can only be characterized, if very broadly, as domesticated animals [→Domestication: Coevolution;→Social History;→History of Emotions;→History of Pets;→History of Animal Slaughter;→History of Veterinary Medicine;→History of Agriculture], some are doing the exact opposite and show how "wild", or "other" animals shape both the imagination of the world around us as well as one's own position in the micro- and macrocosm of life. How baboons, silkworms, bison and possums shape their respective environments is central to understanding also the cultural make-up of societies at large. Thus, the question of "nature out there" does not only concern an →Environmental History that seeks to highlight the co-dependencies of entangled habitats, but also →Global History and →(Post)Colonial History, which address the transfer of ideas on notions of controlling nature as means of solidifying power structures. Similarly, the takes on non-European histories, and particularly on →Australian and Pacific Studies, →African Studies and →(East) Asian Studies complicate and challenge absolute dichotomies of domesticated vs. wild and, implicit therein, of nature vs. culture, and how it is necessary for animal historians to historize such distinctions. With this in mind, a more liminal comprehension of species being has lately been developed in animal studies that also resonates in animal history.[15] Literary and cultural studies scholars have rightfully pointed to the fact that the engaging with (fictional) contact and transition zones helps to illustrate the flexibility of nonhuman animals and their usage for social determinations and differentiations.[16] The ways in which animals take on diverse roles in these negotiation processes, both in terms of their symbolism and in terms of their material being, is a topic of →Political as well as →Cultural History. That these roles can be attributed to real and distinct individual animals is probably nowhere more apparent than when they change their physical state or place, for example from living beings to dead material, as is shown by research in →Material Culture Studies, the →History of Taxonomy/Animal Collections, the →History of Hunting, or even the →History of Experimental Animals and the History of Animal Experiments. That changing symbolical roles likewise can be attributed to spatial relocations of living animals is one of the central motifs of a →History of the Zoo as well as that of a →Diplomatic History, which pay attention not only to the economic value of animals given as gifts, but also to the changing attitudes towards animals per se. Animal histories of this attentive kind have multiplied over the last decade. While the sheer range of topics is now impossible to cover in an introductory chapter such as this, something that is true for both the subjects addressed in this handbook as well as by the field of human–animal history as such, there are nonetheless some framing concepts and narratives that bind these topics together.

---

15 Donaldson, Sue and Will Kymlicka. *Zoopolis: A Political Theory of Animal Rights*. Oxford: Oxford University Press, 2011; Howell, Philip, Aline Steinbrecher and Clemens Wischermann (eds.). *Animal History in the Modern City. Exploring Liminality*. London: Bloomsbury, 2018.
16 Ohrem, Dominik and Roland Bartosch (eds.). *Beyond the Human-Animal Divide: Creatural Lives in Literature, Culture, and History*. London: Palgrave Macmillan, 2017.

Agency

The concept – and challenge – of animal agency is probably at the core of most of the articles presented in this volume, even if not always explicitly referred to, which merely shows how central it is to human-animal studies scholars across all disciplines.[17] Yet for historians especially, the question of what constitutes a historical actor is important as well as ongoing. Agency is often the subject of furious debates within the discipline, where the arrival of novel actors is either seen as a sign of willingness to choose a more integrative and intersectional approach to its subject [→Intersectional Feminist History] or as a dangerous trivialization of traditional actors' agency by becoming analytically arbitrary. In following mainly labor and gender history and thus the proponents of what is more widely received as →Social History, animal historians are routinely looking for clues of what animal agency could look like and how to detect it in sources available to them.[18] As Philip Howell and Hilda Kean have put it, agency is "important in recovering the lives and experiences of the less privileged, the subordinated, the exploited"[19] and it is this heritage that leads animal historians to view it as a tool to show that more is possible than to merely write the history of animals as objects of human appropriation. Moreover, it illuminates how the writing of animal history is seen for many historians as a political act, a necessary and important intervention in discourses on animals' societal standing and recognition, a "histoire engagée" that has similarities with previous attempts of widening spheres of influence, for example of women, by insisting on telling and narrating their histories. In the wake of the widening influence of social theory as well as the rise of posthumanist thought [→Post-Domestication: The Posthuman;→History of Ideas], the concept of agency has changed to be more open to symmetrical, relational and distributed leanings.[20] In the present book, most histories follow the latter ap-

---

[17] See McFarland, Sarah E. and Ryan Hediger (eds.). *Animals and Agency: An Interdisciplinary Exploration*. Brill, 2009; Wirth, Sven, Anett Laue, Markus Kurth, Katharina Dornenzweig, Leonie Bossert and Karsten Balgar (eds.). *Das Handeln der Tiere: Tierliche Agency im Fokus der Human-Animal Studies*. Bielefeld: transcript Verlag, 2015.
[18] Flack, Andrew. 'In Sight, Insane': Animal Agency, Captivity and the Frozen Wilderness in the Late-Twentieth Century. *Environment and History* 22, no. 4 (November 2016): 629–652; Nance, Susan. *Entertaining Elephants: Animal Agency and the Business of the American Circus*. Baltimore, MD: Johns Hopkins University Press, 2013; Pearson, Chris. History and Animal Agencies. In *The Oxford Handbook of Animal Studies*, Linda Kalof (ed.), 204–257. Oxford: Oxford University Press, 2017; Shaw, David Gary. The Torturer's Horse: Agency and Animals in History. *History and Theory* 52, no. 4 (December 2013): 146–167.
[19] Howell, Philip and Hilda Kean. Writing in Animals in History. In *The Routledge Companion to Animal-Human History*, Philip Howell, Hilda Kean (eds.), 3–27. London and New York, NY: Routledge, 2018, 12.
[20] Howell, Philip. Animals, Agency and History. In *The Routledge Companion to Animal–Human History*, Philip Howell, Hilda Kean (eds.), 197–221. London and New York, NY: Routledge, 2018; Roscher, Mieke. Actors or Agents? Defining the Concept of Relational Agency in (Historical) Wildlife Encoun-

proach while remaining firmly rooted in the kind of history that encourages an active involvement with the state of society in the past as well as in the present and future.

Representation

How animals were imagined, depicted, portrayed, narrated, used for fables, employed as tropes, metaphors or allegories to paint a picture of the constitution of (human) societies of the past is probably one of the oldest themes of a history that takes an interest in animals [→History of Ideas]. Take, for example, Keith Thomas' ground-breaking work in *Man and the Natural World*, which explores the question of how changing attitudes towards animals in early modern England were intertwined with changes in the social and infrastructural fabric of the land.[21] With the development of historical animal studies, this approach has not been without its critics. Concentrating on representations has come with the assumption that all that animal history can ever do is to rephrase what other humans in different times have thought of animals without being able to look at the animal beyond its symbolic and anthropomorphized self.[22] Yet, over the last couple of years and specifically within →Cultural History and →Political History an understanding has developed that takes the representation of animals as a starting point for research to interrogate depictions of animals not only with regard to their symbolic value, but also their material implications. Here, source material on biological evolution is read next to cultural texts in ways where culture itself is regarded from the beginning as being co-constituted alongside, with the help of and in co-operation with animals. This is very much in line with themes offered in this handbook that have been influenced particularly by the field of →Visual Culture Studies and Art History and that are traced further here in the entry on the →History of Animal Iconography, both of which highlight the autonomous animal-being behind the image and aim to decipher the relationship between writers, chroniclers or artists and the animals that are the subjects of their depictions.[23]

---

ters. In *Animal Encounters*, Jessica Ullrich, Alexandra Böhm (eds.), 149–170. Stuttgart: J.B. Metzler, 2019.

21 Thomas, Keith. *Man and the Natural World: Changing Attitudes in England 1500–1800*. New York, NY: Pantheon Books, 1983.

22 Fudge, Erica. A Left-Handed Blow: Writing the History of Animals. In *Representing Animals*, Nigel Rothfels (ed.), 3–18. Bloomington, IN: Indiana University Press, 2002.

23 See also Donald, Diana. *Picturing Animals in Britain, 1750–1850*. New Haven, CT and London: Yale University Press, 2007.

Practices

To avoid a representational "trap," many chapters in this book make animal-related practices their thematic focus, whether they are encapsulated in economic structures and trade negotiations to which animals are objects [→Economic History], practices that involve and require the bodily involvement of human and animals, and where the death of an animal is often the likely result [→History of War;→History of Hunting; →History of Animal Fights and Blood Sports], or the making of animal made or inspired objects [→Material Culture Studies]. As such, the concentration on practices also provides new ways of opening up the study of historical power relations by exploring the relationship between knowledge production and the representation of and actions upon the body that is produced by that knowledge.[24] For the practicing of animal history, this means to take seriously the shifts that occur between the semantic typifications and the material realities that are part of certain practices. Whether it is in the realm of →Diplomatic History, →Political History, the →History of Science or the →History of Experimental Animals and the History of Animal Experiments or in a concentration on themes that offer glimpses into the specificities of interchanges between species that only produce a certain knowledge base or symbolic inventory – practices surface as a means of avoiding having to jump to conclusions as to who is the actor in an assemblage. Also, it offers new ways for expanding the microhistoric view on everyday encounters with animals that serve to form certain mentalities and ideas [→History of Emotions].

Matter/Materials

Everyday experiences more often than not are based on material interaction with animals or animal matter, with animal objects or animal made objects. This is reflected by those chapters in this volume that are inspired both by discussions in animal history as well as →Material Culture Studies. Taking the material (animal) body as a vantage point for scrutinizing human–animal relations of the past has become a fruitful way to address topics such as breeds and breeding, mounting and exhibiting animal remains as well as confining live animals in zoos and circuses. [→History of Agriculture;→History of Animal Collections/Taxonomy;→History of the Zoo;→History of Circus Animals].[25] The presentation of animals – dead or alive, in menageries and cabinets of curiosities, as working or as pet animals – defined social textures. Mate-

---

[24] Clever, Iris and Willemijn Ruberg. Beyond Cultural History? The Material Turn, Praxiography, and Body History. *Humanities* 3, no. 4 (December 2014): 546–566, 562.

[25] See also Landes, Joan B., Paula Young Lee and Paul Youngquist (eds.). *Gorgeous Beasts: Animal Bodies in Historical Perspective*. University Park, PA: Penn State University Press, 2012; Raber, Karen. *Animal Bodies, Renaissance Culture*. Philadelphia, PA: University of Pennsylvania Press, 2013; Specht, Animal History after Its Triumph.

rial furnishings and their codification said possibly as much about the individual animal as about their owner or carrier. However, while most of animal history still seems to shy away from taking such a "hands-on" approach to the animal body that builds on an involvement with bones, furs and feathers, zooarchaeology reveals quite literally how enmeshed our pasts are with animal matter [→Pre-Domestication: Zooarchaeology;→Domestication: Coevolution].

Spaces and Places

In addition to animal attentive social and cultural geography that focuses on the places that humans and animals inhabit, share and build together and which conceptualizes these places as "hybrid" spaces,[26] more and more historical works use spatiality as a tool for expanding and at the same time grounding their narratives [→Historical Animal Geography].[27] In this work, encounters are framed in such ways to use them as lenses to uncover the bodily experiences of everyday assemblages of humans and nonhumans that are prone to being tied to spatial arrangements.[28] Urban studies in particular, both historical as well as sociological, has taken up the lived experience of shared spaces and has looked at what changes when certain parameters of geographical and structural compositions are being transformed, for example by the forces of industrialization [→Urban (and Rural) History].[29] Space and place are also central not only to the relationship of humans and animals in history but also to the experiencing of that past. →Public History, for example, takes involvement with everyday life as a starting point for furthering interest in how deeply historical everyday life really is.

Another aspect and spatial dimension more indirectly addressed in this book is that of the predominance of a Western perspective in much of historical animal studies. The relationship of European actors with "others" was, of course, for most of Europe's history defined by a forceful and violent encounter of seemingly very one-directional interspecies relations, justified by an alleged superiority of the European [→(Post)Colonial History;→Global History]. The socio-spatial constellation of what was and is considered European or Neo-European is, however, very much deter-

---

[26] See Gillespie, Kathryn and Rosemary-Claire Collard (eds.). *Critical Animal Geographies: Politics, Intersections and Hierarchies in a Multispecies World*. Abingdon and London: Routledge, 2017; Philo, Chris and Chris Wilbert. *Animal Spaces, Beastly Places. New Geographies of Human-Animal Relations*. London and New York, NY: Routledge 2000.
[27] Wilcox, Sharon and Stephanie Rutherford (eds.). *Historical Animal Geographies*. London and New York, NY: Routledge, 2018.
[28] Crane, Susan. *Animal Encounters: Contacts and Concepts in Medieval Britain*. Philadelphia, PA: University of Pennsylvania Press, 2012.
[29] Atkins, Peter (ed.). *Animal Cities: Beastly Urban Histories*. Farnham: Ashgate, 2012; Holmberg, Tora. *Urban Animals: Crowding in Zoocities*. London: Routledge, 2015.

mined by interspecies relations and at the same time regulates how they were experienced, represented and negotiated.³⁰ What is more, they work both on a material as well as on a semiotic level. We hope that the concentration on the production of specific animal related knowledges provided by many of the chapters in this volume will help us to critically tease out the Western perspective of animal history by simultaneously moving beyond its Eurocentrism. By capturing both the internal and external dimensions of writing about interspecies relationships – structured, for example, around ideas of metropole and periphery, of wild and domestic, of culture and nature, for which the European intellectual tradition so strongly stands [→History of Ideas] – the West is framed here as a set of cultural ideas in itself. By providing global perspectives [→Global History], focusing on non-European places and taking cues from animal related practices in different parts of the world, we reflect the expansion to non-Western animal histories that can be observed in the wider field of animal studies.³¹

3 Methods and Approaches

As an interdisciplinary effort, human-animal studies is characterized by a diverse and ambitious plurality of methods and approaches. Methodological concerns reflect therein a particularly challenging tension. At the point in time when this handbook is published, human-animal studies as a field of scholarship, or the scholarly study of animals and their relations with humans, has reached a crossroads in its own history. On the one hand, it is obvious that animals have gained a firm place as subject matter (and research approach, more on that below) in the humanities and social sciences.³² The explosion of literature on animals and human-animal relations over the past years suggests that their study will not easily disappear again like a fad. Likewise, the perception of animals as subjects, or the very least actors in their own right, including in the natural sciences, is well established, even if this insight is not yet universally recognized. On the other hand, however, this centering of animal subjectivity has not yet produced a clearly defined and delimited field of scholarship. Quite to the contrary, while in the 1980s, 1990s and even the early 2000s it was possible to refer to a human-animal studies community that was primarily defined by a set of scholars that were turning their attention to the role and place of animals in

---

30 See Ham, Jennifer and Matthew Senior (eds.). *Animal Acts: Configuring the Human in Western History*. London and New York, NY: Routledge, 1997.
31 See for example Alves, Abel A. *The Animals of Spain. An Introduction to Imperial Perceptions and Human Interaction with Other Animals, 1492–1826*. Leiden, Boston: Brill, 2011; Deb Roy, Rohan (ed.). Nonhuman Empires. Special issue of *Comparative Studies of South Asia, Africa and the Middle East* 35, no. 1 (May 2015); Norton, Marcy. The Chicken or the Iegue. Human-Animal Relationships and the Columbian Exchange. *The American Historical Review* 120, no. 1 (June 2015): 28–60.
32 Cf. Specht, Animal History after Its Triumph.

human societies, the growth and wide acceptance of the animal as a worthwhile lens for scholarship has frayed the edges of human-animal studies and has seemed to lead the field and its scholarship to lose any clear outlines and definitions. Indeed, rather than human-animal studies solidifying into an academic field, it seems to oscillate between efforts to institutionalize itself as its own field (attempts, however, that have proven much less fruitful than in other fields of research, for example gender studies, which by now is present with professorships in many universities) and forming sub-sections of traditional, established fields – history, sociology, geography, political sciences, for example. This situation, then, underlines the specific methodological challenges of historical animal studies.

Whereas nonhuman animals are present throughout historical research, from a traditional disciplinary point of view historical methods tend to consider animals as objects, things and resources, as furnishings, accessories and extras. In fact, a significant number of animal histories that have been driving the explosion in animal literature in recent years might be just this – histories of how humans deal with and handle animals. While this is a perfectly legitimate undertaking (although it remains questionable whether such histories can ever be sound if indeed at least some animals are conscious, reflexive, self-deciding subjects), these scholars are often not the least bit interested in animals; they are interested in human history, from a human perspective, writing histories with animals inside them, rather than animal histories. An animal history written from a human-animal studies perspective, instead, takes the central impetus of human-animal studies – to consider nonhuman animals as subjects, actors and beings in their own right – and writes history with the active influence of animals on human cultures in mind as well as the history of animals as active subjects. However, given the centrality and pervasiveness of animals in human cultures, this perspective always cuts both ways. Historical animal studies require sound and rigorous understanding of the methods and workings of history as a discipline, while the acknowledgement of the impetus of human-animal studies requires a fundamental challenge to history. (This might also describe human-animal studies' "dilemma", if one wants to call it that, in its struggle to establish itself as its own field.) As a consequence, there are three central dimensions to methods and approaches in historical (and any disciplinary) animal studies: 1) human-animal studies as a method and approach for challenging the anthropocentrism and obliviousness to animals as subjects; 2) the animal as a method and approach to reread the historical constitution of animal–human societies; and 3) the deployment and revision of traditional, disciplinary methods and approaches, that is, methods and approaches that define the scholarly field of history, that help us within human-animal studies to find and trace the animal and their entanglements with humans and their cultures.

One of the main challenges that has been leveled against an animal history, and thus at the effort of historical animal studies to be taken seriously as rigorous scholarship that is conscious of animals as subjects in their own right, is the authorship of sources.³³ The problem here is a deeply intertwined one. Traditionally, the discipline of history privileged (and made in fact exclusive) written sources, which especially in the context of an animal history leads to the problem that animals do not leave behind sources because they do not write human script and language. How, then, can an animal history be anything else than a history of the representation and perception of animals by humans, and thus a human history of animals and their human perception, and therefore ultimately a human history? Animal history shares this challenge with →Environmental History. However, there is precedence for overcoming this problem in the development of history as a scholarly discipline itself. Many human perspectives from the past are equally not visible in documents, from the history of peasants, factory workers, women and slaves, who were illiterate or otherwise deprived of opportunities to write down their history, to indigenous groups who do not record their history on paper, and events, actors and processes for which simply no written documents survive. While the recognition of these perspectives had to be fought for, and methods and theoretical approaches had to be developed to access and uncover them, the limited availability of written sources is no longer considered an unsurmountable barrier in writing the history and accessing the experiences of historical actors, as the chapters in this volume make clear [for more prominent discussions of this issue see →Post-Domestication: The Posthuman;→African Studies;→Environmental History;→Material Culture Studies;→History of Emotions].³⁴

Indeed, what scholars mean probably half the time if not more when they point out the problem of the authorship of sources for animal history is a different issue, one closely connected to the availability of sources but that still faces its own set of methodological challenges: that we are unable to bridge and access the different experiences and perspectives of other animals. *What is it like to be?* has become a familiar shorthand for this problem.³⁵ Here also we find precedents in non-animal-centered historical research, though, as Hilda Kean reminds us.³⁶ As historians we, in fact, constantly try to imagine what it was like to be alive, or this or that person, dur-

---

33 Cf. Baratay, Éric. Building an Animal History. In *French Thinking About Animals*, Louisa Mackenzie, Stephanie Posthumus (eds.), 3–14. East Lansing, MI: Michigan State University Press, 2015.
34 Benson, Etienne S. Animal Writes: Historiography, Disciplinarity, and the Animal Trace. In *Making Animal Meaning*, Linda Kalof, Georgina Montgomery (eds.), 3–16. East Lansing, MI: Michigan State University Press, 2011; Fudge, A Left-Handed Blow; Kean, Hilda. Challenges for Historians Writing Animal–Human History: What is Really Enough? *Anthrozoös* 25, Supplement (2012): 57–72; Swart, Sandra. "But Where's the Bloody Horse?": Textuality and Corporeality in the "Animal Turn". *Journal of Literary Studies* 23, no. 3 (September 2007): 271–292.
35 Fudge, Erica. What Was it Like to be a Cow? History and Animal Studies. In *The Oxford Handbook of Animal Studies*, Linda Kalof (ed.), 258–282. Oxford: Oxford University Press, 2017; Nagel, Thomas. What is it Like to be a Bat? *The Philosophical Review* 83, no. 4 (October 1974): 435–450.
36 Kean, Challenges for Historians; Kean, Hilda. Finding a Man and his Horse in the Archive. In *Animal Biography: Re-framing Animal Lives*, André Krebber, Mieke Roscher (eds.), 41–56. Cham: Palgrave Macmillan, 2018.

ing a time that was markedly different from our own and during which people's experiences and their ways of perception would have differed strikingly from our own present; Walter Benjamin's work on the baroque or the advent of photography and the mechanical reproduction of artworks are profound cases for the historical specificity of perception and how the qualities of experience change.[37] In the case of other animals, of course, the challenge is aggravated by the greater difference in species-specific biology and perception of those that are experiencing. However, the →History of Emotions shows us how potent such an argument can be even in the case of humans. Indeed, if history as an undertaking has to rely on inhabiting the perspectives of others to be possible, we must assume that this is equally possible in the case of other animals, however incomplete, fragile and precarious the results might be.

Both these problems or challenges can still be felt as markers or impulses for the methodological discussions of the individual chapters throughout this volume. Yet the fact that only a few authors make these challenges explicit, highlights how significantly historical animal studies has left these concerns behind and surpassed these challenges as fundamentally calling animal history into question. Just as historians have surmounted the challenges in the context of human histories and actors by reading sources against the grain, trying harder to find sources that express the experiences and challenging the dominance of written sources ever since modern history emerged as its own scholarly discipline in the nineteenth century, animal historians over the past decades have equally pushed the boundaries of and revitalized and innovated historical scholarship by looking for ways to access and represent the active influence and experience of animals in history. Due to the fundamental challenge to our, quite modern (and possibly quite European in its origin) understanding of animals as passive *things*, this is a necessarily interdisciplinary undertaking that has had to include the sciences, although it remains a challenge of how to successfully and complementarily, if critically, implement such perspectives. Three methodological challenges crystalize as centrally driving the methodological considerations and debates across this volume: how to access the perspective of animals in history; how to perceive and conceptualize animal (both individual and species-specific) agency beyond mere comparisons to human forms of agency; and how to write and represent history from a non-anthropocentric perspective. Engaging these challenges has led to a methodological and theoretical pluralism in the field of historical animal studies, one that is difficult to overlook and which the chapters in this volume both testify and provide an introduction to.

Across their variety, there are a number of recurring perspectives that as approaches or practices run through the methodological foundations of an animal history that is more than just a history with animals. The first and most broadly propa-

---

[37] Benjamin, Walter. *The Work of Art in the Age of its Technological Reproducibility, and Other Writings on Media.* Michael W. Jennings, Brigid Doherty, Thomas Y. Levin (eds.). London: The Belknap Press of Harvard University Press, 2008; Benjamin, Walter. *Origin of the German Trauerspiel.* Translated by Howard Eiland. Cambridge, MA: Harvard University Press, 2019.

gated is to read existing, human authored sources anew and question them critically on their animal content, or read them critically to extract an animal perspective from them.[38] A classic example of this would be Harriet Ritvo's *The Animal Estate*, to which references abound in this handbook.[39] The second, responding specifically to the question of *What is it like to be?*, is to draw on ethology as a way of enriching our understanding of the living world and the possible experiences of animals, seeing their behaviors as affected by the environments they live in. Thirdly, animal historians have been exploring and drawing on new, nonwritten sources. The application of these practices, however, are deeply intertwined with the specific historical perspectives and contexts they are used to elucidate, and we leave it to the chapters of this volume to flesh them out.

4 Contents, Gaps, and Missing Links

Despite the comprehensive approach of this handbook, there are still some obvious candidates of fields and topics that should have been included but were not due to either not finding the right author or our inability to fathom a topic in the beginning and only later, by reflecting on this volume, seeing what is missing. In addition, the COVID-19 pandemic as well as other unforeseen circumstances left their mark in so far as articles planned and promised could not be finished in the end. This is true for the history of the body, for example, a field that has shown promising perspectives on how to conceptualize animal–human bonds, both physically and figuratively.[40] Maritime history as an angle that moves beyond earthly surfaces offers new and exciting insights that we would have preferred to have included, but which other handbooks are now left to adopt.[41] Further topics that we have not explicitly included such as the history of animal welfare or conservation and on which there is a plethora of literature[42] appear in other entries such as the →History of Animal Fights and Blood Sports, the →History of Experimental Animals and the History of Animal Experiments or →Environmental History.

As mentioned above, this volume does not try to write a world history of animal–human relations. The regional approaches show glimpses of what that could entail.

---

38 Kean, Finding a Man.
39 Ritvo, Harriet. *The Animal Estate: The English and Other Creatures in the Victorian Age.* Cambridge, MA and London: Harvard University Press, 1987.
40 Eitler, Pascal. Animal History as Body History: Four Suggestions from a Genealogical Perspective. *Body Politics: Zeitschrift für Körpergeschichte* 2, no. 4 (2014): 259–274.
41 Jones, Ryan Tucker. Running into Whales: The History of the North Pacific from Below the Waves. *The American Historical Review* 118, no. 2 (April 2013): 349–377.
42 Kean, Hilda. *Animal Rights: Political and Social Change in Britain Since 1800.* London: Reaktion Books, 1998; Pearson, Susan J. *The Rights of the Defenseless: Protecting Animals and Children in Gilded Age America.* Chicago, IL: University of Chicago Press, 2011.

There is still much more to be done however, including on regions that we have not covered here in more detail, such as South America, North Africa, Eastern Europe and South Asia.⁴³ Furthermore, the histories assembled in this volume show a clear bias towards the early modern and the modern age. Again, this has as much to do with our own background as with the fact that these modern histories have been much more influenced by the theoretical discussions around the concepts brought in by cultural studies in particular that rose to prominence with the animal turn. However, we do see that in the past few years archaeological studies and those focusing on ancient and, to a slightly lesser extent, on medieval histories are thriving,⁴⁴ in particular with their focus on the material side of shared lives.⁴⁵ As such we are hopeful that an animal history that takes seriously the notion of the *longue durée* will develop along more particular histories and those advancing beyond the modern.

5 How to Use this Book

This volume follows a path that winds down from the general to the particular, from the macro to the micro, and from meta-categories to specific topics of animal–human relations. As time and chronology form the major instruments of the historic discipline, the volume opens with Timelines. The chapters in this part do not, however,

---

43 That does not mean, of course, that there are not excellent animal histories of these regions but that they are not as prominent as others. See for example Costlow, Jane and Amy Nelson (eds.). *Other Animals: Beyond the Human in Russian Culture and History.* Pittsburgh, PA: University of Pittsburgh Press, 2010; Derby, Lauren. Bringing the Animals Back in: Writing Quadrupeds into the Environmental History of Latin America and the Caribbean. *History Compass* 9, no. 8 (August 2011): 602–621; Few, Martha and Zeb Tortorici (eds.). *Centering Animals in Latin American History.* Durham, NC: Duke University Press, 2013; Mikhail, Alan. *The Animal in Ottoman Egypt.* Oxford: Oxford University Press, 2014; Mondry, Henrietta. *Political Animals: Representing Dogs in Modern Russian Culture.* Leiden: Brill Rodopi, 2015; Rangarajan, Mahesh and Kalyanakrishnan Sivaramakrishnan (eds.). *Shifting Ground. People, Animals, and Mobility in India's Environmental History.* Oxford: Oxford University Press, 2014; Saha, Jonathan. Milk to Mandalay: Dairy Consumption, Animal History and the Political Geography of Colonial Burma. *Journal of Historical Geography* 54 (October 2016): 1–12; Szczygielska, Marianna. Elephant Empire: Zoos and Colonial Encounters in Eastern Europe. *Cultural Studies* 34, no. 5 (June 2020): 1–22.
44 See for the discussion Armstrong Oma, Kristin and Linda Birke. Archaeology and Human-Animal Studies. *Society & Animals* 21, no. 2 (January 2013): 113–119; Boyd, Brian. Archaeology and Human–Animal Relations: Thinking Through Anthropocentrism. *Annual Review of Anthropology* 46 (October 2017): 299–316; Hill, Erica. Archaeology and Animal Persons: Toward a Prehistory of Human–Animal Relations. *Environment and Society* 4, no. 1 (2013): 117–136.
45 Crane, *Animal Encounters*; McCracken, Peggy. *In the Skin of a Beast: Sovereignty and Animality in Medieval France.* Chicago, IL: University of Chicago Press, 2017; Turner, Nancy K. The Materiality of Medieval Parchment: A Response to 'The Animal Turn'. *Revista Hispánica Moderna* 71, no. 1 (June 2018): 39–67.

align with classical human-centered epochs but introduce, inspired by works such as that of Richard Bulliet,[46] new ways of thinking periodization that are more attentive to animal perspectives. As space and place are equally central to our understanding of history, part one is followed by Regional Approaches. Because Europe and its cultures have been prevalent within the study and field of human-animal history, views from beyond and outside of Europe take center stage here to emphasize in what ways animal history is a field that especially profits from and can contribute to challenging and bypassing a Eurocentric perspective. The third part, Historical Fields, reviews the state of animal history in classical historical subfields that on one hand have been long established and on the other seem to have been battlegrounds for a more inclusive approach to historical writing. Of course, as a result of such battles, the more traditional and more clearly delimited fields are now differentiated in new ways. The handbook reflects this by surveying the animal perspective in more recent Historical Approaches that themselves are defined by their shifting, less clearly denominated structures in part four. The last and fifth part of the handbook finally compiles concrete topics in the History of Human-Animal Interactions. These show, even if they are not exclusive, that the ways in which we narrate human-animal histories always depend on the relation we, as humans, chose or were forced to have with other animals, and that these relations are very much dependent on species characteristics as well as individual traits of both human and nonhuman animals.

Following a composition that has also been used for this introductory chapter, each entry presents the reader with the main topics and themes that are relevant and central to the specific subfield. Under the heading of methods and approaches, the articles then go on presenting roadmaps for how to advance towards the animal and animal–human relationships with regard to the specific topic. This part furthermore addresses methodological difficulties and tries to offer solutions. Some chapters provide us here with new readings of classical methods of historical disciplines and how they can be adapted, read anew or transformed, while others suggest new ways or introduce and adapt methods offered by other disciplines, such as sociology, literary studies or geography. The subsection "Implication(s) of the *Animal Turn*" spells out subsequently how the animal turn has affected work in the specific area. Each entry concludes with a list of suggested further readings, a reminder that this handbook is ultimately an invitation to its users to continue the work of rethinking human-animal relationships in the past and present, placing animals at the center of our shared history.

---

46 Bulliet, Richard W. *Hunters, Herders, and Hamburgers. The Past and Future of Human-Animal Relationships.* New York, NY: Columbia University Press, 2005.

Selected Bibliography

Baratay, Éric. Building an Animal History. In *French Thinking About Animals*, Louisa Mackenzie, Stephanie Posthumus (eds.), 3–14. East Lansing, MI: Michigan State University Press 2013.
Brantz, Dorothee (ed.). *Beastly Natures: Animals, Humans, and the Study of History*. Charlottesville, VA: University of Virginia Press, 2010.
Howell, Philip and Hilda Kean (eds.). *The Routledge Companion to Animal-Human History*. London and New York, NY: Routledge, 2018
Kalof, Linda (ed.). *The Oxford Handbook of Animal Studies*. Oxford: Oxford University Press, 2017.
Kalof, Linda (ed.). *A Cultural History of Animals*. 6 vols. Oxford: Berg, 2011.
Nance, Susan (ed.). *The Historical Animal*. Syracuse, NY: Syracuse University Press, 2015.
Marvin, Garry and Susan McHugh (eds.). *Routledge Handbook of Human-Animal Studies*. Routledge, 2014.
DeMello, Margo. *Teaching the Animal: Human-Animal Studies Across the Disciplines*. New York, NY: Lantern Books, 2010.
Ritvo, Harriet. On the Animal Turn. *Daedalus* 136, no. 4 (Fall 2007): 118–122.
Rothfels, Nigel (ed.). *Representing Animals*. Bloomington, IN: Indiana University Press, 2002.
Walker, Brett L. Animals and the Intimacy of History. *History and Theory* 52, no. 4 (December 2013): 45–67.
Weil, Kari. A Report On the Animal Turn. *differences: A Journal of Feminist Cultural Studies* 21, no. 2 (2010): 1–23.

Part I: **Timelines**

Erica Hill
Pre-Domestication: Zooarchaeology

1 Introduction and Overview

Most of the human past – some 200,000 years – was populated by small bands of people who survived by hunting, fishing and collecting birds, eggs, insects, and plant foods. Wild animals provided meat and marrow, as well as the raw materials for tools, clothing, and shelter. As foragers (or hunter-gatherers, or hunter-gatherer-fishers), these people lived in mobile and semi-sedentary social groups comprised of extended families. Their mobility and unparalleled ability to adapt enabled them to colonize some of the most challenging environments on the planet.

In every region, humans engaged with nonhuman animals, often as prey, but more commonly as cohabitants whom they might conflict with, respect, or ignore. When humans encountered new environments, changing climates, and other humans, their relations with animals changed. Between 20,000 and 15,000 years ago, some humans and animals began to forge new sets of relations, linking their mutual survival in the process of co-domestication[1] and forming the foundations for agriculture and pastoralism [→Domestication: Coevolution; →History of Agriculture]. Some human–animal relations intensified as co-dependencies emerged in more settled life. Agriculture and pastoralism also required the renegotiation of space, as activities such as milking and shearing brought cattle, sheep and goats into much closer proximity with humans and each other. Many non-domesticates – animals that humans hunted, feared, or admired – adjusted to the newly created landscapes of farm and field, encountering roads, structures and humans in increasing numbers.

The availability of "secondary products," such as milk, cheese, and wool, further altered human–animal relations among farmers and herders. As human subsistence became more focused on grains and domesticated animals, time spent hunting and gathering declined. Members of many farming communities continued to hunt in order to supplement and diversify their diets; however, the nature of their relations with non-domesticates shifted [→History of Hunting]. Wild animals were invested with new meanings, and the practice of hunting was deployed for new purposes, such as displaying wealth and power.[2]

---

[1] For discussion of how domestication changed human-animal relations, see Ingold, Tim. From Trust to Domination: An Alternative History of Human-Animal Relations. In *The Perception of the Environment: Essays on Livelihood, Dwelling and Skill*, Tim Ingold (ed.), 61–76. London: Routledge, [1994] 2000; Mlekuž, Dimitrij. The Birth of the Herd. *Society & Animals* 21, no. 2 (January 2013): 150–161.
[2] Hamilakis, Yannis. The Sacred Geography of Hunting: Wild Animals, Social Power and Gender in Early Farming Societies. In *Zooarchaeology in Greece: Recent Advances*, Eleni Kotjabopoulou, Yannis

In many parts of the world, hunter-gatherers encountered farmers and herders, with whom they traded, intermarried, or fought. In some cases, foragers adopted farming and herding. Elsewhere, hunting and gathering persisted until the 1500–1800s, when colonial expansion initiated global demographic and economic change. Despite the impacts of contact, epidemic disease, and acculturation, some foragers continued to rely upon hunting, fishing, and collecting wild foods. Today, their modern descendants selectively hunt and gather, including Yupiit of Alaska, Inuit of Alaska, Canada, and Greenland, Aboriginal Australians, the Aché of Paraguay, and the Hadza of Tanzania.

Since the late nineteenth century, domestication and the "birth of civilization" have been mythologized as watershed events in human history. In contrast, foraging societies without domesticated plants and animals have been considered primitive and lacking in the technologies (e.g., pottery, metal-working) and specialists (e.g., scribes) that distinguished the earliest civilizations of the Near East and Egypt. Instead of documents, foraging peoples recorded their histories through oral narratives and other, unwritten forms of record-keeping. As a result, archaeology, which studies the material remains of human activity, is one of the only ways to reconstruct millennia of human engagement with animals prior to the origins of written history.

Like history, archaeology offers temporal depth to the study of human–animal relations, as well as opportunities to write new (pre)histories[3] of the animals themselves[4] – histories of migration, adaptation, and extinction. The earliest and most enduring material remains of hunter-gatherer lives, animal bones and the stone tools used to hunt and process prey, date human–animal relations to the origins of the genus *Homo*. Therefore, how people *used* animals (human action), rather than how they related to them (human–animal *inter*action), has been the subject of study.

Zooarchaeologists (or archaeozoologists) specialize in the identification, analysis and interpretation of animal bones and teeth from contexts of past human activity. Working primarily in laboratory and museum contexts, zooarchaeologists sort through all of the nonhuman animal remains recovered from site excavation. They seek to identify, using comparative specimens, every bone, tooth, shell, horn or antler to species, determine which skeletal element of the animal body is represented, and record any form of human modification, such as butchery marks, burning, and

---

Hamilakis, Paul Halstead, Clive Gamble and Paraskevi Elefanti (eds.), 239–247. London: British School at Athens, 2003; Borić, Dušan. Theater of Predation: Beneath the Skin of Göbekli Tepe Images. In *Relational Archaeologies: Humans, Animals, Things*, Christopher Watts (ed.), 42–64. New York, NY: Routledge, 2013.

3 "Prehistory" generally refers to that extended time period prior to the development of writing or contact with literate societies. In the Americas, the term has been applied to the entirety of the Native American past prior to 1500. Increasingly, archaeologists are rejecting the term, since it discounts oral narratives and other, non-Western modes of record-keeping. As used here, (pre)history refers to time periods with non-written records.

4 Ingold, From Trust to Domination.

degree of fragmentation. Zooarchaeological analysis may involve thousands of bones from contexts ranging from ancient hearths to graves, middens, and house floors.

At sites where hunter-gatherers camped, lived, hunted, and buried their dead, the majority of an archaeological assemblage may be comprised of animal remains and animal-related artefacts, which also inform us about human-animal prehistories. Artefacts may include animal representations; objects made from animal bodies; and implements used to hunt, capture, process, and transform them [→Material Culture Studies]. Curated animal bones and teeth, used to fashion personal ornaments and amulets, for example, are commonly encountered in human burial contexts. These objects hint at the perceived efficacy and aesthetic, apotropaic, or medicinal values of select species, as well as those characteristics of animals most desired by humans.

Representations of animals, as figurines, decorative elements, in rock art, and a host of other media, may also aid in the reconstruction of human–animal relations [→History of Animal Iconography]. Though imagery presents major interpretive challenges, it highlights those species, both real and imagined, that figured most prominently in the human past. Species of greatest dietary value – as indicated by faunal remains – may differ from those people chose to draw, paint, or sculpt, thereby providing insight into the conceptual domains that animals inhabited.

Zooarchaeology emerged as a specialty in Anglophone archaeology in the mid-twentieth century[5] and by the 1970s had developed an economic focus. Animals were considered resources, objects for human consumption and use. In these functional, utilitarian scenarios, bones were the material remains of past human adaptations to the environment.[6] This focus on subsistence and human adaptation was part of a discipline-wide theoretical trend in archaeology known as processualism. Processualist archaeologists emphasized scientific analysis of animal remains in order to reconstruct human hunting, butchery, and discard behaviors.

By the 1990s, some zooarchaeologists were finding economic interpretations inadequate when animal remains were encountered in contexts unassociated with food debris or middens. In a series of publications on the Iron Age, British zooarchaeologists laid the groundwork for the shift to social zooarchaeology in their studies of problematic deposits of animal bones. Such features – "special," "structured deposits" or "associated bone groups" – were generally considered to be ritual in nature, but defied explanation within the framework of economic zooarchaeology.[7]

---

5 Boyd, Brian. Archaeology and Human–Animal Relations: Thinking Through Anthropocentrism. *Annual Review of Anthropology* 46 (October 2017): 299–316; Reitz, Elizabeth J., and Elizabeth S. Wing. *Zooarchaeology*. Cambridge: Cambridge University Press, 2008.
6 Binford, Lewis R. *Bones: Ancient Men and Modern Myths*. New York, NY: Academic Press, 1981.
7 Grant, Annie. Economic or Symbolic. Animals and Ritual Behaviour. In *Sacred and Profane: Proceedings of a Conference on Archaeology, Ritual and Religion*, Paul Garwood, Robin Skeates, Judith Toms (eds.), 109–114. Oxford: Oxford Committee for Archaeology, 1991; Hill, James D. *Ritual and Rubbish in the Iron Age of Wessex: A Study on the Formation of a Specific Archaeological Record*. Oxford: Tempus Reparatum, 1995.

Deposits of animal parts such as skulls or bird wings in human graves, and the burials of entire animals, both with and without humans, presented similar interpretive dilemmas. Interments of dogs, relatively common in some regions, might plausibly be dismissed as the idiosyncratic deposition of a "prehistoric pet [→History of Pets].[8] However, interments of other species with humans – a large headless wading bird[9] with a man and boy, for example, or the wing of a whooper swan with a woman and infant[10] – suggested that animal remains offered insight on much more than diet and hunting practices.

Twenty-first-century *social* zooarchaeology, as opposed to the processualist *economic* zooarchaeology of the 1970s, acknowledges the centrality of animals to human subsistence whilst recognizing their multiple and complex roles in art, cosmology, and social organization. Recent studies demonstrate the ways in which animals were deployed in the service of politics, religion, and ideology. Certain species have been linked to status and power in feasting rituals, for example, which often leave robust archaeological signatures. Elsewhere, human and animal lives were entangled, often in violent ways, in the contexts of performance and sacrifice.[11] Social zooarchaeologists are also exploring whether and to what extent animals had agency in the past, questioning categories such as "wild" and "domestic," and asking how perceptions of animals changed as some humans adopted agriculture. Social zooarchaeology takes an anthropocentric position on animals, as the purpose of their study is to understand humans, not animals *per se*.[12]

Multispecies archaeology radically departs from both economic and social (zoo) archaeology by shifting from exclusive focus on human-centered prehistories to the ways in which animals, including humans, created, lived, and altered ecosystems in the past. From this perspective of "archaeo-ecology,"[13] nonhuman animals, plants, and microbiota are all potential agents in emergent ecological networks. In highlight-

---

8 See Chapter 7 in Russell, Nerissa. *Social Zooarchaeology: Humans and Animals in Prehistory*. Cambridge: Cambridge University Press, 2012.

9 Gibson Mound 3, a Hopewell site in Illinois, United States. See Parmalee, Paul W. and Gregory Perino. Prehistoric Archaeological Record of the Roseate Spoonbill in Illinois. *Transactions of the Illinois Academy of Science* 63 (1971): 80–85; Perri, Angela R., Terrance J. Martin, Kenneth B. Farnsworth. A Bobcat Burial and Other Reported Intentional Animal Burials from Illinois Hopewell Mounds. *Midcontinental Journal of Archaeology* 40, no. 3 (January 2015): 282–301.

10 Grave 8 at the Mesolithic site of Vebæk, Denmark. See Overton, Nick J. and Yannis Hamilakis, A Manifesto for a Social Zooarchaeology: Swans and Other Beings in the Mesolithic. *Archaeological Dialogues* 20, no. 2 (December 2013): 111–136.

11 Russell, *Social Zooarchaeology*; DeFrance, Susan D. Zooarchaeology in Complex Societies: Political Economy, Status, and Ideology. *Journal of Archaeological Research* 17, no. 2 (June 2009): 105–168; Arbuckle, Benjamin S. and Sue Ann McCarty (eds). *Animals and Inequality in the Ancient World*. Boulder, CO: University Press of Colorado, 2014.

12 Russell, *Social Zooarchaeology*, 5; Gifford-Gonzalez, Diane. *An Introduction to Zooarchaeology*. Cham: Springer, 2018, 5.

13 Birch, Suzanne E. Pilaar. Introduction. In *Multispecies Archaeology*, Suzanne E. Pilaar Birch (ed.), 1–8. New York, NY: Routledge, 2018, 4.

ing these multispecies entanglements, the anthropocentrism of archaeology is implicitly undermined as the human actor yields to study of social, biological, environmental and evolutionary networks. Multispecies archaeology promises new insights on the effects of humans and other invasive species on island ecosystems, for example,[14] on the spread of multi-vector epidemic disease, and on the co-construction of ecological niches in the built environment. A multispecies approach informed by animal studies, moreover, has the potential to highlight types of agency specific to animals, and qualitatively different from those of plants and other biotic actors.

Below, I discuss key themes for zooarchaeologists working within HAS frameworks. These include animal agency and personhood and hunter-gatherer perspectivism, as represented in material culture. Dogs are another dynamic topic of research, particularly their earliest encounters with humans and their transition from commensal species to codependent.

2 Topics and Themes

Agency and Personhood

Agency lies at the foundation of many questions in human–animal studies. When archaeologists and historians attribute agency to animals in their (pre)histories, they implicitly critique anthropocentric views of the past by directing attention to nonhuman actors. For zooarchaeologists, this means rejecting the human subject–animal object dichotomy and recognizing that both humans and animals acted and were acted upon by others on a daily basis. To act as an agent in this context requires neither intentionality nor personhood; rather, an agent is an animal capable of effecting change.

Archaeology detects these agents by identifying sites of human–animal engagement in the past and reconstructing human and animal behaviors and adaptations. For example, landscape change initiated by beavers may have encouraged human settlement in some regions of Europe, changing the population distributions of both species.[15] Animal agency is also at the forefront of debates over domestication, as archaeologists ask to what extent domesticates were agential in their own biological and behavioral transformations. Related questions include when, how, and why humans modified their own lives – their social relations, living arrangements, and daily practices – to accommodate domesticates and reformulate their understanding of "the wild."

---

14 For example, Leppard, Thomas P. Rehearsing the Anthropocene in Microcosm: The Palaeoenvironmental Impacts of the Pacific Rat (*Rattus exulans*) and Other Non-Human Species during Island Neolithization. In *Multispecies Archaeology*, Suzanne E. Pilaar Birch (ed.), 47–64. New York, NY: Routledge, 2018.
15 Coles, Bryony. *Beavers in Britain's Past*. Oxford: Oxbow Books, 2006.

Many scholars are no longer asking whether animals were agents; rather they ask when and where animals exercised agency and how historians and archaeologists might identify it in documents and material culture.[16] In the study of hunter-gatherer ontologies – how people perceived their world and their relations with animals and others – agency has emerged as a key animal attribute.

Persons may be defined as distinct types of human and animal agents – those constructed through social engagement with other persons. All agents are animate and capable of action; however, only some agents are constituted relationally, through interaction with others.[17] Personhood is conditioned by corporeal experience, and by variables such as age, sex/gender, and (dis)ability. Persons may include, for example, a whale who recognizes and avoids a hunter, an elephant matriarch who leads herd members to a watering hole, or a crow who identifies a dangerous human and warns conspecifics.

Hunter-Gatherer Ontologies

Few zooarchaeologists are also ethnographers; however, ethnographers describe many beliefs and practices that archaeologists seek to identify and reconstruct in the past [→Multispecies Ethnography]. Late nineteenth- and twentieth-century ethnographic work has demonstrated that, in general, hunter-gatherers, including many Alaska Natives, Aboriginal Australians, and First Nations peoples of Canada, attributed agency and personhood to many more entities than either agriculturalists or modern Westerners. Minimally, an agent is any entity or creature capable of acting or effecting change, regardless of intention or sentience.[18] Persons, however, are agents who also possess the capacity to behave socially, to *inter*act with others.[19] The distinction between agency and personhood matters because landscapes for many hunter-gatherers were densely populated by a host of agents – animals, plants, topographic features, the dead – only some of whom were persons.

In many Arctic, Subarctic, and Amazonian perspectivist ontological systems, discussed below, "animals" and humans differ primarily in their bodies and behavior – not in their interiority. Like humans, many "animals" have families and social rules; they live in villages, cooperate, and communicate. They do this, however, in bodies

---

16 For an excellent overview of agency, see Nyyssönen, Jukka and Anna-Kaisa Salmi. Towards a Multiangled Study of Reindeer Agency, Overlapping Environments and Human–Animal Relationships. *Arctic Anthropology* 50, no. 2 (January 2013): 40–51.
17 Hill, Erica. Personhood and Agency in Eskimo Interactions with the Other-Than-Human. In *Relational Identities and Other-Than-Human Agency in Archaeology*, Eleanor Harrison-Buck, Julia A. Hendon (eds.), 29–50. Boulder, CO: University Press of Colorado, 2018.
18 Fudge, Erica. What Was It Like to Be a Cow? History and Animal Studies. In *The Oxford Handbook of Animal Studies*, Linda Kalof (ed.), 258–278. Oxford: Oxford University Press, 2017.
19 Hill, Personhood and Agency.

that differ from human ones. Due to this bodily difference, they consume different foods and live in different places – the river, sea ice, or tundra. Their abilities differ from those of humans – humans paddle watercraft on the surface of the water, while seals dive beneath to catch fish. But the interior self of a seal is analogous to the self of a human – and both of those selves are persons.

Animal persons can interact and communicate with humans, though of course they must do so in different ways. For example, in Southwest Alaska, Yup'ik seal-persons, like humans, observe cultural norms and rules for behavior, one of which is to allow themselves to be hunted and consumed by humans. Humans, in turn, are obligated to respect animal persons, accept their bodies as gifts, and treat their remains with respect. Relating to animals as persons may leave material markers that are discernible archaeologically, such as distinctive patterns of bone or body part use and disposal and elaborate mortuary treatment of some species.

Along the Bering and Chukchi seacoasts, human efforts to engage in reciprocal behavior with prey animals were materialized in intentional curation and deposition of sea mammal skulls, identified archaeologically.[20] Ethnohistoric and ethnographic sources indicate that sea mammals refused to offer themselves to humans who trampled their bones and hides, wasted meat, or allowed dogs to chew on their remains. In exchange for the bodies of seal-persons taken in a hunt, Yup'ik families facilitated animal regeneration by collecting seal bones and bladders, ritually thanking the animals, and returning their remains to the sea, as the animals preferred.

In some cases, animal skulls were placed at coastal "meeting points" where animals could see or retrieve them. In Southwest Alaska at a popular hunting site, dozens of beluga skulls were intentionally arranged on shore where other belugas could see them, evidence that human hunters had treated their bones respectfully. Sea mammals demonstrated agency and intentionality in choosing hunters who behaved appropriately, taking part in the reciprocal gift giving that enabled humans to eat and animals to regenerate. Similar treatments of animal remains in other foraging societies – including hanging bones in trees or caching reindeer antlers – suggest that animal agency and personhood were key components of many zoontologies.

Humans and Dogs

Another theme in the archaeology of human–animal relations is the perennial interest in dogs, with whom humans have lived for at least 15,000 years.[21] The unique,

---

[20] Hill, Erica. Animals as Agents: Hunting Ritual and Relational Ontologies in Prehistoric Alaska and Chukotka. *Cambridge Archaeological Journal* 21, no. 3 (October 2011): 407–426; Hill, Erica. The Nonempirical Past: Enculturated Landscapes and Other-than-Human Persons in Southwest Alaska. *Arctic Anthropology* 49, no. 2 (February 2012): 41–47.

[21] Morey, Darcy F. In Search of Paleolithic Dogs: A Quest with Mixed Results. *Journal of Archaeological Science* 52 (December 2014): 300–307.

often ambiguous positions occupied by dogs in many human societies and their status as the first domesticate have led to scrutiny of their osteology, ancestry, and archaeological context.[22] Recently recorded rock art panels in Saudi Arabia dated to about 7000 BCE depict the earliest cooperative hunting by humans and dogs; some of the panels show dogs with leashes, evidence that a particular management regime prevailed.[23]

Elsewhere, dogs played critical roles in human colonization and occupation of particular ecological niches. For example, hunting dogs may have enabled Jōmon foragers to settle temperate deciduous forests in Japan. Without dogs, species such as wild boar and sika deer would have been inaccessible in the dense forest cover. Greater human dependence on dogs starting some 5500 years ago was signaled by increased numbers of dog burials at Jōmon sites.[24]

Dog traction enabled Thule sea mammal hunters to colonize high latitudes and rapidly expand across the North American Arctic between 1200 and 1400 CE. In exchange for provisioning, dogs assisted humans by pulling sleds, dragging watercraft upriver, and serving as pack animals. Human–canine co-dependence, however, did not ensure the well-being of Arctic dogs, many of whom suffered broken ribs, fractured skulls, and tooth removal at the hands of humans.[25] Physical abuse of dogs and other animal cohabitants is in no way restricted to the North American Arctic; however, treatment of Arctic dogs provides a dramatic contrast to the respectful treatment of sea mammals and highlights contingent notions of "care" for animals.[26]

3 Methods and Approaches

Archaeologists depend upon three primary lines of evidence in their efforts to reconstruct human–animal relations: faunal remains, artefacts, and representational im-

---

[22] For example, Germonpré, Mietje, Mikhail V. Sablinb, Rhiannon E. Stevens, Robert E. M. Hedges, Michael Hofreiter, Mathias Stiller and Viviane R. Després. Fossil Dogs and Wolves from Palaeolithic Sites in Belgium, the Ukraine and Russia: Osteometry, Ancient DNA and Stable Isotopes. *Journal of Archaeological Science* 36, no. 2 (January 2009): 478–490; Morey, Darcy F. *Dogs: Domestication and the Development of a Social Bond*. Cambridge: Cambridge University Press, 2010.
[23] Guagnin, Maria, Angela R. Perri and Michael D. Petraglia, Pre-Neolithic Evidence for Dog-Assisted Hunting Strategies in Arabia. *Journal of Anthropological Archaeology* 49 (March 2018): 225–236.
[24] Perri, Angela R. Hunting Dogs as Environmental Adaptations in Jōmon Japan. *Antiquity* 90, no. 353 (October 2016): 1166–1180.
[25] Losey, Robert J., Erin Jessup, Tatiana Nomokonova, Mikhail Sablin. Craniomandibular Trauma and Tooth Loss in Northern Dogs and Wolves: Implications for the Archaeological Study of Dog Husbandry and Domestication. *PLoS ONE* 9, no. 6 (June 2014): e99746; Park, Robert W. Dog Remains from Devon Island, N.W.T.: Archaeological and Osteological Evidence for Domestic Dog Use in the Thule Culture. *Arctic* 40, no. 3 (September 1987): 184–190.
[26] Thomas, Richard. Towards a Zooarchaeology of Animal Care. In *Care in the Past: Archaeological and Interdisciplinary Perspectives*, Lindsay Powell, William Southwell-Wright, Rebecca Gowland (eds.), 169–188. Oxford: Oxbow Books, 2016.

agery. Whether the material under study is a faunal assemblage (i.e., a collection of animal bones, teeth, or shell), artefact, image, or effigy, context – the temporal and spatial situation of the animal or object within a site or feature – is central to interpretation. Animal remains from a midden or a house floor, for example, have different implications than those recovered in a burial. Human and animal interments, as well as contexts such as caches and shrines, are particularly informative on the non-dietary roles animals played.[27]

A major challenge for both archaeology and history is how to write *the animals themselves* into the human past.[28] Below, I highlight three approaches employed in zooarchaeology that seek to reconstruct the spatial, sensory, and experiential components of human-animal relations in the past: animal geography, ethology, and osteobiography. Both animal geography and ethology have their own disciplinary histories [→Historical Animal Geographies]; they have been applied in zooarchaeology in ways tailored to the challenges of ancient animal remains. All three approaches recognize that the subjective lifeworlds of animals themselves are critical components of our multispecies past.

Animal Geographies: Meeting Points

Animal geography explores human-animal spatial engagements – how humans understand, enact, and experience their interactions with animals in different spatial contexts. Animal geography also explores the lived geographies of animals themselves, in both the presence and absence of humans. In the context of zooarchaeology, taking the perspective of animal geography involves reconstructing both where and when humans engaged with animals in the past. Among hunter-gatherers, animal concerns structured spaces ranging from the intimate geographies of fur clothing and animal-tooth ornaments to the choice of site location and housing material. For example, animal migration routes had significant influence on the placement of camp sites, villages, hunting blinds, look-outs, and drive fences. On the coasts of the Bering and Chukchi seas, sites were positioned on points and peninsulas, making migrating walruses and whales visible and accessible. Ideally, sites would also be situated near the nesting cliffs of colonial seabirds and walrus haul-outs. A similar logic applied to the location of sites near waterfowl migration routes, caribou calving grounds, and watering holes.

Archaeologists are often preoccupied with sites of hunting activity, such as "kill" or butchery sites, which leave robust material signatures. However, such sites are

---

[27] For discussion of structured deposits, see Morris, James. *Investigating Animal Burials: Ritual, Mundane and Beyond.* Oxford: BAR Publishing, 2011.
[28] Armstrong Oma, Kristin and Lynda Birke. Guest Editors' Introduction: Archaeology and Human-Animal Studies. *Society & Animals* 21, no. 2 (January 2013): 113–119; Domańska, Ewa. Animal History. *History & Theory* 56, no. 2 (June 2017): 267–287.

only one kind of human–animal "meeting point." Routine tasks for all members of hunter-gatherer societies involved travel, observation, and daily encounters with multiple species, both living and dead. In perspectivist ontologies, relations with animals may continue after death, with the animal person still present in the bones or hide. Dealing with the material remains of animals was therefore a form of social interaction, experienced through skinning, butchering, sharing, eating, and discarding. As bones, teeth, feathers, and hides were extracted and further modified by scraping, tanning, sewing, and wearing, new chains of relations developed, with each body part creating opportunities for human-animal engagement.

Animal teeth and bones fashioned into ornaments are relatively common finds at forager sites. These objects become part of the intimate geography of the human body and establish highly personalized links between humans and specific animals. These kinds of geographies are most accessible in human burials, which preserve specific instances of human–animal engagement. Among the oldest grave goods ever recovered archaeologically are red deer teeth and aurochs horns from Neandertal burial contexts. The antiquity of these objects is evidence that the earliest symbolic behavior in human history involved human–animal relations. Further, like clothing made of animal bodies, the choice to wear certain animal ornaments indicates that display and manipulation of animals were key components of human identity.

In European Mesolithic (9500–5500 BCE) burials, networks of relations with many types of animals – not just those hunted and consumed – are apparent in personal ornaments and accompanying artefacts. At least 18 species were represented by antlers, horns of aurochs and bison, and animal tooth pendants.[29] Larsson has characterized such finds as "compositions" – intentional assemblages of animal affordances structured by cultural norms and personal preferences.[30]

In other contexts, relations with non-mammals received the greatest elaboration in mortuary features. Betts et al.[31] highlight the role of sharks in Archaic and Woodland societies of Eastern North America. Their study suggests that shark teeth mediated relations between humans and these charismatic predators, with whom they competed for prey. Mutual use of marine environments and human knowledge of and respect for sharks made teeth powerful signifiers that facilitated the transfer, sharing, or hybridization of human and animal traits.

Whilst the human body served as a dynamic arena for human–animal engagement, meaningful relations also occurred at the other end of the spatial scale. Hussain and Floss have argued that the landscapes inhabited by Upper Paleolithic peo-

---

[29] Grünberg, Judith M. Animals in Mesolithic Burials in Europe. *Anthropozoologica* 48, no. 2 (January 2013): 231–253.
[30] Larsson, Lars. Tooth-Beads, Antlers, Nuts and Fishes. Examples of Social Bioarchaeology. *Archaeological Dialogues* 20, no. 2 (December 2013): 148–152.
[31] Betts, Matthew W., Susan E. Blair and David W. Black. Perspectivism, Mortuary Symbolism and Human–Shark Relationships on the Maritime Peninsula. *American Antiquity* 77, no. 4 (October 2012): 621–645.

ple during the Aurignacian (42,000–30,000 years ago) were primarily a construct of mammoths.[32] Employing analogies to modern elephant behavior, they describe how herds inscribed trails, altered soil and drainage configurations, and processed vegetation. Many of these impacts on the landscape benefitted human co-habitants, who would have observed and experienced them on a daily basis. The high salience of both living mammoths and their impacts contributed to a material culture in which mammoths figured prominently, even though they were likely neither hunted nor a primary food source. Instead, Aurignacian people materialized their relations with mammoths by using their trails, curating mammoth bones and tusks, and carving objects made of ivory.

Reconstructing Animal Lifeworlds

What is it like to be a bat?[33] What *was* it like to be a cow?[34] Imagining subjective animal lifeworlds, or recovering the animal's point of view, involves exploration of the sensory, emotive, affective, and experiential components of past animal lives. Critical to this pursuit is an understanding of animal biology and behavior, and an appreciation of the species-specific nature of animal engagement with the world. Hunter-gatherers had such understandings, compiled over centuries, even millennia, of experience with some species.

Whitridge, for example, has drawn attention to those abilities of dogs that specially suited them for life among Canadian Inuit, noting their superior sense of smell, spatial reckoning, and sociality.[35] Those capacities situated dogs in past Inuit lives in particular ways – as co-habitants, guides in bad weather, and actors with their own social networks. Mobilizing this sort of ethological information does not preclude appreciation of how the lives of specific dogs, or specific human-dog relations, diverged from those of conspecifics. Like documentary sources, the archaeological evidence suggests that there will always be tension between generalizing animal pasts based on the biology and behavior of particular species and highlighting a particular animal life – the difference between a history of dogs in the Arctic and the detailed study of the dog interred in a log tomb at the Ipiutak site in Northwest Alaska. This

---

32 Hussain, Shumon T. and Harald Floss. Regional Ontologies in the Early Upper Palaeolithic: The Place of Mammoth and Cave Lion in the Belief World (*Glaubenswelt*) of the Swabian Aurignacian. In *Prehistoric Art as Prehistoric Culture: Studies in Honour of Professor Rodrigo de Balbín-Behrmann*, Primitiva Bueno-Ramírez, Paul G. Bahn (eds.), 45–58. Oxford: Archaeopress, 2015.
33 Nagel, Thomas. What Is It Like to Be a Bat? *The Philosophical Review* 83, no. 4 (October 1974): 435–450.
34 Fudge, What Was It Like to Be a Cow?
35 Whitridge, Peter. The Government of Dogs: Archaeological (Zo)Ontologies. In *Relational Engagements of the Indigenous Americas: Alterity, Ontology and Shifting Paradigms*, Melissa R. Baltus, Sarah E. Baires (eds.), 21–40. Lanham, MD: Lexington Books, 2018.

dog, who received elaborate burial treatment comparable to that of few humans, was singled out, perhaps for some notable act or distinctive behavior.[36]

Knowledge of the biological and behavioral capacities of waterfowl enabled Overton and Hamilakis to imagine the sensory experience of hunting, killing and processing whooper swans for meat and bones at Aggersund, Denmark around 4500 BCE.[37] Swan social structure, pair bonding, and seasonal arrival would have been especially salient to human observers. The size and aggression of swans would have made their capture a loud and violent struggle, and their death an affective and emotive experience for both human and avian persons.

Animal capacities also feature in Argent's study of human-horse relations in Iron Age Pazyryk society, c. 400 BCE.[38] Equine pattern recognition, cooperation, and capacity to form emotional bonds made horses valued members of the Pazyryk interspecies community. They were also sacrificed and buried when important people died. Argent imagines the sensory and experiential elements of horse sacrifice; the ritual would have been a chaotic, violent, and emotional event for both human and animal participants as frightened and horrified horses were dispatched by their grieving friends and handlers.

Animal Burial and Osteobiography

One marker of animal personhood is burial, like the Pazyryk horses, in species-specific ways that demonstrate respect for the animal person or in ways analogous to the treatment of deceased humans. Examples of animals buried like humans and in human cemeteries include dogs and wolves around Lake Baikal[39] and a bobcat (*Lynx rufus*) buried wearing a shell necklace in Illinois.[40] Canine burials at the Swedish Mesolithic sites of Skateholm I and II are particularly instructive, as the treatment of dogs in the same cemeteries differed. Several dogs were given individualized burial treatment like that of humans, including being strewn with red ochre and receiving grave goods. Other dogs were apparently killed and then thrown into the graves

---

36 Hill, Erica. The Archaeology of Human–Dog Relations in Northwest Alaska. In *Dogs in the North: Stories of Cooperation and Co-Domestication*. Robert J. Losey, P. Wishart, Jan Peter Laurens Loovers (eds.), 87–104. London: Routledge, 2018.
37 For example, Overton and Hamilakis, A Manifesto for a Social Zooarchaeology.
38 Argent, Gala. Killing (Constructed) Horses: Interspecies Elders, Empathy and Emotion, and the Pazyryk Horses. In *People with Animals: Perspectives and Studies in Ethnozooarchaeology*, Lee G. Broderick (ed.), 19–32. Oxford: Oxbow Books, 2016.
39 Losey, Robert J., Vladimir I. Bazaliiskii, Sandra Garvie-Lok, Mietje Germonpré, Jennifer A. Leonard, Andrew L. Allen, M. Anne Katzenberg and Mikhail V. Sablin. Canids as Persons: Early Neolithic Dog and Wolf Burials, Cis-Baikal, Siberia. *Journal of Anthropological Archaeology* 30, no. 2 (June 2011): 174–189.
40 Perri, Martin and Farnsworth, Bobcat Burial.

of humans, their discard contrasting with treatment of the primary human decedent and of the dogs buried by themselves.[41]

Animal burials may be further investigated using osteobiography, an approach first applied to human remains,[42] which recreates the life and death history of individuals. Osteobiography involves the detailed study of osteology, osteometrics, trauma, disease, and cultural treatment at death of a single animal. Diet and mobility patterns may be reconstructed through isotopic analysis of bone collagen, whilst ancient DNA provides data on the sex of the individual, evolutionary history, and genetic distance from conspecifics. Treatment of the animal during life may also be accessible, as when a significant injury exhibits evidence of human care, or, conversely, when successive healed fractures and blunt force trauma indicate human abuse.[43]

Detailed contextual and osteobiographical study of associated human and animal remains inform us not only about human relations with selected species, but also about the links between specific animals and specific humans. Variables such as sex, social status, and identity may be discernible, giving insight into the nature of personhood, humanity, and animality at particular moments in time and space.

4 Implication(s) of the Animal Turn

As the *animal turn* has gained momentum in the humanities and social sciences, some zooarchaeologists have pushed the discipline further, influenced by thinkers such as Donna Haraway.[44] These scholars are increasingly critiquing anthropocentric (pre)histories and reinterpreting the past as a co-creation of humans and animals – a

---

41 Larsson, Lars. Big Dog and Poor Man: Mortuary Practices in Mesolithic Societies in Southern Sweden. In *Approaches to Swedish Prehistory: A Spectrum of Problems and Perspectives in Contemporary Research*, Thomas B. Larsson, Hans Lundmark (eds.), 211–223. Oxford: British Archaeological Reports, 1989.
42 Saul, Frank P. and Julie M. Saul. Osteobiography: A Maya Example. In *Reconstruction of Life from the Skeleton*, Mehmet Yaşar İşcan, Kenneth A. R. Kennedy (eds.), 237–301. New York: Alan R. Liss, 1989; Stodder, Ann L. W. and Ann M. Palkovich (eds.). *The Bioarchaeology of Individuals*. Gainesville, FL: University Press of Florida, 2012.
43 Bartelle, Barney G., René L. Vellanoweth, Elizabeth S. Netherton, Nicholas W. Poister, William E. Kendig, Amira F. Ainis, Ryan J. Glenn, Johanna V. Marty, Lisa Thomas-Barnett, Steven J. Schwartz. Trauma and Pathology of a Buried Dog from San Nicolas Island, California, U.S.A. *Journal of Archaeological Science* 37, no. 11 (November 2010): 2721–2734; Bendrey, Robin. Care in the Community. Interpretations of a Fractured Goat Bone from Neolithic Jarmo, Iraq. *International Journal of Paleopathology* 7 (December 2014): 33–37; Tourigny, Eric, R. Thomas, E. Guiry, R. Earp, A. Allen, J. L. Rothenburger, D. Lawler and M. Nussbaumer. An Osteobiography of a 19th-Century Dog from Toronto, Canada. *International Journal of Osteoarchaeology* 26, no. 5 (October 2016): 818–829; Thomas, Zooarchaeology of Animal Care; Losey et al., Craniomandibular Trauma.
44 Haraway, Donna J. *The Companion Species Manifesto: Dogs, People, and Significant Otherness*. Chicago: Prickly Paradigm Press, 2003; Haraway, Donna J. *When Species Meet*. Minneapolis: University of Minnesota Press, 2008.

multispecies archaeology, as briefly mentioned above.[45] In its explicit recognition of animal agency a multispecies approach moves beyond both economic and social zooarchaeology to explore issues central to human–animal studies more generally, including: the ontological status of animals in non-Western worldviews; "subjective animal lifeworlds"[46] within multispecies communities; and the spatial, sensory and experiential mechanics of human–animal interaction.

In 2013, following the lead of ethnographers [→Multispecies Ethnography], Hamilakis and Overton[47] advocated an approach that recognized the past as a co-construction of humans and animals. That is, animals have the ability to act, personal life histories, and behavioral and emotional repertoires that introduce complexity to every bio-social system. Like humans, nonhuman animals effect change in the archaeo-ecology of every time and place, past and present. A true multispecies archaeology therefore must consider the roles of animal agents. Historical ecology, paleoepidemiology, and landscape and environmental history [→Environmental History] may all benefit from multispecies perspectives informed by (zoo)archaeology. Archaeology is well positioned to contribute to the deep history of the Anthropocene. Archaeologists have already documented some 50,000 years of continuous ecological change and countless anthropogenic extinctions, a "deep history" that supports the position that the Anthropocene is coeval with the genus *Homo*.[48]

A multispecies archaeology offers the potential to write (pre)histories of the animals themselves, exploring the spatial, sensory and experiential components of subjective animal lifeworlds within multispecies communities. These communities involved networks of relations linking animals to humans and to other animals. Armstrong Oma, for example, has pointed to the complex interspecies relationships that developed among humans, sheep and dogs in Bronze Age Norway around 1500 BCE.[49] Her work draws attention to social spaces in which multiple animal species converged, both with and without human involvement. Dogs, in particular, figure prominently in multispecies pasts, mediating relations among humans, prey animals, predators, and herds. With its materialist orientation, archaeology is uniquely positioned to interrogate the labor of co-dependence that linked humans, dogs, and other animals [→Domestication: Coevolution].

---

45 A term first used in print by Hamilakis, Yannis and Nick J. Overton. A Multi-Species Archaeology. *Archaeological Dialogues* 20, no. 2 (December 2013): 159–173.
46 Argent, Gala. Turns, Tropes and Terminology. Toward an Interspecies (Inter)social. *Archaeological Dialogues* 20, no. 2 (December 2013): 137–143, 142.
47 Hamilakis and Overton, Multi-Species Archaeology.
48 See the special December 2013 issue of *Anthropocene* dedicated to archaeology, especially Braje, Todd J. and Jon M. Erlandson. Human Acceleration of Animal and Plant Extinctions: A Late Pleistocene, Holocene, and Anthropocene Continuum. *Anthropocene* 4 (December 2013): 14–23.
49 Armstrong Oma, Kristin. Sheep, Dog and Man: Multi-Species Becomings Leading to New Ways of Living in Early Bronze Age Longhouses on Jæren, Norway. In *The Farm as a Social Arena*, Liv-Helga Dommasnes, Doris Gutsmiedl-Schümann, Alf Tore Hommedal (eds.), 23–61. Munich: Waxmann, 2016.

These new multispecies narratives demonstrate that the animals themselves mattered; so too did factors such as sex, age, and social status, which structured the life histories of specific humans and specific animals. The subjective lifeworlds of animals also await recovery, making it apparent that knowledge of ethology – and consideration of behavioral changes through time – must inform (zoo)archaeology.

Selected Bibliography

Argent, Gala. Do the Clothes Make the Horse? Relationality, Roles and Statuses in Iron Age Inner Asia. *World Archaeology* 42, no. 2 (April 2010): 157–174.

Armstrong Oma, Kristin. Between Trust and Domination: Social Contracts between Humans and Animals. *World Archaeology* 42, no. 2 (April 2010): 175–187.

Betts, Matthew W., Mari Hardenberg and Ian Stirling. How Animals Create Human History: Relational Ecology and the Dorset–Polar Bear Connection. *American Antiquity* 80, no. 1 (January 2015): 89–112.

Birch, Suzanne E. Pilaar (ed). *Multispecies Archaeology*. New York, NY: Routledge, 2018.

Boyd, Brian. Archaeology and Human–Animal Relations: Thinking Through Anthropocentrism. *Annual Review of Anthropology* 46 (October 2017): 299–316.

Hill, Erica. Archaeology and Animal Persons: Towards a Prehistory of Human–Animal Relations. *Environment and Society: Advances in Research* 4, no. 1 (December 2013): 117–136.

Losey, Robert J., Vladimir I. Bazaliiskii, Sandra Garvie-Lok, Mietje Germonpré, Jennifer A. Leonard, Andrew L. Allen, M. Anne Katzenberg and Mikhail V. Sablin. Canids as Persons: Early Neolithic Dog and Wolf Burials, Cis-Baikal, Siberia. *Journal of Anthropological Archaeology* 30, no. 2 (June 2011): 174–189.

Nyyssönen, Jukka, and Anna-Kaisa Salmi. Towards a Multiangled Study of Reindeer Agency, Overlapping Environments, and Human–Animal Relationships. *Arctic Anthropology* 50, no. 2 (January 2013): 40–51.

Overton, Nick J. and Yannis Hamilakis. A Manifesto for a Social Zooarchaeology: Swans and Other Beings in the Mesolithic. *Archaeological Dialogues* 20, no. 2 (December 2013): 111–136.

Russell, Nerissa. *Social Zooarchaeology: Humans and Animals in Prehistory*. Cambridge: Cambridge University Press, 2012.

Abel Alves
Domestication: Coevolution

1 Introduction and Overview

Humans share the planet with other animals, and by interacting with those animals, both learned behaviors and genetic templates can be altered. Domestication both teaches behaviors, and, through artificial selection, changes biological programming. There are also occasions when animal interactions can be coevolutionary, but true biological coevolution involving humans appears to be rare in domestication.

Charles Darwin once described the evolutionary process as being like an entangled bank – a mutualism in which the interactions of multiple life forms synergistically promote both a healthy ecosystem and the survivability of individual beings because of the healthy ecosystem. Thus, "birds singing on the bushes", "insects flitting about" and "worms crawling through the damp earth" interact with each other to provide for the survival of well-adapted individuals and groups, even though some individuals may die before they can reproduce.[1] By its very nature, evolution is interactive, and, to an extent, always coevolutionary. Properly, however, the term "coevolution" refers to the "complementary evolution of closely associated species" (as in an "arms race" between predator and prey affecting such traits as speed and stealth).[2] It can also be understood as the interaction of genes and learned adaptations, or culture, within a given species.[3] By extension then, "coevolution" may also be the interaction of genes and culture involving more than one species, as in the process of domestication.

To disentangle such a web of life in order to pinpoint exact causality is difficult, and Darwin himself recognized that natural selection might be the preponderant mechanism of evolution, not the only mechanism.[4] Hence, it should come as no surprise that coevolution should also be difficult to untangle, since different animal species with their unique genotypes and phenotypes are interacting. It is not easy to determine that which is innate and that which is learned. However, it can be said that learning and even cultural adaptations are only successful when they build on innate

---

[1] Darwin, Charles. On the Origin of Species by Means of Natural Selection. In *From so Simple a Beginning: The Four Great Books of Charles Darwin*, Edward O. Wilson (ed.), 441–760. New York: W. W. Norton and Company, 2006, 760.
[2] Allaby, Michael (ed.). *The Oxford Dictionary of Natural History*. Oxford: Oxford University Press, 1985, 150.
[3] Durham, William H. *Coevolution: Genes, Culture, and Human Diversity*. Stanford: Stanford University Press, 1991, 37–38.
[4] Durham, *Coevolution*, 24.

templates. Unless placed on an airplane, a pig cannot be culturally constructed to fly, but a symbiotic behavior may be developed where a pig, usually a sow, is taught by humans to root for truffles and not to eat immediately what she finds. In the process of mutual domestication, the human has learned something about porcine behavior, adapting to it, even while the pig must learn something about human wants. Truffle hunting develops from: the biological need for food that pigs and humans share; a sow's apparent attraction to a pheromone found in both the testes and saliva of boars and in the truffles as well; the human social animal's desire to express hierarchy by means of a food deemed a luxury by cultural construction; and the construction of a cross-species set of learned behaviors built on natural behaviors.[5] Since neither the genotype of pigs nor humans has been altered by this learned, cultural interaction, biologists will not call this coevolution in the technical sense, and it probably behooves scholars of human interaction with other animals to shy away from the use of the term coevolution. It is, however, more than appropriate for historians to discuss changes in the behavior of humans and other animals as they interact, and nowhere is this more apparent than in the process of domestication.

Domestication is a function of human dominance, the human desire to nurture and the agency and sociability of nonhuman animals. Although power is unequal in domestication, nonhuman animals, as sentient beings, contribute to the process and change humans to some extent. By contemplating the hierarchical, yet reciprocal, communities we construct with other animals, we explore historical cooperation and what historian Richard Bulliet has described as our domestic and postdomestic worlds. Domestication is culture, moreover, in that, from its very origins, it relies on at least two species learning a set of behaviors allowing for greater interaction. However, there is a thematic discussion found in Juliet Clutton-Brock's now classic 1989 anthology, *The Walking Larder: Patterns of Domestication, Pastoralism, and Predation*, that remains central to the historical study of domestication. To what extent do sentient nonhumans shape the behaviors of humans, even as humans shape the behaviors, and even genotypes, of animals under their dominion? In *The Walking Larder*, Sandor Börkönyi argued, "Domestication is the beginning of a symbiosis that needs at least two partners, and it is simplistic to view it from the side of one of the partners alone."[6] This was countered in the same volume by Pierre Ducos, who stated that domestication is not a symbiosis, and that "it exists because humans (and not

---

[5] Mizelle, Brett. *Pig*. London: Reaktion Books, 2011, 94–95; Claus, R., H. O. Hoppen and H. Karg. The Secret of Truffles: A Steroidal Pheromone? *Experientia* 37, no. 11 (November 1981): 1178–1179; Boyer, Jacques. Truffle Culture in France. *Scientific American: Supplement* 61, no. 1566 (January 1906): 25087–25089.

[6] Börkönyi, Sandor. Definitions of Animal Domestication. In *The Walking Larder: Patterns of Domestication, Pastoralism, and Predation*, Juliet Clutton-Brock (ed.), 22–27. London: Unwin Hyman, 1989, 24.

the animal) wished it."[7] To separate humans and human culture, as Ducos did, from the animal kingdom remains a position taken by some, but, overwhelmingly, an animal studies approach to history builds on themes of animal sentience and agency, interactive interspecies choices, and human dominance in the midst of some benefits accrued by some other animals. It is this tension between the theories of coevolution and the paradigm of domestication that this article tries to highlight.

2 Topics and Themes

Biologists and dog breeders Raymond and Lorna Coppinger have cautioned us to be wary of applying the term coevolution to the interactions of humans and other animals, since in biological terms, coevolution involves genetic changes in both the species involved. However, "Sled dogs are genetically adapted for pulling a sled while the human only has to learn to ride it."[8] Despite this apparent slavery, the Coppingers, in a 2001 book, proposed that dogs chose us. More curious and less nervous wolves, sometime between 135,000 and 12,000 years ago, were attracted to the refuse at human sites of habitation, where they increasingly interacted with humans until they became guards similar to the wandering village dogs who still inhabit many parts of the earth, and, finally, hunting companions – and more [→Pre-Domestication: Zooarchaeology].[9] This runs counter to the earlier hypothesis that humans artificially selected from among the most docile wolf cubs, perhaps after having killed the adults, to create dogs in a foreshadowing of today's genetic engineering. While Dmitry Belyaev's artificial selection of doglike behaviors in silver foxes lends some credence to the older argument, our ability to determine whether dogs chose us, we chose dogs, or the attraction was mutual, remains unknowable.[10] We may have "coevolved" with dogs behaviorally, if not genetically, on many different occasions, and the Coppingers have recently been challenged by a more expansive cultural

---

[7] Ducos, Pierre. Defining Domestication: A Clarification. Translated by Marie Matthews. In *The Walking Larder: Patterns of Domestication, Pastoralism, and Predation*, Juliet Clutton-Brock (ed.), 28–30. London: Unwin Hyman, 1989, 29.
[8] Coppinger, Raymond and Lorna Coppinger. *Dogs: A Startling New Understanding of Canine Origin, Behavior, and Evolution*. New York, NY: Scribner, 2001, 159.
[9] Coppinger and Coppinger, *Dogs*, 283–286, 60–61, 72–81. For the debate as to when exactly dogs evolved, see Morey, Darcy F. In Search of Paleolithic Dogs: A Quest with Mixed Results. *Journal of Archaeological Science* 52 (December 2014): 300–307; Vilà, Carles, Jesús Eduardo Maldonado and Robert K. Wayne. Phylogenetic Relationships, Evolution and Genetic Diversity of the Domestic Dog. *The Journal of Heredity* 90, no. 1 (January 1999): 71–77; Freedman, Adam H., Ilan Gronau, Rena M. Schweizer, Diego Ortega-Del Vecchyo, Eunjung Han, Pedro M. Silva, Marco Galaverni et al. Genome Sequencing Highlights the Dynamic Early History of Dogs. *PLoS Genetics* 10, no. 1 (January 2014): e1004016. doi.org/10.1371/journal.pgen.1004016.
[10] Trut, Lyudmila N. Early Canid Domestication: The Farm-Fox Experiment. *American Scientist* 87, no. 2 (March-April 1999): 160–170.

analysis incorporating knowledge gathered from Amerindian traditions in North America.

Described by early European observers as quite lupine, the dogs of the Great Plains were seen by the First Nations of North America themselves as wolf companions who had chosen to hunt with humans. Referring to a complex of evidence, including the similarity in wolf and dog genotypes and their ability to mate and produce fertile offspring, Raymond Pierotti and Brandy R. Fogg argue, in their 2017 book *The First Domestication: How Wolves and Humans Coevolved*, that wolves chose to join us in the hunt, teaching us more effective hunting methods. According to Pierotti and Fogg, this happened a number of times around the globe, with early dogs probably also continuing to add genetic information from feral wolves through ongoing mating and hybridization.[11] Humans were shaped by wolves, made more cooperative, and, in effect, domesticated by interaction with another species before agriculture. From this mutually beneficial partnership the increasingly dependent pet would develop.

There is evidence for human attachment to nonhuman animals since late Paleolithic and early Neolithic times. There are multiple gravesites where humans were buried with small mammals. Around 12,000 years ago, an elderly person, probably a woman, in northern Israel was buried with her hand on a puppy, which implies that there may have been an affection for the puppy, but also that the puppy may have been "sacrificed" to accompany the woman in death.[12] While not technically biological "mutualism," Linda Kalof, reflecting on the work of James Serpell writes, "The human-companion animal relationship is 'mutualistic,' conferring adaptive benefits on both humans and other animals."[13] Serpell, in his 1986 classic *In the Company of Animals* has shown that the role of the pet, or "proto-pet" was always there. It is not a European cultural formulation of the nineteenth-century Parisian bourgeoisie as some misreadings of Kathleen Kete's 1994 *The Beast in the Boudoir* have made it out to be.[14] Serpell argues that the Barasana people of Colombia have kept small birds, small rodents and dogs as pets, and that in the twelfth century, monkeys kept by Thomas à Beckett were "undoubtedly" pets.[15] The affection shown towards companion animals is cross-cultural and historical, with continuity and change over time and place, but with mutually supportive emotional bonds al-

---

[11] Pierotti, Raymond and Brandy R. Fogg. *The First Domestication: How Wolves and Humans Coevolved*. New Haven, CT: Yale University Press, 2017, 83–104, 143–165.

[12] Davis, Simon J. M. and François R. Valla. Evidence for Domestication of the Dog 12,000 Years Ago in the Natufian of Israel. *Nature* 276, no. 5688 (December 1978): 608–610; Clutton-Brock, Juliet. *A Natural History of Domesticated Mammals*. Cambridge: Cambridge University Press, 1999, 58.

[13] Kalof, Linda. Introduction. In *The Oxford Handbook of Animal Studies*, Linda Kalof (ed.), 1–24. New York, NY: Oxford University Press, 2017, 3.

[14] Kete, Kathleen. *The Beast in the Boudoir: Petkeeping in Nineteenth-Century Paris*. Berkeley: University of California Press, 1994, 3.

[15] Serpell, James. *In the Company of Animals: A Study of Human-Animal Relationships*. Cambridge: Cambridge University Press, 1996, 47–48, 63–64.

ways present [→History of Pets]. Archaeologically and in writing, the sentience of nonhuman animals is demonstrated in their ability to contribute to such bonds, and this is perhaps best captured when the sixteenth-century French author Michel de Montaigne wondered whether he played with his cat, or his cat played with him.[16]

Humans and nonhuman animals interact, communicate and receive mutual biologically based benefits, with humans providing for dogs and cats who, in turn, show positive desires to interact with us, respectively wagging tails and rubbing against us, and with stress-suppressing oxytocin levels increasing in both humans and dogs who are interacting with gazes of affection.[17] Indeed, a number of studies show the physiologically demonstrable benefits of other animals' companionship, and evidence is beginning to accumulate in favor of a possible biological coevolution occurring through this interaction, although much remains to be studied and tested.[18] Anthropologist Pat Shipman even suggests that early humans who were more successful at communication and empathetic understanding with others animals were more successful at the first domestications. Then, these humans accumulated all the benefits provided by hunting dogs, milk cows and wool-bearing sheep – who learned to work with us as we learned to work with them. Although admitting that she cannot prove it, she postulates that "humans who were more successful at handling and living with animals accrued a selective advantage" through a mutual domestication. This means their genes would be passed on to more offspring than those less successful with survival-enhancing relationships with other animals, thus altering the human gene pool.[19] If Shipman is correct, this is true coevolution related to domestication, but the jury remains out.

Leaving aside possible alterations in human genotypes, domestication has always involved cross-species communication. Historically, a number of humans have understood that other animals communicate with each other and with us in meaningful ways. As cited by Erica Fudge, King James I of England, in the early seventeenth century, used his own observations of his dogs to argue for a breakdown of boundaries between so-called instinct and reason. He saw humans and dogs on a continuum, arguing that he had one hunting dog, who when alone in picking up the scent of the prey, returned to the pack and convinced his fellows to follow him by means of "yelling arguments" that could not be "carried on without an exer-

---

16 De Montaigne, Michel. An Apology for Raymond Sebond. Translated by Michael Andrew Screech. In *The Complete Essays*, Michael Andrew Screech (ed.), 489–683. London: Penguin Books, 1991, 505.
17 Nagasawa, Miho, Shouhei Mitsui, Shiori En, Nobuyo Ohtani, Mitsuaki Ohta, Yasuo Sakuma, Tatsushi Onaka, Kazutaka Mogi and Takefumi Kikusui. Oxytocin-Gaze Positive Loop and the Coevolution of Human-Dog Bonds. *Science* 348, no. 6232 (April 2015): 333–336.
18 Shipman, Pat. *The Animal Connection: A New Perspective on What Makes Us Human*. New York, NY and London: W. W. Norton and Company, 2011, 272; Headey, Bruce. Health Benefits and Health Cost Savings Due to Pets: Preliminary Estimates from an Australian National Survey. *Social Indicators Research* 47, no. 2 (June 1999): 233–243.
19 Shipman, *The Animal Connection*, 257–258.

cise of understanding."[20] By the nineteenth century, and *The Expression of the Emotions in Man and Animals*, Charles Darwin continued to argue for a continuum when he stated that animals like cats, dogs and chimpanzees felt affection and communicated this to the subject of their affection by rubbing against a human companion's leg, or, in the case of chimpanzees, embracing.[21]

If domesticated animals participate in the development of human-animal bonds through expressing their own individual feelings and responses based on particular circumstances, they can also reject association, thereby demonstrating their agency. While Chris Pearson is correct to point out that nonhuman animal resistance is not the full-fledged political resistance of humans, it remains resistance nonetheless.[22] In human history, horses, cows and pigs have all escaped at times. In an invidious comparison, Spaniards in the sixteenth and seventeenth centuries applied the term "*cimarrón*," meaning "wild" and "renegade," to runaway slaves and nonhuman animals alike. By escaping from Spanish "*imperio*" (defined as "dominion," "authority" and "territory" in the Spanish Royal Academy's eighteenth-century *Diccionario de Autoridades*), human and nonhuman *cimarrones* proved their agency – that they were fully animate beings and not insensible things.[23] In the eighteenth century, Benedictine priest and proponent of the Enlightenment Benito Feijóo summarized a theme of sentience and agency in Spanish writings about nonhuman animals in his *Teatro Crítico Universal*, arguing that the movement of nonhuman animals was not purely passive like that of a timepiece. He stated that a cat will pause and deliberate about the best path to take to get a hard-to-reach piece of meat.[24] Likewise, cats and dogs will refrain from taking food that is to their liking in front of their owners if they are punished for doing so, and they will show hesitancy before fighting each other.[25]

---

20 King James I of England, cited in Fudge, Erica. *Brutal Reasoning: Animals, Rationality and Humanity in Early Modern England*. Ithaca: Cornell University Press, 2006, 102–103.

21 Darwin, Charles. *The Expression of the Emotions in Man and Animals*. Paul Ekman (ed.). Oxford: Oxford University Press, 1998, 212–213.

22 Pearson, Chris. History and Animal Agencies. In *Oxford Handbook of Animal Studies*, Linda Kalof (ed.), 240–257. Oxford: Oxford University Press, 2017, 250–251.

23 For "*cimarrón*" see Real Academia Española. *Diccionario de autoridades*, 3 vols. facsimile ed. Madrid: Editorial Gredos, 1963–1964, 1: 350. For "*imperio*," 2: 224. The *Diccionario de autoridades* was originally published between 1726 and 1739. For application of the term "*cimarrón*" to cows and pigs, see Relación y descripción de la ciudad de Loxa. In *Biblioteca de autores españoles*, Vol. 184: *Relaciones geográficas de Indias.-Perú, Tomo 2*, Marcos Jiménez de la Espada (ed.), 3 vols. Madrid: Ediciones Atlas, 1965, 296. For the origins of the sixteenth-century application of the term to Amerindians and escaped black slaves, see Arrom, José. Cimarrón: apuntes sobre sus primeras documentaciones y su probable origen. *Revista española de antropología americana* 13. Madrid: Editorial Universidad Complutense, 1983, 47–57. URL: revistas.ucm.es/index.php/REAA/article/view/REAA838311 0047A (December 30, 2019).

24 Feijóo, Benito. Racionalidad de los brutos. In *Biblioteca de autores españoles. Obras escogidas del Padre Fray Benito Jerónimo Feijóo y Montenegro*, Vicente de la Fuente (ed.), 130–141. Madrid: Ediciones Atlas, 1952, 136.

25 Feijóo, Racionalidad de los Brutos, 137–138.

Feijóo argued that nonhuman animals pursue their objectives, and that René Descartes (1596–1650) was wrong to reduce "all their movements to pure mechanism," while the physician Gómez Pereira (1500–1567) was wrong to attribute those movements to "sympathies and antipathies" the animals automatically have with objects, resembling iron when it is attracted to a magnet.[26]

Should nonhuman animals have sentience and agency and the evidence indeed supports this, then domestication is the story of interspecies choices and interactions in a web of unequal power relationships favoring *Homo sapiens*, but sometimes providing some benefits to the other animals involved. In Ming China, when devout Buddhists and some others bought live animals in order to liberate them and prevent them from being turned into meat, human dominance was admitted and confronted with compassion.[27] Feedback loops and multiple forms of interaction, including compassion and the development of emotional bonds, abound in domestication as a form of interspecies cultural construction, but these constructions develop in interaction with the biological conditions that are already there.

Livestock, called *"ganado"* [*"something* gained"] in Spanish, are exemplary of sentient animal beings whom humans attempt to reduce to machines of production. This occurred long before the Concentrated Animal Feeding Operations of the twentieth and twenty-first centuries. Reducing these animals to mere objects was prevalent when the Spanish Jesuit José de Acosta, in 1587, tallied 64,350 cattle hides worth 96,532 pesos being sent from colonial New Spain, now Mexico, to Seville.[28] Likewise, in *Creatures of Empire*, Virginia DeJohn Anderson shows how surpluses in livestock were on the minds of New England farmers of the seventeenth century, and Samuel Maverick, in 1660, reported that "many thousand Neate Beasts and Hoggs are yearly killed" and exported for profit.[29] While David Wheeler of Bradford, Massachusetts could mourn the killing by another colonist of a deer with a red collar he had tamed, livestock slaughtered for export were probably seldom mourned – the sheer number of the slaughter being seen as profitable and praiseworthy.[30] In the analytical terms of James Serpell, humans, across cultures, are given to distinguishing between "pets" and "pigs" (for the slaughter).[31] In the Andes, sociologist Edmundo Morales noted this as well. While in the twentieth century, large work-animals kept by Andeans might be named after plants and flowers, and considered companions

---

[26] Feijóo, Racionalidad de los Brutos, 131.
[27] Smith, Joanna Handlin. *The Art of Doing Good: Charity in Late Ming China*. Berkeley, CA: University of California Press, 2009, 21–25.
[28] De Acosta, José. *Historia naturale, e morale delle Indie* [Natural and Moral History of the Indies]. Translated by Frances M. López-Morillas. Jane E. Mangan (ed.). Durham, NC: Duke University Press, 2002, 231.
[29] Anderson, Virginia DeJohn. *Creatures of Empire: How Domestic Animals Transformed Early America*. Oxford: Oxford University Press, 2004, 151.
[30] Anderson, *Creatures of Empire*, 43–45.
[31] Serpell, *In the Company of Animals*, 12–20.

in life and labor, guinea pigs were raised *en masse* for food and "rarely named."[32] The guinea pig, identified by physical attributes, became a dominated comestible.

Still, some animals declared livestock found ways in very unequal power relationships to express their sentience. In her studies of the interaction of cattle with the Khoikhoi and Dutch in seventeenth-century South Africa, Sandra Swart has noted that Dutch documents complain about the native African cattle's unwillingness to follow the wishes of Dutch drovers when Khoikhoi drovers were present. The Dutch would place European cattle at the front and back of a herd to prevent the dissolution of the herd through the resistance of African cattle who were far more accepting of interaction with dominant Khoikhoi drovers than they were with the Dutch [→African Studies].[33]

With domestication being a complicated and multifaceted process, such human perceptions as the assumed need to "break the spirit of a willful horse" indicates that humans have recognized nonhuman sentience and agency, even when their ultimate goal may be the reduction of another being to the status of object. However, there is also a great deal of evidence that what goes on in domestication is unconscious. In *Hunters, Herders, and Hamburgers*, Richard Bulliet notes, "When the first wild animals were being penned or tolerated as they skulked around the campsite catching mice or scavenging garbage, no one knew that ox-drawn plows, horse-drawn chariots, pinstriped wool suits, frozen yogurt, and Kentucky Fried Chicken were looming in the distant future."[34]

The transition from a predomestic world to a domestic one was not planned strategically, and at least one unplanned case of technical coevolution may have transpired, with the interaction of cows and humans yielding higher percentages of adult lactose absorption in some human populations. It has been estimated that among the pastoralist Tutsi, lactose absorption ranges from 88.2% in Uganda to 92.6% in Rwanda.[35] Among the nomadic Fulani of Nigeria, estimates of lactose absorption have been reported at approximately 80%.[36] In Spain, it is estimated that between 71% and 85.3% of the adult population are capable of absorbing lactose, while 96–97.5% of adult Danes retain sufficient amounts of the enzyme lactase to

---

32 Morales. Edmundo. *The Guinea Pig: Healing, Food, and Ritual in the Andes*. Tucson, AZ: The University of Arizona Press, 1995, 11.
33 Swart, Sandra. Settler Stock: Animals and Power in the Mid-Seventeenth-Century Contact at the Cape, circa 1652–62. In *Animals and Early Modern Identity*, Pia F. Cuneo (ed.), 243–267. Farnham, England: Ashgate, 2014, 255–256.
34 Bulliet, Richard W. *Hunters, Herders, and Hamburgers: The Past and Future of Human-Animal Relationships*. New York, NY: Columbia University Press, 2005, 136–137.
35 Durham, *Coevolution*, 234.
36 Kretchmer, Norman, Ruth Hurwitz, Olikoye Ransome-Kuti, Claibourne Dungy and Wole Alakija. Intestinal Absorption of Lactose in Nigerian Ethnic Groups. *The Lancet* 298, no. 7721 (August 1971): 392–395.

allow them to absorb lactose.³⁷ All these populations, from Africa to Europe, have lived for quite some time with cows, consuming dairy products that would have caused gastrointestinal problems and reduced food options for those adults in their population who could not take advantage of the nutrition found in milk and milk products. Whether a high percentage of lactose absorbers preceded the reliance on domesticated cattle or not, these groups are genetically and biochemically adapted to cattle as some other groups are not, and their cows have multiplied under human dominance.

Under human dominance and domestication, the growth in the sheer number of cows has come at a cost that threatens the biodiversity, variation and intertwining mutualism of Darwin's "entangled bank." According to the Yale School of Forestry and Environmental Studies, cattle ranching currently accounts for 80% of the deforestation rates in the Amazon, and one estimate of earth's biomass has livestock weighing in at 0.1 gigatons and wild mammals at only 0.007 gigatons.³⁸ Given the fact that our livestock species are limited to such animals as bovids, pigs, horses and a few others, biodiversity suffers. In turn, current human cultural desire for conformity has led to the impoverishment of breed types within a species. Science journalist Richard C. Francis reports that central Californian dairy herds of the 1960s were composed of at least three breeds: Guernseys, Jerseys and Holsteins. Today, only Holsteins are commonly found.³⁹ If early modern European thinkers like Descartes and Pereira proposed that nonhuman animals could be thought of as mere stimulus-response machines, today's human practices certainly do reduce them to manipulated objects of our dominance. While an occasional pig may run from the slaughter, other animals are forced to do our bidding at great cost to their ability to express their individual behaviors in overcrowded Concentrated Animal Feeding Operations.⁴⁰ While many cats and dogs might benefit greatly from domestication, domesticated livestock usually experience nothing more than domination today.

Once upon a time, the categories may have been a bit more blurred. John Duncan, in *Travels in Western Africa in 1845 and 1846*, wrote that he saw horses playing with African children, "nibbling their heads" and "licking their faces, as a spaniel

---

37 Durham, *Coevolution*, 235; Storhaug, Christian Løvold, Svein Kjetil Fosse and Lars T. Fadnes. Country, Regional, and Global Estimates for Lactose Malabsorption in Adults: A Systematic Review and Meta-Analysis. *Lancet Gastroenterology & Hepatology* 2, no. 10 (October 2017): 738–746, 743.
38 Cattle Ranching in the Amazon Region. *Global Forest Atlas*. New Haven, CT: Yale School of Forestry and Environmental Studies. URL: globalforestatlas.yale.edu/amazon/land-use/cattle-ranching (November 27, 2020); Bar-On, Yinon M., Ron Phillips and Ron Milo. The Biomass Distribution on Earth. *PNAS* 115, no. 25 (June 2018): 6506–6511, 6508.
39 Francis, Richard C. *Domesticated: Evolution in a Man-Made World*. New York, NY and London: W. W. Norton and Company, 2015, 145–146.
40 Serpell, *In the Company of Animals*, 7–12, 19, 201–202; Imhoff, Daniel (ed.). *CAFO (Concentrated Animal Feeding Operation): The Tragedy of Industrial Animal Factories*. San Rafael, CA: Earth Aware, 2010.

would."[41] While contemporary historian Diana Ahmad notes that animals were overburdened, abused, abandoned and killed on the nineteenth-century North American overland trails, an A. J. McCall, upon reaching his destination in the west, refused to sell a horse named Charley, as he was not a "mercenary slave-trader."[42] Likewise, the young Spanish naval lieutenant, explorer and natural philosopher Antonio de Ulloa (1716–1795) noted that Quito was not only a place where excellent meat might be purchased. It was a city where Amerindian women so loved the chickens they raised that they did not eat them and only sold them with great sorrow and regret if they were in dire need.[43] A city whose population grew through migration in the sixteenth century, Quito was a locus for the accumulation of diverse Amerindian traditions, and while accumulated evidence points to the Eurasian chicken's becoming a less preferred substitute for culturally preferred guinea pig meat among Quechua speakers, there are also sources that tell us of Amerindians who kept chickens as pets and suppliers of ornamental feathers.[44] Like other humans, Amerindians both used and loved nonhuman animals, and Ulloa, who took time to reflect on the behavior of South American crocodilians, demonstrated the human capacity to flit back and forth between viewing other animals as resources or fearsome threats on the one hand and showing real interest in their behaviors and lives on the other.[45]

To once again channel James Serpell, human relationships with domesticated animals can reduce those animals to mere objects and machines or embrace them as boon companions. In fact, research for colonial Mexico (New Spain) by Sonya Lipsett-Rivera and Zeb Tortorici demonstrates the presence of cared-for nonhuman animal companions. Mining the judicial archives of New Spain, Sonya Lipsett-Rivera writes that in 1685 a man named Fernando de Lezcano was denounced to the Inquisition "because he said that his many dogs were his children; he ate and slept with them and showered them with love." Indeed, tales of cruelty toward dogs can be intertwined with accounts of those who would defend the dogs. When a boy named Josef María Villanueva kicked a little dog who had barked at him, he was reprimand-

---

41 Duncan, John. *Travels in Western Africa in 1845 and 1846, Comprising a Journey from Whydah, Through the Kingdom of Dahomey, to Adofoodia in the Interior.* 2 vols. London: Richard Bentley, 1847, vol. 2: 138.
42 Ahmad, Diana L. *Success Depends on the Animals: Emigrants, Livestock, and Wild Animals on the Overland Trails, 1840–1869.* Reno, NV: University of Nevada Press, 2016, 85.
43 De Ulloa, Antonio. *Viaje a la América meridional.* Andrés Saumell (ed.), 2 vols. Madrid: Historia 16, 1990, vol. 1: 512.
44 Powers, Karen Vieira. *Andean Journeys: Migration, Ethnogenesis, and the State in Colonial Quito.* Albuquerque, NM: University of New Mexico Press, 1995, 7–8, 13–43; Morales, *The Guinea Pig*, 13, 62; Seligmann, Linda J. The Chicken in Andean History and Myth: The Quechua Concept of *Wallpa*. Ethnohistory 34, no. 2 (Spring 1987): 139–170, 143; Nordenskiöld, Erland. *Comparative Ethnographical Studies, Vol. 5: Deductions Suggested by the Geographical Distribution of Some Post-Columbian Words Used by the Indians of S. America.* Translated by George Ernest Fuhrken. Gothenburg: Elanders Boktryckeri Aktiebolag, 1922, 9–12.
45 For caimans, see Ulloa, *Viaje a la América meridional*, vol. 1: 269–274.

ed by the animal's human companions, the husband and wife Alexo Hernández and Jacinta Flores. Lipsett-Rivera aptly notes that this couple and others in New Spain considered "their dogs to be part of the household and worth protecting."[46] In comfortable Mexico City households, Zeb Tortorici has found that dogs were subjected to baptisms and wedding ceremonies that satirized the excesses of an imperial hierarchy that saw no separation of theology and politics. However, it seems as though such events were not only satires done at the expense of the animals, and one baptism was never actually performed since the dogs' human, María Dolores, "became concerned that they would be harmed if they were doused in the makeshift baptismal font."[47] The officials of the Inquisition seemingly worried that "these acts challenged the divinely ordained natural and social orders," but they were not worthy of serious punishment in the first instance of offense.[48]

In his study of nonhuman animals in the Ottoman Empire, Alan Mikhail corroborates that the desire to nurture and care for a nonhuman companion transcends restrictive religious and cultural constructions. Rather than just a beast with foul habits and contaminating saliva, the status of the dog has actually been debated in Islam, with one Muslim holy man telling his followers that dogs might be found in heaven.[49] Both positive and negative feelings toward the dog were always there in Islam. In the 1830s, a solitary woman mourned her dog, her only companion, with a Muslim funeral.[50] Emotional bonds existed across species in Ottoman Egypt, and, in today's Cairo, comfortable non-Muslims and Muslims alike will keep dogs as pets in their apartments, even as street dogs are few to be found, having been mostly eradicated in the nineteenth century as nuisances and disease vectors. An older domestic world where street dogs cleared garbage and peasants labored side-by-side with donkeys had been replaced with a world focused on human and mechanical labor, an absence of a myriad of domestic animals in the streets, and some pampered dogs kept as pets at home.[51] In Egypt, with the coming of what has been called modernity, the stage was now set for the debates of what Richard Bulliet has called the postdomestic world.

---

46 Lipsett-Rivera, Sonya. A New Challenge: Social History and Dogs in the Era of Post-Humanism. *Sociedad indiana* (August 2015). URL: socindiana.hypotheses.org/320 (November 27, 2020).
47 Tortorici, Zeb. In the Name of the Father and Mother of All Dogs: Canine Baptisms, Weddings and Funerals in Bourbon Mexico. In *Centering Animals in Latin American History*, Martha Few, Zeb Tortorici (ed.), 93–119. Durham, NC: Duke University Press, 2013, 100.
48 Tortorici, In the Name of the Father, 102.
49 Mikhail, Alan. *The Animal in Ottoman Egypt*. Oxford: Oxford University Press, 2014, 68–77, 83.
50 Mikhail, *The Animal in Ottoman Egypt*, 100–101.
51 Mikhail, *The Animal in Ottoman Egypt*, 53–59, 106.

3 Methods and Approaches

Perhaps unfortunately, a history of domestication must always confront the fact that humans, in their records and artifacts, have demonstrated a persistent cross-cultural tension where domesticates are concerned. We approach them with varying degrees of nurturing care and a lust for domination. We appreciate their individual attributes and agencies, and we also objectify them as machines to be exploited and extensions of our own symbolic thinking to be anthropomorphized. Indeed, this spectrum of attitudes has been fully recognized by major luminaries in the field. Juliet Clutton-Brock argued that although domestication may have started with the human desire to nurture, especially cute baby animals, it also incorporates the view that domesticated animals exist only for the benefit of humanity, leading to contemporary factory farms where "The personality of each individual livestock animal is lost [...] where rows of caged animals are treated like animate vegetables."[52] James Serpell, in turn, identified "a hard-nosed, economic attitude to the exploitation of domestic animals" that is contradicted by the pampering and even "trivializing" of our pets.[53] For Serpell, rather than appreciating animal others in their own right, we make them over into machines or *Ersatz* human babies. Finally, and explicitly transforming our standard methods of categorizing our anthropocentric histories, Richard Bulliet has given us the helpful, overlapping categories of "predomestic," "domestic" and "postdomestic."

In Bulliet's predomestic world, nonhuman animals were independent agents, competitors for food, killers of our own kind, teachers and guides, and prey. Bulliet is quick to point out that as some became our domesticated companions, vestiges of the older worldview were not entirely lost, even as, increasingly, domesticated animals have been reduced to resource-producing machines. In turn, as we, in our postdomestic world, have distanced ourselves from constant interaction with many different species of other animals, the affective uses we make of our pets elicit positive feelings in us, which lead some of us to question the unfeeling, uncaring use we make of factory-farmed cows, pigs and chickens. According to Bulliet, in postdomesticity, "Those who become guilt-ridden about the productive beasts we cannot humanize feel a corresponding yearning to reconnect with the wild animals that our human ways are rapidly driving to extinction."[54] Ultimately, then, just as early nineteenth-century European Romantics escaped into an idealized nature to flee from the pollution and exploitative child labor of industrialization, some of us today try to escape the growing reality of the Anthropocene – an entire planet in service to humanity alone – by finding some refuge in the domesticated and wild animals who remain.

---

[52] Clutton-Brock, Juliet. *Animals as Domesticates: A World View through History.* East Lansing: Michigan State University Press, 2012, 9, 1–3.
[53] Serpell, *In the Company of Animals*, 13, 20.
[54] Bulliet, *Hunters, Herders and Hamburgers*, 35.

In fact, are there any "wild" animals left when they must be protected from human poachers by human rangers and genetically selected for breeding purposes in zoos? Bulliet is quite correct in pointing out just how postdomestic we are as a species, but also that vestiges of our predomestic and domestic attitudes remain.

Methodologically, the very best histories of domestic animals are guided by Serpell's exploited pig to pampered pet spectrum. Thus, Alfred Crosby in his chapter on animals in *The Columbian Exchange* (1972), could recognize both the use that was made of livestock by conquistadores and their propensity to name and individualize the horses with whom they interacted and formed bonds.[55] In *Creatures of Empire*, DeJohn Anderson refined the use of this spectrum with her discussions of Algonquian *manitous*, early modern English folklore traditions attributing religious significance to robins, a deer with a red collar kept as a pet, and English colonists' use of livestock as resource machines who ultimately displaced Amerindians and their uses of land and animals. In short, the methodology of studying domestication must always look at: (a) the exploitative use of domesticated animals; and (b) the companionship we develop with some domesticates. These trends must also be studied over time, as is implicit in Bulliet's three developmental categories, and they must be related to the agency and species-specific behaviors we detect in the domesticated animals we study.

Although there is no question that domesticated animals live in a very unequal power relationship with *Homo sapiens*, they once had land to which they could escape. They could become *cimarrones*, like the wild mustangs of the western United States. It is not only a recalcitrant mule or camel who will refuse to do what we want, and sometimes other animals simply cannot do what we want because of species-specific biological realities that we now ponder changing through genetic engineering, fulfilling the mad dreams of the fictional Dr. Moreau. Historians should not shy away from the evolutionary biology that has shaped us and the other animals of our world, and Helen Cowie's *Llama*, in its review of the evolution and biology of South America's four camelids *(llamas, alpacas, guanacos and vicuñas)*, provides historians with a model of how this may be done. To see domestication and other animals as cultural constructs alone would be the ultimate postdomestic lack of appreciation of who they truly are.

4 Implication(s) of the Animal Turn

On a planet currently shaped by the whims and desires of humanity, the *animal turn* forces us to remember that we did not always live in the Anthropocene's "zoo," and within that human-dominated environment commensal raccoons and foxes, who

---

[55] Crosby, Alfred W., Jr. *The Columbian Exchange: Biological and Cultural Consequences of 1492.* Westport, CT: Greenwood, 1972, 81.

now live in our cities with pigeons and rats, remind us that there is still animal agency beyond our control.⁵⁶ Indeed, when our pets "disobey," they remind us of that as well.

What is transparently apparent is that our domesticated animals, from pigs to pets, continue to shape and define us, even as we continue to reflect on them. If not a centimeter of the planet can escape human impact in the twenty-first century, there is still some "push back" in the form of cooperation and resistance from the other sentient beings who inhabit Earth. In 1998, Sundance and Butch, porcine siblings who came to be known as the Tamworth Two, successfully escaped from a truck taking them to slaughter. After a week on the run, they were recaptured, but having won over the hearts of the British public, they were sent to a sanctuary where they both eventually died after living some thirteen years in a "luxury pen" at the Rare Breeds Centre in Kent. A human propensity to nurture that may very well have been there at the start of domestication saved Butch and Sundance from becoming nothing more than resources.⁵⁷ In our postdomestic world, all facets of *Homo sapiens*' attitudes to other animals exist simultaneously, and the *animal turn* in historical scholarship makes us remember that our very existence on this planet has always been tied to the interactions we have with the other animals we use as resources and embrace as companions. In doing this, we actually return to an earlier form of writing our history. Pliny the Elder, in his *Natural History*, contextualized his discussion of humanity by seeing us as another aspect of nature, influencing such sixteenth-century European writers as José de Acosta to do the same in their intertwining discussions of American natural and cultural history.⁵⁸ Having always interacted with the rest of nature, impacting it and being impacted by it, human history is natural history, and the *animal turn* reminds us of that.

The *animal turn* also reminds us that "the better angels of our nature" are animal. In 2013, Jane Goodall and Rebecca Atencia released the chimpanzee Wounda at the Jane Goodall Institute's Tchimpounga sanctuary. In a video that has been watched millions of times on social media, Atencia first calls Wounda over to embrace her, but the chimpanzee then goes on to embrace Goodall spontaneously.⁵⁹ There is evidence that gratitude, compassion and love are animal expressions. To

---

56 O'Connor, Trent. Commensal Species. In *Oxford Handbook of Animal Studies*, Linda Kalof (ed.), 525–541. New York, NY: Oxford University Press, 2017, 525–526, 533, 535; Francis, *Domesticated*, 78–84.
57 The Last of the Tamworth Two Pigs Dies at Age 14. *BBC News*, May 23, 2011. URL: www.bbc.com/news/uk-england-wiltshire-13503690 (November 27, 2020).
58 For references to Pliny, see Acosta, *Historia naturale*, 99–100, 216, 234. For Pliny's influence on natural history authors of the Americas, like Gonzalo Fernández de Oviedo, see Wagschal, Steven. *Minding Animals in the Old and New Worlds: A Cognitive Historical Analysis*. Toronto: University of Toronto Press, 2018, 121–132.
59 Sullivan, Ashley. Wounda: The Amazing Story of the Chimp Behind the Hug with Dr. Jane Goodall. *Jane Goodall's Good for All News*, November 21, 2017. URL: news.janegoodall.org/2017/11/21/tchimpounga-chimpanzee-of-the-month-wounda/ (November 27, 2020).

see these emotions as somehow only human or supernatural is to either minimize or multiply causality, while ignoring the sensible – the animal – evidence. From Charles Darwin who wrote of the emotions we share with other animals in *The Expression of the Emotions in Man and Animals* to our own interactions with our domesticated pets, it is evident we gain a great deal in an unequal power relationship that is nurturing stewardship. To see this as "something gained," a new connotation of the Spanish word for livestock, "*ganado*," is to go beyond mere resource-driven materialism in an *animal turn* generated from benevolence rather than brutality.

Selected Bibliography

Anderson, Virginia DeJohn. *Creatures of Empire: How Domestic Animals Transformed Early America*. Oxford: Oxford University Press, 2004.

Bulliet, Richard W. *Hunters, Herders, and Hamburgers: The Past and Future of Human-Animal Relationships*. New York, NY: Columbia University Press, 2005.

Clutton-Brock, Juliet. *A Natural History of Domesticated Mammals*. Cambridge: Cambrdige University Press, 1999.

Cowie, Helen. *Llama*. London: Reaktion Books, 2017.

Kalof, Linda, and Brigitte Resl (eds.). *A Cultural History of Animals*. 6 Volumes. Oxford: Berg, 2007.

Mikhail, Alan. *The Animal in Ottoman Egypt*. Oxford: Oxford University Press, 2014.

Mitchell, Peter. *Horse Nations: The Worldwide Impact of the Horse on Indigenous Societies Post-1492*. Oxford: Oxford University Press, 2015.

Serpell, James. *In the Company of Animals: A Study of Human-Animal Relationships*. Oxford: Basil Blackwell, 1986.

Shipman, Pat. *The Animal Connection: A New Perspective on What Makes Us Human*. New York, NY: W. W. Norton and Company, 2011.

Etienne S. Benson
Post-Domestication: The Posthuman

1 Introduction and Overview

The growing attention to animals and human-animal relations among historians that has been evident since the closing decades of the twentieth century, is just one part of a broader scholarly effort to reconsider assumptions about the centrality, distinctiveness, and superiority of humanity over other animals, which have long been taken for granted by historians and other scholars in the humanities and social sciences. Sometimes going under the name 'posthumanism', this broad-ranging reconsideration challenges not only the philosophical underpinnings of historical scholarship but also its everyday practice.[1] To meet this challenge, historians have sought to develop new methods and new sources that make it possible to write histories that are not centered on human subjectivity or human accomplishments – even if they continue to try to shed light on both. In developing these new methods and sources, they have drawn inspiration from a variety of disciplines both adjacent and distant, from anthropology to zoology. This chapter describes some of the most important of those sources of inspiration.

The term 'posthumanism' is both appropriate and problematic for describing this movement, and it has not been adopted by all practitioners of histories that seek to decenter the human. Over the past several decades, the term has appeared in a variety of contexts and with various nuances of meaning, from scholarship on cybernetics and semiotics in literary and media studies to studies of domestication in environmental history. Despite this variation, it can still serve as a useful catchall for an evolving and loosely defined set of scholarly approaches that seek to challenge the human exceptionalism at the core of historical scholarship as it has usually been practiced. As used here, the term does not refer to efforts to escape or exceed the biological human body – an idea better captured by the term 'transhumanism', which represents an elaboration on the idea of human exceptionalism rather than a challenge to it. On the contrary, the posthumanism addressed in this chapter is an effort to shake free of the conceptual shackles that have bound scholars to the idea that reason is a unique property of humanity that makes human history entirely distinct from, superior to, and independent that of other beings on Earth.

The remainder of this chapter outlines some of the key developments in several of the most important fields that have influenced practitioners of posthumanist his-

---

[1] Hayles, N. Katherine. *How We Became Posthuman: Virtual Bodies in Cybernetics, Literature, and Informatics.* Chicago, IL: University of Chicago Press, 1999; Wolfe, Cary. *What is Posthumanism.* Minneapolis, MN: University of Minnesota Press, 2010; Barad, Karen. Posthumanist Performativity: Toward an Understanding of How Matter Comes to Matter. *Signs* 28, no. 3 (March 2003): 801–831.

tory. In particular, it focuses on research in the fields of evolutionary theory, ecology, and psychology that has challenged the uniqueness, superiority, and autonomy of humanity. Rather than describing these developments solely in terms of the internal dynamics of each discipline, this chapter seeks to show how they were shaped by the changing material conditions of human-animal relations over the course of the twentieth century. Although generalizations about such conditions are difficult to make given the wide variations between and within human societies, it is nonetheless possible to identify certain trends whose effects are either global or directly relevant to scholarly production. An example of the former is the growth of capital-intensive animal agriculture and the biotechnology industry; an example of the latter is the shifting emphasis of higher education toward specialization in science, technology, engineering, and medicine and accompanying concerns about the status and role of the humanities. This chapter's concluding section shows how these developments have influenced the emergence of explicitly posthumanist approaches to history.

2 Topics and Themes

Evolutionary Entanglements

It is a matter of conventional wisdom that Charles Darwin's theory of evolution by natural selection challenged Western understanding of humanity's place in nature by showing that *Homo sapiens* emerged as a consequence of the same evolutionary processes that produced all other living beings. True in a limited sense and certainly apparent in retrospect, this claim is nonetheless misleading if it is taken to mean that the publication of Darwin's *Origin of Species* in 1859 was recognized in its own time as a watershed moment in the emergence of a perspective that challenged the superiority or uniqueness of humanity. On the contrary, Darwin's theory was quickly integrated into existing evolutionary frameworks that drew on older theories of the 'great chain of being' or 'animal series' to suggest that, whatever its origins may be, divine or otherwise, humanity was unquestionably at the pinnacle of animal life on Earth [→History of Ideas].[2] Moreover, as the ubiquity of evolutionary justifications for European imperialism and American racial discrimination that identified both 'savage races' and nonhuman animals as distinctly inferior suggests, Darwinian social-evolutionary thought expressly identified certain components of humanity as

---

[2] Lovejoy, Arthur O. *The Great Chain of Being: A Study of the History of an Idea*. Cambridge, MA: Harvard University Press, 1936; Appel, Toby A. Henri de Blainville and the Animal Series: A Nineteenth-Century Chain of Being. *Journal of the History of Biology* 13, no. 2 (1980): 291–319; Tresch, John. The Animal Series and the Genesis of Socialism. In *Of Elephants & Roses: French Natural History, 1790–1830*, Sue Ann Prince (ed.), 194–202. Philadelphia, PA: American Philosophical Society, 2013.

superior to others [→(Post)Colonial History].³ Such teleological and hierarchical thought was not especially prominent in Darwin's own writings, but nor was it entirely absent. In any case, it remained the dominant interpretation of evolutionary theory well into the twentieth century.⁴

Thus, while present-day posthuman thinkers may look back to Darwin as a source of inspiration, argumentation, and evidence, evolutionary theory in Darwin's time was not in any meaningful sense 'posthuman'. It was only with the development of the so-called 'modern synthesis' of genetics and evolutionary theory in the mid-twentieth century by biologists such as Sewall Wright, Theodosius Dobzhansky, Ernst Mayr, and George Gaylord Simpson that a real challenge to the idea of humanity as separate from and superior to the rest of nature was mounted.⁵ These evolutionary theorists did so by showing, theoretically and in some cases also experimentally, that it was possible for extraordinarily complex and adaptive biological features to emerge without the involvement of any designing hand, predetermined goal, or vital force, whether understood in secular or religious terms. Among such complex evolved features were those that supposedly distinguished humanity from other animals, including language, tool use, creativity, foresight, and the capacity for cultural transmission. According to evolutionary biologists involved in the development of the modern synthesis, all of these features were fully capable of being produced by an aimless, contingent evolutionary process that had no inherent values or end goal in sight. That humanity existed at all was a lucky accident, as was the fact that it had thrived to the extent that it could consider itself to be distinct from the rest of earthly life.

Despite its rejection of earlier forms of teleological evolutionary theory that placed humanity at the pinnacle of the hierarchy of species, even the fully contingent and mechanistic form of evolution posited by the modern synthesis allowed for the continuation of an essentially 'humanist' evolutionary narrative. In this new narrative, however, it was not the case that humanity was depicted as the naturally superior outcome of the evolutionary process. On the contrary, it was precisely through the scientific process that had led to the modern evolutionary synthesis itself that humanity had supposedly become capable of directing evolution and re-engineering life – a perspective reflected in the 'transhumanism' of Julian Huxley, who had given the 'modern synthesis' its name in a popular summary published in 1948.⁶ This conviction was only strengthened by the discovery in the early 1950s of the double-helical structure of DNA, which provided a fully mechanistic – and implicitly reengi-

---

3 Kuklick, Henrika. *The Savage Within: The Social History of British Anthropology, 1885–1945*. New York, NY: Cambridge University Press, 1991.
4 Bowler, Peter J. *Evolution: The History of an Idea*. Berkeley, CA: University of California Press, 1989.
5 Smocovitis, Vassiliki Betty. *Unifying Biology: The Evolutionary Synthesis and Evolutionary Biology*. Princeton, NJ: Princeton University Press, 1996.
6 Huxley, Julian. *Evolution: The Modern Synthesis*. London: George Allen & Unwin, 1942.

neerable – means for the faithful reproduction of genetic material.[7] In other words, it was precisely through the recognition of the fact that humanity's existence and properties were just as contingent and based in physical processes as those of other forms of life that humanity could be seen as something different from and superior to those other forms, since it had now become the only known form of life capable of recognizing and reorganizing its own contingency. Thus, a certain kind of 'humanism' or human exceptionalism – that is, the belief that humanity was unique, separate, and superior to other forms of animal life – continued to be authorized by the modern synthesis and by new discoveries in genetics, even as an older form of humanism was undermined.

Building on the discovery of DNA's double-helical structure, the development of the biotechnology industry in the last half of the twentieth century became an important site for this new form of exceptionalism to flourish. In particular, from the 1970s onwards, it linked the development of technical means for altering genetic material to the expansion of neoliberal forms of capitalism premised on the capacity of humanity to reengineer all forms and ways of life for maximum profit.[8]

It was during this period of biotechnology's explosive growth, however, that a countervailing movement developed that is perhaps the first one that truly deserves the name 'posthuman'. Building on the modern synthesis but also rejecting some of its basic premises, this alternative evolutionary 'theory' – perhaps too strong a word for a loose collection of critiques and counterexamples united by a shared sentiment – focused on symbiosis rather than competition and on the entanglement of genes and environments rather than on their separation. Lynn Margulis and James Lovelock for example, working both together and separately, proposed that symbiotic processes dominated evolution on every scale from the microscopic to the planetary.[9] Meanwhile, Richard Lewontin and others argued that genes and their environments had proved to be so closely entangled with each other that the idea that the former were adapted to the latter – as if environments were somehow outside and prior to genes – could no longer be defended.[10]

What made this countervailing research program 'posthuman' even if the term itself was not used by the scientists involved was that it undermined the claims to complete understanding and control that had been made with the help of the modern evolutionary synthesis and biotechnology, thereby removing the residual argu-

---

[7] Kay, Lily E. *Who Wrote the Book of Life? A History of the Genetic Code.* Stanford, CA: Stanford University Press, 2000.
[8] Rabinow, Paul. *Making PCR: A Story of Biotechnology.* Chicago, IL: University of Chicago Press, 1996.
[9] Lovelock, James E. *Gaia, A New Look at Life on Earth.* New York, NY: Oxford University Press, 1979; Margulis, Lynn. *Symbiosis in Cell Evolution: Life and Its Environment on the Early Earth.* San Francisco, CA: W. H. Freeman, 1981.
[10] Lewontin, Richard C. *The Triple Helix: Gene, Organism, and Environment.* Cambridge, MA: Harvard University Press, 2000.

ments for human exceptionalism that they had provided. Research on 'epigenetics', for example, suggested that it was possible for aspects of an organism's internal environment that were not encoded in genes to nonetheless be passed to its offspring in ways that influenced gene expression and helped determine the phenotype of organisms in succeeding generations. One implication was that even if biologists were able to precisely manipulate an organism's genetic code – a longstanding biotechnological dream that seemed to grow closer to reality every year, from the recombinant DNA techniques developed in the 1970s to the CRISPR gene editing method developed in the 2010s – they would still not be able to predict precisely what kinds of organisms that code would help to generate. Consequently, even if humans had developed a better understanding of the operation of processes of evolution and development, the existence of epigenetic mechanisms of inheritance meant that that understanding did not give them complete control over other forms of life, nor did it free them from being subject to the same unpredictable processes.

By the beginning of the twenty-first century, the triumphant narrative of the previous 'century of the gene' – that is, a narrative in which discoveries about the physical mechanisms of genetic inheritance had transformed Darwinian natural selection from one of many speculative evolutionary mechanisms into a complete and convincing explanation for all of life's diversity – had begun to fade, replaced by a more complex and more muddled set of incomplete and overlapping explanations.[11] What made this development an important inspiration for 'posthuman' thought was that it not only challenged the claim that the human species was the result of a teleological process that singled it out as unique and superior, as mid-century genetics and evolutionary theory had done, but also the claim that in the insights and techniques of modern science had transformed an ordinary animal – that is, humanity as the product of Darwinian natural selection working on the products of a deterministic genetic code – into something beyond or above the rest of animal life. Instead, biological research revealed a human species whose knowledge of biological processes, even if they provided powerful new tools for redirecting evolutionary processes, did not allow humanity to completely control the biological life of other species or of its own.

Ecological Interdependencies

In parallel with these developments in evolutionary theory and genetics, and sometimes (but not always) in dialog with them, were shifts in the study of ecology that similarly challenged conventional understandings of human exceptionalism. Ecolo-

---

[11] Keller, Evelyn Fox. *The Century of the Gene.* Cambridge, MA: Harvard University Press, 2000; Keller, Evelyn Fox. *The Mirage of a Space Between Nature and Nurture.* Durham, NC: Duke University Press, 2010.

gy, like evolutionary theory, did not begin its career as an inherently posthumanist discipline. On the contrary, when it was first identified as a distinct field of study in the late nineteenth century, it was seen as compatible with then-common hierarchical models of humanity's relationship to the rest of the living world. Thus, when the zoologist Karl Möbius described the 'living community' as an essential requirement for all living organisms to flourish in an 1877 report on oyster cultivation in the Baltic Sea, he was attacking the idea of the individual organism's autonomy and self-sufficiency but not the idea of an 'animal series' according to which humanity stood at the pinnacle of creation.[12] The lesson he drew from his studies was that humanity's efforts to control nature needed to be carefully tailored to the specific affordances of each place and its living community. This was a radical claim for its time, but it did not challenge the idea that humanity was distinct from and superior to the rest of nature. On the contrary, ecology at its foundations was a science that aimed to refine rather than reject the human domination of nature.

Ecology was also closely intertwined with the theory and practice of European and U.S. imperialism. This can be clearly seen in the work of the U.S. plant ecologist Frederic Clements, who built upon the foundational work of European ecologists such as Möbius and Eugen Warming to develop an elaborate and influential theoretical framework for ecology in the early twentieth century.[13] Clements' theory of ecological succession posited that there was an ideal assemblage of vegetation for each given set of climatic conditions, which he identified as the 'climax', toward which vegetation would automatically progress and where it would remain unless disturbed. The theory explained the possibility of a variety of different vegetative assemblages existing under identical climatic conditions as the result of disturbances (such as fire or flood) that shifted a particular assemblage toward an earlier stage of ecological succession. The human exceptionalism of Clements' model came from the fact that it depicted humanity not as a member of these ecological communities – just one species among many – but rather as one of the external forces capable of disturbing them. Humanity's status as an external disturbance was particular visible, according to this model, when a 'civilized' society colonized a territory that was uninhabited or inhabited only by 'primitive' humans. Thus, for Clements, one of the main functions of ecology was to provide guidance in managing the disturbances to natural ecological succession caused by human action in general and the European colonization of the Americas in particular [→American Studies]. Ecology was, in this sense, quite explicitly a tool of empire, serving to reinforce hierarchies among humans as well as between humans and other animals.

---

[12] Nyhart, Lynn K. *Modern Nature: The Rise of the Biological Perspective in Germany*. Chicago, IL: University of Chicago Press, 2009.
[13] Tobey, Ronald C. *Saving the Prairies: The Life Cycle of the Founding School of American Plant Ecology, 1895–1955*. Berkeley, CA: University of California Press, 1981; Donald Worster. *Nature's Economy: A History of Ecological Ideas*. New York, NY: Cambridge University Press, 1994.

Just as with evolutionary theory, this teleological (and colonial) understanding of ecological succession came under fire in the mid-twentieth century by theorists who emphasized contingency and adaptation in ecological processes and who sought to explain all biological phenomena in terms of physical mechanisms. In ecology, the leading theorists of this sort rallied around an approach known as 'ecosystem ecology', which emerged as one of the dominant varieties of ecology in the decades following World War II with generous support from U.S. government sponsors such as the Atomic Energy Commission.[14] Whereas Clements and other earlier ecologists had sought to demonstrate how 'living communities' developed toward end states that were predetermined by their climatic conditions, ecosystem ecologists explained stable assemblages as the result of circuits of information and control that maintained the organization of the system even as the environment around it changed. While ecosystem ecologists such as the brothers Eugene Odum and H.T. Odum still saw the maintenance of balance as essential, they rejected the idea of an orderly progression from primitive to advanced stages. Instead, they argued, an ecosystem maintained itself in a given state through its own internal organization, which determined its 'goals'. Humanity, from this perspective, was not an external disturbance but rather a species with a particularly important role in the feedback systems that maintained ecosystems.[15]

The parallels with the history of evolutionary theory continue, however. Even if ecosystem ecology dethroned a particular form of human exceptionalism – that is, the idea that humanity was a disturber of balanced ecological communities rather than a member of them – it also simultaneously enthroned a new one based not on the natural order but rather on the human capacity to understand and manipulate that order. With its functionalist analyses of the roles played by different components of the ecosystem in transforming matter and transferring energy, ecosystem ecology lent itself to engineering metaphors such as the concept of 'Spaceship Earth', which called attention to humanity's outsized demands on the planet's limited resources while holding forth the possibility of 'tuning' the Earth's performance in ways that would allow humanity to maintain its dominance (and Western populations to maintain their consumption-intensive lifestyles).[16] Just as genetic engineering seemed to make humanity exceptional by giving it the tools to take evolutionary contingency into its own hands, so did ecosystem ecology promise to allow humanity to reengineer for its own benefit the ecological systems of which it was an integral part. In both cases, human exceptionalism was simply transferred to new foundations.

---

14 Hagen, Joel B. *Entangled Bank: The Origins of Ecosystem Ecology.* New Brunswick, NJ: Rutgers University Press, 1992; Bocking, Stephen. *Ecologists and Environmental Politics: A History of Contemporary Ecology.* New Haven, CT: Yale University Press, 1997.
15 Odum, Eugene P. *Fundamentals of Ecology.* Philadelphia, PA: Saunders, 1953.
16 Taylor, Peter J. Technocratic Optimism, H. T. Odum, and the Partial Transformation of Ecological Metaphor after World War II. *Journal of the History Biology* 21 (1988): 213–244; Höhler, Sabine. *Spaceship Earth in the Environmental Age, 1960–1990.* London: Pickering & Chatto, 2015.

Fundamental to ecosystem ecology's promise was the idea that ecological systems, whether natural or influenced by human action, maintained a certain kind of dynamic equilibrium or balance based on their particular patterns of organization. It was that promise of balance that made the reengineering of earthly ecosystems seem achievable, if no doubt difficult. From the 1970s onwards, this residual 'humanism' was challenged by a shift in focus toward disequilibrium and disturbance as inherent properties of ecological systems rather than as products of occasional disasters or human interventions.[17] According to the new field of disturbance ecology, humanity was neither inherently and necessarily outside the system – a claim already convincingly challenged by ecosystem ecology – nor continently removed from the system through its growing understanding of the feedback processes that governed complex systems that included among living things. On the contrary, a new generation of ecologists began to argue, unpredictable, chaotic, and contingent effects were integral to ecological systems, regardless of whether they included humans and of whether humans had attempted to consciously manipulate them. However massive or sophisticated its influence on the earth might be, in order words, the claim was that humanity remained subject to the same ecological processes that affected other animals.

In the past several decades, these challenges to human exceptionalism in the ecological domain have been extended and intensified with the development of additional evidence and theories showing that human existence depends on myriad entanglements with other living and nonliving entities, none of which seem inherently balanced. These include the discovery of a complex 'microbiome' consisting of bacteria and other microscopic forms of life that not only survive within and upon the human body – in the gut and on the skin, for example – but also have been shown to be essential to various bodily functions, including digestion and immune system function. These discoveries make it clear that 'human nature' is something fundamentally produced by interactions with nonhumans.[18] At the same time, increasing evidence that human reproduction and other biological functions are being reshaped by industrial pollutants, including 'endocrine disruptors' that mimic endogenous hormones, has revealed both the mutability of human nature and the impossibility of completing isolating the body from potential threats.[19] In other words, human bodies are always already 'posthuman' in the sense that they are continually being remade by the organisms and chemical substances they en-

---

[17] Botkin, Daniel B. *Discordant Harmonies: A New Ecology for the Twenty-First Century.* New York, NY: Oxford University Press, 1990.
[18] Thomas, Julia Adeney. History and Biology in the Anthropocene: Problems of Scale, Problems of Value. *American Historical Review* 119 (2014): 1587–1607.
[19] Langston, Nancy. *Toxic Bodies: Hormone Disruptors and the Legacy of DES.* New Haven, CT: Yale University Press, 2010; Murphy, Michelle. *The Economization of Life.* Durham, NC: Duke University Press, 2017.

counter and incorporate. In the face of such findings, fantasies of human uniqueness, separation, and superiority have become harder to maintain.

Cognitive Relations

Late-nineteenth-century social evolutionary thinkers often attributed to the 'lower' animals capacities for cognition, emotion, and complex social relations that exceeded those of later scientists, to the extent that it might be tempting to see their work as presaging later posthumanist attempts to eliminate the divide between humans and other animals. The U.S. anthropologist Louis H. Morgan, for example, argued in the 1860s that the impressive dam-building activities of beavers required a level of foresight and cooperative planning that indicated that they possessed, if not the full range of human reasoning abilities, at least their rudiments.[20] Like the evolutionary and ecological theory of the day, however, the psychological 'posthumanism' of social-evolutionary thought was deeply compromised by its commitment to a hierarchical model of nature in which humans inevitably stood at the forefront of development. Within the framework of this model, the sometimes impressive cognitive or sensory capacities of nonhuman animals – like those of supposedly 'primitive' human groups – did nothing to challenge human (or Western) exceptionalism. However much foresight beavers might develop, thinkers of this period argued, it would always remain inferior to that of progressively advancing humanity.

In the early twentieth century, as the appeal of social-evolutionary theories faded, the apparent cognitive gap between humans and nonhuman animals was widened even further by a mechanistic turn in biology and comparative psychology that sought to belittle claims about animal intentions, experiences, and consciousness as 'subjective' and 'speculative', as opposed to the 'objective' and 'proven' results of experimental science.[21] For example, 'Morgan's Canon', attributed to the British psychologist C. Lloyd Morgan, established as a scientific principle the necessity of avoiding complex explanations for animal behavior when simpler ones would suffice, even when the behavior in question was one that in humans would be attributed to those more complex processes.[22] More broadly, the rise of behaviorism – including the evolutionarily informed version developed as 'ethology' by Konrad Lorenz, Nikolaas Tinbergen, and others – sought to eliminate from comparative psychology all references to an animal's internal psychological states, instead seeking explanations in relationships between stimuli and behaviors that were externally observable and

---

[20] Morgan, Lewis H. *The American Beaver and His Works*. Philadelphia, PA: J.B. Lippincott, 1868.
[21] Pauly, Philip J. *Controlling Life: Jacques Loeb and the Engineering Ideal in Biology*. New York, NY: Oxford University Press, 1987.
[22] Morgan, C. Lloyd. *Animal Behaviour*. London: E. Arnold, 1908.

instrumentally registerable.[23] Taken to its extreme, as it was in the work of the radical behaviorist B.F. Skinner, this approach could even be used to eliminate intentions, experiences, and consciousness from explanations of human behavior.[24] For most psychologists, however, strict behaviorism was appropriate only for nonhuman animals, whose behavior they considered to be governed largely by 'instinct'. Meanwhile, they continued to believe that more complex cognitive processes were necessary to explain the flexibility and diversity of human behavior [→History of Science].

Thus, behaviorism and the ethological study of evolved 'instincts', which were applied in their strictest forms only to nonhuman animals, did little to displace human psychological exceptionalism. A more serious challenge came with the development of systems-theoretic approaches to cognition in the mid-twentieth century, as in the efforts of mathematician Norbert Wiener to explain the behavior of animals, machines, and humans as the result of complexly organized 'cybernetic' circuits of feedback and control.[25] Cybernetics shared with behaviorism and ethology the search for mechanistic explanations of complex behaviors, but it rejected the idea that such behaviors could be adequately explained without reference to internal states. On the contrary, cybernetics assumed that the significance of a given stimulus under given conditions for a particular organism depended precisely on its internal patterns of organization, whose structure and function were amenable to experimental determination. What made cybernetics nascently 'posthuman' was that it did not assume that human cognition had unique properties that created a difference of kind between humans and nonhuman animals rather than simply a difference of degree. Ultimately, however complex they might be or become, cybernetics postulated that human cognitive systems operated according to the same principles as the simplest nonhuman systems.

Human cognitive exceptionalism also received a blow in the mid-twentieth century from new studies that showed that, while many of the late-nineteenth-century claims about animal cognition and emotion were incorrect, differences between animals and humans were much smaller than early-twentieth-century comparative psychologists had assumed in their pursuit of the simplest or 'lowest' explanations of animal behavior. Beginning in the 1950s, for example, Jane Goodall's field studies of chimpanzees demonstrated the existence of tool use, intergenerational learning, individual personality, and complex social relations that endured over time, including shifting alliances and violent intergroup conflicts.[26] These and other studies also showed that various species of nonhuman animals possessed rich communicative

---

[23] Burkhardt, Richard W., Jr. *Patterns of Behavior: Konrad Lorenz, Niko Tinbergen, and the Founding of Ethology.* Chicago, IL: University of Chicago Press, 2005.
[24] Skinner, Burrhus Frederic *The Behavior of Organisms.* New York, NY: Appleton-Century-Crofts, 1938.
[25] Wiener, Norbert. *Cybernetics.* New York, NY: J. Wiley, 1948.
[26] Van Lawick-Goodall, Jane. *Behaviour of Free-Living Chimpanzees in the Gombe Stream Reserve.* London: Baillière, Tindall & Cassell, 1968.

repertoires that, rather than being fixed by instinct and triggered automatically by certain predetermined stimuli, were adaptable to changing circumstances, much like human language.[27] At the same time, studies by psychologists showed that, under ordinary conditions, human reason was limited and biased in various ways. Amos Tversky and Daniel Kahneman's work on cognitive heuristics, for instance, revealed various 'shortcuts' used to make decisions about risks and costs that led to major deviations from the classical model of the rational actor.[28] Humanity, in other words, was much less rational than it had presumed itself to be.

Thus, by the late twentieth century, many animal behaviorists and comparative psychologists were ready to explain human and nonhuman animal behavior as the end result of the same kinds of fundamental processes, while also recognizing that many nonhuman animals demonstrated reason, tool use, communicative flexibility, and even 'culture' and that humans, like other kinds of animals, almost always deviated from the precepts of pure reason in ways both large and small. The gap between humans and other animals was also narrowed by the increasing recognition of the importance to human cognition of emotion and embodiment, two areas where the commonalities between humans and nonhuman animals had long been recognized. The work of the neuroscientist Antonio Damasio, for example, showed that bodily experiences and emotions were essential to various forms of higher human cognition.[29] In other words, it was not merely the case that human cognition was limited by various heuristics and biases, but more fundamentally that human cognition was not possible without emotions and embodiment. In these ways, human and nonhuman animal cognition were drawn ever closer together.

Since the 1980s, the distance between human and nonhuman minds has been further narrowed by models proposing that cognition is distributed between individual organisms and the other objects and organisms in their surroundings, rather than being localized within the individual. J.J. Gibson's psychology of 'affordances', for example, suggests that human cognition and action are both constrained and enabled by the possibilities offered to them by their environments, while Edwin Hutchins' model of 'distributed cognition' suggests that important aspects of the production and transformation of representations take place beyond the bodily boundaries of any one individual.[30] Both of these approaches decenter the individual organism and undermine the claim that certain species have inherent capacities that can be

---

[27] Radick, Gregory. *The Simian Tongue: The Long Debate about Animal Language*. Chicago, IL: University of Chicago Press, 2007; Munz, Tania. *The Dancing Bees: Karl von Frisch and the Discovery of the Honeybee Language*. Chicago, IL: University of Chicago Press, 2016.

[28] Kahneman, Daniel, Paul Slovic and Amos Tversky (eds.). *Judgment Under Uncertainty: Heuristics and Biases*. New York, NY: Cambridge University Press, 1982.

[29] Damasio, Antonio R. *Descartes' Error: Emotion, Reason, and the Human Brain*. New York, NY: Putnam, 1994.

[30] Gibson, James J. *The Perception of the Visual World*. Boston, MA: Houghton Mifflin, 1950; Hutchins, Edwin. *Cognition in the Wild*. Cambridge, MA: MIT Press, 1995.

deemed to be superior or inferior to those of others. Instead, they suggest, cognitive capacities are products of a set of contingent relationships that are subject to change over a range of time scales. One implication is that when humans are thinking in situations that include nonhuman animals, those animals may have profound effects on humans' cognitive processes. Thus, the capacities of humans and nonhuman animals are both determined partly by their circumstances, and human and animal cognition may become entwined with each other. Psychology of this sort leaves little room for the idea that human cognition is rational in a way that makes humans exceptional in relation to other kinds of animals.

3 Implication(s) of the Animal Turn

Much of what has been written by historians about nonhuman animals – including some of the most influential works in the field of animal history – cannot be considered 'posthuman' in the sense discussed in this chapter because they fail to decenter the human. While such histories may be radical in their consideration of subjects previously considered trivial or nonhistorical in nature, they continue to implicitly reinforce human exceptionalism inasmuch as they depict humanity as unique and superior to other animals and as the only true subject of history. Much of the work in the cultural history of animals falls into this category [→Cultural History]. Often bold in its willingness to take human-animal relations seriously as aspects of human history, such scholarship is just as often conventional in its assumption that only humans are capable of generating, understanding, and representing historical change – even if that change is now understood to include nonhuman animals. In work of this kind, animals commonly become 'mirrors of' or 'windows onto' human history rather than historical subjects in their own right.[31] Again, while this represents an important step forward in relation to scholarship that ignores animals entirely or sees them as nothing but material resources available for human exploitation, it falls short of a truly posthuman history that contributes to the dismantling of the exceptional human subject.

One of the reasons why such scholarship continues to reinforce human exceptionalism, sometimes despite its own explicitly professed aim to do otherwise, is that it relies on conventional sources for human history and interprets them in ways that privilege human language, reason, and experience. To use the metaphors of the mirror and window in another way, such works assume that historical animal lives only become visible as reflected in or seen through human eyes – an assumption that raises insurmountable challenges to any attempt to recount a history that is not fundamentally anthropocentric, inasmuch as it transforms potential evidence for

---

31 Mullin, Molly H. Mirrors and Windows: Sociocultural Studies of Human-Animal Relationships. *Annual Review of Anthropology* 28 (1999): 201–224.

how nonhuman animals have shaped and experienced historical change into evidence of nothing more than how humans have viewed animals.[32] This approach of historical scholarship on animals depends on a constricted understanding of what counts as a historical source – namely, textual or nontextual documents that were produced by humans for human purposes and that are seen as reflecting solely human intentions and understandings. It becomes possible to escape the trap of anthropocentric historical scholarship only when the definition of a legitimate historical document is broadened to include material traces of past animal action and when even human-generated documents are understood to arise from embodied interactions between human and nonhuman animals [→Pre-Domestication: Zooarchaeology; →Material Culture Studies].[33]

In addition to requiring a broadened view of historical sources and their interpretation, posthuman history requires adopting a model of historical agency that no longer depends on capacities believed to be unique to the human. While there is a very large body of historical scholarship on the question of whether animals can be said to have agency, much of that scholarship is limited by the fact that it hews to a conventionally 'humanist' definition of agency, according to which history can only be shaped by self-determining, self-conscious agents who are capable of expressing their intentions and experiences in human language [→Social History]. For those who embrace this definition of agency, the debate over animal agency becomes a debate over whether animals of various kinds possess those properties in sufficient degree to qualify as historical agents – a debate which largely consists of historians arguing with each other in inconclusive ways about their disparate interpretations of the shifting findings of biologists. When agency is defined in ways that do not depend on properties unique to the human, however – for instance, as the capacity to change the world through embodied, mutually transformative interactions with other beings who are also pursuing their own desires and interests – new possibilities for posthuman history open up that productively sidestep much of the longstanding debate over 'animal agency'.[34]

Posthuman history also involves a reconsideration of what it means to be human and what kinds of histories humans can make. The developments in the natural sciences described earlier in this chapter provide a set of resources for reconsidering an

---

[32] Fudge, Erica. A Left-Handed Blow: Writing the History of Animals. In *Representing Animals*, Nigel Rothfels (ed.), 3–18. Bloomington, IN: Indiana University Press, 2002.
[33] Benson, Etienne S. Animal Writes: Historiography, Disciplinarity, and the Animal Trace. In *Making Animal Meaning*, Linda Kalof, Georgina Montgomery (eds.), 3–16. East Lansing, MI: Michigan State University Press, 2011; Tortorici, Zeb. Animal Archive Stories: Species Anxieties in the Mexican National Archive. In *The Historical Animal*, Susan Nance (ed.), 75–98. Syracuse, NY: Syracuse University Press, 2015.
[34] Nash, Linda. The Agency of Nature or the Nature of Agency. *Environmental History* 10 (2005): 67–69; Rees, Amanda. Animal Agents. Historiography, Theory and the History of Science in the Anthropocene. *The British Journal for the History of Science Themes* 2 (2017): 1–10.

image of the human inherited from the Enlightenment – one in which reason is identified as a property that is unique to humanity among the many species in the living world and that makes it possible for secular history to have some kind of transcendent meaning that ultimately justifies historical scholarship.[35] If, on the contrary, humanity is caught up in the same evolutionary, ecological, and cognitive webs as all other species, such that its capacities are always fundamentally relational and contingent, then historical narratives predicated on a fixed vision of what makes humanity distinctive from and independent of all other living beings are no longer tenable.

Ultimately, therefore, the move toward a posthuman history is about much more than simply incorporating nonhuman animals into conventional histories. It involves fundamentally rethinking what kinds of sources, methods, and subjects historical scholarship is capable of considering, so that the past in all of its complexity falls within the remit of the historian – that is, a past that cannot be reduced to human representations, intentions, and actions but instead includes the full panoply of interactions among living beings through time. Such a history draws on the insights of the natural sciences, including the biological, ecological, and psychological challenges to human exceptionalism described above, without seeking to replicate the reductionist and universalist explanatory strategies commonly pursued by those sciences. On the contrary, posthuman history continues to draw on history's traditional strengths – its sensitivity to contingency and to the specificities of time and place, its capacity to rigorously explore speculative possibilities in the face of profound evidentiary gaps, and its ability to craft compelling narratives from the overwhelming complexity of the past – even while abandoning history's traditional assumption that only humans can be the proper subjects and agents of history.

Selected Bibliography

Barad, Karen. Posthumanist Performativity: Toward an Understanding of How Matter Comes to Matter. *Signs* 28, no. 3 (March 2003): 801–831.
Emmett, Robert S. and David E. Nye. *The Environmental Humanities: A Critical Introduction.* Cambridge, MA: MIT Press, 2017.
Haraway, Donna. *When Species Meet.* Minneapolis, MN: University of Minnesota Press, 2008.
Kirksey, Eben (ed.). *The Multispecies Salon.* Durham, NC: Duke University Press, 2014.
LeCain, Tim. *The Matter of History: How Things Create the Past.* Cambridge: Cambridge University Press, 2017.
Nance, Susan (ed.) *The Historical Animal.* Syracuse, NY: Syracuse University Press, 2015.
Wolfe, Cary. *What is Posthumanism?* Minneapolis, MN: University of Minnesota Press, 2010.

---

35 Smail, Daniel Lord. *On Deep History and the Brain.* Berkeley, CA: University of California Press, 2008.

Part II: **Regional Approaches**

Andreas Hübner
American Studies

1 Introduction and Overview

In 1980, John Berger, the English cultural thinker, published an essay entitled "Why Look at Animals?"[1] Dedicated to Gilles Aillaud, a French painter recognized for his portraits of animals confined in zoos, Berger's essay provided a critique of capitalism through the lens of human–animal relations. "Why Look at Animals?" reached beyond animal representations, projections, and symbolisms and instead explored places and spaces designated as belonging to animals in modernity. To Berger, zoos represented one such place, a place of confinement and marginalization; elsewhere, he noted, "animals disappear. In zoos they constitute the living monument to their own disappearance."[2] Berger did not stop at the zoo gates; he produced an irredeemable vision of modern capitalist society and engaged in a critique of corporate capitalism, the seeds of which, he thought, had been planted in western Europe and North America. His criticism of zoos, hence, strongly resonated with a critique of American capitalist culture and society.[3] Berger expressed that the "marginalization of animals is today being followed by the marginalization and disposal of the only class who, throughout history, has remained familiar with animals and maintained the wisdom which accompanies that familiarity: the middle and small peasant."[4]

In American studies, animals have appeared everywhere in recent years, as have studies that trace the shape of human–animal relations from a historical perspective. In the 1980s, historical animal studies emerged as a subfield of social, cultural, and environmental history [→Social History; →Cultural History; →Environmental History]. Ever since, the field has taken a transcultural, transatlantic, and transdisciplinary path across academia and society. The scholarship in this area is now abundant. Any overview of historical animal studies faces a serious challenge: A process of canon formation has not been initiated, and scholars have not agreed on a collective set of approaches, methods, and theories.[5] If forced to produce a list of influential

---

[1] Berger, John. *Why Look at Animals?* London: Penguin Books, 2009. Written in 1977, first published in 1980: Berger, John. Why Look at Animals? In *About Looking*, John Berger (ed.), 3–28. New York, NY: Pantheon Books, 1980.
[2] Berger, Why Look at Animals?, 36.
[3] Cf. Berger, Why Look at Animals?, 12.
[4] Berger, Why Look at Animals?, 36.
[5] Cf. Marvin, Garry and Susan McHugh. In it together: An Introduction to Human-Animal Studies. In *Routledge Handbook of Human-Animal Studies*, Garry Marvin, Susan McHugh (eds.), 1–9, 3–5. London: Routledge, 2014; Nance, Susan. Introduction. In *The Historical Animal*, Susan Nance (ed.), 1–16. Syracuse, NY: Syracuse University Press, 2015, 5–7.

https://doi.org/10.1515/9783110536553-008

works, many present a plethora of authors. Virginia DeJohn Anderson, Andrew Isenberg, and Ann Norton Greene – to identify just three – are frequently mentioned.

Anderson, Isenberg, and Greene represent the multitude of scholars who are thoroughly engaged in historical animal studies. In *Creatures of Empire*, Anderson delineates human–animal relations in colonial New England and the Chesapeake region and uncovers different concepts of animals, property rights, and land use among settlers and Native American groups in the seventeenth century.[6] Thus, *Creatures of Empire* places historical animal studies at the heart of American history. Isenberg does likewise.[7] His *Destruction of the Bison* intertwines approaches from various disciplines to discuss the near-extermination of the American bison in the Great Plains. Isenberg not only succeeds in connecting human–animal studies to the history of the American West, but he also offers a compelling case for the significance of human–animal relations in the long nineteenth century. With *Horses at Work*, Greene rewrites the history of industrialization and reminds us of the continuing relevance of animals in modernity. Recognizing animals as sites of social change, she notes that "horses, not steam engines, established the material environment and cultural values that have shaped energy use in the twentieth century."[8]

What follows will pick up the scent of Anderson, Isenberg, and Greene, arguing that the study of human–animal relations is key to our understanding of American history since the early 1600s. In their works, these scholars connect animal histories of bison, horses, and domesticated livestock to overarching topics and themes of American history, such as settler colonialism, the "age of extermination," and the periods of industrialization and modernization. In this respect, they belong to a set of scholars who introduce animal-related topics "as a part of the general history of a given time and place rather than isolating them in peripheral, or even antiquarian, sub-fields."[9]

Notwithstanding, neither Anderson, nor Isenberg and Greene can be characterized as scholars of animal history by profession; rather, their scholarship is informed by cultural, ecological, and environmental history. Still, their explorations are enriched by a keen interest in human–animal studies. The subsequent ideas will show that their work reflects the diversity of methods and approaches applied to dissecting human–animal sociality. In conclusion, this chapter will discuss the impact of the *animal turn* in American studies. Given the rise of the animal rights movement

---

6 Cf. Anderson, Virginia DeJohn. *Creatures of Empire: How Domestic Animals Transformed Early America*. New York, NY: Oxford University Press, 2004.
7 Cf. Isenberg, Andrew. *The Destruction of the Bison: An Environmental History, 1750–1920*. New York, NY: Cambridge University Press, 2000.
8 Greene, Ann Norton. *Horses at Work: Harnessing Power in Industrial America*. Cambridge, MA: Harvard University Press, 2008, 9.
9 Ritvo, Harriet. History and Animal Studies. *Society & Animals* 10, no. 4 (January 2002): 403–406, 405.

in the 1970s, any such discussion needs to observe the shifting attitudes of historians toward the study of animals.[10]

2 Topics and Themes

Daniel Gookin, the seventeenth-century Massachusetts overseer of Indian affairs, is nowadays often overlooked in the histories of colonial New England. If at all, the colonial magistrate is recognized for his *Historical Collections of the Indians in New England* (1674).[11] To the scholar of animal history, however, these *Historical Collections* constitute a valuable source of information and allude to a key topic in the study of early human–animal relations on the North American continent: Native Americans and Europeans held different conceptions of human and animal worlds. Whereas colonial settlers categorically regarded all nonhumans as animals and "relegated animals to a purely physical existence,"[12] Native American traditions encompassed animals as distinct species and placed them at the core of indigenous epistemologies, spirituality, and culture.[13] Any conflict between Native Americans and colonial settlers has to be read against this backdrop, whether it be disputes over livestock, land and trespass disputes, or other legal disputes.

In the field of American history, the formation of native/settler–animal sociality remains a central theme. Again, Gookin's *Historical Collections* offer a point of departure. Upon visiting Hassanamesitt, a "Praying Indian" town in the Boston area, Gookin commented on the Indian way of living: "[It] is by husbandry, and keeping cattle and swine; wherein they do as well, or rather better, than any other Indians, but yet are very far short of the English both in diligence and providence."[14] Historian David Silverman has used this account to help decipher Native Americans' cultural practices. Native Americans, Silverman remarks, "selectively borrowed European cultural practices by carefully weighing outside demands against their own needs and priorities."[15] Regardless of Gookin's pejorative assessment of these practices, it becomes apparent that Native American groups reshaped conceptions of human–animal rela-

---

10 Cf. Ritvo, History and Animal Studies, 404. In this spirit, historian Erica Fudge (A Left-Handed Blow: Writing the History of Animals. In *Representing Animals*, Nigel Rothfels (ed.), 3–18. Bloomington, IN: Indiana University Press, 2002, 6.) once declared that the history of animals is essentially "the history of human attitudes to animals".
11 Cf. Silverman, David J. "We Chuse to Be Bounded": Native American Animal Husbandry in Colonial New England. *William and Mary Quarterly* 60, no. 3 (July 2003): 511–548.
12 Anderson, *Creatures of Empire*, 18.
13 Cf. McHugh, Susan and Wendy Woodward. Introduction. In *Indigenous Creatures, Native Knowledges, and the Arts: Animal Studies in Modern Worlds*, Susan McHugh, Wendy Woodward (eds.), 1–9. Cham: Palgrave MacMillan, 2017, 2. Also cf. Anderson, *Creatures of Empire*, 18.
14 Gookin, Daniel. *Historical Collections of the Indians in New England [...]*. Ann Arbor, MI: Text Creation Partnership, 2004–12, 45, URL: name.umdl.umich.edu/N18748.0001.001 (October 4, 2020).
15 Silverman, We Chuse to Be Bounded, 547.

tions, in practicing both husbandry and Christianity, to counter European colonization.

Any discussion of the colonial period also needs to address the "quintessential American epic": The impact of invasive species on Native American ways of living and environments. The most prominent and debated example of this is the history of Plains Indians and the horses that were introduced to the Americas by Spanish conquistadores.[16] Historian Pekka Hämäläinen has long defined Plains horse culture as a "mixed blessing." Instead of presenting the horse era as the anomaly of ecological imperialism and celebrating Indian equestrianism, as scholars have tended to do in the past, Hämäläinen considers the full complexity of the transformational power of horses in the Great Plains. In a statement that holds true for the impact of any invasive species on the continent, he writes: "[Horses] disrupted subsistence economies, wrecked grassland and bison ecologies, created social inequalities, unhinged gender relations, undermined traditional political hierarchies, and intensified resource competition."[17]

The Plains region is central to another theme of historical animal studies: The "age of extermination." The nineteenth century saw the disappearance of many species, resulting in a hitherto unknown transformation of the American environment. The poster child of this era is the bison, whose once thriving populations were almost extinct by 1890.[18] Scholars have also examined the excessive hunting of other species, most importantly of fur-bearing animals, plume birds, passenger pigeons, and the whale populations in the Atlantic and Pacific Oceans.[19] In their analyses, historians have stressed the killing of animals for profit and emphasized that contemporaries often understood the extinction of species as a means of paving the way for improvement and civilization. On the other hand, it is now well established

---

16 Cf. Hämäläinen, Pekka. The Rise and Fall of Plains Indian Horse Cultures. *Journal of American History* 90, no. 3 (December 2003): 833–862, 833.
17 Hämäläinen, The Rise and Fall, 834.
18 Cf. Hornaday, William T. *The Extermination of the American Bison*. Washington, DC: Government Printing Office, 1889.
19 Cf. Foster, John Wilson. *Pilgrims of the Air: The Story of the Passenger Pigeon*. London: Notting Hill Editions, 2013; Davis, Lance E. and Robert E. Gallman. American Whaling, 1820–1900: Dominance and Decline. In *Whaling and History: Perspectives on the Evolution of the Industry*, Bjørn L. Blasberg, Jan Erik Ringstad, Einar Wexelesen (eds.), 55–65. Sandefjord: Sandefjordmuseene 1993; Davis, Lance E., Robert E. Gallman, and Teresa D. Hutchins. The Decline of U.S. Whaling: Was the Stock of Whales Running Out? *The Business History Review* 62, no. 4 (Winter 1988): 569–595; Minichiello, J. Kent. The Audubon Movement: Its Origins, its Conservation Context, and its Initial Accomplishments. *Journal of the Washington Academy of Sciences* 90, no. 2 (Summer 2004): 30–44; Cf. Richards, John F. *The Unending Frontier: An Environmental History of the Early Modern World*. Berkeley, CA: University of California Press, 2005, ch. 13.

that the near-extinction of bison and plume birds helped spark the American Conservation Movement that began to develop in the 1850s.[20]

The American Conservation Movement reflected the philanthropic tendencies of the Progressive era and responded to the challenges of industrialization and modernization in the late nineteenth century. With social stability at stake, conservationists – and preservationists – turned to flora and fauna to redefine the American nation. Most prominently, historian Frederick Jackson Turner would place "winning a wilderness" at the center of American democracy and culture in his essay "The Significance of the Frontier in American Society."[21] With the closing of the frontier, however, America was abruptly facing an existential identity crisis. In reaction, intellectuals, scholars, and politicians, among them Theodore Roosevelt, began advocating conservationist and preservationist policies. Ever since, institutions, organizations, and other bodies that were formed as a result, such as the US National Park Service, have shaped the understanding of human-environment relations in North America. In recent years, scholars of animal history have paid much attention to this period: Historians have discussed the impact of naturalist thinkers such as Ralph Waldo Emerson, Henry David Thoreau and John Muir on the transformation of human–animal relations. Considering naturalist observations, historians have also revised long-lasting narratives about wilderness and civilization, and they have deconstructed conceptions that categorize nature as stable and static.[22]

Scholars have come to understand human–animal relations as central to numerous (master) narratives of American history. Critically decoding foundational myths, for example, the Frontier and the West, they have reassessed the ways in which societies speak and think about animals, thus contesting the language of animality and human–animal difference that marked westward expansion from the 1600s onward.[23] In this sense, historians such as Dominik Ohrem have stressed the interdependence of human and nonhuman agency in American history and emphasized that "[even] the many historical manifestations of anthropocentrism and human ex-

---

20 Vermont Congressman George P. Marsh first addressed the destructive force of humans on the North American continent in 1847. The American Bison Society, founded in 1905 by William Temple Hornaday (see footnote 18), represents the American Conservation Movement until today, cf. Isenberg, *The Destruction of the Bison*, 4; and cf. Marsh, George P. *Address Delivered before the Agricultural Society of Rutland County, September 30, 1847.* Rutland, VT: Herald Office, 1848.
21 Turner, Frederick J. The Significance of the Frontier in American Society. In *Annual Report of the American Historical Society for the Year 1893.* Washington, DC: Government Printing Office, 1894, 199–227.
22 Cf. Isenberg, The Destruction of the Bison, 12; Taylor, Dorceta E. *The Rise of the American Conservation Movement: Power, Privilege, and Environmental Protection.* Durham, NC: Duke University Press, 2016.
23 Cf. Ohrem, Dominik. The Ends of Man: The Zooanthropological Imaginary and the Animal Geographies of Westward Expansion in Antebellum America. In *American Beasts: Perspectives on Animals, Animality and U.S. Culture, 1776–1920*, Dominik Ohrem (ed.), 245–278. Berlin: Neofelis, 2017, 249.

ceptionalism remain unwillingly but inevitably expressive of a human dependence on animal life."[24]

The remaking of the American West and the near-extermination of bison was but a prelude to the impact of the livestock industry that would slowly transform the Great Plains in the second half of the nineteenth century and the early twentieth century. As bison disappeared from the grasslands of the Plains, the introduction of cattle and pastures gave rise to new animal spaces. "Long drives in Texas, ranches in Wyoming, cattle towns in Kansas, feedlots in Illinois" would soon dominate the Midwest.[25] During these years, Chicago, as many scholars have illustrated, became the nation's slaughterhouse. Here, butchers and meatpackers optimized divisions of human labor in the slaughtering process, promoted mechanization in the meat industry, and overcame problems in refrigeration to successfully market beef consumption in America. The result was what Upton Sinclair criticized in *The Jungle:* The beginning of the animal–industrial complex that forever changed human–animal landscapes [→History of Animal Slaughter].[26]

Scholars have long explored human–animal landscapes and urban–rural understandings of human–animal sociality in the nineteenth and twentieth centuries. They have studied the rise of pets in the city, the contributions of non-domesticated mammals to the urban landscape, and the "replacement" of working animals with technology and machinery in rural and urban spaces [→Rural and Urban Studies].[27]

Of particular interest have been the interconnections of animal and children's welfare organizations. As "dependent beings," animals and children came to be protected from cruelty in Gilded Age America. Both were regarded as "speechless," a trope that "stood for the inability to act physically, legally, and politically on one's own behalf."[28] Historically, animal welfare initiatives can be traced to the colonial period when anti-cruelty laws were initially enacted in the Massachusetts Bay Colony

---

[24] Ohrem, Dominik. A Declaration of Interdependence: American History and the Challenges of Postanthropocentric Historiography. In *American Beasts: Perspectives on Animals, Animality and U.S. Culture, 1776–1920*, Dominik Ohrem (ed.), 9–48. Berlin: Neofelis, 2017, 22.

[25] Cronon, William. *Nature's Metropolis: Chicago and the Great West.* New York, NY: Norton, 1992, 224.

[26] Cf. Sinclair, Upton. *The Jungle.* New York, NY: Penguin Books, 2006. First published in *Appeal to Reason* 482–518 (1905). Coined by Barbara Noske in 1989, the term animal-industrial complex draws attention to the industrialized, institutionalized and commodified exploitation of animals, cf. Noske, Barbara. *Humans and Other Animals: Beyond the Boundaries of Anthropology.* London: Pluto, 1989; Twine, Richard. Revealing the Animal-Industrial Complex: A Concept & Method for Critical Animal Studies. *Journal for Critical Animal Studies* 10, no. 1 (2012): 12–39.

[27] Cf. Benson, Etienne. The Urbanization of the Eastern Gray Squirrel in the United States. *Journal of American History* 100, no. 3 (December 2013): 691–710; Norton, *Horses at Work;* Wilcox, Sharon and Stephanie Rutherford (eds.). *Historical Animal Geographies.* New York, NY: Routledge, 2018.

[28] Pearson, Susan J. *The Rights of the Defenseless: Protection Animals and Children in Gilded Age America.* Chicago, IL: University of Chicago Press, 2011, 23.

in the mid-1600s.²⁹ In 1867, the State of New York passed the first modern anti-cruelty statute. Supported by the American Society for the Prevention of Cruelty to Animals (ASPCA), all states had adopted anti-cruelty legislation by the end of the nineteenth century.³⁰ Historians Susan Pearson and Kimberly Smith have shown that such anti-cruelty legislation largely contributed to the creation of humane societies. Devoted to the welfare of animals and children, these societies laid the groundwork "for later welfare regimes."³¹

Given the series of global conflicts in the nineteenth and twentieth century, scholars of animal history have also turned their attention to the Civil War, the World Wars and the complexities of the Cold War [→History of War]. At their best, these studies come to see the full ambiguity of animal representations and subjectivity in conflict-ridden societies. Once again, horses have received most academic attention in recent years; yet, it has not gone unnoticed that other animals also played central roles.³²

In Civil War America, the envelopes of the soldiers' letters of the US Christian Commission carried a stamp that displayed carrier pigeons, thus portraying the importance of "pigeon messages" to front-line communication.³³ In post-war America, research in dolphin communication and cognition uncovered the "curious double legacy of the modern bottlenose," as both the hippie movement and the US Navy claimed the marine mammal for their objectives. Whereas military scientists and commanders worked to train dolphins for undersea combat, the flower children recognized, as historian Graham Burnett maintains, that "the *Tursiops truncatus* was an erotically liberated, spiritually profound pacifist, intent on saving humans from their materialistic, violent, and repressive lives."³⁴

---

29 Cf. Massachusetts Historical Society (ed.). *Collections of the Massachusetts Historical Society*. Boston, MA: Little and Brown, 1843, 232: A Coppie of the Liberties of the Massachusets Collonie in New England, §92: "No man shall exercise any Tirrany or Crueltie towards any bruite Creature which are usuallie kept for man's use."
30 Cf. Pearson, Susan J. and Kimberly K. Smith. Developing the Animal Welfare State. In *Statebuilding from the Margins: Between Reconstruction and the New Deal*, Carol Nackenoff, Julie Novkov (eds.), 118–139. Philadelphia, PA: University of Pennsylvania Press, 2014, 130.
31 Pearson and Smith, Developing the Animal Welfare State, 119.
32 In their research, many scholars have also explored the history of dogs, for instance cf. Wang, Jessica. Dogs and the Making of the American State: Voluntary Association, State Power, and the Politics of Animal Control in New York City, 1850–1920. *Journal of American History* 98, no. 4 (March 2012): 998–1024.
33 Cf. "Civil War envelope for U.S. Christian Commission showing carrier pigeon with letter," ca. 1861–1865, Liljenquist Family Collection of Civil War Photographs: Civil War Era Photographs, Manuscripts, Ephemera, and Miscellaneous Objects in the Liljenquist Collection, Library of Congress Prints and Photographs Division Washington, DC, Call Number: LOT 14043–6, no. 2.
34 Burnett, David Graham. A Mind in the Water: The Dolphin as Our Beast of Burden. *Orion Magazine* 29, no. 3 (May/June 2010): 38–51, 49. See also Burnett, David Graham. *The Sounding of the Whale: Science and Cetaceans in the Twentieth Century*. Chicago, IL: University of Chicago Press, 2012; Reiss,

At present, human–animal studies resound in every corner of American history; rather than limit their research to the themes and topics outlined above, scholars of animal history now strive to explore all aspects of human–animal interactions on the North American continent since the early 1500s. At the same time, new methods and approaches have taken the stage: What could be considered a history of human ideas about animals in the past has become the field of historical animal studies [→History of Ideas].

3 Methods and Approaches

Historical animal studies have, since their gradual emergence, taken a transcultural, transatlantic, and transdisciplinary path. Scholars of animal history have exchanged ideas and experiences across cultural, spatial, and disciplinary boundaries with cultural thinkers, artists, and researchers from the humanities, social sciences, and natural sciences. The current cultivation of the field within American studies owes much to the groundbreaking contributions of cultural, ecological, and environmental history as well as political activism and societal movements that can be traced back to the early 1960s. In particular, the global environmental movement and the animal rights movement have influenced many historians interested in human–animal relations and introduced ideas of nonhuman emancipation and posthumanism to the field of human–animal studies.

Rachel Carson – marine biologist, ecologist, and environmental activist – is often credited with launching the modern global environmental movement. In 1963, Carson published *Silent Spring*, a book that criticized the excessive use of pesticides. With *Silent Spring*, Carson raised awareness of the dangers lurking behind the employment of DDT (Dichlorodiphenyltrichloroethane) and other insecticides in agriculture and ignited a debate that would shatter the environmental status quo in America. Almost immediately, President John F. Kennedy initiated an investigation into the use of pesticides. Carson was asked to testify before the Senate. During the hearings, she not only lamented the environmental, pesticide-driven pollution of water, soil, air, and vegetation, but also condemned the "contamination of various kinds" that have "even penetrated that internal environment within the bodies of animals and of men."[35]

Diana. *The Dolphin in the Mirror: Exploring Dolphin Minds and Saving Dolphin Lives*. Boston, MA: Houghton Mifflin Harcourt, 2011.

35 Carson, Rachel. Environmental Hazards Control of Pesticides and Other Chemical Poisons, Statement before the Subcommittee on Reorganization and International Organizations of the Committee on Government Operations, Hearing on S. Res. 27, 88th Cong. 1st Sess., June 4, 1963, 206–219, cited in Edward P. Weber (ed.), *Endangered Species: A Documentary and Reference Guide*. Santa Barbara, CA: ABC-Clio, 2016, 100–104, 101. In the long run, the resultant "pesticide report" of the President's Science Advisory Committee brought about two considerable changes in environmental politics: The in-

Writing about the "silencing of spring," Carson further turned her attention to human–animal relations. In a general sense, scholars have shown that she accentuated humans' "moral responsibilities to, and the moral standing of animals."[36] This call for moral values and courage did not constitute an end in itself. To Carson, the absence and extinction of animals served as a prelude to the silencing of ecosystems by chemical contamination. The muting of animals indicated that environmental destruction affected flora and fauna alike.[37] Thus, in Carson's writing, the absence of animals and their silence mark a historical change in ecosystems and the environment. It is this finding that has attracted the interest of scholars of ecological, environmental, and animal history and shaped scholars' thinking about concepts and methods of environmental and human–animal studies.

The "use of animals as indicators" of historical change distinguishes the work of Alfred Crosby, a pioneer in ecological and environmental history whose research was closely connected to the political and intellectual debates sparked by Carson. In the seminal *Columbian Exchange*, published a decade after *Silent Spring*, Crosby explores the impact of the European conquest in the Americas and examines the exchange of plants, animals, and microbes between the continents since 1492. He places animals (and microbes) in the broader context of colonization and ecological imperialism.[38] For Crosby, the success of the early colonizers and conquistadores was essentially tied to their animals: "The migrant Europeans could reach and even conquer, but not make colonies of settlement of these pieces of alien earth until they became a good deal more like Europe [...]. Fortunately for the Europeans, their domesticated and lithely adaptable animals were very effective at initiating that change."[39] Moreover, Crosby anticipates some key ideas of Global and Atlantic history: In *Columbian Exchange*, animals play a central role in the circulation of people, knowledge, capital, goods, and ideas across and around the Atlantic Ocean.[40] The introduction of oxen, for instance, brought about the plough and the farmer, varieties of wheat and barley, and the concepts of extensive agriculture and agro-capitalism.[41] Yet in contrast to later studies of human–animal interactions, Crosby rarely considers animals as historical agents. Animals remain biological entities that simply reflect European colonialism and imperialism.

---

troduction of the Chemical Pesticides Coordination Act and the formation of the Environmental Protection Agency (EPA).
36 Bekoff, Marc and Jan Nystrom. The Other Side of Silence: Rachel Carson's Views of Animals. *Human Ecology Review* 11, no. 2 (December 2004): 186–200, 186–187.
37 Cf. Bekoff and Nystrom, The Other Side of Silence, 187.
38 Cf. Crosby, Alfred. *The Columbian Exchange: Biological and Cultural Consequences of 1492*. Westport, CT: Greenwood, 1972; Crosby, Alfred. *Ecological Imperialism: The Biological Expansion of Europe, 900–1900*. Cambridge: Cambridge University Press, [1986] 2004.
39 Crosby, *Ecological Imperialism*, 172.
40 Cf. Games, Alison. Atlantic History: Definitions, Challenges, and Opportunities. *American Historical Review* 111, no. 3 (June 2006): 741–757.
41 Cf. Crosby, *The Columbian Exchange*, 110.

In American studies, the idea of animals as agents of historical change first emerged in the studies of another ecological and environmental historian, William Cronon. With *Changes in the Land*, Cronon produced a monograph that added plants and animals to the master narrative of colonial America. Cronon understood native–settler relations to function within an interacting, ecological system, and he criticized Crosby for overemphasizing the impact of species transfer. Cronon stressed the significance of human–animal networks across the Atlantic Ocean: "Important as organisms like smallpox, the horse, and the pig were in their direct impact on American ecosystems, their full effect becomes visible only when they are treated as integral elements in a complex system of environmental and cultural relationships."[42]

Cronon can be seen as a forerunner of environmental history and historical animal studies. Concepts such as relational and entangled animal agency, now well established, are inherent in his work. In 2004, Virginia DeJohn Anderson incorporated Cronon's ideas into *Creatures of Empire*. Starting from the premise that domesticated and non-domesticated animals were never fully under human control in colonial New England, she revises the notion that animals were solely looked upon as objects and the property of settlers. Instead, she notes that animals "became a reliable indicator of the tenor" of native–settler relations.[43] In Anderson's understanding, animals caused changes to the land and people's behavior, and animal activities shaped the lives of natives and settlers alike. As a result, natives regarded domestic animals as "being agents of English imperial dominion." Despite this finding, Anderson stops short of approaching the idea of animal agency, identifying livestock as "other kinds of actors […] driven by instinct rather than reason."[44]

Concepts of animal agency would surface in the context of the modern animal rights movement of the 1970s and 1980s. Stimulated by scholars and activists such as Peter Singer, Tom Regan, and Roslind and Stanley Godlovitch who led the "Oxford Group," the animal rights movement adopted traditional animal welfare positions and criticized animal use as morally unjustifiable.[45] Singer, an Australian philosopher who limits his criticism of animal exploitation to a utilitarian perspective, introduced elements of poststructuralism to the debate and therefore enabled "discussions of animals as agents who are not just humanlike subjects or thing-like objects, but actors of a different order."[46] In American studies, some scholars have promoted a critical animal studies perspective and they have integrated animal-

---

42 Cronon, William. *Changes in the Land: Indians, Colonists, and the Ecology of New England.* New York, NY: Hill and Wang, [1983] 2003, 14.
43 Anderson, *Creatures of Empire*, 5.
44 Anderson, *Creatures of Empire*, 3, 230.
45 Cf. Singer, Peter. *Animal Liberation: A New Ethics for Our Treatment of Animals.* New York, NY: Avon Books, 1975; Regan, Tom. *The Case for Animal Rights.* Berkeley, CA: University of California Press, 1983.
46 Marvin and McHugh, In it together, 5.

rights positions and approaches into their work – and scholars have examined the institutions responsible for animal cruelty, experimentation, and exploitation. Only recently, Daniel Bender, a cultural historian of animal studies, has explored the conflicted history of zoos in the United States, dismantled the global politics, economies, empires, and wars key to any historical analysis of institutionalized animal confinement, and argued that animals on display dramatically influenced society's understanding of "faraway places, environments, and peoples."[47] Bender, of course, is by no means bound to the methods and approaches of the 1970s and 1980s. Much of his work is informed by the *animal turn* and responds to Harriet Ritvo's call for challenging settled assumptions about human–animal relations. Like many of his fellow scholars, Bender thoroughly deconstructs histories and narratives of human–animal interactions in American history.[48]

4 Implication(s) of the Animal Turn

In recent decades, the *animal turn* has made an impact within most disciplines of the humanities, the social sciences, and the natural sciences. American studies is no exception to this trend. Since the 1980s, the field has encompassed animals and taken on the deconstruction of anthropocentrism and anthropocentric worldviews.[49] In wider historical discourse, scholars of animal history have questioned narratives of human historicity and animal otherness. To theorize and to deconstruct the human–animal boundary, concepts of animal agency have been introduced and categories of analysis inherent to cultural studies, such as class, gender, race, space, and materiality, have been implemented. The concept of race was, of course, from the beginning inextricably linked to the colonization of the American continent and to the enslavement of African people [→Feminist Intersectionality Studies]. In this regard, racialized practices and cultures of human-animal relations have been discussed by a multitude of studies that explore the entangled histories of African Americans and animals.[50] Applying intersectional approaches, many scholars have focused on the histories and practices of race relations and of analogizing animals with blacks, that is, they have analyzed and deconstructed "attempts to exclude non-

---

[47] Bender, Daniel E. *The Animal Game: Searching for Wildness at the American Zoo.* Cambridge, MA: Harvard University Press, 2016, 4.
[48] Cf. Ritvo, Harriet. On the Animal Turn. *Daedalus* 136, no. 4 (October 2007): 118–122, 122.
[49] Cf. Thomas, Keith. *Man and the Natural World: A History of the Modern Sensibility.* New York, NY: Pantheon Books, 1983; Ritvo, Harriet. *The Animal Estate: The English and Other Creatures in the Victorian Age.* Cambridge, MA: Harvard University Press, 1987.
[50] Cf. Boisseron, Bénédicte. The Animal and African American History. In *Oxford Bibliographies Online* in African American Studies, Gene Andrew Jarrett (ed.), published 27 February 2019. doi.org/10.1093/obo/9780190280024–0065.

white people from the category of human."⁵¹ Following the 2014 shooting of Michael Brown, the operations of the Ferguson Police Department initiated numerous works that, once again, sought to trace the historical connection between race and the repressive use of police- and watchdogs as well as other animals.⁵²

More generally, scholars examine how animals influence, and are influenced by, culture and cultural practices and how animal-related cultures and cultural practices change over time. In an attempt to institutionalize human–animal studies, US-based scholars, such as Kenneth Shapiro, Margo DeMello, and Brett Mizelle, have launched numerous journals, volumes, and monographic series that reflect the *animal turn* in American studies.⁵³

Today, most scholars of American studies agree that animals matter. Animals have shaped rural and urban American environments, they have been in constant interplay with American society, and, dead or alive, animal bodies have come to narrate, authenticate, and conserve American history. Influenced by the works and theories of Bruno Latour and Donna Haraway, scholars of animal history have studied modes and dynamics of human–animal relations and produced concepts of animal agency that emphasize situated practice and interaction.⁵⁴ Nevertheless, to some extent, scholars are struggling to combine animal significance with animal agency.⁵⁵ Few recognize the connection as clearly as David Gary Shaw, a historian of medieval England: "Agency and its acting is a shibboleth of historical significance: if you can't perform it, you can't matter."⁵⁶

---

51 Fielder, Brigitte Nicole. Animal Humanism: Race, Species, and Affective Kinship in Nineteenth-Century Abolitionism. *American Quarterly* 65, no. 3 (September 2013): 487–514, 487.

52 Cf. Boisseron, Bénédicte. Afro-Dog. *Transition* 118 (2015): 15–31, 15; Boisseron, Bénédicte. *Afro-Dog: Blackness and the Animal Question*. New York, NY: Columbia University Press, 2018; Meacham, Sarah H. Pets, Status, and Slavery in the Late-Eighteenth-Century Chesapeake. *Journal of Southern History* 77, no. 3 (August 2011): 521–554; Spruill, Larry H. Slave Patrols, Packs of Negro Dogs and Policing Black Communities. *Phylon* 53, no. 1 (Summer 2016): 42–66, 45.

53 In 1983, Kenneth Shapiro founded Psychologists for the Ethical Treatment of Animals (PsyETA), the forerunner to the Animals and Society Institute (ASI) that also publishes the *Society & Animals* Journal. Margo DeMello has published numerous seminal and introductory works in the field of human–animal studies. Brett Mizelle is co-founder and editor of H-Animal Listserv.

54 Cf. Roscher, Mieke. Geschichtswissenschaften: Von einer Geschichte mit Tieren zu einer Tiergeschichte. In *Disziplinierte Tiere? Perspektiven der Human–Animal Studies für die wissenschaftlichen Disziplinen*, Reingard Spannring, Karin Schachinger, Gabriela Kompatscher, Alejandro Boucabeille (eds.), 75–100. Bielefeld: transcript, 2015, 83–87. See also Latour, Bruno. *Reassembling the Social: An Introduction to Actor-Network-Theory*. Oxford: Oxford University Press, 2005; Haraway, Donna. *The Companion Species Manifesto: Dogs, People, and Significant Otherness*. Chicago, IL: Prickly Paradigm Press, 2003.

55 Historians often argue that "the recognition of historical agency is often synonymous with the recognition of an individual, group or category's historical significance," Rees, Amanda. Animal Agents. Historiography, Theory and the History of Science in the Anthropocene. *British Journal for the History of Science Themes* 2 (2017): 1–10, 2.

56 Shaw, David Gary. A Way with Animals. *History and Theory* 52, no. 4 (December 2013): 1–12, 8.

Performance is central to ideas of animal agency – to build upon the findings of historians Mieke Roscher and André Krebber, performance is central to the "making" of the individual's biography.[57] Accordingly, questions of animal significance, animal agency, animal performance, and animal biography are closely intertwined. Scholars of animal history have recognized this connection in recent years, and they have embarked on the journey of writing animal biographies. In American studies, the biography of the elephant known as Topsy has received most scholarly attention. Born around 1875, Topsy was publicly electrocuted in 1903.[58] After her capture in Southeast Asia by order of international wildlife dealer Carl Hagenbeck, Topsy was sold to Forepaugh and Sells Brothers, a leading institution in the American circus business [→History of Circus Animals]. For almost thirty years, Topsy performed as a circus elephant before she was villainized as a "rogue elephant," that is, as an elephant who harmed and killed people. In his biography of Topsy, activist-scholar Kim Stallwood has deconstructed this narrative. He argues that Topsy's electrocution was not linked to the elephant's "rogue" behavior, but rather to the business interests of her owners. In turn, Topsy's biography exposes the anthropomorphism and anthropocentrism at work in American history.[59]

Topsy's biography opens up yet another field of investigation associated with the *animal turn*. Filmed by employees of Edison Manufacturing Company, the electrocution represents one of the many ways in which animals – dead and alive – have been put on display throughout US history. Historians, informed by the study of visual, literary, and material culture [→History of Animal Iconography;→Material Culture Studies], have analyzed various modes and practices of animal display in films, books, museums, archives, and zoos [→History of Zoos;→History of Animal Collections/Animal Taxonomy]. In sum, their research maintains that "making animals visible" fulfills two primary functions: (1) the display suggests a human–animal proximity that is nostalgically "imagined within a historio-cultural space;"[60] (2) the materiality of the animals is overemphasized to create a sense of historicity that refers to a "real" and "authentic" period and place in the past.[61] Such manipulations become

---

57 Cf. Krebber, André and Mieke Roscher. Introduction: Biographies, Animals and Individuality. In *Animal Biography: Re-Framing Animal Lives*, André Krebber, Mieke Roscher (eds.), 1–15. Cham: Palgrave Macmillan, 2018, 4.
58 Cf. Jacob Blair Smith or Edwin S. Porter. *Electrocuting an Elephant*. Edison Manufacturing Company, January 17, 1903.
59 Cf. Stallwood, Kim. Topsy: The Elephant We Must Never Forget. In *Animal Biography: Re-Framing Animal Lives*, André Krebber, Mieke Roscher (eds.), 227–242. Cham: Palgrave Macmillan, 2018, 246.
60 Thorsen, Liv Emma, Karen A. Rader and Adam Dodd. Introduction: Making Animals Visible. In *Animals on Display: The Creaturely in Museums, Zoos, and Natural History*, Liv Emma Thorsen, Karen A. Rader, Adam Dodd (eds.), 1–15. University Park, PN: Pennsylvania State University Press, 2013, 3.
61 Cf. Thorsen, Rader, Dodd, Introduction, 4–5. See also Harris, Adam D. *Wildlife in American Art: Masterworks from the National Museum of Wildlife Art*. Norman, OK: University of Oklahoma Press, 2009.

most powerful, as Donna Haraway and Linda Kalof have respectively shown, when the technology of taxidermy is used to preserve and display dead animals. Indeed, taxidermied animals are often mounted in museums to perform an "unforgettable gaze" that is fundamental to the experience of the viewer, who, by this means, partakes in a "moment of supreme life."[62]

The works of Haraway and Kalof resonate with John Berger's "Why Look at Animals?" Haraway and Kalof also propose a critical approach towards the representation of animals. In the broader sense of the *animal turn*, rather than restrict their research to a critique of animal marginalization, they further develop our assessment of human–animal relations. Like other scholars, they stress animal performance, materiality, and agency. In the future, it will be the task of scholars of historical animal studies to sharpen our understanding of these analytical categories, especially animal agency, and to explore "new frontiers" of human–animal interaction, both thematically and methodically. In American studies, the examination of so-called "unexpected" animals might prove fruitful. Insects and other invertebrate species warrant further research. Finally, given the dominance of the animal-industrial complex and the rise of human–pet relations in the United States, historians need to enhance our knowledge of how humans have historically transformed animal bodies.[63]

Selected Bibliography

Anderson, Virginia DeJohn. *Creatures of Empire: How Domestic Animals Transformed Early America*. New York, NY: Oxford University Press, 2004.
Bender, Daniel E. *The Animal Game: Searching for Wildness at the American Zoo*. Cambridge, MA: Harvard University Press, 2016.
Benson, Etienne. *Wired Wilderness: Technologies of Tracking and the Making of Modern Wildlife*. Baltimore, MD: Johns Hopkins University Press, 2010.
Boisseron, Bénédicte. *Afro-Dog: Blackness and the Animal Question*. New York, NY: Columbia University Press, 2018.
Greene, Ann Norton. *Horses at Work: Harnessing Power in Industrial America*. Cambridge, MA: Harvard University Press, 2008.
Grier, Katherine C. *Pets in America: A History*. Chapel Hill, NC: University of North Carolina Press, 2006.
Isenberg, Andrew. *The Destruction of the Bison: An Environmental History, 1750–1920*. New York, NY: Cambridge University Press, 2000.
Pearson, Susan J. *The Rights of the Defenseless: Protecting Animals and Children in Gilded Age America*. Chicago, IL: University of Chicago Press, 2011.
Rothfels, Nigel (ed.). *Representing Animals*. Bloomington, IN: Indiana University Press, 2002.

---

[62] Haraway, Donna. *Primate Visions: Gender, Race, and Nature in the World of Modern Science*. New York: Routledge, 1989, 30. See also Kalof, Linda. *Looking at Animals in Human History*. London: Reaktion Books, 2007.
[63] Cf. Specht, Joshua. Animal History after Its Triumph: Unexpected Animals, Evolutionary Approaches, and the Animal Lens. *History Compass* 14, no. 7 (July 2016): 326–336.

Thorsen, Liv Emma, Karen A. Rader and Adam Dodd (eds.). *Animals on Display: The Creaturely in Museums, Zoos, and Natural History.* University Park, PA: Pennsylvania State University Press, 2013.

Sandra Swart
African Studies

1 Introduction and Overview

In 2015, a team of scientists breathlessly announced a fresh fossil find from South Africa's "Cradle of Humankind": a brand-new hominin species! They had discovered the largest collection of hominin fossils from a single species ever found. This visitor from a long-forgotten past gave South Africa a reason to celebrate in a time of economic gloom and socio-political despair. *Homo naledi*, as the scientists baptized the new species, certainly seemed to bear a resemblance to modern humans, with hand and feet akin to ours.[1] But the little creature weighed only about 45 kg and stood 1,5 meters high, and possessed a brain the "size of an orange". At the time, the bones had not been dated, but researchers hoped they would teach us more about the shift between Australopithecus and the *Homo* genus, our direct ancestors.[2] Remarkably, given its relatively primitive form, this hominin evinced complex behavior: the scientists suggested it may even have buried its dead. The world was electrified; international press besieged the scientific team. The nation breathed a sigh of relief: this was cutting-edge modernity uncovering a shared global history of the common roots of humanity buried in our own soil – South Africa was in the international news for positive reasons for once. Yet, almost at once there was a stern disavowal of kinship from some national leaders and public figures. Zwelinzima Vavi, a prominent Marxist trade unionist at the forefront of the labor movement, disowned the discovery. He tweeted to his 300,000 followers: "No one will dig old monkey bones to back up a theory that I was once a baboon. Sorry."[3]

When challenged with the scientific consensus on evolutionary theory, Vavi retorted defiantly: "Then prove that I was a monkey before – please don't bring old baboon bones." At the same time, the president of the South African Council of Churches, an organization renowned for having protested Apartheid, announced that it was insulting to claim that Africans were descended from baboons. In addition, African National Congress MP, Dr Mathole Motshekga, declared that the fossil find was a scheme by the West: "[it's] dangerous to send a message, which says we are related

---

[1] Berger, Lee R., John Hawks, Darryl J. de Ruiter, Steven E. Churchill, Peter Schmid, Lucas K. Delezene, Tracy L. Kivell et al. *Homo naledi, a* New Species of the Genus *Homo* from the Dinaledi Chamber, South Africa. *eLife* 4 (2015): e09560. doi.org/10.7554/eLife.09560.001.
[2] Homo naledi's a youngster at just 250,000 years, say surprised scientists. *Daily Dispatch*, April 28, 2017.
[3] Discovery of new human ancestor Homo naledi sparks racism row in South Africa. *ABC News*, September 17, 2015.

to baboons because many South Africans are still undergoing a healing process [...] This is a backdoor attempt to remind us that we are sub-human[...]."[4]

Many questioned why the extinct species had received an African name: was it a racist way to covertly critique African culture?[5] In fact, the fossil remains had been found at the end of a tapering claustrophobia-inducing corridor of the cave, known as the "dark zone" in a chamber known as *Dinaledi* – or the Chamber of Stars. No sunlight had ever touched this realm.

Indeed, for some this lack of light also seemed a powerful metaphor for the Dark Ages to which the three celebrity fossil-deniers appeared to hark back. Enlightenment values and the authority of science were invoked by outraged South Africans. In the press and on social media, they ridiculed this triumvirate, accusing them of "baboon hysteria". But how to understand these three very different men – a trade unionist, a bishop and a career politician – responding to a long-dead hominid by disavowing any kinship with baboons?

The question goes much deeper: to the core of writing animal history in Africa and how we reach (and fail to reach) the animal 'other' in the past. Moreover, as this chapter will argue, it touches a deep seam running through the African past, informing how humans relate to animals – and, through animals, how humans relate to other humans. To do so, it explores the human-animal lineages of such ideas, arguing that these challenges to the consensus of Darwinian thought are not merely examples of reactionary ignorance. Instead, this chapter investigates the long history that complicates our understanding of our relationship with the other primates in South Africa and the often contradictory but overlapping palimpsestic bodies of knowledge about baboons. It chooses this explosively controversial case study to showcase entanglements of materiality and metaphor in analyzing past human-animal relationships and to show how these are connected to human-human relationships in Africa's past. This case study helps the chapter in its broader remit: to explore the possibilities and constraints in writing animal histories of Africa, by using the social metaphor of the baboon and their lived historical reality in the segregationist and later Apartheid state in order to understand the current socio-political use of the baboon in human society.

After all, African 'animal history' "has remained little studied and elusive [...] roaming the disciplinary deep forest, nibbling at the edges of the conferences and journals, straying into unexpected territories and prone both to local extirpations and bursts of fecundity. But as a living, breathing beast, it has been grazing in full sight of everyone in the historiographical field for as long as African history

---

4 Interview with Modise, Tim. 24 September 2015, Mathole Motshekga: Homo Naledi, human ancestry link offensive – should be rejected. URL: www.biznews.com/transformation/2015/09/24/mathole-motshekga-homo-naledi-human-ancestry-link-offensive-should-be-rejected/ (October 5, 2020).
5 The Sesotho name Naledi was not meant to associate this discovery with African ancestors, just the location where the discovery was made – the Rising Star cave ('naledi' means 'star').

has existed and we locals have often caught and consumed it."⁶ Historians of Africa have enduringly analyzed animals and humans as entangled vectors of change in our shared histories. Yet, in Africa, animal history is still sometimes seen as irrelevant or self-indulgent – which this chapter tries to refute with this case-study. Nevertheless, recently, African animal history has become a more self-conscious sub-disciplinary 'turn'. A useful if solipsistic example of this historiographic shift is found in contrasting two monographs on horses in African history: Robin Law's 1980 *The Horse in West Africa*, in which he stated sternly that he had 'no special affections for these animals', analyzing their function as technologies of change in human societies.⁷ Three decades later, my 2010 *Riding High – Horses, Humans and History in South Africa*, acknowledged (indeed, unabashedly celebrated) horses as historical beings: historical subjects and agents – co-creating multispecies history. Although it flirted with the view from the horses' eyes, it remained the view from the saddle.⁸ Sometimes a truly multispecies history may be attempted.⁹ However, generally when the starting point is the human (but, unlike mainstream histories, the "human" studied in relation to the "animal other"), the term "animal-sensitive history" may be more appropriate. Certainly, historians have embraced this turn – yet there remains an urgent need for research into the shifting histories of local ideas, vernacular culture and indigenous society in shaping ideologies of the animal within human society rather than merely studying their imposition by colonizing outsiders – this chapter tries to do just that.

2 Methods and Approaches

The study of human–animal relations is critical in our understanding of African pasts: with livestock key to both African then subsequent settlers, moving with human diasporas and trade networks.¹⁰ African historians have studied the impact of animals in transforming human lifeways: from hunting economies and transhumant pastoralism, to agribusiness.¹¹ Work on animals comes from environmental his-

---

6 Drawn from Swart, Sandra. Animals in African History. In *Oxford Research Encyclopedia of African History*, Thomas Spear (ed.), published 26 April 2019. doi.org/10.1093/acrefore/9780190277734.013.443.
7 Law, Robin. *The Horse in West Africa*. Oxford: Oxford University Press, 1980.
8 Swart, Sandra. *Riding High – Horses, Humans and History in South Africa*. Johannesburg: Witwatersrand University Press, 2010.
9 In *Riding High*, I tried to tell a multi-species history of horses and humans (a gentle herbivore and small rogue primate) co-fashioning a shared history.
10 For *longue durée* see Smith, Andrew B. *African Herders: Emergence of Pastoral Traditions*. Walnut Creek, CA: AltaMira Press, 2005.
11 This paragraph and others in the essay draw from Swart, Sandra. Writing Animals Into African History. *Critical African Studies* 8, no. 2 (August 2016): 95–108.

tory,[12] philosophy[13] and literary studies.[14] It includes wildlife[15] and domestic animals[16], "vermin"[17],, insects and zoonotics[18], and how animals came between people in the colonial past.[19] Cattle have featured as particularly societally central.[20] African 'animal histories' owe much to studies of conservation, critiques of colonialism, epizootics, livestock farming, pastoralism, hunting and conservation and environmental history, zooarchaeology[21] and some engagement with IKS movements that can be traced back to the early 1990s. Unlike in some other regions mentioned in this volume, the animal rights movement have not influenced many historians.

Since at least the 1960s, African historians have refused to be chastened by those who suggest that the idea of history 'as normal' happens elsewhere and African history is the aberrant Other. Equally, as Southern theorists have contended, the North is often modelled as representative of the "normal course of history" (they have bat-

---

12 Beinart, William. African History and Environmental History. *African Affairs* 99, no. 395 (April 2000): 269–302.
13 Horsthemke, Kai. *Animals and African Ethics*. Basingstoke: Palgrave Macmillan, 2015.
14 Woodward, Wendy. *The Animal Gaze*. Johannesburg: Witwatersrand University Press, 2008.
15 Pooley, Simon. The Entangled Relations of Humans and Nile Crocodiles in Africa, c.1840–1992. *Environment and History* 33, no. 2 (August 2016): 421–454; Draper, Malcolm. Going Native. Trout and Settling Identity in a Rainbow Nation. *Historia* 48, no. 1 (May 2003): 55–94; Carruthers, Jane. Wilding the Farm or Farming the Wild. The Evolution of Scientific Game Ranching in South Africa from the 1960s to the Present. *Transactions of the Royal Society of South Africa* 63, no. 2 (October 2008): 160–181.
16 Jacobs, Nancy. The Great Bophuthatswana Donkey Massacre: Discourse on the Ass and the Politics of Class and Grass. *The American Historical Review* 106, no. 2 (April 2001): 485–507; Bankoff, Greg and Sandra Swart. *Breeds of Empire: The Invention of the Horse in the Philippines and Southern Africa, 1500–1950*. Copenhagen: Nordic Institute of Asian Studies, 2007, 529–549; Van Sittert, Lance and Sandra Swart. *Canis Familiaris – A Dog History of Southern Africa*. Leiden: Brill, 2008.
17 Mazarire, Gerald. The Burrowed Earth: Rodents in Zimbabwe's Environmental History. *Critical African Studies* 8, no. 2 (August 2016): 109–135; Mavhunga, Clapperton. Vermin Beings: On Pestiferous Animals and Human Game. *Social Text* 29, no. 1 (Spring 2011): 151–176.
18 Brown, Karen. Political Entomology: The Insectile Challenge to Agricultural Development in the Cape Colony, 1895 to 1910. *Journal of Southern African Studies* 29, no. 2 (June 2003): 529–549; Beinart, William. Transhumance, Animal Diseases and Environment in the Cape, South Africa. *South African Historical Journal* 58, no. 1 (2007): 17–41.
19 Shadle, Brett. Cruelty and Empathy, Animals and Race, in Colonial Kenya. *Journal of Social History* 45, no. 4 (Summer 2012): 1–20; Swart, Sandra. It is as Bad to be a Black Man's Animal as it is to be a Black Man – The Politics of Species in Sol Plaatje's Native Life in South Africa. *Journal of Southern African Studies* 40, no. 4 (July 2014): 689–705; Steinhart, Edward. *Black Poachers, White Hunters: A Social History of Hunting in Kenya*. Athens, OH: Ohio University Press, 2006.
20 Herskovits, Melville. J. The Cattle Complex in East Africa. *American Anthropologist* 28, no. 4 (October-December 1926): 633–664; Evans-Pritchard, Edward Evan. The Sacrificial Role of Cattle Among the Nuer. *Africa* 23, no. 3 (July 1953): 181–197; Shutt, Allison. The Settlers' Cattle Complex: The Etiquette of Culling Cattle in Colonial Zimbabwe, 1938. *Journal of African History* 43, no. 2 (July 2002): 263–286.
21 Mitchell, Peter. *African Connections: Archaeological Perspectives on Africa and the Wider World*. Walnut Creek, CA: AltaMira Press, 2005.

tled this ever since Hegel pronounced Africa "no part of history").[22] Africanists write with *context* uppermost: "human-animal histories" have a strong sense of place.[23] Histories especially tend towards thick description and studies of local processes. While avoiding the pitfall of the meta-narrative, these can help answer questions of larger causation. Being aware of context, being unabashedly local, does not diminish significance.

Yet, although this chapter contends that it is important to factor in vernacular context and not generalize from the North, it would be unwise to set up too stark a dichotomy between 'African' and 'the Rest' in terms of historiography or methodology. Since the 1960s, invigorated by continental movements towards independence, African history has sometimes declared the distinctiveness of an "African perspective".[24] I would argue strongly against such a flattening description: African history is as eclectic, syncretic and diachronic as that of any other continent. There are variances but not a deep gulf: difference is not dichotomy. Perhaps African historians deploy oral history more. Perhaps they are more used to recovering unwritten pasts. Here, perhaps African history might offer more to animal histories than more text-based historiographies – after all, African historians have long rejected the privileging of written sources, instead adopting the careful and corroborative use of both a diversity of primary sources and multidisciplinary methods, including DNA analysis, paleontology, archaeology, ethnography, anthropology, linguistics. Perhaps they are more attuned to *longue durée* processes hitherto under-represented in animal histories. Thus, this chapter includes a brief engagement with the oldest archive of Africa: rock art of KhoeSan peoples. Perhaps African scholars resist faddishness – the push to conformity – a little better than most. Perhaps they are more skeptical of universalizing meta-narratives[25] and intellectual 'turns' or trends: better able to resist the screw of the 'turn', as it were.

Moreover, animal histories may be reconstructed within the framework of African history's persistent concern with "agency" in writing the history of the suppressed and the silenced. It has resisted the idea that oppressed groups were docile instruments or victims. Yet in trying to give them "voice", historians of Africa have been criticized of speaking *for* the Other.[26] This is a useful precedent: similar doggedness in claiming to view history through the "animals' eyes" is not possible; such

---

22 Thiong'o, Ngũgĩ Wa. *Moving the Centre. The Struggle for Cultural Freedoms.* London: Heinemann, 1993; Connell, Raewyn. *Southern Theory: The Global Dynamics of Knowledge in the Social Sciences.* Sydney: Allen and Unwin, 2007.
23 Swart, Animals in African History.
24 Barbosa, Muryatan Santana. The African Perspective in the General History of Africa (Unesco). *Tempo* 24, no. 3 (September 2018): 400–421.
25 See Cooper, Fred. Africa's Pasts and Africa's Historians. *Canadian Journal of African Studies* 34, no. 2 (2000): 298–336, 327.
26 Rassool, Ciraj. Power, Knowledge and the Politics of the Public Past. *African Studies* 69, no. 1 (March 2010): 79–101, 83–84.

thought experiments are just historiographical versions of an inkblot test, displaying more about the historian's era than that of the animal through whose "eyes" they purport to see. Historians of Africa have, however, deployed methodologies for discussing those that are silenced ("representing" skirts perilously near, as noted above, to ventriloquizing the subject). Moreover, several animal historians have noted that it is time to move beyond "agency" as the central concern.[27] The call to move beyond just "discovering agency in the animal past" parallels a concurrent call in African history to move beyond just discovering (or even simply asserting) agency. As Lynn Thomas has argued: "Too often agency slips from being a conceptual tool or starting point to a concluding argument. For example, in my subfield of African women's and gender history, statements like 'African women had agency' can stand as the impoverished punch lines of empirically rich studies […]".[28] Here animal historians could learn from historians of Africa. Thus, rather than simply asserting or repeatedly establishing that agency exists, we should ask how agency was understood in different periods and what kind of archive and methodology might yield this data.

3 Topics and Themes

As noted, a key characteristic of writing African history is awareness of context.[29] Correspondingly, although all processes of "othering"[30] pivoted on domination, they took local forms and inflections. Apartheid South Africa provided perhaps the most powerful contemporary example of how racial categories could be embedded in shifting power relations. This is pivotal in explaining the power of the monkey metaphor, the ape analogy and simian simile – and why they continue to wound. Very recently, for example, an elderly white woman rekindled a nationwide debate about racism after posting about littering on the beach: "From now I shall address the blacks of South Africa as monkeys as I see the cute little wild monkeys do the same, pick and drop litter." Using the hashtag #RacismMustFall, people expressed their shocked outrage at her, as she went from obscurity to a figure of national loathing – disciplined by the South African Human Rights Commission.[31] It was so serious

---

27 Specht, Joshua. Animal History after Its Triumph: Unexpected Animals, Evolutionary Approaches, and the Animal Lens. *History Compass* 14, no. 7 (July 2016): 326–336.
28 Thomas, Lynn. Historicising Agency. *Gender & History* 28, no. 2 (August 2016): 324–339.
29 Even a cursory survey of African-based book and article titles includes the geographic in the title, as opposed to studies from the global north that frequently assume their case to be generalizable.
30 Borkfelt, Sune. Non-Human Otherness: Animals as Others and Devices for Othering. In *Otherness: A Multilateral Perspective*, Susan Yi Sencindiver, Maria Beville, Marie Lauritzen (eds.), 137–154. Frankfurt a.M.: Peter Lang, 2011.
31 Mutiga, Murithi. South African Woman Faces Criminal Charges Over Racist Tweets. *The Guardian*, June 10, 2016.

a charge that it was considered – by some – to warrant rape and death threats towards her. These calls for direct violence went unpunished – they were considered less harmful socially than the metaphor. Then the Swedish clothing company H&M featured a black child modelling a sweatshirt bearing the label "coolest monkey in the jungle."[32] H&M came under attack by the Economic Freedom Fighters and protestors ransacked the stores. Metaphors matter.

Dehumanization

Indeed, ironically, the compelling urge to dehumanize is all too human. It is not limited to one place, one people or one time. The rhetoric of dehumanization helped rationalize patriarchal social orders, slavery, colonialism, and xenophobic violence globally, including genocide towards Jews in the Holocaust, and Tutsis in Rwanda. Historically, the 'animalized Other' is classified thus in order to create emotional bridges that transfer the 'bad feelings' that accompany the specific species to the human group that the speaker wants to denigrate. Often these are feelings of horror or hatred.[33] When people are stripped of their humanity, they are stripped of their ability to arouse empathy. This permits their abuse without the perpetrator feeling the usual guilt – so it removes inhibitions to violence. Exposure to dehumanizing hate speech numbs ordinary individuals to injustice as it normalizes the abnormal. This is why those that deploy strategies of dehumanization find animalization so useful.

The Making of Metaphors and the Unmaking of People

Dehumanization not only erupts at dramatic flashpoints; it also operates as quotidian social phenomenon. In this context, the dehumanization or infrahumanization of Africans as almost-simian has a long history.[34] Simians have long represented a challenge to drawing an unequivocal boundary between the animal and the human.[35] Prior to the 18th century, the West supposed that all species were unalterable in a

---

[32] Fortin, Jacey. H&M Closes Stores in South Africa Amid Protests Over Monkey Shirt. *New York Times*, January 13, 2018.
[33] Haslam, Nick. Dehumanization: An Integrative Review. *Personality and Social Psychology Review* 10, no. 3 (February 2006): 252–264.
[34] Jahoda, Gustav. *Images of Savages: Ancient Roots of Modern Prejudice in Western Culture*. London: Routledge and Kegan Paul, 1999; Montague, Ashley and Floyd Matson. *The Dehumanization of Man*. New York, NY: McGraw-Hill, 1983; Hund, Wulf, Charles Mills and Sylvia Sebastiani. *Simianization. Apes, Gender, Class, and Race*. Zürich: Lit Verlag, 2016.
[35] Fudge, Erica, Ruth Gilbert and Susan Wiseman. *At the Borders of the Human: Beasts, Bodies and Natural Philosophy in the Early Modern Period*. Basingstoke: Palgrave, 2002, 2–3.

god-ordained immutable natural and socio-political order.[36] The Dutch scientist Petrus Camper (1722–1789) argued that the original man was an ancient Greek, who wandered from his geographic origin and degenerated under the undesirable influence of other climates: so monkeys and apes were degenerate varieties of original man. His paradigm was overturned in 1809, when French naturalist Lamarck (1744–1829) insisted that all beings began from a single spontaneous act of creation before changing through an ill-defined process somehow coupled to the vicissitudes of the environment. This model – capsizing Camper's – conjectured that humans originated from apes, with Africans imagined as the link between the latter and Europeans. By the mid-19th century, polygenesis (the notion of distinct and separate origins of modern "races") was commonly held by intellectuals, many of whom believed in racial hierarchies associating black people with apes. This notion became entrenched in the long nineteenth century, further ensconced by popular misunderstandings of Darwinism – politically useful discourse in colonial spaces like South Africa. The advent of scientific racism further validated the analogy: if they had to concede that all humans were related to the apes, they nonetheless insisted that Africans' consanguinity was much closer. Yet, it must be remembered, during the same period there was also always a strong stream of evolutionary science that explicitly rejected this racist model. In fact, evolutionary theory emphasized the deep similarities between different races, and argued that differences in behavior were the product of culture not biology. Darwinism did not cause racism; misusing Darwinism helped buttress racism.

So, the boundary between people and the other primates has a long history of porosity.[37] On the one side of the spectrum there has been an intellectual tradition that regards the other primates with horror, as a parody or debasement of the human, at their "aping" of man. Yet, there has also been a tradition of ideas of physio-emotional connectivity, a closeness between the two species, evident in one strain of thought from Linnaeus to the twenty first century [→History of Science]. As noted, there is an enduring tradition dating back to the classical period and reinvigorated by travelers' tales that established blurred categories between human and simian "natives" in colonial discourse, as Manfred Pfister argues. If those to be colonized were not "quite human," then they could be missionized, subjugated or even eradicated. This received renewed impetus between the 17th and early-20th centuries, enveloped in colonialism's rhetoric of a "civilizing mission", which compared African wildlife to the "wild" state of African societies (and, as we have seen, animals to Africans themselves) [→(Post)Colonial History]. But this contributed to a strange twist:

> The definition of the savages as sub-human animals dissolved any clearcut distinction between man and animal. It not only put savages at a distance from human beings, it also had, paradoxi-

---

36 Hund, Mills and Sebastiani, Simianization: Apes, Gender, Class, and Race.
37 See Corbey, Raymond. *The Metaphysics of Apes: Negotiating the Animal–Human Boundary.* Cambridge: Cambridge University Press, 2005.

cally, the contrary effect of bringing man and animal closer to each other. If savages are regarded as some higher kinds of animals, certain animals could come to be regarded as some quasi-human species.[38]

So, it became harder to guard the wall between the human and the Other. Simian others were increasingly figuratively deployed as metaphors for the human condition. As biology developed, they became understood as literally related to humans, on the visceral levels of anatomy and (much later) genetics. This further eroded the foundations of the already permeable boundary between humans and others. The border between the groups was always shifting – but patrolled differently at different historical junctures and has moved, sometimes in a non-linear fashion. Over the past century, the status of the other primates, especially that of the great apes, has transformed in the (so-called) Western tradition. This has become evident in genetic analysis, cognitive research and long-term behavioral studies, and demonstrated in conservation programs led by primatologists (in Africa: Jane Goodall, Dian Fossey, and Shirley Strum). Today there is a shared consensus among scientists over the classification of all humans, regardless of our race or nationality, as a species that we now believe originated in Africa and share a common ancestor with other primates. All humans are now believed to be, in a sense, simply very successful African apes.

Thus, there has been a remarkable paradigmatic shift. To explain this mind-shift, there are several excellent accounts of "how we got there", which examine the contours of ape–human relations over time. But these are all from the perspective of the Global North.[39] It is presented as a triumph of Darwinian thought overcoming antiquarian antipathy towards apes and a growing acceptance of the similarity between humans and the other primates – until the apogee reached today with rights of "personhood" being considered in the courts for some primate species. To date, there has been very little analysis of historical thinking about primates from and in Africa, in either academic or popular domain. But the full history is not a simple story.

Certainly, the South African backlash against the fossil-find disconcerted people around the world at a time when evolutionary theory seemed to have global consensus. Oxford University's Richard Dawkins jumped into the South African mêlée with his own tweets on the "We're all African apes" trope, arguing that the fossil denial just "breathes new life into paranoia". Lee Berger, who ran the *Homo naledi* dig, just explained that humans did not descend directly from baboons. What was interesting was the repeated refrain of the "baboon" taint within the discourse, despite no professional scientist ever mentioning baboons in any connection to *Homo naledi*. But

---

38 Pfister, Manfred. Man's Distinctive Mark': Paradoxical Distinctions between Man and his Bestial Other in Early Modern Texts. In *Early Modern Texts in Telling Stories*, Elmar Lehmann, Bernd Lenz (eds.), 17–33. Amsterdam: Grüner, 1992, 28.
39 Corbey, Raymond and Bert Theunissen. *Ape, Man, Apeman: Changing Views Since 1600*. Leiden: Leiden University, 1995. Most scholarship on this is Anglo-American in focus, but occasionally Japan, with its strong history of primatology, is included.

academia equally did not intervene or explain *why* the errors were being made and, moreover, there was hardly any attempt by social scientists and natural scientists to bridge the gap between them in order to explain why people were misunderstanding the fossils so badly. Generally, the exchange of ideas between the biosciences and the social sciences/humanities in this area is still negligible. This chapter tries to address this, as human-animal studies – especially history – can play an important role here.

Fossil Fictions and Frictions?

At certain points in various cultures, apes have been understood as *ur*-human, creatures capable of becoming – through a civilizing process – human. Some ethnic groups, classes, races, and women have been (at various times and places) located by the dominant group as closer to the animals on the man-beast continuum [→Feminist Intersectionality Studies].[40] Like the great apes themselves, these groups have been seen as "almost but not quite human", roaming the range between the rational and irrational, the mind and the body, the articulate and the voiceless. These liminal figures at the "edge of humanness" have been drawn into ideologies of power that needed the "other" to define themselves. Once the idea that humans had "descended from apes" (rather than Darwin's more complicated articulation) gained traction, public imagination became stimulated by evolutionary notions filtering down and the display of the first gorilla in the London Zoo (from 1860) [→History of the Zoo]. They displaced their own anxiety about their newly discovered descent by claiming close kinship between apes and "primitive peoples", thereby dissociating themselves from the near relatedness of either.

But did "ape" or "baboon" mean the same thing for everyone at all times? No. We cannot know if most or even many Africans calqued or borrowed a new meaning of "baboon" from the new colonial dispensation, but it seems African antipathy to baboons came in with the shift to crop farming as baboons were a serious threat to grains, vegetables and small livestock. Some of this history may be reclaimed through oral tradition and rock art [→History of Animal Iconography], which certainly does not suggest a contempt for baboons. Indeed, some groups – like the AmaTola or Ncube – chose baboons as constitutive of their personal identities as their totemic animals. Baboons were given a quasi-human position, replete with agency, in the cosmology of the Bushman/San, who did not suffer crop losses or see baboons as enemies.[41] Baboons were venerated and sometimes imitated because of their associ-

---

40 For pioneering research in the gendered history of primate studies see Haraway, Donna. *Simians, Cyborgs and Women – the Reinvention of Nature*. London: Free Association Books, 1991.
41 Jolly, Pieter. Therianthropes in San Rock Art. *The South African Archaeological Bulletin* 57, no. 176, (December 2002): 85–103; Challis, Sam. Taking the Reins: The Introduction of the Horse in the Nineteenth-Century Maloti-Drakensberg and the Protective Medicine of Baboons. In *The Eland's People:*

ation with root medicines,⁴² believed to render them safe from illness: a belief consequent to millennia of humans watching baboons choose certain plants and avoid poisonous ones. Thus, baboons and humans were not always enemies and their relationships were complex.

There are scores of references to baboons in early travelers' descriptions. There was a racist (but, at this stage, pre-Darwinian) link made between the indigenous population and baboons, in a populist "Great Chain of Being" – a rough-and-ready taxonomy that put Europeans above the others they encountered in their voyages of exploration from the fifteenth century onwards. This was an extension of the long intellectual legacy of the Classical period, premised on the idea that every existing entity had its place in a divinely ordained hierarchy, corresponding with increasing corporeal complexity and mental prowess. *Contra* (later) evolutionary thought, this offered a static view of nature, with all "beings" in a linear, continuous series and all always kept strictly in their relative position on the chain. This supposedly "natural" social hierarchy could be deployed to validate political oppression. But there is something interesting in the early travelers' tales: in the seventeenth and eighteenth centuries, despite the bigotry of the time, there was sometimes a grudging admiration for both indigenous hunter-gatherers and baboons. This was the reluctant respect of those who found southern Africa daunting and could not help but marvel at the survival skills of the inhabitants most like them.

Moreover, to further complicate the metaphor, the baboon insult was also seen (quite early) in some circles as a mark of the backwardness of the *insulter*, rather than the *insulted*. English travelers sometimes noted with repugnance how Boers would animalize black Africans.⁴³ At the same time, ironically, Boers (the progenitors of "Afrikaners") were occasionally likened by English commentators to baboons,⁴⁴ while other English-speakers used it as a derogatory epithet for Africans in the interior. Racial thinking shifted from the ill-defined but functional universalism of the late eighteenth century to a cultural racism, predicated on stereotyping the so-called "native" (or, as we have seen, other ethnicities like Boers) as closer to nature and to the other primates.⁴⁵

The baboon analogy cropped up again in the early twentieth century, but by then it became garnished with the embellishment of "race science". "Scientific racism" helped validate the segregationist state, with theorists arguing that the differences

---

New Perspectives in the Rock Art of the Maloti-Drakensberg Bushmen, Peter Mitchell, Ben Smith (eds.), 104–107. Johannesburg: Wits University Press, 2009.

42 Holleman, Jeremy. *Customs and Beliefs of the /Xam Bushmen*. Johannesburg: Witwatersrand University Press, 2004, 10–13.

43 Castell Hopkins, John and Murat Halstead. *South Africa and the Boer-British War, Volume I*. Toronto: War Book Publishing, 1990, 113.

44 Tangye, H. Lincoln. *New South Africa: Travels in the Transvaal and Rhodesia*. London: Horace Cox, 1896.

45 Stepan, Nancy. *The Idea of Race in Science: Great Britain 1800–1960*. London: Macmillan, 1982.

in socio-political and economic status between whites and blacks were caused by physiological differences.[46] While many nineteenth-century white liberals had faith in the transformative power of education and Christianity, a growing opposition deployed the new race science to buttress a rigid anti-African policy. Many white settlers embraced this metropolitan-inspired race theory, drawing on populist (mis)understandings of Social Darwinism. As Jan Smuts (then on the cusp of becoming Prime Minister) said, "natives have the simplest minds [...] and are almost animal-like in the simplicity of their minds and ways".[47] The state's so-called "expert" on "black tribes", Sir Godfrey Lagden, head of the South African Native Affairs Commission, declared: "A study of the physiognomy of the masses shows a lack of intellect [...] and gives the impression of being not unlike baboons."[48] Segregationist legislation, legitimized by such "race science", was increasingly enacted by the state in the 1910s. The rural paternalism of the early twentieth century was really the polite face of a brutal racism.

So, the animalized metaphor was adopted as a form of policing African behavior. By the 1920s, at least in some places, the epithet "baboon" conveyed a sense of primitivism, uncivilized behavior and – perhaps most significantly – recidivism from "Christian respectability". For example, some missionary preachers lambasted their converts for "baboon behavior". The same "baboon" trope appeared repeatedly in the reaction to perceived assaults on respectability. For instance, the introduction of "Bantu Dancing" into the state's teacher training syllabus in 1948 provoked indignation from those who felt that it would precipitate not only dissolute behavior, but cause outsiders a kind of cruel amusement.[49] Those within the intelligentsia like Sidney Ngcongo feared that not only would it lead to social vulgarity if not licentiousness, but that it would provoke white *schadenfreude* and epicaricacy, a kind of gloating self-satisfaction at African failure to be respectable: "The joy that a civilized [white] man gets when watching Zulus dance, is the same kind of joy he gets when looking at monkeys playing on the trees. [...] Just as he never thinks of improving the monkeys, so it is with the poor African dancing before him."[50]

The baboon came to represent the anathema to the ideals of the *amaRespectables* and other African groups right up until the 1940s. So, as South Africa had Apart-

---

[46] Dubow, Saul. *Scientific Racism in Modern South Africa*. Cambridge: Cambridge University Press, 1995.
[47] Smuts, Jan. C. The White Man's Task, 22 May 1917. In *Plans for a Better World: Speeches of Field-Marshal the Right Honourable J.C. Smuts*, Jan. C Smuts (ed.). London: Hodder and Stoughton, 1942.
[48] Denoon, Donald. *A Grand Illusion: The Failure of Imperial Policy in the Transvaal Colony During the Period of Reconstruction, 1900–1905*. London: Longman, 1973, 100; Burton, David Raymond. *Sir Godfrey Lagden: Colonial Administrator*. PhD diss.: Rhodes University, ZA (1989).
[49] Marks, Shula. Patriotism, Patriarchy and Purity: Natal and the Politics of Zulu Ethnic Consciousness. In *The Creation of Tribalism*, Leroy Vail (ed.), 215–240. Berkeley, CA: University of California Press, 1989.
[50] Excerpts from the *Natal Native Teachers' Journal*, quoted in Marks, Patriotism, Patriarchy and Purity.

heid imposed upon it, for many Africans, part of the performance of respectability lay in distancing oneself from such animalized insults, creating further antipathy towards baboons. A generation later, the political scene was dominated by the Apartheid state's brutal crackdown on dissent, especially following the 1976 protests and the rise of black consciousness. By then, restricting jobs by skin color was actually being eroded. This infuriated the openly segregationist white *Mineworkers'* Union, led by right-wing trade unionist *Arrie Paulus*, who unwisely told a *New York Times* reporter: "You have to know a black. He wants someone to be his boss. They can't think quickly. You can take a baboon and learn [sic] him to play a tune on the piano, but it's impossible for himself to use his own mind to go on to the next step."[51] When this appeared in the press, nation-wide protests were organized by both black and white South Africans.[52] Paulus was charged with *crimen injuria* for "impairing the dignity of the Black people of South Africa or a part thereof".[53] He escaped prosecution altogether: the magistrate decided that his words constituted no offence because not *every* African would be offended. But this was the first time in history that the simian simile was publicly condemned up to the level of the state, the first time that such animalizing analogy received not only international indignation but national condemnation – and attempted redress through the courts.

I will not detail all the quotidian viciousness of all these everyday affronts. The cases are many and persist right up until today – but what is important is that out of the streams of invective, the word "baboon" stood out for people. It retained a strong power to shock and to wound, akin to the "k-word". The cases overwhelm one, even after the transition to a post-Apartheid state in the 1990s (with the first democratic elections in 1994) saw a heightened drive towards ensuring peace through policing "hate speech". Yet, parliament continued to hear a lot of baboon insults.[54] Julius Malema, at the time the ANC's firebrand Youth League President warned a whistle-blower who exposed his alleged corruption in securing state tenders, "It there is a baboon, come explain yourself, you coward. You don't have a face, you bloody ape."[55] Cases kept erupting in ordinary society: social media, at places of work, in incidents of road rage – even children's birthday parties.

Yet, despite the notoriety surrounding each incident, the cases continued. It is also important to reflect on the power of zoomorphism (in this case the baboon insult) as a means of self-authentication by othering – and the danger it poses to the humans harmed in the comparison. However, in zoomorphism, the danger is not only to people but to animals. Baboons have, over time, become heavily targeted creatures: hated, shot at, poisoned and hunted with dogs. So, there is a codicil to the discursive deployment, with the affective arrow pointing the other way, and pri-

---

51 Paulus, Arrie. Blacks Want Bosses. *Daily Dispatch*, September 14, 1979.
52 Paulus, Arrie. Baboon Remark Harmful to Relations. *Dully Dispatch*, August 17, 1980.
53 Paulus, Arrie. Defence Team Seeks Acquittal. *The Citizen*, May 21, 1980.
54 Buti's Baboon Blunder. *The Times*, November 26, 2008.
55 Juju Rants at Illiterate and Uneducated Media. *Times Live*, July 29, 2011.

mates acting as human proxies. Public outrage towards primate invasions is often disproportionate. There has been evidence of extreme violence towards primates in the suburbs, which suggest that the rage is not only about the baboons and monkeys but about societal anxiety more generally. People have suffered through the metaphor, but so have animals. Baboons have stood for too much – more than their small, grey bodies can bear. The derogatory metaphors are bad for individuals and for society – but also for baboons.

4 Implication(s) of the Animal Turn

This chapter explored a controversial and hurtful theme in order to *demonstrate*, rather than merely *assert*, the significance of animal histories in understanding Africa's past – and present. This is partly to refute any claims that adding "species" to the analytical trinity of "race, class and gender" is indulgent or irrelevant [→Feminist Intersectionality Studies]. Because it is pivotal in, for example, demonstrating how animalization and dehumanization be defeated. A key weapon is the writing of history itself. What is clear is that it is not enough to simply point out examples of othering – one needs to explain it ideographically to counter it. One basic tool is empathy, which comes from an understanding of history. An understanding of past deployment shows that social context and shifting knowledge have – by operating together – helped shaped the "baboon metaphor" in popular discourse. The chapter shows how shifting ideas about animals allow humans to express ideas about human identity [→History of Ideas]. As noted at the start of the chapter, the "full history" of the baboon-human relationship is not simple – it shows that the relationship between the two species was non-linear, was geographically and culturally idiographic, and had ramifications external to the relationship between the two. It argues that shifts *within* the culture(s) of the dominant species impacted on the relationship *between* the species. This relationship was influenced by scientists and writers but was also influenced by the knowledge generated by quotidian interactions between ordinary members of both species.

Fundamentally, this chapter has explained why a trade unionist, a politician and a bishop once joined together to disavow any kinship with a baboon. But on a much broader level, it has shown why an animal-sensitive perspective impacts such histories: it wishes to contend that bodies of knowledge about "animals" are not static, and they are diachronic, porous and accretive. In essence, this chapter is both a plea for and a demonstration of the necessity for animal-sensitive histories. A more nuanced story can be told if we include the perspective from a very different historical context. The point is that local idiographic contours powerfully shape the local knowledge held by the public – and this too is fissured by the trajectories of local social history. You cannot just tell a "Global North" story of ape–human relations from ancient Greece to Anglo-American primatology and think you have told the *whole* story. Finally, shifting human understandings of the animal Other impacts not only on the animal but – through them – on other people. Indeed, the human relationship with animals was connected to both how humans see themselves and, further, how they see other humans. Baboon metaphors have been integral as mechanisms of social control. But what is equally important is that "baboon" has meant different things to different people in different periods. If such ideas have changed before, they may change again – which offers hope for a different kind of relationship between humans and baboons.

Selected Bibliography

Beinart, William, Karen Middleton and Simon Pooley. *Wild Things: Nature and the Social Imagination.* Cambridge: White Horse Press, 2013.
Haraway, Donna. *Simians, Cyborgs and Women – the Reinvention of Nature.* London: Free Association Books, 1991.
Hund, Wulf, Charles Mills and Silvia Sebastiani. *Simianization. Apes, Gender, Class, and Race.* Zürich: Lit Verlag, 2016.
Skabelund, Aaron. Animals and Imperialism: Recent Historiographical Trends. *History Compass* 11 (October 2013): 801–807.
Speitkamp, Winfried and Stephanie Zehnle. *African Animal Spaces.* Cologne: Rüdiger Köppe, 2014.
Swart, Sandra. Animals in African History. In *Oxford Research Encyclopedia of African History*, Thomas Spear (ed.), published 26 April 2019. doi.org/10.1093/acrefore/9780190277734.013.443.
Swart, Sandra. *Riding High – Horses, Humans and History in South Africa.* Johannesburg: Witwatersrand University Press, 2010.
Van Sittert, Lance and Sandra Swart. *Canis Africanis – a dog history of Southern Africa.* Leiden: Brill, 2008.
Wylie, Dan. *Death and Compassion: The Elephant in Southern African Literature.* Johannesburg: Witwatersrand University Press, 2018.

Anna Boswell
Australasian and Pacific Studies

1 Introduction and Overview

To consider animal histories in the southern island worlds of Australasia and the Pacific is to grapple with habitats that extend beyond human occupation and with pasts that exceed human memory. These adjacent and interconnected regions began forming roughly 80 million years ago through the fragmentation of the ancient supercontinent of Gondwana. Contemporary polities in the regions comprise the island nations of Australia and Aotearoa/New Zealand as well as archipelagos stretching across Melanesia, Micronesia and Polynesia and the vast ocean that links them. The inherited ecosystems that make up Australasia and the southern Pacific have been described by governments within these regions as constituting their "biological wealth" and "natural capital".[1] Diverse in nature, they comprise scattered atolls, coral reefs, volcanic cones, extensive shorelines, lowland swamps, glacier-scapes, mountain ranges and ancient forests, and encompass desert and tropical zones as well as temperate and sub-Antarctic ones. Extremely high rates of faunal endemism coupled with the regions' geological distinctiveness and remoteness from major continental land masses and centers of population make Australasia and the southern Pacific remarkable cases in island biogeography and evolutionary adaptation. The regions' predominantly aquatic rather than land-based worlds of life and their deep soundings – which have recently revealed Zealandia, a submerged continent[2] – also make them world-leading subjects for studies in marine ecology and ocean-dwelling.

Moreover, within recent human occupation and memory, Australasia and the southern Pacific have become engulfed in upheaval. In large part, this is a consequence of the commodification of their biological wealth and natural capital for state and private gain, pursued by and in the service of colonial regimes [→(Post)Colonial History]. Since the closing decades of the eighteenth century, patterns of indigenous environmental stewardship have been interrupted and suspended by the annexation of territories and extension of British, European, and Euro-American settlement. Colonial and ecological histories in these regions are further complicated by what might be termed the 'secondary' colonial activities of settler governments. In the twentieth century, for instance, Aotearoa/New Zealand colonized Samoa, the

---

[1] See for example New Zealand Government. *The New Zealand Biodiversity Strategy*. Wellington: Department of Conservation and Ministry for the Environment, 2000, 3–4.
[2] See Nace, Trevor. Stunning New Maps Reveal What the Lost Continent of Zealandia Looks Like. *Forbes*, June 29, 2020. URL: forbes.com/sites/trevornace/2020/06/29/stunning-new-maps-reveal-what-the-lost-continent-of-zealandia-looks-like/#2e113db822be (June 30, 2020).

Cook Islands, Niue and Tokelau, while Australasian nations have intensively mined rock phosphates from islands such as Nauru, using them to manufacture chemical fertilizers which have conditioned southern-world soils for agricultural makeover. As a result of colonial histories, habitats across Australasia and the southern Pacific have been destroyed, degraded, fragmented and extensively remade; terrestrial and marine resources have been unsustainably harvested and harnessed; a deluge of introduced exotic species has irrevocably altered food chains and faunal densities; endemic biodiversity has declined; and an avalanche of extinctions has begun to take hold. Indeed, as settler governments in these regions have publicly conceded, human-wrought ecological transformation in Australasia and the southern Pacific in the last 250 years has, by world standards, been catastrophic in its intensity, rapidity and scale.[3] At the same time, fueled by the northern-world imaginary that persists in figuring these regions as pristine South Seas eco-paradises, tourism has burgeoned [→Environmental History].

Amid these contradictions, Australasia and the southern Pacific are tasked with safeguarding their unique biodiversity on behalf of the planet.[4] Yet forceful logics of economic development – which align the 'soft' industries of tourism and conservationism with harder extractive industries such as forestry, agriculture, aquaculture and mining – mean such protection tends to be piecemeal at best, and focused on turning a profit. On an ongoing basis, unpredictable impacts on the regions' biota are produced by forces of nature: tectonic contours make Australasia and the southern Pacific prone to earthquakes, tsunamis, volcanic activity and sea-level shifts, and the regions' landmasses – which span tropical as well as southern latitudes, and are largely low-lying and water-locked – are vulnerable to weather patterns that bring cyclones, monsoonal rainfall, flooding, droughts, erosion, bushfires and blizzards. At the same time, the regions' oceanic ecosystems are being redefined by what Tess Lea terms "the warming acidity of an increasingly seasick environment".[5] In this context, human-wrought planet-wide climate change looms as a horizon of further calamity for the lives supported by these regions.

The very real complexities arising from these combined forces make southern-world regions of immense interest for the study of animal histories. Indeed, because indigenous attachments to place in these regions are directly expressed through relationships with animals, and because economic development in the regions relies so directly on lands and waters and living organisms, Australasia and the southern Pacific are places where deep contestation is emerging over how whole worlds of life are constituted and tended. Not all of the animal studies scholarship produced in or about these regions deals with issues particular to the regions and their histories. In what follows, I focus on topics, themes and approaches that foreground regional

---

3 See NZ Government, *Biodiversity Strategy*, 4–6; Australian Government, *Australia's Biodiversity Conservation Strategy*. Canberra: Natural Resource Management Ministerial Council, 2010, 3.
4 See for example NZ Government, *Biodiversity Strategy*.
5 Lea, Tess. *Darwin*. Sydney: NewSouth Publishing, 2014, 154.

concerns and offer lessons and comparators for animal historiographies emerging elsewhere. As I hope to show, there are many explicit and implicit points of connection with diverse areas covered in the present volume, including →History of the Zoo;→History of Agriculture;→Public History;→American Studies;→Zooarchaeology. While the scholars whose work I discuss tend to be based in southern-world contexts, one does not have to be physically located in Australasia or the Pacific in order to study animal histories associated with these regions. Many materials are available online and the histories in question are not confined within the regions' geographical compass: animals from Australasia and the southern Pacific are found in zoos and museums worldwide; some of the distinctive faunal species associated with these regions – including *kiore* (Pacific rats) and dingoes and *kurī* (Australian and Pacific dogs) – were seafarers from Asia; island states and dependencies may historically have been subject to federal laws enacted on distant continents; markets for commodities that affect southern-world animals stretch across the globe, and so on. Within these regions, too, it is not the case that endemic creatures are the only ones of concern or that their value is commonly agreed; introduced animals such as rabbits and sheep, for instance, take on extraordinary valence when imported to southern-world places.

2 Topics and Themes

Recovery

To a significant degree, work in the area of historical animal studies involves excavating and piecing together knowledges from the past and striving to understand how the past surfaces in and inflects the present. As sketched above, in relation to Australasia and the southern Pacific, the operative timescales are at once expansive, compressed and complexly interwoven. A single species might have an evolutionary history dating back 80 million years; a history of value and meaning in the terms of an indigenous culture dating back 50,000 or 500 years; a history of attempted extermination in the terms of a settler worldview dating back 200 or 250 years (drawing, of course, on a longer northern-world history of human-animal relationships); and a history of esteem as a prized zoo exhibit dating back ten years or less. As the oldest-living sentient inhabitants of Australasia and the southern Pacific, animals endemic to those regions can be understood as carriers of long memory.[6] To attend to the histories and habits of such animals is to recall pasts that are otherwise obscured from view. Animals also bring the past to life through their present-day activities and their interactions with other endemic and newcomer species, and

---

6 Boswell, Anna. The Sensible Order of the Eel. *Settler Colonial Studies* 5, no. 4 (February 2015): 363 – 374.

through their various treatments by humans. Faunal imports – including ancient ones, such as the rats and dogs brought by the first human inhabitants of the southern Pacific, as well as the more recent flood of foreign species brought by European settlers – express relationships that create further layers of overlap. The work of recovery involves attending to entwined short and long histories of place, and to temporal disjunctions, hauntings, paradoxes and willed forgettings of various kinds. It also involves attending to attitudes and logics that may be radically inconsistent across time, and to questions of belatedness, futurity, restoration and starting over.

(Re-)Mapping

The study of animal histories involves piecing together maps and tracks of various kinds. Animals may produce footprints that take the form of archaeological traces, physical evidence of foraging or nesting, and so on. Equally, their traces may be virtual, materializing as imprints in media or legislative records, or in storytelling or image-making traditions. Australasia and the southern Pacific have strong histories of human migration, and settler governments have striven to anchor their newly-constituted nations via strenuous public memory-work in the form of written histories, memorials and exhibitions. For these reasons, observing and tracking animals in southern-world regions may involve disassembling fixed cartographies, geopolitical boundaries, socio-cultural norms and historiographical practices otherwise taken as given.

Through their own mobility as well as through their transportation by humans, animals expose barriers that are defined differently at different junctures. Strict quarantine and biosecurity regimes currently enforced at international ports and airfields in Australasia, for instance, acquire new meaning when considered in light of the coastal breeding depots set up by acclimatization societies in the 1800s to husband foreign species shipped for intentional release into the wild. Animal populations may disperse or recede over time and their status may change as they cross jurisdictional thresholds. While the brushtail possum enjoys state-sponsored protection in Australia, for example, the same species is slated for state-sponsored eradication in Aotearoa/New Zealand – having been introduced by the New Zealand government deliberately and repeatedly in the nineteenth century in order to found a national fur industry. Human-possum relations in Aotearoa/New Zealand are further complicated by the fact that the possum may be understood as a surrogate or proxy for the figure of the settler – as seen in its destructiveness, its territorial invasiveness and its aptitude for establishing abundant populations when transplanted to already-inhabited lands. In completing their migratory journeys, too, southern-world animals may travel across terrains where they are subject to sharply differentiated modes of protection and aggression. Yet because such routes have the capacity to diagram ranges that are not human ones, animal "ichnographies" also produce alternate proximities and patterns of life-in-place.[7] The migratory journey of the *tuna* (or longfin eel), for instance, connects fresh water bodies in Aotearoa/New Zealand with deep ocean trenches off the coast of Tonga, reminding that Aotearoa was once swampy land and that it is still entwined with Pacific homelands.[8] In these ways, place-based knowledges enacted by animals re-center local worlds as their own domains of life, proceeding without reference to conceptual geographies founded on hemispheric thinking, on notions of periphery and metropole, or on hard distinctions between terrestrial and aquatic environments.

Stories told about animals function as oral maps, bringing places and peoples into being and forming the basis for shared attachments and culturally-located worldviews. Such stories are key resources for the animal historian: Australasia and the southern Pacific shed light on how narratological processes form, unfold and falter. The same creature might be 'storied' differently across different locations within these regions, with stories acquiring or requiring divergent inflections as they are handed down [→Multispecies Ethnology]. Indeed, histories of settler occupation in Australasia and the southern Pacific and ongoing indigenous attachments to species and place mean there is no commonly-held sense of what an animal or its story might mean in any given context, and radical reversals might make the stories circulated by settlers in a single location inconsistent across short spans of time. Zoos and museums across the globe have sought southern-world fauna for its exoticism, requiring the animals in question to forge new relationships to place. The types of taxonomic sociality that such collections create – whereby animals are excised from their own ecosystems and thrust into unfamiliar relationships – speak to powerful institutionalized forms of dislocation and displacement [→History of Animal Collections/Animal Taxonomy]. The experience of alienation is not restricted to creatures removed from their homelands, however. As Mark Jerome Walters poignantly puts it in *Seeking the Black Raven*, habitats in places such as Hawai'i have left the animals who originate from there, making these creatures "familiars in a strange land".[9]

Absence/s

Environments in Australasia and the southern Pacific have historically been characterized from afar for their strangeness and depauperization – that is, for what they seemed to lack. Northern-world newcomers alternately marveled or were aghast at what they perceived as the ecosystemic niche-shifts that caused southern-world insects to act like rodents, birds to act like mammals and so on. Once identified,

---

[7] Carter, Paul. *Dark Writing: Geography, Performance, Design*. Honolulu: University of Hawai'i Press, 2009, 166.
[8] See Boswell, Sensible Order.
[9] Walters, Mark Jerome. *Seeking the Black Raven: Politics and Extinction on a Hawaiian Island*. Washington, DC: Island Press, 2006, 3.

such so-called deficiencies functioned as enabling absences, justifying large-scale efforts to stock southern-world places with free-range protein and game, domestic companions, livestock, wild mammals and birds, and service animals for use in desert exploration and agricultural development. Sentimentality and imagined economic functionality have been the dual drivers of these introductions. Among other things, constructed absences have saturated southern-world places with new categories of 'pest': self-sustaining mobs or plagues of camels, rabbits and possums, to name just a few. Several imported gap-fillers have created further gaps, preying on or replacing endemic creatures and/or throwing into jeopardy the ecosystems and economies they are supposed to serve. As an unwitting vector for bovine tuberculosis, for instance, the brushtail possum looms as a direct threat to the national dairy industry in Aotearoa/New Zealand. At the same time, select endemic species – such as the thylacine, crocodile and longfin eel – have been pathologized by settler populations as vermin and willed towards extinction.

What these circumstances more deeply reveal is that settler newcomers lacked foundational knowledge about how to live in southern-world places. They also reveal regimes of value which promote non-indigenous norms as baselines, prioritize industrial expansionism and commercial profitability, and presume authority to overturn existing ways of life and make-over existing environments. As endemic biodiversity has come under strain, and as the long-term effects of unsustainable harvesting of endemic fauna and the ecosystems that support animal and human lives have come to light, absence and erasure have come to signify in new ways in Australasia and the southern Pacific. This reversal has coincided with national identities coming to rely on the animals of the regions – propelled, in part, by mounting tourist demand for encounters with endemic wildlife, and by the global dawn of modern environmental consciousness. Refigured as something slipping or already-slipped, absence is now likely to be construed as a hole in the fabric of the lifeworld: a gap where the thylacine or the Hawai'ian crow ('alalā) once flourished.

From a contemporary standpoint, the history of the acclimatization experiments conducted in Australasia and the Pacific seems shockingly fanciful. Arctic foxes were proposed for introduction in Aotearoa/New Zealand in the late 1800s; moose were shipped there from North America but failed to establish free-living populations, and so on. Yet, as seen in the case of the New Zealand government's recently unveiled Predator Free 2050 campaign, shockingly fanciful future absences are also still being willed. Possums, stoats and rats face state-sponsored nationwide obliteration under this world-leading campaign, whose own publicized lack is sufficient funding to enable development of the gene editing and immunocontraceptive advances that will "silence" these animals in perpetuity.[10] Because their histories are strongly implicated in large-scale biodiversity loss, Australasia and the southern Pacific are also at the global forefront of de-extinction debates. The thylacine, a carniv-

---

10 See Boswell, Anna. Stowaway Memory. *Pacific Dynamics* 2, no. 2 (November 2018): 89–104.

orous marsupial hunted out of existence by Tasmanian pastoralists and renowned as a species whose endling died in a zoo in 1936, beckons as a candidate for resurrection.[11] For the animal historian, such stories focalize difficulties which illuminate processes of national identity-making and the workings of political ecology.

Identity

Across Australasia and the southern Pacific, indigenous worldviews hold that people are indistinguishable from animals. The interconnectedness of everything in the world of life means that humans share ongoing genealogical relationships with animals, who are ancestors and spiritual guardians, and may also be ingested for sustenance and/or as a delicacy. Crocodiles, for instance, feature in the stories, songlines and ancient rock art of several northern Australian Aboriginal clans, and because some believe the spirits of the dead are contained in the bodies of certain large crocodiles, the deaths of such crocodiles cause widespread mourning.[12] Clans, communities and individuals are understood as belonging to certain animals: in Hawai'i, for instance, the name of the 'alalā is bestowed on a class of human orators whose skill is inspired by the bird's own communicative prowess.[13] Because animals supply the coordinates by which people understand themselves and their relationships to the world, identity and orientations to time and place are powerfully informed by animal lore and law. Te Ika a Maui, the landmass subsequently renamed by Europeans as the North Island of Aotearoa/New Zealand, is a fish hauled up by Maui, the trickster demigod figure known across the Pacific. As Alice Te Punga Somerville puts it, "This is a story of firstness: as long as there has been a fish, there have been us [...] We are standing in the mouth of a fish on which our people have always lived [on Matiu/Somes Island], and on this basis, we are Indigenous".[14]

Human-animal intimacies and identification practices work very differently in settler cultures. As Adrian Franklin has noted, early northern-world visitors to Australasia and the southern Pacific historically defined themselves as absolutely different from animals, construing the closeness of connections between indigenous peoples and animals as evidence of these peoples' evolutionary inferiority [→History of

---

11 See Narraway, Guinevere and Hannah Stark. Re-animating the Thylacine: Narratives of Extinction in Tasmanian Cinema. *Animal Studies Journal* 4, no. 1 (January 2015): 12–31.
12 See Pooley, Simon. Invasion of the Crocodiles. In *Rethinking Invasion Ecologies from the Environmental Humanities*, Jodi Frawley, Iain McCalman (eds.), 239–255. London and New York, NY: Routledge, 2014.
13 See Walters, *Black Raven*.
14 Te Punga Somerville, Alice. *Once Were Pacific: Māori Connections to Oceania*. Minneapolis, MN and London: University of Minnesota Press, 2012, xviii.

Ideas].[15] Yet the ability of humans to metamorphose into other animals by means of shape-shifting is a widespread and ingrained trope in Australasian settler cultures.[16] The strong tendency for settlers to self-identify with totemic faunal icons such as the kiwi, wallaby and kangaroo emerged as an expression of eco-nationalist yearning in the early to mid-twentieth century, fueled by the rise of conservationism and tourism. More deeply, however, settler cultures persist in identifying with the introduced animals that most remind them of themselves, creating "anti-animals" such as the stoat, possum and rat in Aotearoa/New Zealand, and the rabbit and fox in Australia.[17] Publicly reviled and targeted for state-sponsored extermination, such animals bear the burdens of settler misgiving. Other disturbing pattern-work is visible in relation to people and animals in southern-world histories. As Carol Freeman explains, the "profound absences" marked by faunal extinctions were expected by settler newcomers to foretell the disappearance of indigenous populations.[18] In some cases, too, faunal emblems are deployed as surrogates in depicting more recent histories of migration. Such characterizations surface in the Australian reality television series *Border Control*, with xenophobic sentiments displaced onto cane toads and wasps.[19]

Emergence / Emergency

For indigenous populations, Australasia and the southern Pacific have long been places of emergent knowledges taught by and through animals. Such teachings are at once practical and figurative. Te Ahukaramū Charles Royal notes, for instance, that the fish (*ika*) is a metaphor for ideas in *te ao Māori* (the Māori world), so that appearances of ika in oral traditions evoke theoretical debate and the pursuit of understanding and innovation.[20] At the same time, knowing fish involves knowing what fish know – about the health of a waterway and its inhabitants, about plenitude and depletion, about currents and tides, about seasons and calendars, about *taonga tuku iho* or ancestral treasures handed down through time – patterning human life so that it is synchronized with the lives on which it depends. Over centuries or perhaps millennia, fishing, hunting and fire management regimes, conducted through trial

---

15 Franklin, Adrian. *Animal Nation: The True Story of Animals and Australia*. Sydney: University of New South Wales Press, 2006, 48.
16 See Simmons, Laurence and Philip Armstrong. *Knowing Animals*. Leiden and Boston, MA: Brill, 2007; Boswell, Anna. Anamorphic Ecology or the Return of the Possum. *Transformations* 32 (2018): 1–18.
17 See Holm, Nicholas. Consider the Possum: Foes, Anti-Animals, and Colonists in Paradise. *Animal Studies Journal* 4, no. 1 (2015): 32–56.
18 Freeman, Carol. *Paper Tiger: A Visual History of the Thylacine*. Leiden and Boston, MA: Brill, 2010, 1; see also Boswell, Dodo.
19 See Frawley, Jodi and Iain McCalman (eds.). *Rethinking Invasion Ecologies from the Environmental Humanities*. London and New York, NY: Routledge, 2014.
20 Cited in Te Punga Somerville, *Once Were Pacific*, 245.

and sometimes error, yielded insights into sustainable practices of harvest and management. The crisis precipitated when the giant flightless moa was hunted to extinction in pre-European Aotearoa, for instance, gave rise to new forms and modes of traditional ecological knowledge encoded as *kaitiakitanga* (or guardianship) and passed on through *whakataukī* or proverbs.[21] Knowledges about times and places for acting in respect of animals and the wider world of life are also embodied in Dreamings, creation stories, dances, songs and prayers. As this begins to make clear, indigenous environmental stewardship in southern-world places is inter-generationally transmitted and modelled on reciprocity: people are charged with observing and taking care of fellow creatures who are, in turn, charged with taking care of them. In a larger sense, then, local understandings in these regions hold that animals and humans and their shared worlds of life are emergent – in a state of continuously "becoming" together.[22]

Emergence and emergency have accelerated in the 250 years since Europeans first made landfall in Australasia and the south Pacific, acquiring new inflections and new urgencies. Reconstruction of lands and waterways, degradation of ecosystems and interference in mnemonic traditions has subjected endemic species to distress, reverse-engineering these places from the point of view of newcomers and amounting to what Deborah Bird Rose terms "death-worlding".[23] As the Predator Free 2050 campaign demonstrates, settler responses to the extreme ecologies created in Australasia and the southern Pacific have, themselves, proven extreme, throwing up new forms of incipient knowledge. In a remarkably short span of time, for instance, newly mobile categories of niche-shifting creature (familiar exotics, state-sponsored serial killers, endemic strangers, pure contaminants, endangered pests) have arisen. Meanwhile, resilience has come to be construed as a harmful attribute in a number of animal populations, heralding the arrival of a "feral future".[24] Refashioned or newly constructed institutions and technologies of crisis (wildlife sanctuaries, predator- and pest-control methods and so on) have also been pioneered in southern-world places as part of ecological disaster-containment efforts in these regions. Instituting new regimes of "violent-care" in the name of conservation, such institutions and technologies have turned exile-at-home into a permanent state for critically endangered endemic creatures.[25]

---

21 See Wehi, Priscilla M., Hēmi Whaanga and Tom Roa. Missing in Translation: Māori Language and Oral Tradition in Scientific Analyses of Traditional Ecological Knowledge. *Journal of the Royal Society of New Zealand* 39, no. 4 (November 2009): 201–204.
22 See Haraway, Donna. *When Species Meet*. Minneapolis, MN and London: University of Minnesota Press, 2008, 244.
23 Rose, Deborah Bird. *Wild Dog Dreaming: Love and Extinction*. Charlottesville, VA and London: University of Virginia Press, 2011, 12.
24 Low, Tim. *Feral Future: The Untold Story of Australia's Exotic Invaders*. Chicago, IL: University of Chicago Press, 2002.
25 Van Dooren, Thom. *Flight Ways: Life and Loss on the Edge of Extinction*. New York, NY: Columbia University Press, 87–122.

Yet catastrophic narratives are not the only ones possible in respect of animal histories in Australasia and the southern Pacific. Indeed, the disrupted ecologies produced by colonial histories in these regions have been figured by indigenous populations as transformative scenes of new understanding, intimacy and potential. Species such as camels and donkeys – deliberately introduced to Australia as valued beasts of burden tasked with constructing the modern settler nation, yet subsequently reviled as unwanted ferals – have been integrated in complex ways into stories of identification and patterns of memory and knowledge, such that they may be considered to belong to country [→Political History].[26] Likewise, counterfactual indigenous histories – such as those telling how possums might have been dealt with by councils of tribal elders prior to European arrival in Aotearoa/New Zealand – suggest the basis for re-remembering or reimagining shared futures.[27] In material-semiotic terms, such imaginings are simultaneously impossible and already-lived.

3 Methods and Approaches

Animals abound – or perhaps fail to abound – in Australasia and the southern Pacific, and animals from these places circulate and dwell elsewhere too. Approaches to studying their histories call for attentiveness: noticing the arrival of a species or discursive category (such as 'the pest') or perhaps noticing its disappearance; detecting something that is in the process of spreading, shifting or fading; intuiting strangeness, contradiction, difficulty or elusiveness. The most prevalent approach taken to studying the animal histories of these regions to date is that of single-species studies or animal biographies. Such studies might focus on an extinct animal, a threatened or critically endangered one, an agricultural one, an introduced pest, an endemic pest or vermin species, or a crossover species which is at once familiar and unfamiliar to settlers and/or indigenous communities in southern-world contexts. While some scholarship focuses on so-called charismatic fauna such as albatrosses and whales, a growing volume devotes itself to species deemed "unloved" by settler cultures, exploring the ethical dilemmas generated by species which are marginalized and/or maligned.[28] Studies may focus on individual animals – as in the case of the 'alalā family tree that tracks through *Seeking the Black Raven*.[29] Shifting the frame a fraction, scholars may elect to focus on decontextualized animals: Harriet Ritvo considers the first platypus taken to London, whose perplexed reception reveals how southern-world places began to expose the shortcomings of the Linnaean

---

[26] Rose, Deborah Bird. Judas Work: Four Modes of Sorrow. *Environmental Philosophy* 5, no. 2 (Fall 2008): 51–66.
[27] See Boswell, Anamorphic Ecology.
[28] Rose, Deborah Bird and Thom van Dooren. Introduction – Unloved Others. *Australian Humanities Review* 50 (May 2011): 1–4.
[29] Walters, *Black Raven*.

taxonomic imaginary;³⁰ Philip Armstrong invites rumination on the armies of sheep which were imported to Aotearoa/New Zealand as an occupying force, yet which also sanctified settlement as a Judeo-Christian spread-event.³¹ Or scholars might focus on institutions such as zoos, sanctuaries, museums and international exhibitions in Australasia and the Pacific or elsewhere, and on the role of southern-world creatures in thwarting conventional rhetorics of display.³²

Further approaches might be inspired by animal-dependent industries such as agriculture, fishing and ecotourism, or on their commercial products – export commodities such as a packaged leg of lamb, or gourmet pet food manufactured from *tuna*, an endemic species that is at risk and in decline.³³ Contemporary phenomena affecting the lives of animals provide prompts in some cases: a disaster like the earthquakes which levelled much of Aotearoa/New Zealand's second largest city in 2011;³⁴ a donkey cull in outback Australia or a pack of dingoes strung from a tree on the fringe of a national capital;³⁵ a rescue effort to assist penguins at a rapidly gentrifying New South Wales beach;³⁶ an investigation of how waterfowl have responded to the new spaces of irrigated agriculture that have replaced their former wetland habitats;³⁷ a human trial of xenotransplantation using an isolated colony of introduced pigs.³⁸ Or a study might be prompted by a chanced-upon archival relic, such as a photo of a faked dodo fossil donated to a New Zealand museum at a time when the country was making global headlines for its own ornithic extinctions.³⁹ As is the case internationally, representational media frequently supply objects of interest. Donna Haraway, for instance, meditates on how the queer paedomorphic sculptures produced by Australian artist Patricia Piccinini re-work the trope of aberrance initially applied to Australian fauna by northern-world observers

---

30 Ritvo, Harriet. *The Platypus and the Mermaid, and Other Figments of the Classifying Imagination*. London and Cambridge, MA: Harvard University Press, 1997.
31 Armstrong, Philip. *Sheep*. London: Reaktion, 2016.
32 See Anderson, Kay. Animals, Science, and Spectacle in the City. In *Animal Geographies: Place, Politics and Identity in the Nature-Culture Borderlands*, Jennifer Wolch, Jody Emel (eds.), 27–50. London and New York, NY: Verso, 1998.
33 See Potts, Annie, Philip Armstrong and Deidre Brown. *A New Zealand Book of Beasts*. Auckland: Auckland University Press, 2013, 1–6; Boswell, Sensible Order.
34 See Potts, Annie and Donelle Gadenne. *Animals in Emergencies: Learning from the Christchurch Earthquakes*. Christchurch, NZ: Canterbury University Press, 2014.
35 See Rose, Judas Work; Rose, *Wild Dog Dreaming*.
36 See van Dooren, *Flight Ways*.
37 See Hobbins, Peter. Invasion Ontologies: Venom, Visibility and the Imagined History of Arthropods. In *Rethinking Invasion Ecologies from the Environmental Humanities*, Jodi Frawley, Iain McCalman (eds.), 181–195. London and New York, NY: Routledge, 2014.
38 Carr, Rachel. 100% Pure Pigs: New Zealand and the Cultivation of Pure Auckland Island Pigs for Xenotransplantation. *Animal Studies Journal* 5, no. 2 (2016): 78–100.
39 Boswell, Anna. Lessons from the Dodo. *Journal of New Zealand Studies* 24 (June 2017): 72–86.

at the same time as they comment on topologies of endangerment and technologies of preservation and reproduction.[40]

In practice, the story of an individual creature or species or event tends to refract several themes or topics at once. Studies may adopt explicitly comparative approaches, within Australasia and the Pacific or beyond these regions. Nicholas Holm, for instance, posits a therianthropic explanation for the possum's radically different treatment on both sides of the Tasman Sea.[41] In Jodi Frawley and Ian McCalman's edited volume *Rethinking Invasion Ecologies*, Ritvo tracks the spread of camels and starlings across the US before locating them in Australia; Morgan Richards follows the cane toad from the sugar plantations of Hawai'i to those of the Gold Coast; Peter Hobbins unravels the historical processes whereby stories of the New Zealand katipo spider "discursively envenomated" public ideas about the Australian redback.[42] Some studies – such as those of the fabricated nineteenth century "Feejee mermaid" – deal with circulating signs of southern-world regions that contain nothing of the regions.[43] And other studies which begin with animals centrally present, such as *Seeking the Black Raven*, turn out to be as much about encounters with the people, texts and sites of history (voyagers, administrators, journals, policies, libraries, archives, graveyards) as they are about creatures.[44] One of the poignant parallels in Walters' study is that as the 'alalā fades from its own story and becomes dependent on the prostheticizing work of sanctuaries, Walters and his readers become dependent on textual prosthetics which transform the 'alalā into a "paper bird".[45]

As this suggests, colonial legacies in Australasia and the southern Pacific have laid down strong documentary records which are of immense value to historical animal studies scholarship. From the time of James Cook's Pacific voyages in the 1760s and 1770s – which were, in part, charged with scientific discovery of economically useful plants and animals – visitors began recording observations [→History of Science]. As they wrote and drew, and as they collected and distributed specimens, these visitors were aware that their activities were historic in that they were placing on the record species and taxa previously unknown to Linnaean classification. One of the key scholarly methodologies in historical animal studies in Australasia and the southern Pacific, then, involves examining archival arks that include voyage documents, early writings by missionaries and colonists (such as letters, reports and transactions and proceedings of scientific institutes, acclimatization societies and so on), and representations made by itinerant and immigrant artists. Embalmed information about animals and human-animal relationships is prominent in these

---

40 Haraway, Donna. Speculative Fabulations for Technoculture's Generations. In *The Multispecies Salon*, Eben Kirksey (ed.). Durham, NC and London: Duke University Press, 2014.
41 Holm, Consider the Possum.
42 See Frawley and McCalman, *Invasion Ecologies*.
43 Ritvo, *Platypus and Mermaid*.
44 Walters, *Black Raven*.
45 See also Freeman, *Paper Tiger*.

sources, as it is in ongoing traditions of scientific writing and in public records and government reports and legislative documents. An especially important suite of historical sources is furnished by Charles Darwin, Alfred Russel Wallace, Charles Elton and Alfred Crosby, who reflect on species evolution, faunal realms and imperial-invasion ecologies. Indeed, through this corpus of early writings it is possible to trace the pivotal role played by Australasia and the southern Pacific in the development of scientific conceptions of the globe as being comprised of distinct biogeographic regions.

If scholarly methods tend to be varied and rich, potential challenges are nevertheless posed by the complexity of the relevant materials and contexts:

First, southern-world places have long-been conceptualized from afar as being unnatural, anomalous, strange and exotic. In a range of ways, the upside-downness and back-to-front-ness (or mirror-world 'australity') that was projected onto these so-called antipodean places from the outside have become lived realities, such that the animal histories furnished by Australasia and the southern Pacific are – by world standards – extraordinary. Yet those who work in the area of southern-world historical animal studies need to be mindful of the conceptual paradigms on which they draw, and the intellectual traditions they privilege. It is important for scholars to strive to understand southern-world animal histories as being extraordinary or remarkable *in their own terms*, and to beware the banal exceptionalist logics and self-exceptionalizing discursive formulations that are characteristically deployed by settler cultures in so-called new world places.[46] Whatever might seem unfamiliar or peculiar from an outside perspective may well be ordinary or 'not-strange' within. For this reason, historical animal studies scholarship needs to be prepared to suspend or invert imported hierarchies and seemingly conventional forms of common sense, and to exhibit reflexive awareness of the models to which it turns and the citational practices it perpetuates or inducts.

Second, while the inherited archival record is strong, it also evidences the absence or possible extinction of certain kinds of paperwork. Many of the faunal specimens first sent to Europe and North America, for instance, have no surviving documented provenance. Historical records are sometimes misleading, too. As Walters points out, the naturalists on board Cook's ships mis-labelled the 'alalā specimens they collected, introducing errors into recorded knowledge at the very moment of the global dispersal of this species.

Third, reliance on the historical archive of written records risks reproducing a bias towards northern-world forms of textuality, and especially European forms. Carol Freeman and Murray Garde take care to show that human-animal knowledges are also transmitted in inscription practices such as rock painting and carving, as well as being documented in oral histories and in a vast range of objects of adorn-

---

[46] See Boswell, Anna. Climates of Change: A Tuatara's-Eye View. *Humanities* 9, no. 2 (May 2020): doi:10.3390/h9020038.

ment and status.[47] Not all animal-related knowledges are available, however, as a matter of public record in southern-world places. Walters, for instance, experiences difficulty in finding keepers of traditional knowledge about the 'alalā, coming to realize that boundaries need to be observed and preserved around sacred species.[48] Likewise, Dreamings and animal-informed cosmological knowledges may be embargoed. In dealing with complex cross-cultural histories, then, scholars need to develop reflexive awareness of how northern-world methodologies might come unstuck or run aground; why contemporary understandings and norms should be contextualized and historicized rather than universalized and back-projected; and how written sources, vocabularies and translation practices work to construct, circumscribe or occlude different knowledge practices.

Fourth, animal histories associated with Australasia and the southern Pacific are dispersed across vast swathes of the planet. They are also diffuse, inviting consideration in a wide array of academic fields – from media, art history and literary studies through to indigenous studies, settler colonial studies, anthropology, gender studies, conservation biology, science and technology studies, and legal studies. Because colonial histories in Australasia and the southern Pacific are recent and subject to radical revisionism as they continue to unfold, animal histories associated with these regions can be tricky to stabilize. Awareness of the interconnectedness of everything in worlds of life in the southern hemisphere and awareness of the co-existence of indigenous and non-indigenous modes of story-telling and knowledge-sharing can make it difficult to isolate stories – to work out what to do about seepage, how to make 'cuts', where to expand or link, where to stop. Materials and stories may also be obliquely recorded or dispersed and fragmented through public policies, songs, proverbs, voyage accounts, tribunal findings and so on. These challenges may require scholars to adopt new practices for working with unfamiliar knowledges, learning to stretch, sieve and "engraft"[49].

Finally, because animal histories associated with Australasia and the southern Pacific are multivalent, analyses drawn on and/or produced by scholars working on topics in this area may not always take the form of conventional academic outputs. Prominent commentators – such as Deborah Bird Rose and Thom van Dooren – have pioneered affectively-attuned practices of writing that combine personal narrative and scholarly critique with philosophical reflection and ethical concern. Such work is at once meditative and urgent, and it is committed to finding new ways to accommodate and account for difficulty. Powerful animal histories have emerged through artistic activities, too: a film about cane toads in Australia or an exhibition of specimens from Patricia Piccinini's queer menagerie might be understood as an

---

47 See Freeman, *Paper Tiger*; Garde, Murray. *Something About Emus: Bininj Stories from West Arnhem Land*. Canberra: Aboriginal Studies Press, 2017.
48 See Walters, *Black Raven*.
49 Boswell, Dodo, 84.

intellectual intervention in the historical study of southern-world animals.[50] In deciding how to convey the stories they seek to tell, scholars may need to consider public impacts and responsibilities – and, in particular, how to make work accessible and responsive to the indigenous communities whose lifeworlds and livelihoods are at stake. Again, animal histories in these southern-world regions offer powerful lessons about how frontiers of knowledge might be known, navigated and re-negotiated.

4 Implication(s) of the Animal Turn

As this chapter has sought to document, the *animal turn* has precipitated an emergent body of vibrant scholarship associated with southern-world places. Creative-critical in nature, this work re-centers historical animal studies from southern perspectives. Attending to place-based knowledges, it addresses how and why animals in southern-world contexts refuse globalized ways of knowing and it acknowledges that local animal histories have difficult and ongoing legacies. A larger implication of this work is that boundaries demarcating what counts as 'animal studies' in these regions begin to morph and/or break down. Indeed, in a number of the publications that I find most helpful for my own reference, animals are only obliquely present, yet they ghost-write historical geographies. Tess Lea, for example, explains how the city of Darwin in the Northern Territory of Australia was shaped by the flight range of the mosquito,[51] while Paul Carter and Philip Armstrong attend to the already-animalized nature of places by examining the chiaroscuro inscriptions which underwrite human designs.[52] Other studies barely contain animals at all, yet they speak in profound ways to understandings of animate worlds and the interconnectedness of all things. Alice Te Punga Somerville, for instance, unfolds a study of relationships between the Pacific and Aotearoa/New Zealand which refers to southern-world animals only in passing yet manifests orientations to place and history unthinkable except in the aqua-spheric terms and webs of past-and-future-relations such animals call forth.[53] And birds take center stage as elders and teachers in historical animal studies published by Mark Walters and Murray Garde, fledging living knowledge systems that are at once ancient, current and impending.[54] Ultimately, then, one of the key challenges posed by Australasia and the southern Pacific is

---

[50] See Richards, Morgan. Cane Toads: Animality and Ecology in Mark Lewis's Documentary Films. In *Rethinking Invasion Ecologies from the Environmental Humanities*, Jodi Frawley, Iain McCalman (eds.), 149–165; Haraway, Speculative Fabulations.
[51] See Lea, *Darwin*.
[52] Carter, *Dark Writing*; Armstrong, Philip. Moa Ghosts. In *A New Zealand Book of Beasts: Animals in Our Culture, History and Everyday Life*, Annie Potts, Philip Armstrong, Deidre Brown, 9–32. Auckland: Amsterdam University Press, 2013.
[53] Te Punga Somerville, *Once Were Pacific*.
[54] Walters, *Black Raven*; Garde, *Something About Emus*.

the need for historical animal studies to develop practices of inquiry that respond to and are able to crystallize the non-linear temporalities of the more-than-animal.

Selected Bibliography

Franklin, Adrian. *Animal Nation: The True Story of Animals and Australia.* Sydney: University of New South Wales Press, 2006.
Freeman, Carol. *Paper Tiger: A Visual History of the Thylacine.* Leiden and Boston, MA: Brill, 2010.
Frawley, Jodi and Iain McCalman (eds.). *Rethinking Invasion Ecologies from the Environmental Humanities.* London and New York, NY: Routledge, 2014.
Garde, Murray. *Something About Emus: Bininj Stories from West Arnhem Land.* Canberra: Aboriginal Studies Press, 2017.
Potts, Annie, Philip Armstrong and Deidre Brown. *A New Zealand Book of Beasts: Animals in Our Culture, History and Everyday Life.* Auckland: Auckland University Press, 2013.
Rose, Deborah Bird. *Wild Dog Dreaming: Love and Extinction.* Charlottesville, VA and London: University of Virginia Press, 2011.
Simmons, Laurence and Philip Armstrong. *Knowing Animals.* Leiden and Boston, MA: Brill, 2007.
Te Punga Somerville, Alice. *Once Were Pacific: Māori Connections to Oceania.* Minneapolis, MN and London: University of Minnesota Press, 2012.
Van Dooren, Thom. *Flight Ways: Life and Loss on the Edge of Extinction.* New York, NY: Columbia University Press, 2014.
Walters, Mark Jerome. *Seeking the Black Raven: Politics and Extinction on a Hawaiian Island.* Washington, DC: Island Press, 2006.

Barbara R. Ambros and Ian Jared Miller

(East) Asian Studies

1 Introduction and Overview

People in East Asia have been writing on animals for as long as they have been writing. The earliest known form of Chinese script is carried on turtle plastron and ox scapula "oracle bones," inscriptions used for divination from at least the Shang dynasty (c. 1766–1122 BCE). Given this history, it is of little surprise that East Asia is home to rich traditions of writing about nonhuman animals. Natural historians, philosophers, clergy, scholars, and others in China, Japan, and Korea have long made animals the subject of consideration. Queries about animals – whether or not to sell livestock, for example – are among the first divinations found on those oracle bones, and the oxen whose scapulae carried script were themselves probably the subject of sacrificial rites. With a four-millennia history that spans from China to the coasts, mountains, and cities of Japan, Korea, and Taiwan, the simplest answer as to how to characterize animal histories in East Asia is to argue the obvious: they were – and are – profoundly diverse.[1]

The scope of animal history in East Asia narrows considerably when we limit our focus, as we have here, to works written in English and shaped by the *animal turn* in Anglophone scholarship, with its focus on questions of agency, companionship, emphasis on a particular set of ethics, and oftentimes normative modern and Western conception of the "animal". One function of reading in the East Asian literatures – like those other "other" provinces of world history – is to stymie the universalizing tendencies of Euro-American histories of the "animal," which can at times take the unitary status of their subject – the categorical fact and cultural fabrication that is "animal" – for granted. The history of East Asian engagements with the living world suggests the risks of rendering diverse histories into subsidiary aspects of a single cosmopolitan, globalizing story akin to world history.[2] Taken as a whole, East Asian animal histories show us how the "animal" at the center of animal history is always best understood as the contingent result of changeful cultural and historical processes.

Note: Subodhana Wijeyeratne and Aaron Van Neste provided research and editorial support for this paper.

1 Elvin, Mark, Three Thousand Years of Unsustainable Development: China's Environment from Archaic Times to the Present. *East Asian History* 6 (December 1993): 7–46.
2 Adelman, Jeremy. What is Global History Now? *Aeon*, March 2, 2017. URL: aeon.co/essays/is-global-history-still-possible-or-has-it-had-its-moment (February 10, 2021).

2 Topics and Themes

The stakes for animal studies are significant in East Asia. Roughly 20% of the world's human population lives in the region, which encompasses an array of environments, from steppe to tropical climates. Even as the oracle bones mark the depth of East Asian animal history, they also tell us that those engagements were often violent. The scale of this trouble is obvious. In 2016 the International Union for Conservation of Nature (IUCN) listed 1,500 plant and animal species in the region as "Critically Endangered" and further noted rapid degradation in key ecologies. Pet-keeping has expanded, tracking the emergence of bourgeois cultures from Meiji Japan (1868–1912) to the People's Republic of China (PRC) today, where the pet industry saw compound growth rates of nearly 50% from 2010 to 2016 alone.[3] The exotic animal trade is robust. The *New York Times* recently noted that even after official efforts to stem trade, at least 20,000 elephants are killed annually to supply tusks to the PRC. The problem is not just terrestrial. Japan and China are the world's third- and first-largest consumers of seafood, respectively, with Japan's impact compounded through preference for top predators such as bluefin tuna that fill sushi platters served to apex predators: humans.[4]

The catalogue of "human-animal interactions" runs blood-red in East Asia, but the turn to understand nonhuman animals as agents of historical processes is recent. Animals have been there, to be sure, but until the closing decades of the 20th century they were often rendered as passive objects or listed as commodities. The reasons for this dismissal are complex, but two dynamics are suggestive. First, intensive crop farming was prioritized in China, Japan, and Korea for centuries, diminishing the relative importance of domesticated animals. Direct contact with living animals was often associated with rural life. As such, animals were segregated from metropolitan centers by markets and class distinctions [→Urban (and Rural) History]. Even after the advent of modern animal husbandry and the elevation of meat-eating as a "civilized" act in Japan, for example, the act of slaughter continued to carry the stigma of pollution used to discriminate against so-called "outcastes," whose occupation-based status was at times linked to service in abattoirs [→History of Animal Slaughter].[5]

---

[3] Wang, Yiqing. Dogs and Cats Have a 'Pet Economy' of Their Own. *China Daily*, October 6, 2018. URL: chinadaily.com.cn/a/201810/06/WS5bb86859a310eff303280d78.html (February 10, 2021).
[4] Larmer, Brook. China's Mixed Messages on the Global Trade in Endangered-Animal Parts. *The New York Times*, November 27, 2018. URL: nytimes.com/2018/11/27/magazine/chinas-mixed-messages-on-the-global-trade-in-endangered-animal-parts.html (October 5, 2020); Pauly, Daniel, Jackie Alder, Elena Bennett, Villy Christensen, Peter Tyedmers and Reg Watson. The Future for Fisheries. *Science* 302, no. 5649 (November 2003): 1359–1361.
[5] Abele, Michael Thomas. *Peasants, Skinners and Dead Cattle: The Transformation of Rural Society in Western Japan, 1600–1890*. PhD diss.: University of Illinois at Urbana-Champaign, USA (2018).

Anthropocentrism shaped the second key dynamic, carrying over into the modern era. As Avenell has observed, the contrast between Japan's strong environmental movement and US dynamics is instructive. Rachel Carson's *Silent Spring* must surely stand as a key work in the field of animal studies, focusing as it does on the devastation wrought by chemical production on birds and fish. In East Asia, in contrast, imaginaries have highlighted the human victims of industrialization. Even as environmentalist discourse developed in Japan at much the same moment that Carson was writing, exploding into consciousness in the 1960s and 1970s, it did so through narratives grounded in an ethics of human pain and damaged human bodies [→Environmental History]. This human-centered imaginary was augmented by longstanding aesthetic preferences – who or what counted as beautiful and natural – that valued landscapes and gardens that were largely devoid of animals. There were, of course, crucial exceptions to such generalizations, but aesthetics have tended to foreground the human and the land, even when dynamics such as hunting and fishing have been practiced widely.[6]

Hunting

In China and Japan, hunting has had a paradoxical cultural position and has received limited scholarly attention. Hunting was generally an elite privilege, but it has also had a marginal status, often being associated with peripheries. As Sterckx explains, in ancient China, royal hunts and ritual killings of wild animals typically occurred in hunting parks where spatial confinement assured a supply of animals. Royal hunts were symbolic acts in which the ruler manifested his privilege of subduing living beings within his realm and as opportunities to express hierarchies among elites.[7] While Han commoner populations were not supposed to hunt, hunting remained vital in borderland ethnic communities as a marker of ethnic identity into the late imperial and modern periods [→History of Hunting; →Political History].[8]

In Japan, hunting also had an important, if ambivalent, status. Both hunting and fishing were closely tied to identity. Archeological studies stress the importance of

---

6 Avenell, Simon. *Transnational Japan in the Global Environmental Movement*. Honolulu, HI: University of Hawai'i Press, 2017; Miyamoto, Ken'ichi. *A Critical History of Environmental Pollution in Postwar Japan*. Tokyo: Iwanami Shoten, 2014.
7 Sterckx, Roel. Attitudes towards Wildlife and the Hunt in Pre-Buddhist China. In *Wildlife in Asia: Cultural Perspectives*, John Knight (ed.), 15–35. London: Routledge, 2004.
8 Bello, David. The Cultured Nature of Imperial Foraging in Manchuria. *Late Imperial China* 31, no. 2 (December 2010): 1–33; Andrade, Tonio. Pirates, Pelts, and Promises: The Sino-Dutch Colony of Seventeenth-Century Taiwan and the Aboriginal Village of Favorolang. *The Journal of Asian Studies* 64, no. 2 (May 2005): 295–321; Koo, Hui-wen. Deer Hunting and Preserving the Commons in Dutch Colonial Taiwan. *The Journal of Interdisciplinary History* 42, no. 2 (Autumn 2011): 185–203; Lynteris, Christos. Skilled Natives, Inept Coolies: Marmot Hunting and the Great Manchurian Pneumonic Plague (1910–1911). *History and Anthropology* 24, no. 3 (August 2013): 303–321.

hunting among the Jōmon people before the advent of rice agriculture (c. 1000BCE–300CE), a period that is generally considered formational.[9] Hunting played a role in pre-modern Japan, too, both as the privilege of the elites and for the procurement of sacrificial offerings for Shinto shrines. This fact challenges the oft-repeated claim that pre-modern Japan was exclusively an agrarian rice culture.[10] Game was probably a larger source of protein than domesticates because oxen and horses were considered essential to agriculture and transportation.[11] Eating them was taboo. Even so, hunting was accorded a marginal status associated with the periphery, such as mountain uplands, where it played an important role in folklore.[12]

Fishing and Whaling

Seafood was the main source of animal protein in pre-modern Japan. Fishing has played such a central role in Japan since the Jōmon period that some archeologists refer to Neolithic inhabitants of the archipelago as "hunter-fisher-gatherers."[13] Fishing remained important after the introduction of rice farming in the Yayoi period (c. 300BCE-300CE). In ancient Japan, coastal communities paid taxes in ocean products, which also served as offerings at shrines. Fishing was largely a male occupation with the exception of female divers (*ama*) who specialized in procuring shellfish, such as

---

[9] Akazawa, Takeru. Regional Variation in Jomon Hunting-Fishing-Gathering Societies. In *Interdisciplinary Perspectives on the Origins of the Japanese*, Keiichi Omoto (ed.), 223–231. Kyoto: International Research Center for Japanese Studies, 1999; Seguchi, Shinji. Landscape Neolithization Among the Hunter-Fisher-Gatherers of Lake Biwa, Central Japan. *Journal of World Prehistory* 27, no. 3 (December 1, 2014): 225–245; Perri, Angela R. Hunting Dogs as Environmental Adaptations in Jōmon Japan. *Antiquity* 90, no. 353 (October 2016): 1166–1180.
[10] Long, Hoyt. Grateful Animal or Spiritual Being. Buddhist Gratitude Tales and Changing Conceptions of the Deer in Early Japan. In *JAPANimals: History and Culture in Japan's Animal Life*, Gregory M. Pflugfelder, Brett L. Walker (eds.), 21–58. Ann Arbor, MI: University of Michigan Center For Japanese Studies, 2005; Grumbach, Lisa. *Sacrifice and Salvation in Medieval Japan: Hunting and Meat in Religious Practice at Suwa Shrine.* PhD diss.: Stanford University (2005); Pitelka, Morgan. *Spectacular Accumulation: Material Culture, Tokugawa Ieyasu, and Samurai Sociability.* Honolulu, HI: University of Hawaii Press, 2015.
[11] Krämer, Hans Martin. Not Befitting Our Divine Country: Eating Meat in Japanese Discourses of Self and Other from the Seventeenth Century to the Present. *Food and Foodways* 16, no. 1 (March 2008): 33–62; Shimizu, Akira. Meat-Eating in the Kōjimachi District of Edo. In *Japanese Foodways, Past and Present*, Stephanie Assmann, Eric C. Rath (eds.), 72–107. Urbana, IL: University of Illinois Press, 2010.
[12] Knight, John. Indigenous Regionalism in Japan. In *Indigenous Environmental Knowledge and Its Transformations: Critical Anthropological Perspectives*, Alan Bicker, Roy Ellen, Peter Parkes (eds.), 151–176. London: Routledge, 2003; Knight, Catherine Heather. *The Bear as Barometer: The Japanese Response to Human-Bear Conflict.* PhD diss.: University of Canterbury, NZ (2007).
[13] Seguchi, Landscape Neolithization.

abalone.[14] Ericson, Heé, and Tsutsui have shown us how fishing expanded exponentially through industrialization and fossil-fueled transport in modern times.[15]

Writing on whales marks the deepest point of Japan's oceanic literature. A wedge issue in Japanese politics akin to that of gun rights in the US, studies of whaling bear on local and international questions. These works range from Arch's examination of shore-based early modern whaling to works on whales as food, resource, and deities from pre-modern to modern times.[16]

Animal Domestication and Livestock

If hunting and fishing are crucial to different aspects of Japanese identity, domestication is associated with Chinese identity. The field of zooarchaeology is small but growing across the region, as are deeper historical studies of animals and humans [→Pre-Domestication: Zooarchaeology].[17] Domestication in China began with dogs and pigs in the Neolithic period (8000–7000BCE), when some species also became victims in sacrificial rites.[18] Livestock were introduced starting in the 3rd and 2nd millennium BCE [→Domestication: Coevolution]. Historians have contended that animal populations began to dwindle from the Han on as crop farming expanded, displacing communities into borderlands that had subsisted on herding or hunting. Compared to Europe, where mixed crop and livestock farming was the norm, animal husbandry

---

[14] Kalland, Arne. *Fishing Villages in Tokugawa, Japan*. Honolulu, HI: University of Hawaii Press, 1995.

[15] Ericson, Kjell. Water Before Fish: Japan's Fundamental Fisheries Survey and the Currents of Empire, 1909–1918. *Zinbun* 49 (March 2019): 11–29; Heé, Nadin. Negotiating Migratory Tuna: Territorialization of the Oceans, Trans-war Knowledge, and Fisheries Diplomacy. *Diplomatic History* 44, no. 3 (June 2020): 413–427; Tsutsui, William. The Pelagic Empire: Reconsidering Japanese Expansion. In *Japan at Nature's Edge: The Environmental Context of a Global Power*, Ian Jared Miller, Julia Adeney Thomas, Brett L. Walker (eds.), 21–38. Honolulu, HI: University of Hawaii Press, 2013.

[16] Morikawa, Jun. *Whaling in Japan: Power, Politics and Diplomacy*. London: C Hurst & Co, 2009; Watanabe, Hiroyuki. *Japan's Whaling: The Politics of Culture in Historical Perspective*. Melbourne: Trans Pacific Press, 2009; Kalland, Arne and Brian Moeran. *Japanese Whaling: End of an Era*. London: Routledge, 2010; Miller, Ian Jared, Julia Adeney Thomas, Brett L. Walker (eds.). *Japan at Nature's Edge: The Environmental Context of a Global Power*. Honolulu, HI: University of Hawai'i Press, 2013; Kishigami, Nobuhiro, Hisashi Hamaguchi and James M. Savelle (eds.). *Anthropological Studies of Whaling*. Osaka: National Museum of Ethnology, Senri Ethnological Studies, 2013; Holtzman, Jon. On Whale: Conundrums of Culture and Cetaceans as Local Meat. *Ethnos* 82, no. 2 (March 2017): 277–297; Itoh, Mayumi. *The Japanese Culture of Mourning Whales: Whale Graves and Memorial Monuments in Japan*. Singapore: Palgrave Macmillan, 2018; Arch, Jakobina K. *Bringing Whales Ashore: Oceans and the Environment of Early Modern Japan*. Seattle, WA: University of Washington Press, 2018.

[17] Kowner, Rotem, Michal Brian, Meir Shahar, Guy Bar-Oz and Gideon Shelach (eds.). *Animals and Human Society in Asia: Historical, Cultural and Ethical Perspectives*. Cham: Palgrave Macmillan, 2020.

[18] Liu, Li and Xiaolin Ma. The Zooarchaeology of Neolithic China. In *The Oxford Handbook of Zooarchaeology*, Hannah Russ, Umberto Albarella, Kim Vickers, Mauro Rizzetto, Sarah Viner-Daniels (eds.), 304–318. Oxford: Oxford University Press, 2017.

was not as central in China, and pastoralism was associated with borderlands that lacked arable land. After the Tang, small-scale crop farming became the norm and, as Bray notes, "the huge flocks of sheep and teams of oxen characteristic of manorial farming likewise disappeared from the scene."[19] Nevertheless, mixed farming that relied on penned livestock rather than on pastures while using draft animals, remained an integral feature. Taboos against beef eating became prevalent from the Song, but mutton and goat remained staples in northern cuisine whereas pork, poultry, and fish were common in the south.[20]

Livestock farming played a marginal role in Japan until the modern era. Domesticated species arrived in Japan from the continent, beginning with dogs between 7300 and 5300 BCE. Livestock were introduced between 300–500 CE. In ancient Japan, livestock was bred on state-operated farms, but beginning in the late Heian, warriors bred horses for warfare. For much of the pre-modern period, horses – like cattle – were also used as draft animals. In the 19th century, Japanese horses and cattle were crossbred with Western breeds to produce modern Japanese breeds of cavalry horses and beef cattle while dairy cattle and new dog breeds were introduced from the West.[21] Warhorses served as crucial biopower in each of Japan's modern wars, notably the Second World War [→History of War].

Despite taboos against the meat of livestock, medieval and early modern warrior culture required leather for military use. Herding, butchering, and tanning cattle and horses fell to stigmatized "outcaste" populations. Even though outcaste status was formally abolished in the late nineteenth century, such labor has retained stigma into contemporary times.[22]

---

19 Bray, Francesca. Where Did the Animals Go? In *Animals Through Chinese History: Earliest Times to 1911*, Roel Sterckx, Martina Siebert, Dagmar Schäfer (eds.), 118–131. Cambridge: Cambridge University Press, 2018, 121.

20 Bray, Where Did the Animals Go?; DuBois, Thomas David. China's Dairy Century: Making, Drinking and Dreaming of Milk. In *Animals and Human Society in Asia: Historical, Cultural and Ethical Perspectives*, Rotem Kowner, Michal Brian, Meir Shahar, Guy Bar-Oz, Gideon Shelach (eds.), 179–211. Cham: Palgrave Macmillan, 2019; Goossaert, Vincent. The Beef Taboo and the Sacrificial Structure of Late Imperial Chinese Society. In *Of Tripod and Palate: Food, Politics, and Religion in Traditional China*, Roel Sterckx (ed.), 237–248. New York, NY: Palgrave Macmillan, 2005.

21 Grass, Noa. A Million Horses: Raising Government Horses in Early Ming China. In *Animals and Human Society in Asia*, Kowner, Brian, Shahar, Bar-Oz, Shelach (eds.), 299–328. Hongo, Hitomi. Introduction of Domestic Animals to the Japanese Archipelago. In *The Oxford Handbook of Zooarchaeology*, Russ, Albarella, Vickers, Rizzetto, Viner-Daniels (eds.), 333–350.

22 Cangià, Flavia. *Performing the Buraku: Narratives on Cultures and Everyday Life in Contemporary Japan*. Münster: LIT Verlag, 2013; Amino, Yoshihiko. *Rethinking Japanese History*. Ann Arbor, MI: University of Michigan Press, 2012; Ambros, Barbara R. Partaking of Life: Buddhism, Meat-Eating, and Sacrificial Discourses of Gratitude in Contemporary Japan. *Religions* 10, no. 4 (April 2019): 279–300.

Sericulture

Sericulture played vital roles across the region. Silk cultivation was a labor-intensive process that combined growing mulberry trees for feed, breeding silkworms, reeling thread, and weaving fabric. Studies of pre-modern sericulture have paid attention to the technical details of silk manufacture as well as human social dynamics involved in production, but human-silkworm interactions are generally treated only as background. In the modern era, however, Onaga has used the study of silkworms in Japan to explore questions of multispecies history.[23]

Evidence suggests that silk was cultivated in China as early as 3000 BCE. The earliest written references to sericulture appear on Shang oracle bones.[24] From the Zhou to the Ming, the state required households to pay taxes in grain and cloth, including silk. The raising of silkworms, reeling, and weaving was typically women's work; tilling fields, including growing mulberry trees, was the domain of men, a division of labor that reinforced Confucian gender norms associating women with the domestic sphere [→Feminist Intersectionality Studies].[25] Zhou dynasty ritual texts mention that the empress and other court women conducted sericulture rites. During the Han, imperial rituals emerged that were dedicated to the inventor of sericulture, but the identity of the first sericulturalist fluctuated over the centuries, eventually settling on Leizu, the wife of the Yellow Emperor. Moreover, silk contributed to the association of women's weaving and wearing the delicate, luxurious fabric with immortals and divinities.[26]

As in China, sericulture and weaving was considered women's labor in Japan, offering women a source of income.[27] Como has demonstrated that ancient Japan incorporated Chinese sericulture rites associated with female immortals and female divinities owing to the influence of immigrant kinship groups from the continent.[28] As silk production was modernized in the modern era, raising silkworms and reeling be-

---

23 Onaga, Lisa A. Bombyx and Bugs in Meiji Japan: Toward a Multispecies History. *Scholar and Feminist Online* 11, no. 3 (2013): 569–604.
24 Kuhn, Dieter. *Science and Civilisation in China, Vol. 5. Chemistry and Chemical Technology, Part IX. Textile Technology: Spinning and Reeling.* Cambridge: Cambridge University Press, 1988.
25 Bray, Francesca. *Technology and Gender: Fabrics of Power in Late Imperial China.* Berkeley, CA: University of California Press, 1997.
26 Kuhn, Dieter. Tracing a Chinese Legend: In Search of the Identity of the First Sericulturalist. *T'oung Pao, Second Series* 70, no. 4–5 (1984): 213–245; Rothschild, N. Harry. *Emperor Wu Zhao and Her Pantheon of Devis, Divinities, and Dynastic Mothers.* New York, NY: Columbia University Press, 2015; Cho, Philip S. The Circulation of Sericulture Knowledge through Temple Networks and Cognitive Poetics in Eighteenth Century Zhejiang. In *Motion and Knowledge in the Changing Early Modern World: Orbits, Routes and Vessels,* Ofer Gal, Yi Zheng (eds.), 11–37. Dordrecht: Springer, 2014.
27 Imai, Shiho. The Independent Working Woman as Deviant in Tokugawa Japan, 1600–1867. *Michigan Feminist Studies* 16 (2002): 117–140.
28 Como, Michael. *Weaving and Binding: Immigrant Gods and Female Immortals in Ancient Japan.* Honolulu, HI: University of Hawaii Press, 2010.

came separate labor. Women working in new factories were initially recruited from wealthy peasant and former warrior families but working conditions for women eventually deteriorated in the 1880s; yet silkworm cultivation continued to be promoted as women's labor.[29]

Pet-Keeping

Pet-keeping in East Asia has only relatively recently begun to receive scholarly attention. While it has largely been associated with the modern era, the breeding and keeping of insects and animals for leisure has a deeper history, particularly among the wealthy. Ornamental goldfish were popular as early as the Southern Song and later across a broad segment of society. Breeding ornamentals grew into a vehicle of experimentation and national pride in the modern era, even under Mao.[30] Ornamental goldfish also became popular in Japan along with exotic birds, fancy mice, cats, and lapdogs. Exotic species and fancy breeds became status symbols, often obtained through trade with the West. Conversely, the evaluation of local breeds was closely linked to imperialism.[31]

Pet-keeping grew exponentially in the late twentieth century. In Japan, scholars generally refer to a "pet boom", which led to new ways of conceptualizing the relationship between humans and their companions. Pets have come to be viewed as family members. In death they have been accorded ritual attention.[32] For those who cannot keep their own pets, services such as cat cafés allow patrons to temporarily experience intimacy with animals – a phenomenon pioneered in Taiwan in 1998 that spread to Japan and elsewhere.[33] This boom has also had its dark side: the proliferation of abandoned and stray companion animals [→History of Pets].[34]

---

[29] Yamasaki, Akiko. Handicrafts and Gender in Modern Japan. *The Journal of Modern Craft* 5, no. 3 (April 2012): 259–274.

[30] Jiang, Lijing. The Socialist Origins of Artificial Carp Reproduction in Maoist China. *Science, Technology & Society* 22, no. 1 (March 2017): 59–77.

[31] Chaiklin, Martha. Exotic-Bird Collecting in Early-Modern Japan. In *JAPANimals: History and Culture in Japan's Animal Life*, Gregory M. Pflugfelder, Brett L. Walker (eds.), 125–160. Ann Arbor, MI: University of Michigan Center for Japanese Studies, 2005; Skabelund, Aaron. Can the Subaltern Bark? Imperialism, Civilization, and Canine Cultures in Nineteenth-Century Japan. In *JAPANimals* Pflugfelder, Walker (eds.), 195–243. Chan, Ying-kit. The Great Dog Massacre in Late Qing China: Debates, Perceptions, and Phobia in the Shanghai International Settlement. *Frontiers of History in China* 10, no. 4 (January 2015): 645–667.

[32] Ambros, Barbara R. *Bones of Contention: Animals and Religion in Contemporary Japan*. Honolulu, HI: University of Hawaii Press, 2012.

[33] Plourde, Lorraine. Cat Cafés, Affective Labor, and the Healing Boom in Japan. *Japanese Studies* 34, no. 2 (June 2014): 115–133.

[34] Iida, Motoharu, director. *Inu to Neko to Ningen to: Dogs, Cats and Humans*. Group Low Position, 2009. DVD; Chan, Ying-kit. No Room to Swing a Cat. Animal Treatment and Urban Space in Singapore. *Southeast Asian Studies* 5, no. 2 (August 2016): 305–329.

Exhibition and Popular Culture

East Asia's greatest animal export may be the Giant Panda. Images of these endangered bears are almost inescapable in the modern world. Their black-and-white forms are the mascot of the World Wildlife Fund (WWF), panda videos are an internet staple, and live pandas rank among the most viewed cultural attractions in the world. As Miller argues for Japan and Songster echoes for China, this cultural ubiquity is not a coincidence. It is the result of political, diplomatic, and biological convergence around a species distinctly ill-suited to captivity, but which humans the world over find captivating. Pandas have long been a source of fascination in Asia and elsewhere, but it was only after the consolidation of the PRC in 1949 that "panda diplomacy" accelerated the bear's evolution into a global icon of Chinese national identity, on the one hand, and symbols of endangered species everywhere, on the other [→Diplomatic History].[35]

The contradictory status of pandas points to the changeful history of animal captivity in modern East Asia. Pandas took on meaning in zoological gardens, whose East Asian history began with the Tokyo's Ueno Zoological Garden in 1882. Over time, East Asian zoos have built on the deep history of animal exhibition in a region where people have long liked to look at captive animals [→History of the Zoo].[36] The English-language literature on Chinese and Korean zoological gardens has not been as extensive as work on Japan, initiated by Kawata, and very little has been published on North or South Korean animal collections.[37]

No event has garnered as much attention as the so-called "Great Zoo Massacre": the mass-mediated slaughter of much of the Tokyo Zoo's inhabitants in 1943, including popular trained elephants. Miller argues that the killings constituted a fascist form of animal sacrifice. Itoh offers a detailed chronicle of the events, which were the stuff of one of postwar Japan's best-selling children's books. *The Faithful Elephants* converted the grim wartime slaughter into a parable of postwar pacifism.[38] Litten focuses on questions of blame and responsibility, a difficult issue when key records were destroyed in the war. Seeley and Skabelund suggest avenues forward for the field as a whole when they underline the importance of multi-lingual work,

---

[35] Songster, E. Elena. *Panda Nation: The Construction and Conservation of China's Modern Icon*. Oxford: Oxford University Press, 2018; Szczygielska, Marianna. Pandas and the Reproduction of Race and Heterosexuality in the Zoo. In *Zoo Studies: A New Humanities*, Tracy McDonald, Daniel Vandersommers (eds.), 211–236. Montreal: McGill-Queen's University Press, 2019; Miller, Thomas and Walker, *Japan at Nature's Edge*.
[36] Ito, Takashi. Flying Penguins in Japan's Northernmost Zoo. In *Zoo Studies*, McDonald, Vandersommers (eds.), 237–262.
[37] Kawata, Ken. Zoological Gardens of Japan. In *Zoo and Aquarium History: Ancient Animal Collections to Zoological Gardens*, Vernon N. Kisling (ed.), 295–330. Boca Raton, FL: CRC Press, 2000; Walker, Sally. Zoological Gardens of India. In *Zoo and Aquarium History*, Kisling (ed.), 251–294.
[38] Tsuchiya, Yukio and Ted Lewin. *Faithful Elephants: A True Story of Animals, People, and War*. Mooloolaba: Sandpiper, 1997.

noting colonial and post-colonial connections in a region where Japanese imperialism sparked the creation of zoos in Korea, Taiwan, and Northeast China.[39]

3 Methods and Approaches

A tension between the material and the symbolic, the historical living animal and the history of belief about animals run through animal studies of East Asia, highlighting key debates over animal agency, especially in the Japan field. While any number of scholars invoke the language of "animal agents", few provide satisfying explications of such arguments, gesturing through jargon rather than demonstrating significance; fewer still ground those assertions in rigorous philosophical, scientific, or historical argumentation. Walker's seminal history of the Japanese wolf and its extermination makes the strongest claims in favor of thoughtful, intentional, emotionally complex nonhuman agency. Reaching back to Darwin, he argues descriptions of wolf behavior can be read as a kind of "text" through which historians might grant canids historical agency. Miller takes up Walker's prompt, arguing that the limits of the archive define the threshold of historians' ability to perceive thoughtful agency in animals, nonhuman and human alike. Here, animals are framed as "actors" capable of shaping events and possessing rich inner lives, but whose thoughtful agency remains beyond the epistemological reach of historians. Ambros, in contrast, draws on work in Buddhist Studies to elevate the idea of animal agents in discussions of rites as well as pet-keeping, noting that animals were understood as active participants in ritual traditions. Marran expounds on this turn to Buddhist thought, noting resonance with Haraway's arguments for the simultaneity of semiosis and materiality to ground claims of agency in a materialist philosophy. Such discussion marks the borders of ongoing work at the interface of animal studies, science studies, and the sciences from scholars such as Arch, who employs marine science in her history of whaling while eschewing more explicitly philosophical approaches.[40]

---

39 Itoh, Mayumi. *Japanese Wartime Zoo Policy: The Silent Victims of World War II*. New York, NY: Palgrave Macmillan, 2010; Litten, Frederick S. Starving the Elephants: The Slaughter of Animals in Wartime Tokyo's Ueno Zoo. *The Asia-Pacific Journal* 7, no. 38 (September 2009): 1–18, 18; Miller, Ian Jared. *The Nature of the Beasts: Empire and Exhibition at the Tokyo Imperial Zoo*. Berkeley, CA: University of California Press, 2013 (errata at URL: scholar.harvard.edu/ianmiller, December 4, 2020); Seeley, Joseph and Aaron Skabelund. Tigers – Real and Imagined – in Korea's Physical and Cultural Landscape. *Environmental History* 20, no. 3 (July 2015): 475–503; Seeley, Joseph and Aaron Skabelund. Bite, Bite against the Iron Cage: The Ambivalent Dreamscape of Zoos in Colonial Seoul and Taipei. *The Journal of Asian Studies* 79, no. 2 (May 2020): 429–454; Henry, Todd A. Ch'anggyŏng Garden as Neocolonial Space: Spectacles of Anticommunist Militarism and Industrial Development in Early South (Ern) Korea. *Journal of Korean Studies* 21, no. 1 (March, 2016): 7–43.
40 Walker, Brett L. *The Lost Wolves of Japan*. Seattle, WA: University of Washington Press, 2005; Ambros, Partaking of Life. Marran, Christine L. *Ecology without Culture: Aesthetics for a Toxic World*. Min-

Discussions of agency and action highlight the particularity of work in Asian Studies. What sorts of creatures counted as special or worthy of distinctive sorts of moral consideration? Such questions come to the fore in analysis of ritual, memorialization, and sacrifice, which cut across time from oracle bones to contemporary laboratory practice and even into modern zoological gardens. Animal bodies played central roles in East Asian ritual traditions. Historically, animals were often seen as active participants in mantic, sacrificial, and memorial rituals. Understood as meriting moral consideration, they were subjects of dietary restrictions and taboos and they were actors in animal release rituals. Some animal bodies were made to mediate communication with the spirit world, as with oracle bones and sacrificial rituals.[41] Others – especially physical anomalies or instances of odd behavior – became vehicles for divination. The symbolic capacity to mediate between spiritual and material realms, signaling affiliation and distinction with the human sphere made the procurement of suitable creatures a priority for elites and imperial officials.

The introduction of Buddhism to East Asia with its rhetoric against killing – even of undesirable pests[42] – led to the development of new ritual technologies in order to replace sacrificial practices and to mitigate the karmic effects of killing animals, from ritual animal releases[43] to memorials for dead animals.[44] These aspects contributed

---

neapolis, MN: University of Minnesota Press, 2017; Arch, *Bringing Whales Ashore*; Miller, *The Nature of the Beasts*.

41 Keightley, David N. *Sources of Shang History: The Oracle-Bone Inscriptions of Bronze Age China*. Berkeley, CA: University of California Press, 1978; Sterckx, Roel. *The Animal and the Daemon in Early China*. Albany, NY: SUNY Press, 2002; Sterckx, Roel. Of a Tawny Bull We Make Offering: Animals in Early Chinese Religion. In *A Communion of Subjects: Animals in Religion, Science and Ethics*, Paul Waldau, Kimberley Patton (eds.), 259–272. New York, NY: Columbia University Press, 2006; Sterckx, Roel. Animal to Edible: The Ritualization of Animals in Early China. In *Animals through Chinese History*, Sterckx, Schäfer, Siebert (eds.), 46–63. Flad, Rowan K. Divination and Power: A Multiregional View of the Development of Oracle Bone Divination in Early China. *Current Anthropology* 49, no. 3 (June 2008): 403–437; Kory, Stephan N. From Deer Bones to Turtle Shells: The State Ritualization of Pyro-Plastromancy during the Nara-Heian Transition. *Japanese Journal of Religious Studies* 42, no. 2 (November 2015): 339–380.
42 Heirman, Ann. How to Deal with Dangerous and Annoying Animals: A Vinaya Perspective. *Religions* 10, no. 2 (February 2019): 113–130; Chen, Huaiyu. The Road to Redemption: Killing Snakes in Medieval Chinese Buddhism. *Religions* 10, no. 4 (April 2019): 249–278.
43 Williams, Duncan. Animal Liberation, Death, and the State: Rites to Release Animals in Medieval Japan. In *Buddhism and Ecology*, Mary Evelyn Tucker, Duncan Williams (eds.), 149–162. Cambridge, MA: Harvard University Press, 1997; Handlin Smith, Joanna. Liberating Animals in Ming-Qing China: Buddhist Inspiration and Elite Imagination. *The Journal of Asian Studies* 58, no. 1 (February 1999): 51–84; Eichman, Jennifer. *A Late Sixteenth-Century Chinese Buddhist Fellowship: Spiritual Ambitions, Intellectual Debates, and Epistolary Connections*. Leiden: Brill, 2016; Yang, Der-Ruey. Animal Release: The Dharma Being Staged between Marketplace and Park. *Cultural Diversity in China* 1, no. 2 (January 2015): 141–163; Ambros, Barbara R. Cultivating Compassion and Accruing Merit: Animal Release Rites During the Edo Period. In *The Life of Animals in Japanese Art*, Robert T. Singer, Masatomo Kawaii (eds.), 16–27. Washington, DC: National Gallery, 2019.

to strong denial of blood sacrifice's history in Japan. Some argue, however, that cattle and horses were sacrificed in ancient Japan or that wild animals served as sacrificial victims in the medieval period.[45] Ambros has shown how sacrificial narratives, if not necessarily actions, became more common alongside increased consumption of meat.[46]

Another extension of the rhetoric against killing, the abstention from meat, has remained a strong marker of Buddhist identity in China and other parts of East Asia.[47] In pre-modern Japan, Buddhist rhetoric against meat also became a strong cultural influence, but there were exceptions to this position, particularly in the esoteric Buddhist traditions and Jōdo Shin Buddhism. In modern Japanese Buddhism, the consumption of meat has become largely accepted, if not normative, along with the Westernization of the Japanese diet.[48] Similar tensions existed in Chinese Buddhism around the use of silk and leather. While the Chinese Buddhist tradition offered moral arguments against silk and the killing of silkworms, the monastic community and the laity remained engaged in the production and use of silk, and eventually developed sacrificial narratives that described silkworms as bodhisattvas who voluntarily gave their lives to alleviate suffering.[49]

4 Implication(s) of the Animal Turn

It is telling that one origin story of the COVID-19 pandemic is a parable of voracious East Asian appetites: the mass consumption of wild species such as bats or pango-

---

[44] Kenney, Elizabeth. Pet Funerals and Animal Graves in Japan. *Mortality* 9, no. 1 (February 2004): 42–60; Nakamaki, Hirochika. Memorials of Interrupted Lives in Modern Japan: From Ex Post Facto Treatment to Intensification Devices. In *Perspectives on Social Memory in Japan*, Yun Hui Tsu, Jan van Bremen, Eyal Ben-Ari (eds.), 44–57. Folkestone: Global Orient, 2005; Veldkamp, Elmer. Memorializing and Remembering Animals in Japan. In *Perspectives on Social Memory in Japan*, Tsu, van Bremen, Ben-Ari (eds.), 58–74. Ambros, *Bones of Contention*.
[45] Como, Michael. Horses, Dragons, and Disease in Nara Japan. *Japanese Journal of Religious Studies* 34, no. 2 (November 2007): 393–415.
[46] Ambros, Partaking of Life.
[47] Kieschnick, John. Buddhist Vegetarianism in China. In *Of Tripod and Palate: Food, Politics, and Religion in Traditional China*, Roel Sterckx (ed.), 186–212. New York, NY: Palgrave Macmillan, 2005; Eichman, *Late Sixteenth-Century Chinese Buddhist Fellowship*; Greene, Eric. A Reassessment of the Early History of Vegetarianism in Chinese Buddhism. *Asia Major* 29, no. 1 (2016): 1–43; Verchery, Lina. Both Like and Unlike: Rebirth, Olfaction, and the Transspecies Imagination in Modern Chinese Buddhism. *Religions* 10, no. 6 (June 2019): 364–374.
[48] Jaffe, Richard. The Debate over Meat Eating in Japanese Buddhism. In *Going Forth: Visions of Buddhist Vinaya*, William M. Bodiford (ed.), 255–275. Honolulu, HI: University of Hawaii Press, 2005.
[49] Young, Stuart. For a Compassionate Killing: Chinese Buddhism, Sericulture, and the Silkworm God Aśvaghoṣa. *Journal of Chinese Religions* 41, no. 1 (May 2013): 25–58; Young, Stuart. Bald-Headed Destroyers of Living Things: Shaping Buddhist Identity in the Silk Cultures of Medieval China. *Asia Major* 30, no. 2 (2017): 27–70.

lins via the so-called wet markets of central China. ("Wet" because of the melted ice, but also by the blood of animal slaughter.) This trouble is compounded by the fact that such narratives were quickly put to racist ends by American political leadership and by outside critics who unselfconsciously echoed long histories associating Asian foodways with the denigration of the East Asian other. This tangle of cultural, ecological, and political dynamics reminds us – yet again – that the question of the animal has always been vexed by issues of race and ethnicity in East Asia, even as the virus reminds us – yet again – of our own common animal natures. The *animal turn* in East Asia has marked the acceleration of efforts to historicize such dynamics, most often through work that has held such fraught histories in tension, staying with the trouble of real animal exploitation and the cultural mischief arising from unstable distinctions between the animal and the human. In the Japan field, Marran, in particular, marks the latest intervention in this discourse, reminding readers of the ways that Japanese nationalists have claimed the mantle of a "green Orientalism," defining particularity through the history of dealings with the natural world. Forthcoming work from Chee and Keck in Chinese Studies, on the other hand, suggests how such scholarship may help to reshape the storyline of COVID-19 into more responsible narratives.[50]

If the most exciting work in the field has historicized the boundaries between animals and people, often drawing on rich philosophies of action or agency drawn from the Euro-American canon, East Asian Studies as such has been far too polite in its dealings with more conventional borders. When Seeley and Skabelund follow tigers across the historical terrain of colonial Korea they sketch an itinerary that confounds the dictates of the national historiographies that dominate the *animal turn* in East Asia.[51] Tracking a creature lauded in Chinese folklore since the time of the oracle bones as it traversed Korean forests under Japanese occupation, the pair point to the next frontier for this field: East Asian animal histories that account for the trans- and international nature of the region. In this context, "old" questions about the status of the nonhuman under Buddhism, for example, take on new, potentially startling implications, offering a different sort of philosophy in response to the very present-day questions of the *animal turn*. Over time, this may become the most significant implication of the *animal turn* in East Asia: the defamiliarization

---

50 Singer, Peter and Paola Cavalieri. The Two Dark Sides of COVID-19. *Project Syndicate*, March 2, 2020. URL: project-syndicate.org/commentary/wet-markets-breeding-ground-for-new-coronavirus-by-peter-singer-and-paola-cavalieri-2020 – 03 (October 8, 2020); Rogers, Katie, Lara Jakes and Ana Swanson. Trump Defends Using Chinese Virus Label, Ignoring Growing Criticism. *The New York Times*, March 18, 2020. URL: nytimes.com/2020/03/18/us/politics/china-virus.html (October 8, 2020); Keck, Frédéric. *Avian Reservoirs: Virus Hunters and Birdwatchers in Chinese Sentinel Posts*. Durham, NC: Duke University Press, 2020; Chee, Liz P. Y. *Mao's Bestiary: Medicinal Animals and Modern China*. Durham, NC: Duke University Press, 2021.
51 Seeley and Skabelund, Bite, Bite against the Iron Cage.

offered by the deep time and occasional difference of that other "other" part of the world.

Selected Bibliography

Ambros, Barbara R. *Bones of Contention: Animals and Religion in Contemporary Japan*. Honolulu, HI: University of Hawaii Press, 2012.
Arch, Jakobina K. *Bringing Whales Ashore: Oceans and the Environment of Early Modern Japan*. Seattle, WA: University of Washington Press, 2018.
Knight, John. *Waiting for Wolves in Japan: An Anthropological Study of People–Wildlife Relations*. Oxford: Oxford University Press, 2003.
Miller, Ian Jared. *The Nature of the Beasts: Empire and Exhibition at the Tokyo Imperial Zoo*. Berkeley, CA: University of California Press, 2013.
Pflugfelder, Gregory M. and Brett L. Walker (eds.). *JAPANimals: History and Culture in Japan's Animal Life*. Ann Arbor, MI: University of Michigan Center for Japanese Studies, 2005.
Skabelund, Aaron. *Empire of Dogs: Canines, Japan, and the Making of the Modern Imperial World*. Ithaca, NY: Cornell University Press, 2011.
Sterckx, Roel, Martina Siebert and Dagmar Schäfer (eds.). *Animals through Chinese History: Earliest Times to 1911*. Cambridge: Cambridge University Press, 2018.
Sterckx, Roel. *The Animal and the Daemon in Early China*. Albany, NY: SUNY Press, 2002.
Sterckx, Roel. *Food, Sacrifice and Sagehood in Early China*. Cambridge: Cambridge University Press, 2011.
Walker, Brett L. *The Lost Wolves of Japan*. Seattle, WA: University of Washington Press, 2005.

Part III: **Historical Fields**

Mieke Roscher
Social History

1 Introduction and Overview

"Animals are everywhere, and there has never been any purely human moment in world history," Susan Nance declared in her introduction to "The Historical Animal" in 2015.[1] But does that mean that we can say that animals are part of society and hence also social agents? Must social history therefore include animals? Is there such a thing as an animal social history? Most proponents of the now "classical" new social history would have resisted such an extension of the social realm. Moreover, social historians like Edward P. Thompson, who can rightly be identified as one of the founders of the field, even turned explicitly to the difference between the human and the animal in order to lend the working class a voice and depict it as a collective with historical agency, in contrast to animals. Ever since Thompson's and others' attempt to enfranchise the "common people" or the "subaltern" by acknowledging them as having a say in our histories, agency has become an integral part of social history's historiographical approach. This approach emphasizes the lives of individual agents, the makers of their own history, above societal structures, but notoriously overlooks the animal element of history.

The new social history brought to life by Thompson concerned itself with the emancipation of history from more traditional historiographical approaches. Special emphasis was placed on the human, while animals were excluded as actors in social history. History was to be devoted to the process by which humans formed and transformed their reality. Structural barriers were, accordingly, overcome in a conscientious act of appropriation of the social world. Thompson was of course not alone in this concentration on the human actor. While there were numerous debates as to how to treat society as an object of research – considerable discrepancies existed between British and American social history, the French *Annales* school (an obvious antecedent of the other approaches), and the German *Gesellschaftsgeschichte* – all of these approaches looked at the position of humans in (human) society.[2] Eric Hobsbawm also referenced the distinction between the human and the animal in his definition of social history: Social history was "the history of societies as well as of

---

[1] Nance, Susan. *The Historical Animal*. Syracuse, NY: Syracuse University Press: 2015, 5.
[2] Cf. Rüter, Adolf J. C. Introduction. *International Review of Social History* 1, no. 1 (1956): 1–7; Burke, Peter. Reflections on the Historical Revolution in France: The Annales School and British Social History. *Review (Fernand Braudel Center)* 1, vol 3/4 (1978): 147–156. Because at its inception animal history adopted little from its influential counterparts in continental Europe, the connections between animal history and the Annales school as well as the *Gesellschaftsgeschichte* of the Bielefeld school are not explored in as much detail.

human society (as distinct from, say, that of apes and ants)."³ With this reference, Hobsbawm sought to address conceptual problems in historiography. He insisted that, when writing history, it was necessary to define the object of investigation and warned that during the process the social structures being dealt with were often represented reductively or in a universal fashion.

Notwithstanding the drawing of these lines between humans as social actors and animals as some sort of naturalized foil, the pursuit of nailing down what constituted historical agency paved the way for a social history informed by animal history as it is pursued today, that is a history of society that does not exclude animals.⁴ Indeed, early discussions in animal history made the question of who and under which circumstances someone became a historical actor their focal point. Agency, it can be surmised, acquired the status of what Walter Johnson termed a "master trope" throughout the field of human-animal studies.⁵ The other focal point in an animal social history has been the re-definition of the social, of social relationships and of interspecies communication. For animal historians the fact that many animals have lived with us as part of an interspecies relationship thus serves as an entry point to gain a more inclusive picture of what past societies might have looked like. When the status of an animal or human as an actor, and the question of what innate ability constitutes an actor, are set aside, then the shared life of the human and nonhuman can become the focus of investigation.

In the following, I will therefore try to first present some form of a genealogical overview of which aspects of animal-human relations have been taken up by social history, with a special focus on the human relation to horses and dogs. It might come as no surprise here that animals appeared in social history before the *animal turn*, yet that they remained mostly representatives of social hierarchies that were not of their own making. In the discussion of theories and methods, I will nevertheless argue that the toolbox provided by the new social history, with its focus on the analysis of practices and actions, can be applied to animal-human relations. Finally, I will outline how the *animal turn* enables the question of the social to be rethought. All in all, this entry takes up the questions posed at the beginning and will debate whether there can be a social history that goes beyond the human.

---

3 Hobsbawm, Eric J. From Social History to the History of Society. *Daedalus* 100, no. 1 (Winter 1971): 20–45, 30.
4 Looking at it from another angle, it is certainly justified to speak of societal animal-human relationships because society continually shapes these relationships.
5 Johnson, Walter. On Agency. *Journal of Social History* 37, no. 1 (October 2003): 113–124, 113. For the application within human-animal studies see McFarland, Sarah E. and Ryan Hediger (eds.). *Animals and Agency: An Interdisciplinary Exploration*. Leiden: Brill, 2009.

2 Topics and Themes

A piece often cited by socio-historically informed animal history is a satirical contribution written in 1974 by a certain Charles Phineas (a pseudonym). With ever more groups emerging as historically oppressed, Phineas asked whether it was not also time for a social history of pets.[6] In contrast to the first generation of new social historians, for whom the animal functioned merely as foil for the human, animals functioned here as an argument against a social history that was too inclusive, that sought to be more than a history of the workers' movement. Phineas' words, likely to his chagrin, were taken literally. Five years later, in the same publication, the *Journal of Social History*, John K. Walton published an article about how changing human-dog relations could serve as a fruitful illustration of how Victorians dealt with political and social upheaval.[7] The articles in the *Journal of Social History* and other works that emerged at the same time – on animal welfare, antivivisectionism,[8] and hunting[9] – very obviously focused on human society. Here, animals served to create a more detailed portrait of past human society, yet the works also opened the door to taking animals seriously as social actors. Along with the history of ideas [→History of Ideas], social history was thus one of the most important sources of methodological and empirical inspiration for animal history, despite or perhaps because of the manner in which it previously delimited the animal.

The history of pets, for example, the social history of the animals with which humans cohabit, both physically and emotionally, was and remains a central theme [→History of Pets]. Asking what the social implications of rising pet ownership at the end of the nineteenth century were and how they coincided with increasingly differentiated classes, John K. Walton still saw (for instance in the structures surrounding dogs in breeders' associations) a social microcosm of Victorian society – the same microcosm that was at play in hot-button topics of the time such as nature conservation and child welfare. This thread was taken up by Harriet Ritvo, who delved

---

6 Phineas, Charles. Household Pets and Urban Alienation. *Journal of Social History* 7, no. 3 (Spring 1974): 338–343. For its reception, see, for instance, Fudge, Erica. A Left-Handed Blow: Writing the History of Animals. In *Representing Animals*, Nigel Rothfels (ed.), 3–18. Bloomington, IN: Indiana University Press, 2002; Shaw, David Gary. A Way with Animals. *History and Theory* 52, no. 4 (December 2013): 1–12. Kete, Kathleen. Animals and Ideology: The Politics of Animal Protection in Europe. In *Representing Animals*, Nigel Rothfels (ed.), 19–34. Bloomington, IN: Indiana University Press, 2002.
7 Walton, John K. Mad Dogs and Englishmen: The Conflict over Rabies in late Victorian England. *Journal of Social History* 13, no. 2 (Winter 1979): 219–239.
8 French, Richard D. *Antivivisection and Medical Science in Victorian Society.* Princeton, NJ: Princeton University Press, 1975; Harrison, Brian. Animals and the State in Nineteenth-Century England. *The English Historical Review* 88, no. 349 (October 1973): 786–820. A social history of animal welfare that considered the animal as social actor was published by Kean, Hilda. *Animal Rights: Political and Social Change in Britain Since 1800.* London: Reaktion Books, 1998.
9 Itzkowitz, David C. *Peculiar Privilege. A Social History of English Foxhunting, 1753–1885.* New York, NY: Harvester Wheatsheaf, 1977.

deeper into how pet ownership correlated with social status in the Victorian era. In contrast to Walton, however, her story was less about the way the working class and the bourgeoisie experienced life in clubs and associations, nor was it so much about aristocratic attempts to breed animals. Rather, it took up changing interactions with very specific animals to analyze foundational transformations, societal conditions, and animal-human relations.

Walton had already made it clear that breeders' associations had implications for various levels of society, opening up discussions that negotiated the authority, creation, and organization of knowledge, power, and hierarchies.[10] However, while offering an important insight, he neglected to attend to the formative role of animals in shaping society. Dogs, Ritvo contrarily argued, lent themselves particularly well to the strategies surrounding the development and maintenance of social hierarchies. The dog "is subject to much greater physical variation than any livestock species (or indeed, as it happens, than any other mammalian species); it therefore offered enormous opportunities for differentiating breeds and for ranking individuals of varying qualities within breeds."[11]

The dog show thus became an event in which the bodies of animals served not only as vessels for ideas about race, but also class, and notions of the social were put into practice. This, of course, required the cooperation of the dogs. The shows put on by the Kennel Club, which began in 1873, featured the cream of the crop, limiting "competition to a carefully screened segment of the canine and, implicitly, the human population."[12] According to Ritvo, it was striking that there was no longer any simple relation between pure-bred animals and the aristocracy. Rather, social belonging was constantly renegotiated at dog shows and was mirrored in fashion and breeding preferences. In this way, the status of particular dogs was subject to change, as was the case with the bulldog, for example. Up until the middle of the nineteenth century, the bulldog was primarily used for fighting and was therefore associated with the leisure activities of the working class. By the end of the century, however, the bulldog had become the "darling of the refined and fashionable."[13] The breed, "ready to take its place in the elaborately graduated hierarchy of canine society,"[14] thus vividly illustrates processes of social transformation and the bourgeois ideal of self-improvement. In this way, canine society stood as a model of how human societies, too, could change, improve, and be remade. The idealized society was one in which there was generally room for improvement.

---

[10] Ritvo, Harriet. *The Animal Estate: The English and Other Creatures in the Victorian Age*. Cambridge, MA: Harvard University Press, 1987.
[11] Cf. Ritvo, Harriet. Pride and Pedigree: The Evolution of the Victorian Dog Fancy. *Victorian Studies*, 29, no. 2 (Winter 1986): 227–253, 235.
[12] Ritvo, Pride and Pedigree, 241.
[13] Ritvo, Pride and Pedigree, 245.
[14] Ritvo, Pride and Pedigree, 252–253.

Yet at the same time, class distinctions were naturalized by taking dog breeds as reference points. In colonial societies and colonies too, according to Robert Gordon, dogs served as "mobile metaphors for understanding gradations."[15] Animals could only provide this function because they were also physically part of the societal system and assigned their place accordingly in the hierarchy. In contrast to colonial accounts [→(Post)Colonial History], in which animals and the colonized population were usually placed on the same level, both deemed equally in need of civilizing,[16] other societies offered hierarchies in which there were finer distinctions. In the Third Reich, dogs were considered part of the *Volksgemeinschaft* and thus were clearly invested with more value than many humans,[17] becoming part of the political propaganda machine. This implies that the animals' role was never static but closely interrelated with the specific social and political fabric [→Political History].

The social historical perspective typically scrutinizes those animals that can easily be portrayed as individuals, such as dogs, characterizing them as "meaning-making figures"[18] that have obviously become part of the human-animal sociality. The history of coevolution, mutual influence and cooperation, at least at first glance, can be told better through these creatures. This is very much in keeping with the methodological approaches of social history, which turns to the biography in order to gain access to the experience of individuals who have suffered most under societal structures.[19] Social history thereby attends particularly to historical upheavals and the effects these have on the formation of society and the individual members of society. For this reason, the industrial revolution and the accompanying processes of urbanization [→Urban (and Rural) History] received most of the attention of social historians, at least those predominantly concerned with British society.

In these accounts, animal history concerns itself with the animals who are closely tied up with these upheavals, above all horses.[20] In this respect, societal structures themselves and how they affect the lives of humans and animals are of more interest than the social lines of demarcation surrounding animal beings. Anna Sewell's novel

---

15 Gordon, Robert J. Fido. Dog Tales of Colonialism in Namibia. In *Social History & African Environments*, Beinart, William McGregor, Joann McGregor (eds.), 240–251. Athens, OH: Ohio University Press, 2018, 248.
16 Cf. Poon, Shuk-Wah. Dogs and British Colonialism: The Contested Ban on Eating Dogs in Colonial Hong Kong. *The Journal of Imperial and Commonwealth History* 42, no. 2 (March 2014): 308–328.
17 Roscher, Mieke. New Political History and the Writing of Animal Lives. In *The Routledge Handbook of Animal-Human History*, Hilda Kean, Philipp Howell (eds.), 53–75. London: Routledge, 2018.
18 Haraway, Donna. *When Species Meet*. Minneapolis, MN: University of Minnesota Press, 2008, 4; see also Swart, Sandra. Dogs and Dogma: A Discussion of the Socio-Political Construction of Southern African Dog Breeds as a Window onto Social History. In *Canis Africanis: A Dog History of Southern Africa*, Lance van Sittert, Sandra Swart (eds.), 267–287. Leiden: Brill, 2008.
19 Cf. Krebber, André and Mieke Roscher (eds.). *Animal Biography. Re-framing Animal Lives*. London: Palgrave, 2018.
20 Clutton-Brock, Juliet. *Horse Power: A History of the Horse and the Donkey in Human Societies*. London: Natural History Museum Publications, 1992.

Black Beauty, a narrative about these demarcations published in 1877, deals both with the treatment of carriage horses and, quite pointedly, the working class.[21] Thus, even in the contemporary literary imagination horses were seen as both being part of the workforce as well as being emblematic for the social stratification accompanying the division of labor. As Gary David Shaw has observed: "Dogs and horses are almost unique in their master-slave/servant relationship with people. It was that sort of hierarchical relationship that until fairly recently characterized many if not most European human social relations as well: slaves, yes, but more typically servants of all kinds, subordinates with whom you worked."[22] Horses were, however, less symbolically laden than dogs because they played a direct role in the value chain and could thus more easily be conceptualized as part of the working class.[23]

F.M.L. Thompson famously described Victorian society as a "horse-drawn society," with its various elements defined by the manner in which they utilized horses.[24] One's social status was determined by how and when one spent time with horses, worked with them, and/or lived with them. But horses were not merely mediators – they were simultaneously part of the social setting. The more targeted breeding of purebred horses preceded that of dogs and was informed by aristocratic ideals, particularly the emergence of bloodlines, as well as narratives of gallant war horses. Because this type of breeding originated in feudal society, it can be described as preindustrial. Taking industrialism as a turning point in history, the transition from early modern to modern times and from the preindustrial to the industrial age as the focal point of the "horse age" (*Pferdezeitalter*), as Reinhart Koselleck did, also allows for an historical understanding of the social realities of the coexistence of humans and horses.[25] Ann Norton Greene has investigated the interdependence of industrial progress, social upheaval, and the impact of horses in America during the industrial revolution, noting that the "horse's historical agency lies in the physical power they produced, and the role of this power in shaping material and social arrangements".[26] With Greene's "Horses at Work," a category was invoked that immediately interested social historians: work and how social configurations are determined by it [→Eco-

---

21 Sewell, Anna. *Black Beauty.* Norwich: Jarrold & Sons, 1877.
22 Shaw, Gary David. The Torturer's Horse: Animals and Agency in History. *History and Theory* 52, no. 4 (December 2013): 146–167.
23 Hribal, Jason. Animals are Part of the Working Class: A Challenge to Labor History. *Labor History* 44, no. 4 (November 2003), 435–453; Hribal, Jason. Animals are Part of the Working Class Reviewed. *Borderlands* 11, no. 2 (2012): 1–37.
24 Thompson, Francis Michael Longstreth. *Victorian England: The Horse-drawn Society: An Inaugural Lecture.* London: Bedford College, 1970.
25 Koselleck, Reinhart. Der Aufbruch in die Moderne oder das Ende des Pferdezeitalters. In *Historikerpreis der Stadt Münster. Die Preisträger und Laudatoren von 1981–2003*, Berthold Tillmann (ed.), 159–174. Münster: Stadt Münster, 2005.
26 Greene, Ann Norton. *Horses at Work: Harnessing Power in Industrial America.* Cambridge, MA: Harvard University Press, 2009, xi.

nomic History]. Up until the middle of the twentieth century, horsepower was the unit by which power was measured on both sides of the Atlantic.[27] The asymmetrical distribution of horsepower, to which elites had more access, had a number of effects on working conditions and workload. How this manifested not only on the macro-level, but also on the micro-level – in other words, how this constellation could become part of the history of everyday life – is an essential question also taken up by social history. In Fernand Braudel's "Civilization and Capitalism," a longitudinal study of the everyday practices in the Mediterranean on its path to industrialization, he places special emphasis on the function of draft animals. Together with the farmers, horses, oxen, donkeys, and mules could be considered part of a set social system: "Plants, animals and people all had a place in it."[28]

A social history of human-animal relations thus also examines how these relations change in different contexts. Established narratives portray the transformation of human-animal relations in the period since the industrial revolution fairly one-dimensionally, with the relative proximity between humans and animals giving way to mutual alienation, with some animals paradoxically becoming an ever more important part of the human emotional landscape [→History of Emotions].[29] The micro-analyses offered by social history, however, are capable of recognizing multifaceted and complex transformations that are equally important and exist simultaneously alongside one another.[30] Subjects that illustrate these complex entanglements include, for example, the social consequences of animal epidemics,[31] animal thefts[32] or the distribution of land and its use,[33] and how animals and humans were affected by these developments.

---

[27] Tarr, Joel A. A Note on the Horse as an Urban Power Source. *Journal of Urban History* 25, no. 3 (March 1999): 434–448.
[28] Braudel, Fernand. *Civilization and Capitalism, 15th-18th Century, Vol. I: The Structure of Everyday Life*. Berkeley, CA: University of California Press, 1992, 117.
[29] A classic example is Berger, John. Why Look at Animals? In *About Looking*, John Berger (ed.), 3–28. London: Vintage, 1980.
[30] Cf. Woods, Abigail. Rethinking the History of Modern Agriculture: British Pig Production, c. 1910–65. *Twentieth Century British History* 23, no. 2 (June 2012): 165–191.
[31] Swabe, Joanna. *Animals, Disease and Human Society: Human-Animal Relations and the Rise of Veterinary Medicine*. London: Routledge, 2002; Woods, Abigail. Flames and Fear on the Farms: Controlling Foot and Mouth Disease in Britain, 1892–2001. *Historical Research* 77, no. 198 (October 2004): 520–542.
[32] Freeman, Michael. Plebs or Predators? Deer-stealing in Whichwood Forest, Oxfordshire in the Eighteenth and Nineteenth Centuries. *Social History* 21, no. 1 (January 1996): 1–21.
[33] Cf. Hoyle, Richard W. and C. J. Spencer. The Slaidburn Poor Pasture: Changing Configurations of Popular Politics in the Eighteenth- and Early Nineteenth-Century Village. *Social History* 31, no. 2 (2006): 182–205.

3 Methods and Approaches

That the new social history saw a veritable boom in the 1970s and 1980s was due to the fact that new historical actors were recognized and subjected to scrutiny. In the process, the materialist impetus behind social history – as the history of everyday life, mentalities, and culture – that drove the previous generation of social historians became increasingly weak. Social history was no longer simply the history of the poor and radical social movements, particularly of socialism,[34] but was expanded to include postcolonial and gender-oriented historiographies that turned to discourse analysis and cultural history.[35] Instead of class, social history now concentrated more on gender, race, material culture, and the environment. The discipline turned to actors whose social lives were overlooked by discourse and thus required different methods of investigation.[36]

Different approaches were also developed: Practices, relations, and interactions acquired a much more prominent role. Agency was interpreted in an increasingly relational and interconnected fashion. Especially for social historically informed animal history, this approach grants animals a historical role, without necessarily assigning them individual identities.[37] Here, agency is interpreted as being distributed in human-animal networks; the stress is therefore on social coproduction. In this way, like material culture studies [→Material Culture Studies], animal history is more connected to sociological approaches, particularly Bruno Latour's actor-network theory, than social historical approaches focusing on individual actors. Animal history focuses on the historical interrogation of the space between material and semiotic categories.[38] Including material dimensions of the past allows for a more

---

34 See Hobsbawm, From Social History to the History of Society, 21.
35 Eley, Geoff. Is all the World a Text? From Social History to the History of Society Two Decades Later. In *Practicing History. New Directions in Historical Writing After the Linguistic Turn*, Gabrielle M. Spiegel (ed.), 35–61. New York, NY and London: Routledge 2005; Cf. Conrad, Christoph. Social History. In *International Encyclopedia of the Social & Behavioral Sciences*, Neil J. Smelser, Paul B. Baltes (eds.), 14299–14306. Oxford: Elsevier 2001.
36 See Poster, Mark. *Cultural History and Postmodernity. Disciplinary Readings and Challenges*. New York, NY: Columbia University Press, 1997, 45. This upheaval by no means occurred without conflict. See Kocka, Jürgen. Losses, Gains and Opportunities: Social History Today. *Journal of Social History* 37, no. 1 (Autumn 2003): 21–28.
37 Rees, Amanda. Animal Agents? Historiography, Theory and the History of Science in the Anthropocene. *British Journal for the History of Science*, 2, no. 1 (September 2017): 1–10; Pearson, Chris. Dogs, History and Agency. *History and Theory* 52, no. 4 (December 2013): 128–145.
38 Roscher, Mieke. Actors or Agents? Defining the Concept of Relational Agency, in (Historical) Wildlife Encounters. In *Animal Encounters*, Jessica Ullrich, Alexandra Böhm (eds.), 149–170. Stuttgart: J.B. Metzler, 2019. On the conflict and relation between sociology and social history, see Braudel, Fernand and Immanuel Wallerstein. History and the Social Sciences: The Longue Durée. *Review (Fernand Braudel Center)* 32, no. 2 (2009): 171–203; Burke, Peter. *History and Social Theory*. Cambridge: Polity Press, 1992.

comprehensive understanding of social practices and historical processes. The analysis of practices is the genuine task of the social historian, as social history has long been concerned with its connection to praxis.³⁹

The influencing of historical processes can thus be interpreted as doing history. In this way, a dynamic concept of history is employed in which it becomes a kind of history in the making, one made also by animals. As with the concept of doing culture, action can be considered "a type of activity that is not connected to intention," but rather includes "an elementary practice that precedes the symbolic condensation of 'actions.'"⁴⁰ Animals can therefore easily be incorporated in this framework, also because praxis and action each show a different set of empirical values. They can be regarded in this way as "*embedded* in the performance of social phenomena."⁴¹

Looking to theories of praxis, which conceptualize the social as becoming-in-doing, a social history of animals or animal-human society can encompass what we do with animals as well as what they themselves do and the material consequences of this. In line with practice theorist Theodor Schatzki an animal history informed by social history requires a "social ontology that depicts social existence as inherently transpiring in nexuses of practices and material arrangements."⁴² He thereby sought to conceptualize nature and technology as forces unto themselves and as part of society, refusing to recognize the artificial division between nature and culture. According to Jürgen Kocka, social history does not accept "that the past can sufficiently be understood as a context of perceptions, experiences, discourses, actions and meanings alone." Instead, he continues, social historians "insist that conditions and consequences, structures and processes have to be taken seriously and brought back in."⁴³ It is thus the material consequences of practices on the animals in animal-human society that animal historians should focus on.

Social history also offers animal history a plethora of approaches to methodology. Its prerogative is – to take up an oft-quoted phrase from Walter Benjamin – "to brush history against the grain," and in this way capture the voices of the historically oppressed. This found expression particularly in the German history of everyday life (*Alltagsgeschichte*) and the microhistory inspired by the *Annales* school,⁴⁴

---

39 Polyakov, Michael. Practice Theories: The Latest Turn in Historiography. *Journal of the Philosophy of History* 6, no. 2 (January 2012): 218–235.
40 Hirschauer, Stefan. Praktiken und ihre Körper. Über materielle Partizipanden des Tuns. In *Doing Culture. Neue Positionen zum Verhältnis von Kultur und sozialer Praxis*, Karl H. Hörning, Julia Reuter (eds.), 73–91. Bielefeld: transcript, 2004, 73 [translation mine].
41 Hirschauer, Praktiken und ihre Körper, 89 [translation mine].
42 Schatzki, Theodore R. Nature and Technology in History. *History and Theory* 42, no. 4 (December 2003): 82–93; Cetina, Karin Knorr, Theodore R. Schatzki and Eike von Savigny (eds.). *The Practice Turn in Contemporary Theory*. London: Routledge, 2005.
43 Kocka, Losses, Gains and Opportunities, 28.
44 Lüdtke, Alf (ed.). *The History of Everyday Life: Reconstructing Historical Experiences and Ways of Life*. Princeton, NJ: Princeton University Press, 1995.

which interprets the social as lived praxis on the smallest possible level.⁴⁵ The traces of this lived praxis "are often erased, hidden, or disguised, either through ideology or through social geographies that separate human and animals, consumers and laborers, urban and rural, pleasure and pain."⁴⁶ Sometimes all that is required to recover this lived praxis is the development of lines of sight that also include animals. In his social history of dogs in colonial Namibia, Robert Gordon, for example, found that in order to see the role of the animal-human relationship comprehensively, "one must infer it from a variety of contextual documents like folktales, poems, proverbs and especially photographs. It is striking how many photographs of settlers have dogs either deliberately posed, or in many cases simply wandering into a scene being photographed."⁴⁷ Instead of determining what constitutes action in advance, a more generous notion of history requires the inclusion of nature and animals. "Human history, accordingly, is a social-natural history: a perpetual development that encompasses the omnipresent and varied active presence of nature in human life."⁴⁸ This, in turn, demands an "amplified understanding of social life."⁴⁹

4 Implication(s) of the Animal Turn

Forty years after Phineas' satire, social history as a discipline has shown itself to be relatively resistant to regarding animals as societal actors and has failed to produce any foundational works analyzing past animal-human societies.⁵⁰ The early works portrayed here "for the most part [have] been concerned with adding in animals to extant human-focused histories", as Hilda Kean rightly criticizes.⁵¹ Nevertheless, the categories utilized by social history and the questions it raises about the influence of race, class, and gender have become central for an animal history informed by social history. In their groundbreaking essay from 2010, Susan Pearson and Mary Weismantel explained that the social should constitute the focus per se, as it would only be this perspective that could help to bridge the gap between anthropomor-

---

45 See also: Pręgowski, Michał Piotr. *Companion Animals in Everyday Life: Situating Human-Animal Engagement within Cultures*. London: Palgrave, 2016.
46 Pearson, Susan J. and Mary J. Weismantel. Does "The Animal" Exist? Toward a Theory of Social Life with Animals. In *Beastly Natures: Animals, Humans, and the Study of History*, Dorothee Brantz (ed.), 17–37. Charlottesville, VA: University of Virginia Press, 2010, 24.
47 Gordon, Dog Tales of Colonialism, 249.
48 Schatzki, Nature and Technology in History, 90.
49 Pearson and Weismantel, Does "The Animal" Exist?, 32.
50 Stephen Mosley reaches a similar conclusion regarding environmental history, which has also failed to receive its due attention from social history, despite itself being informed by it: Mosley, Stephen. Common Ground: Integrating Social and Environmental History. *Journal of Social History* 39, no. 3 (March 2006): 915–933, 924.
51 Kean, Hilda. Animal-Human Histories. In *New Directions in Social and Cultural History*, Sasha Handley, Rohan McWilliam, Lucy Noakes (eds.), 173–190. London: Bloomsbury Publishing, 2018, 175.

phism and anthropocentrism, as well as between the way animals function as symbols and the material experience of them: "Our relationship to animals is [...] neither wholly symbolic nor wholly material; rather it is profoundly social."[52] For this reason, they argue, one must look to social experience, to the manner in which we live with animals. With this appeal, Pearson and Weismantel entered the debate as to whether social history, under the influence of cultural studies, had not lost track of the social and how it should be defined. Indeed, in the field of social history, Patrick Joyce claimed, "the social stands in need of constant theoretical scrutiny and reinterpretation."[53]

In conceptualizing animals as part of the social, Pearson and Weismantel also seize on the urgency of constant re-examination. They go on to argue that, in everyday practice, the subject and object of an act can often not be clearly differentiated and that, for this reason, a broader theory of social life is necessary. This is relevant to animal history insofar as it suggests the volatility of the concepts of subject and society, which always have a historical dimension. For social history, in fact, what counts as a legitimate subject of society has changed radically, and with it what can potentially be perceived as a historical actor.[54] In keeping with Pearson and Weismantel, a history of the social lifeworld of the human-animal relationship must take into account symbolic, economic, and spatial dimensions.[55]

The *animal turn* thus allows animals to be grasped as bound up with the social structures of an interspecies society, which, however, raises a number of new questions. Sandra Swart, for instance, casts new light on the social history of the human-horse relationship: "The 'world the horses made' is still too much a history of their riders. It is still too much the 'world the horses were made to make' (by humans) rather than the 'world they made.'"[56] She concludes that it should be possible, at least to some extent, to write the history of the horse from the horse's perspective. Here, too, the skills of social historians are called upon in order to read actions without interpreting them with only one side or the other in mind. Swart sees no point in seeing every equine action as an act of resistance; rather, she calls for the development of a finer sense of how horses express themselves. A "horsestory" would unite history and natural history and examine shared experiences of humans and horses on the microlevel.[57] She continues, "[t]he social history of the horse–human relation-

---

52 Pearson and Weismantel, Does "The Animal" Exist?, 22.
53 Joyce, Patrick. What is the Social in Social History. *Past & Present* 206, no. 1 (February 2010): 213–248, 224.
54 Shaw, A Way with Animals, 2.
55 Pearson and Weismantel, Does "The Animal" Exist?, 28.
56 Swart, Sandra. The World the Horses Made: A South African Case Study of Writing Animals into Social History. *International Review of Social History* 55, no. 2 (August 2010): 241–263, 252. See also: Swart, Sandra. *Riding High: Horses, Humans and History in South Africa*. New York, NY: New York University Press, 2010.
57 Swart, The World the Horses Made, 257.

ship reveals how its experiences alter in time (and space) and concomitantly so does the social experience of that relationship."[58] It is thus the historically specific relations and contexts that determine how the realm of possibility for animals is formed. When Amanda Rees poses a series of questions, she refers to this realm of possibility and social context: "Could different kinds of horses – warhorses, as compared to pack mules, say – both be seen to experience the same kind of agency? Or does making this distinction itself reflect the reading of human class conflict onto the animal canvas, privileging the warrior over the peasant?"[59] Only by addressing these sorts of questions can the animal-human relationship be adequately incorporated into the organization of the social, according to Pearson and Weismantel, and can the historical changes resulting from this relationship be comprehended.[60]

In this regard, the structural categories offered to us by social history can help us to include animals in history without necessarily introducing species as another category. The relations that influence actors always have an economic dimension, which must be considered in the writing of animal history as social history. This economic dimension illuminates the power relations inherent to human-animal dependencies and highlights the centrality of the animal for economic development as well as the classic Marxist "human-animal relations of production."[61] For this reason, Dorothee Brantz insists on disclosing the corresponding sociocultural, economic, and political circumstances in which the animal-human relationship is situated.[62] The relationship between societal structures and animal-human interactions should therefore be regarded as dialectical.

This dialectic again has two sides to it. The question of how models of society can be applied to animal societies in particular has been of interest to philosophy since antiquity, ethnology since the nineteenth century, and, at least to a degree, sociology since the twentieth century.[63] But does it make sense to utilize the approaches and methods of social history in this context? Sandra Swart has pointed out the challenges in conceptualizing society as going beyond the boundaries of the species, maintaining that doing as much would eventually lead to misinterpreta-

---

[58] Swart, The World the Horses Made, 262–263. For the historian's task of imagining the past, see Kean, Animal-Human Histories, 178.
[59] Rees, Animal Agents, 4.
[60] Pearson and Weismantel, Does "The Animal" Exist?, 26.
[61] Tapper, Richard. Animality, Humanity, Morality, Society. In *What Is an Animal*, Tim Ingold (ed.), 47–62. London: Unwin Hyman, 1988, 52–54.
[62] See Brantz, Dorothee (ed.). *Beastly Natures: Animals, Humans and the Study of History*. Charlottesville, VA: University of Virginia Press, 2010, 3.
[63] For newer approaches, see, for example, Whitehead, Hal. *Analyzing Animal Societies: Quantitative Methods for Vertebrate Social Analysis*. Chicago, IL: University of Chicago Press, 2008; De Waal, Frans and Peter L. Tyack (eds.). *Animal Social Complexity: Intelligence, Culture, and Individualized Societies*. Cambridge, MA: Harvard University Press, 2009.

tions of human society.[64] She takes baboons as an example, a species whose complex collective life has led reputable ethnologists to consider primate societies as characterized by violence and sexual dominance. Notwithstanding, when categories used to approach one species are reductively applied to another, they become problematic. Moreover, these reductive, trans-species models were re-applied to human society, where they were also naturalized: "The baboon model of human nature was based on naturalizing nationalism and aggrandizing aggression – validating the existing social hierarchy of race, class and gender."[65] According to Swart, it was only in the 1980s that this reductive model of dominance was discarded. However, the damage had already been done with regard to the analysis of animal societies. Swart maintained that a social history of animals must take into account "that there is no such thing as the baboon or at least no such thing as a model baboon – organically essential, unchanging and homogenous. They are as historical as we are ourselves."[66]

Brett L. Walker therefore suggests a more promising strategy to do away with essentialism when it comes to looking at animals historically: to stop considering humans as an anomaly and as somehow situated outside nature.[67] It follows that animals should no longer be conceptualized as existing beyond the confines of society. In the Anthropocene, in which animals are influenced by and themselves influence the actions of humans, animals are as subject to evolving societal structures as humans.[68] What could emerge from this is an animal history informed by social history that dissolves the tension between structure and praxis, as well as between text and body. In an animal history driven by fundamental microhistorical research, actors would no longer be as clearly defined, and the social, with all of its asymmetric constellations of power, would be brought to the foreground.

Selected Bibliography

Hribal, Jason C. Animals, Agency, and Class: Writing the History of Animals from Below. *Human Ecology Review* 14, no. 1 (2007): 101–112.
Kean, Hilda. *Animal Rights: Political and Social Change in Britain Since 1800*. London: Reaktion Books, 1998.

---

[64] Swart, Sandra. *The Lion's Historian. Animal Histories from the South*. Inaugural Lecture University of Stellenbosch, delivered on 09 October 2017. Stellenbosch: Stellenbosch University, 6.
[65] Swart, *The Lion's Historian*, 9.
[66] Swart, *The Lion's Historian*, 14.
[67] Walker, Brett L. Animals and the Intimacy of History. *History and Theory* 52, no. 4 (December 2013): 45–67, 48.
[68] Cf. the discussion in Latour, Bruno. *We Have Never Been Modern*. Cambridge, MA: Harvard University Press, 2012.

Kean, Hilda. Animal-Human Histories. In *New Directions in Social and Cultural History*, Sasha Handley, Rohan McWilliam, Lucy Noakes (eds.), 173–190. London: Bloomsbury Publishing, 2018.

Krebber, André and Mieke Roscher (eds.). *Animal Biography. Re-framing Animal Lives*. London: Palgrave, 2018.

Pearson, Susan J. and Mary J. Weismantel. Does "The Animal" Exist? Toward a Theory of Social Life with Animals. In *Beastly Natures: Animals, Humans, and the Study of History*, Dorothee Brantz (ed.), 17–37. Charlottesville, VA: University of Virginia Press, 2010.

Rees, Amanda. Animal Agents. Historiography, Theory and the History of Science in the Anthropocene. *British Journal for the History of Science Themes* 2, no. 1 (July 2017): 1–10.

Ritvo, Harriet. *The Animal Estate: The English and Other Creatures in the Victorian Age*. Cambridge, MA: Harvard University Press, 1987.

Shaw, Gary David. The Torturer's Horse: Animals and Agency in History. *History and Theory* 52, no. 4 (December 2013): 146–167.

Swart, Sandra. The World the Horses Made: A South African Case Study of Writing Animals into Social History. *International Review of Social History* 55, no. 2 (August 2010): 241–263.

Walker, Brett L. Animals and the Intimacy of History. *History and Theory* 52, no. 4 (December 2013): 45–67.

Helen Cowie
Cultural History

1 Introduction and Overview

Animals have long been central to human culture. From stone-age depictions of bison in the caves at Altamira to characters in Disney cartoons, animals have been a constant presence in human art, literature and belief systems. They have shaped how we live, how we work and how we think.

Cultural history is about recovering the mentalities and ideologies of past societies and exploring how their members interpreted and understood the world around them. Animals have often been fundamental to these analyses because of the rituals and taboos surrounding the human-animal relationship. Early cultural histories used animals as a prism through which to deconstruct human beliefs, unpicking the complex meanings attached to a range of human-animal interactions. Some more recent histories, influenced by the *animal turn*, focus more explicitly on the experiences of actual animals in the past (rather than purely their symbolic functions) and pose questions about emotions, agency, and even whether animals themselves can have a 'culture' – that is, a distinct set of ideas, customs or social behaviors that distinguishes one group of individuals from other members of the same species. This chapter explores how cultural historians have used animals to illuminate past beliefs and social practices and examines the methodological approaches they have taken. I draw on examples from the sixteenth to the nineteenth centuries, showing how animals functioned as symbols, victims and cultural actors in their own right.

2 Methods and Approaches

Cultural history is a broad and expanding field. In the nineteenth century, practitioners focused primarily on elite or 'high' culture. Since then, increased emphasis has been placed upon mass or popular culture, and the different ways in which people understood the world around them.[1] Cultural history is primarily about meanings and interpretations. While social historians have traditionally concentrated on quantitative changes, tracing shifting class, race and gender relations [→Social History], cultural historians have tended to adopt a more qualitative approach, using detailed case studies to explore the layers of meaning individuals attached to particular experiences or events. What did people think when they took part in a carnival procession or participated in a food riot? What cultural beliefs shaped the perpetration of vio-

---

[1] Burke, Peter. *What is Cultural History.* London: Polity Press, 2008, 29.

lence against other living beings – human or animal? What did it mean when someone baptized a pet dog, encountered a giraffe for the first time or fed gingerbread to an elephant in a travelling menagerie? Cultural historians seek to tease out the often multi-layered meanings of such events, and to go beyond the material causes of human behavior.

Methodologically, cultural historians have been strongly influenced by developments in anthropology. In the 1960s and 70s, anthropologists conducted sustained observations of non-western societies in an effort to learn about the cultural beliefs that underpinned them. Clifford Geertz, for instance, drew upon observations of cockfighting in Bali to illuminate Balinese conceptions of power and masculinity, employing a technique called thick description to build up layer upon layer of meaning onto the events he witnessed.[2] Cultural historians have imitated this technique, using archival sources to reconstruct past world views. In *The Cheese and the Worms*, Carlo Ginzburg draws upon surviving Inquisition records to recover the somewhat eccentric religious views of a sixteenth-century miller named Menocchio – an approach described as microhistory.[3] In another cultural history classic, *Montaillou*, Emmanuel Le Roy Ladurie uses an earlier set of inquisitorial documents to chart the beliefs and life ways of another group of heretics, the Cathars of twelfth-century France.[4] Literary theory has also impacted on the work of some cultural historians, prompting them to examine written texts more critically and to be more sensitive to language, silences, and the construction of narratives. Central indeed to many cultural histories is the concept of close reading, or reading against the grain, in order to recover voices that had previously been absent or suppressed – including those of animals. The historian Inga Clendinnen has applied this technique to the records of Spanish missionaries in sixteenth-century Yucatán in an attempt to explore the impact of the Spanish conquest on Maya women.[5]

Today, cultural history remains a vibrant and evolving discipline. Cultural histories now exist of almost everything, from emotions to death, football to sex, and crime to chocolate. Cultural history has also informed the work of many social historians who increasingly incorporate cultural factors into their research. Indeed, the boundaries between the two disciplines are increasingly blurred, with many historians combining cultural and material approaches to the past; Kathleen Kete's study of pet-keeping in nineteenth-century Paris, for instance, focuses on the cultural significance of dogs and cats within the bourgeois household, but sets this against the

---

2 Geertz, Clifford. Deep Play: Notes on the Balinese Cockfight. In *The Interpretation of Cultures*, Clifford Geertz, 412–453. New York, NY: Basic Books, 1973.
3 Ginzburg, Carlo. *The Cheese and the Worms: The Cosmos of a Sixteenth-Century Miller.* London: Routledge, 1980.
4 Le Roy Ladurie, Emmanuel. *Montaillou. Cathars and Catholics in a French Village, 1294–1324.* London: Penguin, 1980.
5 Clendinnen, Inga. Yucatec Maya Women and the Spanish Conquest: Role and Ritual in Historical Reconstruction. *Journal of Social History* 15, no. 3 (Spring 1982): 427–442.

backdrop of shifting class relations.[6] The *animal turn* has brought animals further within the remit of cultural historians, posing new questions about human relationships with other species. This, however, represents a shift in emphasis rather than a complete innovation, as animals have been central to cultural history from its inception.

3 Topics and Themes

Cultural history, then, is about reconstructing the mental worlds of past societies. In what follows I focus specifically on the role of animals within cultural history and consider the important place they have assumed in helping us to interpret the attitudes and mentalities of cultures distant in time and place from our own. I address three key areas in which animals have been crucial in this regard: symbolism, religion and art; deviance and social subversion; and cross-cultural encounters.

Symbolism, Religion and Art

One prime area of analysis for cultural historians has been the symbolic function of animals in past societies. Because of their proximity to humans, and their crucial role as suppliers of meat, wool or labor, animals have often featured prominently in a range of human belief systems; as the anthropologist Claude Lévi-Strauss famously expressed it, animals are "good to think" with.[7]

If we start with the theme of religion, we find that animals have often been highly visible in the mythology, scripture and folklore of almost all societies. The Ancient Egyptians worshipped the cat goddess Bastet and mummified dead cats in her honor.[8] The Aztecs and the Maya venerated the jaguar, often wearing jaguar skins into battle and sometimes ingesting the flesh of the animals to imbibe their feline courage.[9] The Incas sacrificed llamas and alpacas to mark key points of the agricultural cycle, while the Ashanti of West Africa prayed to the spider god, Anansi.[10] Through analyzing the multiple depictions of animals in spiritual contexts, histori-

---

6 Kete, Kathleen. *Petkeeping in Nineteenth-Century Paris.* Berkeley, CA: University of California Press, 1994.
7 Lévi-Strauss, Claude. *Totemism.* Translated by Rodney Needham. Boston, MA: Beacon Press, 1963, 89.
8 Malek, Jaromir. *The Cat in Ancient Egypt.* London: British Museum, 1993.
9 Benson, Elizabeth P. The Lord, the Ruler. Jaguar Symbolism in the Americas. In *Icons of Power: Feline Symbolism in the Americas,* Nicholas J. Saunders (ed.), 53–76. London: Routledge, 1998.
10 Cowie, Helen. *Llama.* London: Reaktion Books, 2017, 39–49; Allen, Barbara. *Animals in Religion. Devotion, Symbol and Ritual.* London: Reaktion Books, 2016, 33.

ans have begun to disentangle the belief systems of past human societies, gradually reconstructing their mental worlds.

Animals have also been incorporated into secular rituals, which can likewise be highly revealing. Exploring how animals were perceived in medieval Europe, for instance, Esther Cohen describes the phenomenon of the 'backwards ride' in which criminals were carried to the gallows "riding backwards upon a donkey, a ram (for licentious women), or a sick horse". Sometimes the rider was "forced either to hold on to the animal's tail or lower his face into its anal cleft", deepening his sense of shame.[11] These kinds of shaming punishments played upon the permeability of the human-animal boundary and were inflicted as a deliberate inversion of social norms. In an interesting colonial adaptation of the ritual, Spanish officials in Peru punished disobedient Indians by making them ride naked on the back of a llama – a punishment considered more degrading if the animal selected was piebald in color.[12] Jesuit-trained Andean chronicler Felipe Guaman Poma depicts a royal administrator subjecting a native Peruvian to this humiliation (Fig. 1).

As well as encoding spiritual and ritual meanings, animals have often taken on important symbolic functions as representatives of individuals, cities and nations. They appear on crests, coats of arm, flags, statues and coins and are increasingly seen as part of the national heritage. A bear, for instance, features on the crest of the Swiss city of Bern, commemorating the hunting exploits of a local lord, while a nineteenth-century sculpture by Luis Rochet uses native animals (an anteater, a tapir and a capybara) to represent the natural environment and (romanticized) indigenous heritage of Brazil (Fig. 2). Today, animals continue to act as shorthand for cities, regions and nations, standing in for countries in political discourse and sport.[13] Animals thus form part of personal and national iconographies, serving as visual representation of people and places.

Finally, animals have long made their presence felt in human art and literature, providing essential outlets for human artistic expression [→Visual Culture Studies and Art History]. Whether in fables, fairy tales, novels, poems or films, animals have been central to human storytelling. In some cases, such as George Orwell's *Animal Farm* (1945), animals function largely as surrogate humans, offering a thinly veiled critique of human actions and failings. In other cases, like Anna Sewell's *Black Beauty* (1877), animals assume more individualized characters and become mouthpieces for other members of their own species. Whatever format they take, depictions of animals in books, paintings or films can be highly revealing of our changing relationship with other species and can tell us a lot about how past societies grappled with the human animal divide. As Lorraine Daston and Gregg Mitman ob-

---

11 Cohen, Esther. Animals in Medieval Perceptions. In *Animals and Human Society: Changing Perspectives*, Aubrey Manning, James Serpell (eds.), 59–80. London: Routledge, 1994, 69.
12 Cobo, Bernabe. *Historia del Nuevo Mundo*. Seville: Imprenta de E. Rasco, 1891, 321.
13 Swart, Sandra. The Other Citizens: Nationalism and Animals. In *Handbook for Animal–Human History*, Philip Howell, Hilda Kean (eds.), 31–52. London: Routledge, 2018.

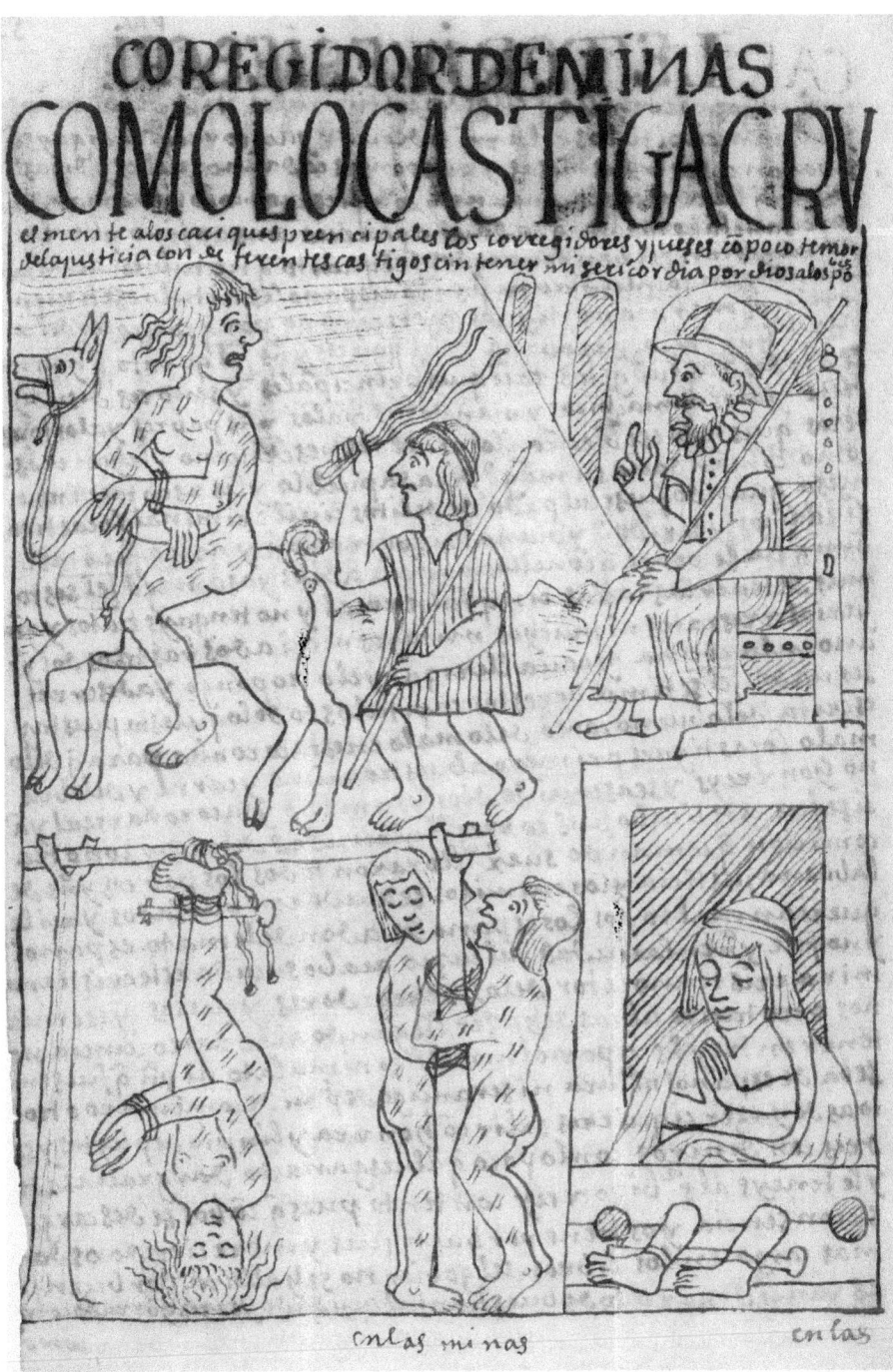

Figure 1: Drawing 211. 'The Administrator of Royal Mines Punishes the Native Lords with Great Cruelty'. From Felipe Guaman Poma, *El Nuevo Corónica y Buen Gobierno* (1615), © Royal Library of Copenhagen, shelfmark GKS 2252 4.

Figure 2: Luis Rochet, detail from 'Equestrian Sculpture of Don Pedro I', Praça de Tiradentes, Rio de Janeiro, 1862. Note the anteater and capybara alongside the idealized figure of a Native American.

serve, "When humans imagine animals, we necessarily re-imagine ourselves, so these episodes reveal a great deal about notions of the human – the 'anthropos' of anthropomorphism".[14]

---

14 Daston, Lorraine and Gregg Mitman. The How and Why of Thinking with Animals. In *Thinking*

Deviance and Subversion

One of the best ways to unpick the complex social codes adhered to by past societies is to examine instances in which those codes were transgressed. Precisely because animals have been so central to human power structures and belief systems they often feature in tales of subversion, sometimes as victims of extreme violence, in other cases as the objects of excessive human affection. By looking at instances in which animals have been treated inappropriately (according to the standards of the place and time), historians have gained a deeper insight into the social norms governing past societies and the (often unspoken) rules according to which they operated.

To understand how animals have become the subject of human taboos, let's turn first to one of the classic texts in cultural history, Robert Darnton's essay, "The Great Cat Massacre". In this engaging essay, Darnton relates a bizarre and disturbing incident that happened in Paris in 1730. Drawing on the sole written account of the episode – a narrative penned by one of the participants, Jerome, many years after the event – Darnton describes how a group of apprentices, tired of being kept awake at night by the sound of cats howling, rounded up the animals, subjected them to a mock trial and executed them by hanging. The boys apparently regarded the massacre as hilarious and re-enacted it many times over subsequent years as a form of entertainment. According to Jerome, they took particular pleasure in disposing of the mistress' grey cat, *la grise*, which they knew to be a much-loved pet.

Killing cats doesn't sound like the height of mirth to the modern reader, so Darnton naturally asks how something we would regard as supremely unfunny, even obscene, could prove so amusing to the apprentice Jerome and his colleagues. He argues that the very fact that we do not "get" the joke in the cat massacre points to a disjuncture between a past society and our own, and that this disjuncture is precisely what merits investigation. Using the anthropological method of thick description, Darnton analyses the multiple layers of cultural meaning attached to the cat massacre, highlighting its function as a covert way of attacking the apprentices' master and mistress, its invocation of the carnivalesque and the longstanding association between cats and witchcraft in early modern France. His detailed unpicking of a strange and violent incident thus illuminates the mental world of eighteenth-century French artisans, with the brutal death of the cats providing a vital entrée into these complex social dynamics.[15]

Another example of someone not 'getting' an animal-related joke features in Zeb Tortorici's analysis of dog baptisms, marriages and funerals in eighteenth century

---

with Animals: New Perspectives on Anthropomorphism, Lorraine Daston, Gregg Mitman (eds.), 1–14. New York, NY: Columbia University Press, 2005, 6.

15 Darnton, Robert. Workers Revolt: The Great Cat Massacre of the Rue Saint Severin. In *The Great Cat Massacre: And Other Episodes in French Cultural History*, Robert Darnton (ed.), 75–104. London: Allen Lane, 1984.

Mexico. Making use, in this case, of Inquisition records from the Catholic Church, Tortorici relates how a party was held in Mexico City in 1770 to celebrate the marriage of two dogs. The animals, specially dressed up for the occasion, were attended by their owners, and later placed in a matrimonial bed. A real priest, Father Toribio Basterrechea, officiated the ceremony, pronouncing the animals man and wife "in the name of the Father and Mother of all Dogs".[16] While the participants appear to have viewed that the canine nuptials were a joke – perhaps a reflection of the esteem in which the dogs were held, perhaps just a diversion for bored Mexican aristocrats to fritter away an afternoon – the Spanish Inquisition, not known for its sense of humor, perceived the incident as unfunny and potentially sacrilegious. As a consequence, Father Basterrechea found himself under arrest and threatened with torture (though all parties were ultimately exonerated with a stiff reprimand).[17]

Attempting to explain the broader significance of this peculiar footnote in the inquisitorial archives, Tortorici traces the wider social and cultural meanings of canine marriages and shows how they subverted both the sanctity of the holy sacraments and the traditional relationship between humans and animals. The most subversive element of the performance, he contends, lay not in the words uttered by the priest, but in treating dogs as humans. As he explains:

> In canine weddings and baptisms, carnivalesque religious rituals veered dangerously close to heresy, but the larger issue at stake was that, at least in the eyes of ecclesiastical authorities, these acts challenged the divinely ordained natural and social orders.[18]

Though very different in tone from the Great Cat Massacre, the marriage of two dogs served a similar purpose in undermining and mocking traditional social hierarchies. In stark contrast to the former incident, however, it may also have reflected genuine affection for the dogs involved, revealing a more complex emotional relationship between humans and animals.

In both of the above cases the perpetrators of animal mis-use were the jokers, deriving pleasure from their abnormal interactions with other species. In other instances, however, those who behaved unnaturally around animals became the butt of the joke for others, their inappropriate relationships exposed to ridicule in biting social satires. In eighteenth-century Britain, for instance, as Ingrid Tague has shown, women who lavished attention on parrots, cats and lapdogs were accused of perverting gender norms by pampering an animal rather than directing their attentions towards a member of the opposite sex, becoming the subjects of vicious cartoons.[19]

---

[16] Tortorici, Zeb. In the Name of the Father and the Mother of All Dogs: Canine Baptisms, Weddings and Funerals in Bourbon Mexico. In *Centering Animals in Latin American History*, Martha Few, Zeb Tortorici (eds.), 93–119. Durham, NC: Duke University Press, 2013, 95.
[17] Tortorici, In the Name of, 100.
[18] Tortorici, In the Name of, 102.
[19] Tague, Ingrid. *Animal Companions: Pets and Social Change in Eighteenth–Century Britain*. Philadelphia, PA: Penn State University Press, 2015, 91–137.

Two centuries earlier, in sixteenth-century France, King Henri III's passion for lapdogs became a symbol of his effeminacy and homosexuality, mirroring the king's predilection for effeminate male favorites. As Juliana Schiesari observes: "The uncontrolled population of tiny dogs serves as a powerful synecdoche of the excess, sterility and general ruin of the kingdom under his rule".[20] Misplaced affection for inappropriate animals could thus function as a vehicle for critiquing un-kingly – or indeed unmanly/unwomanly – conduct.

The case of George IV's giraffe offers a revealing example of how culturally inappropriate relationships with animals could provide fodder for political opponents. A gift to George from the Pasha of Egypt, the giraffe arrived in 1827 and was installed in the king's private menagerie at Sandpit Gate, near Windsor. Unfortunately, the animal was in poor health after its long journey and suffering from a badly swollen knee. George, however, smitten with his new acquisition, devoted every attention to it, visiting the menagerie frequently, ordering that the giraffe be clothed in a blanket to protect it from the cold and arranging for its troublesome knee to be "constantly bathed in salt water".[21] While such attentions might seem touching to the modern reader, contemporaries were less impressed, subjecting George to a barrage of ridicule. One critic, Sir Henry Halford, made a connection between the giraffe's ailments and George's own notorious obesity, informing the Privy Council that "the indisposition of the Giraffe at Windsor has arisen from the animal's loyal sympathy in his Majesty's twinges in his toe, in his late fit of gout".[22] Another commentator, a destitute merchant, wrote a sarcastic letter to the *Liverpool Mercury* in which he accused George of putting the care of his exotic pet above the suffering of his subjects:

> I have perpetually before me the afflicting sight of a wife and six daughters almost heart-broken, not only deprived of the comforts of life to which they have hitherto been justly accustomed, but almost destitute of its necessaries, and that, too, without any prospect of amendment. But this is nothing compared with the misery I have undergone from solicitude for 'that rare animal, the giraffe', which at present appears very properly to occupy most of his Majesty's attention.[23]

A series of satirical cartoons also circulated widely, showing the king riding on the giraffe's back with his mistress (Fig. 3), coddling it in his boudoir and raising up the ailing animal with a specially designed pulley. Like Henry III's lapdogs, George IV's giraffe thus became the focus of opposition for the king's many detractors, and a convenient vehicle through which to attack an already unpopular monarch.

---

20 Schiesari, Juliana. Bitches and Queens. Pets and Perversion at the Court of France's Henri III. In *Renaissance Beasts: Of Animals, Humans and Other Wonderful Creatures*, Erica Fudge (ed.), 37–49. Urbana, IL: University of Illinois Press, 2004.
21 Tuesday's Post. *Ipswich Journal*, May 10, 1828.
22 The Mirror of Fashion. *The Morning Chronicle*, July 15, 1828.
23 General Mourning. *Liverpool Mercury*, August 7, 1829.

As the above examples show, animals often featured in cases of deviance, where social norms were subverted or challenged. For contemporaries, inappropriate behavior towards animals gave the powerless a way of symbolically overturning social hierarchies or mocking social superiors. For historians, decoding these zoological 'jokes' can provide a window onto the sometimes elusive cultural practices of the past. Looking at animals can therefore offer an insight into alien traditions and customs, helping us to reconstruct past mentalities.

Figure 3: William Heath, 'The Camelopard, or a New Hobby', 1827 (color engraving), courtesy of Bridgeman Image.

Cross-Cultural Encounters

A third area of cultural history in which animals have played a prominent role is in cross-cultural encounters [→(Post)Colonial History; →Global History]. When different cultures meet, contrasting attitudes towards animals become particularly apparent. Focusing on these encounters can reveal some of the deepest practices and assumptions at the heart of different societies and bring to the fore key cultural distinctions.

An excellent example of how historians have used animals to chart cultural difference is Virginia DeJohn Anderson's *Creatures of Empire* (2004), which examines animal encounters in colonial North America. Looking at the early years of British settlement in New England and the Chesapeake, Anderson shows how European settlers and Native Americans clashed in their perceptions of the newly-introduced European livestock, sometimes with tragic consequences. The Indians of north-eastern America "conceived of their relationship with animals in terms of balance and reciprocity", while the Christian Europeans viewed animals as servants to be owned, farmed and domesticated.[24] When feral cattle and pigs strayed beyond European settlements and damaged Indian crops, these contrasting conceptions of animals were put to the test, generating conflict, litigation and, occasionally, physical violence. Loose cows and other introduced species thus function as important prisms through which historians can chart ideas about property rights, notions of agricultural improvement and conceptions of the natural world. Here, as Anderson demonstrates, animals were much more than just symbols: they were flesh and blood agents of the conquest – albeit unwitting ones [→American Studies].

In her article, "The Chicken and the Iegue", Marcy Norton builds on DeJohn Anderson's analysis to explore how cultural conceptions may explain the supposed 'failure' of Amerindian societies to domesticate more native species. Focusing on Caribbean and lowland South American Indians, Norton demonstrates that these groups frequently captured and tamed individual animals as pets (e. g. parrots, monkeys, tapirs), but that cultural taboos meant that these animals were never killed or eaten (although wild animals of the same species were). Norton argues that we should not apply European conceptions of pet-keeping and farming to non-European societies, and that other paradigms for relating to animals may exist. As she remarks:

> For many Europeans, the human-animal and hunting/livestock binaries were organizing principles, and those who confused or attempted to cross these boundaries were troubling. For Amerindians [...] the fundamental dividing line was between wild and tame beings. This divide bridged and superseded the human/nonhuman binary, grouping human kin and tamed animals on one side and human enemies and prey on the other.[25]

---

24 Anderson, Virginia DeJohn. *Creatures of Empire. How Domestic Animals Transformed Early America.* Oxford: Oxford University Press, 2004, 7.
25 Norton, Marcy. The Chicken and the *Iegue*. Human–Animal Relationships and the Columbian Exchange. *American Historical Review* 120, no. 1 (February 2015): 28–60.

In this case, unravelling the cultural parameters that undergirded human-animal relations can help to explain why one society domesticated animals for food while another did not.

While DeJohn Anderson and Norton look at the encounters between whole societies with differing customs and values, other historians have explored how individual animals can act as cross-cultural ambassadors, moving between places and receiving differing reactions. In a thought-provoking article, Erik Ringmar assesses the differing receptions of three imported giraffes to explain why Europeans engaged in imperialist ventures in the sixteenth and nineteenth centuries. The first giraffe, presented to Lorenzo de' Medici in 1486, reflected the curiosity of the Renaissance, and arrived in Europe just prior to the age of transatlantic exploration. The second giraffe, presented to Charles X of France in 1827, was by turns a commodity, Orientalist fantasy and scientific specimen, and pre-dated by a couple of decades the French colonization of Algeria. The third giraffe reached China in 1414 from Melinda in East Africa, shortly before the Emperor prohibited all travel overseas by Chinese subjects. It was classified by contemporary scholars as a unicorn – a fantastical but already known being – and interpreted as a good omen. Ringmar concludes that these differing responses reflected broader cultural attitudes towards the exotic and can help to explain why Europeans chose to explore and conquer foreign lands while the Chinese became increasingly inward-looking. He suggests that the giraffe, "an emissary from the unknown", acted as "an empty signifier that force[d] people to reveal their cultural predispositions".[26]

The reception of two globetrotting anteaters in the eighteenth and nineteenth centuries likewise illustrates the cultural connotations of exotic animal acquisition. In this case, the first anteater, a royal gift, arrived in Madrid in 1776 as a present for King Charles III. It was donated by Don Manuel de Basavilbaso, Administrator of Post in Buenos Aires, and housed in the Buen Retiro Menagerie, where it subsisted on "little pieces of bread, minced meat and flour dissolved in water".[27] The second anteater, a commercial speculation, was brought to London from Rio de Janeiro by two German showmen in 1853 and exhibited in a dilapidated shop in Bloomsbury. The novelist Charles Dickens watched it consume "an egg, which it had heard cracked against the wall", "licking the yolk" out of the shell "with its long tongue".[28]

---

26 Ringmar, Erik. Audience for a Giraffe: European Expansionism and the Quest for the Exotic. *Journal of World History* 17, no. 4 (December 2006): 375–397. For a similar study of a cross–cultural animal, see Juan Pimentel's recent book, *The Rhinoceros and the Megatherium*, which charts the reception of a rhinoceros in sixteenth–century Lisbon and a giant ground sloth in eighteenth–century Madrid. See Pimentel, Juan. *The Rhinoceros and the Megatherium: An Essay in Natural History.* Cambridge, MA: Harvard University Press, 2017.

27 MNCN, Fondo Museo, Sección A, Real Gabinete de Historia Natural, legajo 373; De Azara, Félix. *Apuntamientos para la Historia Natural de los Quadrúpedos del Paraguay.* Madrid: Imprenta de la Viuda de Ibarra, 1802, 16.

28 Dickens, Charles. A Brazilian in Bloomsbury. *The Albion: A Journal of News, Politics and Literature*, November 5, 1853.

Like the two European giraffes, both anteaters became cultural icons, generating multiple textual and visual representations. The Madrid anteater was painted from life by the court painter Rafael Mengs (or possibly his apprentice, Franisco de Goya) and described in detail by the soldier naturalist Félix de Azara, who revealed that the species was called *yurumí* ('small mouth') in its native Paraguay and that the locals in that country used anteater fat to "cure sores on horses' [backs]".[29] The London anteater was painted by the Austrian-born artist Joseph Wolf, examined by the comparative anatomist Richard Owen and fêted in an article by the satirical magazine, *Punch*, which compared its body to that of "a German pig", its snout to "a cucumber" and its long tongue to "some very fine surgical instrument that had shot out of its case upon a spring being touched".[30] Dickens speculated that "should it live and get its rights, we shall have ant-bear quadrilles, ant-bear butter dishes, ant-bear paper weights, ant-bear pictures of all sorts, and perhaps a dash of ant-bear in the Christmas pantomime".[31] The anteaters thus functioned by turns as scientific specimens and popular commodities, reflecting the differing outlooks and preoccupations of the people who encountered them. For the Paraguayans they were useful animals with medicinal properties; for the Spanish, they were reflections of royal power; for the British they were above all commercial attractions, as well as subjects for contemporary science. By looking at the reception of particular animals in different locations and social settings we can gain a better understanding of the priorities and world views of distinct (and often competing) cultural groups.

4 Implication(s) of the Animal Turn

The *animal turn* has brought about a shift in the kinds of questions cultural historians ask when studying animals. From using animals primarily as a mirror onto human society, historians have begun to write the histories of animals in their own right, treating them not merely as symbols, but as historical subjects with distinctive and recoverable pasts. This has resulted in the opening up of new areas for investigation and the explicit inclusion of nonhuman animals within established historical contexts.

First, the *animal turn* has encouraged historians to write animals back into history. Sometimes using new sources, more often asking new questions of existing material, cultural historians have begun to highlight the crucial role that animals played in past societies and to trace changing conceptions of other species. A range of historical works now address such topics as the history of zoos and circuses [→History

---

29 Pérez, Ana Victoria Mazo. El Oso Hormiguero de Su Majestad. *Asclepio* LVIII, no. 1 (2006): 286–288; De Azara, Félix. *Apuntamientos para la Historia Natural de los Quadrúpedos del Paraguay*. Madrid: Imprenta de la Viuda de Ibarra, 1802, 66–67.
30 The Fashionable Zoological Star. *Punch*, December 24, 1853.
31 Dickens, A Brazilian in Bloomsbury.

of the Zoo;→History of Circus Animals], the history of pet-keeping [→History of Pets] and the history of animal welfare organizations.³² Many of these pose broader cultural questions, placing animals within the contexts of the industrial revolution, popular imperialism or Renaissance court culture and asking what role animals played within these wider human settings.³³ By putting animals at the center of their analysis, rather than on the periphery, a new generation of cultural historians are highlighting the centrality of other species to human life and probing more deeply their role in shaping past societies.

Another group of works has emerged that puts either individual animals or species under the spotlight and explores their influence on human society. Drawing on the methodology of object histories, or the history of things [→Material Culture Studies], such studies focus on individual animals and trace their life histories, often, like Darnton, moving out from particular incidents to chart broader social trends. Susan Nance, for example, has examined the furor surrounding the sale of the African elephant Jumbo to American showman P.T. Barnum in 1882, describing how the famous pachyderm was celebrated, mourned and commodified on both sides of the Atlantic.³⁴ Samuel Alberti's *The Afterlives of Animals* chronicles the posthumous histories of a range of well-known beasts, from the elephant Maharajah, who walked from Edinburgh to Manchester in 1872, to the lion, Wallace, who fought against six dogs at Warwick in 1825 (Fig.4). Alberti shows how these animals attained new meanings after death, appearing in new settings and reaching new audiences [→History of Taxonomy/Animal Collection].³⁵

Third, influenced by developments in the field of cultural geography [→Historical Animal Geography] and the history of science [→History of Science], cultural historians have started to pay more attention to the spatial dynamics of human-animal interactions. This means looking at animals in a range of different sites and contexts, from the royal palace to the scientific laboratory, the natural history museum to the travelling menagerie, and the city street to the bourgeois bedroom. In his book *At Home and Astray*, cultural geographer Philip Howell highlights the geographical dimensions of dog life in Victorian London, showing how different places and spaces conferred different meanings on its canine inhabitants. Dogs in the home formed

---

32 See, for example, Kete, Kathleen. *The Beast in the Boudoir. Petkeeping in Nineteenth–Century Paris*. Berkeley, CA: University of California Press, 1994; Cowie, Helen. *Exhibiting Animals in Nineteenth–Century Britain. Empathy, Education, Entertainment*. Basingstoke: Palgrave Macmillan, 2014; Amato, Sarah. *Beastly Possessions. Animals in Victorian Consumer Culture*. Toronto: University of Toronto Press, 2015; Davis, Janet. *The Gospel of Kindness*. Oxford: Oxford University Press, 2016.

33 On race, zoo architecture and civic identity in the mid–twentieth–century USA, see, for example, Uddin, Lisa. *Zoo Renewal: White Flight and the Animal Ghetto*. Minneapolis, MN: University of Minnesota Press, 2015.

34 Nance, Susan. *Animal Modernity. Jumbo the Elephant and the Human Dilemma*. New York, NY: Palgrave Macmillan, 2015.

35 Alberti, Samuel (ed.). *The Afterlives of Animals: A Museum Menagerie*. Charlottesville, VA: University of Virginia Press, 2011.

Figure 4: Wallace the lion, Saffron Walden Museum. Photo author.

part of the domestic idyll of the bourgeois world, owned, loved and cared for as part of the family; dogs on the streets represented disorder and disease and were increasingly perceived as a danger to be extirpated. Howell contends that these two processes operated in tandem, creating what he calls a "moral geography of dogs".[36] The cultural history of animals has thus taken an important spatial turn, considering more explicitly how location influenced meaning.

Closely related to the geographies of human-animal relations, other historians have started to explore the sites, nature and meaning of human-animal interactions, asking what these tell us about cross-species connections in the present and in the past. Whether through milking a cow, cuddling a lapdog, riding a horse or stroking a hyena's paw in a menagerie, humans have enjoyed multiple close connections with

---

36 Howell, Philip. *At Home and Astray. The Domestic Dog in Victorian Britain*. Charlottesville, VA and London: University of Virginia University Press, 2015, 176.

animals. By examining these interactions in detail, we can learn about the affection, mutual pleasure and violence inherent in our relationship with other species and recover intimate details of animal encounters in past societies. An early-twentieth-century *Animal Care Journal* for Manchester's Bellevue Zoo, for instance, records how one keeper tickled a tapir in the ear, fed cakes to Daisy the elephant and massaged a giraffe's stiff knee, "apparently to the Giraffe's pleasure as he put his head down to caress me".[37]

A fourth area in which historians have become increasingly interested is the history of the emotions [→History of Emotions]. Did people in the past experience the same range of feelings as we experience today and how did they express those feelings? When extended to animals, the history of the emotions becomes particularly contentious and often invites allegations of anthropomorphism. Do animals experience human-like emotions, and if they do can we interpret them? In recent years, however, a number of cultural historians have taken the view that animals do exhibit emotions, and, drawing on scientific research in the field of cognitive ethology, they have begun to explore the behavioral and emotional worlds of animals in the past.[38] In a study that blends history with zoology, for instance, Nicola Foote and Charles W. Gunnels apply modern biological knowledge about sea lions, doves and tortoises to explain reported behaviors of animals on the Galapagos Islands.[39] In an in-depth exploration of one famous early-modern feline, Sarah Cockram documents that short life of a little cat, or *animalino* (little animal) owned by the Marquess of Mantua, Isabella d'Este, and wonders whether the creature liked being stroked and carried about in her sleeve. Cockram also explores what having this pet may have meant to its owner, Isabella, emphasizing the sensory dimension of the encounter "There is," she suggests, "evidence for Renaissance Italy of the breeding of luxury animals specifically for strokability and of responses to such qualities".[40]

Of course, in practice, it is often difficult to do more than speculate about what animals felt in the past. What we can sometimes do, however, is learn about what previous generations of humans *thought* animals were feeling, which can itself reveal changing cultural attitudes, values and relations – a central component of cultural history. Take, for instance, a 1908 court case, in which the RSPCA prosecuted circus keepers Schreida and Havadia for "cruelly abusing and terrifying an elephant" by

---

[37] *Animal Care Journal*. Belle Vue Gardens. Jennison Collection. Chetham's Library, Manchester, F.5.04, 1 January 1910; 12 July 1912; and 30 November 1909.
[38] Important studies in the field of animal emotions from a scientific perspective include Masson, Jeffrey and Susan McCarthy. *When Elephants Weep. The Emotional Lives of Animals*. London: Vintage, 1996; Bekoff, Marc. *The Emotional Lives of Animals*. Novato, CA: New World Library, 2007.
[39] Foote, Nicola and Charles W Gunnells IV. Exploring Early Human–Animal Encounters in the Galapagos Islands Using a Historical Zoology Approach. In *The Historical Animal*, Susan Nance (ed.), 203–220. Syracuse, NY: Syracuse University Press, 2015.
[40] Cockram, Sarah. Sleeve Cat and Lap Dog. Affection, Aesthetics and Proximity to Companion Animals in Renaissance Mantua. In *Interspecies Interactions. Animals and Humans from the Middle Ages to Modernity*, Sarah Cockram, Andrew Wells (eds.), 34–65. London: Routledge, 2018.

making it slide down a steep chute into a pool of water while performing at the White City in London. In stating the case for the prosecution, lawyer Stuart Bevan emphasized not only the physical pain suffered by the elephant, when Havadia "dug the pointed end of the elephant stick six times into the right cheek of the animal", but also the mental anguish it was thought to have experienced. One witness testified that the elephant "trumpeted loudly and gave vent to shrill cries of pain *and terror*". Another stated that "From the movements of the beast it was suffering the utmost terror, and *evinced the greatest alarm* at the prospect of going down". The magistrate, Mr Garret, convinced by this argument, concluded that the animal, "before it knew what was in front of it [...] could only think that it was plunging down into sheer space" and "must [have been] *greatly frightened*". In this instance, therefore, it mattered not just what happened to the elephant, but what it thought was going to happen to it. The fact that several witnesses assumed the animal was afraid may or may not have reflected the actual state of mind of the elephant, but it certainly reflected their interpretation of it and their judgement that inflicting fear on a fellow creature – particularly a highly intelligent mammal like an elephant – was unacceptable. The elephant's sensory capacity was, indeed, referenced in the court proceedings, with one witness, elephant trainer Charles Miller, testifying that "An elephant was as sensitive as a human being".[41] Here, then, we can see humans endowing animals with a repertoire of emotional responses, and modulating their treatment of the latter accordingly. By unpicking animal emotions in the past – and the human responses they elicited – cultural historians are expanding earlier work on the history of mentalities and making an important contribution to the burgeoning field of the history of emotions.

Finally, the *animal turn* is prompting scientists, and some historians, to ask a controversial but important question: can (nonhuman) animals have culture? When historians talk about culture, either high or low, they have almost exclusively focused on human beings – understandably, given the source material available. The work of several biologists, however, has begun to challenge the idea that culture is something confined to humans, showing that certain other species do indeed have what might be referred to as culture. Primatologists such as Jane Goodall have demonstrated that chimpanzee families possess distinct modes of existence, passing on particular skills or forms of tool use from generation to generation.[42] Recent research on sperm whales off the Galapagos Islands suggests, similarly, that different family groups, or clans, have developed their own distinct set of vocal clicks to communicate, a phenomenon that Marucio Canto attributes to social learning rather than

---

[41] The Elephants at the Franco–British Exhibition, *The Animal World*, November 1908, 246–247. My italics.
[42] See, for instance, Goodall, Jane. *In the Shadow of Man*. London: Weidenfeld and Nicholson, 1988.

shared genetics.[43] Elephant expert Cynthia Moss, meanwhile, recounts how the elephants she observed for fourteen years in Amboseli National Park, Kenya, appeared to mourn their dead and to recognize the skulls and skeletons of their own species. She describes, in one case, how the calf of a dead elephant found the jaw of his mother near her camp and remained with it for a long time "repeatedly feeling and stroking the jaw and turning it with his foot and trunk".[44] Whether such behaviors constitute culture remains open to debate, but they do raise important questions about how we perceive, study and represent other species, both in the present and the past.

Selected Bibliography

Alberti, Sam (ed.). *The Afterlives of Animals: A Museum Menagerie*. Charlottesville, VA: University of Virginia Press, 2011.
Anderson, Virginia DeJohn. *Creatures of Empire: How Domestic Animals Transformed Early America*. Oxford: Oxford University Press, 2004.
Cockram, Sarah and Andrew Wells (eds.). *Interspecies Interactions: Animals and Humans from the Middle Ages to Modernity*. London: Routledge, 2018.
Darnton, Robert. *The Great Cat Massacre: And Other Episodes in French Cultural History*. London: Allen Lane, 1984.
Few, Martha and Tortorici Zeb (eds.). *Centering Animals in Latin American History*. Durham: Duke University Press, 2013.
Fudge, Erica. *Renaissance Beasts: Of Animals, Humans and Other Wonderful Creatures*. Urbana: University of Illinois Press.
Kalof, Linda and Brigitte Resl (eds.). *A Cultural History of Animals*. London: Bloomsbury, 2007.
Norton, Marcy. The Chicken and the *Iegue*. Human–Animal Relationships and the Columbian Exchange. *American Historical Review* 120, no. 1 (February 2015): 28–60.
Ringmar, Erik. Audience for a Giraffe: European Expansionism and the Quest for the Exotic. *Journal of World History* 17, no. 4 (2006): 375–397.
Tague, Ingrid. *Animal Companions. Pets and Social Change in Eighteenth Century Britain*. Philadelphia, PA: Penn State University Press, 2015.

---

43 When Sperm Whales Click, is it Culture? *Agence France Presse*, 8 September, 2015. See also Whitehead, Hal. Social and Cultural Evolution in the Ocean. In *Sperm Whales: Social Evolution in the Ocean*, Hal Whitehead, 316–359. Chicago, IL: University of Chicago Press, 2003.
44 Moss, Cynthia. *Elephant Memories*. Chicago, IL: University of Chicago Press, 2000, 271.

Aaron Skabelund
Political History

1 Introduction and Overview

Political history is typically understood as those aspects of the past which have to do with the formal exercise of power in society, which for most societies means the state. It is primarily concerned with political developments, ideology, movements, organs of government, leaders, and – for democratic systems – with voters and parties. Since the formation of history as a professionalized academic discipline in the late nineteenth century until around 1970, political history told the story of the state, political elites who struggled for control of the state, and inter-state relations, though by that time diplomatic history had become a separate but closely related subfield within the discipline. Political history was, as is often said, a "Grand Narrative" of important (read "political") events conducted by "Great Men." It dominated historiography until history as a discipline began to splinter and fragment into a variety of subfields from the late 1960s onward. Since then, political history has been influenced by a wider concern with society, culture, and non-elites leading to what is sometimes called new political history or the cultural history of politics.

In the last two decades or so, the study of animals, real and symbolic, has begun to influence political history, though its impact is less substantial compared to other subfields of history.[1] Few books, articles, or edited volumes that focus on animals in history can be categorized as political history even in its new forms.[2] Nothing like the six-volume series *A Cultural History of Animals*, which "review[s] [...] the changing roles of animals in society and culture throughout history" from antiquity to the modern age, exists for political history.[3] Even among what little work has been produced that considers relations between animals and political history, few would call it even new political history or the cultural history of politics. Its authors would certainly not call themselves political historians. But politics, political ideology, govern-

---

[1] For another, more theoretical evaluation of the nexus between animal studies and (new) political history, see Roscher, Mieke. New Political History and the Writing of Animal Lives. In *The Routledge Handbook of Animal-Human History*, Hilda Kean, Philipp Howell (eds.,), 53–75. London: Routledge 2018.
[2] For edited volumes, only one chapter in these three recent collections is concerned with political history: Castow, Jane and Amy Nelson (eds.). *Other Animals: Beyond the Human in Russian Culture and History*. Pittsburgh, PA: University of Pittsburgh Press, 2010; Few, Martha and Zeb Tororici (eds.). *Centering Animals in Latin America History*. Durham, NC: Duke University Press, 2013; Nance, Susan (ed.). *The Historical Animal*. Syracuse, NY: Syracuse University Press, 2015.
[3] Kalof, Linda (ed.). *A Cultural History of Animals*, 6 vols. Oxford: Berg, 2007, series preface.

ment institutions, and political leaders are some of its concerns. And regardless, this emerging attention to animals has enriched the study of political history.

This work has helped historians recognize that politics has actually always been concerned with animals. From time immemorial, rulers have mobilized animals to wage war, exploited them for their economic value, used them to stage spectacles of authority, and rhetorically deployed them to animalize human others who they sought dominion over both at home and abroad. Over the centuries, a host of animal-related issues, from agricultural and environmental policies to debates about their welfare, have concerned governments and fueled political debate. This chapter will leave a recounting of many of those influences to other chapters in this handbook that deal with subfields that emerged in the late 1960s such as diplomatic, economic, social, and colonial history [→Diplomatic History;→Economic History;→Social History;→(Post)Colonial History], and to topical chapters that are closely related to politics and government that deal with war, zoological gardens and agriculture [→History of War;→History of the Zoo;→History of Agriculture]. Recent historiographical developments demonstrate that the study of the political past specifically and the study of history more generally are not complete without a consideration of animals. As illustrated by the examples mentioned in this and the other chapters in this handbook, animals have long been implicated, though not by their choice, in the exercise of political and governmental power. That involvement often takes a physical form and sometimes it is only metaphorical, but either way animals serve as pliable, malleable tools for those deploying governmental power, as well as those opposing it.

This chapter will focus on the relationship of animals and politics since roughly the beginning of the twentieth century because from around that time in much of the world the actual use of animals became less important – or at least less visible – to the exercise of power of all kinds including political power. John Berger and Steve Baker have argued that the increasing visibility of animals as symbols, signs, and images in the twentieth century resulted despite, or perhaps because of, an alienation from and the disappearance of actual animals, other than pets, from much of the human world.[4] Motorized vehicles and machines replaced horses, oxen, and other animals as sources of labor and a means of transportation. In war, tanks rather than horses moved cavalries. Although synthetic processes made the use of some animal products a thing of the past, animals as always provide protein and other invaluable dietary and material resources, yet with the rise of industrial agriculture, urbanization, and zoning regulations, people became increasingly physically separated from actual animals. In more explicit political settings, leaders rarely used the spectacle of parading about on a fine steed or their exploits hunting big game to bolster their symbolic power. People, including political leaders, now spend more time with

---

4 Berger, John. Why Look at Animals? In *About Looking*, John Berger (ed.), 1–28. New York, NY: Pantheon, 1980; Baker, Steve. *The Postmodern Animal*. London: Reaktion Books, 2000, 7–25.

pet animals and little time around living animals who serve utilitarian purposes. One might call this new situation that began to take shape in some parts of the world "postdomesticity," as historian Richard W. Bulliet has, but regardless it seems to represent a marked departure from the past [→Post-Domestication: The Posthuman].[5]

Although this chapter approaches this topic from a comparative perspective with consideration from a number of geographic subfields of history, it will draw primarily from examples in the United States, Britain, Japan, and Germany. This is not because American, British, Japanese, or German history in the twentieth century is necessarily representative of wider trends, though they are among the first countries Bulliet identifies as entering the stage of postdomesticity. American presidential "first dogs," which I will discuss, are clearly a particular American phenomenon. Yet they and other examples are inflections of wider trends, and often have been influenced and emulated by developments elsewhere in the world. For example, the use of dogs for police work, which I will also mention, is a story of cross-fertilization among Europe, the United States, and other countries in the late nineteenth and twentieth centuries.

As suggested by those two examples, this chapter will highlight the deployment of canines in the political realm because the use of dogs, both actual and figurative, has been among the most prominent deployment of animals for political purposes in the last century. As the use of animals for practical purposes dropped and the separation of people and animals became more pronounced in the early twentieth century, dogs were among the most widely-adopted pets and given their social nature the most visible.[→History of Pets] They became (and arguably already were), as zoologist James Serpell has suggested, closer to people in affective and symbolic terms, than just about any other animal.[6] Dogs have been ideal partners for political leaders' efforts to humanize themselves and government efforts to inculcate their populations with messages of loyalty or to strike fear in their hearts.

2 Topics and Themes

Historians have increasingly begun to recognize the ways by which government and political institutions, parties, and leaders have used animals, both actual and symbolic, to wield power. This deployment may have included real personal physical interaction with animals through the keeping or hunting of animals, the use of animals for policing, bureaucratic directives dictating the protection or extermination of certain animals, and the symbolic use of animals as mascots, clothing items, and edu-

---

[5] Bulliet, Richard W. *Hunters, Herders, and Hamburgers: The Past and Future of Human-Animal Relationships*. New York, NY: Columbia University Press, 2005, 189–204.
[6] Serpell, James. From Paragon to Pariah: Some Reflections on Human Attitudes towards Dogs. In *The Domestic Dog: Its Evolution, Behaviour, and Interactions with People*, James Serpell (ed.), 245–56. Cambridge: Cambridge University Press, 1995.

cational material. Its purposes are many, multi-faceted, and may include both practical and metaphorical manipulation or just the latter. Here are some topics and themes that illustrate the nexus between animals and politics, some of which historians have begun to explore.

First Dogs

One striking example of the deployment of animals for political ends is what historian Helen C. Pycior calls "the making of the 'first dog,'" the keeping of a pet canine by U.S. presidents for companionship and public relations.[7] More generally, historian Katherine C. Grier has argued that "[p]et animals belonging to the president [...] became community pets" or celebrities by the early twentieth century.[8] Although presidents have kept a variety of animals including cats in the White House, dogs have by far been the most popular and the most publicly prominent. Presidents from George Washington onward kept canines but the nature of that ownership and the purposes it was put to changed. Theodore Roosevelt may have represented the last example of an earlier kind of relationship with animals. He kept a wide variety of animals including dogs, but they were not considered part of his family in the mores of modern pet culture. Roosevelt's most notable and noted interactions with animals were him hunting and killing them in exotic locations in Africa and the American West. After that, historians have suggested, it became "conventional wisdom that the president should have a dog on display, as a symbol of stability, courage, fidelity, and devotion."[9]

Pycior argues specifically that Laddie Boy, an Airedale terrier adopted by President Warren C. Harding, became the first "First Dog."[10] The confluence of several societal trends in the 1920s – the emergence of celebrity culture (including some cinematic canines such as Strongheart and Rin Tin Tin); major technological advances such as the radio and motion picture camera; and the spread of the keeping of dogs more as pets than for practical purposes – all contributed to the celebration of presidential dogs. Harding's thirty-year marriage had produced no children for him and his wife Florence, so Laddie Boy served as a surrogate child to form a "folksy First Family" the public generally embraced. As a longtime newspaperman, Har-

---

[7] Pycior, Helena. The Making of the First Dog: President Warren G. Harding and Laddie Boy. *Society & Animals* 13, no. 2 (July 2005): 109–138.
[8] Grier, Katherine C. *Pets in America: A History.* Raleigh, NC: University of North Carolina Press, 2006, 224.
[9] Derr, Mark. *A Dog's History of America: How our Best Friend Explored, Conquered, and Settled a Continent.* New York, NY: North Point Press, 2004, 267.
[10] Pycior, The Making of the First Dog, 110.

ding understood the power of stories and symbols and used Laddie Boy to create the image of a "simple and homey life, [...] which [...] included a family dog."[11]

Harding's successors in the White House – Calvin Coolidge and Herbert Hoover – kept pet dogs, but the next president Franklin Delano Roosevelt took the political mobilization of his dog to a new level. This may have been in part because FDR was not only one of the most beloved presidents, but one of the most disliked. Early in Roosevelt's third term, his Scottish Terrier Fala became his constant companion, frolicking with the press corps at the White House and elsewhere, travelling to Allied conferences and political events with him, and acting as an endearing ambassador for Roosevelt that served to humanize the president. Roosevelt put Fala to use particularly during his tight fourth presidential election campaign, when the dog accompanied him nearly everywhere on the campaign trail. This led Republicans to also deploy Fala in their criticisms of the president. In response, Roosevelt delivered what became known as the "Fala speech," in which he declared the dog was "furious" about the "libelous statements" leveled at him.[12] Fala's influence was so great that at the height of the campaign Pulitzer Prize-winning journalist John H. Crider of the *New York Times* declared – with only slight exaggeration – that "Fala is no longer just a dog; he is a personage."[13] Fala also served as an unofficial war dog, enlisting as a private in the Dogs for Defense Reserve, an organization that encouraged owners of dogs capable of becoming military dogs to contribute them to the army and owners of dogs not capable of becoming military dogs, like Fala, to make financial contributions to the cause.

The keeping of a dog by U.S. presidents since Roosevelt has continued and has become so much a tradition that Donald Trump, who dislikes dogs and did not have one at the White House, broke another rule by becoming the first president since the nineteenth century to not have a "First Dog." Indeed, presidential dog-keeping has become so much a tradition that just a day after Joe Biden was declared the winner of the election in 2020, the media noted his bringing two German shepherds, Champ and Major, to the executive mansion would be another way he would restore normalcy to 1600 Pennsylvania Avenue.[14] In *A Dog's History of America* (2004), author Mark Derr contrasts the "organic" relationship of earlier presidents and their dogs – such as Roosevelt and Fala – with more recent president-dog relationships – such as that of George H. W. Bush's and George W. Bush's with their dogs – which he characterizes as "self-consciously self-referential," "staged for publicity, [and] a political ploy de-

---

[11] Pycior, The Making of the First Dog, 113.
[12] Pycior, Helena. The Public and Private Lives of 'First Dogs': Warren G. Harding's Laddie Boy and Franklin D. Roosevelt's Fala. In *Beastly Natures: Animals, Humans, and the Study of History*, Dorothee Brantz (ed.), 176–203. Charlottesville, VA: University of Virginia Press, 2010, 195.
[13] Crider, John H. Fala. Never in the Doghouse. *New York Times Magazine*, October 15, 1944: 26.
[14] Morales, Christina. Biden to Restore a White House Tradition of Presidential Pets. *New York Times*, November 8, 2020, www.nytimes.com/2020/11/08/us/politics/biden-dogs.html.

signed to connect to [...] dog lovers in the electorate."[15] This is most certainly a romanticization of the past. As Pycior's work shows, both Harding's and even more so Roosevelt's use of their dogs was politically calculated, and in the latter case even modified to cater to public opinion.[16]

Political Attack Dogs

The "First Dog" may be a distinctive American phenomenon, but it is certainly not without parallels elsewhere. As in the United States, political image makers positioned dogs, both real and imagined, as well as other animals to portray leaders in endearing terms.[17] One of the most famous examples is Adolf Hitler and his German Shepherd dogs, including Blondie. One such story from a German elementary school primer featuring Hitler attempted to show his humanity and kindness by highlighting his supposed concern for youngsters and animals. It told of two little girls using their own allowance money to buy a big sausage for his Shepherd dog, Wolf, and flowers for the Führer on his birthday. After giving them a talk about the importance of the role of women and girls as mothers willing to sacrifice for the nation, Hitler cut off a piece of the wiener, placed it above Wolf's nose, and let him jump for it. Such treatment elicited praise from the children, who were moved to exclaim, "My goodness, Herr Hitler, you really have an excellent way of training dogs!"[18] In addition to showing a more approachable side of Hitler, another message seemed to be that the ideal child would be just as teachable and compliant in executing Hitler's commands as Wolf.

The Nazis manipulated many other animals and animal symbols beyond dogs. Hitler's fondness for German Shepherd dogs like Blondie can be explained in part by the name he gave the first German Shepherd dog he owned: Wolf. Like Max von Stephanitz (1863–1936), who codified the German Shepherd breed and theorized that they were entirely the descendent of ancient wolves whereas all other dogs were recipients of varying amounts of jackal blood,[19] Hitler venerated wolves. He and other National Socialist leaders continually appropriated wolves for symbolic purposes, referring to themselves as wolves and their various headquarters as lairs. As Boria Sax has observed, the "Nazis were constantly involving dogs and wolves

---

15 Derr, *A Dog's History*, 348.
16 Pycior, The Public and Private Lives of 'First Dogs', 195–197.
17 For an analysis of the place of dogs in German politics and society, see Wippermann, Wolfgang and Detlef Berentzen, *Die Deutschen und ihre Hunde: Ein Sonderweg der Mentalitätsgeschichte*. Berlin: Siedler, 1999.
18 Kamenetsky, Christa. *Children's Literature in Hitler's Germany: The Cultural Policy of National Socialism*. Athens, OH: Ohio University Press, 1984, 183.
19 Von Stephanitz, Max. *The German Shepherd in Word and Picture*. Translated by Carrington Charke. Jena: Anton Kämfe, 1923, 18, 22–23.

as models for the qualities they wanted to cultivate: loyalty, hierarchy, fierceness, courage, obedience, and sometimes even cruelty."[20] The Nazis, who made a practice of "imagining the nation in nature,"[21] mobilized the symbolic land, as well as other large predators – ungulates and other animals – for symbolic purposes, but dogs because of their immediate, constant physical presence were among their preferred tools.

Hitler's fascist ally, Mussolini also deployed animals, real and imagined, past and present, to bolster the aura of *Il Duce*. A master of choreographing political spectacle, Mussolini frequently portrayed himself as a modern tamer of wild animals. He made a practice of being photographed "completely at ease visiting cages of lions" – who as it turns out had been defanged – and riding in the back of an automobile with a lion cub in his lap.[22] As part of his effort to effectively link the Roman empire with his fascist regime, he actively incorporated wolf and eagle iconography to inspire national passion. The former symbol was based on a myth of a she-wolf who was said to have raised the future founders of the city of Rome, Romulus and Remus, after the twins were abandoned on the shores of the Tiber Sea by their parents, Mars, the god of war, and Rhea Silvia, a princess.[23] As in Nazi Germany, Imperial Japan, and elsewhere, such imagery was part of fascism's yearning for a pure, indigenous cultural aesthetic grounded in the countryside, nature, and a mythical distant past.

Although they perhaps cannot be placed under the rubric of fascism, other regimes have invoked animals as part of an effort to foster national identity and bolster their political power. Several examples come to mind. After Mobutu Sese Seko seized direct power of the Congo in 1965, he launched the *autheniticité* campaign to garner public support for his regime and construct a new national identity. As an act of decolonization from lingering Belgian colonial influences, Mobutu sought to "move away from borrowed or imposed ideas toward an increased awareness and privileging of indigenous cultural beliefs and values." This campaign began with the renaming of cities, replacing colonial names with "African" ones, most prominently the Congo – both the country and the river – becoming known as Zaïre. Mobuto invoked the leopard as a symbol of the nation and himself by doing such things as creating the National Order of the Leopard as the highest national honorific decoration and frequently wearing a leopard skin head-dress that transformed him into "a personi-

---

20 Sax, Boria. *Animals in the Third Reich: Pet, Scapegoats, and the Holocaust.* New York, NY: Continuum Books, 2000, 66.
21 Lekan, Thomas M. *Imagining the Nation in Nature: Landscape Preservation and German Identity, 1885–1945.* Cambridge, MA: Harvard University Press, 2004.
22 Falasca-Zamponi, Simonetta. *Fascist Spectacle: The Aesthetics of Power in Mussolini's Italy.* Berkeley, CA: University of California Press, 1997, 68; Thompson, Molly. *Mussolini: Italy's Nightmare,* 50 min., A&E Home Video, 1995.
23 See Melograni, Piero. The Cult of the Duce in Mussolini's Italy. *Journal of Contemporary History* 11, no. 4 (October 1976): 221–237, 229–230.

fication of the [national] body politic."²⁴ Elsewhere during the same decades, President Park Chung Hee of South Korea mobilized the Jindo dog as part of a joint state-society campaign to fabricate national identity. His government celebrated cultural practices and environmental objects identified as purely Korean, unstained by Japanese colonial, Chinese, and American influences such as shamanism, mask dancing, and "traditional" music. By designating the Jindo as a natural treasure, the government claimed to be protecting the breed's purity of blood and preventing its "extinction."²⁵ Private fans of the breed suggested the Jindo was especially worth of selection because the dogs had supposedly long been fierce defenders of the Korean nation. They claimed that on the day before the armada of the Japanese warlord Hideyoshi invaded the Korean peninsula in the late sixteenth century, thousands of Jindo had faced the sea and barked viciously in the direction of the Japanese islands. More recently, historian E. Elena Songster has analyzed how the Chinese communist government has used the giant panda to define national identity and as a tool for international diplomacy, especially in the last few decades as the unifying adhesive that communism once served has waned.²⁶ The use of animals to invent and instill political identity is one of the richest veins among recent research.

A contemporary leader who frequently deploys animals for political purposes is Russian president Vladimir Putin. Photographs have shown him, to cite just a few examples, flying in a motorized deltaplane alongside migrating cranes, riding a horse shirtless, and fastening a tranquilized Siberian tiger with a satellite tracking transmitter in the Russian Far East. Many of these photos are clearly staged and intent on portraying Putin as tough, virile, and in control of but still caring about the natural world. Putin has also deployed dogs for political purposes. In 2007, he apparently summoned one of his dogs, a black Labrador named Koni, in an attempt to intimidate German president Angela Merkel during a meeting at Putin's residence in Sochi. As one account relates:

> As the dog approached and sniffed her, Merkel froze, visibly frightened. She'd been bitten once, in 1995, and her fear of dogs couldn't have escaped Putin, who sat back and enjoyed the moment, legs spread wide. 'I'm sure it will behave itself,' he said. [...] Later, Merkel interpreted Putin's behavior. 'I understand why he has to do this – to prove he's a man,' she told a group of reporters. 'He's afraid of his own weakness. Russia has nothing, no successful politics or economy. All they have is this.'²⁷

---

24 Dunn, Kevin C. *Imagining the Congo: The International Relations of Identity.* New York, NY: Palgrave Macmillan, 2003, 111, 113.
25 Chung-ho, Kim. The Chindo Dog: A Proud Korean Breed. *Koreana* 1, no. 1 (Spring 1994): 74–77.
26 Songster, Elena. *Panda Nation: The Construction and Conservation of China's Modern Icon.* Oxford: Oxford University Press, 2018.
27 Packer, George. The Quiet German: The Astonishing Rise of Angela Merkel, the Most Powerful Women in the World. *The New Yorker,* 1 December, 2014.

Journalists have noticed how Putin uses animals in these ways; at some point historians will analyze these physical and metaphorical machinations.

Political (Party) Animals

For a historical figure as prominent and as well studied as Winston Churchill, scholars have paid little attention to the ways in which animals were implicated in his political activities. During Churchill's service as prime minister, the bulldog was not only identified with British national and imperial might, an association as old as the early nineteenth century,[28] but with Churchill himself and by extension the Conservative Party. The reasons for these latter associations may have emerged as early as the mid-1920s. By that time, Churchill had begun, as biographer William Manchester observes, to "resemble the cartoonist's conception of John Bull,"[29] who was originally bovine and had become a human form and the national personification of the United Kingdom in general, and was often shown accompanied by a bulldog, who looked like John Bull, and thus like Churchill. During the Second World War Churchill began to be identified with the bulldog directly, in part because of physical appearance but also due to his obstinate courage in the face of Nazi German aggression. In 1950, as the Conservative Party sought to regain power, it nourished Churchill's identification with the bulldog adopting it as the party's unofficial mascot and staging photo opportunities of Churchill patting the head of a bulldog. Historians, though, have so far had little to say about these developments.

Other political parties and movements across the world have regularly used animal images. Some of these come to be associated with parties by others such as political cartoonists, like the identification of the U.S. Democratic and Republican parties with the donkey and elephant, respectively, and are neither much embraced by the parties nor have much meaning for the public. Others resonate in both ways. Perhaps one of the most powerful were parties that not merely used an animal as an icon but used it in their very name. Black Panther groups emerged across the United States from the mid-1960s. They derived their symbolic power, both for group members and observers because they were closely related to the actual animal. In 1965, civil rights activist Stokely Carmichael helped organize the Alabama Lowndes County Freedom Organization (LFCO) in a county that was 80 percent black but did not have a single black citizen registered to vote. The following year (and just a month before the comic book superhero Black Panther appeared in the *Fantastic Four* series), the party "selected a snarling black panther as their ballot symbol to meet state require-

---

[28] Ritvo, Harriet. Pride and Pedigree: The Evolution of the Victorian Dog Fancy. *Victorian Studies* 29, no. 2 (Winter 1986): 227–253, 250.
[29] Manchester, William. *The Last Lion: Winston Spencer Churchill–Visions of Glory, 1874–1932*. Boston, MA: Little, Brown and Company, 1983, 755.

ments that every political party have a logo due to the high rate of adult illiteracy."[30] They did so, Carmichael recalled, because it was "a bold, beautiful animal, representing the strength and dignity of black demands today." For party members, it was "a positive, brave and militant symbol; however, it struck fear in the hearts of whites and black moderates."[31] Although black panthers – an umbrella term for any big cat such as a leopard or jaguar with a melanistic black coat – were only found in Africa, Asia, and other parts of the Americas, there was the misperception that they might emerge from woods and swamps of the American South or perhaps from the hills or urban streets of California. That is what in part made the logo so powerful that "other young militants around the country found inspiration in the LCFO and its powerful symbol, [and began] forming their own Black Panther Parties," the most famous of which was an Oakland-based group that violently clashed with government authorities.[32]

K-9's Bite Back Too

Besides political leaders and parties, government institutions of all sorts have made use, both practically and symbolically, of animals to bolster the power of the state. This use goes beyond the management of animals for economic purposes. Perhaps the most prominent mobilization of animals by an arm of the state is the use of dogs and (now much less so) horses by policing agencies. Canines have served as sentries, in investigation and search operations, and for crowd control.[33] In these roles, police dogs acted as protectors, deterrents, and enforcers of political and social control. But because they are living beings whose actions could not be entirely controlled by their handlers, police dogs were (and remain) not merely tools but agents of police work, not simply new crime-solving devices but deputies, and active participants rather than weapons. Despite being bred, trained, and handled to serve the police, they were unpredictable. They sometimes did less or went beyond what was expected – by both the police and policed – and this animate volatility, along with their super-human scenting abilities and potential to inflict serious injury or death, made them particularly intimidating extensions of policing power. Physically and figuratively, they lend their masters an air of authority and help to maintain repressive regimes, as a result of the canines' conditioned hostility. There seems to be something particularly empowering in having a creature at one's command to tor-

---

30 Jeffries, Hasan Kwame. *Bloody Lowndes: Civil Rights and Black Power in Alabama's Black Belt.* New York, NY: New York University Press, 2009, 152.
31 Ogbar, Jeffrey O. G. *Black Power: Radical Politics and African American Identity.* Baltimore, MD: John Hopkins University Press, 2004, 75.
32 Ogbar, *Black Power*, 77.
33 Lilly, J. Robert and Michael B. Puckett. Social Control and Dogs: A Sociohistorical Analysis. *Crime and Delinquency* 43, no. 2 (April 1997): 123–147.

ment other people, and something extremely humiliating about being on the receiving end of such action.

Some of the most iconic images from the Civil Rights movement in the U.S.-American South and apartheid South Africa feature police officials such as Alabama Public Safety Commissioner Bull Conner deploying dogs to suppress protests and dissent. They had an unquestionable powerful physical impact, but as symbols, police dogs were tremendously pliable. They were highly effective in intimidating protestors, but their use especially when caught in moving and still photography could backfire and cause a backlash against those who deployed them. Perhaps the best example of this was a photograph of a police dog appearing to attack a young black teenager at a demonstration in Birmingham in 1963, which according to author Diane McWhorter "shifted international opinion to the side of the civil rights revolution."[34] As journalism scholar Meg Spratt and others have observed, civil rights leaders including Martin Luther King Jr. recognized that pliability and sought to take advantage of it.[35]

Such images and anecdotes are familiar but American historians and other scholars have focused little to no analysis on them. Spratt, for example, treats police dogs as just another policing crowd suppression tool not much different than the fire hose that police also used to suppress protests. In recent work, some historians have given greater attention to police dogs. Lance van Sittert, Sandra Swart and Keith Shear have analyzed their use during the apartheid-era South Africa, and Chris Pearson and Binyamin Blum have begun to examine the role of police dogs both in continental Europe and colonial areas in the modern era.[36]

---

[34] McWhorter, Diane. *Carry Me Home: Birmingham, Alabama, The Climactic Battle of the Civil Rights Revolution.* New York, NY: Simon & Schuster, 2001, caption of photo 32.

[35] Spratt, Meg. When Dogs Attacked: Iconic News Photographs and Construction of History, Mythology, and Political Discourse. *American Journalism* 25, no. 2 (Spring 2008): 85–105.

[36] Van Sittert, Lance and Sandra Swart. Canin Familiaris: A Dog History of Southern Africa and Shear, Keith. Police Dogs and State Rationality in Early Twentieth-Century South Africa. In *Canis Africanis: A Dog History of Southern Africa.* Lance van Sittert, Sandra Swart (eds.), 1–34. Leiden: Brill, 2008; Shear, Keith. Police Dogs and State Rationality in Early Twentieth-Century South Africa. In *Canis Africanis: A Dog History of Southern Africa.* Lance van Sittert, Sandra Swart (eds.), 193–216. Leiden: Brill, 2008; Pearson, Chris. Beyond 'Resistance': Rethinking Nonhuman Agency for a More-Than-Human World. *European Review of History* 22, no. 5 (September 2015): 709–725; Pearson, Chris. Between Instinct and Intelligence: Harnessing Police Dog Agency in Early Twentieth-Century Paris. *Comparative Studies in Society and History* 58, no. 2 (April 2016): 463–490; Blum, Binyamin. The Hounds of Empire: Forensic Dog Tracking in Britain and its Colonies, 1888–1953. *Law and History Review* 35, no. 3 (August 2017): 621–665.

Nationalized Dogs: The Teacher's Pet

Some governments have also used environmental designations and breeding regulations as a way to bolster national identity. As part of its veneration of the wolf and attributes associated with wolves, Nazi Germany placed lupines under protection in 1934, a first among modern nations and a gesture that was in fact symbolic since wolves had been extinct within the country's borders for almost a century.[37] Around the same time, the Japanese Ministry of Education's Bureau of Cultural Affairs designated seven dog breeds including the Akita, who government bureaucrats and private enthusiasts deemed to be indigenous and uncorrupted by foreign canine blood, as national natural treasures. Dogs were certainly not the first animals to be given such special status, but they were among the most celebrated, in part because of political and social intimations. Talk about canine purity echoed contemporary rhetoric about the unbroken and undefiled blood ties of the Japanese nation and the imperial line.[38] Likewise, nationalism has played an important role in Korean celebrations of the "Korean" tiger, from South Korean officials' selection of them as the mascot for the Olympic Games at Seoul (1988) and more recently in Pyeongchang (2018) to plans to "reintroduce" them in a spacious arboretum.[39] These – and South Korea's protection of Jindo dogs – are just a few of many examples of the governmental protection and preservation of animals to bolster national identity.

Education officials have also used animals to foster values seen as fundamental to identity. Again, an example from wartime Japan involving dogs is instructive. A year after its Bureau of Cultural Affairs declared the Akita a national natural treasure in 1931, the Ministry of Education learned of an Akita dog Hachikō who seemed to embody loyalty, purity, and bravery – distinctive and superior Japanese attributes that political leaders wanted children to emulate. A newspaper article in 1932 reported that Hachikō appeared each evening outside of a Tokyo train station to await the return of his master, who had died many years earlier in 1925. This led ministry officials to deploy Hachikō as a pedagogical tool to try to instill in imperial subjects, especially children, a sense of loyalty. Working closely with private dog aficionados, they organized a fund- and national identity-raising campaign to construct a statue of him outside of Shibuya Station in 1934. That same year, the Ministry published a story praising Hachikō's fidelity in its moral training textbook for primary school students with the intention of encouraging children to emulate the supposed fealty of Hachikō.[40] He became hugely popular because many children loved dogs, and because his living, breathing reality amplified his symbolic power.

---

[37] Sax, *Animals in the Third Reich*, 84.
[38] Skabelund, Aaron. *Empire of Dogs: Canines, Japan, and the Making the Modern Imperial World.* Ithaca, NY: Cornell University Press, 2011, 98–101.
[39] Seeley, Joseph and Aaron Skabelund. Tigers-Real and Imagined-in Korea's Physical and Cultural Landscape. *Environmental History* 20, no. 3 (July 2015): 475–503.
[40] Skabelund, *Empire of Dogs*, ch. 3.

Around the same time, Education Ministry officials as well as private publishers celebrated the exploits of military dogs who were also portrayed as possessing these same attributes. I argue that as with Hachikō, government and private rhetoric took advantage of the enthusiasm for pet dogs to mobilize youth for war and empire, even as they nurtured the relationships between dogs and youngsters. The ministry published the story of two military dogs who supposedly died bravely during the fighting after the Manchurian Incident in 1931 in a national textbook in 1933. This is just one of many instances of the ministry and private media deploying military dogs, both real and wholly imagined, to instill in youth, boys and girls, to follow or prepare dogs for war.[41] The chapter on military history will consider these issues further, but since war is an extension of politics and these military dogs played a role not only on the battlefront but at home, they deserve our attention here as well.

3 Methods and Approaches: Political History after the Animal Turn

The inclusion of animals in political history shares many of the same problems of writing a history of animals in general. In some ways, it may be even more difficult. Chief among these problems is a lack of documents. Animals, at least real, actual animals, may be less present in the political realm than in other aspects of society, perhaps because of the formal spaces wherein politics is often conducted. Furthermore, animals cannot speak, or at least we cannot understand what they say. Almost every source that might allow the animal to "speak" is in some way mediated by humans. Thus, much animal history is unavoidably the history of human animals toward animals. But, as cultural historian Erica Fudge has suggested, if historians ground their analysis of representation in the material relations with the animal, they can explore both the actual and figurative relationship between people and animals.[42] Getting at the actual animal and its practical relationship with humans, which are representations themselves, is not easy, and inevitably historians have tended towards representation over practice.

To overcome this methodological challenge some historians considering animals in political history have used famous animals, whose interactions with people are well documented, as a way to ground their analysis in material relations. Fame and the resulting documentation are essentially why Pycior argues it is possible to construct a biography of presidential dogs and why historian Amy Nelson asserts Laika and other Soviet space dogs allow her to overcome the "problems of representation [...] at least partially" and "consider material relationship as well as human

---

41 Skabelund, *Empire of Dogs*, ch. 4.
42 Fudge, Erica. A Left-Handed Blow: Writing the History of Animals. In *Representing Animals*, Nigel Rothfels (ed.), 3–18. Bloomington, IN: Indiana University Press, 2002.

representations of those relationships."[43] In my work on the practical and symbolic deployment of canines, I have sometimes focused on well-known animals such as Hachikō to try to deal with this problem and tried to identify sources that give the animal the ability to talk back. In *Empire of Dogs*, I argue that photography, both still and moving, and taxidermy give some animals a voice that allows them to possibly complicate or contest how they are represented.

As part of the *animal turn* in historical studies, scholars' attitudes towards animal historical agency has evolved. While earlier studies stressed animals' inability to "speak back" or contest the ways humans used them, some recent studies, including some in the realm of political history, highlight the ways in which animals subverted attempts at complete control and compelled humans to modify their own behavior. Chris Pearson, for example, argues that the practical ways police and military dogs were deployed and how their actions were described provides evidence that they were "purposeful and capable agents *in their own way.*"[44] It is worth noting that these methodological challenges related to sources, representation and practice, and agency have much in common with trying to retrieve the largely silent and invisible past of certain people, who also shaped history. They too were illiterate and left no documents, though they certainly had a greater degree of agency than nonhuman animals. By this same token, some historians argue that animals are partners, if not equals, in the creation of a co-created history and culture.

Relatively few historians, though, have argued that animals are partners in a co-created politics. Yet, as shown by the examples mentioned in this chapter, animals, both real and symbolic, have performed that role. New approaches that highlight practices, relationships, and materiality, as well as representation, offer the possibility of illustrating that role. President Harding's Airdale terrier Laddie Boy certainly shaped presidential politics, in actual and symbolic terms, through his actions, interactions, and corporeal presence. So too have other canines and animals. An embrace by political history of the promise, and challenges, of the animal has yet to be realized. As that happens these developments will surely deepen and change our understanding of politics as a co-authored historical story of species – human and nonhuman.

Selected Bibliography

Blum, Binyamin. The Hounds of Empire: Forensic Dog Tracking in Britain and its Colonies, 1888–1953. *Law and History Review* 35, no. 3 (August 2017): 621–665.

---

[43] Nelson, Amy. The Legacy of Laika: Celebrity, Sacrifice, and the Soviet Space Dogs. In *Beastly Natures: Animals, Humans, and the Study of History,* Dorothee Brantz (ed.), 204–224. Charlottesville, VA: University of Virginia Press, 2010, 205.

[44] Pearson, Chris. Dogs, History, and Agency. *History and Theory* 52, no. 4 (December 2013): 128–145 [emphasis mine].

Nelson, Amy. The Legacy of Laika: Celebrity, Sacrifice, and the Soviet Space Dogs. In *Beastly Natures: Animals, Humans, and the Study of History,* Dorothee Brantz (ed.), 204–224. Charlottesville, VA: University of Virginia Press, 2010.

Pearson, Chris. Between Instinct and Intelligence: Harnessing Police Dog Agency in Early Twentieth-Century Paris. *Comparative Studies in Society and History* 58, no. 2 (2016): 463–490.

Pearson, Chris. Beyond 'Resistance': Rethinking Nonhuman Agency for a 'More-Than-Human' World. *European Review of History* 22, no. 5 (September 2015): 709–725.

Pearson, Chris. Dogs, History, and Agency. *History and Theory* 52, no. 4 (December 2013): 128–145.

Pycior, Helena. The Public and Private Lives of 'First Dogs': Warren G. Harding's Laddie Boy and Franklin D. Roosevelt's Fala. In *Beastly Natures: Animals, Humans, and the Study of History,* Dorothee Brantz (ed.), 176–203. Charlottesville, VA: University of Virginia Press, 2010.

Sax, Boria. *Animals in the Third Reich: Pets, Scapegoats, and the Holocaust.* New York, NY: Continuum Books, 2000.

Skabelund, Aaron. *Empire of Dogs: Canines, Japan, and the Marking of the Modern Imperial World.* Ithaca, NY: Cornell University Press, 2011.

Van Sittert, Lance and Sandra Swart (eds.). *Canis Africanis: A Dog History of Southern Africa.* Leiden: Brill, 2008.

Heinrich Lang
Economic History

1 Introduction and Overview

Before steam and electric driven technology emerged, the everyday life of human society was dependent upon nonhuman labor. Agricultural production, human alimentation, and transport relied heavily on the work of animals. Animal products and animal work were present in every context of human economy. Human food like bread was produced with the animals' contribution to the cultivation of grain, animals generated foodstuffs like milk, honey or eggs, and even their entire bodies served to provide human nourishment.[1] Most commodities, like human clothes, consisted at least in part of some component of an animal's product.

However, animal contributions to human economy have for the most part not been valued as labor. Historiography on economy, and economics itself, does not discuss animals in this vein, because it only reflects the human perspective. Animals were considered as natural producers, objects, sometimes commodities, which humans could simply dispose of. Even the traders on a market where animal competencies like labor force were negotiated did not at all refer to the animals' work as a particular service. Some of the requested qualities were in debate, of course. The animal's health or sex was always part of its marketing, but the animal was not perceived as a laborer sui generis.

In everyday life, animals were subordinate to human culture, which is especially true for human economic culture. Although various animals – nonhumans and humans – shared a common life and lived in a society in which species met and in which species' lives were essentially entangled, animals are not part of social and economic history despite the essential services animals contributed to the society they lived in. Humans silently claimed to make use of animals by their being superior by creation. Nonhuman animals had to bear the consequences of this concept. Their main payoff for their labor was being fed and housed – and being eaten.

As a matter of fact, a certain layer of social relations is constituted by economic interaction. This can happen on various levels and generate complexities. The ox which pulled a plough worked in an agricultural context capitalizing on his physical power for the cultivation of a field which served eventually both the animals on the farm and the human farmer. The symbiotic existence of the ox and its farmer gave life to an intense and productive relationship. The farmers' concept of cultivating his

---

[1] Edwards, Peter. Domesticated Animals in Renaissance Europe. In *A Cultural History of Animals in the Renaissance*, vol. 3, Bruce Boehrer (ed.), 75–94. Oxford: Berg, 2007; Raber, Karen. From Sheep to Meat, From Pets to People: Animal Domestication 1600–1800. In *A Cultural History of Animals in the Age of Enlightenment*, vol. 4, Matthew Senior (ed.), 73–100. Oxford: Berg, 2007.

owner's field and the physical force of an animal constituted an elementary socioeconomic entanglement which laid the foundation for further economic transactions. The fruit of their labor could be sold at a local market or in the next town which brought income to the producers, if it was not partly handed over to a property-owning entity like the Church or a noble landowner.

Industrialization can be regarded as the watershed moment for animals' economic roles in human economies. Animals' labor force was gradually replaced by machines and technical devices. However, animal products have simultaneously increased to provide for human alimentation despite production being moved out of sight [→History of Agriculture;→Urban (and Rural) History]. Cows', chickens', or pigs' contribution to the production of meat has been translocated to industrial structures outside human urban centers. Slaughter now takes place far away from spaces of consumption [→History of Animal Slaughter]. At the same time as these animals have become invisible in the economic context, the last fifty years saw the emergence of a whole industry targeting companion animals. Companion animals share humans' intimate spheres and create markets for supplies like food, accessories, and medication [→History of Pets].[2]

However, statistics of production that could shed light on the real labor of animals enumerate output in simplified categories like the sheer quantity of milk produced per cow. Sometimes mules are mentioned in the context of transport or oxen as suppliers of meat. This is one reason why, although nonhuman animals' contribution to human economy was essential, economic historians do not consider animals as a social category relevant to humans [→Social History]. Current historiography limits social action to the economic activities of humans, so properly describing animals' contributions to human economy profoundly transforms our understanding of economics.[3]

This chapter, therefore, aims at a reevaluation of economic history from an animals' social perspective. Because the role of nonhuman animals in human economy has clearly yet to be written, the goal of this overview is to provide a possible redefinition of the animals' work as labor and to show how from a relational perspective, one that sees human and animal contributions as inherently entangled, economics as such could be re-read and described. After having looked at topics and themes as presented by economic historiography, the section on methods and approaches will show, how a praxeological turn in history has helped to reevaluate such practices as accounting and bookkeeping and asks how this perspective might be expanded on animals in the economy. Finally, this chapter will discuss animals as partici-

---

[2] Kete, Kathleen. *The Beast in the Boudoir: Petkeeping in Nineteenth-Century Paris*. Berkeley, CA: University of California Press, 1994.
[3] Lang, Heinrich. Tiere und Wirtschaft. Nichtmenschliche Lebewesen im ökonomischen Transfer im Europa der Frühen Neuzeit. In *Tiere und Geschichte. Konturen einer Animate History*, Gesine Krüger, Aline Steinbrecher, Clemens Wischermann (eds.), 241–266. Stuttgart: Franz Steiner Verlag, 2014.

pants in labor markets, and, hence, it will reintroduce nonhuman laborers in various and specific contexts of animals' economy.

2 Topics and Themes

A quick glance at the historiography of economy and economic relations reveals an exclusively human-oriented characterization of economic performances. If animals play a part in economic contexts, it is only in the history of the agricultural sector. The majority of animals who shared their lives with humans lived in rural areas.[4] No wonder then that authors on agricultural economy mention the animal factor quite so often. They above all refer to the production of milk, the supply of meat and eggs, and animal husbandry like the keeping of pigs and livestock breeding. What we get to know, however, are mere figures, like the quantity of milk produced per year or the weight of meat produced from each animal. Even studies of animals at markets, like markets of livestock, only count the animals themselves; few scholars have attempted to quantify the share of livestock on markets in comparison to other agricultural production or to any other kind of commodity.

The Ploughshares of History: Oxen and Horses

In an overview of pre-modern European economy, Paolo Malanima explicitly refers to animal power. With regard to eleventh-century England he writes that approximately 70 percent of "available mechanical energy" was "provided by working animals". In general, the most important species that worked in the agricultural context dragging the plough was the ox. This was partly the case because of the generally higher expenses to keep horses, with the exception cart-horses in the peasant's context.[5] While in some regions of Europe, including Northern France and Northern Germany, horses replaced oxen in good numbers from the later Middle Ages onwards, in the Mediterranean oxen remained the prevalent harnessed agricultural workers, employed until mechanization in the middle of the nineteenth century.[6]

Agricultural production in economic and agricultural histories is described in terms of climatic cycles, development of prices like the price of wheat, and income for peasants and their families. Although the cultivation of any kind of cereal crops is the synergic work of human and nonhuman animals, the contribution of oxen and

---

[4] Dülmen, Richard van. *Kultur und Alltag in der Frühen Neuzeit. Zweiter Band: Dorf und Stadt 16.–18. Jahrhundert*. Munich: C.H. Beck, 1992.
[5] Langdon, John. The Use of Animal Power from 1200 to 1800. In *Economia e energia secc. XIII-XVIII*, Simonetta Cavaciocchi (ed.), 213–221. Firenze: Le Monnier, 2003.
[6] Malanima, Paolo. *Pre-Modern European Economy. One Thousand Years (10th-19th Centuries)*. Leiden: Brill, 2009, 65–69.

horses and the effect on this work on them is never mentioned in that context. The only relevant relationship is the development of prices to the human rural and urban communities. So, the productivity of land is calculated as the agricultural product deriving from the technological parameter, the natural resources (capital) comprising livestock, and the labor of the human worker.[7]

A characteristic of European agriculture always had been "mixed farming", that is, the integration of livestock farming and cereal cultivation.[8] The cultivation of forage in turn fed farm animals. Since the Middle Ages in many European regions like Flanders, parts of Lombardy and England, the traditional three-field rotation was replaced by a system of permanent rotation. Fodder crops did not impoverish the land in the same way as cereals did. Hence, forage and livestock farming gave way to agricultural intensification. The increased production of meat was the outcome of more effectively nourished livestock.[9]

Agricultural histories are much more precise about the role of animals in the context of the human economy than other economic histories. Still, they generally refer to animals in human agriculture solely in terms of livestock productivity.[10] The cultivation of land, for instance, underwent technological changes in the late Middle Ages, when different types of ploughs were in use, pulled either by oxen or by horses. If we take a look at paintings from the Middle Ages to modern times, we will see that animals at work turning furrows seem to be omnipresent. The most recent agricultural history of Germany, for instance, uses paintings to illustrate the changing nature of work but does not discuss the labor of the animals shown on them at all.[11] The technological development involved animal labor force, yet agricultural histories also do not refer to oxen's or horses' contribution to labor when characterizing the production of field crops.[12]

A general observation is significant. Bovine animals were smaller and less heavy in the Middle Ages than their wild ancestors. In the seventeenth century, a cow weighed about 200 kg and produced about 80 kg of meat. A fattened ox instead weighed up to 375 kg. Breeding was important, and oxen or cows were grown to be-

---

[7] Malanima, *Pre-Modern European Economy*, 110–116, 129–131.
[8] Ambrosoli, Mauro. *Scienzati, contadini e proprietari. Botanica e agricoltura nell'Europa occidentale*. Torino: Einaudi, 1992.
[9] Slicher van Bath, Bernard H. *The Agrarian History of Western Europe. A.D. 500–1850*. London: Arnold, 1962.
[10] Achilles, Walter. *Landwirtschaft in der frühen Neuzeit*. Munich: Oldenbourg, 1991; Overton, Mark. *Agricultural Revolution in England. The Transformation of the Agrarian Economy 1500–1850*. Cambridge: Cambridge University Press, 1996, 111–120.
[11] Kießling, Rolf, Friedrich Kießling, Frank Konersmann and Werner Troßbach (eds.). *Vom Spätmittelalter bis zum Dreißigjährigen Krieg (1350–1650)*. Cologne: Böhlau, 2016.
[12] Kießling, Rolf, Frank Konersmann and Werner Troßbach. *Vom Spätmittelalter bis zum Dreißigjährigen Krieg (1350–1650)*, vol. 1 of *Grundzüge der Agrargeschichte*, Rolf Kießling, Frank Konersmann, Werner Troßbach (eds.). Cologne: Böhlau, 2016, 52–84. Ironically, all contemporary (historical) illustrations show animals at work producing field crops in this illuminating text.

come heavier. In some regions milk cows and other cattle were separated. Noble landowners especially maintained breeding regimes. Bovine animals from Friesland were heavier than others and, thus, were favorites for breeding and growing activities. In early seventeenth-century Germany, an average cow produced between 440 and 1600 liters of milk per year, the highest rate was produced by the cows on the wetlands in the domains that belonged to the County of Oldenburg. The quality of pastureland was rather varying. Pigs were welcome producers of meat and there was a correlation between the number of pigs and of cows, because pigs could be fed the by-products from the process of producing milk. A growing number of pigs were sent to forests to feed in the sixteenth century onwards for many practical reasons, including maintaining rights of pasturage. Cattle were also efficient producers of dung which served to fertilize cultivated land.[13]

Urban Economic Animals

In urban everyday life, animals were continuously and extensively present [→Historical Animal Geography].[14] In general, all animals that contributed to human subsistence lived alongside their human owners in the houses or in the gardens. Chickens and ducks provided their human neighbors with eggs and eventually with cheap meat. Pigs that spent their times in urban gardens were fed with human food waste and provided lard and meat in return. Bakers and millers kept hogs, because their rubbish served for the pigs' nutrition. Butchers instead were accompanied by dogs that sometimes were the first consumers of the scraps of meat that came off the carcasses. This dimension of urban economies is hardly ever discussed – we only find some traces in the literature that treat the butcher's dog as a sort of companion animal.[15]

Local histories of butchering and slaughter have not received much scholarly attention. Although the social and economic history of butchers is a subject of many histories[16], what happened in slaughterhouses only caught the attention of historians

---

[13] Kießling, Konersmann and Troßbach, *Vom Spätmittelalter*, 98–108.
[14] Steinbrecher, Aline. Hunde und Menschen. Ein Grenzen auslotender Blick auf ihr Zusammenleben (1700–1850). *Historische Anthropologie* 19, no. 2 (September 2011): 192–220.
[15] Hennebo, Dieter and Alfred Hoffmann. Geschichte der deutschen Gartenkunst, Bd. 1: Gärten des *Mittelalters*. Hamburg: Broschek, 1962; Kaiser, Hermann. Über Kuhställe und Schweinekofen. Formen der Viehhaltung in Nordwestdeutschland. *Zeitschrift für Agrargeschichte und Agrarsoziologie* 56, no. 2 (2008): 11–40.
[16] Schläppi, Daniel. Geschäfte kleiner Leute im Spannungsfeld von Markt, Monopol und Territorialwirtschaft. Regionaler Handel als heuristische Kategorie am Beispiel des Fleischgewerbes der Stadt Bern im 17. und 18. Jahrhundert. In *Praktiken des Handels. Geschäfte und soziale Beziehungen europäischer Kaufleute in Mittelalter und früher Neuzeit*, Mark Häberlein, Christof Jeggle (eds.), 451–475. Konstanz: UVK Verlagsgesellschaft, 2010 (Irseer Schriften N. F. Band 6).

for the period when the buildings began to be moved out of sight in the nineteenth and early twentieth centuries [→History of Animal Slaughter].[17]

Animals' bodies provided a number of further products consumed by humans. For the slaughterhouses it is obvious that the main focus of production was meat. However, bones were extracted and used for the production of soap. Skin was transformed into fabric in the form of leather.[18] In late eighteenth-century Leipzig, a porter would carry a basket full of stinking bowels out of the slaughterhouse to the nearby small town of Markneukirchen in the Vogtland every day, where the rather unappetizing raw material was transformed into gut strings. On the way down to the nearby river Weiße Elster, a rafter brought the gut strings to the market in Leipzig town center, where it was sold to the flourishing music instrument business.[19]

Many agricultural products, like wool, which derived from working animals, stood at the beginning of a production chain. Until the nineteenth century, wool was the most important raw material for the textile industry. Sheep helped to maintain fields and at the same time produced wool, which was shorn by humans to be sold on local and inter-regional markets. The provenance of wool, which corroborated its quality, was decisive for prices. For example, in the fifteenth and sixteenth centuries the wool of Spanish merino sheep was the most coveted supply when exports reached their peak.[20]

Animal Trade and Animal Markets

In agricultural history, there is an ongoing debate on the form of pre-modern economy: on the one hand Rainer Beck describes the pre-industrial peasants' economy as "Naturalwirtschaft" (barter economy) which was not profit-oriented but should be defined as subsistence economy. Some scholars use the expression "Knappheitswirtschaft" (scarcity economy) to explain the specific character of pre-industrial economies.[21] On the other hand, the authors of a recently published history of German ag-

---

[17] Brantz, Dorothee. Die animalische Stadt. Die Mensch-Tier-Beziehung in der Urbanisierungsforschung. *Informationen zur modernen Stadtgeschichte* 1 (2008): 86–100; Regnath, R. Johanna. *Das Schwein im Wald. Vormoderne Schweinehaltung zwischen Herrschaftsstrukturen, ständischer Ordnung und Subsistenzökonomie*. Ostfildern: Thorbeke, 2008.
[18] Nowosadtko, Jutta. *Scharfrichter und Abdecker. Der Alltag zweier unehrlicher Berufe in der Frühen Neuzeit*. Paderborn: Ferdinand Schoeningh, 1994.
[19] I owe this story to Josef Focht, director of the University of Leipzig Museum of Music Instruments.
[20] Klein, Julius. *The Mesta. A Study in Spanish Economic History, 1273–1836*. Cambridge, MA: Harvard University Press, 1920.
[21] Beck, Rainer. *Naturale Ökonomie. Unterfinning: Bäuerliche Wirtschaft in einem oberbayerischen Dorf des frühen 18. Jahrhunderts*. Forschungshefte 11. Munich: Deutscher Kunstverlag, 1986. All terms – barter economy, subsistence economy, and scarcity economy – refer to the model of market equilibrium. They refer to limited resources and the use of kind instead of the medium of money for exchange. I am not entirely sure whether the terms are quite what economic historians mean when

riculture claim that trading animals and animals' products were market-oriented like other commerce. Their main questions are whether or not farmers and peasants produced aiming for surpluses and how agricultural production was connected to market commerce. A barely visible process of monetarization – the growing use of coins instead of kind – and complex credit relations in the rural context are indicators which lead the authors to hypothesize that agricultural markets were established alongside the economic structures of urban development and even were connected to regional and trans-regional markets.[22]

However, on the local level most transactions of animals were non-market trade transfers.[23] In Franconia, Germany, for instance, most rural households that owned some land possessed up to six bovids. The property of livestock and the exchange of owners was mediated by transfers without markets. Cows whose work was the production of milk very often were traded by the peasants' wives. Meat produced by bovine animals mainly provided sustenance for its human owners. On the market level, farm oxen and fattened oxen were sold to intermediaries who vended the cattle or brought them to specialized markets. A majority of these traders of bovine animals were Jews who operated in the local and regional context. This group of traders particularly combined the buying and vending of bovine animals with financial obligations like credits. Another important group of traders were the butchers who travelled quite extensively to acquire cattle, often interacting with traders from other regions.[24]

In the German context, for example, every year thousands of oxen were driven from Hungary and Poland in order to be exchanged on the great cattle markets all over the Reich. Some studies give precise detail on numbers especially about either oxen whose bodies served as significant supply for human alimentation in the large towns like the imperial cities of Swabia, or oxen who became labor forces in the broader sense. The size of the drove varied considerably between some dozens of individuals up to thousands. Over the centuries, some of the routes were well established and connected the regions of breeding with the greater markets and consumers.[25]

---

they describe "early modern"/"pre-modern"/"pre-industrial" economies. Observations from ethnology help historians to make comparisons between historical periods and the present: "The past is a foreign country" (Peter Burke).

22 Kießling, Konersmann, Troßbach, *Vom Spätmittelalter*, 175–181; Prass, Reiner. *Vom Dreißigjährigen Krieg bis zum Beginn der Moderne (1650–1880)*, vol. 2 of *Grundzüge der Agrargeschichte*, Stefan Brakensiek, Rolf Kießling, Werner Troßbach, Clemens Zimmermann (eds.). Cologne: Böhlau, 2016, 24, 50–58.

23 Edwards, Peter. *The Horse Trade of Tudor and Stuart England*. Cambridge: Cambridge University Press, 1988.

24 Schenker, Andreas. *Der Rinderhandel im Hochstift Bamberg in der zweiten Hälfte des 18. Jahrhunderts. Struktur, Entwicklung und die Aushandlung der normativen Rahmenbedingungen zwischen Obrigkeit und Marktakteuren. Region und Stadt in der Vormoderne*. Würzburg: Ergon Verlag, 2020.

25 Lerner, Franz. Die Bedeutung des internationalen Ochsenhandels für die Fleischversorgung deutscher Städte im Spätmittelalter und der Frühen Neuzeit. In *Internationaler Ochsenhandel*

The markets specialized in trading cattle had to provide certain infrastructure like easy accessibility, fences, enclosures, and an administration which supervised and taxed the transfer of property rights on animals. The most important issue was guaranteeing the quality of the transferred animal: making sure that the acquired specimen was healthy and would be able to serve the new owner's purposes. Cameralist literature from the eighteenth century, for example, very explicitly presents the needed knowledge about horses and their presumptive future services.[26] Some scholars argue that the reputation of a seller substituted the risk of failure in quality of the divested bovine animals. Others clearly describe that risk was asymmetrically distributed and some of the traders on markets simply failed.[27]

At the core of cattle markets was the development of prices. The price for meat – which is a very important baseline of economic data – remained relatively stable during the centuries and did not so much follow the price of grain which was much more volatile. Interestingly enough, none of this considers the value of the animals' work. Even when studies of livestock markets or on the transfer of bovine animals talk about the practices of selling and purchasing of animals, they never try to change the perspective to include the animals' view of the markets or their contribution to the human economy.

3 Methods and Approaches

In economic history it is all about quantities, practices, and knowledge. As shown in the sections above, animals are not considered in their own right but appear as part of this numeric fixation. For most scholars, it does not make a difference whether the produced object is a certain sort of cereal or milk or meat, whether the traded good is a piece of woolen cloth or an ox, whether the force employed to pull a cart is provided by a machine, a slave or a horse. In consequence, there remains a methodological gap in grasping the difference between nonhuman animals and objects. However, as

---

(1350–1750). Akten des 7th International Economic History Congress Edinburgh 1978. Beiträge zur Wirtschaftsgeschichte 9, Ekkehard Westermann (ed.), 179–218. Stuttgart: Klett-Cotta, 1979; Grillmaier, Anna-Maria. Ochsen- und Rinderimporte nach Augsburg und Schwaben im Lichte bayerischer Zollrechnungen des 15. und 16. Jahrhunderts. In *Beschaffungs- und Absatzmärkte oberdeutscher Firmen im Zeitalter der Welser und Fugger*, Angelika Westermann, Stefanie von Welser (eds.), 175–203. Husum: Matthiesen Verlag Ingwert Paulsen Jr., 2011.
26 Zedler, Johann Heinrich. *Großes vollständiges Universal-Lexicon aller Wissenschaften und Künste*, 27. Halle, 1741, coll. 1400–1405.
27 Fenske, Michaela. *Marktkultur in der Frühen Neuzeit. Wirtschaft, Macht und Unterhaltung auf einem städtischen Jahr- und Viehmarkt*. Cologne: Böhlau, 2006.

part of a more animal-centered perspective, some recent approaches try to shed new light on the impact of the industrialization of agriculture on the dignity of animals.[28]

The praxeological approach also provides us with further perspectives on economies, namely, to focus on the process of quantification. Based on the connection between heterogenous phenomena and homogenous data, meaning that economic transactions were measured for profit, this perspective tries to tease out what is happening in terms of negotiations: if a seller took some bovine animal to a market, he intended to receive a certain amount of money to meet his expenses and to create a profit on his labor. The sovereign imposed some duty on every bovine animal that passed his territory, so his fiscal administration registered every single individual specimen that was brought along.

Another dimension of a practice-led approach is to look at the interactions between a broad variety of actants that are part of the sociality of economic transfers. Hence, the analysis of the constitution of economic action is a qualitative core feature of economies and the knowledge about economic processes and agencies a key indicator for economic change. For instance, the analyses of markets may help uncover decisive impulses for innovation as information is generally viewed as being central to ongoing economic growth.

The fundamental cultural step towards a quantitatively constituted human economy was made by the invention of monetary evaluation, the validation process.[29] If we want to include a more animal-centered perspective on these processes, we have to look into the practices of accounting and reckoning on the animals' behalf.

I want to illustrate this with an example. In his *Libro Grande Verde* – his Green Ledger bearing the signature C – the bookkeeper of the business company Antonio di Filippo Salviati & Co in Pisa documented an income of 127 *fiorini* from selling "their buffaloes" – 123 in number – to various clients in 1616 and early 1617.[30] The vending of milk, the purchase of cows, and the reconstruction of the cows' house were measured in the same way: the bookkeeper simply wrote down a calculation of sheer numbers to get a balanced result at the end of any account. The ledger of the company, which was named *di banco e di Magona* (banking and Magona business) had to produce balanced accounts.

From the output side as well as from the expenditure side of any form of human business a balance was drawn at the reference to the evaluation for a turnover. The technique which was employed for calculation was accounting. Accounting means processing data in economic transfers. It is related to abstraction and organization. In this context abstraction consisted of a homogenizing perspective on heterogenous

---

28 Petrus, Klaus. Die Verdinglichung der Tiere. In *Tiere Bilder Ökonomien. Aktuelle Forschungsfragen der Human-Animal Studies*, Chimaira – Arbeitskreis für Human-Animal Studies (eds.), 43–62. Bielefeld: transcript, 2013.
29 Orléan, André. *L'Empire de la valeur. Refonder l'économie*. Paris: Le Seuil, 2011.
30 SNS, Archivio Salviati, ser. I, no. 1635: Antonio di Filippo Salviati & Co del Banco e di Magona di Pisa: Libro Grande verde Debitori e creditori segnato C (1616–1628), c. 2/II.

objects and actions. The property rights of an object or service are transferred from the seller to the buyer, and, hence, the transfer is validated via an abstract coordination system: the monetary value. The accountant creates a processed evaluation of homogenous measures in order to represent each transfer's contribution to growth and decrease. Each object and each service is de-contextualized and inserted in a system of abstract space – currencies – and abstract time – spans of periods for drawing sums. Organization in this context refers to a written system of accounts and their transposal to account books. The ledger represents the relations of debts between the company as an abstract entity and the bearers of accounts.[31]

These transactions are the basis for the fundamental human method of organizing economy: The physical world of objects and services is validated within a monetary system of reference. Every traded object or service becomes an economic good through an abstract world of numbers (it is inserted into an abstract world of numbers and is, thus, conditioned to be a good). The account book perceived the *bufalo* at his quality of being monetarily evaluated and, thereby, quantifiable. Therefore, the buffalo appears as an object that was in debit to the company being owned at the expense of the human business company. The *bufalo* had to balance his debit by a credit, that is, he had to produce an income for being sold at a reasonable price. The result of the accountant's work is the subordination of the *bufalo* to the system of debt relations. On the accounting level, there is no difference between a desired object like a piece of gold, the payment service of a banking correspondent or a business friend, and the mentioned *bufalo* – no matter the role the animal as a form of lively capital played for human society.[32]

Antonio and Filippo Salviati & Co. ran a farm outside the town in the countryside. Some cows lived there, producing milk, and in turn they were fed on the meadows leased by the business company of the Florentine family Salviati. The accountant evaluated the cows' work at the expenses for the maintenance of the livestock holds and the income of the milk sold at local markets. From season to season the accountant also documented the cost for the acquisition of new cows. However, the bookkeeper registered the referring monetary effects as growth or decrease of the business volume. Whether a cow was bought, milk was sold, the tenant had to employ a carpenter to restore the stands – the process of accounting transposed every change regarding the property of the company into monetary debit or credit.[33]

The operational scheme of calculation is categorized as human rationality. However, we are only able to observe a homogenizing abstraction that made up the accountant's calculation. Moreover, this mono-dimensional view of economy explains

---

31 Lang, Heinrich. *Wirtschaften als kulturelle Praxis. Die Florentiner Salviati und die Augsburger Welser auf den Märkten in Lyon (1507–1559)*. Stuttgart: Franz Steiner Verlag, 2020, 91–99.
32 Miller, Peter. The Margins of Accounting. *The Laws of the Markets (Sociological Review)* 46, no. S1 (May 1998): 174–193.
33 Goldthwaite, Richard A. The Practice and Culture of Accounting in Renaissance Florence. *Enterprise & Society* 16, no. 3 (April 2015): 1–37.

why the buffaloes' contribution to the Pisan business company was identified with expense and income.[34]

There is, thus, a twofold approach to the perspective on animals and economy presented above: On the one hand it explains how humans began to categorize animals' work and how they constructed an economy and, hence, validated animals' contribution to their economy. On the other hand, it shows the point of reference is the documentation we have at our disposal.

The principles of what we may call 'accounting culture' were also inscribed in agricultural numeracy, because the validation of goods and services of any kind followed the very same practices of transforming objects and action into economic valuables. In many parts of medieval and early modern Europe, the summarizing of agricultural income was already established. The use of numeracy was widespread when it came to keeping accounts of agricultural production. For example, in the context of rural Tuscany, smallholders on the countryside had to submit their tax reports to the Florentine tax officials from 1427 onwards. They had to know the value and the size of their land properties, and they recorded their income from the lands in terms of numerical quantities – referring to the measures of grain, olives and wine. Nowhere did they provide the numbers of the bovine workforce employed on their fields. The tax declarations are empty of livestock.[35]

However, there was a gradual refinement of accounting through the centuries in the agricultural sector. The accountants of farms now used many more categories for each species in order to specify the value of every accounting item. Paralleling the expanding knowledge of breeding and animal husbandry, the accounts became much more differentiated. If we take the *Greifensteiner Oeconomie Rechnung* of 1784 – which was the account of an agricultural factory owned by the Franconian nobleman count Schenk zu Stauffenberg – as a telling example, the most important elaborative detail of evaluating animals are sex, age, and color. The account consisted of a list of income and output by category and of a balance sheet. Categories ranged from horses to male calves, from poult to hen, from rearing pigs to boars, but also from grain, oat, butter to milk. Even the transport of pigs to the places of disposal is itemized.[36]

In economic terms, human accounting creates a concept for valuing production which is the value added, the revenue in short, minus the cost of production, such as the expenditure in raw material, fabric units and so on. It does not reflect the animal's contribution to production by its labor. Animals in agricultural production

---

34 Lang, *Wirtschaften als kulturelle Praxis*, 100–107.
35 Herlihy, David and Christiane Klapisch-Zuber. *Les Toscans et leurs families: une étude du catasto florentin de 1427.* Paris: Presses de la fondation nationale des sciences politiques, 1978; Emigh, Rebecca Jean. *The Underdevelopment of Capitalism. Sectors and Markets in Fifteenth-Century Tuscany.* Philadelphia, PA: Temple University Press, 2009, 86–90.
36 State Archives Bamberg. I would like to thank Andreas Schenker for pointing my attention to this marvelous document.

processes are treated by accounting as if they were machines: cost per unit, the oil you have to add to run the machine, and the replacement of components from time to time. On the market, an animal's value was the price the purchaser had to pay for a commodity. In that sense, the surplus value in agricultural production processes which basically derived from animals' labor by either force or reproduction is essentially ignored in accounting, no matter if entrepreneurial accounting, state's accounts or tax reports are considered.

Taking a praxeological approach to economic history means that we have to re-read economic development and practices. In fact, the praxeological approach re-opens the fundamental debate on labor in general which characterizes the implication(s) of the *animal turn* on economic history.

4 Implication(s) of the Animal Turn

Economic history that takes the *animal turn* seriously has yet to be written. The previous section on quantifying animals in economic history through human accounting comprises a qualitative perspective as well. Analyzing the way of categorizing the value of animals by monetary evaluation, however, sheds light on the transformation of the animals' contribution to production and productiveness into sheer numbers of generated commodities. In fact, until industrialization human production was totally dependent on animals working, even if that labor was only seen in terms of human nutrition (eggs, meat, etc.).

The transformative capacity of human accounting paved the way for a rational culture of subordination through abstract calculation (reckoning). However, this very process is decisive, because it prescribes what is submitted to monetary evaluation. While human work appears as services paid, the work of animals does not. In fact, most reproductive action is not considered as service and remains without pay.

Since most economic historians follow neoclassical models, they hardly refer to economic formation at all. The neoclassical theory of economy shifted attention from the relationship of labor and capital to the marginal utility concept of markets. In consequence, the neoclassical theorists put aside the discussion of the impact of labor to production or even the exploitation of labor, prominent in David Ricardo's and Karl Marx's works. In the context of industrialization, of course, the neoclassical models must explain the allocation of labor and capital from factor markets – in theory they do it on the bases of rational choice, perfect competition, and a market equilibrium. This approach totally excludes any kind of non-utility-oriented behavior and it defines labor as an exclusively human, if not to say rational, factor to production.[37] Moreover, Marx, who describes labor as the essential dimension to create a

---

37 Cf. Mazzucato, Marianna. *The Value of Everything. Making and Taking in the Global Economy*. Mil-

surplus value in the process of production, reduces the concept of labor to a male-human activity. In this sense, economic historians generally do not consider the non-human contribution to agricultural production and to markets.[38] The animals' impact on agricultural output and market proceedings is not specified if not simply neglected.

The previous section also examined the statistic (numeral) framework of human economy. The implication of these observations widens the historian's perspective beyond the constitution of human economy. In essence, it helps to look back at a time before the conditioning of human economy by accounting. On a theoretical level, animal historians must therefore leave behind the neoclassical models and take a praxeological approach to reset the production boundaries and to reconstruct the contribution of labor to production processes in practical detail.[39]

Animals have been part of human life, and their force, like their productivity, contributed to human economy throughout time. Thus, looking at animals before accounting means that the social life, the everyday symbiosis of humans and animals, and the synergetic interrelation of animals in the pre-industrial processes of production and commerce constitutes a new perception of an interspecies economy. That is, animals should be considered as laborers, which makes them part of a labor market – even if this market can only be described by asymmetric terms like the labor force of indentured servants.[40]

When it came to the industrial production of meat, for instance, the mere fact that animals contributed with their service to the increase of output, was irrelevant. Of course, feeding livestock required the more or less voluntarily participation of the animals at the reproduction processes, and in the end the animals even gave their entire bodies to the production of meat. However, accounting simply balances cost, for example the material input such as the purchase of a calf and the lease on land, by revenue, that is the sale of the fed animals' bodies. In consequence, animal historians will need to re-write the history of the consumption of meat. As such, the story of the slaughterhouse has to start with the concept of a human society

---

ton Keynes: Penguin Random House UK, 2019, 40–47 on David Riccardo, 47–56, 62–69 on neoclassical theory.

38 Rosen, Aiyana and Sven Wirth. Tier Ökonomien. Über die Rolle der Kategorie Arbeit in den Grenzziehungen des Mensch-Tier-Dualismus. In *Tiere Bilder Ökonomien. Aktuelle Forschungsfragen der Human-Animal Studies*, Chimaira – Arbeitskreis für Human-Animal Studies (eds), 17–42. Bielefeld: transcript, 2013.

39 Nisly, Jadon. Er kömme von seinem Viehe nicht hinweg. Mensch-Nutztier-Beziehung in einem volksaufklärerischen Mustergut (1782–1795). In *Tiere Nutzen. Ökonomien tierischer Produktion in der Moderne*, Lukasz Nieradzik, Brigitta Schmidt-Lauber (eds.), 88–104. Innsbruck: Studien Verlag, 2016. Much of our vision of the "use" of animals is due to the Enlightenment turn in the late eighteenth century, when authors described animals as useful objects to humans.

40 Hribal, Jason. Animals, Agency, and Class. Writing the History of Animals from Below. *Human Ecological Review* 14, no. 1 (June 2007): 101–112.

which keeps animals in order to eat them. Every indicator of production will have to be reconsidered with regard to the animals' impact.

In pre-industrial production we might even describe the animals' contribution to agricultural production by labor in numbers. In economic terms the price of an agricultural good like grain or meat could be valued by the cost of raw material input and the value added, that is, the selling price is the revenue minus production cost. This gap, which is generally considered as gain of selling agricultural products, in fact equals the labor cost of the non-waged labor animals. This amounts to a description *ex negativo* for being absent from real labor markets. When prices of agricultural products increased due to growing or decreasing demand, production output[41] shows the value of non-waged labor of animals. Hence, we have to compare prices of grain and meat for statistical reasons.

The preindustrial economy which could be described as pre-capitalist agricultural production, however, is fairly different from the emerging agricultural business of the nineteenth century.[42] The prices for grain, for example, were not prices fetched on capitalist markets. Governments frequently interfered and, thus, prices for grain were especially subject to regulation. The contributed labor of animals could be explained in a very hypothetical way: if a landowner had had to look for laborers on a labor market to do the animals' work in agricultural production the same way governments had to seek for mercenaries in military affairs, any kind of crops would have become far too expensive for everyday life. Animals offered their work force for production without wages.

Taking a collaborative perspective could change our view on the animals' competences and even historically reevaluate the animals' history of co-habitation and co-operation in economic processes. Although not reflecting on these issues, Mario Ludwig indicates a way forward by examining "animals' jobs" like the very traditional work in hunting provided by falcons or by the truffle pig's job.[43]

To sum up, to allow the *animal turn* to take hold in economic history means to re-read economic production. Firstly, we have to analyze the conventions which defined the animals' labor as neglectable. Secondly, we have to re-interpret agricultural history from its early documentation onwards. Thirdly, we shall re-validate the industrial agricultural production from the nineteenth century onwards to the present day. What does this mean in detail? For once it means to apply the simple work of any historian, that is, to read documents carefully and to understand what was going on. If we put together accounting and visual evidence like paintings, we could observe a more socially dense agricultural sector driven by the animals' force: humans who drove the plough, bovines that dragged the plough. It also means looking at

---

[41] Fara, Andrea, Donatella Strangio and Manuel Vaquero Pineiro (eds.). *Oeconomica. Studi in onore di Luciano Palermo.* Viterbo: Sette Città, 2016.
[42] Beck, *Naturale Ökonomie.*
[43] Ludwig, Mario. *Tierische Jobs. Verblüffende Geschichten aus der Tierwelt.* Darmstadt: wbg Theiss, 2019.

what happens to the basic economic sector where most animals had been living before industrialization: If we change the perspective and allow new economic actors onto the scene a more inclusive economic history might be fathomable.

Selected Bibliography

Beck, Rainer. *Naturale Ökonomie. Unterfinning: Bäuerliche Wirtschaft in einem oberbayerischen Dorf des frühen 18. Jahrhunderts*. Forschungshefte 11. Munich: Deutscher Kunstverlag, 1986.
Brantz, Dorothee. Die animalische Stadt. Die Mensch-Tier-Beziehung in der Urbanisierungsforschung. *Informationen zur modernen Stadtgeschichte* 1 (2008): 86–100.
Chimaira – Arbeitskreis für Human-Animal Studies (eds.). *Tiere Bilder Ökonomien. Aktuelle Forschungsfragen der Human-Animal Studies*. Bielefeld: transcript, 2013.
Edwards, Peter. Domesticated Animals in Renaissance Europe. In *A Cultural History of Animals in the Renaissance. A Cultural History of Animals*. Bruce Boehrer (ed.), 75–94. Oxford and New York, NY: Bloomsbury, 2007.
Kießling, Rolf, Frank Konersmann and Werner Troßbach. *Vom Spätmittelalter bis zum Dreißigjährigen Krieg (1350–1650)*, vol. 1 of *Grundzüge der Agrargeschichte*, Stefan Brakensiek, Rolf Kießling, Werner Troßbach, Clemens Zimmermann (eds.). Cologne: Böhlau, 2016.
Lang, Heinrich. Tiere und Wirtschaft. Nichtmenschliche Lebewesen im ökonomischen Transfer im Europa der Frühen Neuzeit. In *Tiere und Geschichte. Konturen einer Animate History*, Gesine Krüger, Aline Steinbrecher, Clemens Wischermann (eds.), 241–266. Stuttgart: Franz Steiner Verlag, 2014.
Malanima, Paolo. *Pre-Modern European Economy. One Thousand Years (10th–19th Centuries)*. Leiden: Brill, 2009.
Nieradzik, Lukasz and Brigitta Schmidt-Lauber (eds.). *Ökonomien tierischer Produktion in der Moderne*. Innsbruck: Studien Verlag, 2016.
Overton, Mark. *Agricultural Revolution in England. The Transformation of the Agrarian Economy 1500–1850*. Cambridge: Cambridge University Press, 1996.
Prass, Reiner. *Vom Dreißigjährigen Krieg bis zum Beginn der Moderne (1650–1880)*, vol. 2 of *Grundzüge der Agrargeschichte*, Stefan Brakensiek, Rolf Kießling, Werner Troßbach, Clemens Zimmermann (eds.). Cologne: Böhlau, 2016.
Raber, Karen. From Sheep to Meat, From Pets to People: Animal Domestication 1600–1800. In *A Cultural History of Animals in the Age of Enlightenment, vol. 4*, Matthew Senior (ed.), 73–100. Oxford: Berg, 2007.

Nadir Weber
Diplomatic History

1 Introduction and Overview

The history of diplomacy, that is, of the peaceful conduct of relations between political entities, has seen a significant renewal in recent years.[1] Leaving behind the narrow focus on 'great men' and high politics, historians have analyzed in depth the transnational networks which ensured political communication and rediscovered diplomacy as a site of symbolic interaction and intercultural exchange. However, the *animal turn* has had a relatively peripheral effect on the field. Some aspects, especially the use of animals as diplomatic gifts, have gained growing attention in the last years, but these studies usually lack a systematic perspective that takes into account the specificities of animal lives and agency. Furthermore, many other aspects of the presence of animals in diplomatic history such as their role in diplomatic negotiations or representation have remained understudied. This chapter aims to broaden the discussion by discerning different aspects of an animal history of diplomacy. The first part focuses on the roles of nonhuman animals in diplomatic gift-exchange, political representation, and international treatises. The second part discusses methodological approaches that could be adopted for writing a history of animals in diplomacy.

2 Topics and Themes

Animals as Diplomatic Gifts

Gifts always formed an essential part of the political relations among rulers or state leaders because they conveyed mutual esteem and the willingness to cooperate.[2] Since antiquity, living animals or the material remains of animals such as furs, ivory, or hunting trophies were popular choices in the diplomatic exchange of gifts, especially in the case of intercultural encounters. In his *History of Alexander*, Quintus Curtius Rufus reports that an embassy sent by the Indian Mallians after a military defeat in 326/5 BC brought with them "lions of unusual size and tigers,

---

[1] For a brief introduction, see, for example, Dover, Paul M. and Hamish Scott. The Emergence of Diplomacy. In *The Oxford Handbook of European History, 1350–1750. Vol. II: Cultures and Power*, Hamish Scott (ed.), 663–695. Oxford: Oxford University Press, 2015, 663 on the term of diplomacy.
[2] See, most recently, Biedermann, Zoltán, Anne Gerritson and Giorgio Riello (eds.). *Global Gifts. The Material Culture of Diplomacy in Early Modern Eurasia*. Cambridge: Cambridge University Press, 2018.

https://doi.org/10.1515/9783110536553-017

both animals tamed to gentleness, the hides of great lizards and tortoise shells"³ as gifts to the Macedonian ruler. In 801, an Indian elephant called Abul Abbas found its way to Aachen as a gift from the Abbasid Caliph Hārūn Ar-Rašīd to Emperor Charlemagne.⁴ In 1514, another elephant called Hanno accompanied the Portuguese ambassador Tristão da Cunha during his mission to Pope Leo X de' Medici. During the early modern era, a number of Ceylonese elephants (Fig. 1) as well as other large exotic mammals followed Hanno's path, revealing the importance of animals for the diplomatic relations of maritime powers.⁵ On a more regular basis, horses, dogs, falcons, or game animals were exchanged between early modern rulers. Animals thus formed an essential part of the international gift-economy that linked dynastic centers in and beyond Europe [→Global History].⁶

From the nineteenth to the twenty-first centuries, living animals continued to play important roles in diplomatic encounters that became more and more a stage for public relations. In the 1820s, Muḥammad 'Alī Bāshā, the self-proclaimed Khedive of Egypt and Sudan, sent several giraffes as diplomatic gifts to Paris, London and Vienna, where they received enormous attention from spectators on the road and in printed media.⁷ During and shortly after the Second World War, in 1943 and 1947, the Australian government gained public goodwill in Great Britain and the United States by sending platypuses.⁸ From 1958 onwards, dozens of Chinese

---

3 Lewis, Sian and Lloyd Llewellyn-Jones (eds.). *The Culture of Animals in Antiquity: A Sourcebook with Commentaries.* Milton Park, New York: Routledge, 2018, 460. On the practices of gift-exchange in Roman diplomacy, including the transfers of various animal species such as horses, mules, birds, and wild animals used for animal combats, see Necheava, Ekaterina. *Embassies – Negotiations – Gifts. Systems of East Roman Diplomacy in Late Antiquity.* Stuttgart: Franz Steiner, 2014, ch. 4.
4 See Hack, Achim Thomas. *Abul Abaz. Zur Biographie eines Elephanten.* Badenweiler: Wissenschaftlicher Verlag Bachmann, 2011.
5 Bedini, Silvio. *The Pope's Elephant.* London: Penguin Books, 2000; Jordan Gschwend, Annemarie. *The Story of Süleyman. Celebrity Elephants and Other Exotica in Renaissance Portugal.* Zürich: Pachyderm Production, 2010; Simões, Catarina. Non-European Animals and the Construction of Royalty at the Renaissance Portuguese Court. In *Animals and Courts. Europe, c. 1200–1800*, Mark Hengerer, Nadir Weber (eds.), 55–77. Berlin and Boston, MA: De Gruyter, 2020.
6 Weber, Nadir. Lebende Geschenke. Tiere als Medien der frühneuzeitlichen Diplomatie. In *Medien der Außenbeziehungen. Von der Antike bis zur Gegenwart*, Peter Hoeres, Anuschka Tischer (eds.), 160–180. Cologne: Böhlau, 2017; see also Bayreuther, Magdalena. Pferde in der Diplomatie der Frühen Neuzeit. In *Materielle Grundlagen der Diplomatie. Schenken, Sammeln und Verhandeln in Spätmittelalter und Früher Neuzeit*, Mark Häberlein, Christoph Jeggle (eds.), 227–256. Konstanz and Munich: UVK, 2013; Reindl-Kiel, Hedda. Dogs, Elephants, Lions, a Ram and a Rhino on Diplomatic Mission: Animals as Gifts to the Ottoman Court. In *Animals and People in the Ottoman Empire*, Suraiya Faroqhi (ed.), 271–285. Istanbul: Eren, 2010; Weber, Lebende Geschenke.
7 See, among other studies, Allin, Michael. *Zarafa: A Giraffe's True Story, from Deep in Africa to the Heart of Paris.* New York, NY: Walker, 1998; Baratay, Éric. The Giraffe's Journey in France (1826–7): Entering Another World. In *Animal History in the Modern City. Exploring Liminality*, Clemens Wischermann, Aline Steinbrecher, Philipp Howell (eds.), 91–104. London: Bloomsbury Publishing, 2018.
8 Cushing, Nancy and Kevin Markwell. Platypus Diplomacy: Animal Gifts in International Relations. *Journal of Australian Studies* 33, no. 3 (September 2009): 255–271.

Figure 1: In 1562, a young Indian elephant bull had been gifted by Catherine of Austria, Queen of Portugal, to her grandson Carlos of Austria, Prince of Asturias and heir-apparent of Spain. In the following year, the elephant – guided by his Indian mahout, here represented as a 'blackamoor' – made his way via the Netherlands to the Imperial Court in Vienna, where he was integrated into the menagerie of Emperor Maximilian II. (Broadsheet with elephant passing Antwerp in 1563, engraved by Jan Mollijns I (woodcut with hand-color and letterpress; detail). Courtesy of the British Museum, London / © Trustees of the British Museum, Prints and Drawings, No. 1928,0310.97 (Creative Commons).

giant pandas made their way to Western zoos as diplomatic gifts or loans, helping to re-integrate the People's Republic of China into the international system.[9] European states were not only the receivers in China's Panda diplomacy, but also the presenters of animal diplomatic gifts: in January 2018, the riding horse Vésuve de Brekka was gifted to China's President, Xi Jinping, by the French President Emmanuel Macron. Drawing great media attention when they arrive, and often displayed in zoos for years, especially rare and exotic animals proved to be perfect means of what became known as public diplomacy: political communication that addresses not only state leaders, but also the public of foreign countries.[10]

Animal gifts share some characteristics with other types of diplomatic gifts, but also carry their specificities due to their status as living beings.[11] Throughout the past, living animals were supposed to elicit the recipient's benevolence due to their rarity, beauty, or use. In general, large mammals that fulfilled two or all three of these characteristics were favored, with variations in the choice of animals depending on the local fauna and breeding expertise, and the preferences of the rulers or states for whom the gift was destined. Additionally, these animals were frequently accompanied by precious objects such as ornamented harnesses or coaches [→Material Culture Studies]. In contrast to non-living gifts, animals could significantly increase their value as a gift by performing well or drawing emotional attention. Elephant Hanno not only became especially popular among the common people in Rome, but also gained the personal favor of Pope Leo X. The Pope visited Hanno almost daily and sent his personal doctor to cure the animal when he fell ill. Accordingly, some contemporary observers described the elephant as the Portuguese king's best ambassador.[12]

However, gifting animals went hand in hand with specific risks. More than other diplomatic gifts, animals could also be disapproved of by recipients precisely because of their agency and mortality. Ganda, a rhinoceros who was supposed to follow Hanno as an animal diplomat to Rome, died in early 1516 in a shipping accident before reaching the Pope's court. While this animal at least 'survived' in contemporary descriptions and sketches, one of which served as a model for Dürer's famous print, many other animals that died in transit or shortly after their arrival only caused frustration at the receiving court. For instance, a letter from Onolzbach informed the Danish court in June 1734 that the five gifted gyrfalcons from Iceland had immediately died after their arrival at the court of Ansbach, and that this had "heavily affected" the Margrave, who was a passionate falconer and "great lover of these birds."[13] Fur-

---

9 On China's Panda diplomacy, see Hartig, Falk. Panda Diplomacy: The Cutest Part of China's Public Diplomacy. *The Hague Journal of Diplomacy* 8, no. 1 (January 2013): 49–78.
10 Hartig, Panda Diplomacy, 51–52.
11 See Weber, Lebende Geschenke.
12 Bedini, The Pope's Elephant, ch. 4.
13 The Danish National Archives, Rigsarkivet, 303 Overjægermesteren, 333.638, Lit. K, *Obristfalkenmeister* von Pöllnitz to the Danish *Obristjägermeister* Gram, Onolzbach, 18 June, 1734.

thermore, animals that arrived healthy could also thwart the donor's intentions of expressing friendship or symbolizing wealth and power. Thus, an elephant from Congo who was sent to the French court in 1668 caused annoyance in Versailles when he damaged the newly established menagerie and attacked several visitors. A tamed female 'tiger' (probably a leopard) who was sent to the same court by a Moroccan diplomatic mission in 1682 even became the subject of ridicule when she was mortally beaten by a cow in an animal combat that was held in Vincennes in front of the courtly public.[14] Gifting animals was thus a risky business. Because they attracted public attention and evoked emotions, they could be perfect ambassadors, but also quite the opposite.

Animals in Diplomatic Representation

Some species traditionally played important roles in diplomatic representation as participants in ceremonial encounters or as symbols for rulers and states, or in both roles at the same time. In public ceremonies, pre-modern Eurasian rulers often appeared on the back of horses or elephants, which underlined their elevated position in society and dominion over their subjects.[15] When rulers in medieval Europe met each other, horse-related practices such as helping a high guest to descend from the saddle were an eminent means to express hospitality and relations in power and rank. This was also true for the public entries of royal ambassadors, who were regarded as embodiments of their rulers. Around 1600, the use of state coaches began to replace the direct physical contact of rulers or diplomats with their horses for reasons of comfort, privacy, and distinction. However, this transformation strengthened rather than weakened the symbolic role of the animals. For while the ruler or his representatives were now partly or fully hidden behind doors and curtains, the six or even eight carefully trained horses of equal size and color that were hitched up to the carriage became the most visible sign of the passenger's rank and power.[16] It is thus not surprising that horses were among the first victims when diplomats quarreled over precedence. In the notorious coach incident between the Spanish and the French missions that took place in the streets of London on 30 Sep-

---

14 Examples taken from Weber, Lebende Geschenke.
15 Roche, Daniel. *La gloire et la puissance. Essai sur la distinction équestre*. Paris: Fayard, 2011, especially 209–245.
16 See Álvarez, Alejandro López. Some Reflections on the Ceremonial Image of the Kings and Queens of the House of Habsburg in the Sixteenth and Seventeenth Centuries. In *A Constellation of Courts. The Courts and Households of Habsburg Europe 1555–1665*, René Vermeir, Dries Raemaekers, José Eloy Hortal Muñoz (eds.), 267–321. Leuven: Leuven University Press, 2014; Bayreuther, Magdalena. Royal Equipage on its Way: Carriages and Court Ceremonial in Eighteenth-Century Munich. In *Animals and Courts: Europe, c. 1200–1800*, Mark Hengerer, Nadir Weber (eds.), 241–264. Berlin and Boston, MA: De Gruyter Oldenbourg, 2020.

tember 1661, the suite of the Spanish "killed three of the French coach-horses and several men."[17] Interpreted as an attack on his royal body, Louis XIV hereafter threatened war and finally received an apology from the Spanish king that was soon glorified as the latter's acceptance of French precedence in Europe.[18] Up to the eve of the First World War, when the 'Age of the Horse' came to an end,[19] equines prominently participated in diplomatic encounters.

Another prominent animal-related practice in diplomatic relations was hunting, that is, the chasing and killing of more or less free-moving animals – often with the support of other animals [→History of Hunting].[20] The invitation of foreign rulers or their representatives to hunts expressed hospitality and the will to cooperate. It would be misleading, however, to interpret such hunts as a merely informal part of diplomacy. In the late medieval and especially the early modern period, hunts were highly ritualized practices through which princes displayed their abilities as military leaders and rulers, and their control over the territory and the lives of its inhabitants. In some cases, these practices persisted or were even re-invented in the nineteenth and twentieth centuries. For example, the *chasses présidentielles* in France, to which members of the diplomatic corps were regularly invited, were only abolished in 2010. In any case, hunts were supposed to impress foreign guests, and their success always depended on the behavior of the animals that were involved: trained dogs, hawks, cormorants, or cheetahs that served as hunting assistants; and a great variety of game animals – from carefully protected red-legged partridges to red deer and royal tigers – that were destined to die in front of hunting party. While a coordinated chase and attack of a grey heron by the king's gyrfalcons high up in the sky may have been seen as a powerful sign of sovereignty, it was less so when the raptors decided to fly away. The same is true for incidents during which human participants of hunts were wounded or even killed by a furious deer or boar.

The use of animals as symbols for rulers or states, in contrast, was much easier to control than the behavior of living beings [→Political History].[21] In the medieval

---

17 Pepys, Samuel. *The Diary. A New and Complete Translation, vol. 2: 1661.* London: Bell, 1970, 188 (30 September 1661). According to Pepys (189), the Spanish had prepared themselves in advance "to outwit" their rivals "in setting upon the French horses and killing them, for by that means the French were not able to stir."
18 On Louis XIV's 'war of coaches', see Tessier, Alexandre. Des carrosses qui en cachent d'autres. Retour sur certains incidents qui marquèrent l'ambassade de Lord Denzil Holles à Paris, de 1663 à 1666. In *L'incident diplomatique XVIe-XVIIIe siècle*, Lucien Bély, Géraud Poumarède (ed.), 197–240. Paris: Pedone, 2010, 239.
19 Raulff, Ulrich. *Farewell to the Horse. The Final Century of Our Relationship.* Warwickshire: Allen Lane, 2018.
20 For an overview, see Allsen, Thomas T. *The Royal Hunt in Eurasian History.* Philadelphia, PA: University of Pennsylvania Press, 2006, especially 262. However, diplomatic hunts do not seem to have been studied systematically until now.
21 See Baker, Steve. *Picturing the Beast: Animals, Identity and Representation.* Manchester: Manchester University Press, 2001, especially 33–76.

and early modern periods, ruling dynasties in Europe typically chose lions or eagles as heraldic symbols for their dynasties, while Asian rulers associated themselves with tigers, elephants, or dragons – more or less 'real' animals that stood for specific virtues, such as majesty and force. In the nineteenth and especially in the twentieth century, related to the concept of national sovereignty, the association of countries with species became more regionalized. The kangaroo and koala thus served as representatives for Australia, and the giant panda – enormously popular worldwide since the World Wildlife Fund had chosen the animal for its logo – became the emblematic animal of China. In such discourses, the (supposed) natural behavior of these animals was also related to the behavior or goals of the states they represented. The largely vegetarian giant panda, for instance, was propagated as a symbol of China's supposedly peaceful and self-sufficient politics on the international stage. At the same time that the panda was chosen as a motif for the official gold and silver coins (Fig. 2), the Chinese government also began to intensify its endeavors to protect this endangered species by creating national parks and intensifying its panda breeding programs – with remarkable success: in 2016, the giant panda was downgraded from 'endangered' to 'vulnerable' on the global list of species at risk of extinction.[22] Thus, the identification of states with certain animals could have a direct impact on the development of animal populations, as these emblematic species were not only likely to be chosen as diplomatic gifts, but also to be put on the agenda of wildlife preservation by those states.

Animals in International Treatises

As resources or threats, as mobile sources of irritation or creatures deserving protection, animals were constant issues of diplomatic negotiations in the past. The documents resulting from such negotiations – agreements or formal treatises between rulers or states – usually represent an anthropocentric perspective depicting animals as pure objects, not as agents of diplomacy. However, they were often related to practices of interspecies interaction, and at least a part of the negotiations was caused by the specific behavior of certain animals or animal populations – especially by their mobility across political borders that became more and more strictly regulated

---

[22] This downgrade was also presented as a major success by the World Wildlife Foundation who had cooperated with the Chinese government for many decades; see URL: www.worldwildlife.org/stories/giant-panda-no-longer-endangered, published 4 September 2016 (March 28, 2019) and the associated information provided by the WWF. However, the announcement of the International Union for *Conservation* of Nature (IUCN) was largely criticized as coming too soon by Western media. On the history of China's environmental diplomacy, see, for example, McBeath, Jerry and Bo Wang. China's Environmental Diplomacy. *American Journal of Chinese Studies* 15, no. 1 (April 2008): 1–16, 15 on the conservation of pandas.

Figure 2: From 1983 onwards, the People's Republic of China coined silver and gold coins with the Giant panda on the reverse side and the Temple of Heaven on the front side. The number and positions of the pandas represented on the coin vary every year, but they are generally shown as peaceful animals and associated with their favored meal, bamboos. Many other states also regularly use 'typical' wild animals as motifs for their gold and silver coins, from kangaroos, koalas, and kookaburras (Australia), to sea turtles (New Zealand), springboks (South Africa), or bald eagles (United States of America). Being distributed worldwide, these coins and the animals represented on them take part in the states' image policy and public diplomacy. Official 10 Yuan Silver Coin of the People's Republic of China (back side), 1985. Courtesy of Silber.de (www.silber.de) / © Silber.de, Adeos Media GmbH, Laichingen.

during the early modern and modern periods and thus demanded specific international regulations.

As in other cases, the animals' place in international treaties varies significantly among different species. Not only regarding the diplomatic gift exchange, but also regarding treaties between rulers or states, one may speak of a "horse diplomacy" that took shape very early in certain areas. For instance, in the eleventh and twelfth centuries, special treaties were concluded between Song China and Tibet to guarantee the regular export of horses from Tibet to China.[23] Also in pre-modern Europe, the supply of horses was a regular part of alliances or reparations. The medieval ruler

---

23 Sen, Tansen. *Buddhism, Diplomacy, and Trade. The The Realignment of Sino-Indian Relations, 600–1400.* Honolulu, HI: University of Hawaii Press, 2003, 218.

Otto I (r. 936/62–973), for instance, repeatedly enforced other rulers to deliver him certain amounts of horses.[24] The 1709 alliance between the kings of Prussia, Denmark, and Poland stipulated that in the case that one of the partners was attacked they should assist each other with 6,000 men, 1,500 of which should come on horse.[25] Finally, horses were also mentioned in the first international treatises of Humanitarian Law – but rather as objects regarding property rights of soldiers than as subjects of their own [→History of War].[26]

It seems reasonable to assume that not only the number of animals, but also the variation of species evoked in treatises increased in accordance with the territorialization of rulership and the formalization of international cooperation. Already in the early modern period, fishing rights and other animal-related activities were mentioned in negotiated agreements between rulers or states on behalf of the interest of their subjects or semi-independent trading companies. For instance, the treaty of Utrecht (1713) that ended the Spanish War of Succession also included regulations of the fisheries in Northern America.[27] In the same way, the treaties of Paris (1783) and London (1794) that defined the borders between the United States and British Canada included special articles that conceded transboundary mobility to Native Americans and their animal goods (especially beavers) in order to secure the flourishing trade of the Montreal fur companies.[28] Furthermore, authorities of different states also increasingly cooperated among borders due to the uncontrolled mobility of certain animal and bacteria species or viruses. The rinderpest, for instance, had already provoked the coordination of simple protective measures in the early modern period and led to complex international campaigns in the twentieth and early twenty-first centuries.[29] Regarding "invasive" species, the Convention on Biological Diver-

---

[24] In the case of Eberhard of Franconia, who had refused to render homage to Otto in 937, he additionally ordered Eberhard's auxiliaries to bear dead dogs to the king's city of Magdeburg – a dishonouring penalty that reinforced Eberhard's opposition. Cf. Weinfurter, Stefan. Ein räudiger Hund auf den Schultern: Das Ritual des Hundetragens im Mittelalter. In *Die Welt der Rituale. Von der Antike*, Claus Ambos, Stephan Hotz, Gerald Schwedler, Stefan Weinfurter (eds.), 213–219. Darmstadt: Wbg Academic in Wissenschaftliche Buchgesellschaft, 2005.
[25] See Alliance of Cölln an der Spree, July 15, 1709, art. 4, online edition at URL: www.ieg-friedensvertraege.de (January 21, 2018) and numerous other examples in the same database.
[26] Nowrot, Karsten. Animals at War: The Status of Animal Soldiers Under International Humanitarian Law. *Historical Social Research* 40, no. 4 (December 2015): 128–150, 136. Art. 4 of the annex of the Hague convention of 1907 stipulates "that prisoners of war have the right to keep all their personal belongings, 'except arms, horses, and military papers'."
[27] See Dorsey, Crossing Boundaries, 691–692. Mentioning regulations on fisheries in Northern America in the Treaties of Utrecht (1713) and Paris (1763, 1783).
[28] See Carlos, Ann and Frank Lewis. *Commerce by a Frozen Sea. Native Americans and the European Fur Trade*. Philadelphia, PA: Pennsylvania University Press, 2010.
[29] Knab, Cornelia. Infectious Rats and Dangerous Cows: Transnational Perspectives on Animal Diseases in the First Half of the Twentieth Century. *Contemporary European History* 20, no. 3 (August 2011): 281–306; McVety, Amanda Kay. *The Rinderpest Campaigns: A Virus, its Vaccines, and Global Development in the Twentieth Century*. Cambridge: Cambridge University Press, 2018.

sity signed on the United Nations Earth Summit in 1992 requested the member nations to "prevent the introduction of, control or eradicate those alien species which threaten ecosystems, habitats, or species."[30]

In the twentieth century, the protection of animals increasingly became an issue of international negotiations of its own. Although limited in their effect, these negotiations and the resulting treaties should be seen as a new historical phenomenon because, arguably for the first time in history, governmental and nongovernmental actors claimed to negotiate not only about, but also on behalf of animals.[31] In the second half of the century, NGOs such as the World Wildlife Foundation (founded in 1961), the International Fund for Animal Welfare (founded in 1969), or Greenpeace (founded in 1971) were eager to put endangered species on the international political agenda. In contrast to earlier attempts to protect fish populations from over-fishing or endangered mammals in Africa from expanding agriculture (in order to secure game animals for colonialist hunting activities), these organizations contributed to a process during which the preservation of wild species became increasingly recognized as an ecological and political goal in itself.[32]

The Convention on International Trade in Endangered Species of Wild Fauna and Flora (CITES), first signed in Washington, DC in 1973, was gradually ratified by almost all states on the globe. Its implementation is overseen by an international institution with a bureau in Geneva and regular meetings between delegates of the member states. This was certainly a milestone in the history of animal protection, even if its implementation lies in the hands of sovereign states and therefore remains a considerable challenge.[33] Simultaneously, the protection of 'domesticated' animals was also increasingly regulated on an international level, especially in Europe. In 1971, the European Convention for the Protection of Animals During International Transport came into effect (and was extended in 1989 and 2006), followed by conventions regarding the protection on animals kept for farming purposes (1978), for slaughter (1982), for experimental and other scientific purposes (1991) and for animal pets (1992). Hence, new debates about animal rights and the globalization of politics contributed to an increasing regulation of human-animal relations through international conventions.

---

[30] Baskin, Yvonne. *A Plague if Rats and Rubbervines. The Growing Threat of Species Invasion.* Washington, DC: Island Press, 2002, 7 (quote of article 8 h of the convention).
[31] On the rise and limits of these "animal advocacy" politics from a philosophical point of view, see Wissenburg, Marcel and David Schlosberg (eds.). *Animal Politics and Political Animals.* Basingstoke: Palgrave Macmillan, 2014.
[32] See Cioc, Mark. *The Game of Conservation. International Treaties to Protect the World's Migratory Animals.* Athens, OH: Ohio University Press, 2009.
[33] See Reeve, Rosalind. *Policing International Trade in Endangered Species. The CITES Treaty and Compliance.* London: Earthscan, 2002; Liljeblad, Jonathan. *The Convention on International Trade of Endangered Species. Local Authority and International Policy.* New Orleans, LA: Quid Pro Books, 2014; and on the influence of NGOs on diplomacy, Winter Roeder, Larry. *Diplomacy, Funding and Animal Welfare.* Berlin and Heidelberg: Springer, 2011.

3 Methods and Approaches

Social and Cultural History of Diplomacy

Which methods can help us to better understand and contextualize the rich empirical evidence on the roles of animals in diplomatic history? As already stated above, the classical state-centered approach to the history of international relations has been supplemented – and challenged – by new perspectives on the history of diplomacy inspired by social and cultural history in recent years. New aspects such as the different social roles of diplomats, their engagement as transcultural mediators, the materiality and mediality of diplomacy, and the importance of gender in structuring political communication are the focus of new, innovative studies, many of which center on the period from the fifteenth century to the early twentieth, the formative phase of classical diplomacy.[34] This extension in subjects and methods facilitated the consideration of animals as part of diplomatic history. The increasing interest in gift exchanges as a constitutive element of early modern diplomacy resulted in several studies on animal gifting, by now the best studied aspect of animals' roles in diplomatic history. However, most of these studies still tend to see animals as a part of material culture rather than as living beings with their own agencies. At the same time, the existing studies interested in the lives and representations of exotic animals that came to the courts often tend to neglect diplomatic contexts and sources.

The most promising way to bring animals into the new social and cultural history of diplomacy seems to be a combination of methods established in the respective field with concepts developed in animal studies. A broad media-centered approach to foreign relations that reflects not only the mediality of written sources, but also of material artefacts such as gifts and practices of bodily interaction, can be extended to nonhuman living beings if it considers the specificities of the involved animals' agency or interagency [→Cultural History].[35] In cases such as the gifted elephants mentioned above, this perspective would include the close reconstruction of their former lives and conditions of food and housing, but also their close relation-

---

[34] See, among others, Windler, Christian. Diplomatic History as a Field for Cultural Analysis. Muslim-Christian Relations in Tunis, 1700–1840. *The Historical Journal* 44, no. 1 (March 2001): 79–106; Watkins, John (ed.). Toward a New Diplomatic History. Special issue of *The Journal of Medieval and Early Modern Studies* 38, no. 1 (2008); Mösslang, Markus and Torsten Riotte (eds.). *The Diplomats' World. The Cultural History of Diplomacy, 1815–1914*. Oxford: Oxford University Press, 2008; Rudolph, Harriet and Gregor Metzig (eds.). *Material Culture in Modern Diplomacy from the 15th to the 20th Century*. Berlin and Boston, MA: De Gruyter Oldenbourg, 2016; Hoeres, Peter and Anuschka Tischer (eds.). *Medien der Außenbeziehungen von der Antike bis zur Gegenwart*. Cologne: Böhlau, 2017; Hennings, Jan and Tracey A. Sowerby (eds.). *Practices of Diplomacy in the Early Modern World c. 1410–1800*. London and New York, NY: Routlegde 2017.
[35] See reflections in Weber, Lebende Geschenke.

ships to human mahouts who accompanied them through all their journeys without being mentioned in most of the contemporary sources.

Environmental History of Diplomacy

Another approach that addresses the animals' role in diplomatic history has been developed by environmental historians who traced the origins and developments of modern environmental diplomacy.[36] In their introduction to a special issue of the journal *Diplomatic History*, published in 2008, Kurkpatrick Dorsey and Mark Lytle state that most international competitions about resources in the twentieth century can be analyzed through the lens of environmental history. Besides tracing the growing importance of conservational issues in global diplomacy, historians should also pay attention to the unintentional impacts of bilateral or multilateral agreements on environments.[37] In the last years, a number of case studies that situated themselves between the two fields of diplomatic history and environmental history have shown how diplomats, scientists, commercial actors, and non-government organizations interacted on issues of environment including animals [→Environmental History].[38]

The approach of environmental diplomacy allows us to situate debates on salmons, whales, or polar bears in a broader context of changing attitudes of human societies towards "nature" and its inhabitants, and the impacts of these changing attitudes on political decision-making. Dorsey's book on *Whales and Nations*, for instance, traces the shift from a primarily resource-oriented approach in the whaling debates of the first half of the twentieth century, which led to the creation of the International Whaling Commission (IWC) in 1946, to a more conservationist tendency in the second half of the century, which culminated in acts that entirely prohibited whaling in the 1980s.[39] However, the environmental diplomacy approach also has its limitations. It focuses on wild animals rather than on animals in general, and attends less to the question of animal agency than animal historians would do. Furthermore, it is questionable if one could speak of any environmental diplomacy before the twentieth century, while animals, as we have seen, played important roles in

---

[36] For environmental diplomacy, see, for example, Susskind, Lawrence E. and Saleem H. Ali. *Environmental Diplomacy. Negotiating More Effective Global Agreements*. Oxford: Oxford University Press, 2015.
[37] Dorsey, Kurkpatrick and Mark Lytle. Introduction. *Diplomatic History* 32, no. 4 (Forum: New Directions in Diplomatic and Environmental History) (August 2008): 517–518.
[38] See, for example, Wadewitz, Lissa K. *The Nature of Borders: Salmon, Boundaries, and Bandits on the Salish Sea*. Seattle, WA: University of Washington Press, 2012; Dorsey, Kurkpatrick. *Whales & Nations: Environmental Diplomacy on the High Seas*. Seattle, WA: University of Washington Press, 2013. For an overview on recent research, see Dorsey, Crossing Boundaries; Bsumek, Erika Marie, David Kinkela and Mark Atwood Lawrence (eds.). *Nation-States and the Global Environment. New Approaches to International Environmental History*. Oxford: Oxford University Press, 2013.
[39] Dorsey, Whales & Nations.

diplomatic encounters long before. Thus, environmental history certainly adds important elements to our picture of animals' role in diplomacy in modern and contemporary history, but a history of animals in diplomacy is not reducible to the history of environmental diplomacy.

Political Ecology and Posthuman International Relations

The recent debates about enlarging the history of diplomacy coincide with attempts in other disciplines to reframe their vision of politics and international relations when considering environmental issues and the role of animals. One of them, political ecology, an approach originating in discussions among geographers and political scientists, has developed into an interdisciplinary field of academic research in the last decades.[40] Even though the approach rarely seems to have been applied by diplomatic historians until now, works such as Bruno Latour's *Politics of Nature*, originally published in 1999, offer concepts that may help us to integrate animals more directly into the analysis.[41] In his book, Latour aims to set the ground for a renewed political ecology by integrating nonhuman actors (animals, things) as stakeholders, and not as passive objects, into processes of political decision-making. Although they are represented by human 'spokespersons' which also follow their own interests, they affect the organization of the 'collective' through chains of associations. Latour's critique of ontological differences and his insistence on the role of micro-interactions between heterogeneous elements help to raise new historical questions: Which chains of associations link individual whales and their observed behavior with the conclusion of international conventions against whaling, for instance? How did the negotiations result in the creation of new – and often highly artificial – categories that lumped together different species – with substantial effects on their status and lives? And to what extent did not only scientists and activists, but also state leaders and diplomats represent themselves as spokespersons for animals?

The methodological discussions instigated by actor-network theory and posthumanism are also felt within the field of International Relations. Scholars criticize the anthropocentric orientation of the discipline and make a point of establishing a new field of 'posthuman international relations' that opens up the discipline for animals and other nonhumans.[42] Inspired by the works of Latour, Donna Haraway, and Isa-

---

40 See, for example, Perreault, Tom, Gaving Bridge and James McCarthy (eds.). *The Routledge Handbook of Political Ecology*. London and New York, NY: Routledge, 2015; Briant, Raymond L. (ed.). *The International Handbook of Political Ecology*. Cheltenham and Northampton, MA: Edward Elgar, 2015.
41 Latour, Bruno. *The Politics of Nature: How to bring Sciences into Democracy*. Cambridge, MA: Harvard University Press, 2004.
42 Cudworth, Erika and Stephen Hobden. *Posthuman International Relations: Complexity, Ecologism and Global Politics*. London: Zed Books, 2011; Cudworth, Erika and Stephen Hobden (eds.). *Posthuman Dialogues in International Relations*. London and New York, NY: Routledge, 2018; Eroukhmanoff,

belle Stengers, they put forward a new concept of diplomacy and the diplomat and argue that the practices of negotiation and mediation associated with diplomacy do not necessarily require "goodwill, togetherness, the sharing of a common language, or an intersubjective understanding."[43] Instead, they propose a model that deals flexibly with a plurality of heterogeneous agents, interests, and representations. If and to what extent this posthuman vision of diplomacy will be a useful analytical tool to study diplomatic encounters of the past remains an open question. At any rate, the discussion calls for an intensified dialogue between historians and scholars of other disciplines who are interested in the study of the animals' place in diplomacy.

4 Implication(s) of the Animal Turn

Despite these new perspectives that are being developed, the *animal turn* has not yet fully arrived in the history of diplomacy and international relations. In a recent handbook on the field, there is still no chapter devoted to animals, nor do they appear in the text.[44] To which extent the integration of nonhuman animals will transform our general understanding of diplomatic history thus remains to be seen. In any case, the perspective of historical human-animal studies would surely contribute to a broader, more complex, and more vivid picture of political relations across borders in past times. In order to get there, more historical research is needed both on the level of case studies and on the level of generalizations. While the relatively numerous but rather unrelated studies on the roles of animals as diplomatic gifts would profit from a comparative analysis based on the methods of human-animal studies, other topics such as the role of animals in diplomatic ceremonial or negotiations are still to be discovered. The inclusion of animals and 'their humans' does not merely promise to shed light on new aspects to diplomatic history – rather, it has the potential to change our very understanding of the subject. By dint of a combination of various sources, a history of diplomacy that consistently includes animals into its analysis would leave behind for good a narrow view that only concentrates on states and their representatives. It would present past diplomacies as an outcome of complex interactions between men and women of various ranks, depending on historically variable types of media, resources, temporal and spatial limitations, and relations with other animate beings.

Clara and Matt Harker (eds.). *Reflections on the Posthuman in International Relations: The Anthropocene, Security and Ecology*. E-International Relations Publishing, 2017. URL: www.e-ir.info/wp-content/uploads/2017/09/Reflections-on-the-Posthuman-in-IR-E-IR.pdf (December 13, 2018).
43 Stengers, Isabelle. *Cosmopolitics, vol. I: Posthumanities*. Minneapolis, MN: University of Minnesota Press, 2010, 29; cf. Cornago, Noé. Diplomatic Knowledge. In *The SAGE Handbook of Diplomacy*, Costas M. Constantinou, Pauline Kerr, Paul Sharp (eds.), 133–146. Los Angeles, CA: SAGE Publications, 142.
44 Gofas, Andreas, Inanna Hamati-Ataya and Nicholas Onuf (eds.). *The SAGE Handbook of the History, Philosophy and Sociology of International Relations*. Los Angeles, CA: SAGE Publications, 2018.

Just as much as the *animal turn* is an untapped potential for the history of diplomacy, historical human-animal studies may profit from taking diplomatic history more seriously. For instance, the developments in the international legislation and the practice of global animal protection, having crucial impacts on the categorization and evolution of animal populations, cannot be understood without considering diplomacy. Furthermore, diplomacy was often a vital factor in the history of human-animal relations of more distant times, especially regarding the mobility of animals. Many species traveled to distant territories for the first time as diplomatic gifts, and the intensification of diplomatic and commercial relations on both a European and a global scale from the sixteenth century onwards led to a more regular international exchange of animals, with sometimes far-reaching effects on local ecosystems. Finally, the rich sources of diplomatic history – thousands of letters, diaries, memoirs, and other papers – have proven to offer valuable insights for various topics that were not directly related to diplomacy, including elite interactions at court, gender relations, or intercultural contacts. It is therefore a promising enterprise to read them from 'an animal point of view.'

Selected Bibliography

Bodson, Liliane (ed.). *Les animaux exotiques dans les relations internationales. Espèces, fonctions, significations*. Liège: Université de Liège, 1998.

Cushing, Nancy and Kevin Markwell. Platypus Diplomacy: Animal Gifts in International Relations. *Journal of Australian Studies* 33, no. 3 (August 2009): 255–271.

Dorsey, Kurkpatrick. Crossing Boundaries. The Environment in International Relations. In *The Oxford Handbook of Environmental History*, Andrew C. Isenberg (ed.), 688–715. New York, NY: Oxford University Press, 2014.

Dorsey, Kurkpatrick. *Whales & Nations. Environmental Diplomacy on the High Seas*. Seattle, WA: University of Washington Press, 2013.

Hartig, Falk. Panda Diplomacy: The Cutest Part of China's Public Diplomacy. *The Hague Journal of Diplomacy* 8, no. 1 (January 2013): 49–78.

Hengerer, Mark and Nadir Weber (eds.). *Animals and Courts: Europe, c. 1200–1800*. Berlin and Boston, MA: De Gruyter Oldenbourg, 2020.

Leira, Halvard and Iver B. Neumann. Beastly Diplomacy. *The Hague Journal of Diplomacy* 12, no. 4 (December 2016): 337–359.

Pieragnoli, Joan. Les animaux et la diplomatie française XVIe-XIXe siècles. *Revue d'histoire diplomatique* 127, no. 3 (2013): 213–222.

Weber, Nadir. Lebende Geschenke. Tiere als Medien der frühneuzeitlichen Diplomatie. In *Medien der Außenbeziehungen. Von der Antike bis zur Gegenwart*, Peter Hoeres, Anuschka Tischer (eds.), 160–180. Cologne: Böhlau, 2017.

Julia Hauser
Global History

1 Introduction and Overview

The field of global history emerged during the 1990s as a successor to other approaches to writing the history of the world such as universal and world history, and as a partial critique of them. Pre-enlightenment attempts at writing the history of the world strongly tied in with theology and were aimed at making sense of the history of creation.[1] Animals and plants were deemed subordinate to humans in the great chain of being, with humans considered the crowning glory of creation. Then again, not all humans were created equal from the perspective of universal history. Members of oral cultures outside Europe, supposedly representing the "childhood" of humankind were considered part of natural history rather than of its human counterpart.[2] From the late eighteenth century onwards, an increasing number of European scientists, therefore, considered members of these cultures as occupying a liminal zone between humans and animals.[3]

In the course of the Enlightenment, natural history began to split off from "human" history, with the latter taking an increasingly secular turn. New approaches to world history focused on mankind alone while excluding nature and those parts of humanity allegedly closer to it. In this kind of historiography that evolved along with nineteenth-century imperialism, the idea of progress came to be considered central to history, and animals and members of allegedly "primitive" cultures as being outside of it.[4] It was only with the development of "big history" since the 1970s that nature re-entered narrations of the history of the world.[5]

---

[1] Ferreyrolles, Gérard. On the History of Universal History. In *Universal History and the Making of the Global*, Hall Bjørmstad, Helge Jordheim, Anne Régent-Susini (eds.). New York, NY and London: Routledge, 2019, 12–23.
[2] Fabian, Johannes. *Time and the Other. How Anthropology Makes its Object*. New York, NY: Columbia University Press, 1983.
[3] Bourke, Joanna. *What it Means to be Human. Reflections from 1791 to the Present*. London: Virago, 2011.
[4] Conrad, Sebastian. *What is Global History*. Princeton, NJ: Princeton University Press, 2016, 27–28.
[5] McNeill, William and John Robert McNeill. *The Human Web: A Bird's-Eye View of World History*. New York, NY: Norton, 2003; Spier, Fred. *The Structure of Big History. From the Big Bang Until Today*. Amsterdam: Amsterdam University Press, 1996; Christian, David, Cynthia Stokes Brown and Craig Benjamin. *Big History: Between Nothing and Everything*. New York, NY: McGraw Hill Education, 2014. The animal nature of humans is rarely stressed in these works. For an exception, see Yuval Noah Harari, who, nonetheless, emphasizes the alleged superiority of humans. Harari, Yuval Noah. *Sapiens. A Brief History of Humankind*. London: Harvill Secker, 2014.

https://doi.org/10.1515/9783110536553-018

Present-day global history challenges the Eurocentrism underlying both universal and world history. Inspired by both the postmodernist critique of metanarratives and postcolonial studies, its aim is to critically re-examine established grand narratives.[6] Inter alia, global historians do so through a critique of the colonial archive, reading it against the grain, consulting non-Western sources or bringing in anthropological perspectives, thus distancing themselves, to some extent, from early postcolonial debates on whether the subaltern can speak.[7]

Re-centering existing narratives and adding new voices, therefore, are central aims of global history – an approach that echoes that of human-animal studies. Despite what its name suggests, global history is not necessarily a matter of scale: macrohistory is not the only mode in which it can be pursued.[8] It may also be concerned with global connectivity or processes beyond the nation.[9] At the very least, it is history written with a global consciousness and/or one that questions Eurocentric ways of writing history.[10] As a consequence, global history can even take on the form of microhistory, following the trajectories of individual goods or actors across the globe.[11] Yet while global history is openly critical of Eurocentrism, it has rarely questioned the centrality of humankind to history. Animals, consequently, have rarely figured as historical actors in global history to date.

---

6 Conrad, Sebastian and Shalini Randeria. Einleitung. Geteilte Geschichten – Europa in einer postkolonialen Welt. In *Jenseits des Eurozentrismus. Postkoloniale Perspektiven in den Geschichts- und Kulturwissenschaften*, Sebastian Conrad, Shalini Randeria (eds.), 32–73. Frankfurt/M.: Campus, 2003; Hunt, Lynn. *Writing History in the Global Era*. New York, NY: Norton, 2014.

7 Spivak, Gayatri Chakravorty. Can the Subaltern Speak. In *Marxism and the Interpretation of Culture*, Cary Nelson, Lawrence Grossberg (eds.). Chicago, IL: University of Chicago Press, 1988.

8 Conrad, Sebastian and Andreas Eckert. Globalgeschichte, Globalisierung, multiple Modernen: Zur Geschichtsschreibung der modernen Welt. In *Globalgeschichte. Theorien, Ansätze, Themen*, Sebastian Conrad, Andreas Eckert, Ulrike Freitag (eds.), 7–54. Frankfurt a. M.: Campus, 2007, 27.

9 Freitag, Ulrike and Achim von Oppen. Introduction: Translocality: An Approach to Connection and Transfer in Area Studies. In *Translocality. The Study of Globalising Processes from a Southern Perspective. Studies in Global Social History 4*, Ulrike Freitag, Achim von Oppen (eds.), 1–21. Leiden and Boston, MA: Brill, 2010.

10 Davis, Natalie Zemon. Global History, Many Stories. In *Eine Welt – Eine Geschichte? 43. Deutscher Historikertag in Aachen, 26. bis 29. September 2000*, Verband der Historiker und Historikerinnen Deutschlands e.V., May Kerner, Aachener Organisationsbüros (eds.), 373–380. Munich: Oldenbourg Wissenschaftsverlag, 2001.

11 Ghobrial, John-Paul A. The Secret Life of Elias of Babylon and the Uses of Global Microhistory. *Past & Present* 222, no. 1. (February 2014): 51–93; Medick, Hans. Turning Global? Micro-History in Extension. *Historische Anthropologie* 24, no. 2 (August 2016): 241–251; Bertrand, Romain and Guillaume Calafat. La microhistoire globale: affaire(s) à suivre. *Annales Histoire Sciences Sociales* 73, no. 1 (March 2019): 3–18; For practical examples of this kind of global microhistory, see Ghobrial, Secret Life; Sachsenmaier, Dominic. *Global Entanglements of a Man who Never Travelled. A Seventeenth-Century Chinese Christian and His Conflicted Worlds*. New York, NY: Columbia University Press, 2017; Davis, Natalie Zemon. *Trickster Travels. A Sixteenth-Century Muslim Between Worlds*. New York, NY: Hill and Wang, 2006.

Human-animal studies is a strongly interdisciplinary field that brings together scholars trained in the study of literature, history, philosophy, cultural studies, sociology, anthropology, and ethology. Just like global history, it is a fundamentally critical project. It came into being along with the animal rights movement from the late 1970s, thus criticizing the assumption of hierarchies between human and nonhuman animals.[12] Human-animal studies also drew inspiration from posthumanist tendencies in philosophy which questioned the categorical divide between human and other beings.[13] At the same time, scholars concerned with human-animal relations also incorporate feminist and postcolonial approaches, emphasizing and examining intersections between modes of oppression based on species, race, and gender.[14] Just like feminist, postcolonial, and, indeed, later on, scholars of global history, researchers in animal history aim at integrating previously ignored voices [→(Post)Colonial History; →Feminist Intersectionality Studies].[15] Even more so than global history, they are faced with the problem that these voices often are not readily accessible (or beyond human comprehension).

Global history and human-animal studies, therefore, converge in their attempt to question centrisms – global history challenging Eurocentrism, animal history criticizing anthropocentrism – and of bringing in previously ignored voices, with many scholars engaged in both fields considering their respective endeavors a political project. Yet despite this apparent common ground in terms of approaches, both fields have seen little overlap so far.[16]

This chapter, therefore, mainly considers studies focusing on animals that are of interest to global history, whereas only some of their authors actually consider themselves as belonging to this field. It first introduces topics and themes of existing research before reflecting on methodology and approaches. Finally, since the *animal*

---

[12] Roscher, Mieke. Human-Animal Studies. *Docupedia-Zeitgeschichte*, January 25, 2012. URL: docupedia.de/zg/Human-Animal_Studies?oldid=125461 (July 5, 2019).
[13] Posthumanist scholars, therefore, also criticize the concept of the humanities as too narrowly focused on the human. See, for instance, Anderson, Kay J. and Colin Perrin. Up from the Ape: Colonialism, Craniometry and the Emergence of Anatomical Humanism. In *Challenging (the) Humanities*, Tony Bennett (ed.), 1–17. Melbourne: Australian Scholarly Publications, 2013.
[14] Taylor, Nik and Richard Twine. Introduction. Locating the Critical in Critical Animal Studies. In *The Rise of Critical Animal Studies: From the Margins to the Centre*, Nik Taylor, Richard Twine. (eds.), 1–17. Abingdon: Routledge, 2014, 4.
[15] Skabelund, Aaron. Can the Subaltern Bark? Imperialism, Civilization, and Canine Cultures in Nineteenth-Century Japan. In *JAPANimals: History and Culture in Japan's Animal Life*, Gregory E. Pflugfelder, Brett E. Walker (eds.), 194–243. Ann Arbor, MI: Centre for Japanese Studies, University of Michigan, 2005; Krebber, André and Mieke Roscher. Introduction: Biographies, Animals, and Individuality. In *Animal Biography: Re-framing Animal Lives*, André Krebber, Mieke Roscher (eds.), 1–15. Cham: Palgrave Macmillan, 2018; Skabelund, Aaron. A Dog's Life: The Challenges and Possibilities of Animal Biographies. In *Animal Biography*, Krebber, Roscher (eds.), 83–102.
[16] This lack of overlap may already be observed in postcolonial studies. Armstrong, Philip. The Postcolonial Animal. *Society & Animals* 10, no. 4 (December 2002): 413–419; Chagani, Fayaz. Can the Postcolonial Animal Speak? *Society & Animals*, 24, no. 6 (2016): 619–637.

turn has impacted global history only marginally so far, it makes some suggestions for a fruitful cooperation of historical animal studies and global history.

2 Topics and Themes

A number of studies, written well before the emergence of global history, have analyzed the role of animals in processes of global integration in a *longue durée* perspective, focusing on how animals helped transform societies. Fernand Braudel, in his history of the Mediterranean in the age of Philipp II, emphasized the relevance of transhumance and inverse transhumance to economy, demography, and political change in the region. Even though animals inevitably had a role in this process, however, Braudel was far more interested in the movement of those humans that accompanied them.[17] Richard Bulliet looked at how the camel came to replace the wheeled chariot, thereby transforming societies in the Middle East more largely.[18] In a more recent book, Bulliet examines how the domestication of animals, and then the post-domestic era, shaped the behavior of humans towards animals across the world from the emergence of homo sapiens to present times.[19] In both cases, however, animals are not considered actors in their own right but rather resources for human use and consumption [→Post-Domestication: The Posthuman]. Alfred Crosby looked at the circulation of germs, plants and animals in the history of European expansionism. Settler colonialism, or so Crosby argued, would have been impossible without the animals brought from Europe, with some of them breading and spreading far beyond human control, sometimes even posing a threat to colonizers themselves. Crosby thus did not just accord importance, but also some degree of agency to animals. His narrative, however, was strictly diffusionist: according to him, only European animals spread successfully in other parts of the world.[20]

The Zoo's Global Reach

Most studies on animals in global history, however, focus on the nineteenth and twentieth centuries when, at the height of imperialism, nature was exploited more systematically than ever, and new practices of conservation were developed. One im-

---

[17] Braudel, Fernand. *The Mediterranean and the Mediterranean World in the Age of Philip II*. Glasgow: Fontana Press, 1986, 82–105.
[18] Bulliet, Richard W. *The Camel and the Wheel*. Cambridge, MA: Harvard University Press, 1975.
[19] Bulliet, Richard W. *Hunters, Herders, and Hamburgers: The Past and Future of Human-Animal Relationships*. New York, NY: Columbia University Press, 2005.
[20] Bulliet, *Hunters, Herders, and Hamburgers*, 171–194. Crosby had already put forward this argument at greater length in an earlier work: Crosby, Alfred W. *The Columbian Exchange*. Westport, CT: Greenwood, 1972, 64–121.

portant topic in this context is the history of zoological gardens which emerged as public spaces around the middle of the nineteenth century [→History of the Zoo]. In bringing exotic animals to metropoles around the world, zoological gardens were a mirror of globalization and imperialism. They were also intended to demonstrate human rule over nature and the progress of science.

Scholars concerned with the history of zoos stress the unevenness of the exchange in which non-European animals were transferred to zoos around the world while Western animals were hardly sought after.[21] To some extent, this argument may result from the rather Eurocentric focus of existing research on zoo history. Studies focus almost exclusively on Asian and African animals in zoological gardens in Europe and North America.[22] However, the dynamics of the exchange also reflect the asymmetry of power central to the colonial encounter.

Nigel Rothfels takes a critical look at the history of Hagenbeck's zoo in Hamburg, arguing that Hagenbeck, whose zoo was the first to dispense with barred small-scale enclosures, was not just merely a friend of animals, but first of all someone who profited from selling them, a business that involved a great degree of violence towards both humans and animals.[23] Takashi Ito examines how London Zoo helped popularize and disseminate scientific knowledge, thus setting the zoo apart from earlier types of venue for spectacles that involved animals and the demonstration of human mastery over 'nature' in the metropole. His analysis, however, goes beyond the colony-metropole trajectory by stressing that animals were often acquired through transimperial networks.[24] Daniel Bender, in his work on zoos, intends to look at how the exhibition of supposedly wild animals in American zoos made Africa appear both close and distant. While his book does not delve deeply into this aspect, it does shed light on the business behind the acquisition of animals, emphasizing the degree to which traders were dependent on African helpers, their knowledge, and goodwill. Bender does not, however, question hierarchies between human and animal actors, and hardly examines the agency of the latter.[25] While none of these studies explicitly inscribe themselves into global history, at least Rothfels' and Bender's study may be subsumed under this label since they examine how global connections shape the local and indeed ideas of the nation as epitomized by the zoo.

---

21 However, Alan Mikhail shows that Western animals did figure in some zoological gardens outside Europe. Mikhail, Alan. *The Animal in Ottoman Egypt*. Oxford: Oxford University Press, 2014, 450, 452–455.
22 The only exceptions to date are Mikhail, *The Animal in Ottoman Egypt* and Miller, Ian Jared. *The Nature of the Beasts. Empire and Exhibition at the Tokyo Imperial Zoo*. Berkeley, CA, Los Angeles, CA and London: University of California Press, 2013.
23 Rothfels, Nigel. *Savages and Beasts. The Birth of the Modern Zoo*. Baltimore, MD: Johns Hopkins University Press, 2002.
24 Ito, Takashi. *London Zoo and the Victorians, 1828–1859*. Suffolk: Boydell and Brewer, 2014.
25 Bender, Daniel. *The Animal Game: Searching for Wilderness at the American Zoo*. Cambridge, MA and London: Harvard University Press, 2016.

Conservation as a Global Project

While the transport of animals from other parts of the world was central to the concept of the zoo, national parks were conservation projects created in situ. Since the inauguration of Yellowstone in 1872, the concept of the national park came to be gradually appropriated in all parts of the world. In reaction to the rapid exploitation and destruction of nature in a globalizing capitalist economy, national parks aimed at conserving 'pristine' nature. As the term implies, they were tied closely to the project of nation building, aiming at protecting and replenishing flora and fauna deemed nationally specific.

Bernhard Gissibl, Sabine Höhler and Patrick Kupper stress that "[n]ational parks are more adequately understood as 'transnational parks': globalized localities that owe their establishment to transnational processes of learning, pressure, support and exchange."[26] They consider national parks as just one instance of how processes of globalization do not obliterate the national but rather are often played out on the scale of the nation state.[27] In his own contribution, Gissibl shows how a colonial focus on megafauna in Africa (where concepts of conservation partly appropriated approaches taken from US national parks) helped develop concepts of conservation in Germany to which local megafauna were central.[28]

Conservation also became a topic in international diplomacy after World War I [→Diplomatic History;→Environmental History]. Anna-Katharina Wöbse examines the concept of global conservation (*Weltnaturschutz*) as negotiated in the League of Nations, looking, inter alia, at the preservation of endangered fauna. According to Wöbse, animals in these debates were often framed as "universal creatures, standing in for all helpless beings."[29] Even if the League of Nations eventually failed in its attempts to promote animal welfare, she considers its efforts a precursor to modern environmental diplomacy.[30] Kurkpatrick Dorsey studies international efforts to limit whaling from World War I to the present, arguing that debates around the issue were determined by the aspects of sustainability, sovereignty, and science.[31]

---

[26] Gissibl, Bernhard, Sabine Höhler and Patrick Kupper. Introduction. Towards a Global History of National Parks. In *Civilizing Nature. National Parks in Global Historical Perspective*, Bernhard Gissibl, Sabine Höhler, Patrick Kupper (eds.), 1–29. New York, NY and Oxford: Berghahn, 2012, 2.
[27] Gissibl, Höhler and Kupper, Introduction, 7.
[28] Gissibl, Bernhard. A Bavarian Serengeti: Space, Race and Time in the Entangled History of Nature Conservation in East Africa and Germany. In *Civilizing Nature*. Gissibl, Höhler, Kupper (eds.), 102–121.
[29] Wöbse, Anna-Katharina. *Weltnaturschutz. Umweltdiplomatie in Völkerbund und Vereinten Nationen 1920–1950*. Frankfurt a.M.: Campus, 2012, 26.
[30] Wöbse, *Weltnaturschutz*; see also Wöbse, Anna-Katharina. Globale Kreaturen. Tiere in der internationalen Politik des frühen 20. Jahrhunderts. In *Tierische Geschichte. Die Beziehung von Mensch und Tier in der Kultur der Moderne*, Dorothee Brantz, Christoph Mauch (eds.), 304–324. Paderborn: Schöningh, 2010.
[31] Dorsey, Kurkpatrick. *Whales and Nations. Environmental Diplomacy on the High Seas*. Seattle, WA and London: University of Washington Press, 2014.

Globalization, urbanization and the rise of living standards in the industrializing world brought with them even more immediate forms of threat than extinction: a rising demand for meat and other animal products in Europe, North America, and the colonies. Felix Schürmann looks at how the global business of whaling shaped African coastal societies,[32] whereas Chris Otter and Wilson J. Warren analyze the rising consumption of meat around the world during the nineteenth and twentieth centuries, and its consequences for human society.[33] Most of these studies do not focus on living animals. Instead, they look at animals as resources for humans.

Diseases, Wars and Microhistory

From the mid-nineteenth century onwards, rising demand for animal products entailed a steep increase in the transportation of animals. This, in turn, helped promote the global spread of epizootics. Increasing human mobility around the globe likewise had an impact on the spread of other illnesses in which animals served as intermediate hosts, and to an unprecedented need for developing strategies against such illnesses.

Andrea McVety examines the global attempts at containing rinderpest. A "largely imperial disease since the nineteenth century," rinderpest spread with the expansion of global trade. The development of a vaccine against the disease, McVety argues, may be seen as an "act of global consciousness".[34] While she shows that the vaccine was developed in the context of World War II, and deeply entangled with wartime hostilities including apprehensions of biological warfare, it would later be used by international organizations to ensure food security.[35]

The study of the spread of animal-transmitted diseases allows for a nuanced analysis of human-animal relations. In James Webb's global history of malaria from prehistory into the twentieth century, parasites and other animals play a central role. The main actors in his study are the parasites causing the disease. But as Webb shows by drawing on research from both the humanities and the sciences, the spread of malaria was complicated by a range of other factors including human-animal relationships. Agricultural societies in marshlands were exposed to a higher risk of infection due to the prevalence of the host of the disease, the anopheles mosquito, in

---

32 Schürmann, Felix. *Der graue Unterstrom: Walfänger und Küstengesellschaften an den tiefen Stränden Afrikas, 1770–1920*. Frankfurt a.M.: Campus, 2017.
33 Otter, Chris. Planet of Meat: A Biological History. In *Challenging (the) Humanities*, Tony Bennett (ed.), 33–49. Melbourne: Australian Scholarly Publications, 2013; Warren, Wilson J. *Meat Makes People Powerful: A Global History of the Modern Era*. Iowa City, IA: University of Iowa Press, 2018.
34 McVety, Andrea Kay. *The Rinderpest Campaigns*. Cambridge: Cambridge University Press, 2018, 6.
35 McVety, *The Rinderpest Campaigns*.

these regions, whereas pastoral societies were less threatened by malaria infections since cattle acted as a natural barrier in the process of transmission.[36]

The most direct threat to animal, as well as to human life, were wars [→History of War]. While there is a growing strand of research on animals in wars, the topic has hardly been researched from a global history perspective. One exception is Gregg Bankoff's article on Asian fauna during the Cold War. Focusing on zones of war in Southeast Asia, he emphasizes that "animals and their habitats […] were very much the victims of superpower rivalries and the conflicts they generated."[37] Although he finds that there were both losers and winners of the war, particularly in demilitarized zones, he stresses that the overall consequences of the Cold War on animal life were disastrous, with some species being able to flee while others were nearly or fully extinguished.[38] Bankoff's article is innovative in that it considers animals not just as victims but also as subjects endowed with agency.

It is in relation to two further topics – the global history of species and the global biographies of individual animals – that animals appear, to a greater extent, as actors in their own right. Monographs in this field often verge on public history [→Public History]. Ulrich Raulff's *Farewell to the Horse* looks at the gradual separation of humans and horses during the nineteenth and twentieth centuries. Raulff's book certainly offers a history that moves animals into the foreground.[39] With only occasional glances at Asia, Africa, and Oceania, and largely focusing on France, Germany, Britain, and the United States, Raulff's book is not, strictly speaking, an example of global history, although it offers a starting point for it.[40] Philip Armstrong's engaging book about sheep shows how fundamental they, "the most routinely overwritten of animals," were to human existence globally, providing wool, parchment, and meat. Armstrong also stresses that characteristics associated with sheep varied greatly historically as well as culturally. He does not, however, just focus on human perspectives, but also integrates ethological research and reflects on how conditions particularly since the industrial revolution affected the lives of sheep.[41]

---

36 Webb, James L. R. *Humanity's Burden: A Global History of Malaria*. Cambridge: Cambridge University Press, 2009.
37 Bankoff, Greg. A Curtain of Silence. Asia's Fauna in the Cold War. In *Environmental Histories of the Cold War*, John Robert McNeil, Corinna Unger (eds.), 203–226. Cambridge: Cambridge University Press, 2010, 203.
38 Bankoff, A Curtain of Silence.
39 Raulff, Ulrich. *Farewell to the Horse: The Final Century of our Relationship*. London: Allen Lane, 2017.
40 See Osterhammel, Jürgen. Ein Ritt nach Pferdensien. *DIE ZEIT* 46 (November 2015): 61. In his own global history of the nineteenth century, Osterhammel digresses on the role of animals, but not to an extent as to make them leading actors. Osterhammel, Jürgen. *The Transformation of the World. A Global History of the Nineteenth Century*. Princeton, NJ: Princeton University Press, 2014, 213–214, 229–230, 385–388, 654.
41 Armstrong, Philip. *Sheep*. London: Reaktion Books, 2016.

Another genre in which global history and animal history overlap is the field of animal biography. Animal biographies have largely been written about seemingly extraordinary animals that lend themselves to individualization more easily. Silvio Bedini is concerned with the life of the Indian white elephant given to Pope Leo X by King Manoel II in order to win papal support for Portuguese maritime expansion. Bedini looks at the elephant's agency to a limited extent, particularly his resistance towards being an object of attraction, but above all at its role in diplomacy, at the papal court and its posthumous representation.[42] With greater attention to local as well as ethological knowledge about elephants but in a more literary manner, Radhika Subramaniam considers the life of the elephant Abul Abbas, who Harun al-Rashid gifted as a present to Charlemagne.[43] Glynis Ridley reconstructs the travels of an Indian rhinoceros through eighteenth-century Europe.[44] Eric Baratay likewise follows the traces of animals with global lives, yet mainly focuses on their sojourns in France.[45] Given their subject matter, these studies have a potential of contributing to global history. Ultimately, however, they fail to do so either because they are not written for an academic audience or because they do not actively reflect the global context.

As this brief overview reveals, then, animal and global history are still very much separate fields, either because animals are nearly invisible in global history, or because animal histories do not sufficiently take into account global contexts, entanglements and processes of globalization. Which methods and approaches could be employed to bring both fields closer to each other?

3 Methods and Approaches

As scholars like Dominic Sachsenmaier and Sebastian Conrad stress, there is not just one global history. Instead, and despite some degree of exchange, it has taken distinct trajectories in Asia, the United States and Europe.[46] This brief overview mainly focuses on global history in Europe and the US. Particularly, when it first evolved as an approach, historians suggested that global history should take a comparative perspective, an approach that had been developed well before the emergence of global

---

42 Bedini, Silvio. *The Pope's Elephant*. Lanham, MD: Rowman & Littlefileld, 1998.
43 Subramaniam, Radhika. The Elephant's I: Looking for Abu'l Abbas. In *Animal Biography*. Krebber, Roscher (eds.), 207–226.
44 Ridley, Glynis. *Clara's Grand Tour: Travels with a Rhinoceros in Eighteenth-Century Europe*. London: Atlantic Books, 2004.
45 Baratay, Eric. *Biographies animales. Des vies retrouvées*. Paris: Le Seuil, 2017.
46 Sachsenmaier, Dominic. *Global Perspectives on Global History: Theories and Approaches in a Connected World*. Cambridge, New York, NY and Melbourne: Cambridge University Press, 2011; Beckert, Sven and Dominic Sachsenmaier (eds.). *Global History, Globally: Research and Practice Around the World*. London: Bloomsbury Academic, 2018.

history.⁴⁷ Reminiscent of, and partly inspired by, S.N. Eisenstadt's concept of "multiple modernities," the comparative design allowed for a large-scale focus, yet also came with certain problems inherent to historical comparison as such.⁴⁸ Comparison relied on fixed entities, which were juxtaposed with the aim of finding either contrasts or similarities. The problem with comparison is that it is susceptible to essentialism, considering the respective units of investigation as homogeneous and entirely separate entities, thus ignoring diversity, contradictions, and conflicts of interest within and mutual influence between the parties concerned.⁴⁹

It was because of these epistemological shortcomings of the comparative method that the related concepts of *histoire croisée*, entangled history, on the one hand, and shared/divided history [*geteilte Geschichte*], on the other hand were developed, concepts that were not just influential in global history, but also in related fields like the history of colonialism.⁵⁰ Rather than comparing allegedly distinct entities, these concepts focus on analyzing relations between them. In contrast to the older concept of cultural transfers, however, they did not assume that these relations were unidirectional.⁵¹ Instead, they acknowledged their highly dynamic and mutual character. An important inspiration for entangled history was Ann-Laura Stoler's and Frederick Cooper's earlier suggestion that metropole and colony should be investigated in terms of a "single analytic field" in order to expose the "tensions of empire," that is, the contradictions, breaks, and discontinuities at the heart of colonialism.⁵²

Proponents of entangled history have argued since that these dynamics caused the parties involved to become entangled with each other to such an extent as to render the terms metropole and colony, and the hierarchy implied by them, nearly obsolete.⁵³ Thus global history and related approaches no longer rely on a static con-

---

47 Osterhammel, Jürgen. Sozialgeschichte im Zivilisationsvergleich. Zu künftigen Möglichkeiten komparativer Geschichtswissenschaft. *Geschichte und Gesellschaft* 22 (April-June 1996): 143–164.
48 Eisenstadt, Shmuel Noah. Multiple Modernities. *Daedalus* 129, no. 1 (Winter 2000): 1–30.
49 Werner, Michael and Bénédicte Zimmermann. Beyond Comparison: Histoire Croisée and the Challenge of Reflexivity. *History and Theory* 45, no. 1 (February 2006): 30–50.
50 Freitag, Ulrike and Achim von Oppen. *Translokalität als ein Zugang zur Geschichte globaler Verflechtungen*. (ZMOProgrammatic Texts, 2). Berlin: Zentrum Moderner Orient (2005). URL: nbn-resolving.org/urn:nbn:de:0168-ssoar-427594; see also Conrad and Randeria, Geteilte Geschichten.
51 Espagne, Michel and Michael Werner. Deutsch-französischer Kulturtransfer im 18. und 19. Jahrhundert. Zu einem neuen interdisziplinären Forschungsprogramm des C.N.R.S. *Francia* 13 (1985): 502–510.
52 Stoler, Ann Laura and Frederick Cooper. Between Metropole and Colony. Rethinking a Research Agenda. In *Tensions of Empire. Colonial Cultures in a Bourgeois World*, Ann Laura Stoler, Frederick Cooper (eds.), 1–58. Berkeley, CA, Los Angeles, CA and London: University of California Press 1997, 4.
53 Cf. Fischer-Tiné, Harald. *Pidgin-Knowledge: Wissen und Kolonialismus* (Perspektiven der Wissensgeschichte). Zürich: Diaphanes, 2013.

cept of culture, but instead stress its fluidity.[54] Moreover, global history has moved away from focusing on exchanges between one metropole and its periphery, underlining instead the often transcolonial dynamics of appropriation.[55] The concept of shared/divided history, by contrast, stresses that entanglement did not translate into congruence. While colonizers and colonized might share part of their pasts, this very past could also divide them because of the hierarchies and violence that characterized it, with quite different cultures of remembrance as a result.[56] From that perspective, the concept of shared/divided history seems very much applicable to the equally violent global history of human-animal relations – except that the aspect of remembrance, as far as animals are concerned, is likely to remain elusive to historians.

Global history is both an object of research and a perspective, but not yet a closed field. As a consequence, there is disagreement as to whether or not it ought to be limited to a certain period. While some scholars argue that it should be confined to the period from the nineteenth century to the present, considering this the epoch when globalization took place, other historians point to the importance of global connections in earlier periods.[57] In terms of space, it overlaps with colonial history, translocal and transnational history. Global history does not necessarily translate into analyzing the history of the world as a whole. Indeed, such an approach would render difficult one of the central demands of global history, the challenging of received master narratives. While there are indeed studies of global history from a macrohistoric perspective, global history can just as well look at global entanglements on a smaller scale.[58] Inspired by the work of anthropologists Sidney Mintz and Arjun Appadurai, it can trace the circulation of goods or ideas.[59] It could also ask how the national is constituted through global entanglements.[60] Finally, a growing

---

[54] For an introduction to the concept of cultural translation, see Bachmann-Medick, Doris. *Cultural Turns. New Orientations in the Study of Culture*. Berlin and Boston, MA: De Gruyter Oldenbourg, 2016, 181–185.
[55] Cf. Beckert, Sven. *Empire of Cotton: A Global History*. New York, NY: Knopf, 2004.
[56] Conrad and Randeria, Geteilte Geschichten, 17–22.
[57] Cf. Mazlish, Bruce. *The New Global History*. New York, NY: Routledge, 2006. Sachsenmaier, *Global Entanglements*.
[58] Cf. Bayly, Christopher A. *The Birth of the Modern World, 1780–1914. Global Connections and Comparisons*. Malden, MA: Blackwell, 2005; Osterhammel, *The Transformation of the World*.
[59] Mintz, Sidney. *Sweetness and Power. The Place of Sugar in Modern History*. New York, NY: Sifton, 1985; Appadurai, Arjun. *The Social Life of Things. Commodities in Cultural Perspective*. Cambridge: Cambridge University Press, 1986. For a notable example from global history inspired by such approaches, see Beckert, *Empire of Cotton*.
[60] See, for instance, Bender, Thomas (ed.). *Rethinking American History in a Global Age*. Berkeley, CA: University of California Press, 2002.

number of studies combines global history and microhistory, including biographical approaches.[61]

Sources are the main challenge in global history. Often historians are confronted with a relative scarcity of sources beyond the colonial archive where non-European actors may emerge only between the lines. Bringing their voices in, however, is central to any kind of global history that wishes to challenge Eurocentrism. Faced with an actual lack of sources, researchers may be forced to read colonial sources against the grain or analyze processes of silencing in the colonial archive.[62] Global history also confronts historians with a far wider range of language requirements than national history. If scholars resort to primary sources in English or their own first language only, they run the risk of reproducing Eurocentric perspectives.[63] To a certain extent, they may circumvent this risk by entering into dialogue with neighboring disciplines such as anthropology or area studies. But working across the disciplines is central to global history even if researchers are engaged in writing syntheses based on primary sources only.

Even if global history, therefore, has largely neglected animals so far, scholars with a background in global history are well familiar with the necessity of bringing new voices into history, and the problems to be solved in accessing and, literally and metaphorically, translating them.

4 Implication(s) of the Animal Turn

Both global history and animal history, then, are fields very much characterized by working across the disciplines. They also share the aim of bringing in new, and previously underrepresented, actors and challenging established narratives. Consequently, both would profit from joining forces. Some steps in this direction have already been taken. Scholars from global history have turned to the globalization of practices of conservation, exploitation, to researching the history of certain species and the lives of individual animals. But a more serious cooperation of global history and human-animal studies requires some methodological changes.

Scholars in global history would gain new insights by accepting animals as historical actors. They can learn much from debates on whether or not animals are endowed with agency; indeed, these debates will strike a familiar note with cultural

---

[61] Cf. Green, Nile and James L. Gelvin (eds.). *Global Muslims in the Age of Steam and Print*. Berkeley, CA: University of California Press, 2014; Zimmerman, Andrew. *Alabama in Africa: Booker T. Washington, the German Empire, and the Globalization of the New South*. Princeton, NJ: Princeton University Press, 2010. For the discussion on microhistory and global history, see also references in fn. 10 and 11 of this article.
[62] Stoler, Ann Laura. *Along the Archival Grain: Epistemic Anxieties and Colonial Common Sense*. Princeton, NJ: Princeton University Press, 2009.
[63] Pernau, Margrit. *Transnationale Geschichte*. Göttingen: Vandenhoeck & Ruprecht, 2011.

historians specialized in the history of other socially marginalized actors [→Cultural History;→Social History].[64] In learning to approach animals as historical actors, they could cooperate not just with colleagues from humanities and social sciences, but also from the "hard" sciences.[65] Scholars from human-animal studies, on the other hand, should pay more attention to the historical dimension, and take into account cultural specificities as much as historians need to stop considering animals as just one homogeneous entity.

Topics worth researching include the question of how disciplines and practices like ethology and breeding were shaped by the hybrid knowledge resulting from cultural encounters. Appropriations and rejections of Darwin's theory of evolution around the globe might be another fruitful topic for further research.[66] While some scholars have claimed that contempt for animals, and a sense of superiority over them has been a universal phenomenon historically, this blanket statement warrants further research as well.[67] Along with this, the drawing of human-animal boundaries in cultural encounters is a topic that deserves more attention. Finally, it would be interesting to investigate the history of globalization from an animal perspective, both in terms of micro and of macro history: what did globalization mean to animals, or particular species, and how, and how unevenly, did it impact them?

To be sure, both animal history and global history are highly specialized fields. Projects that bring both scholarly traditions together, therefore, might be easier to realize not in terms of one-person efforts but actual teamwork. While this will inevitably complicate discussions, especially where humanities and sciences need to be brought together, it is a worthwhile endeavor. Both fields share some of their central concerns, aiming to bring previously unheard actors into history and challenging established narratives. Both are relatively marginal in mainstream academia. Joining forces can only make both fields stronger, and the current necessity for explaining,

---

64 See the excellent recent work on the fluidity of human-animal boundaries in the context of colonialism: Bourke, *What it Means to be Human*; Sivasundaram, Sujit. Imperial Transgressions: The Animal and Human in the Idea of Race. *Comparative Studies of South Asia, Africa and the Middle East* 35, no. 1 (May 2015): 156–172; Anderson and Perrin, Up from the Ape.
65 Among the historians arguing for a cooperation of history and the sciences, see Baratay, Eric. Pour une histoire éthologique et une éthologie historique. *Études rurales* 189 (2010): 91–106; Otter, Planet of Meat.
66 There are already some studies on the appropriation of evolutionary theory in the Middle East and India where it became central to Hindu nationalism: Elshakry, Marwa. *Reading Darwin in Arabic, 1860–1950*. Chicago, IL: University of Chicago Press, 2014; Nanda, Meena. Madame Blavatsky's Children: Modern Hindu Encounters with Darwinism. In *Handbook of Religion and the Authority of Science*, James R. Lewis, Olav Hammer (eds.), 279–344. Leiden: Brill, 2012.
67 Mason, James B. Misothery: Contempt for Animals and Nature, its Origins, Purposes and Repercussions. In *The Oxford Handbook of Animal Studies*, Linda Kalof (ed.), 135–151. Oxford: Oxford University Press, 2017. For a somewhat contrary position, see the contributions in Waldau, Paul and Kimberley Patton (eds.). *A Communion of Subjects. Animals in Religion, Science, and Ethics*. New York, NY: Columbia University Press, 2006, although many of them lack a historical dimension.

and finding solutions for, the environmental crisis shows that the benefit arising from this would not merely be a strategic one.

Selected Bibliography

Bankoff, Greg. A Curtain of Silence. Asia's Fauna in the Cold War. In *Environmental Histories of the Cold War*, John Robert McNeil, Corinna Unger (eds.), 203–226. Cambridge: Cambridge University Press, 2010.
Bedini, Silvio. *The Pope's Elephant*. Lanham, MD: Rowman & Littlefileld, 1998.
Bourke, Joanna. *What it Means to be Human. Reflections From 1791 to the Present*. London: Virago, 2011.
Bulliet, Richard W. *The Camel and the Wheel*. Cambridge, MA: Harvard University Press, 1975.
Bulliet, Richard W. *Hunters, Herders, and Hamburgers: The Past and Future of Human-Animal Relationships*. New York, NY: Columbia University Press, 2005.
Crosby, Alfred W. *Ecological Imperialism: The Biological Expansion of Europe, 900–1900*. Cambridge: Cambridge University Press, 1986.
Crosby, Alfred W. *The Columbian Exchange*. Westport, CT: Greenwood, 1972.
Ridley, Glynis. *Clara's Grand Tour: Travels with a Rhinoceros in Eighteenth-Century Europe*. London: Atlantic Books, 2004.
Subramaniam, Radhika. The Elephant's I: Looking for Abu'l Abbas. In *Animal Biography: Re-framing Animal Lives*, André Krebber, Mieke Roscher (eds.), 207–226. Cham: Palgrave Macmillan, 2018.
Warren, Wilson J. *Meat Makes People Powerful: A Global History of the Modern Era*. Iowa City, IA: University of Iowa Press, 2018.
Webb, James L. R. *Humanity's Burden: A Global History of Malaria*. Cambridge: Cambridge University Press, 2009.
Wöbse, Anna-Katharina. Globale Kreaturen. Tiere in der internationalen Politik des frühen 20. Jahrhunderts. In *Tierische Geschichte. Die Beziehung von Mensch und Tier in der Kultur der Moderne*, Dorothee Brantz, Christoph Mauch (eds.), 304–322. Paderborn: Schöningh, 2010.

Joanna Dean
Public History

1 Introduction and Overview

One might expect public history to be richly peopled with animal life, but it is usually only the human story that is on display in the museums, historic sites, archives, and art galleries.[1] Nonhuman animals tend to be relegated to a passive role; they serve as backdrop to human drama, as commodity, as specimen or as symbol. The truly historical animal – the individual animal as a subject of its own life – is beginning to appear on the margins: in artistic provocations, in small museums, in temporary exhibits, and in flourishing blogs and twitter accounts. It can be found in the archives, if we know how to look for it, wherever its path crosses that of the human animal. It emerges most frequently when members of the public fashion their own meanings out of the past.

The National Council on Public History defines the field broadly as "the many and diverse ways that history is put to work in the world."[2] Their definition points to a history that is, first, "put to work," and second, "in the world," and so sets public history apart from academic history. And while we might quibble with the assumption that academic history idles cloistered in the ivory tower, it is true that public history is at work in many arenas: it is the history disseminated by historians employed in museums, by producers of television and film documentaries, by archivists, by the antiquarians who run historical societies, and by the enthusiasts who volunteer at heritage sites. Public history by definition speaks to a wider audience than academic history. The historical animal is harder to define: It is the animal who is subject of its own life. One might say that it is the animal about whom a story can be told. As a consequence, this chapter will follow various animal stories in the halls of public history, beginning with Freckles, the transgenic spider goat, and concluding with Jumbo the elephant. It will pick up their traces in museums and art collections, in other forms of public display as well as social media.

---

[1] Hilda Kean has remarked on the absence of animals in key journals like *The Public Historian*. See Kean, Hilda. Public History and Heritage: A Fruitful Approach for Privileging Animals. In *The Routledge Companion to Animal–Human History*, Hilda Kean, Philip Howell (eds.), 76–99. London and New York, NY: Routledge, 2018, 77.
[2] What is Public History? *National Council on Public History*. URL: ncph.org/what-is-public-history/about-the-field/ (May 1, 2019).

2 Topic and Themes

We begin in the animal barns at the Canada Agriculture and Food Museum in Ottawa where a working herd of Holstein cows, a team of Clydesdale horses, and an assortment of other heritage livestock entrance the public, especially the under-ten crowd. For a few years, two transgenic spider goats were unexpected inclusions. "These goats look like goats, act like goats, but have been genetically modified to produce milk containing spider silk proteins," explained the text panel next to their stall. A silk spinning gene inserted into the goats by Nexia Biotechnologies in Montreal meant that their milk contained a silk protein that could be extracted and spun into a biomaterial with twice the tensile strength of steel. The silk protein is used for military purposes (the project was supported by the Canadian Department of National Defense and the United States Army) and, the museum was at pains to point out, such useful items as medical sutures.³ The official messaging went awry when anthropologist Beatriz Oliver visited with her small children. "Seeing the transgenic goats gave me a horrible, sick feeling inside," she said. "I just think it's wrong." The Canadian Biotechnology Action Network took up the issue, urging the public to lobby that "the museum should not be turned into a transgenic zoo to promote the biotech industry," and in 2013 the goats disappeared from view.⁴

A spider goat named Freckles appeared that same year in another museum, Pittsburgh's Center for PostNatural History. The Center is dedicated to the intersection between culture, nature and biotechnology. Their slogan – "That was then. This is now" – might be a direct riposte to the nostalgic undertone of the Ottawa museum's heritage displays. Freckles was related to the Ottawa goats; she was donated by Dr. Randy Lewis of Utah State University, who had acquired the Nexia herd in 2008 when the Montreal firm went out of business. Ottawa's goats were alive, which was part of the aversion felt by Oliver. The Center's goat was not. The Center made her taxidermied form the focus of their display.⁵

---

3 Spider Goats Display Angers Ottawa Professor. *CBC News*, March 29, 2012. URL: www.cbc.ca/news/canada/ottawa/spider-goats-display-angers-ottawa-professor-1.1137229 (May 1, 2019). For the connection to the defense industry see GM Spider Silk Spun by Canadian company. *CBC News*, January 17, 2002. URL: www.cbc.ca/news/technology/gm-spider-silk-spun-by-canadian-company-1.319533 (May 1, 2019).
4 Genetically Modified Goats on Display at Ottawa's Experimental Farm. *Canadian Biotechnology Action Network*, March 27, 2012. URL: cban.ca/1315/ (May 1, 2019). For further background on the Canadian display see Anderson, Peter. Are All Forgotten Friends Worthy of Memory? The Public History of Biotechnology in Canada. In *Creating Heritage: Unrecognised Pasts and Rejected Futures*, Thomas Carter, David C. Harvey, Roy Jones, Iain J.M. Robertson (eds.), 177–190. London and New York, NY: Routledge, 2020.
5 For their description of Freckles, see Center for Postnatural History. Biosteel™ Goat. *Specimen*, August 2014. URL: www.postnatural.org/Specimen-Vault/Biosteel-Goat (December 10, 2020).

The Center for PostNatural History traffics in discomfort. The horror felt by Oliver at the Ottawa Museum was akin to the fascination felt by visitors to the Center. They are invited to consider a variety of animals transformed in disturbing ways by human technology: glowing *E. coli* bacteria, transgenic mosquitoes, giant pumpkins, a laboratory mouse, atomic rats, a Silkie chicken and genetically altered *GloFish*. The Center was founded when Richard Pell, associate professor of art at Carnegie Mellon University, observed that "the organisms most commonly used in science were absent from natural history collections."[6] His specimens point to the absence of manipulated organisms in museums of natural history and the Center demonstrates how these institutions have obscured the vexed history of human relations with other animals. Because his genetic novelties are not visibly different from other animals – Freckles might be any goat – Pell's exhibits rely upon story telling. Each creature is displayed with its own evolutionary and cultural story; each is, indisputably, a historical animal.

Following Animals in Museums and Exhibitions

Pell is not the only critic to draw attention to the absence of the animal in museums of natural history, where the animal is everywhere, but nowhere. Artists Bryndís Snæbjörnsdóttir and Mark Wilson did similar work by surveying the history of all the taxidermied polar bears in the United Kingdom. Bringing the polar bears and their histories together in an exhibit, *Nanoq: Flat Out and Bluesome* (2001–2006), they made generic specimens into individuals with a past.[7] Another artist, Kate Foster, brought attention to these same absences in 2003 by seeking out the story behind one specimen, a hen harrier, *Circus cyaneus*. She returned the harrier from the taxonomic isolation of a museum exhibit to its place of capture in order to trace its biogeography: "tracing and placing it within the life worlds of human-animal encounter, cohabitation and estrangement which brought it to and remain its current situation as a representational scientific object."[8] In effect, she was giving the speci-

---

[6] Pell, Richard and Lauren B. Allen. Bringing Postnatural History into View. *American Scientist* 103, no. 3 (May–June 2015): 224. See also, Pell, Richard W., Emily Kutil and Etienne Turpin. PostNatural Histories. In *Art in the Anthropocene: Encounters Among Aesthetics, Politics, Environments and Epistemologies*, H. Davis, E. Turpin (eds.), 299–316. London: Open Humanities Press, 2015. Pell was apparently not aware of the museological history of spider goats in Ottawa.
[7] Snæbjörnsdóttir, Bryndís and Mark Wilson. *Nanoq: Flat Out and Bluesome: A Cultural Life of Polar Bears*. London: Black Dog Publishing, 2006. For their 2019–2022 project see: Snæbjörnsdóttir, Bryndís and Mark Wilson. Snæbjörnsdóttir/Wilson Receive Major Funding from Icelandic Research Fund for 3-Year Project 2019–2022. URL: snaebjornsdottirwilson.com/projects/visitation/snaebjornsdottir-wilson-receive-major-funding-from-icelandic-research fund-for-3-year-project-2019-2022/ (December 5, 2020).
[8] Foster, Kate, Merle Patchett and Hayden Lorimer. The Biogeographies of a Hollow-Eyed Harrier. In *Afterlives of Animals, A Museum Menagerie*, Samuel J. M. M. Alberti (ed.), 110–133. Charlottesville, VA:

men a past. She then returned the harrier to the Hunterian Zoology Museum with this history.⁹ As she and co-authors Merle Patchett and Hayden Lorimer observe, the story told was necessarily partial, and the personhood granted the harrier was incomplete. They argue, on Deleuzian terms, that this telling was more meaningful in part because of its open-endedness. "Personhood is recast as a constellation of events and encounters where animal and human lives lose neat beginnings and endings."¹⁰

Historians have also begun to trace the cultural history of museum specimens [→History of Animal Collections/Animal Taxonomy]. Samuel Alberti, Rachel Poliquin, Liv Emma Thorsen and Dolly Jørgensen remind us that taxidermied animals are cultural constructions, as well as scientific models.¹¹ Dioramas have come under particular scrutiny. As Karen Wonders has shown, they are predicated upon a kind of timelessness, and they create an illusion of life that is central to naturalizing the biogeographies of nations.¹² Stories that would pin the dioramas to a particular place and time undermine this message. They also raise questions about how the animals came to be in the museum. Jørgensen examines the muskoxen in the diorama at the Biologiska museet, Stockholm, and identifies them as those killed in Gustaf Kolthoff's 1900 collecting expedition to East Greenland. She tells the story that is missing from the diorama, quoting from Kolthoff's graphic account of muskox slaughter, and situating the animals in a place, a time and a community:

> As soon as [the muskoxen] became aware of us, they stood immediately in a protective ring, and calves took their place among the elderly. So the magnificent animals lowered their heads threatening to turn against us, and when we reached them at good shooting range, I asked Kjell to shoot first. He hit a bull in the forehead, who fell instantly.¹³

The backstory – the history – of the animals on display makes the animals individuals. It also turns attention to their deaths. We are squeamish about the deaths necessary for museum display. Of all the many muskoxen on display in Scandinavian

---

University of Virginia Press, 2011, 115; see also Foster, Kate and Hayden Lorimer. Disposition – A Hollow-Eyed Harrier, Displaced and out of Time. *Antennae* 20 (Spring 2012): 117–122.

9 They were exhibited in the UK as part of the exhibitions *Disposition* at the University of Glasgow in 2003 (Scottish Arts Council) and *The Animal Gaze* at Plymouth College of Art in 2009.

10 Foster, Patchett and Lorimer, Biogeographies, 115.

11 Alberti, Samuel J. M. M. (ed.). *Afterlives of Animals, A Museum Menagerie*. Charlottesville, VA: University of Virginia Press, 2011; Poliquin, Rachel. *The Breathless Zoo: Taxidermy and the Cultures of Longing*. Philadelphia, PA: Pennsylvania State Press, 2012; Thorsen, Liv Emma. *Elephants Are Not Picked from Trees: Animal Biographies in the Gothenburg Museum of Natural History*. Aarhus: Aarhus University Press, 2014.

12 Wonders, Karen. Habitat Dioramas and the Issue of Nativeness. *Landscape Research* 28, no. 1 (January 2003): 89–100.

13 Translated and cited in Jørgensen, Dolly. Migrant Muskox. *The Return of Native Nordic Fauna* (blog), December 13, 2013. URL: dolly.jorgensenweb.net/nordicnature/?p=1303 (May 27, 2019). Original quote in Kolthoff, Gustaf. *Til Spetsbergen och Nordöstra Gråonland år 1900*. Stockholm: Fr. Skoglunds Förlag, 1901, 179.

natural history museums, the only one with a name or a personal story identified by Jørgensen was a calf, also called Kjell, at the Muskoxcentrum in Härjedalen.[14] His death was an accidental one, and the human involved was not a hunter but a hero: a veterinarian who tried to save Kjell.

Similar backstories have been told of other dioramas. Harry Snyder's bloody account of his expedition to capture the bison in Canada's Nature Museum points to an international intrigue of sportsmen, wealth and conservation politics in the collection of specimens. Carl Akeley's hunt for the gorillas mounted in a diorama in the American Museum of Natural History was used by Donna Haraway to position the iconic display in a masculine eugenic moment. Hornaday's pursuit of the Buffalo Group at the Smithsonian is similarly explored by Hannah Rose Shell. The accounts of their deaths make these animals individuals: historical animals whose lives are entangled with humans.[15] These stories however remain external to the displays, and the average viewer views the animals without being aware of their lives or their deaths.

In 2012, Finnish visual artist Terike Haapoja and author Laura Gustafsson launched an ambitious project to build a Museum of the History of Others: "a huge museum dedicated to tell the stories of all those species that doesn't [sic] have their own museums or histories, i.e. all the non-human species."[16] They started with one exhibit, the *Museum of the History of Cattle*, on display in Helsinki in 2013. Their aim was a radical one, stated modestly:

> Speaking on behalf of someone else is always a possessive gesture. We can't and don't want to claim that we know what the history of cattle has been like and how they have experienced it. However, we do know for sure that they have been present and that there is a yet untold perspective to our common history.[17]

---

14 Jørgensen, Dolly. Museum Menageries. *The Return of Native Nordic Fauna* (blog), October 7, 2013. URL: dolly.jorgensenweb.net/nordicnature/?p=1186 (May 24, 2019); Jørgensen, Dolly. Naturalized National Identities: Migrant Muskoxen in Northern Nature. YouTube video, 28:30, of a lecture at the RCC Lunchtime Colloquium, December 12, 2013, posted by Rachel Carson Centre, December 13, 2013. URL: youtu.be/3NvbmikWnGo (December 10, 2020).
15 Snyder, Harry M. *Snyder's Book of Big Game Hunting.* New York, NY: Greenberg, 1950; Haraway, Donna. Teddy Bear Patriarchy: Taxidermy in the Garden of Eden, New York City, 1908–1936. *Social Text* 11 (Winter 1984): 20–64; Shell, Hannah Rose. Soul in the Skin: William T. Hornaday and the Buffalo Group, 1896–1996. In *The Extermination of the American Bison [1889]* by William Temple Hornaday, vii–xxii. Smithsonian: Smithsonian Institution Scholarly Press, 2002. Hornaday's Buffalo exhibit is now at the Museum of the Northern Great Plains, in Montana.
16 Jach, Aleksandra. History of Others: Interview with Laura Gustafsson and Terike Haapoja. *The Anthropocene Index*, December 12, 2015. URL: theanthropoceneindex.com/article/32-History-of-Others.-Interview-with-Laura-Gustafsson-and-Terike-Haapoja (December 10, 2020).
17 Gustafsson, Laura and Terike Haapoja. Introduction: How to Write the History According to Cattle. *Gustafsson&Haapoja*. URL: www.gustafssonhaapoja.org/how-to-write-a-history-according-to-cattle/ (December 10, 2020).

Just as Kate Foster argued that the incompleteness of the personhood of the harrier was itself meaningful, these artists conclude that the impossibility of speaking for the cattle is itself a statement: "With this in mind, we can attempt to imagine and give shape to the space that lacks their voice, the space from which they, even now, witness our world."[18]

Haapoja and Gustafsson have not yet succeeded in their larger project for a Museum of the History of Others. Nor have attempts to build an animal history museum with a critical animal studies perspective succeeded. Despite the backing of such eminent animal rights figures as legal scholar Steven Wise, the Animal History Museum, conceived for the purpose of "understanding and celebrating the animal human bond,"[19] is in 2020 on hold for lack of funding. Smaller museums dedicated to specific breeds abound, especially for dogs, cats and horses. Celebratory in nature, these museums are usually made up of collections of art and memorabilia and rarely take a critical or historical position.[20]

Temporary displays in larger institutions have been more successful in sharing a historical perspective. London is a leading venue. The Imperial War Museum hosted *Animal's War* in 2006, just as the field of animal studies was opening. *Making Nature: How We See Animals* (2016) at the Wellcome Collection was explicitly historical.[21] Hannah Velten, author of *Beastly London: A History of Animals in the City* (2016), acted as historical consultant for the Museum of London's 2019 immersive exhibit, *Beasts of London*.[22] Art galleries also tend to ignore the historical animal, even when it is in plain sight, as Hilda Kean has observed. She points instead to small exhibitions, like the Edwin Landseer exhibition at the National Gallery.[23]

There is no lack of documents and artefacts to support such exhibits, but they are often buried in the archives and storerooms. Susan Nance examines an archival photograph, with a dog front and center, in which the caption ignores the animal,

---

18 Gustafsson and Haapoja, Introduction.

19 Mission Statement. *Animal History Museum*. URL: animalhistorymuseum.org/about/mission-statement/ (December 10, 2020).

20 There are numerous museums around the world focused on the dog, such as the American Kennel Club's Museum of the Dog, now in New York City, or on dog memorabilia, such as the Dog Collar Museum in Leeds Castle, Kent England, or on a specific breed, such as the Dackelmuseum in Passau Germany, the Akita Dog Museum in Odate, Japan, or the Musée et Chiens du Saint-Bernard, in *Martigny, Switzerland, or the National Bird Dog Museum in Grand Junction, Tennessee*. The horse is also popular with the International Museum of the Horse at the Kentucky Horse Park and the Le musée du Cheval, which, as its English name – The Living Museum of the Horse – suggests, includes living horses at Domaine de Chantilly, in France. A special exhibit, "The Horse," circulated from Chicago's Field Museum, the American Museum of Natural History and then the Canadian Museum of History, between 2008 and 2016.

21 Making Nature: How We See Animals. *Wellcome Collection*, December 1, 2016 to May 21, 2017. URL: wellcome.ac.uk/press-release/making-nature-how-we-see-animals (May 28, 2019).

22 Beasts of London. *Museum of London*, April 5, 2019 to January 5, 2020.

23 Kean, Hilda. Unimaginative Cat Exhibition at the British Library. *Hilda Kean* (blog), December 29, 2018. URL: hildakean.com (May 23, 2019).

and asks: "how do we find new evidence of nonhuman life in anthropocentric archives, when mostly we are trained to edit animals out of our analysis?"[24] The editing out of the nonhuman is a political act, and Nance calls upon archivists to pay more attention to the multispecies world. Her work has inspired at least one archivist to draw attention to the animals in their collection. In 2019, the Archives of Ontario (Canada) mounted an onsite and online exhibit, *Animalia: Animals in the Archives*.[25]

Trailing and Tracking Animal Life in the "Wild"

Indigenous people see animals differently, but their way of thinking has only recently informed museum practices. When the skeleton of an 18 year old orca, known to science as J34, was offered to the members of the *shíshálh* Nation who operate the *tems swiya* museum in Sechelt, British Columbia, they gave him a name: *Kwentens ?e te sinkwu* [guardian of the sea]. At the opening, they celebrated the cultural ties of his pod to their community and recalled their shared history in the Strait of Georgia. As *shíshálh* Nation chief Warren Paull observed of orca, "They are part of our family. It has huge, huge significance for us."[26] Orca only became individuals for the rest of us when Michael Bigg developed a method for photo-identifying them, and when we grew to care about captive orca held in aquariums.[27] In 2019, when orca J35 lost her calf, and carried its body for over two weeks, the entire world mourned with her, because we knew her history and the history of her dwindling J pod.

Tracking equipment has allowed science to follow living wild animals and so give meaning to their lives. It made it possible to tell the life story of the Yellowstone Park wolf known as 832AF. When she wandered outside the park and was shot by a hunter, her death was covered by the *New York Times*. The detailed notes kept by naturalists who followed, scoped, and photographed 832AF informed the telling of her biography by Nate Blakeslee in *American Wolf: A True Story of Survival and Obsession in the West*.[28] Biologist and artist Colleen Campbell tracked 89 bears in Bow River

---

24 Nance, Susan. Introduction. In *The Historical Animal*, Susan Nance (ed.), 1–17. Syracuse, NY: Syracuse University Press, 2015, 4; see also Tortorici, Zeb. Animal Archive Stories: Species Anxieties in the Mexican National Archive. In *The Historical Animal*, Nance (ed.), 75–98.
25 Jay Young, personal communication. See also Animalia: Animals in the Archives. *Archives of Ontario, Ministry of Government and Consumer Services*, January (2019). URL: www.archives.gov.on.ca/en/explore/online/animalia/index.aspx (April 26, 2019).
26 Woodrooffe, Sophie. Opening Ceremony for Whale Exhibit Highlights Culture and Teamwork. *Coast Reporter*, April 15, 2019. URL: www.coastreporter.net/news/local-news/opening-ceremony-for-whale-exhibit-highlights-culture-and-teamwork-1.23791872 (May 23, 2019).
27 Colby, Jason Michael. *Orca: How We Came to Know and Love the Ocean's Greatest Predator*. Oxford: Oxford University Press, 2018.
28 Schweber, Nate. Famous Wolf is Killed Outside Yellowstone. *The New York Times*, December 8, 2012. URL: www.nytimes.com/2012/12/09/science/earth/famous-wolf-is-killed-outside-yellowstone.

watershed, in British Columbia, from 1994–2004, and reflected upon this work in an art exhibit, *Eastern Slopes Grizzly Bears: Each One is Sacred*. She followed Bear 16 as he became habituated to humans and was exiled to the Calgary Zoo, where he became beloved by visitors as *Skoki*, but, as Campbell notes, neutered and dead to the wild population of bears. In a blog on Campbell's exhibit, historian Tina Loo recounted the life of another bear, 148, observing: "We don't think of wild animals like grizzly bears as having a history, but they do, and not just collectively and evolutionarily, but as individuals."[29]

Radio collar studies informed Mahesh Rangarajan's thought-provoking historical study of the lions in Gir Forest, Gujarat, India. Rangarajan argues that, in addition to having a history, the lions have a *sense* of history. Their cohabitation with humans was learned over time and so reflects the emergence of a specific lion culture: "These lions exhibit a capacity (which humans had liked to think was theirs alone) of perhaps remembering and analyzing events and then passing on that knowledge to younger members of the prides."[30] He reflects: "Although it is going too far to endow the lions with historical consciousness, Gir's lions clearly do have memory of memories" [→Multispecies Ethnography].[31]

Pursuing the Social Media Animal

The historical animal has a lively presence in the independent world of social media. Jonathan Saha, for example, uses historical film footage, archival documents and photographs to explore the role of the Burmese elephant in British colonialism in his blog, *Colonizing Animals*.[32] Like other scholarly blogs, such as *Pet Histories*, and Chris Pearson's *Sniffing the Past: Dogs and History*, Saha's reaches a wide public.[33] Blogs by Hilda Kean and Dolly Jørgensen draw attention to the absence of the

---

html (December 5, 2020); Blakeslee, Nate. *American Wolf: A True Story of Survival and Obsession in the West*. Portland, OR: Broadway Books, 2018.
29 Loo, Tina. Bow Valley and People Without a History. *Network in Canadian History and Environment*, January 10, 2018. URL: niche-canada.org/2018/01/10/the-bow-valley-and-people-without-a-history/ (May 23, 2019); Campbell, Colleen. *Eastern Slopes Grizzly Bears: Each One is Sacred*. Exhibition at Whyte Museum of the Canadian Rockies, Banff, AB, Canada, October 21, 2017 to Jan 28, 2018; Campbell, Colleen. Grizzly Bear #16 a.k.a. Skoki. *Wild Lands Advocate* 18, no. 1 (February 2010): 20–21. URL: albertawilderness.ca/wp-content/uploads/20100200_ar_wla_grizzly_16_colleen_campbell.pdf (December 5, 2020).
30 Rangarajan, Mahesh. Animals with Rich Histories: The Case of the Lions of Gir Forest, Gujarat, India. *History and Theory*, 52, no. 4 (December 2013): 109–127, 126.
31 Rangarajan, Animals with Rich Histories, 109.
32 Saha, Jonathan. Colonizing Animals: A Blog about Beasts, Burma and British imperialism. URL: colonizinganimals.blog/ (May 23, 2019).
33 *Pet Histories: The AHRC Pets and Family Life Project Blog.* URL: pethistories.wordpress.com/ (May 23, 2019); Pearson, Chris. *Sniffing the Past: Dogs and* History (blog). URL: sniffingthepast.word-

historical animal in much public history. Facebook groups, like Horse History Corral, bring together amateurs and historians in a shared love of animals past. Twitter has arguably had the greatest influence among practitioners. In 2018, Adam Koszary, the social media editor of the Museum of English Rural Life, tweeted an archival photograph of a massive Exmoor Horn ram with the caption "look at this absolute unit." (An "absolute unit" is internet jargon for a large muscular person.) It was retweeted 31,000 times and built the museum a twitter following of over 131 thousand, sparking an exchange with entrepreneur Elon Musk, who replaced his avatar with that of the sheep. In 2019 Koszary taunted his betters in the museum world – "hey @britishmuseum give us your *best duck*" – and a lively battle of duck images from major institutions around the world ensued.[34]

Social media is more responsive than most arenas of public history; it is quicker to respond to new historiographies, and it brings professionals in more immediate contact with the public. It is also more ephemeral. Blogs are abandoned and even official twitter feeds lose their authors. In 2019 Adam Koszary left the MERL to join The Royal Academy of Arts (and was briefly lured to Tesla). In his reflections on his work at the Academy, he noted the tensions between the individual voice, the institutional voice, and the internet. He described the efforts he had made to cultivate a "stable and successful tone that reflects the RA without being boring. I hope."[35] It remains to be seen whether social media will retain its critical edge as it is absorbed in the institution.

Chasing the Genomes: Exhibiting Rare Breeds

The historical animal makes a living appearance in collections of "rare breeds." Or does it? Rare breed organizations preserve living examples of the breeds abandoned by modern agriculture. For all their materiality, many of these breeds are the stuff of myth: as Harriet Ritvo and others have observed, heritage breeds, such as the Chillingham Cattle, the Dartmoor Pony, the Malay Chicken, or the Berkshire Pig, become identified with particular places and appeal to popular nostalgia for a rural past.[36] The heritage farms that display these animals are susceptible to the charge that

---

press.com/ (December 5, 2020); Kean, Hilda. *Hilda Kean* (blog). URL: hildakean.com/?page_id=943 (May 23, 2019).
34 Gerken, Tom. Elon Musk Switches with Museum of English Rural Life. *BBC News*, April 18, 2019. URL: www.bbc.com/news/blogs-trending-47975564 (May 23, 2019).
35 Koszary, Adam: Christ What a Weird Year. Post on *medium.com*, February 25, 2020. URL: medium.com/@adamkoszary/christ-what-a-weird-year-42c670d42aba (December 10, 2020).
36 Ritvo, Harriet. Race, Breed and Myths of Origin: Chillingham Cattle as Ancient Britons and Counting Sheep in the English Lake District: Rare Breeds, Local Knowledge and Environmental History. In *Noble Cows and Hybrid Zebras: Essays on Animals and History,* Harriet Ritvo (ed.), 132–156. Charlottesville, VA: University of Virginia Press, 2010.

David Lowenthal levelled at other aspects of the heritage movement: that it "exaggerates and omits, candidly invents and frankly forgets, and thrives in ignorance and error."[37] The farm animal remembered in these places is stripped of its bloody past and practical purpose. Its history is at most a partial one.

Zoos (to which heritage farms owe a great deal) have similarly invoked a mythical heritage animal [→History of the Zoo]. Nigel Rothfels has demonstrated that the *Przewalski* horse, described by numerous respected zoos as the last truly wild horse, is a modern creation. As he shows, the modern breed was developed from as few as twelve wild animals captured in Mongolia at the turn of the twentieth century. In the effort to reduce phenotypic variability, breeders obscured the resemblance of the *Przewalski* horse to the Mongolian domestic horse (and obfuscated the connections between the two) and created a breed with curious resemblance to the primeval horses depicted in cave art. These are not rescued representatives of a remnant population, Rothfels concludes, but "creatures that have at some significant level been made by Western and now international expectations."[38]

Before we dismiss these invented traditions, however, we should consider the work done by emotions. Laurajane Smith calls for us to respect the agency of visitors to create their own meaning in heritage sites. She argues, contra Lowenthal, that this nostalgia is not necessarily a reactionary backward glance:

> it is quite clear that museums and heritage sites are places where people go to feel and indeed they are arenas where people go to 'manage' their emotions [...] permissible places for people to not only feel particular emotions, but to work out or explore how those emotions may reinforce, provide insight or otherwise engage with aspects of the past and its meaning for the present.[39]

Smith's point translates easily to animal history: heritage animals allow the public to work through the complicated mourning of a lost relationship with the animal world described so eloquently by John Berger.[40] Dolly Jørgensen has examined the emotional work done by the exhibition of the last members of an extinct species.[41]

---

37 Lowenthal, David. Fabricating Heritage. *History and Memory* 10, no. 1 (Spring 1998): 5–24.
38 Rothfels, Nigel. (Re)Introducing the Przewalski's Horse. In *The Ark and Beyond: The Evolution of Zoo and Aquarium Conservation* Ben A. Minteer, Jane Maienschein, James P. Collins (eds.), 77–89. Chicago, IL: University of Chicago Press, 2018, 88.
39 Smith, Laurajane and Gary Campbell. The Elephant in the Room: Heritage, Affect, and Emotion. In *A Companion to Heritage Studies*, William Logan, Mairead Nic Craith, Ullrich Kockel (eds.), 443–460. Oxford: Wiley Blackwell, 2015.
40 Berger, John. Why Look at Animals? In *About Looking*, John Berger (ed.), 1–26. London: Bloomsbury, 1980.
41 Jørgensen, Dolly. After None: Memorialising Animal Species Extinction Through Monuments. In *Animals Count: How Population Size Matters in Animal-Human Relations*, Nancy Cushing, Jodi Frawley (eds.), 183–199. Abingdon and New York, NY: Routledge, 2018; Jørgensen, Dolly. *Recovering Lost Species in the Modern Age: Histories of Longing and Belonging.* Cambridge, MA: MIT Press, 2019.

A story from my own Animal History class might serve as a demonstration of how heritage animals engender emotions, and how these emotions can undermine curatorial intentions [→History of Emotions]. Canada's Agriculture and Food Museum acquired a heritage pig, a Lacombe, about five years after they divested themselves of the transgenic goat. The Lacombe had been listed as "critically endangered" by Heritage Livestock Canada, whose chair congratulated the museum on the purchase: "We look forward to the continued production and display of these beautiful farm animals that will tell their own story of agricultural history in Canada."[42] The Lacombe pig, however, was no traditional farm animal; the Lacombe had been created by government scientists between 1947 and 1957 for commercial production. In this respect, the Canadian Lacombe pigs resemble the "fascist pigs" described by Taigo Saraiva: both were technoscientific organisms, bred for modernity.[43] By 2018, the Lacombe pig no longer served the needs of industry; there were only seven young females remaining when the museum stepped in. Most visitors to the farm are oblivious to the breed's history. Even university students sent to study the animal, armed with Saraivo's analysis, get caught up in the livingness of the animal; the sentience, the consciousness, and the emotive draw of the lumbering sow and offspring are more compelling than the technoscientific past.

3 Methods and Approaches

Commemorating

The emotional response to animals is channeled in popular public memorials, and more privately in pet cemeteries.[44] Statues celebrate the faithfulness of dogs who held vigil for their masters, such as Japan's Hachikō, Scotland's Greyfriars Bobby, and Australia's Dog on the Tuckerbox and Eureka dog. They celebrate the service to mankind of dogs like Balto, who delivered diphtheria antitoxin to a remote Alaska town, or Stubby, who was a mascot in the Second World War. And they celebrate companions to the famous, such as Islay, Queen Victoria's Cairn terrier.[45] The most

---

42 The Canada Agriculture and Food Museum Proudly Welcomes Heritage Lacombe Pigs. *Ingenium* (blog), posted by The Canada Agriculture and Food Museum, August 27, 2018. URL: ingeniumcanada.org/newsroom/canada-agriculture-and-food-museum-proudly-welcomes-heritage-lacombe-pigs-farm (December 10, 2020).
43 Saraiva, Tiago. *Fascist Pigs: Technoscientific Organisms and the History of Fascism*. Cambridge, MA: MIT Press, 2016.
44 Howell, Philip. A Place for the Animal Dead: Pets, Pet Cemeteries and Animal Ethics in Late Victorian Britain. *Ethics, Place & Environment* 5, no. 1 (March 1, 2002): 5–22.
45 Kean, Hilda. An Exploration of the Sculptures of Greyfriars Bobby, Edinburgh, Scotland and the Brown Dog, Battersea, South London, England. *Society & Animals* 11, no. 4 (January 2003): 353–373; Kean, Hilda. Public History and Two Australian Dogs: Islay and the Dog on the Tucker Box. *ACH: The Journal of the History of Culture in Australia* 24–25 (2006): 135–162; Kean, Hilda. Balto, the Alaskan

controversial of the dog memorials was one erected by antivivisectionists to remember the suffering of a laboratory animal, the "Old Brown Dog." Erected in 1906, the memorial was protested by rioting medical students, and removed at night under police guard in 1910, only to be reinstated in 1985, and again in 1994. The animals in the statues often stand in for human actors: the Eureka dog represents the men killed in Australia's foundational Pikeman's Protest, and as Coral Lansbury has shown the "old brown dog" channeled the fears of women and workers.[46] Aaron Skabelund has observed of Hachikō that there is a great deal of mythmaking in these histories. Was Hachikō really waiting for his master, or had he come to rely on handouts at the train station? Was he a pure Japanese dog, or perhaps mixed with foreign breeds? The ambiguities only underscore the power of memory and emotion in shaping memorials, and the intensity of public feeling.[47] As Hilda Kean observes, "It is not simply historians who create history."[48]

Animals are particularly effective channels for our feeling about the horror of war [→History of War]. Canada's Peace Tower includes a 1927 representation of the animals that served in the First World War, including reindeer, pack mules, carrier pigeons, horses, dogs, canaries and mice, with the inscription, "The Tunneller's Friends: The Humble Beasts that Served and Died." In 2012, at the urging of a veteran, the *Animals in War* memorial was mounted in a neighboring park: a sculpture of a dog with plaques to horses and mules. (Britain's Animals in War memorial was erected in 2004 in Hyde Park, London.) Horses are popular; a number of exhibits on war horses were inspired by the extraordinary success of the play and film, *War Horse*.[49] The C.A.V. Barker Museum of Veterinary History, at University of Guelph, drew upon a rich collection of veterinary material to mount the exhibit, "Canadian War Horse," at a local museum.[50]

Dog and His Statue in New York's Central Park: Animal Representation and National Heritage. *International Journal of Heritage Studies* 15, no. 5 (July 2009): 413–430. See also White, Richard. The True Story of Gundagai's Dog on the Tuckerbox: Tourists, Truth, and the Insouciance of Souvenirs. *Journeys* 17, no. 2 (January 2016): 115–136.

46 Lansbury, Coral. *The Old Brown Dog: Women, Workers, and Vivisection in Edwardian England*. Madison, WI: University of Wisconsin Press, 1985.

47 Skabelund, Aaron. *Empire of Dogs: Canines, Japan, and the Making of the Modern Imperial World*. Ithaca, NY: Cornell University Press, 2011, ch. 3.

48 Kean, Public History and Heritage, 86.

49 Cf. Gardiner, Juliet. *The Animals' War: Animals in Wartime from the First World War to the Present Day*. London, Portrait, in association with the Imperial War Museum, 2006. Exhibition held from July 14, 2006 to April 22, 2007. URL: www.iwm.org.uk/collections/item/object/80030554 (December 5, 2020).

50 Exhibited at Wellington County Museum and Archives, UK, from December 2016 to April 2017. See Cox, Lisa. Veterinary Artifacts and the Use of Material History. In *The Historical Animal*, Nance (ed.), 99–117.

Refashioning: Jumbo's Second Life

The public also refashions the museum animal, giving it new meaning. Here we can turn to the story of Jumbo, the massive African elephant who won the hearts of Londoners while at the London Zoo from 1865 to 1882, and then went on to win over Americans during his three difficult years with the Barnum and Bailey travelling circus. In 1885, he died crossing the railway tracks in St. Thomas, a small town in Ontario, Canada. His death was the beginning of an extensive "after life": over the next 130 years bodily traces of Jumbo surfaced in museums, archives, and colleges, while the idea of Jumbo, like his name, circulated freely in popular memory and memorabilia.[51] His heart was sold for $40 to a professor at Cornell University. Already decomposing by 1897, the heart had disappeared by the 1940s, although the specimen jar continued to be displayed in a faculty office in the Cornell Veterinary College; it resurfaced in a 2011 online exhibit, *Animal Legends*.[52] Engraved sections of his tusk were distributed to the wives of three circus entrepreneurs.

Jumbo's defining feature was his size, and the most prized relics were his massive skin and skeleton. These were preserved by the young taxidermist Carl Akeley, who added a few inches in the process. (As Barnum said: "It will be a grand thing to take all advantage possible in this direction. Let him show like a mountain.") The new Jumbo, or rather Jumbos (the skeleton and skin were mounted separately), were unveiled in 1886 at a banquet (guests were reported to have been served a dessert gelatinized by ground up tusk) and put on tour for three years.[53] Jumbo's taxidermied self was then displayed at Tufts University. Students put coins in his trunk for good luck, and eventually they transformed Jumbo from natural history specimen into university mascot.[54] When the taxidermied form was destroyed in a fire in 1975, his ashes were placed in a peanut butter jar on the athletic director's desk, where they still serve as a token of good luck. A bronze sculpture was commissioned to mark the 125th anniversary of the year of his death, when his tail and ashes were

---

[51] For a list of Jumbo's body parts, and their possible locations, see Shoshani, Sandar Lash, Jeheskel Shoshani and Fred Dahlinger Jr. Jumbo: Origin of the Word and History of the Elephant. *Elephant* 2, no. 2 (June 1986): 86–122. doi.org/10.22237/elephant/1521732022. They draw upon Goodwin, George G. Whatever Became of Jumbo. *Natural History* 61, no. 1 (1952): 45–46; McClellan, Andrew. P.T. Barnum, Jumbo the Elephant, and the Barnum Museum of Natural History at Tufts University. *Journal of the History of Collections* 24, no. 1 (March 2011): 45–62, 52.

[52] Cornell's Animal Celebrities: Jumbo's Heart. Part of online exhibition *Animal Legends: From the Trojan Horse to Godzilla*, Cornell University, created 2011. URL: rmc.library.cornell.edu/AnimalLegends/exhibition/cuanimalcelebs/jumbo.html (December 10, 2020).

[53] See George G. Goodwin to John J. Murphy, November 14, 1955. A photocopy of the letter is in: Shoshani, Jumbo, 115.

[54] The account of Jumbo's status at Tufts is drawn from Mcclellan, Andrew. Jumbo the Mascot. *Tufts University*. URL: www.tufts.edu/about/jumbo (December 10, 2020).

placed, alongside memorabilia, such as soda bottles, popcorn bags, matches, puzzles, and toys, in the exhibit, *Jumbo: Marvel, Myth and Mascot*.[55]

Jumbo's skeleton was displayed in the Biology of Mammals Hall at the American Museum of Natural History where it also took on unintended meaning. It (or he) served as a "type specimen" for the species *Loxodonta Africana rothschildi* from 1907 to 1969.[56] It then cycled in and out of storage, taking on emotive power over time. Even mammologists broke from their scientific script in 1985, displaying him privately in their department next to the interdepartmental freezer; as the senior mammalogy technician, Helmut Sommer, explained: "Jumbo was an old friend."[57] In 1993, he surfaced in an exhibit, *Jumbo: The World Famous Elephant*, promoting elephant conservation. A quote from the exhibit coordinator is revealing: "It's an important specimen in our collection," she said, emphasizing the science. Then she shifted awkwardly to his emotive presence: "People ask us all the time: Where's Jumbo? What happened to the skeleton? It's nothing macabre at all. People obviously still love it."[58] Her use of the impersonal "it" stands in contrast to the emotional response of the public. In 2018 Jumbo's bones served to draw attention to the life lived: DNA, isotope and forensic analysts hired by television broadcasters in the United Kingdom and Canada confirmed that Jumbo lived a traumatic and stressful life.[59]

Jumbo also lives on in St. Thomas. As Christabelle Sethna has observed, his freakish death gave the small town a useful notoriety, fueling a "fetishist connection" to the African elephant.[60] On the centenary of his death the city mounted a life size Jumbo, constructed of steel and cement, and the local museum put on a display of Jumbo memorabilia, including the tusk section originally owned by Barnum's wife,

---

55 Bambrick, Gail. The Glory that was Jumbo. *Tufts Now*, September 22, 2014. URL: now.tufts.edu/articles/glory-was-jumbo (December 10, 2020). The exhibition was at the Koppelman Family Gallery in the Aidekmann Arts Centre from September to December 2014.
56 It has since been established that there are only two species, and he was a large example of *Loxodonta africana africana*.
57 Shoshani, Jumbo, 100, and photograph in Figure 12; Sommer spoke about the move to the *New York Times*: Collins, Glenn. Barnum's Jumbo is Back in Museum's Center Ring. *New York Times*, January 22, 1993.
58 Collins, Barnum's Jumbo.
59 For clips and broadcast dates of the British documentary, see Dunleavy, Stephen, director. David Attenborough and the Giant Elephant. *BBC*, December 10, 2017. TV Feature. URL: www.bbc.co.uk/programmes/b09jcxrj (December 10, 2020); for comments see Jumbo the Life of an Elephant Superstar. *Canadian Broadcasting Corporation*, January 7, 2018. URL: www.cbc.ca/natureofthings/episodes/jumbo-the-life-of-an-elephant-superstar (May 24, 2019).
60 Sethna, Christabelle. The Memory of an Elephant: Savagery, Civilization and Spectacle. In *Animal Metropolis: Histories of Human-Animal Relations in Urban Canada*, Joanna Dean, Darcy Ingram, Christabelle Sethna (eds.), 29–56. Calgary: University of Calgary Press, 2016.

exhibited along with a section of Jumbo's ear and the contents of his stomach at his death.[61]

Just as there were many material remains of Jumbo, there were many historical Jumbos, serving the various purposes of the past: offering some the opportunity to simply love Jumbo (and through him, perhaps, all elephants), and offering others the opportunity to indulge in the horror of his death and macabre spectacle of his remains.

4 Implication(s) of the Animal Turn

The *animal turn* is still relatively new. Large public institutions move slowly, and nonhuman animals are relegated to a passive role in most museums; they serve as specimen, as backdrop to human drama, or as symbol. The historical animal appears in the margins of these institutions: in the provocations of artists, in the hidden holdings of archives, in the fast-moving world of social media, in the public memorials and in public re-fashionings of exhibits. It is on these margins that the emotional work is done: public memorials allow us to work through the loss of the relationship with the animal. These exhibits also help us to process the complicated feelings of the Anthropocene. The message of the transgenic goat or an extinct species is a disturbing one of new human power to alter nature.

Selected Bibliography

Alberti, Samuel J. M. M. (ed.). *The Afterlives of Animals: A Museum Menagerie*. Charlottesville, VA: University of Virginia Press, 2011.
Despret, Vinciane. From Secret Agents to Interagency. *History and Theory* 52, no. 4 (December 2013): 29–44. doi.org/10.1111/hith.10686.
Kean, Hilda and Philip Howell (eds.). *The Routledge Companion to Animal–Human History*. London and New York, NY: Routledge, 2018.
Kean, Hilda. Balto, the Alaskan Dog and His Statue in New York's Central Park: Animal Representation and National Heritage. *International Journal of Heritage Studies* 15, no. 5 (September 1, 2009): 413–430.
Kean, Hilda. Public History and Two Australian Dogs: Islay and the Dog on the Tucker Box. *Australian Cultural History (ACH): The Journal of the History of Culture in Australia* 24–25 (January 2006): 135–162.
Nance, Susan (ed.). *The Historical Animal*. Syracuse, NY: Syracuse University Press, 2015.
Rangarajan, Mahesh. Animals with Rich Histories: The Case of the Lions of Gir Forest, Gujarat, India. *History and Theory* 52, no. 4 (December 2013): 109–127.

---

61 Forest, Ben. Jumbo the Elephant Featured in New Elgin Museum Exhibit. *St. Thomas Times-Journal*, June 16, 2014. URL: www.stthomastimesjournal.com/2014/06/16/jumbo-the-elephant-featured-in-new-elgin-county-museum-exhibit/wcm/d26deb37-679e-1949-7a99-615206505435 (May 24, 2019).

Ritvo, Harriet. *Noble Cows and Hybrid Zebras: Essays on Animals and History.* Charlottesville, VA: University of Virginia Press, 2010.

Rothfels, Nigel. (Re)Introducing the Przewalksi's Horse. In *The Ark and Beyond: The Evolution of Zoo and Aquarium Conservation,* Ben A. Minteer, Jane Maienschein, James P. Collins (eds.), 77–89. Chicago, IL: University of Chicago Press, 2018.

Dorothee Brantz
Urban (and Rural) History

1 Introduction and Overview

Bees are urbanizing. A recent Bavarian petition to save bees – and by extension other insects – garnered more than 1.7 million signatures in just two weeks.[1] In an unprecedented move, the Bavarian government turned this petition into an expedited law within a month. Of course, a law is just a first step and it remains to be seen how bees will actually be protected. Many cities now house beehives. For instance, in the city of Berlin, some 600 beekeepers are hosting several million bees in backyards, on balconies, rooftops, and in urban parks.[2] In the city and state of Berlin and Brandenburg, over 400 different kinds of wild bees have been recorded.[3] Bees offer a telling example of how animals traverse the boundaries of the urban and the rural. And they are by no means the only animals. Countless insects, birds, mammals, and reptiles have moved to urban areas across the world. In some places, the species diversity is said to be larger in cities than in the surrounding countryside.[4] There are lions in Nairobi, coyotes in Los Angeles, American river crabs in Berlin, and sparrows in Dubai. Animals who are able to adapt to urban living environments might stand a better chance to withstand the anticipated massive species extinction.

From a contemporary and historical perspective, it is worthwhile – indeed indispensable – to gain a better understanding of the rural-urban connections between humans and animals. This is what this chapter seeks to do. One of its central aims is to question the traditional nature-culture dichotomy that has undergirded most thinking about the urban. In the urban context, this dichotomous thinking has contributed to a very instrumental view of the place of animals in cities or led to ignoring their presence altogether. In recent years, our understanding of the constituting ele-

---

[1] The "Save the bees" petition was the most successful citizen-led political campaign in Bavarian history. Zeit online. Rettet die Bienen erfolgreichstes Volksbegehren in Bayern. *Die Zeit*, Feb 14, 2019. URL: www.zeit.de/wissen/2019-02/volksbegehren-rettet-bienen-rekord-beteiligung-bayern (October 4, 2019).
[2] www.berlin.deutschland-summt.de (August 13, 2020).
[3] Freund, Bärbel. Lang lebe die Königin: Gespräch mit dem Biologen Benedikt Polaczek. *Der Freitag – die Wochenzeitung*, January 23, 2004. URL: www.freitag.de/autoren/der-freitag/lang-lebe-die-konigin (October 4, 2019).
[4] It is important to note that while the number of species might be greater in cities, the actual number of animals within a species might be quite small, e.g. there might only be one pair of birds of a particular species.

https://doi.org/10.1515/9783110536553-020

ments of cities has changed substantially.[5] A growing number of scholars is questioning the nature/culture boundary that appeared to be cemented in modernist thinking. As we will see, political ecologists, geographers, environmental historians, and not least, scholars in the field of HAS have opened up new perspectives on the urban as a transgressive field that invites inter- and transdisciplinary perspectives. Indeed, this perspective is also contributing to a more critical view of one of the foundations of urban theory, namely the supposed separation of the urban and the rural. This is not to say that we have already reached our goals. For instance, even in the widely transgressive literature of actor-network theory, which pays a lot of attention to nonhuman actors including infrastructures and technological networks, animals are often not mentioned even though they play a crucial role in many of these networks.[6] My chapter will trace the rural-urban boundaries and what they have meant for the development of human-animal relations throughout history [For the rural see also →History of Agriculture;→Environmental History].

2 Topics and Themes

Animals IN the City and the Countryside

Cities are animal-rich environments. There are, of course, the dogs, cats, rabbits, guinea pigs, birds, mice, rats, reptiles, fish and whatever else people keep as pets. Practices of pet-keeping have differed throughout time and across cultures [→History of Pets]. For instance, while dogs have been kept as domestic animals in the western hemisphere for centuries, in Asian and middle Eastern cultures, for the most part, this has not been the case. In addition to those animals whom people have brought to the city and into their homes on purpose, there are also countless species who have moved alongside human settlements without invitation. Rats and mice as well as insects and parasites ranging from bed bugs to fleas and roaches thrive in the close proximity of humans and have certainly profited from the growing urbanization of the world. Then there are the horses, donkeys, elephants, llamas, camels etc. who were all used for transport and labor. Moreover, one should not forget all of the 'wild' animals who have inhabited cities – the mammals, birds, fish, insects etc. Individual species might have changed over time, but their continuous presence proves that cities are and always have been animal-rich spaces. Indeed, I would argue that cities could only develop as they did through the intricate interrelation-

---

[5] Haumann, Sebastian, Martin Knoll and Detlev Mares. Urban Environmental History as a Field of Research. In *Concepts of Urban Environmental History,* Sebastian Haumann, Martin Knoll, Detlev Mares (eds.), 9–20. Bielefeld: transcript, 2020.
[6] Farías, Ignacio and Thomas Bender. *Urban Assemblages: How Actor-Network Theory Changes Urban Studies.* London: Routledge, 2010; Belliger, Andrea and David Krieger (eds.). *ANTthology: Ein einführendes Handbuch zur Akteur-Netzwerk-Theorie.* Bielefeld: transcript, 2006.

ships between humans and animals. These interrelationships might have been harmonious or violent; humans might have recognized and appreciated the presence of animals or abhorred it; they might have fed them or killed them. But no matter whether humans have paid attention to nonhuman creatures or not, animals have played a vital role in the development of cities across time.

Several points are important here. First, cities would not exist without animals; hence, the history of animals in urban space is as old as the history of urbanization – reaching back several millennia. Second, the existence of animals in the city is closely related to the presence of humans and the interaction of the two. However, this does not mean that animals inhabit cities only as a consequence of human will, planning, or power. Certainly, many animals were purposefully brought to cities as resources, for work, pleasure, experimentation – in short, to serve humans. Yet an even greater number arrived for other reasons or even in spite of human desires and intentions. In other words, while urban animals might not necessarily exist independent from humans, they might live in contrast to human intentions. Hence, the history of urban human-animal relations provides us with countless stories exposing the myth of human dominance and omnipotence in the urban. Third, many animals defy and transgress the categorizations that humans have assigned to them as pets, wild animals etc. For instance, dogs can be pets, strays, or work animals. They can be cherished or feared. Sometimes they serve as food. As Clemens Wischermann, Aline Steinbrecher, and Philip Howell have argued, animals are "liminal creatures," who exist betwixt and between the epistemological categorizations humans have created for them.[7] They often transgress the supposed boundaries between nature and culture, the urban and rural, pre-modern and modern as well as between the so-called human and nonhuman.

The same holds true for the rural. In rural societies across the globe, animals were vital for agricultural production, protection, and as natural resources. All of these engagements were, of course, culturally specific. While mules were prevalent as means of transportation in Latin America, camels were preferred in northern Africa and the Middle East. In southern India and Sri Lanka, by contrast, people often relied on elephants. Similarly, with livestock – pigs were kept in China and across Europe, but not in the Middle East [→Global History]. Geographic characteristics also shaped preferences. In maritime cultures, fish naturally played a significant role. Agricultural production in the Swiss Alps revolves around cows, while in Peru llamas are more common. Hence, in rural areas, human-animal relations developed in a close nexus between geography and culture, in many ways more explicitly so than in urban areas. Generally, what we call countryside is a more diversified space that includes human settlements and agricultural production, but it also

---

7 Wischermann, Clemens, Aline Steinbrecher and Philip Howell (eds.). *Animal History in the Modern City: Exploring Liminality.* London: Bloomsbury Academic, 2018.

reaches far beyond it. As such, the rural is a less bounded space than the urban and, hence, more difficult to tackle as an area of investigation.

Animals AND the Urban and Rural

Until relatively recently animals did not play much of a role in urban history. Urban historians sometimes mention animals in their accounts of urban life, but usually just in passing and without greater detail. As I have argued elsewhere, cities and animals were not even considered a viable topic for environmental historians, and when environmental historians began to study cities, they were predominantly interested in pollution, water supply, and green spaces.[8] Animals first entered urban history via the field of cultural history and the then emerging field of human-animal studies in the early 1990s. Given that the history of HAS has often dealt with urban phenomena, such as the keeping of pets, the creation of zoos, the practice of vivisection, and the establishment of public slaughterhouses, one wonders why animals have not become more present in urban history much sooner [→History of the Zoo;→History of Animal Slaughter].[9] One reason might be that these HAS accounts did not, for the most part, foreground the urban dimension. The city often functioned primarily as a stage where activities unfolded, but not as a focus of analysis in and of itself. Interestingly, urban geography has paid more attention to the role of urban animals [→Historical Animal Geography].[10]

In rural history, animals are frequently mentioned, although usually in relation to landscapes or as part of agricultural production, and only rarely as actual agents in daily life. Similar to the development of urban history as a field, rural history shifted to social perspectives in the 1970s and, following the linguistic turn in the late 1980s, to cultural history. Both of these perspectives were heavily centered on human notions of class, and later gender, language, and culture. If one compares rural and urban historical perspectives, animals seem to be assigned more to the countryside and rural ways of living in close proximity to nature and agricultural production. Among other things, this linkage became manifest through notions of time. The rural is usually associated with naturalized time – day and night, the seasons, human and animal life cycles. Whereas the rural tended to be connected to cy-

---

8 Brantz, Dorothee. The Natural Space of Modernity: A Transatlantic Perspective on (Urban) Environmental History. In *Historians and Nature: Comparative Approaches to Environmental History*, Ursula Lehmkuhl, Hermann Wellenreuther (eds.), 195–225. Oxford: Berg Publishers, 2007.
9 Roscher, Mieke. Darf's ein bisschen mehr sein? Ein Forschungsbericht zu den historischen Human-Animal Studies. *H-Soz-Kult*, December 12, 2016. URL: www.hsozkult.de/literaturereview/id/forschungsberichte-2699 (October 10, 2019).
10 Philo, Chris and Chris Wilbert (eds.). *Animal Spaces, Beastly Places: New Geographies of Human-Animal Relations*. London: Routledge, 2000; Wolch, Jennifer R. and Jody Emel (eds.). *Animal Geographies: Place, Politics, and Identity in the Nature-Culture Borderlands*. London: Routledge, 1998.

clical notions of time, the urban – particularly during the modern period – was increasingly equated with the more rationalized temporalities of human work and leisure. Cities were meant to be functional, consisting of planned and managed spaces and evolving industrialized productions that increasingly replaced animal labor with machines. In this set-up, animals seemed to have disappeared from urban thinking and from historical accounts. Nevertheless, they remained a viable force in daily urban practices and representations.

Having sketched out the role of animals in the rural and urban contexts, we have to ask – what does the actual historiography look like? Looking at the literature about animals in the history of cities and the countryside gives the impression that this history primarily took place in Europe and North America and, moreover, that it is primarily a modern story. As we know from other fields of history, this impression is more a reflection on the state of the field rather than historical actuality. However, this is not meant to assign greater significance to any particular time period or geographic area. Surely in years to come, research foci will evolve so that the next edition of this handbook will hopefully be more global and diachronic.

The Ancient Period

Animals were, of course, part of the earliest human civilizations where many species were first domesticated in different cultures [→Domestication: Coevolution].[11] Once domesticated they served as the main means of transport, an important source of food, and frequently as material for clothing and other human goods. It is still difficult to discern a specifically urban or rural perspective with regard to the use and treatment of animals, which is, however, not so much a result of the actual indistinguishability of the two realms but rather a reflection of the existing historiography. Up to now there is not much literature on explicitly urban or rural human-animal related topics, but ancient human-animal relations can be gleaned from more general texts on animals in the ancient world or from the history of agriculture.[12] Two particular topics that have been studied are horses and animals in entertainment.[13] Human-animal relations are primarily documented through zooarcheology, art and

---

[11] Clutton-Brock, Juliet. *Animals as Domesticates: A World View Through History*. East Lansing, MI: Michigan State University Press, 2012.
[12] Arbuckle, Benjamin S. and Sue Ann McCarty. *Animals and Inequality in the Ancient World*. Boulder, CO: University Press of Colorado, 2015; Campbell, Gordon Lindsay (ed.). *The Oxford Handbook of Animals in Classical Thought and Life*. Oxford: Oxford University Press, 2014; Howe, Timothy. *Pastoral Politics: Animals, Agriculture, and Society in Ancient Greece*. Claremont, CA: Regina Books, 2008; Kitchell, Kenneth, Jr. *Animals in the Ancient World from A to Z*. New York, NY: Routledge, 2013.
[13] Hyland, Anne. *The Horse in the Ancient World*. Westport, CT: Praeger, 2003; Jennison, George. *Animals for Show and Pleasure in Ancient Rome*. Philadelphia, PA: University of Pennsylvania Press, 2005.

remnants of material culture such as pottery and sculptures [→Pre-Domestication: Zooarchaeology;→Material Culture Studies]. Other important sources are references in ancient texts and inscriptions as well as remains that are recovered through archeological work at ancient sites.[14] Based on these sources it is difficult to distinguish between pets and work animals, but animal mummies indicate that cats and dogs were frequently held as pets and buried along with their owners.[15] Much work remains to be done to discern a more explicitly urban or rural perspective on ancient human-animal relations, but I am sure we can look forward to more research to emerge on this soon.

Medieval and Early Modern Attitudes

The Middle Ages were characterized by a sharp division between cultural spaces and what was considered wilderness.[16] At the same time, most people did not draw a clear distinction between domestic and wild animals. For the most part, people still lived in close proximity to animals but not necessarily with clear notions of ownership. Most farm animals, especially pigs, still roamed freely, often in the forest.[17] Others such as oxen, cows, donkeys and – starting in the fourteenth century – horses were used as draft animals.[18] Sheep were kept for wool production. Most importantly, animals served as source of food and material resources to make clothing and tools.[19] On most farms, goats, chicken, ducks, and geese were kept for food. Physically, medieval farm animals were generally smaller and lighter. Some animals were newly domesticated during the Middle Ages, most notably rabbits.[20] Hunting was another significant use for horses and dogs, especially in royal courts where they served as

---

14 Newmyer, Stephen Thomas. *Animals in Greek and Roman Thought: A Sourcebook*. London: Routledge, 2011.
15 Donalson, Malcolm Drew. *The Domestic Cat in Roman Civilization*. Lewiston, NY: Edwin Mellen Press, 1999; Menache, Sophia. Netherworld Envoy or Man's Best Friend? Attitudes Towards Dogs in the Ancient World. In *Routledge Handbook of Human-Animal Studies*, Gary Marvin, Susan McHugh (eds.), 114–123. London: Routledge, 2014; Ikram, Salima (ed.). *Divine Creature: Animal Mummies in Ancient Egypt*. Cairo: American University in Cairo, 2005.
16 For a good overview, see Pascua, Esther. From Forest to Farm and Town: Domestic Animals from ca. 1000 to ca. 1450. In *A Cultural History of Animals in the Medieval Ages*, vol. 2, Brigitte Resl (ed.), 81–102. Oxford: Berg, 2007; Zeuner, Frederick E. *A History of Domesticated Animals*. London: Hutchinson, 1963.
17 Wiseman, Julian. *The Pig: A British History*. London: Duckworth, 2000, 3.
18 Hyland, Anne. *The Horse in the Middle Ages*. Stroud: Sutton Publishing, 1999; Langdon, John. *Horses, Oxen and Technological Innovation: The Use of Draught Animals in English Farming from 1066–1500*. Cambridge: Cambridge University Press, 1986.
19 Grant, Anne. Animal Resources. In *The Countryside of Medieval England*, Grenville Astill, Anne Grant (eds.), 149–187. Oxford: Blackwell, 1988.
20 Bond, James. Rabbits: The Case for Their Medieval Introduction to Britain. *Local History* 18 (1988): 53–57.

a symbol of prestige. The group of hunted animals included a broad range of mammals and birds [→History of Hunting].

The Middle Ages were a period of hardship and crisis. Two of the most formidable experiences were the constant threat of famine and the recurrence of plague. Especially the latter was closely associated with animals, particularly rats.[21] But animal pests were also a serious threat for harvests, which, in turn, often had devastating effects on farm animals, making them weak and more susceptible to epizootics of different kinds. Generally, farm life was still characterized by much communal living of humans and animals. Only the wealthy could afford to have separate stables for animals.

According to Esther Pascua, during the medieval period human-animal relations witnessed a shift from the forest to the farm and from the farm to the urban market.[22] In other words, certain animals were increasingly incorporated in formalized human practices in the countryside and the city. As some scholars have shown, animals were a notable part of medieval town life.[23] Interestingly, while medieval towns were usually clearly separated from the surrounding countryside by walls and moats, town and country dwellers – both human and animal – were still closely connected. On some levels, the boundaries between humans and animals were still ephemeral. One indicator was the legal system, which did not draw a clear distinction between human and animal accountability. Consequently, pigs could be put on trial for murder or chicken for not laying enough eggs.[24] In early modern cities, many animals, among them dogs, pigs, chickens, goose, cats, and others, roamed freely.[25] Some rules and ordinances to regulate the presence of certain animals in cities went back to the fifteenth century, but many binding rules only emerged in the eighteenth century.[26] For instance, the keeping of pigs was increasingly banned. Other animals became objects of pleasure, for example in exhibition fights like bear baiting as well as travelling and courtly menageries [→History of Animal Fights and Blood Sports;→History of Circus Animals].

Pet-keeping became increasingly popular in the eighteenth century throughout Europe – particularly in the more protestant north – even if this pet-keeping cannot

---

21 Burt, Jonathan. *Rat*. London: Reaktion Books, 2006.
22 Pascua, From Forest to Farm, 82.
23 Choyke, Alice M. and Gerhard Jaritz (eds.). *Animaltown: Beasts in Medieval Urban Space*. Oxford: BAR Publishing, 2017; O'Connor, Terry. Animals in Urban Life in Medieval to Early Modern England. In *The Oxford Handbook of Zooarchaeology*, Umberto Albarella, Mauro Rizzetto, Hannah Russ, Kim Vickers, Sarah Viner-Daniels (eds.), 214–229. Oxford: Oxford University Press, 2017.
24 Dinzelbacher, Peter. Animal Trials: A Multidisciplinary Approach. *The Journal of Interdisciplinary History* 32, no. 3 (Winter 2002): 405–421; Fischer, Michael. *Tierstrafen und Tierprozesse: Zur sozialen Konstruktion von Rechtssubjekten*. Münster: Lit Verlag, 2005.
25 Steinbrecher, Aline. Eine Stadt voller Hunde: Ein anderer Blick auf das frühneuzeitliche Zürich. *Informationen für Moderne Stadtgeschichte* 2 (2009): 26–40.
26 Steinbrecher, Aline. Fährtensuche: Hunde in der frühneuzeitlichen Stadt. *Traverse* 3 (2008): 45–59, 46.

necessarily be equated with today's practices of keeping animals in the home.[27] For instance, cats were kept as companions, but they were also viewed as very useful members of a household, especially when it came to warding off mice and rats. At the same time, cats, especially the dead cadavers of free roaming cats, were frequently viewed as sources of danger that could cause illnesses.[28] Dogs were another type of animal that was quite prevalent in early modern cities.[29] Some were kept as guard dogs, others also to fight rats. Most dogs served as a sign of affiliation with a specific social class. Especially very small and very large dogs were regarded as signifiers of wealth and leisure. By the eighteenth century, dogs had become quite popular. In the 1780s, the human-dog ratio of Vienna was eight to one.[30] The city of London already discussed the introduction of a dog tax as a supplemental source of municipal revenue.[31]

Particularly the work of Erica Fudge has taught us a great deal about the role of animals in the Renaissance and early modern period, and even though her work does not focus on the urban and rural per se, readers can find much about the general attitudes towards animals living in the vicinity of humans and hence, in cities and the countryside.[32] For a more explicit treatment of the country and the city, one might turn to Keith Thomas' *Man and the Natural World*, which still offers one of the most impressive accounts of life in early modern Britain.[33] During the Enlightenment, the Western intellectual tradition was in a period of radical transitions, which also affected human attitudes towards animals. On the one hand, the increasing emphasis on reason as a central means of knowledge production shifted the study of animal behavior increasingly towards scientific investigations, many of which took place in urban laboratories [→History of Experimental Animals and the History of Animal Experiments]. At the same time, this rationalization of animal be-

---

[27] Tague, Ingrid. *Animal Companions: Pets and Social Change in Eighteenth-Century Britain*. University Park, PA: Pennsylvania State University Press, 2015.
[28] Hengerer, Mark Sven. Stadt, Land, Katze: Zur Geschichte der Katze in der Frühneuzeit. *Informationen für Moderne Stadtgeschichte* 2 (2009): 13–25.
[29] Steinbrecher, Aline. Tiere und Raum. Verortung von Hunden im städtischen Raum der Vormoderne. In *Tiere und Geschichte. Konturen einer Animate History*, Gesine Krüger, Aline Steinbrecher, Clemens Wischermann (eds.), 219–240. Stuttgart: Franz Steiner, 2014.
[30] Laichmann, Michaela. *Hunde in Wien: Geschichte eines Tieres in der Großstadt*. Vienna: Stadt- und Landesarchiv, 1998, 4.
[31] Tague, Ingrid. Eighteenth-Century English Debates on a Dog Tax. *The Historical Journal* 51 (2008): 901–920.
[32] Fudge, Erica. *Quick Cattle and Dying Wishes: People and Their Animals in Early Modern England*. Ithaca, NY: Cornell University Press, 2018; Fudge, Erica (ed.). *Renaissance Beasts: Of Animals, Humans, and Other Wonderful Creatures*. Urbana, IL: University of Illinois Press, 2004; Fudge, Erica. *Brutal Reasoning: Animals, Rationality and Humanity in Early Modern England*. Ithaca, NY: Cornell University Press, 2006.
[33] Thomas, Keith. *Man and the Natural World: Changing Attitudes in England, 1500–1800*. London: Penguin, 1983; Velten, Hannah. *Beastly London: A History of Animals in the City*. London: Reaktion Books, 2016.

havior, most famously embodied in Descartes' dictum that animals operated like clocks without sense or reason, also raised staunch criticism. Some Enlightenment thinkers, most notably Rousseau, advocated a more emotion-centered engagement with animals [→History of Ideas]. In cities as well as in the countryside, the Enlightenment laid the foundation for the paradoxical attitudes towards animals that reigned throughout the modern period.

The Modern Era

The modern era was a period of rapid transformation. The astounding and often dramatic effects of industrialization, technological innovation, scientific discoveries, and the general rise of (bourgeois) mass society became particularly apparent in cities. Modernist thinking continued along the trajectory of Enlightenment humanism, but it also pushed the idea of reason towards rationalization with all its promising and harrowing implications. Nineteenth-century cities were built on the energy and food resources of animals, which scholars have traced from numerous perspectives.[34] Animals were exposed to growing exploitation but also to a widening emotional and ethical engagement. Moreover, while some animals were removed from the urban landscape, others were purposefully brought to the city. Not surprisingly, dogs and horses received special attention from scholars trying to understand the place of animals in (urban) public space.[35] Horses were of particular interest with regard to questions of labor and mobility.[36]

Technological and scientific advances led to a stark redefinition of the general attitudes towards animals. One example was the theory of evolution that reconfigured the species boundaries between man and certain animals [→History of Sci-

---

34 Atkins, Peter (ed.). *Animal Cities: Beastly Urban Histories*. Farnham: Ashgate, 2012; Brantz, Dorothee. Die animalische Stadt: Die Mensch-Tier-Beziehung in der Urbanisierungsforschung. *Informationen für Moderne Stadtgeschichte* 2 (2008): 86–100; Wischermann, Clemens (ed.). Tiere in der Stadt. Special issue of *Informationen für Moderne Stadtgeschichte* 2 (2009); Wischermann, Clemens, Aline Steinbrecher and Philipp Howell (eds.). *Animal History in the Modern City: Exploring Liminality*. London: Bloomsburg, 2019.
35 McKenzie, Kirsten. Dogs and the Public Sphere: The Ordering of Social Space in the Early Nineteenth-Century Cape Town. In *Canis Africanis: A Dog History of Southern Africa*, Lance van Sittert, Sandra Swart (eds.), 91–111. Leiden: Brill, 2008; Pearson, Chris. Stray Dogs and the Making of Modern Paris. *Past & Present* 234, no. 1 (February 2017): 137–172.
36 McShane, Clay and Joel Arthur Tarr. *The Horse in the City: Living Machines in the Nineteenth Century*. Baltimore, MD: Johns Hopkins University Press, 2007; Greene, Ann Norton. *Horses at Work: Harnessing Power in Industrial America*. Cambridge, MA: Harvard University Press, 2008; Pooley-Ebert, Andria. Species Agency: Comparative Study of Horse-Human Relationships in Chicago and Rural Illinois. In *The Historical Animal*, Susan Nance (ed.), 148–165. Syracuse, NY: Syracuse University Press, 2015.

ence].³⁷ The growing impact of science manifested itself in the rise of veterinary medicine as a professional discipline as well as in the expanding practice and critique of vivisection [→History of Veterinary Medicine; →History of Experimental Animals and the History of Animal Experiments]. None of these were explicitly urban practices, but much of this work took place in cities. More explicit urban places of science were the emerging natural history museums and zoological gardens of the nineteenth century where the display of living and dead animals was undergirded by a mixture of scientific and cultural ideologies.³⁸ Indeed, both of these institutions were closely related to the rise of modern metropolises. Zoological gardens attested to the bourgeois meshing of education and entertainment and how imperial ambitions were brought back to the city [→History of the Zoo]. As such they underlined the close linkage between metropolis and colony and how animals were instrumentalized to serve imperial ambitions.³⁹ In a somewhat odd inversion of the city and countryside, in the course of the twentieth century, zoos also became harbingers of conservation and the protection of species that had become extinct in the wild. Animal science was about knowledge production, but these productions were closely tied to regimes of consumption, particularly as consumer cultures expanded in the second half of the twentieth century.

Production and consumption usually entail destruction. In the case of animals this means killing; and indeed, modernity, in many ways, evolved through the mass-killing of animals.⁴⁰ Zoo exhibits entailed the death of countless animals killed during imperial expeditions. More readily apparent was the mass-killing of livestock for food in the newly established urban (public) slaughterhouses [→History of Animal Slaughter].⁴¹ Other events linking animals and mass death were the repeated wars of the nineteenth and particularly twentieth century. Throughout the world, many different species of animals were engaged in the different theaters of war

---

37 Hochadel, Oliver. Darwin in the Monkey Cage: The Zoological Garden as a Medium of Evolutionary Thinking. In *Beastly Natures: Animals, Humans and the Study of History*, Dorothee Brantz (ed.), 81–107. Charlottesville, VA: University of Virginia Press, 2010.
38 Brantz, Dorothee. Metropolitan Natural Histories: Inventing Science, Building Cities, and Displaying the World. In *Science in the Metropolis*, Mitchell Ash (ed.), 25–42. London: Routledge, 2020.
39 Gißibl, Bernhard. *The Nature of German Imperialism: Conservation and the Politics of Wildlife in Colonial East Africa*. New York, NY: Berghahn, 2016; Krüger, Gesine. Das koloniale Tier: Natur-Kultur-Geschichte. In *Wo ist Kultur. Perspektiven der Kulturanalyse*, Thomas Forrer, Angelika Linke (eds.), 73–94. Zurich: vdf, 2014; Speitkamp, Winfried and Stephanie Zehnle (eds.). *Afrikanische Tierräume: Historische Verortungen*. Cologne: Köppe, 2014.
40 The Animal Studies Group. *Killing Animals*. Urbana, IL: University of Illinois Press, 2006.
41 Brantz, Dorothee. *Slaughterhouse Cities: Paris, Berlin, and Chicago, 1780–1914*. Baltimore, MD: Johns Hopkins University Press, forthcoming; Lee, Paula Young (ed.). *Meat Modernism and the Rise of the Slaughterhouse*. Durham, NH: New Hampshire University Press, 2008; Nieradzik, Lukasz. *Der Wiener Schlachthof St. Marx: Transformation einer Arbeitswelt zwischen 1851 und 1914*. Göttingen: Vandenhoeck & Ruprecht, 2017; Pacyga, Dominic. *Slaughterhouse: Chicago's Union Stockyard and the World It Made*. Chicago, IL: University of Chicago Press, 2015.

[→History of War].⁴² Finally, the modern period also witnessed a massive war on all of those animals variously identified as pests, parasites, and threats to human health.⁴³ War was brought upon them in the name of a civilizing ideology and aided by modern technologies and the science of chemistry.⁴⁴ Clearly, the rise of mass-society also entailed massive extinction underlining the contradictory nature of modernity.

As in centuries past, the urban and rural remained closely interlinked, but industrialization added a new level of interdependence extending industrial style production to agriculture, particularly in the decades after World War II.⁴⁵ Moreover, the distance between cities and countryside broadened because as globalization expanded, animals from far flung places gained in significance while those close by, in the surrounding countryside, lost appeal. Zoos, circuses but also books and films brought lions closer, but cows and pigs moved further and further out of sight. To many urbanites, the countryside became a place of longing, where one spent one's vacations, cherished the landscape, or maybe went for a hunt, but it was not really an actual living space anymore. The study of human-animal relations in an urban and rural context has helped to illuminate the material dimensions of these engagements and how they have changed over time.

3 Methods and Approaches

As stated above, the study of urban (and rural) human-animal relations started with cultural history [→Cultural History].⁴⁶ In line with the linguistic turn, scholars paid increasing attention to how human-animal relations were articulated in written texts. Pioneering works like Keith Thomas' *Man and the Natural World* and Harriet Ritvo's *Animals Estate* showed that the human understanding of society and its environment was often filtered through the discursive engagement with animals. Just a few years later, works like Kathleen Kete's *The Beast in the Boudoir* and Louise Rob-

---

42 Hediger, Ryan (ed.). *Animals and War: Studies of Europe and North America*. Leiden: Brill, 2013; Kean, Hilda. *The Great Cat and Dog Massacre: The Real Story of World War II's Unknown Tragedy*. Chicago, IL: University of Chicago Press, 2017.
43 Biehler, Dawn Day. *Pests in the City: Flies, Bedbugs, Cockroaches, and Rats*. Seattle, WA: University of Washington Press, 2013; Sullivan, Robert. *Rats: Observations on the History and Habitat of the City's Most Unwanted Inhabitants*. New York, NY: Bloomsbury, 2004.
44 Russell, Edmund. *War and Nature: Fighting Humans and Insects with Chemicals from World War I to Silent Spring*. New York, NY: Cambridge University Press, 2001.
45 Schrepfer, Susan and Philip Scranton (eds.). *Industrializing Organisms: Introducing Evolutionary History*. New York, NY: Routledge, 2004; Settele, Veronika. *Revolution im Stall: Landwirtschaftliche Tierhaltung in Deutschland, 1945–1990*. Göttingen: Vandenhoeck & Ruprecht, 2020.
46 In this section, I will primarily talk about the urban realm because scholars have made explicit arguments about how to study urban animals, but not really about rural creatures even though many arguments also hold true for them.

in's *Elephant Slaves and Pampered Parrots* examined the rise of bourgeois Parisian culture and class differences through the cultural engagement with pets. Ever since, a culturally centered perspective has been a central approach in much of the scholarly literature on human-animal relations. This cultural approach worked well because there were ample printed sources ranging from literary sources and poetry to pamphlets, newspapers, diaries, and memoires to discern the role of animals in cities and the countryside. Moreover, this cultural approach provided a somewhat safe entryway to the study of urban animality because it remained clearly focused on humans and how they talked about animals. For the most part, animals appeared as a discursive medium that could be deconstructed and critically analyzed with the conceptual tools of the linguistic turn.

Soon, however, social historians tried to get closer to the daily (urban) practices and interactions between humans and animals [→Social History]. Books like Clay Mc Shane and Joel Tarr's *The Horse in the City* explored how class relations permeated the engagement between humans and animals, in this case how different Bostonians lived and worked with horses. My own research on the history of nineteenth-century slaughterhouses in Paris, Berlin, and Chicago examined how the changing presence of livestock and everyday practices of slaughter played a crucial role in the metropolitan transformation of these cities. In order to get a better sense of animal activities, these studies, in addition to the textual records mentioned above, also turned to some of the more classical sources of social history, e.g., statistics for a sense of the numbers and types of animals, to manuals, reports, and trade publications to understand daily practices, to veterinary records to examine how animal bodies and behaviors were explained, and to police reports to uncover problems. Of course, the actual animals remained allusive because even in these sources they did not speak, but scholarly attention was shifting from discourses about towards the actual (daily) lives of animals. This shift also reemphasized the material dimensions of the urban, which was in line with a concurrent growing interest in urban environmental history [→Environmental History].[47]

The growing curiosity about urban animal behaviors also raised another question – that of animal agency. While animals appear in most urban scholarship primarily as material objects or resources, in recent years, a growing number of HAS scholars insists that animals should be studied for their own sake and as purpose-driven actors in urban and rural assemblages. In addition to claims about agency, the notion of liminality, as pointed out above, has provided a new direction to the study of urban human-animal relations. Focusing on the transgressive nature of human-animal bonds, this perspective questions long held ideas about dichotomies,

---

[47] Brantz, Dorothee. Animals in Urban Environmental History. In *Concepts of Urban Environmental History*, Sebastian Haumann, Martin Knoll, Detlev Mares (eds.), 191–202, Bielefeld: transcript, 2020.

taxonomies, and boundaries of all kinds in order to gain a fresh and more inclusive understanding of how animals have contributed to (human) societies.[48]

Another key approach centers on animal spaces and the question where creatures are found in cities, to which spaces they have been assigned and for what purpose. Scholars from diverse fields including geography, sociology, history, and cultural studies, to name but a few, are investigating places of contact, conflict, and exhibition to identify the transcultural spaces different animals co-inhabited and how they contributed to the formation of transspecies ecologies, or as Jennifer Wolch has called it – a zoöpolis.[49] As examples of such zoöpolises (even if authors are not calling them that), scholars are publishing histories of cities centered on the role animals have played in them.[50] Others are writing about specific animals in these cities.[51]

In general, cities certainly offer rich fields of investigation. Up to now, more emphasis has been given to domesticated animals like dogs, cats, elephants, horses, livestock and to animal-centered institutions – zoos, aquariums, circus, natural history museums, slaughterhouses – than to the more aloof and wild spaces where most animals live – in the underground, in rivers, up in the air etc. This can be in some part attributed to the availability of sources, but perhaps also to the more implicit understanding that most urban animals live in the service of humans, which is of course not true. A growing interest in the urban wilderness is paying tribute to a new, more inclusive, understanding of the urban as a much more diversified and more-than-human environment. A budding number of urban wildlife guides are at-

---

48 Wischermann, Clemens. Liminale Leben(s)räume: Grenzverlegungen zwischen urbanen menschlichen Gesellschaften und anderen Tieren im 19. und 20. Jahrhundert. In *Urbane Tier-Räume*, Thomas Hauck, Stefanie Hennecke, André Krebber, Wiebke Reinert, Mieke Roscher (eds.). 15–31. Berlin: Reimer, 2017.
49 Bull, Jacob, Tora Holmberg and Cecilia Asberg (eds.). *Animal Places: Lively Cartographies of Human-Animal Relations*. London: Routledge, 2018; Hauck, Thomas, Stefanie Hennecke, André Krebber, Wiebke Reinert, Mieke Roscher (eds.). *Urbane Tier-Räume*. Berlin: Reimer, 2017; Holmberg, Tora. *Urban Animals: Crowding in Zoocities*. London: Routledge, 2015; Wolch, Jennifer. Zoöpolis. *Capitalism, Nature, Socialism* 6, no. 2 (February 1996): 21–47.
50 Brown, Frederick L. and Paul Sutter. *The City Is More Than Human: An Animal History of Seattle*. Seattle, WA: University of Washington Press, 2016; Dean, Joanna, Darcy Ingram and Christabelle Sethna (eds.). *Animal Metropolis: Histories of Human-Animal Relations in Urban Canada*. Calgary: University of Calgary Press, 2017; McNeur, Catherine. *Taming Manhattan: Environmental Battles in the Antebellum City*. Cambridge: Harvard University Press, 2015; Miltenberger, Scott A. Viewing the Anthrozootic City: Humans, Domesticated Animals, and the Making of Early Nineteenth Century New York. In *The Historical Animal*, Susan Nance (ed.), 261–272. Syracuse, NY: Syracuse University Press, 2015; Sabloff, Annabelle. *Reordering the Natural World: Humans and Animals in the City*. Toronto: University of Toronto Press, 2001.
51 Benson, Etienne. The Urbanization of the Eastern Gray Squirrel in the United States. *Journal of American History* 100, no. 3 (November 2013): 691–710; Sax, Boria. *City of Ravens: The Extraordinary History of London, the Tower and its Famous Ravens*. New York, NY: Penguin, 2012.

testing to this expanding interest.[52] Several of these publications also look at historical developments. In addition, these contemporary publications are generating future historical records. Research collaborations with (urban) ecologists are proving very valuable with regard to such inter- and transdisciplinary endeavors that move beyond the traditional boundaries of historical scholarship.

4 Implication(s) of the Animal Turn

Asking about the implication for the *animal turn,* it seems clear that a more ecological vision of the urban will contribute to a broadening of vision about what the urban entails. Urban ecologists are helping us to understand the biological dimensions of this animate story, but it is up to historians to put this in a historical context.[53] Working together, we might help to refocus discourses about the urban and the role of humans and animals in it from traditional perspectives of the 'man-made city' to a more encompassing notion of urban biotopias. This notion would offer more room for ideas and practices of co-habitation, of acknowledging animals as neighbors who have shared our urban and rural environments for centuries and who are being massively threatened by climate change and species extinction. To be sure, such biotopias are not necessarily peaceful environments. Quite the contrary, they give rise to many conflicts between humans and animals, but also between animal species sharing the same space.

In conclusion, the work that has been done in this field has contributed a great deal to taking animals more seriously as actants in the production of urban and rural spaces. No longer do animals serve just as a projection screen for human history and its representations. They have emerged as independent and powerful entities worth being studied in and of themselves in order to gain a more differentiated understanding of how cities have developed across time and in different places. Urban HAS offers a rich potential for inter- and transdisciplinary work that critically links theory and practice, history and the present, as well as human and nonhuman spaces and temporalities. Much remains to be explored about multiethnic attitudes towards animals and the inherent cultural specificities and their distinct socio-natural contexts. As urban societies diversified, so did attitudes towards animals and animal products. Debates about methods of slaughtering animals, about which animals made acceptable pets, or which were treated as pests differed not only across cultures but also

---

52 Exemplary for this genre, see Ineichen, Steffen. *Die wilden Tiere in der Stadt: Zur Naturgeschichte der Stadt.* Zurich: Waldgut, 1997; Kegel, Bernhard. *Tiere in der Stadt: Eine Naturgeschichte.* Munich: DuMont, 2013; Lotzkat, Sebastian. *Landflucht der Wildtiere: Wie Wildschwein, Waschbär, Wolf und Co. unsere Städte erobern.* Hamburg: Rowohlt, 2016; Van Horn, Gavin. *City Creatures: Animal Encounters in the Chicago Wilderness.* Chicago, IL: University of Chicago Press, 2015.
53 Schilthuizen, Menno. *Darwin Comes to Town: How the Urban Jungle Drives Evolution.* London: Quercus, 2018.

within urban societies. Many of these differences and their effects on urban living remain to be explored empirically and conceptually. Overall, we should be working towards new notions of kinship and the establishment of an urban multispecies society that builds on the philosophy of posthumanism or even the nonhuman turn.[54] Again, it is geographers and anthropologists who are already working on the contemporary dimensions of this [→Multispecies Ethnography]. Historians could contribute to this dialogue by adding a deeper temporal perspective as well as a close reading of the historical complexities that have led to the present we inhabit together.

Selected Bibliography

Atkins, Peter (ed.). *Animal Cities: Beastly Urban Histories*. Farnham: Ashgate, 2012.
Brantz, Dorothee (ed.). *Beastly Natures: Animals, Humans, and the Study of History.* Charlottesville, VA: University of Virginia Press, 2010.
Choyke, Alice M. and Gerhard Jaritz (eds.). *Animaltown: Beasts in Medieval Urban Space.* Oxford: BAR Publishing, 2017.
Hauck, Thomas, Stefanie Hennecke, André Krebber, Wiebke Reinert and Mieke Roscher. *Urbane Tier-Räume*. Berlin: Reimer 2017.
Holmberg, Tora. *Urban Animals: Crowding in Zoocities*. London: Routledge, 2015.
Kete, Kathleen. *The Beast in the Boudoir: Petkeeping in Nineteenth-Century Paris*. Berkeley, CA: University of California Press, 1995.
Lee, Paula Young (ed.). *Meat Modernism and the Rise of the Slaughterhouse*. Durham, NH: University of New Hampshire Press, 2008.
McShane, Clay and Joel Arthur Tarr. *The Horse in the City: Living Machines in the Nineteenth Century.* Baltimore, MD: Johns Hopkins University Press, 2007.
Philo, Chris and Chris Wilbert (eds.). *Animal Spaces, Beastly Places: New Geographies of Human-Animal Relations*. London: Routledge, 2000.
Ritvo, Harriet. *The Animal Estate: The English and Other Creatures in the Victorian Age.* Cambridge, MA: Harvard University Press, 1987.
Wischermann, Clemens, Aline Steinbrecher and Philipp Howell (eds.). *Animal History in the Modern City: Exploring Liminality.* London: Bloomsbury, 2019.

---

54 Grusin, Richard (ed.). *The Nonhuman Turn*. Minneapolis, MN: University of Minnesota Press, 2015; Wolf, Cary. *What is Posthumanism*. Minneapolis, MN: University of Minnesota Press, 2009.

Mitchell G. Ash
History of Science

1 Introduction and Overview

In his uncompleted utopian text *New Atlantis* (1627), Francis Bacon wrote about the creation of new animal species in an imaginary "Solomon's House": "We find means to make commixtures and copulations of different kinds; which have produced many new kinds, and them not barren, as the general opinion is [...] Neither do we do this by chance, but we know beforehand, of what matter and commixture what kind of these creatures will arise."[1] Present-day readers often find that this passage presages the making of creatures like the genetically modified "*OncoMouse*" (patented in 1988) or the cloning of the famous sheep "Dolly" in 1996. In historical perspective, however, Bacon looked not only into the future, but back to the Lord's injunction in the book of Genesis: "fill the Earth and subdue it, and have dominion over the fish of the sea and the birds of the heavens and over every living thing that moves upon the earth."[2]

Bacon's statement that the purpose of "Solomon's House" was "the enlarging of the bounds of human empire to the effecting of all things possible"[3] was thus a logical extension of the ancient Biblical mandate. Yet his vision also anticipates a fundamental transformation of the meaning of knowledge itself from philosophical understanding of the world to the systematic manipulation of nature which began in the seventeenth century but came to full realization only in the twentieth and twenty-first centuries. This transformation is unthinkable without the exploitation of animal bodies and cooperation with living animals.

This chapter outlines the roles of the natural sciences in the history of animal-human relations. It begins with a summary of the history of animals in science, which includes both experiments on animals [→History of Experimental Animals and the History of Animal Experiments] and other approaches. There follows a selective discussion of recent literature on science with animals, focusing on the impact of

---

Note: Portions of this chapter have been translated by the author from Ash, Mitchell G. Tiere und Wissenschaft. Versachlichung und Vermenschlichung im Widerstreit. In *Tiere und Geschichte. Konturen einer Animate History*, Gesine Krüger, Aline Steinbrecher, Clemens Wischermann (eds.), 267–291. Stuttgart: Franz Steiner Verlag, 2014.

1 Bacon, Francis. *New Atlantis*. In *The Advancement of Learning and New Atlantis*, (The World's Classics 93), Thomas Case (ed.), 235–302. London: Oxford University Press, 1969, 291 [first published in 1627].
2 Genesis 1:28, English Standard Version.
3 Bacon, *New Atlantis*, 288.

the *animal turn* in this field. A consistent theme will be the persistent ambivalence between the treatment of animals as things and the tendency to humanize some privileged animal species – an ambivalence characteristic of animal-human history in general.

2 Topics and Themes

The history of science with animals has not been limited to experiments on animals, for three reasons: First, laboratory experiments did not become an important mode of doing science with animals until the second half of the nineteenth century; until that time, the methods of natural history were predominant. Second, even after animal experiments acquired paradigmatic significance, natural history research in museums continued, and observation of living animals' behavior in the field became more common and systematic.[4] Third, in the twentieth century, the new discipline of ethology produced groundbreaking experimental knowledge of animal behavior without murder or vivisection. To capture this complexity, the following discussion is divided into three parts: (1) the rise of the natural history tradition and animal experimentation from the early modern period until the nineteenth century; (2) scientific animal study without murder – from Darwin to ethology; (3) continued ambivalence in the late twentieth century, including the use and redesign of model organisms alongside field studies of birds, fish and mammals.

Animals as Research Objects from the Early Modern Period to the Nineteenth Century

The modern history of experiments with animals begins with the work of physician William Harvey on the circulation of the blood.[5] His most important study, entitled in English *Movement of the Heart and Blood in Animals*, appeared in 1628, only one year after Bacon's *New Atlantis*.[6] Harvey established that the blood in animal and human bodies circulates continuously, and that arterial and venous blood were not different substances, but the same fluid. To prove these claims, Harvey dissected the hearts of cold- and warm-blooded animals to discover whether their anatomical features allowed blood to circulate. He also experimented on his own body and on living ani-

---

[4] For examples, see Kuklick, Henrika and Robert E. Kohler (eds.). Science in the Field. *Osiris* 11 (1996).
[5] Guerrini, Anita. *Experimenting with Humans and Animals. From Galen to Animal Rights*. Baltimore, MD: Johns Hopkins University Press, 2003, especially 23–25, 28–33.
[6] Harvey, William. Movement of the Heart and Blood in Animals: An Anatomical Essay. Translated by William J. Franklin. In *The Circulation of the Blood and Other Writings*, Andrew Wear (ed.), 1–111. New York, NY: Everyman's Library, 1993 [first published in 1628].

mals, mainly rabbits, to derive conclusions about the movement of the blood in arteries and veins. These were experiments in the modern sense, because they involved planned variations of the situations to be studied and the empirical proof of hypotheses or disproof of counterclaims. The research tradition Harvey founded, linking anatomical studies and physiological experiments on animals and humans, continues to this day.

Animals were also used to demonstrate laws of physical science, most famously in Robert Boyle's proof of the existence of a vacuum in the 1660s. Boyle refuted Descartes' claim that a vacuum was logically impossible by placing small animals – rabbits or doves – in the receiver of the air pump he had designed and removing the air, after which the animals expired. Thus, animal deaths were integral to his experimental design, which was not necessarily so in Harvey's experiments. Animal experiments were severely criticized at the time for interfering impermissibly in God's creation; and vivisection was condemned as a training ground for cruelty.

Predominant in this period, however, was the presentation of stuffed and mounted animal bodies in so-called "cabinets of curiosity" maintained by noblemen.[7] Exotic animals were also dissected after they died, for example at the French court menagerie in Versailles and in Paris [→History of the Zoo].[8] By the eighteenth century the careful classification and display of animal bodies in natural history museums had come to the fore [→History of Animal Collections/Animal Taxonomy]. These exhibited animals were and are seldom alive, but rather stuffed by taxidermists or arranged and portrayed by artists who specialized in such work [→Visual Culture Studies and Art History].[9] The prominence of these collections in court and noble culture was based on a common status principle: the more and the more exotic the species, the better. The scientific potential of such displays had been stated by Bacon in his major work, *Novum Organum* (1620), where he sketched the program for what he called "a natural and experimental history."[10] By this he did not mean anything resembling evolution, but rather the systematically ordered presentation of all kinds of natural objects, including the stars and planets, minerals, plants and animals and also human phenomena, as exemplars of an unchanging Creation.

The *Systema naturae* (1735–1758) of Swedish naturalist Carl Linnaeus can be seen as a realization of this program. In Linnaeus' system, the animal kingdom was part of a natural order organized in a strict hierarchy, beginning with the animal,

---

7 Grigson, Caroline. *Menagerie: The History of Exotic Animals in England, 1100–1837.* Oxford: Oxford University Press, 2016.
8 Guerrini, Anita. *The Courtiers' Anatomists: Animals and Humans in Louis XIV's Paris.* Chicago, IL: University of Chicago Press, 2015.
9 On collaboration of artists and natural historians, see Daston, Lorraine J. and Peter Galison. *Objectivity.* New York, NY: Zone Books, 2007, pt. 2. On taxidermy see Poliquin, Rachel. *The Breathless Zoo: Taxidermy and the Cultures of Longing.* University Park, PA: Pennsylvania State University Press, 2012.
10 Bacon, Francis. *Novum Organon.* Translated and edited by Peter Urbach and John Gibson. Chicago, IL: Open Court, 1995, 297–323 [first published in 1620].

plant and mineral kingdoms and proceeding to phyla, classes, orders, genera and species. Linnaeus' enduring innovation was binomial classification (with genus and species), which applied not only to animals, but also to plants and minerals.[11] Different from Linnaeus' system was that of the Comte de Buffon, head of the Royal Botanical Garden in Paris, who regarded Linnaeus' system as arbitrary and based his classifications of plants and animals in *Histoire naturelle* (1749–1786) not on selected anatomical features, as Linnaeus did, but on functional criteria such as the ability to breed. Buffon utilized drawings of animals to illustrate a continuous progression of forms opposing Linnaeus' emphasis on separately created species, thus marking the transition to a modern historical conception of nature.[12] Because the fixation and preservation of so-called "type" (exemplary) and other specimens was considered necessary in order to secure and stabilize natural knowledge, Liv Emma Thorsen writes that natural history worked with "animal matter" rather than animals.[13]

Natural histories, for example studies of insects (a category that included more types of creatures than it does today) by Maria Sibylla Merian and René de Réaumur, also incorporated narratives of naturalists' encounters with animals. Describing animal behaviors and their development from eggs to larvae to adults presented nature in motion. By showing how researchers had acquired their specimens, they certified that their knowledge was valid; and by including the researchers themselves, they also showed others how to do natural history.[14] Réaumur and others also performed experiments with animals, such as bees and caterpillars.[15]

Science in natural history museums was made possible by worldwide collecting, for example in the voyages of Captain James Cook with naturalists on board in the mid- to late eighteenth century, of Charles Darwin on the *Beagle* (1831–1836), or of the Austrian frigate *Novara* (1857–1859) in the nineteenth century, as well as regional efforts by natural history societies in cooperation with museum experts.[16] Collecting

---

[11] URL: en.wikipedia.org/wiki/Carl_Linnaeus. (October 29, 2019). See also Gibson, Susannah. *Animal, Vegetable, Mineral? How Eighteenth-Century Science Disrupted the Natural Order*. Oxford: Oxford University Press, 2015.
[12] Spary, Emma C. *Utopia's Garden: French Natural History from Old Regime to Revolution*. Chicago, IL: University of Chicago Press, 2000.
[13] Thorsen, Liv Emma. Animal Matter in Museums: Exemplifying Materiality. In *The Routledge Companion to Animal–Human History*, Hilda Kean, Philip Howell (eds.), 171–193. London and New York, NY: Routledge, 2019.
[14] Terrall, Mary. Narrative and Natural History in the Eighteenth Century. *Studies in History and Philosophy of Science* 62 (April 2017): 51–64.
[15] Terrall, Mary. Experimental Natural History. In *Worlds of Natural History*, Helen Anne Curry, Nicholas Jardine, James A. Secord, Emma C. Spary (eds.), 170–184. Cambridge: Cambridge University Press, 2018.
[16] Anderson, Katherine. Natural History and the Scientific Voyage. In *Worlds of Natural History*, Helen Anne Curry, Nicholas Jardine, James A. Secord, Emma C. Spary (eds.), 304–318. Cambridge: Cambridge University Press, 2018.

was also closely linked with hunting, for example the licensed "scientific hunting" of Annie Alexander for the U.S. Biological Survey in the late nineteenth and early twentieth centuries.[17]

In the nineteenth century, exotic animal displays in natural history museums and zoos became part and parcel of urban life [→History of the Zoo;→Urban (and Rural) History]. Natural history museums expanded to new locations and opened their doors to wider publics, while new societies for zoology, ornithology, ichthyology and other fields brought together experts and laypeople.[18] Observations of laypeople in zoos contributed to the acceptance of evolutionary theory.[19]

Laboratory experiments on animals acquired greater prominence during the second half of the nineteenth and the first third of the twentieth century, first in physiology and then in microbiology. Prominent examples of this fundamental shift are the work of physicist and physiologist Hermann Helmholtz on the speed of nervous impulses in the 1850s, performed on the freshly vivisected leg muscles of common pond frogs (*Rana rana*),[20] and experiments on animal nutrition and diseases carried out with rabbits and other species by physiologist Claude Bernard in the same period. Both men believed that biological laws were based on those of physics and chemistry. However, while Helmholtz treated organic processes as mechanical operations and regarded nerves as analogous to telegraph wires, Bernard saw animal organs as locations of physiological processes to be investigated in order to draw clinically relevant conclusions about such processes in human bodies.[21] In his *Introduction to the Study of Experimental Medicine* (1865), for example, Bernard presented his proof that the mechanism of death following poisoning with *Curare* functioned through the nerves rather than the heart, and his discovery that the sugar in animal blood came from the animal itself and not, as was then thought, from its food.[22] Bernard aggressively defended vivisection against public attacks, despite the pain this caused to the animals, by claiming that "great numbers" of animals had to die in order to advance medical knowledge.[23]

---

[17] Kohler, Robert E. *All Creatures: Naturalists, Collectors, and Biodiversity*. Princeton, NJ: Princeton University Press, 2006, 129.
[18] Nyhart, Lynn K. Publics and Practices. In *Worlds of Natural History*, Helen Anne Curry, Nicholas Jardine, James A. Secord, Emma C. Spary (eds.), 335–348. Cambridge: Cambridge University Press, 2018.
[19] Hochadel, Oliver. Observing Exotic Animals Next Door: Scientific Observations at the Zoo. *Science in Context* 24, no. 2 (June 2011): 183–214.
[20] Olesko, Kathryn M. and Frederick Holmes. Experiment, Quantification and Discovery. Helmholtz's Early Physiological Researches, 1843–1850. In *Hermann von Helmholtz and the Foundations of Nineteenth-Century Science*, David Cahan (ed.), 83–95. Berkeley, CA: University of California Press, 1993.
[21] For background see Guerrini, *Experimenting with Humans and Animals*, ch. 4.
[22] Bernard, Claude. *Introduction to the Study of Experimental Medicine*. Translated by Henry Copley Greene. New York, NY: Henry Schuman 1949, 157–158, 163–164 [first published in 1865]. URL: archive.org/details/b21270557/ (October 31, 2019).
[23] Bernard, *Introduction*, 99.

Another role for animals in laboratory science can be seen in the microbiological work of Louis Pasteur, Robert Koch and others in the 1880s. In these studies animals acted as "living laboratory vessels,"[24] that is, as media for the testing of hypotheses about the causal role of so called "microbes" in human diseases as well as the effectiveness of medications designed to eliminate or suppress them. A prominent example of this approach is the work of German physician Paul Ehrlich, known today as a founder of modern chemotherapy.[25] In his Institute for Experimental Therapy in Frankfurt am Main, Ehrlich and his coworkers carried out animal experiments using chemical dyes as tracers, in order to learn whether treatment with particular chemicals supported the production of "antibodies" – a term Ehrlich invented. The white mice they bred and selected for experimental study were "storehouses for micro-organisms,"[26] but contrary to Bernard's defense of killing they were not sacrificial victims whose deaths were accepted as the price of scientific success. Rather, if they died Ehrlich attributed this to poor care or a flawed experimental design.

The rise of systematic laboratory research with animals led to a fundamental change in animal-human relations in the sciences, which was not limited to murder for the sake of science. By the turn of the twentieth century, lab-based breeding of warm-blooded animals reached factory scale in Ivan Pavlov's laboratory in St. Petersburg.[27] However, Pavlov's dogs, like Ehrlich's mice, were cared for and carefully observed for months before they were vivisected for research purposes.

Animal Research Without Vivisection or Murder – From Darwin to Ethology

The killing thought to be required to preserve so-called "type" (exemplary) specimens in museums or to carry out laboratory experiments presupposed that there was a never-ending supply of usable creatures. Extinctions of entire species due to human overreach began to be acknowledged only grudgingly in the mid-nineteenth century.[28] In contrast, research on the adaptation of exotic animals to cold European climates was undertaken to keep the animals alive. So-called "acclimatization" soci-

---

24 Guerrini, *Experimenting with Humans and Animals*, 98.
25 For the following see Hüntelmann, Axel C. Füttern und gefüttert werden. Versorgungskreisläufe und Nahrungsregimes im Königlich Preußischen Institut für experimentelle Therapie, ca. 1900 bis 1910. *Berichte zur Wissenschaftsgeschichte* 35, no. 4 (November 2012): 300–321.
26 Hüntelmann, Füttern und gefüttert werden, 306.
27 Todes, Daniel P. *Pavlov's Physiology Factory: Experiment, Interpretation, Laboratory Enterprise.* Baltimore, MD: Johns Hopkins University Press, 2001.
28 Freeman, Carol. Extinction, Representation, Agency: The Case of the Dodo. In *Considering Animals: Contemporary Studies in Animal-Human Relations*, Carol Freeman, Elisabeth Leane, Yvette Watt (eds.), 153–168. Farnham: Ashgate, 2011.

eties were established in Paris in 1854, and later in London, Moscow, Berlin, Melbourne, Calcutta and other cities.[29] These institutions sought to domesticate exotic breeds for their feathers, fur or meat, and also to improve the survival rate of existing zoos. These goals were at best partially achieved, but the Paris institution still exists today.

From the natural history tradition, and enabled in large part by world-wide expansion of field research and collecting, emerged the theory of evolution, formulated in greatest breadth by Charles Darwin. Darwin's significance for the role of science in the history of animal-human relations cannot be overestimated; put very simply, he abolished once and for all the boundary between humans and other animals. He accomplished this in three ways, exemplified in three of his major books.

In *The Origins of Species by Natural Selection and the Preservation of Preferred Races in the Struggle for Life* (1859), Darwin took a giant step toward overcoming the exceptional status of human beings by claiming that his theory of evolution was valid for all organisms, but in doing so he did not reduce humans to nonhuman animals. Rather, he claimed that population pressure was the driving force of the "struggle for life" for human and nonhuman animals alike, and plants as well.[30]

In *The Descent of Man and Selection in Relation to Sex* (1871), Darwin overcame the animal-human boundary by arguing that all primates, including humans, are descended "from a hairy, tailed quadruped, probably arboreal in its habits, and an inhabitant of the old world."[31] To account for the allegedly unique cultural capabilities of humans, such as language, morals and religion, Darwin postulated a kind of cultural evolution. For each of these capabilities he supposed a natural basis – in the case of language, simple signaling systems like those of birds or monkeys.

Finally, in *The Expression of the Emotions in Animals and Men* (1872),[32] Darwin analyzed animal and human emotional expressions with the same methods, without reducing the emotional behavior of humans to that of animals or vice versa. He based this approach on the claim that nervous excitations exerted the same direct influence on human and animal bodies; as an example, he cited the tendency to shrink back in fear in the face of a threat. On the other hand, physical expressions of different species could be interpreted as expressions of similar emotions.

---

29 Osborne, Michael. *Nature, the Exotic and the Science of French Colonialism*. Bloomington, IN: University of Indiana Press, 1998, ch. 3.
30 Darwin, Charles. *On the Origin of Species by Means of Natural Selection or the Preservation of Preferred Races in the Struggle for Life* (new edition, revised and augmented). New York, NY: Appleton, 1861, 63 [first published in 1859]. URL: darwin-online.org.uk/converted/pdf/1861_OriginNY_F382.pdf (October 29, 2019).
31 Darwin, Charles. *The Descent of Man and Selection in Relation to Sex* (new edition, revised and augmented in one volume). New York, NY: Appleton 1889, 609 [first published in 1871]. URL: darwin-online.org.uk/converted/pdf/1889_Descent_F969.pdf (29 October 2019).
32 Darwin, Charles. *The Expression of the Emotions in Man and Animals*. New York, NY: Appleton 1897 [first published in 1872]. URL: darwin-online.org.uk/converted/pdf/1897_Expression_F1152.pdf (October 29, 2019).

Thus, for Darwin, continuity between humans and nonhuman animals meant not only the reduction of human behavior to that of supposedly "lower" species, but also the observation of capabilities supposed to be unique to humans in nonhuman species. Perhaps that is one reason why Darwin remains a positive reference for animal rights advocates, despite his support for eugenics.

The implications of evolutionary theory for animal-human relations are still controversial. On the one hand, a line can be drawn from Darwin to radical behaviorism, which denied any sort of consciousness to animals.[33] The methodological foundation for this was Conwy Lloyd Morgan's "canon," formulated in 1894, which forbade explaining animal behaviors by reference to "higher" mental processes, if they could also be explained by referring to processes "lower" on the evolutionary scale.[34] On the other hand, George John Romanes, who claimed to be a Darwinist, argued in *Animal Intelligence* (1883) that animals possessed consciousness.[35] Leonard Trelawny Hobhouse did the same under the title *Mind in Evolution* in 1901 on the basis of research on problem-solving with apes at the London Zoo.[36] However, in another book with the title *Animal Intelligence* published in 1898 and expanded in 1911, American psychologist Edward Thorndike tried to refute Romanes on the basis of experiments which showed in his view that animals' problem solutions were based on random trial and error.[37]

Yet another way of doing science with animals emerged early in the twentieth century, when ethologists carried out close observations and experimental studies of animal behavior in natural environments. This was a countermovement to the artificial control of animal behavior in laboratories. Ethology and primatology, which emerged at the same time, suggest that science with animals is possible without vivisection or murder. The most prominent examples of ethological research are studies by Konrad Lorenz and Nikolaas Tinbergen with birds and Karl von Frisch's studies of honeybee communication in the 1920s and 1930s.[38]

Lorenz began with extensive studies of companionship behavior in crows.[39] In this work he discovered and named the phenomenon of "imprinting," by which

---

[33] Boakes, Robert. *From Darwin to Behaviorism: Psychology and the Minds of Animals*. Cambridge: Cambridge University Press, 1984.
[34] Fitzpatrick, Simon and Grant Goodrich. Building a Science of Animal Minds. Lloyd Morgan, Experimentation, and Morgan's Canon. *Journal of the History of Biology* 50, no. 3 (August 2017): 525–569.
[35] Romanes, Geroge John. *Animal Intelligence*. New York, NY: Appleton 1883.
[36] Hobhouse, Leonard Trelawny. *Mind in Evolution*. London: Macmillan 1901.
[37] Thorndike, Edward. *Animal Intelligence: An Experimental Study of Associative Processes in Animals*. New York, NY: Macmillan, 1898; Thorndike, Edward. *Animal Intelligence: Experimental Studies*. New York, NY: Macmillan, 1911.
[38] Burkhardt, Richard W. *Patterns of Behavior: Konrad Lorenz, Niko Tinbergen, and the Founding of Ethology*. Chicago, IL: University of Chicago Press, 2005, chs. 3–4; Munz, Tania. *The Dancing Bees: Karl von Frisch and the Discovery of the Honeybee Language*. Chicago, IL: University of Chicago Press, 2016.
[39] Burkhardt, *Patterns of Behavior*, ch. 3.

he meant the fixation of instinctive behavior on the creature who induced it, whether this was the "natural" object of such behavior or, as in these studies, the researcher. With Dutch naturalist Tinbergen, he then conducted experiments on the egg rolling behavior of geese published in 1938.[40] They found that when eggs or egg-like objects had fallen out of the nest or were placed next to it, mother geese reacted with two behaviors: pushing the egg with their beaks, and additional movements suited to keeping the eggs rolling in a straight line toward the nest.[41] Lorenz argued that these were inherited behavioral schemata that could be released by different external stimuli, but could not be interpreted as chains of reflexes. The studies took place in the open air; the animals were neither vivisected nor killed and dissected afterward, but continued to live on the Lorenz family estate where they had been raised. As a basic point of method Lorenz stated that one must love the animals in order to learn what they have to teach us. This meant that one had to grasp their entire repertoire of behaviors before performing experiments with them. He also named the creatures he studied; one of them, the grey goose Martina, became world famous.[42]

Sympathetic as such experiments may appear to be, it must also be noted that Lorenz joined the Nazi party in 1939, and then presented a crude analogy between "decadent behaviors" in domestic animals and "degeneration" in supposedly "over-civilized" humans; properly biological selection programs should therefore favor "natural" [*arteigene*] over "degenerate" types.[43] He expressed similar views on the heritability of behavior long after 1945, for example in his book *On Aggression* (1963).[44] Tinbergen, who had been in the Dutch resistance during the war, took a rather different path. The research tradition he established at Oxford in the 1940s avoided theoretical speculation or analogies from animals to humans.[45] This is true also of Karl von Frisch's experimental studies of honeybee communication. Despite their different political views, all three scientists shared a commitment to studying animal perception and behavior in relation to the animals' own perceived environments. This was a decisive step beyond both objectification and anthropomorphism; yet the use of human-derived terminology persisted, for example von Frisch's references to honeybee "dancing".

---

40 Burkhardt, *Patterns of Behavior*, 205–208.
41 Films of the experiments or recreations of them can be viewed on YouTube under the keywords "egg rolling experiment," or "Fixed Pattern Behavior".
42 Munz, Tania. My Goose Child Martina: The Multiple Uses of Geese in the Writings of Konrad Lorenz. *Historical Studies in the Natural Sciences* 41, no. 4 (Fall 2011): 405–456.
43 Burkhardt, *Patterns of Behavior*, ch. 5.
44 Lorenz, Konrad. *On Aggression*. Translated by Marjorie Kerr Wilson. New York, NY: Harcourt Brace & World, 1963.
45 Tinbergen, Nikolaas. *The Animal in its World. Explorations of an Ethologist*. Cambridge, MA: Harvard University Press, 1972.

A Persistent Ambivalence in the Late Twentieth Century

Of course, ethological experiments did not displace invasive methods [→History of Experimental Animals and the History of Animal Experiments]. Experiments with fruit flies (*Drosophila melanogaster*) by geneticists led to planned modifications of this model organism, first by transplanting organs and later by transfusing fluids into fly larvae or feeding them with chemicals.[46] Procedures like these eventually led to the genetically modified "onco-mouse" in the later twentieth century.[47] Such organisms can be patented in the US since the 1980s, and are offered for sale on the internet.[48] All this has encountered political opposition, some of it violent. Efforts by animal rights activists to intervene into or even prevent laboratory science with animals continue in some respects the tradition of the anti-vivisection movement of the nineteenth century; increasing state regulation and academic self-regulation have not reduced the activists' commitment.[49]

Alongside all this, naturalistic field work with animals has persisted and even intensified, driven by ecological perspectives and a commitment to studying animals in their own habitats, with minimally invasive methods. Examples are field studies of migratory birds, fish, whales and dolphins. The use of marking and tracking techniques in this work exemplifies technology in the service of scientific understanding, rather than technoscientific manipulation.[50] However, the best-known examples of science with animals based on intensive animal-human contact come from primatology, the study of monkeys and anthropoid apes.

The roots of primatology go back to nineteenth-century studies of apes in zoos. After 1945, field work with apes in Africa and Asia intensified; by the 1970s the majority of doctorates in the field were being awarded to women.[51] The high involvement of women scientists apparently affected the choice of species to be studied (more peaceful Bonobos rather than aggressive Pavians), but male primatologists also had human-like relations with apes. Work in this field has stimulated an intense

---

[46] Kohler, Robert E. *Lords of the Fly: Drosophila Genetics and the Experimental Life*. Chicago, IL: University of Chicago Press, 1994, ch. 7.
[47] Endersby, Jim. *A Guinea Pig's History of Biology: The Plants and Animals Who Taught us the Facts of Life*. London: William Heinemann, 2007, ch. 12.
[48] The author received an unsolicited advertisement for transgenic mice from the firm Cyagen Biosciences in Santa Clara, California by E-mail on 15 January 2014.
[49] On the origins and history of the anti-vivsection and animal welfare movements in Britain, see Roscher, Mieke. *Ein Königreich für Tiere. Die Geschichte der britischen Tierrechtsbewegung*. Marburg: Tectum Wissenschaftsverlag, 2009.
[50] Benson, Etienne. *Wired Wilderness. Technologies of Tracking and the Making of Modern Wildlife*. Baltimore, MD: Johns Hopkins University Press, 2010.
[51] Haraway, Donna J. *Primate Visions: Gender, Race, and Nature in the World of Modern Science*. New York, NY and London: Routledge, 1989, especially pt. 3.

debate on whether to assign human capabilities such as toolmaking, language, or even cooperation and conflict, to nonhuman animals.[52]

The recent period has seen new approaches to animal-human interactions in the sciences. Only two examples can be mentioned here. In the 1970s, efforts were made by biologists and seismologists in California to coordinate research on laboratory and farm animals to establish whether they behaved abnormally before earth tremors, as folk tradition has long held, and thus might act as seismic sensors.[53] More recently, efforts have been made to integrate ethnology and climate science in studies of human-animal disequilibrium in the Arctic, to determine whether recently observed rapid decreases in animal populations are more appropriately analyzed at regional and local scales, rather than being caused by global climate change. Chief investigator Igor Krupnik speaks here not only of human or animal, but also of "habitat agency".[54]

3 Methods and Approaches

A wide variety of methods has been employed to study this topic, ranging from archival research, close reading and contextual interpretation of historical scientific texts that are standard procedure in history of science,[55] to approaches from social and cultural history that focus on the history of exotic animal display in broader cultural contexts, including the history of zoological gardens and more recently of museum displays [→Cultural History;→History of Ideas;→History of the Zoo;→Material Cultural Studies;→Public History].[56] Pioneering work on animals by cultural historians includes examples from the sciences, but is not limited to this field.[57] Due to the so-called "practical turn" since the 1990s, scholarship in the history of science has shifted from grand theories to detailed studies of research practices. The history of

---

[52] De Waal, Frans. *Chimpanzee Politics*. London: Jonathan Cape, 1982; Radick, Gregory. *The Simian Tongue. The Long Debate about Animal Language*. Chicago, IL: University of Chicago Press, 2008.
[53] Aronova, Elena. Earthquake Prediction, Biological Clocks, and Cold War Psy-ops: Using Animals as Seismic Sensors in 1970s California. *Studies in History and Philosophy of Science* 70 (August 2018): 50–57.
[54] Krupnik, Igor. Arctic Crashes: Revisiting the Human–Animal Disequilibrium Model in a Time of Rapid Change. *Human Ecology* 46, no. 5 (October 2018): 685–700. doi.org/10.1017/9781316481066.
[55] Curry, Helen Anne, Nicholas Jardine, James A. Secord and Emma C. Spary (eds.). *Worlds of Natural History*. Cambridge: Cambridge University Press, 2018.
[56] On museum displays see Thorsen, Liv Emma, Karen A. Rader and Adam Dodd (eds.). *Animals on Display: The Creaturely in Museums, Zoos and Natural History*. University Park, PA: Pennsylvania State University Press, 2013.
[57] Ritvo, Harriet. *Noble Cows and Hybrid Zebras: Essays on Animals and History*. Charlottesville, VA: University of Virginia Press, 2010.

zoological gardens, including scientific research in zoos, is increasingly being written as a history of animal-human relations.[58]

Recent efforts to include animals in the history of science have been inspired in part by feminism and poststructuralism [→Feminist Intersectionality Studies]. The work of Donna Haraway remains foundational in this respect, precisely because it reaches far beyond the history of science.[59] Further examples of feminist and poststructuralist approaches involving animal-human relations are Londa Schiebinger's essay on how Linnaeus came to call mammals by that name,[60] Sara Jansen's study of the history of entomology and the battle against insect pests as a history of maleness,[61] and Marga Vicedo's study of psychologist Harry Harlow's work on primate mothering.[62] Political historians have also entered the mix, with emphasis on the history of the anti-vivisection movement and on science with animals in dictatorships [→Political History]. Notable examples of the latter include Boria Sax's work on the Reich Animal Protection Law of 1933, the first legislation worldwide that limited (but did not abolish) experiments with animals,[63] work on the attempts by German zoo directors Lutz and Heinz Heck to reverse breed prehistoric animals such as the aurochs,[64] and studies of agricultural animal research in Fascist regimes.[65]

In sum, a wide range of approaches to historical scholarship has been mobilized to include animals in the history of science, and work on animal-human history more broadly conceived also touches on science with animals. However, studies on these topics are a minority in each subdiscipline, and until very recently the perspective has often been that of the human researchers rather than that of the animals.

---

[58] See Ash, Mitchell G. Zoological Gardens. In *Worlds of Natural History*, Helen Anne Curry, Nicholas Jardine, James A. Secord, Emma C. Spary, (eds.), 418–432. Cambridge: Cambridge University Press, 2018 and the literature cited there.

[59] Haraway, *Primate Visions*; Haraway, Donna. *Simians, Cyborgs, and Women – the Re-Invention of Nature*. New York, NY: Routledge 1990.

[60] Schiebinger, Londa. Why Mammals are Called Mammals: Gender Politics in Eighteenth-Century Natural History. *The American Historical Review* 98, no. 2 (April 1993): 382–411.

[61] Jansen, Sara. *Schädlinge. Geschichte eines wissenschaftlichen und politischen Konstrukts 1840–1920*. Frankfurt a. M. and New York, NY: Campus Verlag, 2003.

[62] Vicedo, Marga. Mothers, Machines and Morals: Harry Harlow's Work on Primate Love from Lab to Legend. *Journal of the History of the Behavioral Sciences* 45, no. 3 (Summer 2009): 193–208.

[63] Sax, Boria. *Animals in the Third Reich. Pets, Scapegoats and the Holocaust*. New York, NY and London: Continuum, 2000, especially ch. 11, 110–123 and Appendices 1 and 2.

[64] Driessen, Clemens and Jamie Lorimer. Back-Breeding the Auerochs: The Heck Brothers, National Socialism and Imagined Geographies for Non-Human *Lebensraum*. In *Hitler's Geographies: The Spatialities of the Third Reich*, Paolo Giaccaria, Claudio Minca (eds.), 138–159. Chicago, IL: University of Chicago Press, 2016.

[65] Saraiva, Tiago. *Fascist Pigs: Technoscientific Organisms and the History of Fascism*. Cambridge, MA: MIT Press, 2016, chs. 4 and 6.

4 Implication(s) of the Animal Turn

In the debate on animal-human relations, scientific knowledge frequently plays a role. Often, however, the portrayal of science in such contexts appears to be based on stereotypes rather than serious historical study. The father figures of this master narrative are Bacon and Descartes. Bacon is indicted on the basis of his comparison of scientific research with hunting or his metaphor of experiments as the torture of nature, drawing upon the practice of criminal court judges in the examination of alleged witches.[66] Descartes stands accused because of his supposed claim – actually made by epigones – that animals are soulless machines, incapable of feeling pain;[67] Descartes himself ascribed sensation to animals, including the ability to feel pain.[68] Contrary to such one-sided polemics, human attitudes toward animals have always vacillated between objectification and humanization [→History of Ideas], and both attitudes have been present in the sciences as well. The focus on invasive laboratory experiments obscures the actual variety of human-animal relationships in science.

The history of science has long been concerned with knowledge of animals, but until quite recently work in this area was focused mainly on human activities, even when animals were involved.[69] More explicit inclusion of nonhuman animals in science is evident in the literature on model organisms. This began decades ago with studies of frogs (Genus *Rana*, many species) and other animals as "martyrs" or "heroes" of research [→History of Experimental Animals and the History of Animal Experiments]. More recent studies focus on model organisms as privileged research objects and as organizing centers for communities of scientific practice. Examples include the fruit fly (*Drosophila melanogaster*) and the nematode worm *Coenorhabditis elegans*.[70] Both the choice of species and the transferability of research results from model organisms to others have been discussed often, most profoundly by Georges Canguilhem.[71] Important for the history of animal-human relations is the criticism that research with model organisms focuses on a limited number of highly standar-

---

66 Merchant, Carolyn. *The Death of Nature: Women, Ecology, and the Scientific Revolution.* New York, NY: Harper & Row, 1980, ch. 7.
67 See, for example, Singer, Peter. *Animal Liberation: A New Ethics for our Treatment of Animals.* New York, NY: New York Review, 1975; Regan, Tom. *The Case for Animal Rights.* Berkeley, CA: University of California Press, 1983.
68 Guerrini, *Experimenting with Humans and Animals,* especially 33–37.
69 See, for example, Nyhart, Lynn K. *Biology Takes Form: Animal Morphology and the German Universities.* Chicago, IL: University of Chicago Press, 1995; Kohler, *All Creatures.*
70 Kohler, *Lords of the Fly*; Ankeny, Rachel A. The Natural History of *Coenorhabditis elegans* Research. *Nature Reviews Genetics* 2, no. 6 (June 2001): 474–479.
71 Canguilhem, Georges. Experimentation in Animal Biology. Translated by Stefanos Gerulanos and Daniela Ginsburg. In *Knowledge of Life,* Paola Morati, Todd Meyers (eds.), 3–24. New York, NY: Fordham University Press, 2008.

dized species;[72] this suggests that biologists are more likely to study artefacts of their own technologies than animals in their natural habitats. The *animal turn* has also influenced media studies in history of science; scientists' films of animal behavior have been interpreted as allowing viewers to see animal agency directly.[73]

Recent work has cited Bruno Latour's actor-network-theory in support of regarding animals as "actants", or even subjects. An example is Axel Hüntelmann's study of the mice in Paul Ehrlich's laboratory. Hüntelmann claims to see animal agency in the mice's instinctive refusal to accept poison or dyestuffs as nourishment,[74] and argues that animals' deaths were a form of resistance. However, such behavior took place in human-built laboratories, and did not alter the power relations there. It seems questionable to speak of "agency" or of a spontaneously created animal "society" in such cases. Since the animals and the scientists were dependent on one another in such settings, a kind of cooperation existed, but this was hardly a symmetrical relationship.[75] Like human slaves, these animals could extract themselves from their implicit "contracts" only by escaping, falling ill, or dying.

As suggested above, historical work on science with animals outside the laboratory has expanded. Examples include work on the history of zoological field research,[76] on natural history museums as places for research on animals,[77] and the history of zoo biology.[78] In 2007 Jim Endersby claimed in a book title to investigate the history of biology from the animals' point of view.[79] Unfortunately, the study actually considers the suitability of certain animals for particular research topics from the point of view of researchers, and does not attribute agency to animals. Bernd Hüppauf has devoted a cultural historical monograph to the frog, in which, however, the animal appears as an abstract entity with no life or agency of its own.[80]

Of course, the variety of ways of doing science with animals does not deny the predominance of laboratory research. The mass breeding of animals for laboratories, the view of animals and their bodies as energy consuming machines, and the persis-

---

[72] Ankeny, Rachel A. Historiographic Reflections on Model Organisms: Or How the Mureaucracy may be Limiting our Understanding of Contemporary Genetics and Genomics. *History and Philosophy of the Life Sciences* 32, no. 1 (January 2010): 91–104. doi.org/10.2307/23335054.
[73] For examples, see Munz, The Dancing Bees, 74; Winter, Alison. Cats on the Couch: The Experimental Production of Animal Neurosis. *Science in Context* 29, no. 1 (March 2016): 77–105.
[74] Hüntelmann, Füttern und gefüttert werden, 311.
[75] Krebber, André. Washoe: Das Subjekt in der Tierforschung. In *Philosophie der Tierforschung 3: Milieus und Akteure*, Martin Böhnert, Kristian Köchy, Matthias Wunsch (eds.), 187–220. Freiburg and Munich: Verlag Karl Alber, 2018.
[76] Mitman, Gregg. When Nature is the Zoo: Vision and Power in the Art and Science of Natural History. *Osiris* 11 (1996): 117–143.
[77] Nyhart, Publics and Practices.
[78] Ash, Zoological Gardens, 426–427 and the literature cited there.
[79] Endersby, *A Guinea Pig's History of Biology*.
[80] Hüppauf, Bernd. *Vom Frosch. Eine Kulturgeschichte zwischen Tierphilosophie und Ökologie*. Bielefeld: transcript, 2011.

tence of machine models of animal behavior in cybernetics all testify to the treatment of animals as technoscientific objects. The natural history tradition persists, though it is underfunded and engaged in a Sisyphean task of finding and classifying the millions of species estimated to exist, only a fraction of which have actually been described or classified. Tens of thousands of "lost species" lie in museum collections awaiting discovery and naming.[81] But the numbers of animals bred for today's laboratories far exceed those studied in ethology or primatology. The exploitation of animals in laboratory science is now hedged about by a web of administrative regulations; as a result, the availability of animals for scientific research is no longer unlimited. However, whether stronger regulation and the professionalization of animal care have in fact served the wellbeing of animals remains in dispute.

In this chapter I have attempted to show how the sciences have gained new impulses through varied ways of working with animals. Not only exploitation and murder of animals, but also cooperation with them has influenced the creation of scientific knowledge. Darwin's experiments with his dogs,[82] the relations of birds like Martina with Konrad Lorenz, or the warm-hearted interactions of primates and the people who study them show how far such collaborations have gone. Donna Haraway describes laboratory animals as "working companions."[83] Yet even when laboratory or zoo animals displayed resistance or cooperation, and thus agency, they rarely went beyond the roles assigned to them by scientists. Research on animal behavior that employs human-sounding concepts like cognition and intelligence, but gives them entirely new, animal-centered meanings, has become better established in recent years.[84] Nonetheless, the ambivalence between objectification and humanization of animals' characteristic of animal-human relations in general persists in scientific research with animals, and is unlikely to be overcome any time soon.

Selected Bibliography

Burkhardt, Richard W. *Patterns of Behavior: Konrad Lorenz, Niko Tinbergen and the Founding of Ethology*. Chicago, IL: University of Chicago Press, 2005.

Curry, Helen Anne, Nicholas Jardine, James A. Secord and Emma C. Spary, (eds.). *Worlds of Natural History*. Cambridge: Cambridge University Press, 2018.

Endersby, Jim. *A Guinea Pig's History of Biology: The Plants and Animals who Taught us the Facts of Life*. London: William Heinemann, 2007.

---

[81] Kemp, Christopher. *The Lost Species: Great Expeditions in the Collections of Natural History Museums*. Chicago, IL: University of Chicago Press, 2017.

[82] Townshend, Emma. *Darwin's Dogs. How Darwin's Pets Helped Form a World-Changing Theory of Evolution*. London: Francis Lincoln Limited, 2009.

[83] Haraway, Donna J. *The Companion Species Manifesto. Dogs, People and Significant Otherness*. Chicago, IL: University of Chicago Press, 2003.

[84] See, for example, Safina, Carl. *Beyond Words: What Animals Think and Feel*. New York, NY: Henry Holt and Company, 2015.

Guerrini, Anita. *Experimenting with Humans and Animals. From Galen to Animal Rights*. Baltimore, MD: Johns Hopkins University Press, 2003.
Haraway, Donna J. *Primate Visions: Gender, Race, and Nature in the World of Modern Science*. New York, NY and London: Routledge, 1989.
Munz, Tania. *The Dancing Bees: Karl von Frisch and the Discovery of the Honeybee Language*. Chicago, IL: University of Chicago Press, 2016.
Nyhart, Lynn K. *Modern Nature: The Rise of the Biological Perspective in Germany*. Chicago, IL: University of Chicago Press, 2009.
Rader, Karen A. *Making Mice: Standardizing Animals for American Biomedical Research 1900–1955*. Princeton, NJ: Princeton University Press, 2004.
Radick, Gregory. *The Simian Tongue. The Long Debate about Animal Language*. Chicago, IL: University of Chicago Press, 2008.
Thorsen, Liv Emma, Karen A. Rader and Adam Dodd (eds.). *Animals on Display: The Creaturely in Museums, Zoos and Natural History*. University Park, PA: Pennsylvania State University Press, 2013.

André Krebber
History of Ideas

1 Introduction and Overview

The challenges of a history of ideas in relation to animals start with the attempt to identify its subject matter. Rather than 'animal' simply forming a subsection of human ideas, the historical emergence of ideas and thinking in concepts themselves seems already intricately and irreversibly bound up with nonhuman animals. Paleolithic cave paintings with their pronounced animal imagery would suggest as much; the vastness of animal symbolism and metaphors across human societies and cultures outlasts such origins [→History of Animal Iconography]. Approaching the topic from within the field of history of ideas proves little more successful. Browsing the six volumes of the *New Dictionary of the History of Ideas* reveals no single entry for "Animal," but its pages teem with animals and the thinking about them, from "Animism" to "Behaviorism" and "Cartesianism" to "Organicism" and "Wildlife."[1] Similarly, while Joyce Chaplin asks in her 2016 Arthur O. Lovejoy Lecture as published in the *Journal of the History of Ideas* whether the nonhuman can speak, one remains pressed to find much prominent reference to animals in the journal's volumes.[2] And are ideas not after all the matter of people anyways, reserved for humans and demarcating precisely what is not animal? Thus, Theodor W. Adorno and Max Horkheimer open the fragment "Man and Human" in their generative work *Dialectic of Enlightenment*:

> Throughout European history the idea of the human being has been expressed in contradistinction to the animal. The latter's lack of reason is the proof of human dignity. So insistently and unanimously has this antithesis been recited by all the earliest precursors of bourgeois thought, the ancient Jews, the Stoics, and the Early Fathers, and then through the Middle Ages to modern times, that few other ideas are so fundamental to Western anthropology.[3]

Here, the history of ideas is problematized as an attempt to distinguish the human from and raise him above the animal, and represents a history of marking out

---

Note: I want to thank Tom Tyler (University of Leeds, UK) for our discussions on ideas and their translation over the years and in response to this chapter in particular.

1 Horowitz, Maryanne C. (ed.). *New Dictionary of the History of Ideas*, 6 vols. Detroit, MI and New York, NY: Thomson-Gale, 2005.
2 Chaplin, Joyce E. Can the Nonhuman Speak? Breaking the Chain of Being in the Anthropocene. *Journal of the History of Ideas* 78, no. 4 (October 2017): 509–529.
3 Horkheimer, Max and Theodor W. Adorno. *Dialectic of Enlightenment: Philosophical Fragments*. Translated by Edmund Jephcott. Stanford, CA: Stanford University Press, 2002, 203–204.

what is specifically human. Indeed, rather than identify and determine a specifically animal quality, the history of ideas seems to pay testimony to attempts to separate something "human" from animals.[4] Discourses in human-animal studies, by contrast, are built on the attempt to challenge the very claim of human exceptionalism, of which the capacity to form ideas or concepts is probably the last stubbornly flickering beacon that promises to remain the most resistant to surrender.

The relationship between animals and ideas as well as between the history of ideas and human-animal studies hence proves deeply conflicted yet inextricably interwoven.[5] This chapter aims to chart this path by tracing lines of confluence and identify areas of cross-fertilization between human-animal studies and the history of ideas. Thereby, an outline of what an animal history of ideas might encompass, look like and be able to achieve will gradually emerge.

The fields' subject matters make both the history of ideas and human-animal studies inherently interdisciplinary and indeed highly fluid areas of inquiry. Rather than closed in and delimited by neat borders around their subject matter or methodological comport, they push outwards, and have a history of continuously reorganizing, reframing and restructuring themselves as scholarly fields.[6] As a consequence, I engage with arguments and lines of inquiry in the following, instead of foregrounding disciplinary conventions. To identify areas of overlap and cross-contamination or contagion as well as conflict, I probe the relationship between the two fields along the following broad lines of inquiry: a) the challenge to the idea of the animal by the field of human-animal studies, b) ideas that have proven particularly prominent within human-animal studies, and c) the history of the thinking on animal consciousness. As there is no established subfield of an animal history of ideas with working methods and approaches, I take up and discuss specific methodological challenges for an animal inclined historian of ideas in the respective section of this chapter. The chapter finishes with reflections on repercussions of an *animal turn* for the history of ideas.

---

[4] For example the definition in Burrow, John W. Intellectual History in English Academic Life: Reflections on a Revolution. In *Palgrave Advances in Intellectual History*, Richard W. Whatmore, Brian Young (eds.), 8–24. Houndmills: Palgrave Macmillan, 2006, 11.
[5] Cf. Fudge, Erica. What Was it Like to be a Cow? History and Animal Studies. In *The Oxford Handbook of Animal Studies*, Linda Kalof (ed.), 258–278. New York, NY: Oxford University Press, 2017, 259.
[6] Grafton, Anthony. The History of Ideas: Precept and Practice, 1950–2000 and Beyond. *Journal of the History of Ideas* 67, no. 1 (January 2006): 1–32; Kelley, Donald R. *The Descent of Ideas: The History of Intellectual History*. Adershot: Ashgate, 2002; Krebber, André and Mieke Roscher. Spuren suchen, Zeichen lesen, Fährten folgen. In *Den Fährten Folgen: Methoden Interdisziplinärer Tierforschung*, Forschungsschwerpunkt Tier–Mensch–Gesellschaft (eds.), 11–28. Bielefeld: transcript, 2016.

2 Topics and Themes

As a scholarly field of inquiry, the history of ideas is engaged with human ideational expressions and perceptions of the world, including in particular the reconstruction of how this thinking and the content of ideas changes. For a long time, this work was steeped in efforts to revive the thoughts and conceptual essences of past authors on a predominantly intellectual level, that is, to identify the purely intellectual content of ideas uncoupled from any material and socio-cultural context this intellectual work was situated in. Moreover, historians of ideas focused on ideas and concepts that were perceived as great, universal and timeless in so far as they were deemed trans-historically important. Their work was akin to what has come under scrutiny as the history of grand narrative and great white men. In the case of the history of ideas, it was actually both, ideas and the intellectuals that dealt with them, which were made to embody these qualities of individual greatness.

Since roughly the 1970s and in response to internal and external forces acting on the field, scholars of ideas have increasingly come to show how ideas and their emergence are intertwined with, predicated by and impacting upon actual, lived cultural and social contexts. This development was part of a wider shift in the twentieth century that saw a growing problematization of the teleological tendencies in nineteenth century historiography and the increasing acknowledgement of the marginalized, local and particular in history. The emergence of first social and women's history and later cultural and gender history illustrate this process. In the history of ideas, the shift is also manifest within the field's recasting as intellectual history, which emerged as a fusing of the history of ideas and cultural history and stresses the cultural, social and material dimensions of ideas, although there was always a group, if long radiating less prestige, that focused on philosophizing's social underpinnings.[7]

From this perspective, human-animal studies represents itself a potentially monumental shift in the history of human thinking about animals, one that in turn drives a recovery of the historical diversity of animal ideas. As such, it presents as much an empirical player in the history of ideas as it proves implicated in studying the history of ideas. As a consequence, it seems unsurprising that early pioneers in the historical study of animals such as Harriet Ritvo, Nigel Rothfels, Erica Fudge, Donna Haraway and Lorraine Daston have explored perceptions of animals in history and made visible the incongruities within our thinking and ideas of animals; some of these scholars are not by accident also renowned within the field of intellectual history. Indeed, it has been pointed out innumerable times that all our engagements with animals would always already be nothing more than an exploration of our ideas of animals, and that we cannot gain access to the animal beyond our own ideas. Much of the early work in animal history as well as in human-animal studies more generally, accordingly, has taken the form of history of ideas or engaged with thinking about an-

---

[7] Grafton, History of Ideas; Kelley, *Descent of Ideas*.

imals, or at least relates to such an endeavor, as has been criticized by a recent group of animal historians who emphasize the social and political history of human-animal relations [→Social History;→Political History].[8]

Challenging the Modern Idea of the Animal

The period most consistently identified across the literature in which a human-animal studies perspective started to shift our ideas of animals are the 1970s and 1980s. Often, the watershed moment for how we perceive and understand animals and the impact they exert on societies and cultures is traced back to either Peter Singer's *Animal Liberation* (1975), John Berger's "Why Look at Animals" (1977/80) or Keith Thomas' *Man and the Natural World* (1983).[9] There are, of course, many more, as seminal texts tend to materialize and pay especially poignant testimony to moods and suspicions of a historical moment, while recording, between their lines, what has been occupying people's minds during that historical phase.[10] Rather than relegating human-animal studies back to either one of these works, indeed all three can be regarded as emblematic for different tendencies and streams of inquiries that seem to dominate and shape the texture of the field to this day: Singer for a philosophically tinged cultural and especially ethical revaluing of animals, Berger for a critique of our cultural and visual representations of animals, and Thomas for the reconsideration of lived human-animal relationships. Albeit neither mutually nor generally exclusive, these three perspectives appear to represent the central threads that make up the pattern of the fabric that is human-animal studies as a field and can certainly be regarded as central coordinates for a re-exploration of our thinking about the animal.

The early modern and Victorian periods seem to hold particular importance for both our ideas of animals as well as our current imagination, as engagement with these periods has proven especially productive for the reevaluation of our perception of animals. Erica Fudge begins her introduction to *Brutal Reasoning* by clarifying that

---

8 Kean, Hilda. *Great Cat and Dog Massacre: The Real Story of World War Two's Unknown Tragedy*. Chicago, IL: University of Chicago Press, 2017; Roscher, Mieke. New Political History and the Writing of Animal Lives. In *The Routledge Handbook of Human-Animal History*, Hilda Kean, Philipp Howell (eds.), 53–75. London: Routledge, 2018.
9 Berger, John. Why Look at Animals? In *About Looking*, John Berger, 1–26. London: Writers and Readers, 1980; Singer, Peter. *Animal Liberation: A New Ethics for our Treatment of Animals*. New York, NY: Avon Books, 1975; Thomas, Keith. *Man and the Natural World: Changing Attitudes in England 1500–1800*. London: Allen Lane, 1983.
10 Cf. Marvin, Garry and Susan McHugh. In it Together: An Introduction to Human-Animal Studies. In *Routledge Handbook of Human-Animal Studies*, Garry Marvin, Susan McHugh (eds.), 1–9. London and New York, NY: Routledge, 2014, 4.

this book began its life as an attempt to examine what early modern English writers believed about the reasoning capacity of animals in the period before the appearance of Descartes' 'beast-machine' hypothesis. I wanted to see if any aspects of this hypothesis, that declared the automatism of animals and the absolute distinction of the human from the animal, could be traced in early modern English culture prior to the appearance of *Discourse on the Method* (1637).[11]

Indeed, no other idea has held more sway as reference of demarcation in the ethical reconsideration of the animal in human-animal studies than Descartes' machine-metaphor. Describing animals as mechanistically animated and comparing them to intricately manufactured automata, he argued that they would not recognize and experience pain as such. Because of this reasoning, his status as founder of modern philosophy and the notoriety of horrendous vivisections on living animals during his time and subsequently, Descartes has evolved into the foremost modern animal abuser.[12] [→History of Experimental Animals and the History of Animal Experiments] Yet Fudge's historical study makes visible that the history of mechanistic ideas is more complex and that there is much more nuance involved than references to Descartes as the first modern 'animal abuser' imply. This even seems to be true for Descartes himself, who struggled throughout his life to present a definitive mechanistic theory of animal behavior, no matter how much he was convinced that they really were intricately and masterfully constructed apparatuses.[13]

Where *Brutal Reasoning* traces the borderlands of human and animal, asking what being human and animal was made to mean in the early modern period, Harriet Ritvo's study *The Platypus and the Mermaid* turns to the malleability of animal classifications in the Victorian period in relation to the needs and demands of specific cultural contexts. "Butchers and artists, farmers and showmen, all displayed distinctive taxonomies in their work, although they seldom bothered to articulate them theoretically. Scientific systematizing was similarly polymorphic."[14] Neither a history of ideas nor an intellectual history in the narrow sense, the book provides a cultural historical examination of the interplay and disagreements between quotidian and scientific approaches to classification of the creaturely world. But by chart-

---

[11] Fudge, Erica. *Brutal Reasoning: Animals, Rationality, and Humanity in Early Modern England*. Ithaca, NY: Cornell University Press, 2006, 1.
[12] For example, Ryder, Richard D. *Animal Revolution: Changing Attitudes Toward Speciesism*. Oxford: Berg, 2000, 52–53. This *perception* of Descartes is also reflected by his defenders, for example Gaukroger, Stephen. *Descartes: An Intellectual Biography*. Oxford: Oxford University Press, 2004, 3; Leiber, Justin. Descartes: The Smear and Related Misconstruals. *Journal for the Theory of Social Behaviour* 41, no. 4 (April 2011): 365–376.
[13] Cottingham, John. A Brute to the Brutes: Descartes' Treatment of Animals. *Philosophy* 53, no. 206 (October 1978): 551–559; Harrison, Peter. Descartes on Animals. *The Philosophical Quarterly* 42, no. 167 (April 1992): 219–227; Krebber, André. Raising the Memory of Nature: Animals, Nonidentity and Enlightenment Thought. PhD diss.: University of Canterbury, NZ (2015), ch. 3.
[14] Ritvo, Harriet. *The Platypus and the Mermaid and Other Figments of the Classifying Imagination*. Cambridge, MA and London: Harvard University Press, 1997, xii.

ing the ways in which diverse constituencies tried to make sense of animals, both on their own and in relation to their cultural functions, Ritvo complicates our assumptions about how animals were thought of in the eighteenth and nineteenth centuries, providing an example of the focus on lived relationships in human-animal studies.

Prominent Ideas

Both Ritvo's and Fudge's work present influential and paradigmatic examples of the challenging of the modern idea of the animal in human-animal studies, as well as the opening up of the plethora of animal ideas that have been making their rounds in human cultures for centuries. While the modern and early modern period has received primary attention, neither the middle ages nor classics have been left out completely.[15] What these explorations bring to light, in some instances much to their own surprise, is the diversity and sophistication of the ideas about animals that existed throughout history. The various functions of these ideas, in different cultural contexts, sit uncomfortably with both our late modern ideas of the animal as well as with our ideas about other historical periods. To some degree, then, an animal history of ideas initiates an unlearning of a specifically modern European perspective that perceives animals as resources and mere exemplars of their species, and 'premodern' periods and cultures as less sophisticated in their knowledge production and thinking. Nowhere does this become more apparent than in Claude Lévi-Strauss' suggestion that certain animals were chosen as totemic symbols in indigenous cultures not because they were 'good to eat' but because they were 'good to think.' Indeed, rather than discovering a connection between totems and food source – a deeply Western European interpretation to begin with, where people seem to find it incomprehensible to think significantly of animals in any other terms than food – Lévi-Strauss pointed out that the differing characteristics of species and types of animals and their physical appearances and behaviors makes the natural world an ideal reservoir for mediating human experiences.[16]

Thinking with and through animals is for the very same reason not restricted to indigenous cultures.

> Cartoons and animated feature films show the adventures of Bambi, Mickey Mouse, and the Road Runner to rapt audiences; countless pet owners are convinced that their dogs and cats un-

---

[15] For example, the contributions in Campbell, Gordon L. (ed.). *The Oxford Handbook of Animals in Classical Thought and Life*. Oxford: Oxford University Press, 2014; Crane, Susan. *Animal Encounters: Contacts and Concepts in Medieval Britain*. Philadelphia, PA: University of Pennsylvania Press, 2012; Sælid Gilhus, Ingvild. *Animals, Gods and Humans: Changing Attitudes to Animals in Greek, Roman and Early Christian Ideas*. London: Routledge, 2006; Steel, Karl. *How Not to Make a Human: Pets, Feral Children, Worms, Sky Burial, Oysters*. Minneapolis, MN and London: University of Minnesota Press, 2019.

[16] Lévi-Strauss, Claude. *Totemism*. Translated by Rodney Needham. London: Merlin Press, 1964.

derstand them better than their spouses and children; television wildlife documentaries cast the lives of elephants and chimpanzees, parrots and lions, in terms of emotions and personalities that appeal to viewers around the world.[17]

Unsurprisingly then, the concept of anthropomorphism has remained a central point of reference and been substantially explored, criticized and, especially in more recent times, carefully revalidated among human-animal studies scholars.[18] While these explorations do not primarily engage with the history of anthropomorphic thinking, and instead with its validity and deficiencies, historical perspectives are nonetheless incumbent in these works, and they would provide an ideal starting point to develop a more coherent look at the history of anthropomorphism as an idea. Our current difficulty of coming to terms with both our own agency and that of the nonhuman world would make this a very worthwhile and timely undertaking.

Animal Consciousness

Two central challenges are intricately tied up with such anthropomorphisms: the legitimacy of attributing mental experiences to animals on the one hand, and how we might be able to access the experiences of animals on the other. They converge around the question of animal minds and consciousness, and what the qualities and specific characteristics of various animals might be in this respect. The discourse around this is far from new. Even when Descartes published his *Discourse on Method* (1637), the philosopher Henry More (1614–1687), who in general was much in favor of Descartes' work, wrote that he

> turns not with abhorrence from any of your opinions so much as from that deadly and murderous sentiment which you professed in your *Method* [...] the sharp and cruel blade which in one blow, so to speak, dared to despoil of life and sense practically the whole race of animals, metamorphosing them into marble statues and machines. [...] why deny that they [parrots and magpies] are quite aware of what they want, viz., the meal which by this device [their ability to imitate the human voice] they acquire from their masters?[19]

---

17 Daston, Lorraine and Gregg Mitman. The How and Why of Thinking with Animals. In *Thinking with Animals: New Perspectives on Anthropomorphism*, Lorraine Daston, Gregg Mitman (eds.), 1–14. Chichester, NH: Columbia University Press, 2007, 1.
18 See, for example, Mitchell, Robert W., Nicholas S. Thompson and H. Lyn Miles (eds.). *Anthropomorphism, Anecdotes, and Animals*. Albany, NY: State University of New York Press, 1997; Crist, Eileen. *Images of Animals: Anthropomorphism and Animal Mind*. Philadelphia, PA: Temple University Press, 1999; Tyler, Tom. *Ciferae: A Bestiary in Five Fingers*. Minneapolis, MN: University of Minnesota Press, 2012, 50–64; Daston, Lorraine and Gregg Mitman (eds.). *Thinking with Animals: New Perspectives on Anthropomorphism*. Chichester, NH: Columbia University Press, 2007.
19 Letter from More to Descartes from 11 December 1648, in Cohen, Leonora D. Descartes and Henry More on the Beast-Machine—A Translation of their Correspondence Pertaining to Animal Automatism. *Annals of Science* 1, no. 1 (1936): 48–61, 50.

With the demise of behaviorism and the rise of human-animal studies as a new field of inquiry in the second half of the twentieth century, these debates took on new zeal, and there is a large and widely developed discourse around the inner lives of animals – from their capacity to reason to their emotions – that conjure up all sorts of historical positions, perspectives and validations [→History of Emotions;→ Post-Domestication: The Posthuman].[20]

A significant amount of recent work on the mental capacities of other animals is motivated by advances in neurosciences and debates in the philosophy of mind, and thus occupied with either general notions of consciousness or, in more particular terms, human mindfulness in demarcation from the animal mind.[21] The human-animal studies-side, by contrast, points to the consciousness and complex mental self-activity of animals, including their emotional lives and mental processing capacities.[22] A particularly fruitful tendency in relation to the history of ideas in this respect is a return to late nineteenth and early twentieth century ethological and psychological research, before behaviorism's post-war rise to prominence, alongside a pronounced reconsideration of Russian scholarship of that period [→History of Science].[23] An animal history of ideas reveals here, how the forgetting of the inner lives of other animals might truly very well be a thing only of the first half of the twentieth century, even if it was already prefigured by the mechanism of Descartes and other early modern thinkers. At the same time, such a reevaluation of the mindfulness of animals challenges our understanding of perception, consciousness, thinking, imagination, reasoning and so forth more generally, defying human exceptionalism while facilitating a species-critical, or species-specific, reengagement with these concepts and ideas.

Alongside these reconsiderations of the mental lives of animals surface two final ideas that have proven central to human-animal studies and thus also become central in an animal history of ideas: agency and personhood. The reflection on these terms is not restricted to theories of mind and the challenge they provide for denying agency and personhood to animals. Quite to the contrary, the concepts have been ap-

---

20 Eitler, Pascal. The Origin of Emotions: Sensitive Humans, Sensitive Animals. In *Emotional Lexicons: Continuity and Change in the Vocabulary of Feeling 1700–2000*, Ute Frevert (ed.), 91–117. Oxford and New York, NY: Oxford University Press, 2014; Wild, Markus. *Die anthropologische Differenz: Der Geist der Tiere in der Frühen Neuzeit bei Montaigne, Descartes und Hume*. Berlin and New York, NY: De Gruyter Oldenbourg, 2006.
21 See, for example, the entries on "Consciousness" and "Mind" in Horowitz, *New Dictionary*, vol. 2: 440–444 and vol. 4: 1457–1461.
22 Especially Allen, Colin and Marc Bekoff. *Species of Mind: The Philosophy and Biology of Cognitive Ethology*. Cambridge, MA: MIT Press, 2000; Griffin, Donald R. *Animal Minds: Beyond Cognition to Consciousness*. Chicago, IL and London: University of Chicago Press, 2001; Gross, Aaron and Anne Vallely (eds.). *Animals and the Human Imagination: A Companion to Animal Studies*. New York, NY: Columbia University Press, 2012.
23 For example, Mondry, Henrietta. *Political Animals: Representing Dogs in Modern Russian Culture*. Leiden: Brill, 2015, especially 309–362.

proached primarily from the perspective of an empirically driven social and cultural history, while their substance has been discussed largely within the context of providing animals with legal rights to their lives and wellbeing [→Cultural History; →Social History; →Pre-Domestication: Zooarchaeology].[24] But as the debate on the mental capacities of animals shows, any assessment of individuals as agents, and therefore, by extension, as persons, is inevitably tied up with the conception of agency – and who is considered commanding agency – that is employed in any given historical moment and place. Because personhood has been perceived as something quite different across historical periods and cultural contexts,[25] the recovery and acknowledgement of the agency of the oppressed has been the backbone of the recognition of ignored social groups as having a history or being central to the study of history in the twentieth century, making the consideration of an animal agency an extension of this history as an idea. As animal historians attempt to describe and account for the agency of their historical subjects, a history of the concept promises to make visible its precariousness while offering opportunities to expand the concept of agency beyond the rational human self.[26] The question of animal agency thus makes visible how an animal history of ideas is inextricably connected with the social, political and material reconsideration of animals as historical agents.

3 Methods and Approaches

Like human-animal studies, the history of ideas is an inherently interdisciplinary venture. Ideas are intricately bound up with the forms and formats they are expressed in and the ways in which they are handed down in history, and thus far from just the provenance of, or a neatly staked out subfield of history.[27] Ideas materialize as much in philosophical texts and written form as in vernacular literature and visual objects. As a consequence, the history of ideas fuses the history of philosophy

---

24 For the latter, see, for example, Hutton, Christopher. *Integrationism and the Self: Reflections on the Legal Personhood of Animals*. New York, NY: Routledge, 2019; Andrews, Kristin, Gary Comstock, G.K.D. Crozier, Sue Donaldson, Andrew Fenton, Tyler M. John, L. Syd. M. Johnson, Robert C. Jones, Will Kymlicka, Letitia Meynell, Nathan Nobis, David Peña-Guzmán and Jeffrey Sebo. *Chimpanzee Rights: The Philosopher's Brief*. Abingdon and New York, NY: Routledge, 2019; Cochrane, Alasdair. *Should Animals Have Political Rights*. Cambridge, MA: Polity, 2020.
25 Cf. the entry "Idea of the Person" in Horowitz, *New Dictionary*, vol. 4: 1740–1743; Johnson, Walter. On agency. *Journal of Social History* 37, no. 1 (October 2003): 113–124.
26 Pearson, Chris. History and Animal Agencies. In *The Oxford Handbook of Animal Studies*, Linda Kalof (ed.), 240–257. New York, NY: Oxford University Press, 2017; Rees, Amanda. Animal Agents. Historiography, Theory and the History of Science in the Anthropocene. *Britsh Journal of the History of Science Themes* 2 (July 2017): 1–10.
27 Young, Brian. Introduction. In *Palgrave Advances in Intellectual History*, Brian Young, Richard W. Whatmore (eds.), 1–7. Houndmills: Palgrave Macmillan, 2006, 11.

with the study of literature, cultural history with philology, the history of science with visual studies, without being limited to these relations.

During the 1960s and 1970s, the rise of social and cultural history alongside the widespread influence of New Criticism in literary theory instigated a set of substantial methodological challenges to the history of ideas that are still grappled with today, and in which much of the recent study of human-animal interactions is implicated as well. The recognition of the lives and experiences of people outside the confined circles of the powerful as worthy of inquiry was almost by definition an antidote to the practices of a traditional history of ideas, geared as it was towards the work of *acknowledged* philosophizing intellectuals, few of whom would have descended from the huddled masses (even though this is not to say that there were not any thinkers worth to engage with among the latter!). As ideology, ideas were increasingly perceived with suspicion, and everyday material practices became privileged in the study of history over self-referential exercises in idealist abstraction. The new attention to the inner coherence and logic of texts and the growing acceptance of the constitutive role of language in the making of ideas, revealed in turn a certain naivety or insensitivity in the textual analyses of the history of ideas.[28] This critique was furthered in the theoretical, constructivist debates of the 1980s and 1990s, which highlighted hermeneutical and interpretative challenges for humanities scholarship at large.[29]

The criticisms led to a reevaluation of the approaches, methods and practices in the study of the history of ideas that emphasized social context alongside textual construction. This meant an overwhelming shift to the particularity, discontinuity and limitation of ideas to specific historical, discursive contexts. In other words, rather than thinking of an idea as a more or less stable, if changing, entity over time, ideas became grounded and tied to the socio-historical contexts in which they were put to use. As a consequence, the focal point of studies in the history of ideas shifted from the relationship between content and idea (or concept), that is, what ideas were expressing, to the socio-material contexts they were referenced, produced and cultivated in.[30] Today, however, the pendulum seems to be swinging back again, striking a more leveled note that recovers the strength of a focus on individual concepts while also considering the specific contexts from which they emerge, are used in and developed without falling back behind the theoretical insights of the past 40 years.[31]

---

[28] McMahon, Darrin M. The Return of the History of Ideas. In *Rethinking Modern European Intellectual History*, Darrin M. McMahon, Samuel Moyn (eds.), 13–31. New York, NY: Oxford University Press, 2014, 15–17.
[29] Grafton, History of Ideas, 22.
[30] Grafton, History of Ideas.
[31] McMahon, Return of the History of Ideas; Gordon, Peter E. Contextualism and Criticism in the History of Ideas. In *Rethinking Modern European Intellectual History*, Darrin M. McMahon, Samual Moyn (eds.), 32–55. New York, NY: Oxford University Press, 2014.

Human-animal studies emerged against the same background, and the same theoretical debates also significantly shaped its methodological repertoire, as diverse and multifarious as it is.[32] Indeed, Robert Darnton's list of new methodological concepts that the history of ideas faced in 1980: "*mentalité*, episteme, paradigm, hermeneutics, semiotics, hegemony, deconstruction [...],"[33] rings largely true with the new study of human-animal relations as well. The most talked about methodological maneuver in this respect has been arguably Derrida's deconstruction of the singular *animal* as a container for a seemingly incongruous diverse *animals*.[34] This primary impulse – to open up the nonhuman to more nuanced analysis – is felt across the work that is pursued in human-animal studies, and thus deconstruction as a reference and method is deployed, however faithfully to Derrida, to reveal the restrictions of our conceptual knowledge. However, writing in 1807, none other than Hegel noted that "just as when I say: *all animals* the word cannot pass for a zoology, just as obvious is that such words as the divine, the absolute, the eternal, etc., do not pronounce what is contained in them."[35] Derrida's observation accordingly does not seem as new and revelatory of modernity as contemporary human-animal studies scholars like to suggest, even if it may prove new and enlightening to them in the twenty-first century. On the one hand then, the forgetting of the incongruity between idea and phenomenon, material diversity and its apprehension in thought might be a development more recent than the early modern Enlightenment that is usually held responsible for it, while on the other hand the awareness and, more importantly, negotiation of this tension, even in relation to animals, is much more present in the history of ideas than often presumed.

In the spirit of questioning the coherence of "the animal" and the tracing of the changing construction of animals, human-animal studies scholars have turned to (among other fields, but the following are particularly relevant for both the historical and philosophical study of ideas) ethnography and anthropology in order to access alternative ways of thinking and conceptualizing animals [→Multispecies Ethnography]. While these attempts are rarely pursued from a historical point of view themselves, they have proven highly productive in a number of historically-focused animal studies, from social and cultural history to urban and rural history and (post) colonial history [→Social History;→Cultural History;→Urban (and Rural) History;→(Post)Colonial History]. What makes ethnography and anthropology so po-

---

32 Krebber and Roscher, Spuren suchen, 17.
33 Darnton, Robert. Intellectual and Cultural History. In *The Past Before Us: Contemporary Historical Writing in the United States*, Michael G. Kammen (ed.), 327–354. Ithaca, NY and London: Cornell University Press, 1980.
34 Derrida, Jacques. The Animal that Therefore I am (More to Follow). Translated by David Willis. In *The Animal That Therefore I Am*, Marie-Louise Mallet (ed.), 1–51. New York, NY: Fordham University Press, 2008.
35 Hegel, Georg Wilhelm Friedrich. *Phänomenologie des Geistes. Werke, Bd. 3*, Eva Moldenhauer, Karl M. Michel (eds.). Frankfurt a.M.: Suhrkamp, 1986, 24 [emphasis and translation mine].

tent is their capacity to take seriously cosmologies outside of hegemonical modern Western reason. As this includes specific access to and appreciation of alternative, non-philosophical modes of thinking and exploring animals, they will no doubt prove just as productive for an animal history of ideas.

The recent recognition of the possibilities and potentialities of reading history through objects, or the effect of what is widely termed as the "material turn" on the historical study of ideas, converges likewise productively with human-animal studies [→Material Culture Studies]. On the one hand, this new awareness or sensibility in the history of ideas makes animals more directly its subject matter. This possibility is exemplified within a growing body of work that looks at human alterations of animal physical appearances as a manifestation of human ideas of beauty, but also productivity and usefulness, as it becomes apparent in the history of breeding of bovines and dogs for example [→Economic History; →History of Agriculture].[36] On the other hand, such materialist perspectives open up new sources and new ways of reading source materials in the study of history across numerous historical subfields.[37] One aspect that has particularly been made use of in this respect is the consideration of taxidermied animals/objects and their display in museum exhibitions as well as how these intersect with the construction of knowledge through both their configuration and their arrangement in exhibits [→History of Animal Collections/Animal Taxonomy].[38] Although not necessarily focusing on and making use of their material to explore the ideas associated and implied within them, these studies are saturated with insights and approaches to thinking animals in ways that also prove fruitful for an animal history of ideas. Here in particular, the latter blends with Foucault's genealogical approach to history and its situation within the body, human and nonhuman, as he exemplified so fascinatingly in *Les mots et les choses*.[39]

---

[36] For example, Landes, Joan B., Paula Young Lee and Paul Youngquist (eds.). *Gorgeous Beasts: Animal Bodies in Historical Perspective*. University Park, PA: Pennsylvania State University Press, 2012; Whiston, Kate. Conference Report: The Ideal Animal: How Images of Animals and Animals Were Created, 02.06.2016–03.06.2016, University of Kassel, Witzenhausen. *H-Soz-Kult*, June 8, 2016. URL: www.hsozkult.de/conferencereport/id/tagungsberichte-6646 (July 31, 2020).

[37] Grafton, History of Ideas, 26–28.

[38] Landes, Joan B. Animal Subjects: Between Nature and Invention in Buffon's Natural History Illustrations. In *Gorgeous Beasts: Animal Bodies in Historical Perspective*, Joan B. Landes, Paula Young Lee, Paul Youngquist (eds.), 21–40. University Park, PA: Pennsylvania State University Press, 2012; McGhie, Henry A. Images, Ideas, and Ideals: Thinking with and about Ross's Gull. In *Animals on Display: The Creaturely in Museums, Zoos and Natural History*, Liv Emma Thorsen, Karen A. Rader, Adam Dodd (eds.), 101–127. University Park, PA: Pennsylvania State University Press, 2013.

[39] For more perspectives on the relevance of Foucault not just for historical animal studies see Chrulew, Matthew and Dinesh Joseph Wadiwel (eds.). *Foucault and Animals*. Leiden: Brill, 2017; Eitler, Pascal. Animal History as Body History: Four Suggestions from a Genealogical Perspective. *Body Politics* 2, no. 4 (2014): 259–274; Roscher, Mieke. Animals as Signifiers – Re-reading Michel Foucault's *The Order of Things* as Genealogical Working Tool for the Historical Human-Animal Studies. In *Beyond the Human-Animal Divide: Creatural Lives in Literature, Culture, and History*, Dominik Ohrem, Roland Bartosch (eds.), 189–214. London: Palgrave Macmillan, 2017.

In addition to Foucault's work, the exploration of which has just really begun recently in human-animal studies, there are a number of other approaches and theorists that have seen limited but growing use in the study of an animal history of ideas. Donna Haraway's work, of course, has had a far-reaching and fundamental, if not foundational impact on human-animal studies. Yet her adoption of practices of competitive dog training in *When Species Meet* have strained her relationship with some human-animal studies scholars, since it has been received as glossing over moments of appropriation in human-animal relationships.[40] Such reasonable criticism should not overshadow, however, the conceptual elements on the conjoined production of knowledge that are as much part of her explorations in *When Species Meet* as the practical elements of a mutual interspecies world-making. This requires us, of course, to think of ideas and knowledge as something not just human, thus opening up the history of ideas to an interspecies world of knowledge. Haraway's early work on the history of primatology still remains a productive source for rethinking the history of ideas from the perspective of science and feminism that has been superseded by the success, and theoretical allure, of her later work.[41] The work of the first generation of critical theorists, finally, from Adorno to Horkheimer, Herbert Marcuse and Walter Benjamin, provides not only an original point of departure of thinking about animals, but their writings are themselves rich with critical references to the history of animal ideas and thus promise to be highly productive yet remain generally underappreciated in animal history more broadly.

4 Implication(s) of the Animal Turn

From the perspective of the history of ideas, the recent discovery of nonhuman animals as agents in their own subjective rights may be not as monumental a turn as is claimed sometimes in human-animal studies discourses. The historical study of our thinking about, with and of animals and animality sensitizes us instead for the historical contentiousness of the idea of the animal and all that is tied up with it, as John Berger pointed out some 40 years ago.[42] Thus, the history of animal ideas reveals a continuous effort to bring the animal under epistemological control, although even the identification of such a trend might be informed more by contemporary coordinates of the animal discourse than its previous constellations. Such qualification echoes Benjamin's *Theses on the Philosophy of History*, which conceives history as a

---

40 Haraway, Donna. *When Species Meet*. Minneapolis, MN and London: University of Minnesota Press, 2008; critically see, for example, Weisberg, Zipporah. The Broken Promises of Monsters: Haraway, Animals and the Humanist Legacy. *Journal for Critical Animal Studies* VII, no. II (January 2009): 22–62.
41 Haraway, Donna. *Primate Visions: Gender, Race and Nature in the World of Modern Science*. London and New York, NY: Routledge, 1989.
42 Berger, Why Look at Animals?

rubble heap of oppression that is coherently perceived only, if crucially, from the perspective of the victors. In contrast, Benjamin carefully plots every historical moment as a ground for competing positions, which have to be remembered through reading history against the story of the victorious.[43]

Ideational sensitivity of late towards the agency of animals thus seems to be more of a rediscovery, or even just a re-acknowledgement rather than a new discovery. The forgetfulness of the self-determination of the animal, in turn, appears to be a much more recent event of the early modern and modern period, or even just of the twentieth century. The central task of an animal history of ideas is to shed further light on this question and the causes for such forgetting and remembering. Why and what makes us remember the animal in their self-minded idiosyncrasy now? And how do we make sure we do not fall back into the same history that simply tries to deal with this remembrance by subjugating such self-mindedness to a rigid, human idea? Indeed, precisely in making us aware of the ambiguity of the animal, a challenge rises to the history of ideas as an inherently anthropocentric or even anthropologic venture. Notwithstanding the differences between humans and other animals, the history of ideas actually points to its own animal qualities. At the same time though, the contentiousness of the animal itself, in analogy to the plurality of heterogeneous models of nature in the history of science, "is maybe just an expression of the limitations of our cognitive ability to determine the object definitively."[44] As such, it should be clear that an animal history of ideas equally would not put the case for a neurological determinism in which animals are driven by instincts rather than shaping their own desires. Pushed further in light of current reevaluations of animal mind and consciousness, the case of animals then throws open the question if and how far the concept of the idea itself has to be realigned with recent developments in animal studies[45]; discussions on embodied consciousness might set a precedent here by distancing thinking, and with it, ideas, from its cerebral contraction.

What comes into view as content for an animal history of ideas, then, could prove potentially decisive for both the history of ideas and human-animal studies. Brushing our thinking historically against the animal allows us to recognize the central organizing force of the animal and animality for and within our thoughts. Yet at the same time, this awareness undercuts the novelty of human-animal studies, placing it within a long, contentious and contradictory history of human animal thinking. Likewise, it reveals how an anthropocentric history of ideas that is reinforced by dismissing the consideration of ideas for not dealing with real animals represents the

---

[43] Benjamin, Walter. Theses on the Philosophy of History. In *Illuminations: Essays and Reflections*, Hannah Arendt (ed.), 253–264. New York, NY: Schocken Books, 2007.
[44] Gloy, Karen. *Das Verständnis der Natur. Bd. 1: Die Geschichte des wissenschaftlichen Denkens*. Munich: C.H. Beck, 1995, 225 [translation mine].
[45] See, for example, Kaufman, Allison B. and James C. Kaufman (eds.). *Animal Creativity and Innovation*. San Diego, CA, Waltham, MA and Oxford: Academic Press, 2015.

view of a human victor, who increasingly, however, comes to choke on this victory. Ideas matter in this history as they mediate our experience and thereby shape and influence the ways we interact with nonhuman animals. But more than that, attention to animals moves us from a mere history of animal ideas that traces the changes in human thinking about animals to an animal history of ideas, that reveals the mutual interdependency of human and animal ideas and wherein the anthropocentric concept of ideas and the subject that holds them crumbles.[46]

Selected Bibliography

Adamson, Peter and G. Fay Edwards (eds.). *Animals: A History*. Oxford: Oxford University Press, 2018.
Berkowitz, Beth. Animal. In *Late Ancient Knowing: Explorations in Intellectual History*, Catherine M. Chin, Moulie Vidas (eds.), 36–57. Berkeley, CA: University of California Press, 2015.
Böhnert, Martin, Kristian Köchy and Matthias Wunsch (eds.). *Philosophie der Tierforschung 1: Methoden und Programme*. Freiburg and Munich: Verlag Karl Alber, 2016.
Calarco, Matthew. *Thinking Through Animals: Identity, Difference, Indistinction*. Stanford, CA: Stanford Briefs, 2015.
Dauler Wilson, Margaret. Animal Ideas. *Proceedings and Addresses of the American Philosophical Association* 69, no. 2 (November 1995): 7–25.
Derrida, Jacques. *The Beast & the Sovereign*, 2 vols. Translated by Geoffrey Bennington. Michel Lisse, Marie-Louise Mallet, Ginette Michaud (eds). Chicago, IL and London: University of Chicago Press, 2011 & 2017.
Fudge, Erica. *Perceiving Animals: Humans and Beasts in Early Modern English Culture*. Urbana, IL: University of Illinois Press, 2002.
Sorabji, Richard. *Animal Minds & Human Morals: The Origins of the Western Debate*. London: Duckworth, 2001.
Tyler, Tom. *Ciferae: A Bestiary in Five Fingers*. Minneapolis, MN: University of Minnesota Press, 2012.
Weil, Kari. *Thinking Animals: Why Animal Studies Now?* New York, NY: Columbia University Press, 2012.

---

46 Cf. Adorno, Theodor W. On Subject and Object. In *Critical Models: Interventions and Catchwords*, 245–258. New York, NY: Columbia University Press, 2005.

Part IV: **Historical Approaches**

Anna-Katharina Wöbse
Environmental History

1 Introduction and Overview

Environmental history and animal history seem to be natural born partners: *The Encyclopedia of World Environmental History* (2004) lists entries ranging from Acid Rain and Animal Rights to Zebra Mussels and Zoos.[1] Environmental historians have emphasized the impact of animals as environmental actors and factors on politics, economics, culture and human society in general.[2] Many of their studies have incorporated animals in the grand narratives of the fundamental changes in our relations toward the natural world. Recent accounts on environmental history mention human-animal studies as a self-evident subject of research.[3] Both approaches share a genuine interest in de-centering humans as the driving force of history and in gaining new insights by shifting the focus of their scholarly attention to the biosphere and the creatures sharing it. Both consider the many ambivalent and hybrid relations of humans with nonhuman animals. Both emphasize the reciprocal character of their relations. Yet there is a fundamental difference: animal history mainly focuses on animal-human relations; environmental history's interest is directed to trilateral rather than bilateral relations. This encompasses the relations between animals, humans and the environment as such. Environmental history builds on the assumption that humans depend on and interact with their natural environment. It frames our being in the world as cultural and natural actors. In order to understand the shared space, the environment constitutes, environmental historians drew closer looks at those who surround us. Ecology fostered the notion that the planet presents a spatial frame for various forms of life, including animals, plants, archaea, fungi, protists and bacteria.

This chapter offers a brief overview of how the field of environmental history evolved, where it drew inspiration from, and how it contributed to integrating animals into historical writing and research. It turns spotlights on some of its topics and themes and explains some central approaches and methods. The last section in-

---

[1] Krech, Shepard, John Robert McNeill and Carolyn Merchant. *Encyclopedia of World Environmental History*. London: Routledge, 2004.
[2] Guerrini, Anita. Deep History, Evolutionary History and Animals in the Anthropocene. In *Animal Ethics in the Age of Humans*, Bernice Bovenkerk, Jozef Keulartz, (eds.), 25–37. Cham: Springer, 2016, 27–28.
[3] Arndt, Melanie. Environmental History. *Docupedia-Zeitgeschichte*, August 23, 2016. URL: docupedia.de/zg/Arndt_environmental_history_v3_en_2016 (August 23, 2019); Freytag, Nils. Nature and Environment. *Europäische Geschichte Online (EGO)*, September 7, 2016. URL: www.ieg-ego.eu/freytagn-2016-en (January 3, 2018).

vestigates how the *animal turn* and the many ways of "thinking through the animal" might significantly change the ways contemporary environmental history conceptualizes its approach and frames future human environmental relations.

2 Topics and Themes

Animals are among a myriad of features in the physical environment that have always constrained and conditioned the human experience. Vice versa, humans have constrained and conditioned the animal experience. The relations of human animals with their environment over time is the core business of environmental historians: They are interested in human interactions with the natural environment at particular times and in particular places. To write the history of such interactions, environmental historians draw on social, political, economic, and intellectual history, the history of science, and the roots of environmental ideas and ethics.

In the founding phase of environmental history, there had been a strong focus on issues such as pollution, changing patterns of land use, energy systems, or on cultural concepts of wilderness. Over the years, the field expanded tremendously and was cross-fertilized by input drawn from outside the historical discipline like anthropology, ecology and behavioral science. The scope of investigation seems almost unlimited. The following paragraphs look at some topics that have been at the center of environmental historians' research and have a high relevance for animal history as well.

Environmental History of Resources and Energy Systems

Human-animal relations are fundamentally organized along metabolism chains and the use of energy. Human societies depend on the input of energy and resources of all kinds. At the same time, their use of regenerative and fossil energy has had a massive impact on animals themselves. In his pioneering masterpiece on the environmental history of Chicago, William Cronon, showed how human and animal societies interacted in terms of metabolism and energy systems and in which ways the city reached out to its hinterland and changed its topography and ecology [→Urban (and Rural) History].[4] Wild animals and livestock traversed the environmental history of the metropolis and were crucial for supplying energy and resources, power, calories and raw material: Cronon uses draft animals, hogs, cattle, and bison to explain the changing relations and energy regimes between humans, animals and the natural environment. The cities' demand for leather and meat seemed insatiable. The

---

4 Cronon, William. *Nature's Metropolis: Chicago and the Great West.* New York, NY and London: Norton, 1991.

process of industrialization reached out to determine the fate of the bison that provided the skins necessary for transmission belts. European livestock supplanted the bison in the Great Plains prairies as they turned from grassland to farmland, which fed cattle that would go to the urban slaughterhouses.[5] The growing demand for resources, energy and space for human expansion changed not only human-animal but also animal-environment relations. The multi-layered fabric of relations between humans, livestock, wildlife and ecosystems proves to be highly relevant for understanding today's biophysical contradictions of industrial agro-food systems [→History of Agriculture;→American Studies].[6]

The exploitation of animal resources was not restricted to the terrestrial sphere: Since the sixteenth century, whaling evolved as a commercial endeavor on an ever-growing scale. Especially the rather slowly swimming baleen whales were hunted for their fat, which fed lamps with cheap oil and literally 'enlightened' modern societies of the nineteenth century. While the oil was used for illumination and lubricant, the baleen of the whales (their filter-feeder system made of keratin) provided the raw material for anything flexible that modern consumer societies longed for like umbrella ribs, fishing poles, whips and springs. Some baleen whales of the Northern hemisphere, like the right whale, went almost extinct due to the ever-growing market and demands.[7] At the beginning of the twentieth century, the whaling practices and technical infrastructure changed fundamentally due to the 'fossil fuel revolution.' Steam-driven vessels and harpoons armed with dynamite enabled industrial whaling fleets to hunt and process larger and fast-swimming whales in Antarctica. Cheap whale oil substituted for vegetable oils and provided the raw material for margarine, washing powder and explosives. Thus, fossil-fuel-driven machinery fueled the fast and unrestricted exploitation of global whale populations. Yet while industrial whaling sent out thousands of ships and men to 'harvest' the mammals, only little was known of the biology and 'lifestyle' of whales. When Herman Melville published the novel *Moby Dick* in 1851, the narrator Ishmael, after musing about the chance of extinction of the species, had still assumed that the whale was "immortal in his species, however perishable in his individuality."[8] At the turn of the nineteenth to the twentieth century, however, it became conceivable that whale hunting might indeed exhaust the "riches of the sea." The whaling industry had become global – there was no retreat for whales left. While whales were generally perceived as a maritime source for oil, their impending extermination started to slowly change their status in the eyes of

---

[5] Isenberg, Andrew. *The Destruction of the Bison. An Environmental History, 1750–1920.* Cambridge and New York, NY: Cambridge University Press, 2000, 193.
[6] Weis, Tony. *The Ecological Hoofprint: The Global Burden of Industrial Livestock.* London: Zed Books, 2013.
[7] Tonnessen, Johann Nicolay and Arne Odd Johnsen. *The History of Modern Whaling.* London: C. Hurst and Canberra: Australian National University, 1982; Ellis, Richard. *Men and Whales.* New York, NY: Knopf, 1991.
[8] Melville, Herman. *Moby-Dick.* New York, NY: Harper and Brothers, 1851, 514.

scientists and politicians. In 1861, the widely-read historian of the French revolution, Jules Michelet, who would inspire the *longue durée* approach of the French Annales school, described the destructive effects of human action on terrestrial and marine creatures through the ages and explored the interdependence of environmental and social relationships. Michelet challenged the classic division between nature and society and explored the limits of living resources – like the blue whale.[9] Anticipating overexploitation, he envisioned an international protection regime. In the 1920s, the League of Nations discussed conservation schemes to make whaling more sustainable. The Argentinian law expert, José Léon Suárez, reported to the League the need for action. Using the whales' own history, he argued that the seas were not only a commons of humanity but a complex community. Suarez combined moral with economic and ecological aspects and mapped the oceans and its aquatic life as a multifaceted set of relations:

> The wealth constituted by the creatures of the deep is not fixed in the sense of being confined to one region or latitude but varies from year to year according to the biological, physical and chemical circumstances affecting the plankton among which they live.[10]

According to him, the existing migration habits of the oceanic fauna had to be taken into account when drafting any future jurisdictional maritime law: "the biologico-geographical solidarity" of migrating species "should find its counterpart in a legal solidarity in the sphere of international law in which we are working."[11] He suggested a new approach towards the international governance of the shared heritage and its creatures. Suárez's stunning interpretation of the high seas as a commons and the cetaceans' role in it represented a rather exceptional point of view, however. It would take many more decades to attribute individuality to cetaceans and understand their specific cultural history.[12] Since 1986, most whale species are protected by an international moratorium. Today, biologists point to disrupted traditions among whale communities due to the extinction of populations by industrial whaling. In the run of the twentieth century, whales would turn from a material into a moral resource.[13]

Another example of resource depletion and the ways human action changed the lives and environmental settings of animals is the history of fishing. In his book *Cod*

---

9 Michelet, Jules. *La Mer.* Paris: Hachette, 1861.
10 Suárez report in League of Nations Archives: C.P.D.I.28, Geneva, 8th January 1928: 4.
11 League of Nations Archives, C.P.D.I.28, 2.
12 Wöbse, Anna-Katharina. *Weltnaturschutz: Umweltdiplomatie in Völkerbund und Vereinten Nationen, 1920–1950.* Frankfurt a. M.: Campus Verlag, 2012, 171–245; Dorsey, Kurk. *Whales and Nations: Environmental Diplomacy on the High Seas.* Seattle, WA: University of Washington Press, 2014; Cioc, Mark. *The Game of Conservation: International Treaties to Protect the World's Migratory Species.* Athens, OH: Ohio University Press, 2009.
13 Epstein, Charlotte. *The Power of Words in International Relations: Birth of an Anti-Whaling Discourse.* Cambridge, MA: MIT Press, 2005.

– *A Biography of the Fish That Changed the World*, Mark Kurlansky surveys European and North American history from the point of a fish that once thrived by the millions in Northern seas and went eventually commercially extinct almost everywhere due to 1000 years of hunting.[14] The culture and economies of the littoral communities and neighboring states of the Northern Atlantic using the fish stocks depended in part on the overabundance of cod. In the long run, however, the communities closely related to cod fishing did not manage to find any sustainable way of living with the fish. With technologies such as steam and later internal combustion engines to power vessels and frozen food compartments aboard ships, larger nets, and better navigation and tracking equipment, the capacity to catch fish became seemingly limitless, which ultimately led to the collapse of the population. The species was heavily affected not only by the enormous capacity to fish, but also by the destruction of their habitat as bottom-trawling destroyed entire ecosystems. Moreover, even after cod fishing saw restrictions in the 1990s, human interference continues to affect the lives of cod profoundly. Cod is just a proxy for thousands of species whose histories have been changed by upsetting animal-animal relations. In the case of cod, stable food chains were disturbed by removing one of the top predators from the maritime ecosystem, which resulted in cascading effects throughout the trophic levels.[15] While humans have literally incorporated this species by consuming it – cod provides the fish in the traditional dish fish and chips – human understanding of fish histories and their ecology remains limited.

Environmental Movements and the Protection of the Biosphere

Environmental history as a discipline is an offspring of the environmental movement as it evolved in the late 1960s. When in the 1970s the fledging sub-discipline sneaked into the halls of mainstream history, it looked like something completely new – and provocative.[16] The natural world had become a subject for historical investigation long before. Any historian interested in the natural world would inevitably encounter animals crossing humans' paths and ways. There is a long tradition in history writing considering nature as an element determining human culture and vice versa [→History of Ideas]. History has always encompassed stories of human use and exploitation of the environment – including animals. Environmental history, however, resonated with the recent rise of environmental movements and the growing establishment of environmental politics. It applied principles of the civil rights movement like democracy, social justice and nonviolence to environmental issues and en-

---

14 Kurlansky, Mark. *Cod – A Biography of the Fish that Changed the World*. London: Vintage, 1999.
15 Frank, Kenneth T., Brian Petrie, Jonathan A. Fisher and William C. Leggett. Transient Dynamics of an Altered Large Marine Ecosystem. *Nature* 477, no. 7362 (July 2011): 86–89.
16 Isenberg, Andrew. Introduction: A New Environmental History. In *Handbook of Environmental History*, Andrew Isenberg (ed.), 1–22. Oxford: Oxford University Press, 2014.

couraged historians to explore the power relations concerning the exploitation of nature. That implied reframing nature (and animals) not only as an object but also as an agent for changing human history. Moreover, historians became increasingly interested in the history of green activism and thinking. Early initiatives to protest against certain forms of exploitation of nature turned out to be gateways to discover changes in the relational setting of humans and their natural environment. Protecting wild animal species was at the heart of many such initiatives. One of the first modern preservation projects targeted the preservation of birds. Setting out to stop hunting birds for the use of feathers in fashion, the Royal Society for the Protection of Birds formed in 1889 and triggered networks of collaboration all over Europe. Looking at birds through the perspective of preservation organizations allows for an understanding of how the life of birds changed in times of global markets and mass consumerism as they turned into a precious commodity [→Economic History]. Activists not only campaigned to end the feather-trade business but also emphasized responsibility for securing reserves for their feathered friends and encouraged the installation of nest boxes and bird tables. Humans started to account for the loss of habitats that modern societies caused.[17] Birds were drawn into the neighborhood of humans to find a more reliable environment than they had to deal with out in the wild.

In the course of the twentieth century, the overexploitation of wildlife, the expansion of industry, infrastructure, agriculture and tourism changed the living conditions of animals fundamentally. Loss of fauna became a widely experienced phenomenon. When in 1962 the American biologist Rachel Carson published her book *Silent Spring*, she referred to this experience – and expanded it to a dystopian scene. Her book was to become the master narrative for describing deeply troubled human-environment and human-animal relations.[18] Her masterly arrangement of the story explored the results of an unrestricted use of modern pesticides.[19] DDT killed the insects the 'feathered friends' depended on. As DDT accumulated in the food chain, humans were at high risk themselves. Carson referred to historical experiences and drew on observations of ordinary people who had watched the changes in human animal relations due to the excessive application of chemicals in their own backyards. By connecting ecological knowledge with everyday experiences, she demonstrated how closely bound human-environment relations actually were. Insects were much more than just a nuisance: They were pollinators, food for the beloved singing birds and played a crucial part in the web of life, on which humans relied. Carson presented the disturbing imaginary of a human environment that lacked not only the companionship and the ecological service of insects and singing

---

17 Chansigaud, Valérie. *Des Hommes et des Oiseaux: Une Histoire de la Protection des Oiseaux.* Paris: Delachaux et Niestlé, 2012.
18 Carson, Rachel. *Silent Spring.* Boston, MA: Houghton Mifflin Harcourt, 1962.
19 Kinkela, David. *DDT and the American Century. Global Health, Environmental Politics and the Pesticide that Changed the World.* Chapel Hill, NC: University of North Carolina Press, 2011, 9.

birds but also the physical integrity of humans. By putting complex and entangled human-animal relations in very concrete terms, her book became a global bestseller[20] and heralded a fundamental change in political thinking about the environment. In the early 1970s, books like the *Limits to Growth*,[21] and the United Nations conference on the *Human Environment* put this new political approach towards the natural world into global terms [→Global History].[22] The rise of ecological knowledge and interrelatedness very much resembled a "cultural shock," as the French philosopher Gilles Clement put it. It sparked a new interest in rewriting not only the place of humans in the world, who became one species among others,[23] but also the place of animals in human societies' both past and present.

Integrating animals into environmental history writing reflected the field's interdisciplinary outlook. In 1988, Donald Worster defined environmental history broadly as "the interactions people have had with nature in past times"[24] – and animals could easily be herded under such an umbrella-like definition. The use and control of nature, resources, spaces, or technology was meaningful for understanding politics and the human past – and future. Almost any of those early environmental histories that considered and included animals in their narratives shared their interest in power relations.[25]

Animals Changing (Human) Environments

From the outset, environmental history had been interested in the ways animals change the human environment and vice versa. The historian and geographer Alfred Crosby applied natural science to his studies of colonial history and explored the many ways in which plants, animals and microbes had radically altered human lives and societies.[26] He introduced the idea of European biopolitics to a broader historical discourse. In his ground-breaking book *Ecological Imperialism*, he devoted a chapter to animals to illustrate the success of environmental manipulation as essential for European expansion to temperate zones [→(Post)Colonial History]. He fol-

---

20 Gersdorf, Catrin. Tiere und Umwelt. In *Tiere. Kulturwissenschaftliches Handbuch*, Roland Borgards (ed.), 224–229. Stuttgart: J.B. Metzler, 2016.
21 Meadows, Donella H., Dennis L. Meadows, Jorgen Randers and William W. Behrens. *The Limits to Growth*. New York, NY: Universe Books, 1972.
22 Macekura, Stephen J. *Of Limits and Growth. The Rise of Global Sustainable Development in the Twentieth Century*. Cambridge and New York, NY: Cambridge University Press, 2015.
23 Clément, Gilles. *Gärten, Landschaft und das Genie der Natur*. Berlin: Matthes und Seitz, 2015, 17–18.
24 Worster, Donald. *The Ends of the Earth: Perspectives in Modern Environmental History*. New York, NY: Cambridge University Press, 1988, VII.
25 Radkau, Joachim. *Nature and Power. A Global History of the Environment*. Cambridge: Cambridge University Press, 2008.
26 Crosby, Alfred W. *Ecological Imperialism*. Cambridge: Cambridge University Press, 1986.

lowed European cattle, horses and honeybees to North America, Australia and New Zealand, where domesticated animals not only stabilized the hegemonic power but also had a deep impact on existing animal relations. By adding new species to ecosystems, humans radically changed biota.[27] Subsequently, drawing attention to the agency of animals has become a central focus of environmental history, which increasingly became interested in biological connectedness. A telling example of such complex interrelatedness is provided by a seemingly uncharismatic mollusk: The oyster is not necessarily among the most popular species as far as human-animal studies is concerned. The case of the European oyster, however, tells a long history of a close relationship. For thousands of years, oysters were a resource of protein for coastal communities along the shores of the North Sea. In the nineteenth century, oysters turned into a fashionable food staple and millions of oysters were consumed in European capitals. New harvesting techniques and marketing strategies emerged. Machine-driven vessels accelerated the destruction of already overharvested oyster beds in France, Great Britain and Germany. In 1877, when investigating the potential for creating artificial oyster banks, the German zoologist Karl Möbius found that the oyster reefs teemed with life and coined the term "biocenosis" for describing interacting organisms living in a habitat.[28] His plea to stop overexploitation went unheard, and in the 1920s no more oyster reefs could be found in German waters. In the 1980s, Pacific oysters 'escaped' cultivated oyster-beds and started to populate the Wadden Sea followed by a fierce debate on so-called alien species. Humans had to adopt to the animals, as they had razor sharp shells and interfered with the recreational use of the shallow sands. Recently, a European network, the Native Oyster Restoration Alliance (NORA) has been launched aiming at reinforcement and restoration of the native European flat oyster.[29]

The ecological and cultural history of the disappearance and return of the animal demonstrates how attitudes and understanding of the close relationship between oysters and humans have changed over time. Today, the oysters' role as architects of reefs, which are essential for marine biodiversity, water filtration and coastal protection, surpasses their nutritional value.[30] The environmental history of the mollusks echoes the reframing of seas as a shared sphere, where human needs on the

---

[27] Anderson, Virginia DeJohn. *Creatures of Empire: How Domestic Animals Transformed Early America*. Oxford: Oxford University Press, 2004; Gissibl, Bernhard. *The Nature of German Imperialism. Conservation and the Politics of Wildlife in Colonial East Africa*. New York, NY and Oxford: Berghahn, 2016; De Bont, Raf. Eating Game: Proteins, International Conservation and the Rebranding of African Wildlife, 1955–1965. *British Journal for the History of Science* 53, no. 2 (June 2020): 183–205.

[28] Möbius, Karl August. *Die Auster und die Austernwirthschaft*. Berlin: Wiegandt, Hempel und Parey, 1877.

[29] Native Oyster Restoration Alliance. URL: noraeurope.eu (April 2, 2019).

[30] Gercken, Jens and Andreas Schmidt. *Current Status of the European Oyster (Ostrea edulis) and Possibilities for Restoration in the German North Sea 2014*. Bad Godesberg: Bundesamt für Naturschutz Skripte, 2015.

one hand and interests of other life forms on the other have to be considered and constantly negotiated.

3 Methods and Approaches

Environmental history, as Harriet Ritvo acknowledged, surely helped to open the field of historical investigation to including any feature of the natural world.[31] However, although animals played a central part in epic environmental tales about wilderness, extinction, colonialism or agricultural development, environmental historians tended to keep them in a rather vague status of anonymity and abstraction. That might be one reason why Ritvo, who provided one of the first fundamental analyses of the intellectual potential of human-animal history, found herself occasionally reproached for having adopted an eccentric stance by looking at animals in terms of cultural history.[32] Whereas the agency of the animal and the relation between animals and humans are at the center of research for animal historians, the environmental historian will be inclined to include the environment as an actor. Environmental historians explore the trilateral rather than the bilateral setting of the relationship. They are especially interested in the many interactions between the animal, its environment and the human being: The environment shapes the relationship between humans and animals, animals shape the relationship between environment and humans, humans shape the relationship between animals and the environment. At the same time, both nonhuman animals and human animals are themselves part of the environment.

The plurality of topics correlates with the plurality of methods and approaches. Depending on which aspect of human-environmental relation is studied, scholars apply a specific set of methods and instruments. Even if interested in the history of perception, an environmental historian should know about the biological or geographical setting of the field of interest. One such important aspect of the dynamic relations are physical and biological processes, including biotic factors like flora and fauna as well as viruses and bacteria. Environmental history aims at considering all abiotic actors as well as processes like climatic change, soil composition, hydraulic forces or atmospheric compounds.[33] To decipher the complexity of relations between humans and nature it is essential to overcome the confines of pure historical readings [→Multispecies Ethnography]. Environmental historians should be familiar

---

31 Ritvo, Harriet. *The Animal Estate. The English and Other Creatures in the Victorian Age.* Cambridge, MA: Havard University Press, 1987.
32 Ritvo, Harriet. *Noble Cows and Hybrid Zebras. Essays on Animals and History.* Charlottesville, VA: University of Virginia Press, 2010, 1.
33 Krech, Shepard, John McNeill and Carolyn Merchant. Introduction. In *Encyclopedia of World Environmental History*, Shepard Krech, John McNeill, Carolyn Merchant (eds.), vol. 1: ix-xv. London: Routledge, 2004, xi.

not only with historical methods and instruments but also with the methodologies and tools of the natural sciences.[34]

As noted above, environmental history has been inspired by ecology. Modern ecology evolved in the early and mid-1900s and examines the interactions among organisms and their environment and seeks to explain the many relations that signify life on earth. Ecological sciences provide ideas and tools for the analysis of the webs of life and focus on the complex and ever-changing interrelations every organism maintains. Since environmental historians are interested in the mutual conditionality between humans and their environment and diverse sets of relations, they have to consider ecological knowledge. The quite simple and yet fundamental insight that no organism – including humans – is self-contained, was decisive for challenging anthropocentrism and the concept of human superiority and omnipotence.[35] If we are ignorant of the biological needs and behavior of the animals we study, we will not be able to thoroughly understand their actions, their behavior, their outreach or the interest some of them have in us. This also involves considering the science of evolution in environmental and animal history [→Domestication: Coevolution].[36]

Ecologists, vice versa, increasingly draw on historical and philosophical findings and promote "hermeneutic science" to understand the ecological dimension of changing human-animal relations over time.[37] In recent years, human-fish relations have been subject to closer reading and scholars have begun to explore not only the economic but also the emotional and conceptual dimensions of these interactions. Ethology also provides historians with new understandings and approaches. In his book *What a Fish Knows*, Jonathan Balcombe demonstrated not only the uniqueness and diversity of fishes (Balcombe emphasizes the necessity of speaking of fishes to overcome the perception of the collective singular fish reducing the animals to a mere resource), their ability to feel, think, plan and decide, but he also reflected on our human ignorance of the animals' individuality.[38] As one of the essential prerequisites for writing about the history of human-animal relations is to consider all living and habitat conditions of the species involved, it seems decisive to know about their specific environments and the many ways they adapt to them. Interdisciplinary approaches help tremendously to overcome the traditional nature-culture binary in historical writing.

---

34 Winiwarter, Verena and Martin Knoll. *Umweltgeschichte. Eine Einführung.* Cologne: Böhlau, 2007.
35 Warde, Paul, Libby Robin and Sverker Sörlin. *The Environment. A History of the Idea.* Baltimore, MD: Johns Hopkins University Press, 2018.
36 Russel, Edward. *Evolutionary History: Uniting History and Biology to Understand Life on Earth.* New York, NY: Cambridge University Press, 2011.
37 Soentgen, Jens. *Ökologie der Angst.* Berlin: Matthes und Seitz, 2018, 115; Reise, Karsten. *A Natural History of the Wadden Sea. Riddled by Contingencies.* Wilhelmshaven: Waddenacademie and Common Wadden Sea Secretariat, 2013.
38 Balcombe, Jonathan. *What a Fish Knows: The Inner Lives of Our Underwater Cousins.* New York, NY: Scientific American, 2016.

Another discipline that has always been a close ally is geography. Environmental historians are interested in spatial aspects of history – why, where and how did societies react and adapt to certain environmental settings? Geography offers a set of instruments like mapping and using GIS technology that make spatial aspects more concrete.[39] Among the most significant features of animals are their dynamics and patterns of migration. Thus, mapping is an essential tool for visualizing, documenting and understanding changing historical contact zones and conflicts along their routes. Moreover, when it comes to migrating animals, be it birds of passage, fish swarms or cattle driven to slaughterhouses, knowledge of transnational or diplomatic history becomes indispensable [→Historical Animal Geography;→History of Animal Slaughter;→Diplomatic History].[40]

As animal history struggled with identifying and nailing down the right set of sources, environmental history had to face the challenge of finding historical material for a period in which the term "environment" was not even coined. Like so many new approaches before it, environmental history set out to revisit the sources already known: Documents produced by governments and administrations, registries and tax dossiers, printed speeches, journals and newspapers, letters and manuscripts, diaries and memoirs, oral history, photographs and artefacts. Aside from material and written sources, visual material is just as revealing for deciphering the human relations towards nature and animals in the past [→History of Animal Iconography]. Studying photographs and paintings can tell us a lot about the daily life of animals and the fundamental changes in their environments caused by humans – and vice versa. Looking at visual documentary of industrial towns at the beginning of the twentieth century reveals, for instance, the extent to which animals were present in the daily life of urban environments.

The environmental historian Etienne Benson provided an intriguing example of applying methods and tools mentioned above when he looked at the grey squirrels omnipresent in modern American cities. By using biological findings and a variety of historical sources such as newspaper reports, scientific studies, historical photographs and diaries, he explored the many ways in which human efforts fostered urban squirrel populations, described changes in the urban landscape and urban planning, and studied "the squirrels' effort to adapt and thrive."[41] His study revealed the potential of integrating animal history and environmental studies perspectives. Benson revisited the green infrastructure of the nineteenth-century towns and examined them through the squirrel's lens. Public parks, private attics, human food and power lines provided a secure environment for the squirrels, even though it had little in common with the woodlands where they came from.

---

39 Gregory, Ian N., and Alistair Geddes. *Toward Spatial Humanities: Historical GIS and Spatial History.* Bloomington, IN: Indiana University Press, 2014.
40 See Philipp Howells article on Historical Animal Geographies in the present volume.
41 Benson, Etienne. The Urbanization of the Eastern Gray Squirrel in the United States. *The Journal of American History* 100, no. 3 (December 2013): 691–710, 692.

Compared to other areas of historiography, however, environmental historians identified a collection of historical material usually ignored by most of their colleagues: the outdoor archive of landscapes. They found historical material not only in the archives but in the countryside, too. Donald Worster, for example, encouraged students and researchers to "get out of doors altogether, and to ramble into fields, woods, and the open air. It is time we bought a good set of walking shoes, and we cannot avoid getting some mud on them."[42] The list of fields to be visited would soon be extended to industrial areas, sites of natural catastrophes, scenes of environmental crime, urban wilderness, stables and hunting grounds. Thus, environmental historians and animal historians alike should train their multi-sensual skills, get a good set of walking shoes and use binoculars, microscopes, field books and apps. Lately, the history of sound has become a new field of research for comprehending the ways animals communicate and what their silencing means for our soundscapes and environments.[43] Watching animals and getting in contact with them is central for understanding the many ways in which they have changed the human environment and in which we are changing theirs.

4 Implication(s) of the Animal Turn

Environmental history played an active part in paving the way for an *animal turn* as its "vigorous growth has helped direct the attention of other kinds of historians toward animals."[44] It challenged the concepts of human exceptionalism by explaining the many ways humans interact with the natural world. Yet, as Harriet Ritvo aptly observed in 2002, "animals ordinarily have not been among the most prominent concerns of environmental historians." When historians wrote about animals, only few showed any interest in the status of animals as historical actors.[45] Animals entering the narratives of environmental history were presented as objects rather than subjects. This has changed significantly over the last two decades. When the German Historical Institute in Washington organized an international conference in 2005 entitled "Why Look at Animals?," this sounded more like an appeal than a question.[46] Today, it is a given that conferences of environmental historians offer a variety of

---

42 Worster, Donald. Doing Environmental History. In *The Ends of the Earth*, Donald Worster (ed.), 289–308. New York, NY: Cambridge University Press, 1989, 289.
43 Krause, Bernie. *The Great Animal Orchestra: Finding the Origins of Music in the World's Wild Places*. London: Profile, 2012.
44 Ritvo, Harriet. Animal Planet. *Environmental History* 9, no. 2 (April 2004): 204–220.
45 Ritvo, Animal Planet, 206.
46 Brantz, Dorothee (ed.). *Beastly Natures: Animals, Humans and the Study of History*. Charlottesville, VA: University of Virgina Press, 2010; Brantz, Dorothee and Christoph Mauch (eds.). *Tierische Geschichte. Die Beziehung von Mensch und Kultur in der Geschichte der Moderne*. Paderborn and Munich: Schöningh, 2010.

panels on animal history. Meanwhile, environmental histories not considering animals might run the risk of being perceived as incomplete. However, the *animal turn* has another dimension beyond simply making animals more visible. Animals are increasingly seen as actors that historians think of and with. Animals make environmental history more concrete and easier to narrate. Despite the human-animal divide, animals represent a constant in our myths and imaginations. They provide protagonists the audience can relate to. Moreover, the *animal turn* has helped to acknowledge the fact that animals are not only representatives of a species but might be individuals with biographies and very specific agency.[47]

One of the most important impacts animal history has had on environmental history is the challenge and the opportunity to not restrict the reading of animals to an important factor in the history of humankind but to also consider their genuine, very specific influence on humans. Animals "permeate our history and we theirs: tug at the threads and our stories, woven as they are into the same tightly knit tapestry, will not untangle."[48] Thus, animal and environmental history, united by the claim to go beyond the scope of human-only histories, are facing the challenge of meeting the requirements that multispecies studies hold in store. While there "is no human in isolation, no form of human life that has not arisen in dialogue with a wider world,"[49] the same goes for animals – and all the other myriad life forms on the planet. Human and nonhuman animals are "embedded in metabolic and symbiotic relations" with microbes, fungi and plants.[50] The concept of the commons as representing natural resources required by all life forms provides a helpful frame to examine these shared spheres.

There is no reason for insisting on drawing a clear line between the sub-disciplines. They tread on shared ground and both environmental history and animal history pay tribute to the configuration of multispecies studies by offering trajectories that express interdependence in more concrete terms.[51] The recent growth of interest in environmental humanities, which seek to understand and study humanity and its history as part of a larger living system demonstrates that we need the interdisciplinary and multi-perspective approach to help us cope with the diverse and complex relations of dependency and interdependency that signify life on this planet. The

---

47 Krebber, André and Mieke Roscher (eds.). *Animal Biography. Re-framing Animal Lives.* Cham: Palgrave Macmillan, 2018.
48 Walker, Brett. Animals and the Intimacy of History. In *A New Environmental History*, Andrew Isenberg (ed.), 52–75. Oxford: Oxford University Press, 2014.
49 Van Dooren, Thom, Eben Kirksey and Ursula Münster. Multispecies Studies. Cultivating Art of Attentiveness. *Environmental Humanities* 8, no. 1 (May 2016): 1–23, 14.
50 Benson, Etienne S., Veit Braun, Jean M. Langford, Daniel Münster, Ursula Münster and Susanne Schmitt. Introduction. In *Troubling Species. Care and Belonging in a Relational World* (RCC Perspectives: Transformations in Environment and Society 2017/1), The Multispecies Editing Collective (eds.), 5–10. Munich: Rachel Carson Center for Environment and Society, 2017, 6.
51 Tsing, Anna. Unruly Edges: Mushrooms as Companion Species. *Environmental Humanities* 1 (2012): 141–154.

specific strength of environmental history, however, lays not only in its interest in ecological data and relatedness but also in its focus on the concepts of power and politics that determine the shifting relations between humans and the environment.[52] Environmental history is not limited to explaining changing narratives or shifts in awareness. Rather, it includes 'real' material and physical change. The history of oil pollution, for instance, is not limited to explaining how oiled seabirds turned into icons of disturbed human-nature relations in the era of mass communication but also explores the physicality of floating oil, follows it back to the sources and connects it to the change of energy systems.[53]

Both environmental and animal history have contributed fundamentally to a discourse which calls for a more-than-human approach when studying the current planetary crisis. The concept of the so-called Anthropocene, which refers to the massive human environmental impact and might signify a new geological age, encapsulates the fundamental environmental and planetary crisis that cannot be solved by scientific means alone. The concept, however, has suffered from a somewhat anthropocentric and paternalistic undertone.[54] Multidisciplinary approaches help to move beyond a one-dimensional reading of human-nature relations and to forgo the primary focus on disciplines "for a common effort in which the relevance of human action is on par with the environmental aspect."[55] The planet as such has to be taken into account, considered, according to Bruno Latour, as a sovereign agent.[56] Modern societies seem to suffer from a growing environmental amnesia. Animal and environmental history offer some of the essential means to understand and explain the complexity of sweeping shifts of the biosphere.

Selected Bibliography

Balcombe, Jonathan. *What a Fish Knows: The Inner Lives of Our Underwater Cousins*. New York, NY: Scientific American, 2016.

Benson, Etienne. The Urbanization of the Eastern Gray Squirrel in the United States. *The Journal of American History* 100, no. 3 (December 2013): 691–710.

Brantz, Dorothee. *Beastly Natures: Animals, Humans, and the Study of History*. Charlottesville, VA: University of Virginia Press, 2010.

---

[52] Radkau, Joachim. *Die Ära der Ökologie. Eine Weltgeschichte*. Munich: C.H. Beck, 2011, 12.

[53] Wöbse, *Weltnaturschutz*.

[54] Haraway, Donna and Jason W. Moore (eds.). *Anthropocene or Capitalocene? Nature, History, and the Crisis of Capitalism*. Oakland, CA: PM Press, 2016; Emmett, Robert, and Thomas Lekan (eds.). *Whose Anthropocene? Revisiting Dipesh Chakrabarty's 'Four Theses'*. (RCC Perspectives: Transformations in Environment and Society 2016/2). Munich: Rachel Carson Center for Environment and Society, 2016. doi.org/10.5282/rcc/7421; Haraway, Donna. *Staying with the Trouble. Making Kin in the Chthulucene*. Durham, NC: Duke University Press, 2016.

[55] Sörlin, Sverker. Environmental Humanities: Why Should Biologists Interested in the Environment Take the Humanities Seriously? *Bioscience* 62, no. 9 (September 2012): 788–789.

[56] Latour, Bruno. *Down to Earth: Politics in the New Climatic Regime*. Cambridge: Polity Press, 2018.

Isenberg, Andrew. *A New Environmental History*. Oxford: Oxford University Press, 2014.
Krebber, André and Mieke Roscher. *Animal Biography. Re-framing Animal Lives*. Cham: Palgrave Macmillan, 2018.
Krech, Shepard, John Robert McNeill and Carolyn Merchant (eds.). *Encyclopedia of World Environmental History*, 3 vols. London: Routledge, 2004.
Latour, Bruno. *Down to Earth: Politics in the New Climatic Regime*. Cambridge: Polity Press, 2018.
Van Dooren, Thom, Eben Kirksey and Ursula Münster. Multispecies Studies. Cultivating Art of Attentiveness. *Environmental Humanities* 8, no. 1 (May 2016): 1–28.
Warde, Paul, Libby Robin and Sverker Sörlin. *The Environment. A History of the Idea*. Baltimore, MD: Johns Hopkins University Press, 2018.
Wöbse, Anna-Katharina. *Weltnaturschutz. Umweltdiplomatie in Völkerbund und Vereinten Nationen, 1920–1950*. Frankfurt a. M.: Campus Verlag, 2012.

Philip Howell
Historical Animal Geographies

1 Introduction and Overview

The field of animal geography is concerned with the geography of human-animal relations.[1] Historical animal geography is therefore "the exploration of how spatially situated human-animal relations have changed through time".[2] More precisely, however, historical animal geography has been located at the intersection of three sub-disciplines, taking in not only animal geography, but also historical animal studies and historical geography.[3] We are then led to ask what is distinctive about historical animal geography, especially in relation to animal history. My answer to this question is twofold, focusing first on the principal *Topics and Themes* of animal geography: the material and cultural placing of animals, spatial relationships, the role of animals themselves in place-making. The second concerns the wider conceptual framing of animal geography, discussed in *Methods and Approaches*, where I argue that historical animal geographies cover more than geographies of animal-human relations in the past tense: historical animal geographies emphasize animals' contributions to the shaping over time of places, landscapes and environments, and, in conjunction with 'more-than-human' perspectives, call into question the anthropocentric definition of history.

2 Topics and Themes

For the contemporary formulation of animal geography, the all-encompassing concept of *place* is clearly the dominant motif. This means both the *placing* of animals by human beings and the less obvious role of animals themselves in the processes of *place-making*. The contrast has become well-recognized in animal geography, captured in part through Chris Philo and Chris Wilbert's formative discussion of 'animal spaces' and 'beastly places', but more explicitly in the distinction between 'animal geography' and 'animals' geographies'.[4]

---

[1] For an introduction, see Urbanik, Julie. *Placing Animals: An Introduction to the Geography of Human-Animal Relations*. Lanham, MD: Rowman & Littlefield, 2012.
[2] Rutherford, Stephanie and Sharon Wilcox. A Meeting Place. In *Historical Animal Geography*, Sharon Wilcox, Stephanie Rutherford (eds.), 3–9. Abingdon: Routledge, 2018, 5.
[3] Rutherford, Wilcox, A Meeting Place.
[4] Philo, Chris and Chris Wilbert (eds.). *Animal Spaces, Beastly Places: New Geographies of Animal-Human Relations*. London: Routledge, 2000; Hodgetts, Timothy and Jamie Lorimer. Methodologies

Place

Place is "the most fundamental idea that emerges from the body of work that is animal geography".[5] This is more than just the matter of location in space or spatial relations, because place must be understood as a fundamentally social, intersubjective (even interspecific) phenomenon.[6] As "a way of seeing, knowing, and understanding the world", place is also an embodied experience.[7] The role of place in the perception and experience of the world has long been emphasized by 'humanistic' geographers, drawing deeply on the phenomenological tradition.[8] What role animals have in this phenomenology of place, landscape and environment is something of a moot point, however: in a classic work, Yi-Fu Tuan pays his dues to the achievements of nonhuman animals and what we share with them, but the *historical* attachment to place is not one of them: "People have history; other creatures do not".[9] Heidegger is famously categorical in this regard: humans have the place- and world-making privilege that he calls *Dasein*, whilst all other animals are merely "captivated" by their environment, not dispossessed entirely but poor in world compared to human beings, deprived of place and even of space in any philosophically significant sense: "Animals do not experience space *as space*".[10] The avenue to a less anthropocentric approach to the role of place in all animals' lives has not been closed off, and the work of anthropologists such as Tim Ingold on the importance of a "dwelling" framework have become very influential in animal studies, including Geography.[11] Still, to a significant extent animal geography's forwarding of the importance of place had to diverge from the core *humanistic* perspectives of such earlier phenomenological work.

---

for Animals Geographies: Cultures, Communication and Genomics. *Cultural Geographies* 22, no. 2 (March 2014): 285–295.
5 Urbanik, *Placing Animals*, 184.
6 For a helpful review, see Withers, Charles W. J. Place and the Spatial Turn in Geography and in History. *Journal of the History of Ideas* 70, no. 4 (October 2009): 637–658.
7 Cresswell, Tim. *Place: An Introduction*. Chichester: Wiley Blackwell, 2015, 18.
8 For current philosophical thinking, see Casey, Edward S. *The Fate of Place: A Philosophical History*. Berkeley, CA: University of California Press, 1998.
9 Tuan, Yi-Fu. Humanistic Geography. *Annals of the Association of American Geographers* 66, no. 2 (June 1976): 266–276, 272; Tuan, Yi-Fu. *Space and Place. The Perspective of Experience*. Minneapolis, MN: University of Minnesota Press, 1977, 4.
10 The quotation is from the Zollikon Seminars, July 1964, 16. URL: lchc.ucsd.edu/MCA/Mail/xmcamail.2017–07.dir/pdfMkwe9JQh53.pdf (April 29, 2018).
11 See Ingold, Tim. *The Perception of the Environment: Essays on Livelihood, Dwelling and Skill*. London: Routledge, 2002; Johnston, Catherine. Beyond the Clearing: Towards a Dwelt Animal Geography. *Progress in Human Geography* 32, no. 5 (October 2008): 633–648.

Placing Animals: Material and Physical

The focus on place in animal geography was concerned with the domination of nonhuman animals by human beings rather than their species-specific or individual experience of the world. As it developed in the 1990s and the new millennium, animal geography considered the diverse ways in which nonhuman animals have been 'penned in' by human beings and their societies. For convenience, we may think of this general theme as the *placing of animals*, with nonhuman animals themselves rendered somewhat passive if never irrelevant. Given the material power of humans over other animals – for instance in the industrial livestock industry – it is easy to see why we should pay such attention to the modern production of the place of animals [→History of Agriculture]. Historical animal geography has contributed to this analysis of animal exploitation in many ways: analyzing the historical geographies of slaughterhouses and associated sites of animal killing, of animal breeding for the food industry, and of food consumption and its regulation [→History of Animal Slaughter].[12]

Placing Animals: Cultural and Imaginary

These practices are subtended, however, by the imaginative and ideological placing of animals as separate from and inferior to human lives and practices, something that has a longer historical significance. Animal geographers have consistently argued that the cultural understanding of the place of nonhuman animals is inseparable from the material, physical spaces that animals inhabit: "A distinction from animals becomes a way of ordering, regulating, controlling and exploiting them".[13] The critical distinction in Greek antiquity between *zoe* (the life that is common to all creatures) and *bios* (the higher form of life supposedly lived by human beings as 'political animals') does not map out straightforwardly into the familiar modern animal/human distinction, but it does become fundamental to what we as humans are allowed and authorized to 'do' to other animals.[14] There is a strong argument that the philosophical and political placing of animals (at least in Western culture) under-

---

[12] A few examples: Atkins, Peter J. The Glasgow Case: Meat, Disease and Regulation: 1889–1924. *Agricultural History Review* 52, no. 2 (2004): 161–182; Driessen, Clemens and Jamie Lorimer. Back-Breeding the Aurochs: The Heck Brothers, National Socialism and Imagined Geographies for Nonhuman *Lebensraum*. In *Hitler's Geographies: The Spatialities of the Third Reich*, Paolo Giaccaria, Claudio Minca (eds.), 138–157. Chicago, IL: University of Chicago Press, 2016; Laxton, Paul. This Nefarious Traffic: Livestock and Public Health in Mid-Victorian Edinburgh. In *Animal Cities: Beastly Urban Histories*, Peter Atkins (ed.), 107–171. London: Routledge, 2012.
[13] Elden, Stuart. Heidegger's Animals. *Continental Philosophy Review* 39, no. 3 (July 2006): 273–291, 284.
[14] See Agamben, Giorgio. *The Open: Man and Animal*. Translated by Kevin Attell. Stanford, CA: Stanford University Press, 2004.

writes our sovereign authority over nonhuman beings, and so folds back onto human life and the oppression of certain types and classes of human beings.[15] For instance, Kay Anderson's important contribution to zoo history and geography examines that institution's iconic role in mapping out the cultural boundaries of the animal and the human, emphasizing the disciplinary act of enclosure that presents the 'animal' to the human gaze.[16] Alternatively, we might prefer to see general 'biopolitical' processes take priority, so that treatment of other animals depends on the cultural significance of the emergence of a falsely universal discourse of 'Man' which merely disguises the brutal abjection of persons deemed to be less than human.[17] Whatever the precise derivation, there is an obvious additional responsibility for historical animal geography in analyzing the cultural placing of animals by human beings and the instrumental relations it endorses and legitimizes.

Animal Spaces: Processes of Inclusion and Exclusion

Animal geography focuses on the place-specificity of animal-human relations, very obviously in the identification of *animal spaces*, a broad conception that takes in animal landscapes (also known as 'animalscapes', or even 'beastscapes').[18] Historical geographers have been well represented in these key discussions: understandably so, since it is hard to imagine how we might understand either the imaginative geographies of animality or their localized manifestations without historical contextualization. The historical geographer Chris Philo focused our attention for instance on the 'socio-spatial' processes by which animals and animality have been constructed in the modern world, using the removal of animal slaughterhouses from the center of modern, western cities as his exemplar.[19] Philo's concern was with the historical geography of the mass production of meat, but in equal measure with the cultural responses to the killing and dismemberment of animals. The physical/material geographies of animal-human relations are inseparable from cultural/imaginative/discursive geographies. This argument for the expulsion or abjection of animals, animality, and associated 'beastliness' is only part of the story, however, for animals are

---

15 See Wadiwel, Dinesh Joseph. *The War Against Animals*. Leiden: Brill, 2015.
16 Anderson, Kay. Culture and Nature at the Adelaide Zoo: At the Frontiers of Human Geography. *Transactions of the Institute of British Geographers* 20, no. 3 (1995): 275–294.
17 For a starting point, see Schuller, Kyla. *The Biopolitics of Feeling: Race, Sex and Science in the Nineteenth Century*. Durham, NC: Duke University Press, 2017.
18 See Flack, Andrew J.P. Lions Loose on a Gentleman's Lawn: Animality, Authenticity and Automobility in the Emergence of the English Safari Park. *Journal of Historical Geography* 54 (October 2016): 38–49; Matless, David, Paul Merchant and Charles Watkins. Animal Landscapes: Otters and Wildfowl in England 1945–1970. *Transactions of the Institute of British Geographers* 30, no. 2 (July 2005): 191–205.
19 Philo, Chris. Animals, Geography and the City: Notes on Inclusions and Exclusions. *Environment and Planning D: Society and Space* 13, no. 6 (December 1995): 655–681.

nevertheless brought into the city, not only as dead flesh but as living beings, such as pets or companion animals, as working and assistance animals, not to mention the impossibility of purifying the city of wild and feral animals [→Urban (and Rural) History]. As the historical geographer Peter Atkins puts it, there can be no 'Great Separation' between human and animals in western modernity until at least the late nineteenth and early twentieth centuries, and even then, it can hardly be considered remotely complete.[20]

In the most prominent thematic emphasis, animal geographers consider the simultaneous *inclusion* and *exclusion* of nonhuman animals in human worlds. This focus takes its cue from the work of geographers on place identity and the ability of the powerful and the privileged to define the presence of others as 'out of place'.[21] The wider argument in geography about the normative politics of place is here extended to attitudes and treatments of nonhuman animals.[22] I list this theme separately, however, not only because of its pre-eminence, but also because it forms something of a bridge between the human domination of animals and the role of animals themselves in the process of place-making. Social and spatial 'transgression' by nonhuman animals attests not only to the judgements of the powerful but also the actions of the least privileged. Philo insisted that nonhuman animals not only 'endure' but also themselves 'influence' the shared landscapes that are produced. Geographers have continued to argue that the desire to place animals and animality (and keep them there) has been undercut by the necessary limitations of human power over other animals. In a recent text, Tim Cresswell notes that "Part of the process of defining places has been the gradual shifting of what animals belong and which are considered transgressive", and that the behavior of nonhuman animals themselves, often alongside equally recalcitrant human subjects, contributes in a fundamental way to such transgression.[23]

Towards Animals' Geographies

Such studies of the complex and ambiguous placing of animals have become familiar themes in animal geography, with many historical examples. It is worth emphasizing that place is not simply a synonym for the 'local', in the vernacular sense of being spatially confined, isolated, and uniquely singular. We should understand pla-

---

20 Atkins, Peter. Introduction. In *Animal Cities: Beastly Urban Histories*, Peter Atkins (ed.), 1–17. Farnham: Ashgate, 2012, 1–2.
21 Cresswell, Tim. *In Place/Out of Place: Geography, Ideology, and Transgression*. Minneapolis, MN: University of Minnesota Press, 1992.
22 Withers, Place and the Spatial Turn, 658.
23 Cresswell, *Place*, 187.

ces not as singular points – locations or locales – but rather as 'constellations'.[24] So, for instance, slaughterhouses cannot be conceived without thinking of the spatially-extended networks of meat production, retail, and consumption. Nor can zoos be treated in isolation, given the movement of nonhuman animals sourced from the 'wild' or their putative role in global animal conservation.[25] Animal landscapes are *places*, for sure, but they are also *spaces* in this extended and networked sense, linking the human and the animal in characteristically complex ways. Geographers have been prominent in pointing out how interrelated, interconnected, and inseparable are places and spaces. Robert Wilson, concerned with the lack of interest in the more-than-local and more-than-human spatial practice of animal migration, puts it this way:

> A truly animal *geography* – rather than animal ethics or animal history – must attend to how reshaping the landscape has affected animals and how people have, in turn, created spaces to display animals, such as zoos and aquaria, and to conserve them, such as national parks and refugia. Closely aligned with such attention is how animal migration is disrupted or aided by all these changes.[26]

Wilson here sets animal geography at odds with animal history, something to which I will return in the remainder of this chapter. But he is also keen to insist that animals contribute to these geographies. This is not a new idea: early on, animal geographers recognized that by focusing only on discursive or social constructions of animals, or their domination by human beings and societies, we ignore their lived experiences, actions, and agency. But it is clear that we need more than ever to explore not merely animal spaces but also the 'beastly places' nonhuman animals forge for themselves, "reflective of their own 'beastly' ways, ends, doings, joys and sufferings", even if inevitably entangled with human beings and interests.[27] Julie Urbanik is entirely representative of contemporary animal geographers when she argues that we need to "move closer to the animals themselves as individual, subjective beings".[28]

---

24 Driver, Felix and Raphael Samuel. Rethinking the Idea of Place. *History Workshop Journal* 39 (Spring 1995): v–vii.
25 See Whatmore, Sarah and Lorraine Thorne. Wild(er)ness: Reconfiguring the Geographies of Wildlife. *Transactions of the Institute of British Geographers* 23, no. 4 (December 1998): 435–454; see also Braverman, Irus. *Zooland: The Institution of Captivity*. Stanford, CA: Stanford University Press, 2012.
26 Wilson, Robert M. Mobile Bodies: Animal Migration in North American History. *Geoforum* 65 (October 2015): 465–472, 471.
27 Philo, Chris and Chris Wilbert. Animal Spaces, Beastly Places: An Introduction. In *Animal Spaces, Beastly Places: New Geographies of Human-Animal Relations*, Chris Philo, Chris Wilbert (eds.), 1–35. London: Routledge, 2000, 14.
28 Urbanik, *Placing Animals*, 186.

3 Methods and Approaches

How we do this, with the added complexity of considering animals' agency in place-making *historically*, is another question altogether. We can now turn from these themes in animal geography to the less straightforward questions of methods and approaches. In this section, with Wilson's words in mind, I am most concerned to draw out what I see as the elements that effectively distinguish historical animal geography from animal history.

From the Spatial History of Animals to Historical Animal Geographies?

We might begin by insisting that any meaningful historical animal geography must be more than a spatial history of animals. The keen current interest in 'spatial history' looks to new forms of visualization in order to reorient history towards an engagement with space, as a corrective to the historians' instinctual focus on time.[29] For proponents, historical mapping stimulates our thinking as well as reminding us not to ignore the dimension of space. These projects have profitably been extended to animals, using directories and other sources to reveal the presence of nonhuman animals in urban history, for instance.[30] Ranging further afield, in a manner that takes us to global and transnational history, and beyond [→Global History], we can also consider Ben Schmidt's beautiful visualizations of the history of American whaling, which are equally suggestive of the benefits of combining spatial and animal history.[31] We might equate the spatial history of animals with these various attempts at *mapping animal history*. Useful as this work is, however, and fully accepting their preliminary status, even the best examples cannot help but reproduce the assumption that historical animal geography is really about placing animals on maps and in timelines, at most interpreting their distributions, leaving the historiographical heavy lifting to the historical professionals. Typically, such work uses the geography involved merely to 'raise' questions, stimulate thought and analysis,

---

[29] See for instance White, Richard. What is Spatial History? *Stanford University Spatial History Project*, February 1, 2010. URL: web.stanford.edu/group/spatialhistory/cgi-bin/site/pub.php?id=29 (December 5, 2020).

[30] For examples, see Animal City. *Stanford University Spatial History Project*. URL: web.stanford.edu/group/spatialhistory/cgi-bin/site/project.php?id=1047 (December 5, 2020); Kheraj, Sean. Exploring the Geography of Urban Animals in Nineteenth-Century Toronto. *NICHE. Network in Canadian History and Environment*, August 19, 2012. URL: niche-canada.org/2012/08/19/exploring-the-geography-of-urban-animals-in-nineteenth-century-toronto/ (April 29, 2018).

[31] Schmidt, Benjamin. Reading Digital Sources: A Case Study in Ships Logs. *Sapping Attention*, November 15, 2012. URL: sappingattention.blogspot.co.uk/2012/11/reading-digital-sources-case-study-in.html (April 29, 2018).

rather than being the focus in its own right. It 'reveals' historical relations (in the environmental historian Richard White's words), rather than (say) such history 'revealing' geographical relations.³²

This sounds defensive, but there is to the geographer's ear the distinct sense that such spatial histories reinforce what has long been critiqued as "cartographic reason", the modernist fetishization of absolute or abstract space.³³ This would be to reproduce the ideology associated with much of what the philosopher and Marxist critical theorist Henri Lefebvre called "representations of space", meaning (for example) the spatial practice conceived of planners and bureaucrats, scientists and surveyors, rather than his alternatives of lived and perceived space.³⁴ All this is familiar to geographers, but it is necessary to rehearse this argument because so little of it seems to make it through to such spatial history, despite the lip service. For geographers, place is never merely "space with a history", to use the words of Paul Carter in his influential early essay.³⁵ What is lost is the sense of space not only as 'socially produced' but also politically contested, in ways that enroll other species in complex, dynamic and contested forms of place-making.

For the historical animal geographers who emerged in the 1990s, the animating impulse was, by contrast, to give due weight to the ways in which ideas and ideologies about space shaped relations between humans and (other) animals, whilst promoting the recognition that these were historically, geographically, and culturally specific – and moreover that these practices took little account of the complexity of animal-human relations, including the resistance or recalcitrance of animals themselves. Some of this argument has already been advertised, but there is much excellent work still to accomplish. Responding to the concentration on animals in the city, for instance, historical animal geographers have tackled the neglected countryside and the all too easily simplified understanding of the rural past: Carl Griffin has focused on the complex emotional relations between working people and other animals, offering a powerful critique of ideas of capitalism and modernity as he does so.³⁶ The contribution of animals to the process of place-making in seemingly 'natural' or (better) 'unbuilt' environments has been analyzed by the historical geographer Jonathan Peyton.³⁷ Historical geographers have also considered the intersecting his-

---

32 White, What is Spatial History?, 6.
33 See for instance Olsson, Gunnar. *Abysmal: A Critique of Cartographic Reason*. Chicago, IL: University of Chicago Press, 2010.
34 Lefebvre, Henri. *The Production of Space*. Oxford: Blackwell, 1991.
35 Carter, Paul. *The Road to Botany Bay: An Exploration of Landscape and History*. Minneapolis, MN: University of Minnesota Press, 2010, xxiv. This was subtitled *An Essay in Spatial History* in the 1987 UK edition.
36 See for instance Griffin, Carl J. Animal Maiming, Intimacy and the Politics of Shared Life: The Bestial and the Beastly in Eighteenth- and Early Nineteenth-Century England. *Transactions of the Institute of British Geographers* 37, no. 2 (April 2012): 301–316.
37 Peyton, Jonathan. Imbricated Geographies of Conservation and Consumption in the Stikine Plateau. *Environment and History* 17, no. 4 (November 2011): 555–581; Peyton, Jonathan. A Strange

torical geographies of race and slavery in the modern period, tracing the violent politics of white supremacy in the historical geography of, for example, dogs and horses.[38] Still others have begun to tease out the connections between animals and the nature of imperialism and colonialism, war, the military [→Animals and War], geopolitics.[39] My own work, on the place of the dog in Victorian Britain, was animated by the desire to understand the ambiguous place of animals in the modern city, and the historical context in which some animals became (more or less) welcome and others became (more or less) excluded.[40]

From Historical Animal Geography to Historical Animals' Geographies?

All the same, there is a nagging argument about the limitations of current animal geography in general, and historical animal geography in particular, something that may be instructive for animal historians considering the distinctiveness of animal geography. It may be that historical animal geography has focused too much on the 'cultural' or 'socio-cultural' placing of animals rather than animals' own geographies. For animal geography as a whole, Timothy Hodgetts and Jamie Lorimer have argued that until recently the ordering of animals, rather than their contribution to place-making, has taken center stage. This is surely understandable, given the methodological constraints and biases involved, especially in historical geographical studies, but the result is that "Geographers now know a lot more about animal spaces; but relatively less about beastly places – or what we here refer to as *animals'*

Enough Way. An Embodied Natural History of Experience, Animals and Food on the Teslin Trail. *Geoforum* 58 (January 2015): 14–22.

38 Nast, Heidi J. Pit Bulls, Slavery, and Whiteness in the Mid- to Late-Nineteenth Century U.S. Geographical Trajectories, Primary Sources. In *Critical Animal Geographies: Politics, Intersections, and Hierarchies in a Multispecies World*, Kathryn Gillespie, Rosemary-Claire Collard (eds.), 127–145. Abingdon: Routledge, 2015; Lambert, David. Runaways and Strays: Rethinking (Non)Human Agency in Caribbean Slave Societies. In *Historical Animal Geographies*, Sarah Wilcox, Stephanie Rutherford (eds.), 187–200. Abingdon, Oxford: Routledge, 2018.

39 On imperialism and colonialism see Greer, Kirsten. Zoogeography and Imperial Defence: Tracing the Contours of the Nearctic Region in the Temperate North Atlantic, 1838–1880s. *Geoforum* 65 (October 2015): 454–464; Mateer, Jennifer. Rebel Elephants: Resistance Through Human–Animal Partnerships. In *Historical Animal Geographies*, Sarah Wilcox, Stephanie Rutherford (eds.), 123–133. Abingdon, Oxford: Routledge, 2018; Saha, Jonathan. Milk to Mandalay: Dairy Consumption, Animal History and the Political Geography of Colonial Burma. *Journal of Historical Geography* 54 (October 2016): 1–12. On war see Forsyth, Isla. A Bear's Biography: Hybrid Warfare and the More-Than-Human Battlespace. *Environment and Planning D: Society and Space* 35, no. 3 (August 2016): 495–512; Howell, Philip. The Dog Fancy at War: Breeds, Breeding, and Britishness 1914–1918. *Society & Animals* 21, no. 1 (January 2013): 546–567.

40 Howell, Philip. *At Home and Astray: The Domestic Dog in Victorian Britain*. Charlottesville, VA: University of Virginia Press, 2015.

geographies".⁴¹ There is an additional complaint about historical work, for with its reliance on historical sources and methods historical animal geographies run the risk of losing the distinctive 'liveliness' of animals' geographies: Jamie Lorimer and Sarah Whatmore have asserted that "the standard methods of historical geography – which search for discursive meaning in assorted texts – are not wholly sufficient – and indeed run the risk of 'deadening' the practices being examined".⁴² Above all, the risk of reproducing the dualism of culture and nature is insistently raised. Robert Wilson, in making his case for the importance of animal mobility in nineteenth- and twentieth-century North America in particular, argues for more than an ethological recognition of animals' spatiality, for it is impossible to disentangle animals' geographies from human interventions: "we have modified not just the bodies of animals and transformed the landscape they cross. We have also modified animal migration itself, so much so that animal migration is a type of socionature hybrid".⁴³ In using this language, sometimes adverting to 'socionatures' or 'naturecultures', contemporary geographers indicate a refusal to be constrained by the modern Western categories that endorse, amongst other things, the distinction between 'human' and 'animal'.⁴⁴

One thing that is immediately obvious in such work is the desire to forward a "transpecies spatial theory".⁴⁵ We can sum up the present situation as a collective endorsement of a *more-than-human* geography in which spaces, places and landscapes are understood as the work of other agents than human beings alone.⁴⁶ This perspective cannot be considered as synonymous with animal geography, and not only because 'more-than-human' actors/agents include far more than nonhuman animals; 'more-than-human' approaches do not need to focus on living things, let alone animals. All the same, the implications are important to animal geography, not least because destabilizing the nature of what being

> 'human' means leads us to consider an understanding of human-animal interactions as modes of relating in which the principal agents and expertise are 'more-than-human'. Humans and an-

---

41 Hodgetts, Lorimer, Methodologies for Animals' Geographies, 286.
42 Lorimer, Jamie and Sarah Whatmore. After the King of Beasts: Samuel Baker and the Embodied Historical Geographies of Elephant Hunting in Mid-Nineteenth-Century Ceylon. *Journal of Historical Geography* 35, no. 4 (October 2009): 668–689, 675. For a response to some of these criticisms, see Howell, Philip and Hilda Kean. The Dogs that Didn't Bark in the Blitz: Transpecies and Transpersonal Emotional Geographies on the British Home Front. *Journal of Historical Geography* 61 (July 2018): 44–52.
43 Wilson, Mobile Bodies, 471.
44 See Whatmore, Sarah. *Hybrid Geographies: Natures Cultures Spaces*. London: Sage, 2002.
45 Bolla, Andrea K. and Alice J Hovorka. Placing Wild Animals in Botswana: Engaging Geography's Transspecies Spatial Theory. *Humanimalia* 3, no. 2 (Spring 2012): 56–82.
46 See Greenough, Beth. More-Than-Human Geographies. In *The Sage Handbook of Human Geography*, vol. 1, Roger Lee, Noel Castree, Rob Kitchin, Vicky Lawson, Anssi Paasi, Chris Philo, Sarah Radcliffe, Susan M. Roberts, Charles Withers (eds.), 94–119. London: Sage, 2014.

imals are understood to become what they are through situated and embodied interactions, rather than being determined in advance.⁴⁷

The hybrid and heterogeneous spatial practices revealed by this putative 'more-than-human' geography are seen to co-produce not merely animality but also, inevitably, humanity itself. It is logically impossible therefore to isolate a purely '*human* geography' – or any other discipline in the 'humanities' or 'social' sciences for that matter – without accepting the constitutive role of nonhuman animals. At the very least we should enthusiastically accept that "nonhumans play a much larger role in human identity formation, landscape practices, and political conflicts than has heretofore been recognized".⁴⁸

How might historical animal geographers respond to these advances? Since we should never simply assert 'more-than-human' agency, but rather explore "the way that agency is differentially constructed or understood in time and place", the historical geographer might be seen to have a special mission rather than special disabilities.⁴⁹ A number of recent contributions certainly show that historical animal geographers can contribute to an animal geography alert and alive to animals' own agencies. With cities in mind, for instance, Dawn Day Biehler's extraordinary account of 'pests' in twentieth-century American cities offers up a very different take on urban geography and history, using ethological and related scientific accounts of animal behavior, properly situated in a historically informed politics of knowledge, alongside reconstructions of the animals' point of view, imaginatively and instructively revealing those animals' environments and affordances.⁵⁰ Peta Tait's discussion of the modern, western circus argues that the 'exotic geographies' of wildness and danger, which have been central to animal performances from the nineteenth century onwards, invoke dynamic and complex emotional responses, to which circus animals themselves contributed: in her words, these performing animals "trans/act social spaces of emotions", a phenomenon which is essentially prior to the anthropomorphic staging of animals in the scripts of cultural representations; in her work, animals are literally actors, even if the scripts are human ones [→History of Circus Animals].⁵¹ Both these examples accept the legitimacy of a 'responsible anthropomorphism', given the methodological failures of an irresponsible anthropocentrism.⁵²

---

47 Lorimer, Jamie and Krithika Srinivasan. Animal Geographies. In *The Wiley-Blackwell Companion to Cultural Geography*, Nuala C. Johnson, Richard Schein, Jamie Winders (eds.), 333–342. Chichester: John Wiley, 2016, 336.
48 Urbanik, *Placing Animals*, 185.
49 Buller, Animal Geographies I, 309.
50 Biehler, Dawn Day. *Pests in the City: Flies, Bedbugs, Cockroaches, and Rats*. Seattle, WA: University of Washington Press, 2013.
51 Tait, Peta. *Wild and Dangerous Performances: Animals, Emotions, Circus*. Houndmills: Palgrave Macmillan, 2012, 5.
52 Johnston, Beyond the Clearing.

In all such work, nonhuman animals are posited as distinctive actors in the geographies and landscapes that are produced, never losing sight of "the agency and capacities of nonhuman animals as animals: with geographies and lives of their own".[53] Such accounts recognize animals' agency, but they narrate it as 'networked' into the agency of human beings, in specific socio-technic and multispecies 'assemblages'.[54] Though my own work on dogs was not conceived in quite these terms, I have also argued that spatial practices (in my case such mundanities as dog walking) represent a coordination with the interests of human beings that refuses to be reduced to those human interests alone.[55] A somewhat different strategy is to enliven *everything*, to extend the courtesy of agency to all sorts of actors, including animals, but also plants, objects, things and material of all kinds.[56] Some 'more-than-human' historical geographies have approached the beastly agency of animals by incorporating them into socio-technical networks or assemblages of various kinds: even 'dead' animals may be 'enlivened' in this way by attention to the cultural and physical mobility and affective agency of such objects as tiger heads and skins and other animal products.[57]

Animal Geography as Imperfect Histories?

These 'more-than-human' approaches extend to historical animal geographies, then, for all that history seems to impose greater obstacles in narrating *animals' geographies*. But there is another advantage in considering the historical geography of nonhuman animals. What I have in mind is the fact that much excellent historical work is carried out by animal geographers who would not necessarily see themselves as historical animal geographers. Animal geography where it considers historical themes often avoids a stadial approach to history (that is, history divided into neat periods or stages, inevitably organized around anthropocentric priorities), resisting the urge to present the historical past as definitively over. Instead of narrating animal history in the past historic or *preterite* tense (appropriate for completed actions), we could and perhaps should make more use of the *imperfect* past tense (relating to ongoing

---

53 Garlick, Ben. Osprey Involvements: Historical Animal Geographies of Extinction and Return, PhD diss.: Edinburgh University (2017), abstract. URL: www.era.lib.ed.ac.uk/handle/1842/25507 (April 2018).
54 See Howell, Philip. Animals, Agency and History. In *The Routledge Companion to Animal–Human History*, Hilda Kean, Philip Howell (eds.). London and New York, NY: Routledge, 2018.
55 Howell, Philip. Between the Muzzle and the Leash: Dog-Walking, Discipline, and the Modern City. In *Animal Cities*, Peter J. Atkins (eds.), 221–241. New York, NY: Routledge, 2016.
56 See the much-cited Bennett, Jane. *Vibrant Matter: A Political Ecology of Things*. Durham, NC: Duke University Press, 2009.
57 See Patchett, Merle. Tracking Tigers: Recovering the Embodied Practices of Taxidermy. *Historical Geography* 36 (2008): 17–36; Poliquin, Rachel. *The Breathless Zoo: Taxidermy and the Cultures of Knowledge*. University Park, PA: University of Pennsylvania Press, 2012.

processes, actions and states).⁵⁸ Arguably, historical animal geographies tend towards the 'history of the present' – not the terrible historiographical sin of presentism, nor following Foucault's genealogical approach to the letter, so much as recognizing the need for an effective history which is still ongoing, in which we, with others, are in *media res*. In this fully self-conscious history, "Knowledge of past events is deemed valuable only because we also suppose that it leads us to understand ourselves better as creatures of the present".⁵⁹ Historical animal geography, along with other varieties of animal history, recognizes that we are indeed all creatures of the present. If I can be allowed a further conceit, we should not as animal historians wait for the owl of Minerva to alight before philosophy and history can begin; we should be prepared to write animal histories mid-flight. I accept that much good animal history already does this, but there is a special need in animal history to refuse the dead certainties associated with the competing conception of the 'historical animal' [→History of Ideas].⁶⁰ To give just one example, we can cite animal geographies that address the significance of colonial regimes in order to bring out the place of animals in the colonial *present*. Unlike some accounts of North American colonial animal history, for instance, in which animals like wolves are definitive victims (if not passive ones) of European settlement, elbowed aside by the 'triumph' of colonization and penned into shrinking reservations, we can consider the ways in which closely-related animals like the coyote have extended their range following colonial settlement, adapting to urban and suburban modernity in impressively proactive ways, sometimes in the form of unanticipated hybrids such as the 'coy-wolf'.⁶¹ Thinking more generally, it is impossible to consider animals in colonial and imperial history without thinking of the continuing significance of history in the contemporary framing of global wildlife management and conservation [→(Post)colonial History].⁶²

The nature and significance of history itself becomes implicated in the playing out of historical animal geographies, then. Here, history becomes part of the animal stories we have to tell. We are more than merely storytellers on behalf of the nonhu-

---

58 Consider Melillo, Edward D. Global Entomologies: Insects, Empires, and the Synthetic Age in World History. *Past and Present* 223, no. 1 (April 2014): 233–270. Melillo considers in the same breath the short, fragile lives of insects and the durability of their products and influence, neither easy to capture with conventional historical approaches to scale and periodization.
59 Auxier, Randall E. Foucault, Dewey, and the History of the Present. *The Journal of Speculative Philosophy* 16, no. 2 (2002): 75–102, 89.
60 Nance, Susan (ed.). *The Historical Animal*. Syracuse, NY: Syracuse University Press, 2015.
61 Compare Smalley, Andrea L. *Wild by Nature: North American Animals Confront Colonization*. Baltimore, MD: Johns Hopkins University Press, 2017; with Rutherford, Stephanie. The Anthropocene's Animal. Coywolves as Feral Cotravelers. *Environment and Planning E: Nature and Space* (2018). doi.org/10.1177/2514848618763250.
62 See Barua, Maan. Bio-Geo-Graphy: Landscape, Dwelling, and the Political Ecology of Human–Elephant Relations. *Environment and Planning D: Society and Space* 32, no. 6 (December 2014): 915–934.

man animals who cannot by nature take on this role, vital responsibility as this is.[63] One of the things that history and historical geography can do is to bring out – as high theory perhaps cannot – "the vital place of representation in human life, while moving from any narrow sense of representation as distanced and detached".[64] But history does not belong to professional historians alone, and if we are to fulfil the promise of animal history we may need to write a history that is more inclusive and much less magisterial: animal geography, with its modest emphasis on 'co-production', might just help with writing this kind of history, accepting the knotty entanglement of animals' and humans' geographies, and the significance of how we *inherit* histories.[65]

4 Implication(s) of the Animal Turn

The most important implication of the *animal turn* in Geography has clearly been the sustained attention to the geographies of nonhuman animals ("bringing the animals back in" was the rallying cry).[66] This agenda has been diluted by the development of broader 'more-than-human' perspectives, and indeed challenged by it, since, even for sympathetic critics, "the exclusion of humans from the category animals – and thus the necessity of a distinct non-human animal geography – reveals the persistence of the discipline's humanist history".[67] All the same, the move from the *animal turn* to the 'more-than-human turn' produces opportunities for animal geographers, historical animal geographers amongst them – such as a focus on less obvious species, on animal products, and on more complex relationships between humans, other animals, technology and material culture, and forms of representation. Methodologically, the *animal turn* is wholly consistent with current approaches, in contributing to the ongoing 'reanimating' of a previously anthropocentric 'cultural' geography, allowing a dialogue with research and topics previously sequestered as natural science or physical geography, and at least promising the abandonment of the dualism of culture and nature itself.[68] Importantly, for historical geographers and historians, we are not talking about an epochal shift. Rather,

---

[63] Kean, Hilda. Challenges for Historians Writing Animal–Human History: What is Really Enough? *Anthrozoös* 25, Supplement (August 2012): 57–72.
[64] Matless, Merchant, Watkins, Animal Landscapes, 193.
[65] Barua, Maan. Encounter. *Environmental Humanities* 7, no. 1 (May 2015): 265–270, 265.
[66] Wolch, Jennifer and Jody Emel. Bringing the Animals Back In. *Environment and Planning D: Society and Space* 13, no. 6 (December 1995): 631–636.
[67] Lorimer, Srinivasan, Animal Geographies, 339.
[68] Wolch, Jennifer, Jody Emel and Chris Wilbert. Reanimating Cultural Geography. In *Handbook of Cultural Geography*, Kay Anderson, Mona Domosh, Steve Pile, Nigel Thrift (eds.), 184–206. London: Sage, 2003.

Instead of a nature-culture dualism that has been eroded by historical events, a broad movement across the social sciences and social theory has advanced the claim that the world has always been comprised of entanglements between humans, creatures and agents of all kinds.[69]

In some ways we are returning to a more holistic engagement with the vast majority of the world that is not and has never been exclusively human; this seems appropriate enough given the history of Geography, with its dialogue between 'natural' and 'human' sciences, and it might in this sense be easier for historical animal geographers to argue their case than for animal historians.[70] The *animal turn* has also raised, as it does everywhere, insistent ethical questions. In Geography these revolve around the question of how we might "more *justly* share space" with nonhuman others.[71]

The study of animals within historical geography has followed the broad direction of animal history as it has emerged in the last few years: that is, a movement from the marginal to being closer to the center of concerns; an impetus initially from a concern with human culture and society, giving way to a focus on nonhuman animals in their own right; the growing importance of methodological, theoretical, and also political arguments to the nature of our research; the arrival at the present status of at least some acceptance and tolerance from the wider disciplines. Nevertheless, it is worth emphasizing the specific development of animal geography and what distinguishes it from other historical studies of animals. In the first place we have argued that the themes of place and place-making, though far from exclusive, are characteristically geographical. Historical work has been necessary as well as exemplary, for the material and cultural placing of animals is a long-term phenomenon, the animal spaces that have developed being historically and geographically specific, exhibiting characteristic but at the same time particular processes of inclusion and exclusion. Much of this historical work is shared with animal history more generally, but the literature on space and place is particularly rich, to the extent that it resists easy incorporation into a 'spatial history' where animals are simply mapped as a prolegomenon to historical interpretation and analysis. Historical animal geography has developed in a distinctive manner, first in the theoretical and methodological currency of cultural geography before turning more recently the emphasis on a 'more-than-human geography'. Particularly important here is the attention to animals' geographies, something that appears to problematize the methods of historical geography (and history) but within which historical animal geography may not only flourish but be granted a special purpose.

---

[69] Ginn, Franklin. *Domestic Wild: Memory, Nature and Gardening in Suburbia*. Abingdon: Routledge, 2017, 5.
[70] For the successive waves of animal geography since the nineteenth century, see Urbanik, *Placing Animals*.
[71] Gillespie, Kathryn and Rosemarie-Claire Collard. Introduction. In *Critical Animal Geographies: Politics, Intersections and Hierarchies in a Multispecies World*, Kathryn Gillespie, Rosemarie-Claire Collard, 1–16. Abingdon: Routledge, 2015.

Selected Bibliography

Atkins, Peter (ed.). *Animal Cities: Beastly Urban Histories*. Farnham: Ashgate, 2012.
Buller, Henry J. Animal Geography I. *Progress in Human Geography* 38, no. 2 (April 2014): 308–318.
Howell, Philip. *At Home and Astray: The Domestic Dog in Victorian Britain*. Charlottesville, VA: University of Virginia Press, 2015.
Johnston, Catherine. Beyond the Clearing: Towards a Dwelt Animal Geography. *Progress in Human Geography* 32, no. 5 (October 2008): 633–649.
Lorimer, Jamie, and Krithika Srinivasan. Animal Geographies. In *The Wiley-Blackwell Companion to Cultural Geography*, Nuala C. Johnson, Richard Schein, Jamie Winders (eds.), 333–342. Chichester: John Wiley, 2016.
Philo, Chris and Chris Wilbert (eds.). *Animal Spaces, Beastly Places: New Geographies of Human–Animal Relations*. London: Routledge, 2000.
Planhol, Xavier de. *Le Paysage Animal: Une Zoogéographie Historique*. Paris: Fayard, 2004.
Urbanik, Julie. *Placing Animals: An Introduction to the Geography of Human-Animal Relations*. Lanham, MD: Rowman & Littlefield, 2012.
Wilcox, Sharon and Stephanie Rutherford (eds.). *Historical Animal Geographies*. Abingdon: Routledge, 2018.
Wolch, Jennifer and Jody Emel (eds.). *Animal Geographies: Place, Politics, and Identity in the Nature-Culture Borderlands*. New York, NY: Verso, 1998.

Aritri Chakrabarti
(Post)Colonial History

1 Introduction and Overview

In the ever-expanding field of postcolonial studies, scholars like Partha Chatterjee and Dipesh Chakrabarty have championed the ideas of displacing the standards and standpoints offered by Europe or the West in order to decolonize, among others, the academic disciplines of history and political philosophy, both of which are marked by a strong colonial inheritance.[1] The term 'provincializing' has thus become a methodological gateway or standard by itself for representing cultural nuances and alternate possibilities which lie beyond the grand structures or metanarratives of the West, spatially located as they are in the decolonized world. Taking Chakrabarty's questions seriously, we should therefore ask whether the field of animal-human history is decolonized yet? Jonathan Saha, a historian working on animal-human histories in colonial Burma, has emphatically questioned the dearth of animal-human history outside of European or Anglo-American academia while setting the stage for more-than-human histories by incorporating shared trajectories of sensory history, animal geographies and postcolonial theory.[2]

This essay pays homage to Jonathan Saha's trailblazing blog *Colonizing Animals* and mirrors the ongoing quest for finding the 'alternate' histories of brute beasts embedded in the (post)colonial world.[3] It aspires to map some of the most noteworthy developments in the world of historical studies which are spatially, temporally and thematically connected with colonialism or post-eighteenth century imperialism on the one hand and with nonhuman animals on the other. I place the term 'animals' at the end of this formulation for two reasons. Firstly, it underscores the centrality of such categories as empire/colonies or colonizer/colonized in most of the existing historical narrative vis-à-vis animals. Secondly, it can also highlight the undeniable and conspicuous contemporaneity of the practice of inserting animals as fledgling entry-points into the already established realm of post(colonial) history-writing. Here I intend to examine what the *animal turn* has accomplished so far within the corpus of colonial/imperial historiography. I also hope to tease out the prospects and pitfalls of invoking animals through multiple thematic as well as methodological

---

[1] Chatterjee, Partha. Whose Imagined Community. *Millennium* 20, no. 3 (March 1991): 521–525; Chakrabarty, Dipesh. *Provincializing Europe: Postcolonial Thought and Historical Difference*. Princeton, NJ and Oxford: Princeton University Press, 2008.
[2] Saha, Jonathan. Among the Beasts of Burma: Animals and the Politics of Colonial Sensibilities, c. 1840–1940. *Journal of Social History* 48, no. 4 (Summer 2015): 910–932, 914–915.
[3] For excellent resources and framing, see Saha, Jonathan. About. *Colonizing Animals*. URL: colonizinganimals.blog/about/ (October 6, 2020).

approaches already present in the field. Ultimately, I question whether modern histories of empire have been explicitly responsive to nonhuman animals as de facto historical actors.

2 Topics and Themes

As Harriet Ritvo argued, the historical study of nonhuman animals is neither a new practice nor one entirely fueled by the ideological forces of the *animal turn*.[4] Economic histories, mostly agrarian histories, have been partially attentive to animals from the very beginning. Newer fields, such as the history of science, technology and medicine and environmental history, have been more willing to focus on nonhuman animals. Even prior to the *animal turn*, histories of European imperialism and colonialism have palpably wrapped animals into their fold. Two prominent trends still co-exist in this latter field. One provides a broader historical narrative where animals appear as submerged subjects. In the other, animals are articulated as entrypoints or categories of historical analysis in themselves. While these animal-centered (post)colonial histories are still few in number, they have significantly and critically foregrounded theories of representation, agency, affect and/or embodiment.

'Nature', 'Wilderness' and Animals in Shifting Environs

A number of historical works have illustrated the ways in which intellectual ferments impacted human perceptions of nature, which included animals, humans and other life-forms alike since the emergence of modern European imperialism. Historians concerned with the imperial metropoles or the colonies have supplemented each other by unearthing the variegated aspects of the 'age of empire,' when novel regimes of bureaucracy, science, technology, medicine, wildlife management and agriculture rose into prominence in different corners of the world. These new schemes of social and cultural reordering inside the colonized locales are defined as the products of long-term histories of interactions, transactions and resistances that cannot be read through human activities alone. To rehabilitate and coalesce such insights, the field of environmental history has proven to be a lively and dynamic interlocutor for connecting the human and more-than-human worlds under the influence of colonialism and imperialism [→Environmental History]. If we consider the larger historiography of the environment and ecologies of the colonialized world, animals were not initially singled out for historical analysis, but steadily the broad frameworks

---

4 Ritvo, Harriet. History and Animal Studies. *Society & Animals* 10, no. 4 (January 2002): 403–406, 403; see also Ritvo, Harriet. On the Animal Turn. *Daedalus* 136, no. 4 (Fall 2007): 118–122, 118.

of ecological analysis like those of 'wilderness,' 'forests' or 'natural landscape' have been used more frequently as the strongholds for histories of animals.

Scholarship on the imperial/colonial production of natural history disciplines that had linkages with the bewildering domains of colonial forestry has crucially engaged with the intellectual histories of species-oriented imaginations. These works chart the repercussions of the larger histories of the power-knowledge nexus under colonialism. It is certainly interesting to note how the notions of human race and animal taxonomies were visibly entangled in the minds of various imperial naturalists detouring the colonies. Such histories can be found in Meena Radhakrishna's work on colonial ethnography that reflected the evolutionist approaches towards animals in the narratives of imperial natural history, further making inroads into the colonial biographies of 'tribes' in British India.[5] Closely connected to this, Sujit Sivasundaram's work presents the colonial derivations of the human-animal divide through the making of scholarly disciplines such as phrenology, anatomy, anthropology and psychology.[6] Sivasundaram has also voiced the increasing need to widen the historical vision of postcolonial studies towards nonhuman animals in order to locate the critical overlaps and transgressions of human and nonhuman agencies.[7] The 'anthropomorphic' reading of the colonial archive concerning Indian birds, as done recently by Saurabh Mishra, draws heavily on the interdisciplinary field of animal studies and sheds light on the making and unmaking of colonial natural history and scientific tracts.[8]

Have the histories of colonial wildlife as a whole expressed concern for such overlaps? In order to answer this question, it is important to look for the major themes that have signified this field. The intermingling accounts of colonial hunting, leisure activities and conservation of 'game animals' have accommodated a large number of historical works. Studies on the wildlife of colonial Africa focusing on the fissures and liaisons of the human-animal relationship are some of the best instances of such scholarship [→History of Hunting;→African Studies]. For example, William Beinart's classic take on the hunting narratives of the British Empire based on the predatory slaughter of 'game animals' in Eastern and Southern Africa provided a wide-ranging gateway to colonial ecological histories.[9] Such histories have been enriched further by the works of historians such as Edward Steinhart and Lance van Sittert, who have focused on the thriving cultures of animal poaching

---

5 Radhakrishna, Meena. Of Apes and Ancestors: Evolutionary Science and Colonial Ethnography. *Indian Historical Review* 33, no. 1 (January 2006): 1–23.
6 Sivasundaram, Sujit. Imperial Transgressions: The Animal and Human in the Idea of Race. *Comparative Studies of South Asia, Africa and the Middle East* 35, no. 1 (May 2015): 156–172.
7 Sivasundaram, Imperial Transgressions, 171–172.
8 Mishra, Saurabh. History Writing, Anthropomorphism, and Birdwatching in Colonial India. *History Compass* 15, no. 8 (August 2017): e12404. doi.org/10.1111/hic3.12404.
9 Beinart, William. Empire, Hunting and Ecological Change in Southern and Central Africa. *Past & Present* 128, no.1 (August 1990): 162–186.

in the forest lands of colonial Kenya and South Africa respectively.[10] The culture of trophy hunting that still persists in the African continent, as a legacy of colonial practices of indiscriminate animal slaughter, has also been critically analyzed as a complex and superlative expression of imperial masculinity and anthropo-chauvinism by Angela Thompsell, who has, in turn, touched upon wider transnational and cross-cultural consequences of animal-conservation.[11]

The shifting contours of forests and the continuities of the hunting practices or *Shikar* of precolonial, colonial and postcolonial India have found their places in some of the most crucial historical pieces on South Asia.[12] The threads connecting the cultures of official animal hunting in India are best explored in Julia Hughes' accounts of the spectacular hunting practices within the native Princely States of colonial northwestern India under the umbrella of the British bureaucracy.[13] The scholarly emphasis on the impact of imperial or royal hunting in India, moreover, has been thoroughly destabilized by Ezra Rashkow. He has underlined the vulnerability and marginalization of the aboriginal populations, who were essentialized as 'hunting tribes' as a result of deeper penetration of the British imperial power into the remote nooks of wilderness in India.[14]

Animal-centric histories have more prominently appeared in the narratives demonstrating the contests over megafauna conservation in colonial India. A large body of work has been dedicated to this theme. Naturalists and environmentalists like Raman Sukumar, Divyabhanusingh, Raman Kumar, Ghazala Shahabuddin or Mahesh Rangarajan have been actively associated with this corpus through their accounts of the long-term histories of wild animals of South Asia.[15] Rangarajan's ground-breaking work on the histories of lion-human relationships and conservation policies productively speculates about how the lions of Gir forest transmit memories

---

10 Steinhart, Edward I. *Black Poachers, White Hunters: A Social History of Hunting in Colonial Kenya*. Oxford: James Currey Publishers, 2006; Van Sittert, Lance. Bringing in the Wild: The Commodification of Wild Animals in the Cape Colony/Province c. 1850–1950. *The Journal of African History* 46, no. 2 (July 2005): 269–291.
11 Thompsell, Angela. *Hunting Africa*. London: Palgrave Macmillan, 2015, especially 12–41.
12 The term *Shikar*, meaning prey or the act of hunting, has its etymological origins in Persian and Urdu but has entered multiple Indian origin languages.
13 Hughes, Julia E. *Animal Kingdoms: Hunting, the Environment, and Power in Indian Princely States*. Cambridge, MA: Harvard University Press, 2013; see also Pandian, Anand S. Predatory Care: The Imperial Hunt in Mughal and British India. *Journal of Historical Sociology* 14, no. 1 (March 2001): 79–107.
14 Rashkow, Ezra D. Making Subaltern Shikaris: Histories of the Hunted in Colonial Central India. *South Asian History and Culture* 5, no. 3 (April 2014): 292–313.
15 Sukumar, Raman. *The Asian Elephant: Ecology and Management*. Cambridge: Cambridge University Press, 1989; Divyabhanusinh. *The End of a Trail: The Cheetah in India*. Oxford and New York, NY: Oxford University Press, 2002; Divyabhanusinh. *The Story of Asia's Lions*. Mumbai: Marg Publications, 2008; Kumar, Raman and Ghazala Shahabuddin. Effects of Biomass Extraction on Vegetation Structure, Diversity and Composition of Forests in Sariska Tiger Reserve, India. *Environmental Conservation* 32, no. 3 (September 2005): 248–259; Rangarajan, Mahesh. *India's Wildlife History: An Introduction*. Delhi: Permanent Black, 2005.

of their humanly experiences across centuries.[16] The issue of megafauna conservation in India has also been touched upon by Vijaya Ramadas Mandala's research that focuses on the paradoxes emanating from the categories of 'threatening' and 'endangered' animals in British imperial governance, which proved to be perennially 'selective' in its conservationist agenda.[17] Varun Sharma and Neera Agnimitra's long-term take on tiger conservation, which often has been marred by the protracted discourse of the 'endangered' in colonial and postcolonial India, furthermore engages critically with ideas of tokenism, selective appropriation and expediency-orientation that tend to be associated with animal-conservation.[18]

Tigers and elephants are the most frequently figured animals in the wildlife histories of European empire. One can think of the work of Jamie Lorimer and Sarah Whitemore on the affective and embodied encounters of elephant hunting in colonial Ceylon through their study of the deeply equivocal emotions and sensibilities of the British official hunters.[19] On Southeast Asia, Peter Boomgaard's book on the Malay world chronicles the long-term and co-constitutive histories of human-tiger encounters combining the interwoven experiences of rulers, hunters and enchanters at both macro and micro levels [→(East) Asian Studies].[20] Important histories such as these have nonetheless perpetuated a certain epistemic exoticization of the 'wild' animals in the fabric of imperial/colonial history. What about the animals of 'mundane' lives, then? Have they been represented sufficiently in this field of history given their ineluctable presence in the realms of 'everyday'?

Burdened Beasts: Laying out the Lively Commodities

A copious amount of literature has dealt with what can be termed as social and cultural histories of livestock, consisting of such topics as animal breeding and upkeep, and their shared trajectories with agrarian, industrial and military histories. Agricultural histories have shown the most enduring ties by rehabilitating the cattle in their

---

16 Rangarajan, Mahesh. Animals with Rich Histories: The Case of the Lions of Gir Forest, Gujarat, India. *History and Theory* 52, no. 4 (December 2013): 109–127; see also Rangarajan, Mahesh. Region's Honour, Nation's Pride: Gir's Lions on the Cusp of History. In *The Lions of India*, Divyabhanusinh (ed.), 252–261. New Delhi: Black Kite, Permanent Black, 2008.
17 Mandala, Vijaya R. The Raj and the Paradoxes of Wildlife Conservation: British Attitudes and Expediencies. *The Historical Journal* 58, no. 1 (March 2015): 75–110.
18 Sharma, Varun and Neera Agnimitra. Making and Unmaking the Endangered in India (1880–Present): Understanding Animal-Criminal Processes. *Conservation and Society* 13, no. 1 (July 2015): 105–118.
19 Lorimer, Jamie and Sarah Whitemore. After the King of Beasts: Samuel Baker and the Embodied Historical Geographies of Elephant Hunting in Mid-Nineteenth-Century Ceylon. *Journal of Historical Geography* 35, no. 4 (October 2009): 668–689.
20 Boomgaard, Peter. *Frontiers of Fear: Tigers and People in the Malay World, 1600–1950*. New Haven, CT: Yale University Press, 2001.

fold. Histories of equine and canine domestications are also frequently studied topics in imperial histories. Looking closely, it becomes clear that those animals who were significantly connected to the empire's administrative, social, economic, and cultural set-up as facilitators of mobility, communication and drafting have been given priority of place in livestock histories.

Given the focus on the categories of 'labor' or 'work' in (post)colonial studies, it is quite puzzling to find few historical works that have closely dealt with livestock as a potent laboring force. Despite the continuous emphasis on the histories of political economy, institutions, society and culture, the symbiotic histories of human–animal labor are thoroughly underappreciated in colonial history. Heeral Chhabra's recent analysis of the figurations of animal labor in the legal regime of colonial India is an exception that proves the rule.[21] Notwithstanding this lack, it is important to underline one of the major issues that has been handled relatively well by livestock histories, namely the problem of animal breeding. This topic touches upon the affective concerns of desire and discontent regarding the bodily engineering of animals in order to maximize their laboring capacities.

Studies of breeder-livestock encounters show how the animal body became a perennial site of colonial anxieties. The perplexing ideas of 'breed' and 'breeding' involve by and large conscious human intervention in the reproductive lives of domesticated animals [→Domestication: Coevolution;→History of Agriculture]. These notions became further complicated in the colonial context through the emergence of techno-scientific discourse. The kaleidoscopic histories of colonial breeding culture show active networks of dissemination of animal genes and scientific 'inventions' of breeds, which travelled across global empires following the transaction of livestock commodities [→Global History]. The edited volume by Greg Bankoff and Sandra Swart furnishes such diverse species-centric histories of horse breeding in colonial Southeast Asia and Africa, including such countries as Thailand, Indonesia, the Philippines and South Africa.[22] In the field of South Asia's history of equine culture, Saurabh Mishra's contribution locates the British East India Company's horse-breeding operations within a complex network of animal trade and pastoralism, touching the inception of veterinary science and medicine inside the military realms of colonial India.[23] More recently, James Hevia has looked at the administrative aspects of the reliance of colonial government on laboring animals for sustaining col-

---

[21] Chhabra, Heeral. Animal Labourers and the Law in Colonial India. *South Asia Research* 39, no. 2 (June 2019): 166–183.
[22] Bankoff, Greg and Sandra Swart (eds.). *Breeds of Empire: The Invention of the Horse in Southeast Asia and Southern Africa 1500–1950.* Copenhagen: NIAS, 2007; see also Bankoff, Greg. A Question of Breeding: Zootechny and Colonial Attitudes Toward the Tropical Environment in the Late Nineteenth-Century Philippines. *Journal of Asian Studies* 60, no. 2 (May 2001): 413–437.
[23] Mishra, Saurabh. The Economics of Reproduction: Horse-Breeding in Early Colonial India, 1790–1840. *Modern Asian Studies*, 46, no. 5 (September 2012): 1116–1144.

onial warfare.[24] Hevia indulges in an in-depth analysis of the massive scale of animal deployment – camels, mules, donkeys and horses – within the nineteenth-century military conflicts in northern India and Afghanistan and highlights the suffering and exploitation of pack animals, creatures exploited and killed due to faulty and callous implementations of military policies and infrastructure.

Despite not being considered as livestock, canine histories of colonialism figure prominently, as dogs occupy a special status in society and culture as companion or service animals. These histories of canine-breeding also track social-cultural hierarchies within the colonized world. Following such impulse, Sandra Swart and Lance van Sittert's edited book *Canis Africanis* gives a species-centric overview of the history of dog domestication. Sifting through the examination of breeding cultures, disease-control and social hierarchies surrounding breeds in South Africa from the earliest rudiments to the end of the twentieth century, dogs emerge as potent historical agents of the South African past in this volume.[25] Along similar lines, Aaron Skabelund's generative work on the relations between canine culture and imperialism revolving around the fascination and discontent associated with dog breeds in Japan has recovered dogs of Japan as barking 'subalterns' through their presence in photographs and taxidermy.[26]

Several notable works have investigated cattle-breeding in colonial India and South-East Asia. In the Indian context, Brian Caton's detailed account of the connected histories of horse- and cattle-breeding in the government stud-farms shows how imperial ambitions and displeasures regarding the 'public cattle' were articulated and permeated through agrarian and military histories of the British Empire.[27] The British imperial perceptions of milk consumption and inventions of dairy breeds are discovered to be entangled in the spatial imagination of colonial Burma in the work of Jonathan Saha, who connects the histories of cattle breeding with the historical geographies of South and South-East Asia.[28] Similar transnational and trans-colonial networks of cattle breeding are also explored by Peter Boomgaard, who has explored the social and cultural histories of animal breeding, including that of the water-buffaloes in the traditional agrarian history of Indonesia and that of cattle,

---

24 Hevia, James. *Animal Labor and Colonial Warfare*. Chicago, IL and London: University of Chicago Press, 2018.
25 Swart, Sandra and Lance Van Sittert (eds.). *Canis Africanis: A Dog History of Southern Africa*. Leiden: Brill, 2007.
26 Skabelund, Aaron. *Empire of Dogs: Canines, Japan and the Making of the Modern Imperial World*. Ithaca, NY: Cornell University Press, 2011, 13–16.
27 Caton, Brian. The Imperial Ambition of Science and Its Discontents: Animal Breeding in Nineteenth Century Punjab. In *Shifting Ground: People, Animals, and Mobility in India's Environmental History*, Mahesh Rangaraja, Kalyanakrishnan Sivaramakrishnan (eds.), 132–154. New Delhi: Oxford University Press, 2014.
28 Saha, Jonathan. Milk to Mandalay: Dairy Consumption, Animal History and the Political Geography of Colonial Burma. *Journal of Historical Geography* 54 (October 2016): 1–12.

which rose to prominence under Dutch colonialism through thriving cattle-trading networks that Dutch Indonesia maintained with colonial southern India.[29]

Historians have cherished an active interest not only in 'livestock' or fully domesticated animals such as horse, cattle, dogs or sheep, but also in animals like elephants who were predominantly classified as 'wild' yet were regularly captured and tamed for laboring purposes in the process of forest-based accumulation of colonial capitalism. The social contests, ecological factors and cultural negotiations of this process have been explicated by Natasha Nongbri in her study of colonial Northeast India, where elephants were consistently perceived and used as strategic natural resources for empire building.[30] Applying a theoretical lens informed by Donna Haraway, Jonathan Saha's focused and nuanced research on working elephants in Imperial Burma's teak industry has defined elephant labor as 'undead' or lively capital.[31] As Saha explains, elephants were crucial and indispensable historical agents in colonial Burma thanks to their physical abilities, intelligence and corporeality. [32]

The question of livestock upkeep in terms of fodder assurance, housing and caregiving has also been touched upon by works in the South Asian context. For example, Laxman Satya's work on the region of Berar in colonial India in the second half of the nineteenth century teased out the details of the detrimental environmental policies of the British Empire on the once-thriving culture of cattle pastoralism. During famines, inadequate colonial veterinary administration further augmented the ongoing ecological crisis that was causing severe starvation and suffering among the cattle.[33] Saurabh Mishra has also analyzed how the colonial state emerged as an exploitative agent causing huge cattle mortalities during famines in British Western India in the last decade of the nineteenth century. He notes, of course, that it was the livestock who primarily bore the brunt of unwholesome relief measures. [34]

---

29 Boomgaard, Peter. The Age of the Buffalo and the Dawn of the Cattle Era in Indonesia, 1500–1850. In *Smallholders and Stockbreeders: Histories of Foodcrop and Livestock Farming in Southeast Asia*, Peter Boomgaard, David Henley (eds.), 257–282. Leiden: Kitlv Press, 2004.
30 Nongbri, Natasha. Elephant Hunting in Late 19th Century North-East India: Mechanisms of Control, Contestation and Local Reactions. *Economic and Political Weekly* 38, no. 30 (July-August 2003): 3189–3199.
31 Saha, Jonathan. Colonizing Elephants: Animal Agency, Undead Capital and Imperial Science in British Burma. *The British Journal for the History of Science Themes* 2 (April 2017): 169–189; Haraway, Donna. *When Species Meet*. Minneapolis, MN: University of Minnesota Press, 2013; Barua, Maan. Lively Commodities and Encounter Value. *Environment and Planning D: Society and Space* 34, no. 4 (January 2016): 725–744.
32 Saha, Colonizing Elephants, 173–174.
33 Satya, Laxman D. *Ecology, Colonialism, and Cattle: Central India in the Nineteenth Century*. New Delhi: Oxford University Press, 2004.
34 Mishra, Saurabh. Cattle, Dearth and the Colonial State: Famines and Livestock in Colonial India, 1896–1900. *Journal of Social History* 46, no. 4 (Summer 2013): 989–1012.

Colonial Veterinary Medicine and Animals

Since veterinarians are considered to be situated at the margins of the larger realm of medical professions, the wide-ranging historiography of imperial/colonial health and medicine considers veterinary histories as auxiliary spin-offs [→History of Veterinary Medicine]. Thanks to global awareness of zoonotic diseases and biohazard concerns, the resurgence of ecological activism and the proliferation of environment-themed scholarship, veterinary history has increasingly gone mainstream in recent years though and is steadily emerging as an important field for chronicling the institutional matrix of colonial and imperial power and their decolonized legacies. Animals have started to appear in tandem with narratives that are primarily focused on exploring the nature of colonial state sponsorship of veterinary interventions. Many studies have explicated the colonial regimes of epizootic encounters, eradication and the rehabilitation of the affected livestock. Given the abundant historical literature on colonial (human) health and medicine, new veterinary histories can incorporate concerns about animals by unravelling the entangled nature of human and animal health. For example, Saurabh Mishra's comprehensive attempt to chart the colonial history of the veterinary medicine in British India declares itself as a reappraisal of the 'colonial medicine' as practiced by Shula Marks.[35] Apart from sharing its path with human public health and medicine, the trajectory of veterinary history has been mostly connected with agrarian and military history. As of now, the emphasis has been placed by and large on human-centric themes in veterinary history, such as the institutional development and evolution of the profession of veterinarians under the aegis of the colonial government.

For South Asia, Diana K. Davis has provided a detailed comparative analysis of environmental policies of the colonial state and the effects of colonial veterinary administration in French North Africa and British India.[36] So far, the most significant contribution to South Asian veterinary history has come from Saurabh Mishra, who has argued that the focus on able-bodied horses for the military establishment shaped the orientation of colonial veterinary practices, leading to a neglect of the disease-stricken, impoverished 'public cattle' that were the marker of agrarian prosperity in India.[37] A different scenario can be located in colonial Bengal in Samiparna

---

[35] Mishra, Saurabh. Beasts, Murrains, and the British Raj: Reassessing Colonial Medicine in India from the Veterinary Perspective, 1860–1900. *Bulletin of the History of Medicine* (Winter 2011): 587–619, 589; see also Marks, Shula. What is Colonial About Colonial Medicine? And What Has Happened to Imperialism and Health? *Social History of Medicine* 10, no. 2 (August 1997): 205–219.

[36] Davis, Diana K. Brutes, Beasts and Empire: Veterinary Medicine and Environmental Policy in French North Africa and British India. *Journal of Historical Geography* 34, no. 2 (April 2008): 242–267.

[37] Mishra, Beasts, Murrains and the British Raj.

Samanta's work.[38] She has thrown light on the centrality of cattle in colonial Bengal through the outbreaks of bovine epizootics fueled by the ideologies of animal protectionism and the anxieties of Bengali elites regarding cattle health, sanitary conditions and dietary practices.[39] Works such as these reflect a growing body of studies dedicated to the shared field of environmental and veterinary histories.

Karen Brown has highlighted the cross-disciplinary borrowings between environmental and veterinary histories and animal studies through her work on the 'local' healing knowledge-systems in colonial societies across Asia, Africa and Latin America.[40] One of the most crucial resources on the global histories of veterinary medicine is the anthology edited by Karen Brown and Daniel Gilfoyle as it presents fourteen case-studies reflecting topical and geographical diversities of livestock farming and disease-control across continents and colonized worlds, from Africa and South-East Asia to Australia, New Zealand and the Caribbean Islands.[41] Rinderpest and anthrax have also attracted historical research for the virulence and grave impact they inflicted on cattle husbandry and colonial agriculture.[42] Deadly equine diseases like surra and its colonial encounters have been taken into historical account by scholars such as William-Clarence Smith in the context of South-East Asia and by James Hevia on South Asia.[43] Current scholarship largely treats colonial veterinary medicine as a 'tool of empire' which in myriad ways contested or negotiated with 'local' and 'indigenous' animal-healing knowledge often viewed as 'subversive.' Unfortunately, animals themselves have become more or less sidelined as mute objects even in these compelling histories of a hybridized knowledge-power complex.

---

[38] Samanta, Samiparna. Dealing with Disease: Epizootics, Veterinarians and Public Health in Colonial Bengal, 1850–1920. In *Medicine and Colonialism: Historical Perspectives in India and South Africa*, Poonam Bala (ed.), 75–88. London and New York, NY: Routledge, 2015.

[39] Samanta, Samiparna. Cattle, Cruelty, Cow-Doctors: Examining Animal Health in Rural Bengal, 1850–1920. In *Tilling the Land: Agricultural Knowledge and Practices in Colonial India*, Deepak Kumar, Bipasha Raha (eds.), 214–235. New Delhi: Primus Books, 2016.

[40] Brown, Karen. Environmental and Veterinary History – Some Themes and Suggested Ways Forward. *Environment and History* 20, no. 4 (November 2014): 547–559.

[41] Brown, Karen and Daniel Gilfoyle (eds.). *Healing the Herds: Disease, Livestock Economies and the Globalization of Veterinary Medicine*. Athens, OH: Ohio University Press, 2010.

[42] Ballad, Charles. The Repercussions of Rinderpest: Cattle Plague and Peasant Decline in Colonial Natal. *The International Journal of African Historical Studies* 19, no. 3 (1986): 421–450; Gilfoyle, Daniel. Veterinary Research and the African Rinderpest Epizootic: The Cape Colony, 1896–1898. *Journal of Southern African Studies* 29, no. 1 (March 2003): 133–154; Gilfoyle, Daniel. Anthrax in South Africa: Economics, Experiment and the Mass Vaccination of Animals, c. 1910–1945. *Medical History* 50, no. 4 (October 2006): 465–490; Spinage, Clive A. *Cattle Plague: A History*. New York, NY and London: Kluwer Academic/Plenum Publisher, 2003.

[43] Clarence-Smith, William. Diseases of Equids in Southeast Asia, c.1800–c.1945: Apocalypse or Progress. In *Healing the Herds,* Karen Brown, Daniel Gilfoyle (eds.), 129–145. Athens, OH: Ohio University Press, 2010; Hevia, *Animal Labor,* 218–249.

3 Methods and Approaches

But how can we find the 'flesh-and-blood' or corporeal aspects of the history of colonial veterinary medicine?[44] David Arnold quite famously attempted to look at the 'bodily' aspects of colonialism by engaging with the history of public health and human epidemic encounters in colonial India by finding and situating the human 'bodies' and looking at the history of dissection and laboratory medicine.[45] Notwithstanding the general paucity of historical works on colonial veterinary medicine, this area of inquiry opens up the possibility for nurturing the sensory or tactile encounters of human-animal relationship through the unraveling of the shared aspects of institutional or economic colonialism. Furthermore, the study of colonial veterinary medicine can explore the imbricated patterns of the corporeal and material without being bound by defining animals as 'wild' or 'domesticated' because it destabilizes the usual impassive separation between institutional or 'scientific' histories.

Cruelty, Welfare and Colonial Conundrums

The incessant calls for a 'humane' treatment of animals, dissemination of consumption-ethics from both dietary and laboring registers and the clarion calls for banishing animal cruelty are perhaps the most salient and contentious nodes of the *animal turn*. Keeping these features in mind, it would certainly be interesting to see how imperial/colonial historiography has tackled such issues in different cultural contexts. From a mere quantitative assessment, their appearances are relatively few, but they are noteworthy for the diversity they bring. On a general note, such works have opened avenues for more-than-human intellectual or emotional histories of the 'colonial' kind. In most instances, animals are projected as crucial backdrops upon which the historical narratives of 'humane societies,' 'native responses' and 'elite reflections' have been formulated in order to signal the 'discursive' nature of colonial discourse regarding 'violence' against animals.

In colonial Africa, a discourse of indigenous 'cruelty' helped shape highly problematic imperial propaganda promoting racism against the Black population, the latter being perpetually vilified for their alleged 'savagery' against livestock. Amidst the century-old perennial violence and cruelty at both inter-species and intra-species levels, the white European community in colonial Kenya resorted to the usual implementation of a 'civilizing mission' by establishing the East African Society for the Prevention of Cruelty to Animals. This organization aimed to disseminate 'kindness' and

---

[44] Swart, Sandra. "But Where's the Bloody Horse?": Textuality and Corporeality in the "Animal Turn". *Journal of Literary Studies* 23, no. 3 (September 2007): 271–292.
[45] Arnold, David. *Colonizing the Body: State Medicine and Epidemic Disease in Nineteenth-Century India*. Berkeley, CA and London: University of California Press, 1993.

'mercy' among the 'violent' and 'savage' tribes, who were perceived to be exceptionally violent when it came to the matter of either female genital mutilation or treating the brute beasts. Brett Shaddle argues that 'animals' became gradually and ultimately the sole fulcrum of such uneven dealings, which included intellectual violence symptomatic of colonial dominance.[46] In colonial Zimbabwe, by contrast, Alison Shuttle has revealed how white settler colonialist reasoning sought to limit the amount of livestock-herding by the African population through coercive legislations on one hand while expressing conspicuous anxieties regarding the impending doom that resulted from the legalized destocking on the other.[47] Although Shuttle has unraveled the infamous 1938 cattle-culling of Zimbabwe in this context, her description emphasizes the human-centric analysis of cultural and social tensions revolving around animal control and contested notions of etiquette more than the discussion of the cruelty inflicted on the animal subjects. A similarly complex historical account of human-cattle encounters can be found in Sandra Swart's take on South African politician and intellectual Sol Plaatje's invocation of species discourse. She locates Plaatje's personal reflections and contemplation on animal suffering in his famous commentary on the most widely contentious legislation of the Natives Land Act 27 of 1913. The act was dubbed as heinous not only to the human population for its overtly racist agenda and materially detrimental components but was also seen as excruciatingly 'cruel' to the livestock, as it made both humans and livestock co-recipients of banishment, displacement and deprivation.[48]

There are some important scholarly interventions following these themes in the context of British colonialism in Asia. In the South Asian context much of the discussion on animal cruelty debates revolve around the figure of the 'sacred' cattle as seen in earlier works, for example, Marvin Harris' examination of the theological-moral universe of the Indian cattle embedded in the domain of animal husbandry and ecology.[49] Researchers such as Gyanendra Pandey or C.S Adcock have also emphasized the ways in which the 'sacred cow' became the bulwark of the Hindu Right Movement in India under the aegis of British colonialism.[50] More conscious and expressive concerns for animal-centric historical narratives arrived with Samiparna Samanta's

---

[46] Shaddle, Brett L. Cruelty and Empathy, Animals and Race, in Colonial Kenya. *Journal of Social History* 45, no. 4 (Summer 2012): 1097–1116.
[47] Shutt, Allison K. The Settlers' Cattle Complex: The Etiquette of Culling Cattle in Colonial Zimbabwe, 1938. *The Journal of African History* 43, no. 2 (July 2002): 263–286.
[48] Swart, Sandra. It is as Bad to be a Black Man's Animal as it is to be a Black Man – The Politics of Species in Sol Plaatje's Native Life in South Africa. *Journal of Southern African Studies* 40, no. 4 (July 2014): 689–705.
[49] Harris, Marvin. India's Sacred Cow. *Human Nature* 1, no. 2 (February 1978): 28–36.
[50] Pandey, Gyanendra. Rallying Round the Cow: Sectarian Strife in the Bhojpuri Region, c. 1888–1917. In *Subaltern Studies II: Writings on South Asian History and Society,* Ranajit Guha (ed.), 60–129. New Delhi and New York, NY: Oxford University Press, 1983; Adcock, Cassie S. Sacred Cows and Secular History: Cow Protection Debates in Colonial North India. *Comparative Studies of South Asia, Africa and the Middle East* 30, no. 2 (September 2010): 297–311.

visceral exploration of Calcutta slaughterhouses as sites in a more-than-human spatial history concerned with a strong colonial anti-cruelty discourse, at the helm of which stood the Calcutta Society for the Prevention of Cruelty to Animals (CSPCA) that opposed 'barbaric' methods of butchering [→History of Animal Slaughter].[51] The trajectory of the same organization is explored in a slightly different context by Pratik Chakrabarti in the context of early twentieth-century colonial Calcutta.[52] Chakrabarti's work explores how the state-ordained culture of bacteriological experiments in the British Indian laboratories became a triumphant promotion of the authoritative benevolence of British medical governance reeking of 'civilizational discourse,' in which animals took center stage as exploited subjects and resources despite the strong but short-lived anti-cruelty movement that had potent connections with contemporary Hindu sentiments surrounding cow-protection activities.[53]

Looking at the larger scene of British colonialism in Asia, Shuk-Wah Poon's work throws light on the disputatious state legislation banning dog-meat consumption in colonial Hong Kong, where the taboo against dog-meat was carefully constructed by instilling fear of rabies contagion among the colonized population in the 1950s as well as by imposing a culture of keeping dogs as pets worthy of 'wellbeing.'[54] From Nurfadzilah Yahaya's research on meat consumption and on the discourse of 'humane' culling of animals, we get to know about the Malay Archipelago which also witnessed a cultural-ethical clash of sensibilities, as Muslim butchers there were denigrated and penalized for their 'inhumane' method of butchering.[55] Such works reveal the generally ambivalent attitude towards different species of animals within the colonial context that cut across social-political hierarchies.

While current scholarship has mostly been concerned with the positioning of animals within culinary culture, social customs and scientific discourses, greater scope remains for more nuanced and entangled histories of violence, cruelty and care, especially with reference to animal-labor regimes and hunting histories that cross species boundaries.

---

51 Samanta, Samiparna. Calcutta Slaughterhouse: Colonial and Post-Colonial Experiences. *Economic and Political Weekly* 41, no. 20 (May 2006): 1999–2007.
52 Chakrabarti, Pratik. Beasts of Burden: Animals and Laboratory Research in Colonial India. *History of Science* 48, no. 2 (June 2010): 125–151.
53 Chakrabarti, Beasts of Burden.
54 Wah-Poon, Shuk. Dogs and British Colonialism: The Contested Ban on Eating Dogs in Colonial Hong Kong. *The Journal of Imperial and Commonwealth History* 42, no. 2 (November 2014): 308–328.
55 Yahaya, Nurfadzilah. The Question of Animal Slaughter in The British Straits Settlements During the Early Twentieth Century. *Indonesia and the Malay World* 43, no. 126 (May 2015): 173–190.

4 Implication(s) of the Animal Turn

The serious consideration of nonhuman animals in imperial and (post)colonial historiography is a recent development, although animals were objects of inquiry within postcolonial history before the so-called *animal turn*. That said, (post)colonial studies still needs to redress its elision of nonhuman animals to enhance its diverse and culturally specific scholarly engagements. Philip Armstrong has called for a disciplinary alliance between postcolonial and animal studies, one that formulates "sharp, politicized, culturally sensitive, up-to-the-minute local histories that animals and their representations have played – or been made to play – in colonial and postcolonial transactions."[56] Yet while postcolonial studies has made major contributions to historical research on empires and colonialism, a certain skepticism and suspicion still exists about allegedly 'unattended' domains of political and cultural subjecthood.[57] It has also been critically argued that the totalizing notions attached to anything 'colonial' can create sharp myopias that create obstacles to localizing long-term continuities and changes in history.[58] Animals, as the latest heir to the realm of the 'hitherto neglected', have started to be reconsidered and ideas about an animal subaltern have thus been (re)invoked by a few scholars in animal history.[59] Rohan Deb Roy's provocative neologism of 'nonhuman subaltern' testifies to this recent intervention and in the search for animal histories helps to intersect the territories of actor-network theory, subaltern studies and imperial histories that have so far by and large steered clear of each other.[60] While invoking 'subalternity' in tandem with animals might create a certain speculative or preconceived asymmetry, which in turn might cause unintentional essentialization and further alienation of nonhumans, as historians, we must be aware of such limitations that could undermine the enabling contributions of a focus on nonhuman agency. Nevertheless, despite such unease, histories of imperialism and colonialism can no longer ignore the hitherto underrepresented and underexplored contributions of nonhuman animals.

---

[56] Armstrong, Philip. The Postcolonial Animal. *Society & Animals* 10, no. 4 (January 2002): 413–419, 416.
[57] Eaton, Richard. (Re)imag(in)ing Otherness: A Postmortem for the Postmodern in India. *Journal of World History* 11, no. 1 (Spring 2000): 57–78; Sarkar, Sumit. The Decline of the Subaltern in Subaltern Studies. In *Reading Subaltern Studies: Critical History, Contested Meaning and the Globalization of South Asia,* David Ludden (ed.), 400–442. London: Anthem Press, 2002.
[58] Sarkar, Sumit. Orientalism Revisited: Saidian Frameworks in the Writing of Modern Indian History. *Oxford Literary Review* 16, no. 1–2 (July 1994): 205–224.
[59] Skabelund, *Empire of Dogs,* 13–16; Rajamannar, Shefali. *Reading the Animal in the Literature of the British Raj.* New York, NY: Palgrave Macmillan, 2012, 1–15.
[60] Roy, Rohan Deb. Nonhuman Empires. *Comparative Studies of South Asia, Africa and the Middle East* 35, no. 1 (May 2015): 66–75.

Selected Bibliography

Ahuja, Neel. Postcolonial Critique in a Multispecies World. *Publications of the Modern Language Association of America (PMLA)* 124, no. 2 (March 2009): 556–563.
Armstrong, Philip. The Postcolonial Animal. *Society & Animals* 10, no. 4 (January 2002): 413–419.
Bankoff, Greg and Sandra Swart (eds.). *Breeds of Empire: The Invention of the Horse in Southeast Asia and Southern Africa 1500–1950*. Copenhagen: NIAS, 2007.
Boomgaard, Peter. *Frontiers of Fear: Tigers and People in the Malay World, 1600–1950*. New Haven, CT: Yale University Press, 2001.
Brown, Karen and Daniel Gilfoyle (eds.). *Healing the Herds: Disease, Livestock Economies, and the Globalization of Veterinary Medicine*. Athens, OH: Ohio University Press, 2010.
Hevia, James. *Animal Labor and Colonial Warfare*. Chicago, IL and London: University of Chicago Press, 2018.
Mishra, Saurabh. History Writing, Anthropomorphism and Birdwatching in Colonial India. *History Compass* 15, no. 8 (August 2017): e12404. doi.org/10.1111/hic3.12404.
Rangarajan, Mahesh. Animals with Rich Histories: The Case of the Lions of Gir Forest, Gujarat, India. *History and Theory* 52, no. 4 (December 2013): 109–127.
Roy, Rohan Deb. Nonhuman Empires. *Comparative Studies of South Asia, Africa and the Middle East* 35, no. 1 (May 2015): 66–75.
Saha, Jonathan. Among the Beasts of Burma: Animals and the Politics of Colonial Sensibilities, c. 1840–1940. *Journal of Social History* 48, no. 4 (Summer 2015): 910–932.
Swart, Sandra and Lance Van Sittert (eds.). *Canis Africanis: A Dog History of Southern Africa*. Leiden: Brill, 2007.

Dominik Ohrem
Feminist Intersectionality Studies

1 Introduction and Overview

Feminism has been and continues to be a guiding force in the field of animal studies. Viewed from a longer historical arc, feminist perspectives have significantly shaped the conceptual and political grounds from which critiques of essentialist conceptions of human-animal difference and the structural violence of human-animal relations could be articulated. Recent studies have addressed this aspect in more direct ways by emphasizing the partially intersecting histories of feminism and animal advocacy and by identifying salient ethical, political, and philosophical interconnections – "intimate familiarities," as Lynda Birke puts it[1] – between feminist perspectives on the one hand and those informing the field of animal studies on the other. Feminism not only participates in broader endeavors to rethink human relations to a multiplicity of "earth others,"[2] but to think the human itself as a relational creature whose very existence is bound to a meshwork of biosocial relations, beginning with our own more-than-human bodies. At the same time, feminist perspectives, in combination with fields like critical race and postcolonial studies, can help us engage with the blind spots of such post-anthropocentric endeavors, challenging us to ask what may be concealed in and through the very act of attempting to rethink the "the human" [→Post-Domestication: The Posthuman].

Does such a perspective necessarily presuppose some kind of coherent species collective? And, if so, how can it ever account for the internal differentiality of the human, especially with regard to global and societal power relations? Here, in particular those feminist perspectives situated outside the white mainstream of feminist thought offer an important corrective that is crucial to the legitimacy of post-anthropocentric approaches. This chapter focuses on one particularly influential concept that is subject to ongoing debate among feminists regarding its meanings and foundations: intersectionality. As I want to suggest here, intersectional perspectives should be seen as a vital ingredient of any mode of post-anthropocentric critique that is aware of the analytical and political intricacies that grappling with the (d)elusive notion of "the human" and often no less fraught ideas about "human"-"animal" relations entails.

---

[1] Birke, Lynda. Intimate Familiarities. Feminism and Human-Animal Studies. *Society & Animals* 10, no. 4 (January 2002): 429–436.
[2] Plumwood, Val. *Feminism and the Mastery of Nature*. London: Routledge, 1993, 137.

2 Topics and Themes

Intersectionality and (Historical) Animal Studies

While the emergence of intersectionality is commonly traced to the work of Black feminist legal scholar Kimberlé Crenshaw, the development of intersectional perspectives can be seen within a longer historical arc that includes such nineteenth-century figures as abolitionist and women's rights activist Maria Stewart and scholar and educator Anna Julia Cooper.[3] Given its broad reception, it might seem odd that there is no clear consensus on what intersectionality actually is: it has been termed, among other things, a theory, a framework, a method, a "leading feminist paradigm," or a "broad-based knowledge project."[4] Such definitional ambiguities aside, intersectionality comprises a number of key elements that many if not most scholars can agree on.

Historically, intersectionality is strongly grounded in Black feminist activism and social justice projects that sought to make visible the plight of marginalized people (of color) in the US and elsewhere, while at the same time articulating a critique of the implicit whiteness of mainstream feminism's "Woman" and its inability to take into account the experiences of women of color – experiences that were often obscured by a reductive and universalizing preoccupation with gender. These unique historical experiences of women of color with the "interlocking systems of oppression"[5] that affected their lives highlight the interwovenness of intersectionality's commitments to both scholarly analysis and sociopolitical transformation – its importance both as an "analytical strategy" and as a "critical praxis" employed in the pursuit of social justice projects.[6]

Seeking to move beyond the myopia of mainstream feminism, intersectionality not only positions itself against what is often referred to as "single-axis thinking" – attempts, that is, to explain historical and contemporary forms of social injustice that focus on only one vector of difference and power (such as gender) – but also against approaches in which the complexity of social relations and identities is conceived of in terms of a cumulative or additive combination of such vectors that con-

---

[3] See, for example, May, Vivian M. Historicizing Intersectionality as a Critical Lens: Returning to the Work of Anna Julia Cooper. In *Interconnections: Gender and Race in American History*, Carol Faulkner, Alison Marie Parker (eds.), 17–49. Rochester, NY: University of Rochester Press, 2012.
[4] Zack, Naomi. *Inclusive Feminism: A Third Wave Theory of Women's Commonality*. Lanham, MD: Rowman & Littlefield, 2005, 1; Collins, Patricia Hill. Intersectionality's Definitional Dilemmas. *Annual Review of Sociology* 41, no. 1 (2015): 1–20, 3.
[5] The idea that "major systems of oppression are interlocking" was formulated in a 1977 paper by the Bostonian Combahee River Collective. See Combahee River Collective. A Black Feminist Statement. In *The Second Wave: A Reader in Feminist Theory*, Linda J. Nicholson (ed.), 63–70. New York, NY: Routledge, 1997, 63.
[6] Collins, Intersectionality's Definitional Dilemmas, 3.

tinues to treat them as separate and separable categories instead of taking into account their co-constitutive interwovenness. It is not enough, in other words, to argue for a perspective that includes, say, race "plus" gender but to ask how race is always already gendered (and vice versa) in the very process of its social articulation. Importantly, however, the broad reception of intersectionality should not distract us from the fact that, as Anna Carastathis argues, truly intersectional thinking is still a challenge: it "urges us to grapple with [...] our entrenched perceptual-cognitive habits of essentialism, categorial purity, and segregation" rather than offering a "determinate resolution" to these problems.[7]

It is worth noting that intersectional perspectives are historical per se in that they demand careful attention to the complex histories from which social inequality and structural (dis)advantages have emerged. If a crucial task of intersectional perspectives consists in "unsettling dominant imaginaries," as Vivian May suggests, this task entails an intervention in established forms of historical narrative and memory, a confrontation with a "patently racist, sexist, and homophobic intellectual, narrative, and visual archive," and an attention to "submerged histories, disregarded forms of knowing, and long-forgotten or misinterpreted examples of agency and resilience."[8] Such an understanding of intersectional historiography surely resonates with how many historians of human-animal relations see their own work in the face of an overwhelmingly anthropocentric historical archive. But what is the specific relevance of intersectional perspectives for animal studies, and how have intersectionality's insights been implemented by those working in this and related fields? As Patricia Hill Collins explains, one significant result of intersectionality's broad reception is the formation of "a dynamic assemblage of interpretive communities, each of which has its own understanding of intersectionality and advances corresponding knowledge projects."[9] Animal studies scholars and activists can be understood as constituting one such interpretive community, even though certain tensions exist between the academic mainstream of animal studies and the pronouncedly activist field of critical animal studies (CAS). While this is not the place to weigh in on these debates, it is true that CAS has most visibly adopted intersectional perspectives and also taken to heart the important premise that the analytical import of intersectionality cannot be uncoupled from its explicitly political outlook as a social justice project or, to use Collins term, a "critical praxis" (the crux then perhaps being what counts as critical praxis and based on what criteria).

Richard Twine's discussion of intersectionality, for example, identifies "an implicit acknowledgement of intersectionality [...] throughout animal studies" and treats intersectionality as a key concept of CAS, arguing that its traditionally anthro-

---

[7] Carastathis, Anna. *Intersectionality: Origins, Contestations, Horizons*. Lincoln, NE: University of Nebraska Press, 2016, 4.
[8] May, Vivian M. *Pursuing Intersectionality: Unsettling Dominant Imaginaries*. New York, NY: Routledge, 2015, 53–54, 59.
[9] Collins, Intersectionality's Definitional Dilemmas, 3.

pocentric emancipatory focus "can in *some* senses be broadened to include the more-than-human."¹⁰ In turn, animal studies and posthumanist critiques of essentialist conceptions of human-animal difference resonate with and complement feminist, anti-racist, and other kinds of analytics/politics that grapple with the ways in which the notoriously malleable figure of "the human" has been subject to differential constructions within historically shifting fields of power relations. "The human-animal dualism," as Twine puts it, "percolates through intra-human categories of difference, contributing to their sense of fixity."¹¹ A similar line of argument can be found in the work of Maneesha Deckha, who engages with the subtle or overt ways in which 'species' intersects with other vectors of difference and power. Regarding the discursive nexus of racialization-animalization that has shaped much of western history, Deckha points to the ways in which "[t]he metaphor of 'the human female/dark/poor body as *x* animal' operates at the intersections of multiple Othering discourses, including those that subordinate animals," and argues that "uproot[ing] the dynamics of oppression" requires us to take into account how species permeates not only human-animal but also intrahuman relations.¹²

The intersections of sex and species have been explored most prominently by Carol Adams. What Adams calls the "sex-species system" is defined by pervasive associations between carnivorism and hegemonic (hetero-)masculinity and the objectification, rhetorical or material dismemberment into "choice parts," and "consumption" of female and animal bodies. The sex-species system, Adams explains, "ensures that men have access to feminized animal bodies and animalized female bodies,"¹³ underscoring that an ostentatious disregard for the well-being of nonhuman creatures should be understood as a recurring, if not integral, element of what is now often referred to as "toxic masculinity." Adams argues that species as such "is gendered (animals are feminized) and gender, that is, woman, who carries gender identification, is animalized. Man transcends species; woman bears it. So do the other animals."¹⁴ While, as Adams' own work has shown, this no doubt holds true in many contexts, we also need to pay specific attention to the variety of historically situated relationships between different groups of humans – as will be discussed in the next section, "Man" and "Woman" are themselves intersectional configurations that cannot be understood with regard to gender alone – and (to use

---

10 Twine, Richard. *Animals as Biotechnology: Ethics, Sustainability and Critical Animal Studies*. London: Earthscan, 2010, 9, 10.
11 Twine, *Animals as Biotechnology*, 10–11.
12 Deckha, Maneesha. The Salience of Species Difference for Feminist Theory. *Hastings Women's Law Journal* 17, no. 1 (Winter 2006): 1–38, 31.
13 Adams, Carol J. After MacKinnon: Sexual Inequality in the Animal Movement. In *Critical Theory and Animal Liberation*, John Sanbonmatsu (ed.), 257–276. Lanham, MD: Rowman & Littlefield, 2011, 270.
14 Adams, Carol J. *The Pornography of Meat*. New York, NY: Continuum, 2003, 149.

the three-pronged western categorization) the wide range of domesticated, wild, and feral creatures.

With intersectionality's key aspects in mind, it is not too difficult to imagine that a post-anthropocentric angle on intersectionality comes with a number of challenges. Intersectional perspectives that include issues of species in their analytic purview can help us work through the fissures and asymmetries that shape the topography of the human, offering new perspectives on, for example, intersectional histories of race or dis/ability. Things get trickier, however, if we intend to use intersectionality to address animals as intersectional subjects in their own right. Here, we need to think very carefully about the extent to which an intersectional approach can be brought to bear on (historical) animal realities, and what modifications or translations intersectionality undergoes in this process. What, for example, could a micro-level analysis of intersectional experience look like with regard to the experiences of nonhuman beings? As Jason Wyckoff writes:

> When we essentialize the experiences of human beings, we tend to generalize from a set of experiences that are familiar and assume (problematically, of course) that those experiences are the essence of the dimension of social identity being described ('woman', 'working class', etc.). In the case of animals, we cannot do this.[15]

This is not an exclusively historiographic problem but ties into broader epistemological issues of animal alterity, the accessibility of animal minds, and the specter of anthropomorphism frequently invoked to discredit any kind of writing that endows animals with complex forms of interiority [→History of Ideas]. And while it would be disingenuous to downplay the analytic challenges posed by interspecific differences, we should not forget that the notion of experience and the historical task of "reconstructing" past experiences have been problematized even with regard to anthropocentric frameworks.[16] Moreover, we might ask whether micro-level intersectional analyses are necessarily tied to the concept of experience at all or whether they can be approached with a range of different concepts (e.g. bodily practice) that might lend themselves more readily to post-anthropocentric perspectives but still allow us to address the ways in which intersectional power works across different scales of social reality.

Animals assume different locations within or vis-à-vis the anthropocentric social systems they are a part of or (forced to) interact with. In the case of domesticated animals, established forms of utilitarian categorization largely determine whether the destination of an animal of a given species is, pointedly put, a factory farm or a human family [→History of Pets]. Thus, while domesticated animals as a whole

---

[15] Wyckoff, Jason. Analysing Animality: A Critical Approach. *The Philosophical Quarterly* 65, no. 260 (June 2015): 529–546, 537.
[16] See Joan W. Scott's influential Scott, Joan W. The Evidence of Experience. *Critical Inquiry* 17, no. 4 (Summer 1991): 773–797.

are relegated to a subordinate sphere, we can still discern a range of structural (dis)advantages that vary significantly with different species, ranging from the fundamentally terminal existence of "livestock" to the protection and care that companion animals are usually afforded with today. If we want to approach animal lives and identities in intersectional terms, we need to begin from the assumption that human-animal relations are fundamentally shaped by the ways in which animals are gendered, racialized or otherwise marked – including through categories such as utility, charisma, or native-/invasiveness that are specifically employed in hierarchizations of nonhuman species and determine their protection from, or exposure to, violence. Domesticated creatures might be particularly relevant to intersectional analysis because of their deep integration into human social worlds.

Perhaps one of the most fruitful approaches consists in demonstrating how both humans and animals have been exploited through their incorporation into a broader intersectional power structure, as has been the case, for example, with the abuse and mistreatment of vulnerable immigrant workers and the industrial-scale destruction of animals in the turn-of-the-twentieth-century Chicago meatpacking industry, a dual exploitation that seems all the more tragic because of the antagonistic relationship in which both groups have been placed [→History of Animal Slaughter]. Viewed from this angle, the Chicago stockyards emerge as a space marked by the oppression of both "animalized humans" (racialized working-class immigrants represented as "inferior stock" by turn-of-the-century racists and eugenicists) and "animalized animals" (animals whose species membership precludes them from protection against human violence or directly marks them for death), as a space in which human degradation is intertwined with animal suffering.[17]

Post-anthropocentric perspectives may thus serve to broaden intersectional analytics by showing that "differentiated response[s] to animals [are] influenced by multiple axes of difference similar to how differential responses to Othered humans are structured,"[18] and, indeed, that we can often discern an intricate, mutually constitutive correspondence between these two forms of "response." Here, however, a crucial question might be what exactly we mean by "similar" – and perhaps we should be cautious with assertions of structural and/or moral equivalence regarding forms of human and animal oppression. This is in no way to depreciate the political significance of animal oppression but rather, and firstly, to remind us that the desire to establish a sense of moral equivalence in the service of animal advocacy might

---

[17] The terms "animalized humans" and "animalized animals" are part of the typology suggested by Wolfe, Cary and Jonathan Elmer. Subject to Sacrifice: Ideology, Psychoanalysis, and the Discourse of Species in Jonathan Demme's *Silence of the Lambs*. Boundary 22, no. 3 (Autumn1995): 141–170, 146–147. On the Chicago meatpacking industry, see Malay, Michael. Modes of Production, Modes of Seeing: Creaturely Suffering in Upton Sinclair's *The Jungle*. In *American Beasts: Perspectives on Animals and Animality in U.S. Culture, 1776–1920*, Dominik Ohrem (ed.), 123–149. Berlin: Neofelis, 2017.
[18] Deckha, Maneesha. Intersectionality and Posthumanist Visions of Equality. *Wisconsin Journal of Law, Gender and Society* 23 (2008): 249–268, 258.

lead us to simply assume structural equivalence instead of treating it as an object of historical inquiry. Secondly, equivalence is an issue that needs to be handled carefully in a dialogical rather than impositional manner, involving the perspectives of those groups of humans whose experiences of oppression are involved in such arguments, lest we risk creating rifts, or deepening existing ones, between different interpretive communities rather than fostering alliances between them. Equivalence, in other words, is an issue for which a sensitivity to the politics of historiography is no less important than the cogency of historical argument.

Intersectionality is not limited to any specific range of topics but offers a broad lens for the analysis of social (power) relations and identities. While not always explicitly intersectional, a growing number of works in historical animal studies have taken up intersectionality's key analytical focus on the complex synergies between vectors of difference and power. Such works not only complement historiographies of, for example, race and gender by integrating animality and human-animal difference as important elements of analysis, but also broaden the purview of intersectionality to include nonhuman creatures. To mention just two examples from the US-American context: Keridiana Chez discusses how turn-of-the-twentieth-century idea(l)s of domesticity were troubled by anxieties about the supposedly pathological relationship between women and cats, anxieties that were expressed through a "complex discursive-imaginary apparatus linking felinity with femininity to the detriment of both." Chez's analysis opens up a perspective on how (in this case, companion) animal identities were co-shaped by dominant gender norms and how this "gendering of the nonhuman" in turn affected the social location of the group(s) of people discursively and/or materially associated with the animals in question.[19] Combining the perspectives of settler colonial and animal studies, another recent article, by Diné scholar Kelsey Dayle John, develops the concept of "animal colonialism" to highlight "the interconnected nature of Indigenous nonhuman animals, peoples, and lands, and the ways these relationships [...] are tangled with oppressions confronted by various disciplines."[20] John does not expressly situate her article in an intersectional framework, but her focus on settler colonialism's "multifaceted and interlocking forms of oppression" and her definition of animal colonialism as a phenomenon informed by heteropatriarchy, racism and other "forces that connect, intersect, and overlap in complex ways" clearly echo the language of intersectionality.[21]

---

[19] Chez, Keridiana. Man's Best and Worst Friends: The Politics of Pet Preference at the Turn of the Twentieth Century. In *American Beasts: Perspectives on Animals and Animality in U.S. Culture, 1776–1920*, Dominik Ohrem (ed.), 175–200. Berlin: Neofelis, 2017, 177, 200.
[20] John, Kelsey Dayle. Animal Colonialism: Illustrating Intersections between Animal Studies and Settler Colonial Studies Through Diné Horsemanship. *Humanimalia* 10, no. 2 (Spring 2019): 42–68, 42.
[21] John, Animal Colonialism, 57, 42.

Western Man and the History of Slavery

Instead of broaching a number of different topics, I would now like to focus on a historiographic perspective that offers an analytic lens for a wide range of historical topics and contexts – my discussion here will deal with US slavery – and, I hope, brings into focus some of the potentials of combining intersectional and post-anthropocentric approaches: the figure, or, in cultural theorist Sylvia Wynter's terms, "genre" of Man. Wynter's interdisciplinary work is concerned with the ways in which "the human" has been conceptualized and, specifically, with how one particular configuration or "genre" of humanness – (western) Man – has attained a hegemonic status, relegating alternative forms of being human to the subordinate domain of "Man's Human Others."[22] The reason why, in conjunction with more overt processes of colonialist-racist violence, the figure of Man has been so effective in colonizing the topography of the human is precisely because it is positioned less in a straightforwardly dominative way but draws upon notions of human universality: Man, that is, "overrepresents" itself as "the generic, ostensibly supracultural human" – and as if its interests were congruent with those of humanity as such.[23] For Wynter, decolonizing the domain of the human, and perhaps even securing the future of the planet itself, thus means working towards overcoming Man and its universalist pretensions.[24]

Wynter's work resonates with key tenets of intersectionality. Man is not characterized in any primary sense by gender, race, or any other specific vector but is a thoroughly intersectional figure that requires complex analysis and historicization, and one way of thinking about genres of the human is by analyzing them as intersectional configurations emerging from asymmetrical relations of power. Wynter's work encourages us, as historians, to develop a critical awareness of the ways in which the rhetoric of human universality has worked towards the perpetuation of Man, thus constituting treacherous analytical and political ground. In light of this, a crucial task of post-anthropocentric historiography consists in rethinking the human "as an index of a multiplicity of historical and ongoing contestations [...] rather than take 'the human's' colonial imposition as synonymous with all appearances of 'human.'"[25] I would argue that a more comprehensive picture of the historical emergence of Man becomes possible once we take into account how Man is positioned vis-à-vis a collective of both human and nonhuman or, to use the broadly inclusive term

---

22 Wynter, Sylvia. Unsettling the Coloniality of Being/Power/Truth/Freedom: Towards the Human, After Man, its Overrepresentation – An Argument. *CR – The New Centennial Review* 3, no. 3 (Fall 2003): 257–337, 313.
23 Wynter, Unsettling the Coloniality, 288.
24 Wynter, Unsettling the Coloniality, 260.
25 Jackson, Zakiyyah I. Animal: New Directions in the Theorization of Race and Posthumanism. *Feminist Studies* 39, no. 3 (2013): 669–685, 681.

Wynter employs in one of her earlier essays, "Ontological Others"[26] who demarcate frequently overlapping and strategically inconsistent domains of alterity and inferiority.

Wynter's theoretico-historical discussion of the genre of Man opens up a range of perspectives for intersectional post-anthropocentric historiography that combines the insights of fields like critical race, gender, and postcolonial studies with those of animal studies. One area of analysis where such an approach proves useful is what Wynter calls "the tropology of Africa/The Negro"[27] and concomitant constructions of Black humanity. Throughout Western history, Blackness has often served to indicate what Achille Mbembe refers to as "a certain litigious figure of the human,"[28] the epitome of those forms of human otherness contradistinctively sustaining the exalted position of Man. As Mbembe writes with regard to the nineteenth-century US-American context, white Americans frequently questioned whether one could in fact find among Black people

> the same humanity, albeit hidden under different designations and forms [...] Could one detect in their bodies, their language, their work, or their lives the product of human activity and the manifestation of subjectivity [...] a presence that would authorize us to consider each of them, individually, as an alter ego?[29]

This dubiously (in)human humanity of Black people was not only inseparable from but in fact actively produced by the institution of chattel slavery. One of the key tropes of pro-slavery fantasies of plantation domesticity suggested that enslaved Black people willingly embraced their subordinate place in the hierarchical order of things. As irrational but docile domesticated creatures supposedly benefiting from the strict but "benevolent" supervision and instruction of white "masters," pro-slavery discourse commonly incorporated enslaved Black humans into a racialized anthro-patriarchal framework in which white Man as the avatar of rational, civilized manhood – a term that could denote not only adult maleness or virility but also the general state of being human – was ordained by Nature or Providence to preside over a host of supposedly lesser beings, including Black humans and domesticated animals, often collapsing the very distinction between them.

Scholars of slavery have emphasized the harrowing experiences of enslaved Black women, in particular their systematic sexual abuse and exploitation for the purposes of "slave breeding." As Harriet Jacobs recounts in *Incidents in the Life of a Slave Girl*, "[w]omen are considered of no value, unless they continually increase

---

26 Wynter, Sylvia. On Disenchanting Discourse: Minority Literary Criticism and Beyond. *Cultural Critique* 7 (Autumn 1987): 207–244, 217.
27 Wynter, On Disenchanting Discourse [emphasis removed].
28 Mbembe, Achille. *Critique of Black Reason*. Translated by Laurent Dubois. Durham, NC: Duke University Press, 2017, 49.
29 Mbembe, *Critique of Black Reason*, 85.

their owner's stock. They are put on a par with animals."[30] The intersectional configuration of race, animality, gender, and sexuality that shaped Black men's lives, on the other hand, underwent a momentous shift after emancipation, as long-standing anxieties about the supposed hypersexuality of Black men in combination with the challenge to white authority brought about by the abolition of slavery coalesced into the widespread image of the "black beast" rapaciously lusting after white women. A 1901 article by George T. Winston both recapitulates and exemplifies this white supremacist phantasm and the causal linkage it creates between Black freedom and Black masculinity's devolution into a state of unrestrained animality:

> In slavery [the Black man] was like an animal in harness; well trained, gentle and affectionate; in early freedom the harness was off, but still the habit of obedience and the force of affection endured [...]. In Reconstruction came a consciousness of being unharnessed, unhitched, unbridled and unrestrained. The wildest excesses followed [...]. [N]ow, when a knock is heard at the door, [the white woman] shudders with nameless horror. The black brute is lurking in the dark, a monstrous beast, crazed with lust. His ferocity is almost demoniacal. A mad bull or a tiger could scarcely be more brutal.[31]

Winston's article also highlights the ways in which intersectional dynamics could shift significantly according to the specific demands associated with the defense of the white supremacist status quo. In the antebellum order of things, which Winston euphemistically refers to as a harmonious state of "extended and constant social intercourse" supposedly grounded not in a false sense of equality but in honest mutual affection between enslaved Black people and their white "masters," the former were still seen "as individuals, as human beings,"[32] paradoxically suggesting that a recognition of Black people's humanity and personhood in fact relied on their conforming to their imposed role as "animals in harness."

Slave narratives show that enslaved Black people were acutely aware of their troubling proximity – bodily/spatially and in terms of social position and ethical consideration – to the domesticated animals with whom they shared an existence at the bottom of the plantation hierarchy. As Frederick Douglass recounts his experiences at a "slave auction,"

> moral and intellectual beings, in open contempt of their humanity, leveled at a blow with horses, sheep, horned cattle and swine! Horses and men – cattle and women – pigs and children – all holding the same rank in the scale of social existence; and all subjected to the same narrow inspection, to ascertain their value in gold and silver [...]. Personality swallowed up in the sordid idea of property! Manhood lost in chattelhood! [...] Our destiny was now to be fixed for life, and

---

30 Jacobs, Harriet Ann. *Incidents in the Life of a Slave Girl, Written by Herself.* Boston, MA: Published by the Author, 1861, 76; on 'slave breeding,' see Smithers, Gregory D. *Slave Breeding: Sex, Violence, and Memory in African American History.* Gainesville, FL: University Press of Florida, 2012.
31 Winston, George T. The Relation of the Whites to the Negroes. *The Annals of the American Academy of Political and Social Science* 18, no. 1 (July 1901): 105–118, 114, 109.
32 Winston, The Relation of the Whites, 105, 107.

we had no more voice in the decision of the question, than the oxen and cows that stood chewing at the hay-mow.³³

Later, after the sadistic slaveholder and "negro breaker," Edward Covey, has instructed Douglass to work with a pair of unbroken oxen, he notes in his own situation "several points of similarity" with that of the bovines: "They were property, so was I; they were to be broken, so was I,"³⁴ Douglass writes, a kind of human-animal comparison that moves beyond superficial representational analogy into the existential dimension of shared embodied experience – the experience of living property from whom labor was to be forcefully extracted and on whom similar forms of violence – whippings and lashings – were exercised to establish submission.

Beyond the general exploitation of domesticated animals, slavery also featured more specific forms of animal subjugation, in particular the brutalization of bloodhounds, who were systematically abused through beatings and starvation to prepare them for their occupation of "slave-catching." A powerful abolitionist image and the bane of every self-emancipated Black person, the bloodhound opens up a perspective on how slavery and other systems of white supremacist oppression – the abolitionist potency of the bloodhound image was connected to the earlier use of these animals against Native Americans – can be viewed through a trans- and interspecies intersectional lens.³⁵

Can we identify an undercurrent of critique in Douglass' narrative that exceeds the insistence on human dignity in terms of an inclusion of Black humanity into the liberal humanist fold of Man? What exactly is the implication of Douglass' recognition of the "several points of similarity" between his own plight and that of the oxen? Is this assessment "only" supposed to convey or evoke a sense of outrage that "moral and intellectual [human] beings" like him are treated – "broken" – like draft animals? Or does it perhaps (also) convey a broader critique of the degradation of living beings of all kinds through the "sordid idea of property" and its ruthless manifestation in the system of slavery?

While Douglass' imagery frequently alludes to the exploitation of both human and animal life, the key point of his writing is to condemn the ways in which slavery works to "imbrute" Black humans – an experience that, at one particularly desperate point, makes him long for an actual "escape" into animality by exchanging "my

---

33 Douglass, Frederick. *My Bondage and My Freedom*. New York, NY: Miller, Orton and Mulligan, 1855, 175.
34 Douglas, *My Bondage and My Freedom*, 212.
35 Campbell, John. The Seminoles, the Bloodhound War, and Abolitionism, 1796–1865. *The Journal of Southern History* 72, no. 2 (May 2006): 259–302; Fielder, Brigitte. Black Dogs, Bloodhounds, and Best Friends: African Americans and Dogs in Nineteenth-Century Abolitionist Literature. In *American Beasts: Perspectives on Animals and Animality in U.S. Culture, 1776–1920*, Dominik Ohrem (ed.), 153–173. Berlin: Neofelis, 2017.

manhood for the brutehood of an ox."³⁶ Here again, different meanings of "manhood" – and the nuances of its denial – are at play, giving expression to Douglass' experience of slavery. His repeated association with the oxen occurs not only for reasons of his forced interactions with them on Covey's plantation but also because the animals' existence resonates with his own experience: it is not just any kind of "brutehood" or an undifferentiated notion of animality that permeates his narrative but that of domesticated work animals deprived of bodily autonomy, toiling under the yoke and the looming threat of punishment, should they show any sign of resistance. From this perspective, we might say that both Douglass' manhood *and* the oxen's brutehood are "lost in chattelhood" as a mode of existence in which a being's life is entirely dictated by interests that are not, and violently conflict with, their own.

There is also a notably gendered element in Douglass' contrasting of his endangered manhood with the brutehood of oxen, creatures who have been literally "emasculated" to make them more docile. That Douglass imagines exchanging his humanity (and masculinity) for the brutehood of an *ox* thus speaks to the seeming hopelessness of his particular situation – hungry, weak, and exposed to the elements, after he has escaped from the plantation to avoid the consequences of a physical altercation with Covey – and his ambivalent relationship with these animals: a relationship that fluctuates between what Ralph Acampora terms "symphysical"³⁷ resonance and a desire for ontological dissociation, between a sense of shared bodily suffering among fellow living beings and the necessity to distance himself, as a Black (hu)man whose humanity was constantly being assailed through the discursive and material practices of slavery, from animalkind in order to denounce slavery's monstrous degradation of fellow *human* beings. But, as Douglass points out in his speech, "What to the Slave is the 4th of July?," despite the countless ways in which white society questioned Black humanity, the "manhood of the slave" was already conceded in and through the very necessity of ensuring its denial, for example in the form of anti-literacy laws that prohibited teaching enslaved people to read and write:

> When you can point to any such laws, in reference to the beasts of the field, then I may consent to argue [instead of simply *asserting*] the manhood of the slave. When the dogs in your streets, when the fowls of the air, when the cattle on your hills, when the fish of the sea, and the reptiles that crawl, shall be unable to distinguish the slave from a brute, then will I argue with you that the slave is a man!³⁸

This brief discussion already points to some of the questions that present themselves when we look at texts like Douglass' from a post-anthropocentric intersectional

---

36 Douglass, *My Bondage and My Freedom*, 430, 235.
37 Acampora, Ralph R. *Corporal Compassion: Animal Ethics and Philosophy of Body*. Pittsburgh, PN: University of Pittsburgh Press, 2006, 76.
38 Douglass, *My Bondage and My Freedom*, 443.

angle.[39] Read together, Douglass' narrative and Winston's article convey a sense of how intersections of species with race and other vectors shaped Black people's experiences under slavery – including the ways in which their relationships with animals were affected by these experiences – and also informed constructions of Blackness in white supremacist dispositives as a "*distinct* humanity – one whose very humanity was (and still is) in question."[40] How might we connect Mbembe's notion of a distinct Black humanity with the concept of "dehumanization" often employed to describe the workings of extreme forms of oppression such as slavery? Firstly, dehumanization arguably implies a demotion from what is assumed to be a shared default status of humanity that does not, in fact, exist. Conceptualizations of humanity, that is, should not be understood as resulting from a process by which the vector of species becomes "inflected by" or "charged with" ideas about race or gender; rather, the epistemic interpenetration of these and other vectors is crucial to the very emergence of always constitutively intersectional modalities of the human. Secondly, the way in which Blackness is positioned in contradistinction to white- and anthro-supremacist Man is not so much as a void of humanity but in the form of specific modalities of alter- or parahumanity[41] – configurations of the human in which humanity itself is continually in doubt and at stake rather than simply (made) absent.

3 Implication(s) of the Animal Turn

As Philip Armstrong and Laurence Simmons state confidently in a 2007 volume, the effects of the "animal turn" are "comparable in significance to the 'linguistic turn' that revolutionized humanities and social science disciplines from the mid-twentieth century onwards."[42] But while the idea of an *animal turn* may well be an appropriate way to describe these transformations, such "turn talk" may also indicate some problematic tendencies. As Gary Wilder argues in an essay on the politics of academic turns, the latter are often marked by a trajectory "from optic to topic," a development through which "the analytic openings that were created by [these] turns [are] foreclosed" and "new optics [are] transformed into routine research topics."[43] As a result,

---

39 On animals and animality in Douglass' writing, see Boggs, Colleen Glenney. *Animalia Americana: Animal Representations and Biopolitical Subjectivity.* New York, NY: Columbia University Press, 2013, ch. 2; Jackson, Zakiyyah I. Losing Manhood: Animality and Plasticity in the (Neo)Slave Narrative. *Qui Parle: Critical Humanities and Social Sciences* 25, no. 1–2 (Fall/ Winter 2016): 95–136.
40 Mbembe, *Critique of Black Reason*, 85.
41 See Allewaert, Monique. *Ariel's Ecology: Plantations, Personhood, and Colonialism in the American Tropics.* Minneapolis, MN: University of Minnesota Press, 2013.
42 Simmons, Laurence and Philip Armstrong. Bestiary: An Introduction. In *Knowing Animals*, Laurence Simmons, Philip Armstrong (ed.), 3–24. Leiden: Brill, 2007, 3.
43 Wilder, Gary. From Optic to Topic: The Foreclosure Effect of Historiographic Turns. *The American Historical Review* 117, no. 3 (June 2012): 723–745, 723.

the critical dynamics of these phenomena – including their internal tensions and contradictions that in part account for their vitality – are relinquished in favor of academic marketability or survive as watered-down versions.

Do animals and animality inhabit the world of academia in optical or topical form? More important than an answer to this question (which would probably be: both) is keeping the question itself relevant. Even though animal historiography is already being practiced for some time now, its practitioners should cultivate a certain skepticism about any kind of methodological or political consensus regarding what exactly animal history should do, be, and look like [→History of Ideas]. Perhaps we should thus refuse to answer in any definite way the question that makes up the subtitle of Hilda Kean's article on animal historiography – "what is really enough?"[44] – and instead use this question to think, as Carastathis suggests with regard to intersectionality, the writing of animal history as an ongoing challenge. A broader task of animal studies, then, lies in its commitment to sharpening "a critical lens like that offered by race or feminist theory"[45] – a lens, however, that should not be understood as merely "like" but interlaced *with* these other lenses, yielding a complex, multifaceted optics.

Twine's assessment of an "implicit acknowledgement" of intersectionality in animal studies is a good way to look at how works in the field have been shaped by the concept's interdisciplinary travels without always explicitly adopting an intersectional framework. But to what extent has intersectionality studies reacted to the *animal turn*? In a 2013 article, Deckha notes that "despite all [...] debates about intersectionality's purpose, one boundary has remained certain: the anthropocentric focus of even recent literature in this area."[46] While Nina Lykke points in a 2010 book to an "emerging field of feminist studies of human-animal relations," she notes that "the feminist discussion of nonhuman actors runs parallel to, rather than being integrated with, explicit feminist theorizing of intersectionality."[47] A look at more recent monographs on intersectionality might be a decent enough indicator on whether this (still) holds true: the monographs by Carastathis and Hancock that I have drawn upon here testify to the richness of the debate surrounding intersectionality, but they do not acknowledge, defend, critique, or otherwise engage with the anthropocentric purview of the concept, nor do they contain any reference to works in animal studies

---

44 Kean, Hilda. Challenges for Historians Writing Animal-Human History: What Is Really Enough? *Anthrozoös* 25, Supplement (April 2012): 57–72.
45 Gross, Aaron S. Introduction and Overview: Animal Others and Animal Studies. In *Animals and the Human Imagination: A Companion to Animal Studies*, Aaron Gross, Anne Vallely (eds.), 1–23, 4. New York, NY: Columbia University Press, 2012.
46 Deckha, Maneesha. Animal Advocacy, Feminism and Intersectionality. *DEP – Deportate, Esuli, Profughe* 23 (2013): 48–65, 49–50.
47 Lykke, Nina. *Feminist Studies: A Guide to Intersectional Theory, Methodology and Writing*. London: Routledge, 2010, 80, 81.

or ecofeminist scholars like Plumwood, whose work can be regarded at the very least as what Ange-Marie Hancock calls "intersectionality-like thought."[48]

However that may be, in closing I think that two points are worth emphasizing: first, while it is certainly true that species has not been a prominent aspect of intersectional analysis in comparison with the more extensive debates around the triad of race-class-gender, recent discussions have been less centered on the validity, primacy or general relevance of specific categories but on the broader political and ontological implications of intersectionality and its, in part problematic, reception. Moreover, as Hancock points out, a historical perspective may show us "that categories that originally obtained significant attention have fallen out of favor for reasons that are not intellectually defensible."[49] By adopting a longer historical arc that allows us to trace intersectionality-like forms of thought, we might be able to gain insights into how species interacted with more frequently discussed vectors to shape the intersectional experiences of historical actors. Second, if we understand intersectionality studies not as a field but, following Collins, as an assemblage of interpretive communities, a more strategically useful angle than criticizing the omissions of "intersectionality theory" would be to ask how these communities can "mobiliz[e] the language of commonality (however provisional or tentative that commonality might be)" in order to form of a broader coalitional project.[50]

Selected Bibliography

Adams, Carol J. and Lori Gruen. *Ecofeminism: Feminist Intersections with Other Animals and the Earth*. New York, NY: Bloomsbury, 2014.

Ahuja, Neel. Postcolonial Critique in a Multispecies World. *Publications of the Modern Language Association of America (PMLA)* 124, no. 2 (March 2009): 556–563.

Carastathis, Anna. *Intersectionality: Origins, Contestations, Horizons*. Lincoln, NE: University of Nebraska Press, 2016.

Deckha, Maneesha. Animal Advocacy, Feminism and Intersectionality. *DEP – Deportate, Esuli, Profughe* 23 (2013): 48–65.

Hancock, Ange-Marie. *Intersectionality: An Intellectual History*. New York, NY: Oxford University Press, 2016.

Hovorka, Alice J. Feminism and Animals: Exploring Interspecies Relations Through Intersectionality, Performativity and Standpoint. *Gender, Place & Culture* 22, no. 1 (January 2015): 1–19.

Kim, Claire Jean. *Dangerous Crossings: Race, Species, and Nature in a Multicultural Age*. New York, NY: Cambridge University Press, 2014.

Plumwood, Val. *Feminism and the Mastery of Nature*. London: Routledge, 1993.

Twine, Richard. *Animals as Biotechnology: Ethics, Sustainability and Critical Animal Studies*. London: Earthscan, 2010.

---

[48] Hancock, Ange-Marie. *Intersectionality: An Intellectual History*. New York, NY: Oxford University Press, 2016, 24.

[49] Hancock, *Intersectionality*, 199.

[50] Nash, Jennifer C. Re-Thinking Intersectionality. *Feminist Review* 89, no. 1 (June 2008): 1–15, 4.

Kit Heintzman
Material Culture Studies

1 Introduction and Overview

Historians use the term *material culture* to refer to the study of tangible objects that can be used to understand the past in ways that are distinct from textual, visual, and aural sources. Examples of textual sources commonly used by historians include letters, philosophical treatises, and novels; visual sources include paintings, posters, and movies; and audio sources include recordings of speeches, archived radio programs, and music. Textual, visual, and audio sources are, of course, also constituted from material, but using a book as a material source rather than a textual source changes the kinds of questions the historian asks. For example, a historian wanting to know how a literary author's ideas changed over time might read multiple works written by that same author and other texts in circulation at the moment. Historians

Figure 1: Smithsonian Institution Archives, Record Unit 7181, Box 1, Folder: Exhibits central file 1893, 1896–97, 1900–1901, 1903–1904. Image ID: SIA2012–7944. What appears to be a tiger sauntering through Washington, D.C. is a just finished piece of taxidermy for the Smithsonian's Natural History Museum.

https://doi.org/10.1515/9783110536553-028

in such cases would examine these works *textually,* that is, analyzing the prose and the ideas in the text itself. Books, however, are more than reservoirs for content. When a historian wants to know how the prestige of that author changed over time, the historian might look at the book's different material properties of different editions of the book, such as the bindings, paper quality, and size (cut in folio, quarto, octavo, or duodecimo). Physical details like these can be used to estimate how expensive a book was to produce and, therefore, the cost and value of ownership compared to other literary works available at the time. It treats the book as a sensory object, one with weight, finger feel, and the historically constrained aesthetics of typeface. Material culture is, first and foremost, an analytic approach to answering historical questions that lie outside of the written record. Material culture studies are often used in conjunction with other forms of analysis. It is one of many means of including sensory data and of building an experiential aspect to the study of history.[1] Material culture documents the interactions between humans and other animals.

Museums of natural history, comparative anatomy, and veterinary medicine are the sites most commonly chosen by historians using material culture to "get to the/ an animal" [→History of Animal Collections/Animal Taxonomy;→History of Veterinary Medicine]. These venues contain charismatic specimens such as the tiger seen above, and, if the historian is lucky, they might also find a set of corresponding institutional *accession records* that describe how the physical specimen arrived at the museum and how scientists used it for study.

Natural history, comparative anatomy, and veterinary museums are obvious sites for the animal historian's research, because these museums are sure to have animal remains on display. Remains that pointedly retain a sense of animality. However, historians of animals working with material culture may over-rely on animal remains that still retain strong semblance of the animals from which they came. Though the bodies of other animals preserved in natural history museums may highlight individual animals – who merit their own biographies and histories – these settings also elide the multifarious ways that animal matter has been used in everyday life [→Public History]. *Any* museum – of famous historical figures, of great revolutions, of anthropology, of art, of musical instruments, of science, of printmaking, of the military, of textiles, of national or municipal history – will have animal remains among their collections.

Of the once-living animal-derived biomass now housed in museums, undoubtedly the vast majority exists in bits and pieces as parts of human-manufactured objects, not as anatomical specimens. Historians would do well to notice the animal remains present in the inks, dyes, clothing, musical instruments, and furniture of the past, because the everyday objects we have made from other animals document intrahu-

---

[1] Edwards, Elizabeth, Chris Gosden and Ruth B. Phillips. *Sensible Objects: Colonialism, Museums and Material Culture.* Oxford: Berg, 2006.

man and multispecies relations [→Multispecies Ethnology]. For example, many of the ancient manuscripts housed in archives and rare book collections were written with iron gall ink.[2] Iron gall ink is derived from "gall nuts", tannin-rich growths which develop on oak trees after gall wasps lay their eggs within the trees' leaf buds. To make this ink and to record the early history of human philosophy, humans relied on a multispecies parasitic process between insects and plants. Another animal-dependent color, Tyrian purple, was a mollusk-shell-derived dye so precious that Phoenician merchants charted settlements and seafaring voyages around the habitats of the requisite mollusks in the ancient world. By the sixth century BC, Persian king Cyrus the Great codified that certain patterns using this purple could only be worn by royalty.[3] These mollusk shells were highly profitable to the seafarers who collected them, integrated into anthropocentric legislation, and shaped human migration patterns.

Historians of animals have emphasized the importance of "this animal" or "these animals", believing that literary critics' and philosophers' references to "the animal" is a problematic transhistorical abstraction of the lives of individual creatures and herds.[4] It is nearly impossible, however, to recover an individual animal from the many material objects created out of their parts. This kind of objectification of animals, by making them into stuff, may make some historians and animal studies scholars uncomfortable. Material culture thus troubles some of the individualist and agential trends in animal history and instead directs scholars to structural factors that shaped the history of human-animal relations [→History of Ideas]. It took approximately 250,000 mollusk shells to make a single ounce of Tyrian purple dye.[5] (For the curious, in 2019, a reputable global pigment supplier sells "Tyrian Purple, genuine" for about $4000 USD per gram. There are just over 28.3 grams in an avoirdupois ounce.) Trying to extract the history of an individual mollusk from a piece of royal attire is not only futile, but it ignores one of the critical ways that this particular piece of attire documents multispecies relations. Animal individuality is obscured by the process of turning their matter into things. How much this bothers an author hinges on the epistemic values they project onto "individuality".[6] But one can still see an object dyed with Tyrian purple as an archive of how humans depended on animals for commerce and in policing social hierarchies. We miss the chance to

---

[2] Rabin, Ira. Ink Identification to Accompany Digitization of Manuscripts. In *Analysis of Ancient and Medieval Texts and Manuscripts: Digital Approaches*, Tara L. Andrews, Caroline Macé (eds.), 293–308. Turnhout: Brepols, 2014.
[3] Elliot, Charlene. Purple Pasts: Color Codification in the Ancient World. *Law & Society* 33, no. 1 (February 2008): 173–194.
[4] Haraway, Donna. *When Species Meet*. Minneapolis, MN: University of Minnesota Press, 2008.
[5] Jacoby, David. Silk Economics and Cross-Cultural Artistic Interaction: Byzantium, the Muslim World, and the Christian West. *Dumbarton Oaks Papers* 58 (2004): 197–240; Ball, Philip. *Bright Earth: Art and the Invention of Color*. Chicago, IL: University of Chicago Press, 2001, 199–200.
[6] Allen, Paula Gunn. *The Sacred Hoop: Recovering the Feminine in American Indian Traditions*. Boston, MA: Beacon Press, 1986, 149.

tell these histories if we focus only on individuals, such as the tiger in Figure 1. When we expand our historical investigations to include those animals preserved in fragmented and reconstituted forms, new stories of human-animal relationships arise.

2 Topics and Themes

Noticing Animals in Everyday Objects of the Past

The *longue durée* understanding of human use of animal remains – beyond using animals as a food source – stretches back to toolmaking in the Paleolithic Era; humans have been making use of animal matter and transforming it through craftspersonship longer than we have been domesticating them [→Pre-Domestication: Zooarcheology;→Domestication: Coevolution].[7] Archeologists have found evidence that silk manufacturing could date as far back as 8500 years ago, leather clothing has been preserved for more than 5000 years, and wool fibers have survived for more than 3500 years.[8] These animal remains structured the experience of the ancient and early modern world when they were used, providing shelter, clothing, decoration, weaponry, and tools. Whether hunted or scavenged, humans tracked down other animals dead and alive to extract matter from them long before we began to breed them for such purpose.

Human political, economic, technological, and geographic conditions all shape which species of animals are used. These conditions shape many animals' experiences of births and deaths and the life-course in between, what their parts are made into, and the manufacturing processes used on the materials post-extraction. Human political, economic, technological, and geographic conditions shape humans' perceptions about what we need and thus what we make and how.

---

[7] Tuniz, Claudio and Patrizia Tiberi Vipraio. *Homo sapiens: una biografia non autorizzata*. Rome: Carocci, 2015; Soressi, Marie, Shannon P. McPherron, Michel Lenoir, Tamara Dogandžić, Paul Goldberg, Zenobia Jacobs, Yolaine Maigrot et al. Neanderthals Made the First Specialized Bone Tools in Europe. *PNAS* 110, no. 35 (August 2013): 14186–14190. doi.org/10.1073/pnas.1302730110.

[8] Pinhasi, Ron, Boris Gasparian, Gregory Areshin, Diana Zardaryan, Alexia Smith, Guy Bar-Oz and Thomas Higham. First Direct Evidence of Chalcolithic Footwear from the Near Eastern Highlands. *PLoS ONE* 5, no. 6 (April 2010): e10984. doi.org/10.1371/journal.pone.0010984; Gong, Yuxuan, Li Li, Juzhong Zhang and Hao Yin. Biomolecular Evidence of Silk from 8,500 Years Ago. *PLoS ONE* 11, no. 12 (December 2016): e0168042. doi.org/10.1371/journal.pone.0168042; Strand, Eva Andersson, Karin Margarita Frei, Margarita Gleba, Ulla Mannering, Marie-Louise Nosch and Irene Skals. Old Textiles – New Possibilities. *European Journal of Archaeology* 13, no. 2 (August 2010): 149–173.

Partitioning Animals

Material culture can bring a historian's attention to animal *matter*. Here, I focus on particular processes of bodily fragmentation. The list of animal parts and their by-products that humans have made things out of is extensive, including: antlers, beaks, bladders, blood, bones, claws/nails/talons, eggs, exoskeletons, eyes, fangs/teeth/tusks, feathers, fibers/fur/hair, flesh/meat, gut, honey, hooves, milk, scales, shells, sinews, skins/hides, urine, venoms, wax, and other excrement, excretions, and secretions. Identifying the material nature of animals and connecting animal parts with their sources are the first steps necessary to recognize the diverse roles that animals have played in an anthropocentric history.

Using material culture in the history of animals requires consideration of the language we use to describe animal parts. For example, walruses, elephants, rhinoceroses, narwhals, and warthogs all have "tusks", but tusks themselves are a sub-classification for teeth. Such word choice is about more than specificity. It is political. It might seem syntactically neutral to associate "hair" with humans and "fur" with other mammals, even though fur and hair are materially the same, but these associations have also been used to produce and perpetuate hierarchies in theories of human difference. For instance, it was not uncommon for Enlightenment savants and naturalists to use their word for "fur" in their travelogue descriptions of African persons' hair.

Humans have long made things out of each other as well, but examples of using human remains feel exceptional. Devoted followers made relics out of Saint Peter's bones and Napoleon Bonaparte's hair; early modern pharmacists depended on human urine and mummies for their curatives; the halls of ethnographic museums are replete with human remains. One does not make a relic of just anyone and human bodies displayed in ethnographic museums reflect long and enduring colonial power structures. These examples show that when humans do make things out of each other, the political details of which person, what kind of person, why was the object made, and why was the object made of *this* sit at the forefront of scholarly minds. Because making objects out of other animals' materials is so normal, historians often forget to ask about the political decisions regarding which animals and what animal parts got turned into particular objects. Creating things from animal materials is so normal it easily becomes invisible.

3 Methods and Approaches

What Am I Looking At and What Can I Do With It?

Each of the following objects is a part of the history of humans making things of animals and thus a part of animal history. Some are more surprising than others. They are products of human labor using materials from elephants, lac insects, porcupines, sheep, silkworms, and other unspecified mammals (including, probably, cattle). I have not listed the animal products beside the objects because realizing how difficult it is to see an animal in an everyday thing is an important exercise. Doubtlessly, this task would be easier if one had the object in front of them, but, even then, it could take years of training and expensive equipment to ascertain which species of animals were instrumental to these objects' production.

The particular animals used in the four example objects – which each functions as case study – impact how these objects feel, smell, sound, and reflect light. This is one of the ways that the study of material culture facilitates experiential history of human centered objects as well as the extraction of animal parts. The following cases direct historians toward a path of material culture as a means of writing multispecies histories.

Wool: Finding the Animals Within These Objects Is Not Just Identifying Matter

This section specifically addresses Figure 2's object from the *DDR Museum* in Berlin, Germany. The museum inventory describes it as a small, empty bottle with a cover made of grey knitted wool. At first glance, such inventories tell us very little about the objects' cultural significance, who made them how, and how and by whom they were used. Yet, even minor information can be used to reconstruct histories

Figure 2: Small bottle with cover, 1949–1990, *DDR Museum*, Germany

about these everyday items and the social, political, economic, and environmental conditions that shaped their construction.

After World War II, the Allies divided Germany into four military occupation zones, and the north-eastern portion affiliated with the communist Eastern Bloc was called the German Democratic Republic (colloquially East Germany) from 1949 until reunification in 1990. The GDR prioritized two things that reduced government interest in livestock breeding: a more efficient use of grain for food and a new intensive use of synthetic materials in commodities.[9] Organic materials are more vulnerable to climate, disease, and global markets. The GDR, like many post-war nations, hoped that the development of synthetics would make them more self-sufficient and less vulnerable to the vagaries of organic products – like wool, cotton, and silk – whose origins in fragile organisms mean that the sources sometimes die before their parts can be harvested. Historian Eli Rubin has coined the term "synthetic socialism" to describe the GDR's turn to chemical manufacturing of materials, such as plastic. Wool, like other organism-derived products, became increasingly scarce under these new economic strategies.[10] As the number of sheep in the fields diminished, so too did the wool products in peoples' homes. The bottle and its wool cover are therefore a bit of an anomaly in the GDR's world, a vestige of – at the time – outdated reliance on organic manufacturing.

Because the GDR's industrial policies had moved away from organic materials to synthetic ones, owning everyday objects made from organic materials was considered aberrant and transgressive. It was so controversial that the GDR's secret police, the Stasi, kept records of it when they searched peoples' homes.[11] The soft wool cover on the glass bottle would have been a dangerous object to keep in one's home, so we might wonder why someone kept it at all. The cover is both obviously animal-like and clearly also not a representation of a particular animal. One could hardly imagine that it was a functional bottle, given the leg-and-tail-like placement of the decorative pom-poms, though its lid does open. It seems more the kind of object that would be placed on a shelf rather than used to carry anything.

Some pieces of material culture leave behind detailed stories. Others are like letters found in an archive of private correspondence with no signature, no address, and no date. The material may leave clues about what it meant, but certainty can prove quite difficult. Hedging in descriptions – *could have been, may have,* and *possibly* – are more common in the writing of archeologists than that of historians, but it is imperative to temper claims according to the limits of the available information that our sources provide.

---

9 Freeman, James V. Agricultural Reorganization in the German Democratic Republic: 1965–1980. GeoJournal 7, no. 1 (January 1983): 59–66, 60, 63; Brezinski, Horst. Private Agriculture in the GDR: Limitations of Orthodox Socialist Agricultural Policy. Soviet Studies 42, no. 3 (July 1990): 535–553.
10 Rubin, Eli. *Synthetic Socialism: Plastics and Dictatorship in the German Democratic Republic.* Chapel Hill, NC: University of North Carolina Press, 2008, 155.
11 Rubin, *Synthetic Socialism*, 114, 154, 221.

Rubin used oral histories to document the subversive feelings of former inhabitants of the GDR about keeping organic artifacts from their lives before socialism in their homes up until the fall of the Berlin wall. There are no accession records associated with this bottle to tell us whether or not it fits into that narrative. Perhaps its *wooliness* was wholly independent of the owner's attachment to the object, which may have been more personally related to where it came from. We do not know who owned it or what it meant to them. We can, however, use it to ask questions about what it felt like to live in a world where sheep were increasingly scarce, not because of foot-and-mouth disease, but because the state feared economic dependence on animal bodies. The bottle reflects a moment when new ideas about animal vulnerability and technological solutions changed industrialism and where owning wool became dangerous for humans living under a surveillance state. A historian trying to understand agriculture and the rise of plastic in the GDR would miss something if they structured their research solely around financial papers, government correspondences, and farm plots. Using the wool-wrapped bottle to historicize the relationship between industry and private property helps tie large structural forces to the quotidian, non-essential objects that filled one's home. The bottle can help historians see and feel the high stakes of whimsy.

There is no singular sheep to be recovered in the history of this bottle. That is not how wool gets made, and even the projection of "a sheep" onto this bottle misses some of the gaps between systems of craftspersonship and manufacturing and logics of consumption. However, sheep were a necessary condition of this bottle's production as it was made, the feel of it in one's hand, and its political significance to the changing agricultural landscape of the GDR.

Quills, Wool, and Silk: When and How Animal Products are Made Reveal Multispecies Economic and Political Systems

Figure 3's tobacco pouch comes from Austria's *Weltmuseum,* and, like the previous example, it also contains wool. However, wool means something very different in these two cases. Like before, the museum's inventory records tell very little – that it was acquired from the "H. A. Ward collection" in 1879, is attributed to Dakotan peoples, and was made from wool, glass beads, dyed porcupine quills, leather, sinews, and silk. Though an in-depth explanation of indigenous epistemologies of the material world exceeds this chapter, indigenous scholars note that assumed divisions between the animate and inanimate are but one way of perceiving the world, one that became dominant in the Americas through the violence of empire [→American Studies].[12]

---

12 Todd, Zoe. An Indigenous Feminist's Take on the Ontological Turn: Ontology is Just Another Word for Colonialism. *Journal of Historical Sociology* 29, no. 1 (March 2016): 4–22; TallBear, Kim. Beyond the

Figure 3: Tobacco Pouch, before 1879, *Weltmuseum* Wien, Austria

The pouch's diverse materials stitch together histories of human and animal migration, in life and after death. The *Weltmuseum* acquired the tobacco pouch in 1879 from an American settler and naturalist along with six other items claimed to come from Dakota Territory.[13] The history of the region with which the pouch is associated suggests that it was likely produced by Mdewakanton, Wahpekute, Sisseton, or Wahpeton peoples, all important to the mid-nineteenth-century fur trade.[14] Within the *Weltmuseum,* this pouch and its multispecies transoceanic stories are framed through

---

Life/Not Life Binary: A Feminist-Indigenous Reading of Cryopreservation, Interspecies Thinking, and the New Materialisms. *Cryopolitics: Frozen Life in a Melting World*, Joanna Radin, Emma Kowals (eds.), 179–202. Cambridge, MA: MIT Press, 2017.
13 Kohlstedt, Sally Gregory. Henry A. Ward: The Merchant Naturalist and American Museum Development. *Journal of the Society for the Bibliography of Natural History* 9, no. 4 (April 1980): 647–661.
14 McCrady, David G. *Living with Strangers: Nineteenth-Century Sioux and the Canadian-American Borderlands.* Toronto: University of Toronto Press, 2009, xv.

a European curatorial vision of colonial relations.[15] In trying to get to the sheep, porcupines, silkworms, and other nonhuman animals that made this particular pouch possible we must look at the intra-human economic and political conditions that shaped animal migrations.

Studying what the pouch is made of quickly reveals post-contact multidirectional trade and Sioux integration of foreign materials into their craftspersonship. The pouch combines materials made from animals indigenous to the North American continent, such as the porcupine, as well as introduced ones, such as domesticated sheep. Intercontinental trade of live animals like sheep was complemented by intercontinental trade in animal products, like silk.

The introduction of European livestock into the Americas was a fundamental part of the colonial process, and livestock were part of settler-indigenous economic relations from early sixteenth-century contact.[16] While tobacco pouches long predate colonial contact, *this pouch* could not have been made *as it was made* pre-colonization. Therefore, historians can use *this pouch* to reconstruct circumstances in which Sioux introduced animal materials from local and intercontinental sources into material culture craftspersonship. The inclusion of wool and silk into the longer history of Sioux quillwork can serve as a reminder that adaptations in artistry run parallel to the introduction of new materials.

The wool in the tobacco pouch is a legacy of the early colonization of the continent, when Europeans and their imported animals changed stolen land with new hooves, excrement, and teeth.[17] The wool industry, and sheep with it, pushed westward beyond the East Coast port cities, bringing them closer and closer to Dakotan land.[18] Silk products tell another angle. Unlike the successful importation of livestock, eighteenth-century British attempts to introduce sericulture (silk farming) into colonial America repeatedly failed.[19] There was still no real silk industry to speak of in the nineteenth-century United States or Dakotan Territory, so textile traders seeking to expand international textile trade networks shipped silk over when

---

[15] Lonetree, Amy. *Decolonizing Museums: Representing Native America in National and Tribal Museums*. Chapel Hill, NC: The University of North Carolina Press, 2012.

[16] Francisconi, Michael Joseph. *Kinship, Capitalism, Change: The Informal Economy of the Navajo, 1868–1995*. New York, NY: Garland Publishing, 1998, 36–37.

[17] Crosby, Alfred W. *Ecological Imperialism: The Biological Expansion of Europe, 900–1900*. Cambridge: Cambridge University Press, 1986; Anderson, Virginia DeJohn. *Creatures of Empire: How Domestic Animals Transformed Early America*. Oxford: Oxford University Press, 2006.

[18] Ensminger, M. Eugene. *Sheep and Goat Science*. Danville, IL: Interstate Publishers, 2002; Matz, Brendan. Crossing, Grading and Keeping Pure: Animal Breeding and Exchange Around 1860. *Endeavour* 35, no. 1 (March 2011): 7–15.

[19] Anya Zilberstein. *A Temperate Empire: Making Climate Change in Early America*. Oxford: Oxford University Press, 2016, 114–115.

silkworm acclimation projects failed.[20] The pouch being bound with silk shows how some indigenous actors incorporated the materials of global industries.

We may also understand the pouch as an artifact of the shifting relations between Indigenous peoples in the Americas and European settlers. It is unfortunate that we do not know how H. A. Ward acquired it, as the answer to such a question – whether by purchase, theft, or as a gift – could better illuminate the pouch's political significance under empire.[21] Nevertheless, that the pouch arrived in Vienna's *Weltmuseum* allows us to complicate our thinking about those colonial migrations that brought sheep and silk with settlers to already human inhabited land. The pouch's presence in the *Weltmuseum* shows how Europeans collected such objects to showcase Indigenous-"foreignness" even when the object required European materials to be made as it was. Animals domesticated within Europe become "exotic" specimens in their afterlives when filtered through the imperial gaze upon craftspersonship. The Austrian museum uses this piece to represent Indigenous knowledge and craft of the New World, but from a multispecies perspective, the object mingles transoceanic bodies. In this respect, the pouch not only records a history about Sioux integration of global materials into their own material culture, but also how Europeans repatriated their own materials, like wool, via Indigenous peoples' of the Americas material culture. Scholars of nonhuman animals working in the context of colonialism and empire must acknowledge the enduring politics of the access to these items facilitated by their collection in European museums. In the words of Wampum author Karen Coody Cooper, "some of the losses American Indian communities sustained during years of uncertainty and diminishment of their political power" are the losses of material culture, more easily accessed by researchers with funds that travel than the ancestors of the humans who made these objects.[22]

Ivory: Human Ideas About Animals, Nature, and the Materials we Derive from Them are Historically and Culturally Contingent

The ivory diptych sundial shown in Figure 4 was made around 1730 by the Karner family in Nuremberg, a city in present-day Germany deeply connected to the ivory diptych sundial business since the late sixteenth century.[23] The early modern Africa-Europe ivory trade shaped global economies and local markets as part of the deep history of exploitation and violence where it was sold alongside slaves and

---

20 Field, Jacqueline, Marjorie Senechal and Madelyn Shaw. *American Silk, 1830–1930: Entrepreneurs and Artifacts.* Lubbock, TX: Texas Tech University Press, 2007, xxi.
21 McCrady, *Living with Strangers,* xv.
22 Cooper, Karen Coody. *Spirited Encounters: American Indians Protest Museum Policies and Practices.* Lanham, MD: AltaMira Press, 2008, 84.
23 Lloyd, Steven A. *Ivory Diptych Sundials, 1570–1750.* Cambridge, MA: Harvard University Press, 1992, 35.

gold.²⁴ As the slave trade grew, ivory became more available in Europe. As more African resources became commodified, coastal African and Indian transcontinental trade expanded and so did the exposure to colonial contact [→Global History;→(-Post)Colonial History;→African Studies].²⁵

The makers of the ivory sundial in 1730s Nuremberg were probably unconcerned about the lives of animals and the ivory trade's entanglement with slavery. These artisans doubtlessly cared more about ivory's material properties than where it came from or how it was acquired. The material's physical characteristics were prized for carving fine details, and coming from an imported rarity further raised its social and economic value.²⁶

Today, ivory is associated with the threat of African elephant extinction, but the artisans who produced this sundial and the traders who imported the ivory to Europe would not have connected their work to another species' demise. This object comes

Figure 4: Miniature Diptych Sundial with Case, circa 1730, *Collection of Historical Scientific Instruments*, USA.

---

24 Feinberg, Harvey M. and Marion Johnson. The West African Ivory Trade during the Eighteenth Century: The "… and Ivory" Complex. *The International Journal of African Historical Studies* 15, no. 3 (1982): 435–453.
25 Alpers, Edward A. *Ivory & Slaves: Changing Pattern of International Trade in East Central Africa to the Later Nineteenth Century.* Berkeley, CA: University of California Press, 1975, 63.
26 Gouk, Penelope. Historical Introduction to Nuremberg Diptych Sundial. In *Ivory Diptych Sundials, 1570–1750,* Steven A. Lloyd (ed.), 33–44. Cambridge, MA: Harvard University Press, 1992, 35.

from a time before Europeans worried about extinction. In early modern Europe, the suggestion that the earth could "run out" of a type of creature or plant threatened the Christian idea that God had crafted a perfect world with each creature serving a purpose in the Great Chain of Being.[27] When European Christians stopped finding an animal in a region, it was assumed that more of these creatures could be found somewhere else or that they could reoccur at another time.[28] This belief encouraged unfettered approaches to resource extraction and trade. In turn, preconceived notions about how nature works – whether extinction contradicts a theistic ideal that nature is complete or whether it signals humans' harmful effects on other species – informed how individuals related to everyday objects.

Hunting was also only one of several methods of acquiring elephant ivory, along with tipping (cutting the tips off a live elephant's tusks) and scavenging.[29] This can be difficult to understand in our present moment, but ivory was rarely acquired through mass slaughter in the sixteenth and seventeenth centuries. Where documentation exists for contemporaneous hunts, descriptions of ivory extraction are often accompanied by commentary on local consumption of the elephant meat. Small in scale and practiced locally, those hunts looked nothing like dramatic twentieth century images of herd massacres used to lobby support for a global ban of elephant ivory sales.[30]

The choice of particular animal matter to create objects such as the sundial endangered the lives of other tusked animals around the globe as well. Competing ivory markets abounded in the eighteenth century as commercial viability grew [→Economic History]. In the 1760s, Russian fur traders began harvesting walrus tusks in an attempt to break into the industry.[31] When they discovered one of the largest summer breeding homes for walrus herds in 1786 in the Alaskan Pribilof Islands, "large scale slaughter" ensued.[32]

As with wool, archival conditions rarely exist that would allow a historian to uncover which elephant or even which port the ivory in this sundial came from. Yet the sundial's material and timing can direct us to the historically unique patterns of global commerce, harvesting practices, ideas about nature that conditioned its existence, and its impact on elephants and other tusked species. The 1730s era in which

---

[27] Barrow, Mark. *Nature's Ghosts: Confronting Extinction from the Age of Jefferson to the Age of Ecology*. Chicago, IL: University of Chicago Press, 2009, 23.
[28] Keller, Vera. Nero and the Last Stalk of *Silphion:* Collecting Extinct Nature in Early Modern Europe. *Early Science and Medicine* 19, no. 5 (January 2014): 424–447.
[29] Chaiklin, Martha. Ivory in World History – Early Modern Trade in Context. *History Compass* 8, no. 6 (June 2010): 530–542, 534.
[30] Lindsey Gillson and William Keith Lindsay. Ivory and Ecology—Changing Perspectives on Elephant Management and the International Trade in Ivory. *Environmental Science & Policy* 6, no. 5 (October 2003): 411–419.
[31] Jones, Ryan Tucker. *Empire of Extinction: Russians and the North Pacific's Strange Beasts of the Sea, 1741–1867.* Oxford: Oxford University Press, 2014, 129.
[32] Jones, *Empire of Extinction*, 129.

this particular item was made was marked by a decline in the social status of Nuremberg's ivory craftspersonship.[33] While ivory would retain its rare and luxurious status, the objects that it was made into were slowly being replaced by models for time keeping. The sundial represents the last generation of its kind, centuries before worries about the last remaining elephants would attract international attention.

Shellac: Recognizing Animals in Animal-Derived Products has Shaped Consumption Practices

Of course, for many, animal-derived products are not associated with cruelty or fears about extinction. In some cases, that a material comes from an animal becomes symbolic of virtuous organic and natural consumption. This is increasingly true for shellac in the postwar era, but was not the case in its prewar use.

Shellac is an insect-derived resin most commonly associated with wood finishing. Like ivory, shellac's use dates back thousands of years, but it became a global commodity in the sixteenth century, when India exported the material to China, Japan, and Venice.[34] The resin experienced an economic boom in the late nineteenth century when it began being used globally in the production of 78-rpm records such as seen in Figure 5.

As of 2017, there were 99 identified species of lac insect, which subsist of the sap of more than 400 species of plants.[35] Lac insects feed on trees and then secrete a resinous substance that is collected for manufacture. Only a fraction of female lac insects and their food choices have been commercialized, and India remains one of the principal exporters. Cultivation requires attention to the multispecies conditions of production in relation to the insect's body and its ecosystem. Manufacturers, animal morphologists, and ecologists have sought deeper knowledge of lac insects and their environs, turning knowledge of a multispecies world into human manipulated industry. Whether the knowledge of the participation of nonhuman animals in the production process makes it to consumers depends on historically contingent moral connotations of the artificial-vs-natural distinctions of material culture.

"The immediate predecessor to mouldable plastics," historian Edward Melillo explains, "shellac underwrote a vast array of material culture."[36] Ubiquitously deployed to coat furniture and fruit, demand for shellac expanded after it became the standard for 78-rpm phonogram disc production in 1896.[37] Between 1895 and

---

[33] Gouk, Historical Introduction, 35.
[34] Melillo, Edward D. Global Entomologies: Insects, Empires, and the Synthetic Age in World History. *Past & Present* 223, no. 1 (April 2014): 233–270, 240.
[35] Sharma, Kewal K. Lac Insects and Host Plants. In *Industrial Entomology*, Okmar (ed.), 157–180. Springer: Singapore, 2017.
[36] Melillo, Global Entomologies, 242.
[37] Melillo, Global Entomologies, 242.

Figure 5: 78-rpm record Kid Kord, 1930s, *Museum of Obsolete Media*, USA.

1900, India exported 8000 tons of shellac; after the substance's integration into the music industry, exports rose to 20,000 tons from 1910–1915.[38] In 1920, South East Asia exported more than 23 million USD worth of shellac, more than half of which came from Kolkata; the equivalent of nearly 300 million USD in 2020.[39]

Demand for shellac declined in the post-war period due to the World War II rationing regimes, which had propelled a search for a new material to replace it, and its eventual obsolescence in the music industry.[40] Today, ecologists, chemists, and agriculturalists working for the Indian Institute of Natural Resins and Gums ride on the recent turn against synthetics. They advocate that the lac industry facilitates forest and species protections that are necessary to the preservation of regional biodiversity.[41]

There is no reason to believe that music fans listening to these records in the progressive era associated their distinct crackle with the insects from which it was de-

---

38 Ghosh, Amitabha. Orientalism and Technology: A Case Study of Introduction of Voice-Recording in India. In *Studies in History of Sciences*, Santimay Chatterjee, M. K. Dasgupta, Amitabha Ghosh (eds.), 225–263. Calcutta: Asiatic Society, 1997, 232.
39 Ghosh, Orientalism and Technology, 243; *Pacific Ports Manual, vol. 7.* Los Angeles, CA: Pacific Ports, 1921, 267.
40 Melillo, Global Entomologies, 243.
41 Sharma, Kewal K., Anil K. Jaiswa and K. K. Kumar. Role of Lac Culture in Biodiversity Conservation: Issues at Stake and Conservation Strategy. *Current Science* 91, no. 7 (October 2006): 894–898.

rived, but today, when shellac is used in new products, its insect manufacturers are sometimes found at the heart of marketing it as a natural organic product. Melillo argues that growing concerns about the toxicity of synthetic materials such as plastic and nylon has caused a contemporary resurgence in artisanal and early industrial uses of insect-derived materials such as shellac. Anxieties about the healthfulness and danger of everyday objects shape the consumption and marketing of those objects, including whether the animal from which it comes is obscured or emphasized in a social imaginary.

Twenty-first-century resurgent demand for shellac and lac insect products has provoked a consumer-based advocacy to recognize the product's use of insect secretions as a virtue. In a recently published polemic against plastic, an ecologist described the production of shellac thus:

> Shellac comes from the insect *Kerria lacca* (also known as the lac bug). Specifically, shellac comes from its butt in the form of a secretion, which is also called lac. Lac is used as a coating for candies, like jelly beans, and in cosmetics, like lipsticks. And while eating or applying secretion from the rear end of an insect to your lips might strike you as disgusting, it's benign, and far less disgusting and disturbing than the chemicals modern plastics are capable of leaching into your food and drink.[42]

The author expects readers to project experiences of bodily disgust regarding defecation onto the bodies of insects, despite the fact that the glands from which lac is secreted are not anus-analogous. The reader is instructed to substitute their disgust for *natural* bodily processes with a disgust regarding *synthetic* manufacturing processes. That shellac comes from an animal is meant to make it inherently desirable in this post-synthetic commercial wave. The shellac record and its subsequent commercial demise and transformation into historical curiosity serves as a reminder of how historically-specific such a desire is. The 78-rpm record becomes an archive of lac insects' cultivation and industrial use, and the record's crackle is a reminder of how an animal's body shaped the human material and sensory world.

4 Implication(s) of the Animal Turn

Humans have not just collected animal remains in natural history museums and pet cemeteries. We have collected them in our homes, offices, closets, garages, sheds, refrigerators, and storefronts. Animals make up a large portion of the texture of daily life, today and in the past. The material culture of animals compels our attention because it has shaped the everyday practices and environments that humans have occupied. At first glance, very little is revealed by knowing which kinds of animals were

---

42 SanClements, Michael. *Plastic Purge: How to Use Less Plastic, Eat Better, Keep Toxins Out of Your Body and Help Save the Sea Turtles.* New York, NY: St. Martin's Griffin, 2014, 20.

used in certain objects, but this is a common problem for most new information a historian works with. Attention to material culture is a starting point that leads to its own genre of questions. For historians of animals, it can lead to understanding how humans have lived with, depended upon, and manufactured animal remains. It is a way of seeing animal matter in environments that the creatures never experienced in life and also of considering how conditions of extraction set animal life courses, and in some cases the conditions of their death. Animals shape and are shaped by human social, economic, and political systems. We would do well to take notice.[43]

Analyzing what humans have made out of animals is a study in power relations across species and between people. Technological, agricultural, and geopolitical contexts shape which animals we make objects out of and what we make with them. How humans think about other animals impacts which substances we derive from them, how we do it, and how we market the results. Trade, capitalism, and property shape the production of objects. Humans' thoughts about their particular, individual, regional, and national roles in the world always shapes the manufacturing chain (artisanal and industrial) from live animal to animal product.

Selected Bibliography

Alberti, Samuel J. M. *The Afterlives of Animals*. Charlottesville, VA: University of Virginia Press, 2011.
Appadurai, Arjun. *The Social Life of Things: Commodities in Cultural Perspective*. Cambridge, VA: Cambridge University, 1986.
Daston, Lorraine. *Things that Talk: Object Lessons from Art and Science*. New York, NY: Zone Books, 2004.
Gerritsen, Anne and Giorgio Riello. *Writing Material Culture History*. London: Bloomsbury, 2015.
Gosden, Chris and Chantal Knowles. *Collecting Colonialism: Material Culture and Colonial Change*. Oxford: Berg, 2001.
Hodges, Henry W. M. *Artifacts: An Introduction to Early Materials and Technology*. London: J. Baker, 1964.
Hoskins, Janet. *Biographical Objects: How Things Tell the Stories of People's Lives*. New York, NY: Routledge, 1998.
Pluskowski, Aleksander. *Breaking and Shaping Beastly Bodies: Animals as Material Culture in the Middle Ages*. Oxford: Oxbow, 2007.
Roos, Anna Marie. Object Biographies and Interdisciplinarity. *Notes and Records* 73, no. 3 (September 2019): 279–283.
Woodward, Ian. *Understanding Material Culture*. London: Sage, 2007.

---

43 Tsing, Anna. *The Mushroom at the End of the World: On the Possibility of Life in Capitalist Ruins*. Princeton, NJ: Princeton University Press, 2015.

Silke Förschler
Visual Culture Studies and Art History

1 Introduction and Overview

Historical representations of animals serve as demonstrations of mimetic ability and impartment of visual knowledge. Approaches to animal images in art history investigate their production processes, iconographies, particularities of the medial and artistic realization, and their historical reception. With the help of visual culture studies, the scope can be widened from artistic representations to scientific and popular culture images of animals. Issues of hierarchies and power symmetries in the relationship of animals and humans, as posed by human-animal studies, are complemented by inquiries into the relations between images and power, and into evidence production, by visual culture studies. Animals in images are currently receiving increasing attention in exhibition catalogues and general works. Light is being shed from the perspectives of art history and the history of science on individual researchers and artists who have placed animals at the center of their work and in doing so shaped images of animals.[1] In the following, images of animals from the early modern period will be presented as an example of how animals in images have subsequently been handled for centuries. Around 1700 images of animals mediated both the exotic and the anatomical. The evidence demonstrated in images points towards a hierarchically systematized approach to nature. At the same time, animals in images clarify what and how nature is and was representable in the first place.

Historical animal-human relations can be read in images of animals. Representations of animals and animal surfaces were a constant part of scientific diagrams in the early modern period, and they are found in magnificent illuminations and large-size paintings. During this time, living and dead animals, parts of animals and animal materials became visible in still lifes, allegories and hunting scenes as well as in artistic and nature-historical drawings: they were placed center stage, as mediated knowledge and with a determined subject arrangement. The visualization of animals was therefore both constructive and aestheticizing. Complex thought processes in early modern natural history were translated with the help of animals in pictorial form. At the same time, modes of representation emerged in images which influenced the practices of natural history.

---

[1] Schmidt-Loske, Katharina. *Die Tierwelt der Maria Sibylla Merian: Arten, Beschreibungen und Illustrationen.* Marburg: Basilisken Presse, 2007; Ausstellungskatalog Staatliche Kunsthalle Karlsruhe: *Von Schönheit und Tod. Tierstillleben von der Renaissance bis zur Moderne.* Heidelberg: Kehrer, 2011; Degueurce, Christophe and Delalex, Hélène (eds). *Beautés Intérieures. L'Animal à Corps Ouvert.* Paris: Réunion des musées nationaux – Grand Palais, Les éditions Rmn-Grand Palais, 2012.

Staffan Müller-Wille defines the "classic" period of natural history between two prominent works: Carl Linnaeus' tenth edition of his *Systema naturæ* (Stockholm 1758) and Charles Darwin's *On the Origin of Species* (London 1859).[2] However, the central models and pictorial inventions of a nature-historical animal aesthetic emerged before this period in scientific illustration, painting and prints. Discussions of the representability of living and dead animals as well as animal remnants such as horns, teeth and shells prepared the way for the establishment of the animal in artistic and scientific imagery. Remnants of animals and animal parts offered a large spectrum of possibilities to aestheticize and order nature in different ways [→Material Culture Studies]. Living or dead, as a whole or in parts, in portrait or dissected, moving or conserved on the image bearer, animals reveal approaches to nature as well as relationships between nature and culture. Practices of dissection were demonstrated on and with animals. The use of the animal as a material for artistic work, as for example sheep skin in the manufacture of parchment for book illumination, presupposed its killing.

2 Topics and Themes

Exotic animals were a favorite subject of early modern art and science. They were visually recorded, collected and processed scientifically. At the same time, their mode of representation in images was negotiated. The idea of *ad vivum* is central here, that is rendering the most lifelike depiction possible of exotic animals. An important example is the illuminated manuscript *Mira calligraphiae monumenta*, which came into existence over a period of thirty years, in two phases of production carried out by various hands. Georg Bocskay created the font sample in 1561 and 1562 under the commission of Emperor Ferdinand I of Habsburg, while Joris Hoefnagel illuminated the work roughly thirty years later for Ferdinand's nephew Rudolf II, and it eventually became available in two volumes. The first volume performs calligraphy as an art form. Hoefnagel integrates animals, flowers, plants and fruits in the empty spaces on each page. In the second volume, the letters of the alphabet are presented and described individually. Hoefnagel illuminated the pages at the end of his creative period, in the 1590s.

Pictorial approaches to early modern conceptions of nature become clear through the topology of the animals represented. Two things are central here: on the one hand, the use of a technically skilful style, whose mimetic faculty has a history in Walter Benjamin's sense and therefore must be understood as historically spe-

---

[2] Cf. Müller-Wille, Staffan. Verfahrensweisen der Naturgeschichte nach Linné. In *Akteure, Tiere, Dinge. Verfahrensweisen der Naturgeschichte*, Silke Förschler, Anne Mariss (eds.), 109–124. Cologne: Böhlau, 2017.

cific.³ On the other hand, the gleaming parchment made from sheep skin is essential as an image bearer to the mode of representation.

In contrast to rule-based *florilegia*, which made flowers available for usability and a context of use, the central characteristic of Hoefnagel's illuminations is the immediate living presence of the objects represented. This impression is evoked by the animals such as flies, butterflies, beetles, caterpillars, lizards, frogs, toads, snails, spiders and scorpions, which appear to move on the pages. In this manuscript, Joris Hoefnagel created on parchment with the help of the aesthetic, an internal connection between the animals represented. This is made tangible in an iconographic and medium-specific way.

Based on *trompe l'oeil*, a glance at Hoefnagel's dynamic visualization style makes clear how his studies from nature transcended the 'scientific naturalism'⁴ attested to Hoefnagel by Ernst Kris in 1927. Hoefnagel creates a subtle play between extant lines, *naturalia* and space. He crosses the line drawn by Bocskay, not only threading a campion underneath it, but also turning the parchment into a cuff for the stalk (Fig. 1). In the three-cornered blank space to the left, Hoefnagel places the flowers of the campion directly under the lines as *trompe l'oeil* slits, as well as positioning an open mussel beneath Bocskay's line. As the visual equivalent of the curved end of the picture-crossing line, Hoefnagel paints a yellow spotted ladybird with an imaginary eight instead of the actual six legs. The shadowing indicates an almost hovering relationship to the paper. An individual dimension of experience of nature-historical things enables a further optical illusion. A consequence of part of the stalk being guided through the paper onto the reverse side is that the act of turning the page gains particular importance (Fig. 2). The truth of the objects depicted is not only to be perceived through visual impression, but also through the practice of turning the page. Once more the medium-specific approach to the parchment opens up an individual sense of visualization.

The encounter between the represented and the artist in the creative process is not the only relationship that enables animals in images to be composed as living things. To deepen the quality of the presence, the reader has to be active. The spatial presence of animals and nature-historical objects occurs through their perspective-oriented representation on parchment. To the extent that animals and *naturalia* are connected to the parchment, they are freed from it. As Alfred North Whitehead set out in his lecture on "lifeless nature," published in 1983 in his book *Modes of Thought:* In the sixteenth century, 'common sense' about nature consisted of conceiving it as compiled of permanent things. For the early modern conception of nature, things were understood as individual realities, which demonstrated particular characteristics, such as form, movement and color. However, these individual realities

---

3 Benjamin, Walter. Über das mimetische Vermögen. In *Gesammelte Schriften II.I*, Rolf Tiedemann, Hermann Schweppenhäuser (eds.), 210–213. Frankfurt a.M.: Suhrkamp, 1980.
4 Kris, Ernst. Georg Hoefnagel und der wissenschaftliche Naturalismus. In *Erstarrte Lebendigkeit. Zwei Untersuchungen*, Ernst Kris (ed.), 11–25. Zürich: Diaphanes, 2012.

also have connections to one another through spatial relationships. The movement of animal bodies is for Whitehead a stabilizing connection.[5]

If one applies Whitehead's natural descriptions to Hoefnagel's illuminations, the relatedness of objects to each other constitutes an aesthetic relation. Characteristic of this relation is a dynamic which is able to go beyond nature by means of art. Van Mander describes the relationship between art and nature in the dedication of his *Schilder-Boeck*, published in 1604. Art can surpass the appearances of nature as an organizing principle of creation. Only the former has the capacity to imitate objects of nature and achieve their qualities, at the same time leaving out indecorous, banal or ugly objects. In contrast to nature, art has a choice, while nature only generates.[6] Van Mander's category of *naer het leven* or "from the life" describes a mode of representation which is oriented towards a seen object being as true to nature as possible.[7] On the pages of the *Mira Calligraphiae*, the pictorial formula of *naer het leven*[8] functions through the relationship between artist and object and simultaneously ties the reader to uses and practices. The vitality of the animals on the pages perfects the aesthetic ordering of nature.

Artistic representations of exotic animals were evidenced in their reception with the help of the term "from the life." In visual representations of animals for scientific purposes, however, the representation of exotic animals was hotly debated. This is evidence of the highly charged environment in which animal images were situated. Their representation had to measure up to an artistic mimesis, or a mimesis which was able to convey scientific results.

A question often posed in the early modern period was how knowledge about exotic animals comes into existence and can be dispersed. The agenda of the Royal Society, as its first chronicler Thomas Sprat laid down in 1667, consisted in the discovery of unknown creatures, in order to categorize their position in the chain of being and simultaneously make their knowledge useful to humans.[9] In

---

5 Cf. Whitehead, Alfred North. *Modes of Thought*. Chicago, IL: Touchstone 1938; Whitehead, Alfred North. *Denkweisen*. Translated by Stascha Rohmer. Frankfurt a.M.: Suhrkamp 2001, 161–162.
6 Van Mander, Karel. *Het Schilder-Boeck* [Lives of the Illustrious Netherlandish and German Painters]. Haarlem: Passchier Wesbusch, 1604, n.p.
7 Van Mander, *Het Schilder-Boeck*, fol. 9 r/v., fol. 294 v.
8 On this see, for example, the texts of Swan, Claudia. Ad vivum, naer het leven, from the Life: Defining a Mode of Representation. *Word & Image* 11, no. 4 (1995): 353–372; Bakker, Boudewijn. Au vif – naar 't leven – ad vivum: The Medieval Origin of a Humanist Concept. In *Aemulatio: Imitation, Emulation and Invention in Netherlandish Art from 1500 to 1800, Essays in Honor of Erik Jan Sluijter*, Anton W. A. Boschloo, Jacquelyn N. Coutré, Stephanie S. Dickey, Nicolette C. Sluijter-Seijffert (eds.), 37–52. Zwolle: Waanders, 2011; Felfe, Robert. Naer het leven. Eine sprachliche Formel zwischen bildgenerierenden Übertragungsvorgängen und ästhetischer Vermittlung. In *Ad Fontes! Niederländische Kunst des 17. Jahrhunderts in Quellen*, Claudia Fritzsche, Karin Leonhard, Gregor J. M. Weber (eds.), 165–196. Petersberg: Imhof, 2013.
9 Sprat, Thomas. *The History of the Royal Society of London for the Improving of Natural Knowledge*. London: Martyn 1667, 110.

Figure 1: Georg Bocskay 1561–1562, and Joris Hoefnagel, illumination added 1591–1596. Maltese Cross, Mussel, and Ladybird, Watercolors, gold and silver paint, and ink on parchment, 16,6x12,4 cm, *Mira calligraphiae monumenta*, Ms. 20, fol. 37, The J. Paul Getty Museum, Digital image courtesy of the Getty's Open Content Program.

Figure 2: Joris Hoefnagel. Trompe l'Œil Stem of a Maltese Cross. 1591–1596. Watercolors on parchment, 16,6x12,4 cm, *Mira calligraphiae monumenta*, Ms. 20, fol. 37v, The J. Paul Getty Museum, Digital image courtesy of the Getty's Open Content Program.

the 1688 *Memoirs for a Natural History of Animals*, it was expressly stated that while travelers to distant lands could provide descriptions of animals, there was no way to verify their reports:

> But it may be said that there is not found any certainty in these Histories, nor in these Relations. The Materials, of which the Authors have composed their Works, being for the most part desective and layd on sandy Foundations, it may be truly said that the great Structure which they have afterwards build thereon, with so curious a Symmetry, has no real Solidity.[10]

It likewise becomes clear in the minutes of the Royal Society of Science and the *Académie des sciences* that nature-historical practices for the investigation of animals are related to reflections on internal organization.

The precise illustration of exotic animals was especially challenging according to contemporary opinion. It was believed that the true essence of science was demonstrated by means of the exact representation of animals from distant lands. Just as much care had to be taken in the reproduction of their appearance as in the depiction of anatomical facts. Both were equally unknown and should be accurate in every detail.[11]

The modes of representation which corresponded to the required style of aestheticized science can be set out by means of the page on the chameleon in Claude Perrault's *Description anatomique d'un caméléon*, published in Paris in 1669 and translated into English in 1688 (Fig. 3). In the lower part of the diagram, a chameleon can be seen in profile against a landscape in the background. In the upper part of the page, the tableau is overlaid by a further visual plane. A square piece of paper is fixed under the edge of the picture with two nails. Its lower corners are turned inwards, so that its materiality clearly emerges. The piece of paper bears indexed drawings of the organs and complete skeleton. As in a collection, the various parts are presented against a white background. Signs of the different levels of reality in the image are prerequisite for a successful combination of the various stages. The extant knowledge of the respective animal can be tapped into through this internal opportunity for comparison. This arrangement of the image makes the parameters of early modern knowledge production clear. The condition for the obvious establishment of an animal in relation to aesthetic and presentational modes seems to consist in providing both an exterior and an interior view. With both comes the promise of complete comprehension of the animal.

The endeavor to completely capture an exotic animal visually was also present in the eighteenth century, when the presence of an Indian rhinoceros in Europe re-

---

10 Perrault, Claude. *Memoirs for a Natural History of Animals. Containing the Anatomical Descriptions of Several Creatures, Dissected by the Royal Academy of Science.* London: Printed by Joseph Streater and are to be sold by T. Basset, J. Robinson, B. Aylmer, Joh. Southby, and W. Canning, 1688, n. p., preface.
11 Perrault, *Memoirs for a Natural History of Animals.*

Figure 3: Claude Perrault. Description anatomique d'un caméléon. Edition originale chez Frédéric Léonard. Paris 1669, p. 50, Bibliothèque nationale de France, Paris.

leased a flood of images (Fig. 4). Clara was the second rhinoceros on the European continent after a nameless one in the sixteenth century which also attained great fame. Dürer "counterfeited [*abconterfect*]"[12] or depicted the nameless rhinoceros in a woodcut. In May 1515, this rhinoceros was given as a present from Sultan Muzaffar II., ruler of Cambay, to Alfonso d'Albuquerque, Regent of Portuguese India, for the menagerie of Manuel I., king of Portugal in Lisbon. On its way to Rome, where it was to be given to Pope Leo X. and where it never arrived, the rhinoceros was also to be displayed in Marseille.[13] Its exoticism attracted the attention of very different social groups. Generally, the presence of exotic animals in Europe was a result of colonial and economically motivated networks, which spread across the world from the early modern period.

Between February and April 1749, the rhinoceros Clara could be admired as an exhibition piece in the marketplace of Saint-Germain. Jean-Baptiste Oudry sketched

Figure 4: Jean-Baptiste Oudry. Rhinoceros. 1749, Oil on canvas, 310x456 cm, Staatliches Museum Schwerin. Berswordt-Wallrabe, Kornelia von (ed.), Oudrys gemalte Menagerie. Porträts von exotischen Tieren im Europa des 18. Jahrhunderts (Katalog der Ausstellung Staatliche Museen Schwerin), Munich 2008, 143.

---

12 The term means the same as "after the life", "near haet leaven" or "ad vivum". Cf. Ridley, Glynis. *Clara's Grand Tour: Travels with a Rhinoceros in Eighteenth-Century Europe.* New York, NY: Grove Press, 2005, 18.
13 Cf. Dackerman, Susan. Dürer's Indexical Fantasy: The Rhinoceros and Printmaking. In *Prints and the Pursuit of Knowledge in Early Modern Europe,* Susan Dackerman (ed.), 164–171. Cambridge, MA: Yale University Press, 2011.

the exhibited animal with black crayon on blue paper as preparation for his intended painting.¹⁴ As in the subsequent painting, the rhino is seen in profile, so that its height and breadth are apparent. In 1743, Oudry was engaged as a professor of the Académie Royale de Peinture et de Sculpture in Paris and became famous primarily for his hunting paintings commissioned by Ludwig XV, and his tapestries of the *chasses royales de Louis XV*. Up to this point, animal motifs were known to Oudry chiefly through the courtly context of Versailles. In hunting scenes, he visualized animals simultaneously as companions, particularly dogs and horses, and as hunted prey, particularly hares, birds and wolves. He made hunting still lifes in which he represented the slain animals as though they were still alive after their recent death.

Oudry's theoretical statements on art show that he grappled in particular with the possibilities of specific means of painting in order to represent the fur or the feathers of individual animals in a lifelike manner. He sent Prince Friedrich the manuscript of his lectures on color and painting, which he had delivered on 7 June 1749 and 2 December 1752 at the *Académie royale de peinture et de sculpture* in Paris, complete with a personal dedication and in a magnificently bound edition.¹⁵ In the second volume, he described his practice of "peindre au premier coup," which he had developed for the presentation of very ephemeral surfaces, for the lustre and vividness of fur or the feathers of dead animals, to give the impression of freshness and vitality. To do this, Oudry applied a thin layer of color onto the canvas at high speed, and then subsequently carefully formed the layers of varnish; an illusion of "beau terminé" could thus eventually be achieved.¹⁶

Clara was not the first animal Oudry portrayed. Previously he had memorably transformed Ludwig XV's hunting dog into an individual painting, with the name of the dog inscribed on the lower edge of the painting. What was new was for the artist to draw an exotic animal in captivity and later render it life-sized on canvas. As a portrait, the rhinoceros became part of the large-scale menagerie series which Oudry made at the commission of François Gigot de la Peyronie, a doctor and member of the *Académie royale de chirurgerie*.

Oudry made 45 paintings of living animals from the menagerie. In the collective display of the paintings, a whole menagerie can be imagined as an interior. Hung in an inner room, the large-sized paintings create an illusionary space devoted to animals and nature. At the same time, the paintings invoke the knowledge of exotic animals as status symbols in noble menageries from the sixteenth century. The various animal pens in the menagerie at Versailles which are organized around a central pa-

---

14 The *Studie des holländischen Rhinozerosses* measures 27.6 x 44.4 cm and is now located in the British Museum, Inv. 1918–15.7.
15 The dedication in the first volume reads: "DÉDIÉ A.S.S. LE PRINCE FREDERIC PRINCE héréditaire DE MECKLENBOURG. 1750 Par son très humble et très obeissant serviteur J. Bte Oudry." The two bound manuscripts containing Oudry's handwritten text are now in the University Library in Rostock, Mss.var. 74 and Mss.var 75.
16 Second Oudry manuscript, Mss.var 75, 55, 58–59.

vilion make clear the captivity of the exotic animals. Their exposition in specially designed, radial architectures in the court of Versailles transformed Oudry's series on the canvas; in an altered medium the open-air architecture becomes the interior. In Oudry's menagerie paintings, in contrast, one imagines the individual animals as free. Their bondage is, however, illustrated by the fact that the landscape represented here did not correspond with the habitat of the animals. Formal similarities in the composition of Oudry's pictures are also present in the arrangement of the massive animal body in the foreground, as in a human portrait. The background is barely developed. This approach emphasizes the status of the animals as an individual being and creates a closeness between animal and observer. The landscape relegated to use as background only has a supporting function and serves to accentuate the animal's characteristics and features. In the representation of the animal body and the specific features of the species, the color and surfaces of skin, fur or plumage are clearly emphasized.[17] This *mise en scène* obviously can also be explained with the help of the pictorial ideal of "après nature," also known as "after the life." In addition, Oudry wrote to Duke Christian Ludwig II's chamberlain on 25 March 1750, "Ce sont les principaux animaux de la ménagerie du Roy que j'ai tous peints d'après nature par ordre de Sa Majesté et sous la direction de Mr. De la Pe(y)ronie [...]."[18]

In all these examples, the focus is on how the exotic animal becomes an image, and how it is depicted. Broadly speaking, from the perspectives of art history and visual culture studies, images of animals make statements about the historical circumstances of the production of an image. This enables us to reconstruct the specific situation of the human-animal relation and the historically specific image evident in the representation of animals.

---

[17] Cf. Berswordt-Wallrabe, Kornelia von. Die Immanenz der Farbe und der Bildraum im Werk von J.-B. Oudry. In *Jean-Baptiste Oudry, Jean-Antoine Houdon: Vermächtnis der Aufklärung. Ausst.-Kat. Sammlung Staatliches Museum Schwerin*, Kornelia von Berswordt-Wallrabe (ed.), 25–29. Schwerin: Staatliches Museum, 2000.

[18] "These are the most important animals of the royal menagerie. I painted them all after the life, by order of his Majesty and under survey of Mr. De la Peyronie" [tranlsation mine]. Schwerin, Landeshauptarchiv, Älteres Aktenarchiv, 2.12–1/26, Hofstaatssachen, Kunstsammlungen, Angebote und Erwerbungen, No. 109, fol. 11r–11v, supplement to the letter from Oudry to T.J. Caspar of 25 March 1750, quoted in Frank, Christoph. Künstlerisch-Fürstliche Beziehungen. Neue Erkenntnisse zu Jean-Baptiste Oudry und dem Hof von Mecklenburg-Schwerin. In *Oudrys gemalte Menagerie. Porträts von exotischen Tieren im Europa des 18. Jahrhunderts, Ausst.-Kat. Staatliche Museen Schwerin*, Kornelia von Berswordt-Wallrabe (ed.), 31–55. Munich: Deutscher Kunstverlag, 2008, 53.

3 Methods and Approaches

Animal Material

Art history focusing on animal materials investigates animal specimens, animal *naturalia* in collections, and animal materials used in the production of works of art. In addition, the iconography of painted materials in an image can be interpreted by the meaning of real animal materials, like the horn of the rhino. Material iconography is an established concept in the analysis of works of art.[19] Petra Lange-Berndt in particular has researched the use of taxidermy in art.[20] Giovanni Aloi and Jessica Ullrich have recently highlighted how important it is to consider animal materials as part of an artwork. In his essay, "Animal Studies and Art: Elephants in the Room", Aloi argues that the materiality of an artwork is an important trace for making statements about human-animal relationships. In paints, made from animals for book illumination and oil painting, as well as in brush bristles and parchment, Aloi sees dead animals as materials of work and art in connection with their representation in images as living.[21] Ullrich attributes an actor status to the materiality of animals in the production of the meaning of the work of art.[22] In general, these claims can be classed as part of the search for new concepts as they emerge in the course of the material turn to theorize the connection between material waywardness on the one hand and the entanglement of bodies, natures and meanings on the other.[23] For the parchment that Hoefnagel used, this means considering the specific materiality of the sheep skin as a condition for the lifelike representation of animals [→History of Animal Collections/Animal Taxonomy; →Material Culture Studies].

Carlo Ginzburg's conception of the "evidential paradigm" provides one approach to taking into consideration animal materiality, be it sheep skin as a substrate, the organs of the dissected chameleon or the horn of the rhino.[24] Ginzburg's concern

---

[19] Wagner, Monika. *Das Material der Kunst. Eine andere Geschichte der Moderne.* Munich: C.H. Beck, 2001; Hackenschmidt, Sebastian, Dietma Rübel and Monika Wagner (eds.). *Lexikon des künstlerischen Materials. Werkstoffe der modernen Kunst von Abfall bis Zinn.* Munich: C.H. Beck, 2002; Rübel, Dietmar, Monika Wagner and Vera Wolff. *Materialästhetik. Quellentexte zur Kunst, Design und Architektur.* Berlin: Reimer, 2005.
[20] Lange-Berndt, Petra. *Animal Art. Präparierte Tiere in der Kunst 1850–2000.* Munich: Schreiber, 2009.
[21] Aloi, Giovanni. Animal Studies and Art: Elephants in the Room. *Antennae: The Journal of Nature in Visual Culture* (March 2015): 1–27.
[22] Ullrich, Jessica. Tiere und Bildende Kunst. In *Tiere. Kulturwissenschaftliches Handbuch*, Roland Borgards (ed.), 195–215. Stuttgart: J.B. Metzler, 2016, 210.
[23] Cf. Maran, Timo. Semiotization of Matter. A Hybrid Zone Between Biosemiotics and Material Ecocriticism. In *Material Ecocriticism*, Serenella Iovino, Serpil Oppermann (eds.), 141–154. Bloomington, IN: Indiana Press, 2014.
[24] Ginzburg, Carlo. *Clues, Myths and the Historical Method.* Baltimore, MD: John Hopkins University Press, 1989.

is to develop a method of interpretation which is based on what is termed the incidental. With the help of Morelli and Freud, Ginzburg explains his fundamental interest: "By this method, details usually considered of little importance, even trivial or 'minor,' provided the key for approaching higher aspects of the human spirit."[25] An otherwise unattainable reality can be grasped precisely in the elements which evade control through consciousness, and its "infinitesimal traces."[26] He compares this capacity for knowledge with the capacity of the hunter to reconstruct animal traces in mud, broken twigs, pieces of excrement, tufts of hair, caught feathers and lingering scents. The characteristic of hunting knowledge lies in constructing a coherent sequence of events which are not directly tangible from what seems worthless at first sight. In this way, the imagination and the absence of knowledge are recognized as parts of the evidential paradigm.

Ginzburg here carries out a back-door critique of Western scholarly anthropocentrism. He sees a purification as having been at work in the process of understanding texts from the time of Galileo. This means that a process took place which removed the relevance of implicit aspects which were not considered to belong to the subject-matter. To recognize specific features of pictures (for example dating, the difference between the original and a forgery and the work of an artist's hand), it is necessary to use clues. These are found precisely in the places where the purification (or simplification) of the relationship between subject-matter and image has not yet been carried out. For with the image, it is precisely a case of investigating the individual aspects without producing any strict scientific results in the sense of applying physical and mathematical methods.[27] Painting as a system of culturally contingent signs is perfectly suited to demonstrate how the involuntary can be detected from symptoms. Developed through the practices of the hunt, it is therefore on the one hand a question of being able to read the signs of animals to identify a wider interpretative space. For this, it is important and indeed indispensable to pursue incalculable elements intuitively, and to connect with the knowledge that is rooted in the senses, which, as Ginzburg writes, "binds the human animal closely to all animal species."[28] On the other hand then, part of the evidential paradigm is a furtive nestling up to the animal, and in the best case even a process of "becoming-animal."[29]

---

25 Ginzburg, *Clues*, 101.
26 Ginzburg, *Clues*, 101.
27 Cf. Ginzburg, *Clues*, 112–117.
28 Ginzburg, *Clues*, 125.
29 Deleuze, Gilles and Félix Guattari. *A Thousand Plateaus: Capitalism and Schizophrenia 2*. Translated by Brian Massumi. Minneapolis, MN: University of Minnesota Press, 1987, 39.

Varieties of Mimetic Modes of Representation

How things are presented in an image is important for an art historical analysis. Art history as animal history thus inquires how the animal is represented in the image, and what kind of statement is being made about the relation between humans and animals by the mode of representation. On a formal level, the nature-historical aesthetic of the represented animals exemplifies the ideal of mimesis. This ideal comprises the idea of following *ad vivum, après nature, naer het leven* or *after the life* in the production process of the artwork. Eye-witnessing is also part of the reproduction of a mimetic animal morphology and the representation of animal surfaces. A naturalistic style as a technique of representation can be attributed to natural history in general.[30] According to W.J.T. Mitchell, a striving for mastery over nature comes with the claim of truth to life in painting. This endeavor is comparable to that of the hunter who subordinates the game to his purposes in the hunt. Painting is thought of as so controlling and violent that it domesticates and lovingly cares for the wild.[31] In the attempt at a detailed realization of the seen, a new effort can be recognized to realize the visible animal. In such various contexts of image formation in scientific academies and the workshops which served European courts, pictorial invention toyed with mimetically precise detail. The partially hyper-realistic representations mediated trust both in the seen and also in the importance of visibility for nature history in general. Kusukawa attaches this historically typical understanding of the object and its representation to the concept of the "counterfeit," as it was used, for example, by Albrecht Dürer in connection to his 1515 rhinoceros print. In Kusukawa's opinion, this concept of "counterfeit," "after the life" or *"near het leafen"* describes the ideal of representing an object or an act as immediate and living. In this kind of presentation, the process of looking and observing can be made simultaneously concrete and visible.[32] The creation of a collective understanding of visibility is cast in yet another light by Ludwig Fleck. "We look with our own eyes, we see

---

[30] Summers, David. *The Judgment of Sense: Renaissance Naturalism and the Rise of Aesthetic.* Cambridge: Cambridge University Press, 1987, 3; Snyder, Joel. Picturing Vision. *Critical Inquiry* 6, no. 3 (Spring 1980): 499–526; for a careful approach to "naturalism", cf. Kemp, Martin. Taking it on Trust: Form and Meaning in Naturalistic Representation. *Archives of Natural History* 17, no. 2 (June 1990): 127–188; Ackerman, James. Early Renaissance Naturalism and Scientific Illustration. In *The Natural Sciences and the Arts, Aspects of Interaction from the Renaissance to the 20th Century,* Allan Ellenius (ed.), 1–17. Uppsala: Almquist & Wiksell International, 1985.
[31] Mitchell, William John Thomas. *Picture Theory. Essays on Verbal and Visual Representation.* Chicago, IL: University of Chicago Press, 1994, 333.
[32] Kusukawa, Sachiko. *Picturing the Book of Nature. Image, Text and Argument in Sixteenth-Century Human Anatomy and Medical Botany.* Chicago, IL: University of Chicago Press, 2012, 8–25, especially 20.

with the eyes of a collective body."[33] Understood in this way, 'visual styles' of the animal are always part of a historical dispositive.

The Scaling of Nature as Image/Space

Through the way in which an animal is placed both within an image and in an exhibition space, it is given importance, or denied it. Art history as animal history examines the relationships between animals and observers that is established by practices of placement. In cabinets of curiosities and natural history collections, as well as display collections and menageries, the relations between objects and viewers were rehearsed, positions were given to animals, *naturalia* and observers. The embedding of dead and living animals occurred in spatial constellations as much as in pictorial practices. Pursuing early modern spatial conceptions of nature is of particular importance in the light of the egregious temporalization of nature in the modern period.[34] In descriptions of the transformation of natural history into the modern life sciences, emerging attempts at historicization are commonly understood as a "break in the chain of being," since the traditional techniques of arranging objects of knowledge in space were no longer possible for reasons of capacity.[35] It follows that the relation between animals and the space surrounding them, be it in still lifes, cabinets of curiosity or allegories, has greater importance than we would imagine from our "present bias [*Gegenwartsbefangenheit*]."[36] Spatial relations should not be interpreted as monocausal or unidirectional. Instead, we can establish how particular techniques of presentation have led to particular modes of representation, or vice versa – that there are modes of representation which have influenced the techniques of animal presentation. The process of interlacing different spaces is essential for the conception of an order of natural history. European expansions and travel in all directions around 1700 were fundamental to this process, as were the collections of art and natural products that were enlarged by these expansions.[37] In exhibition spaces as well as within images, the animal is thus being turned into an object in a collection. The placement in space leads to a trained, distanced attitude of

---

[33] Fleck, Ludwik. To Look, to See, to Know. In *Cognition and Fact: Materials on Ludwik Fleck*, Robert S. Cohen, Thomas Schnelle (eds.), 129–151. Dordrecht: D. Reidel Publishing Company, 1986, 134.
[34] Foucault, Michel. Questions on Geography. In *Power/Knowledge: Selected Interviews and Other Writings, 1972–1977*, Colin Gordon (ed.), 63–77. New York, NY: Pantheon, 1980, 70.
[35] Lepenies, Wolf. *Das Ende der Naturgeschichte*. Frankfurt a.M.: Suhrkamp, 1986, 16–17.
[36] Lübbe, Hermann. Begriffsgeschichte und Begriffsnormierung. In *Die Interdisziplinarität der Begriffsgeschichte*, Gunter Scholtz (ed.), 31–41. Hamburg: Meiner, 2000, 41.
[37] For example Hans Sloane, see Chakrabarti, Pratik. Sloane's Travels. A Colonial History of Gentlemanly Science. In *From Books to Bezoars. Sir Hans Sloane and his Collections*, Alison Walker, Arthur Macgregor, Michael Hunter (eds.), 71–79. London: The British Library Publishing Division, 2012.

observation by the spectator, and the animal becomes a building block in the history of nature.

4 Implication(s) of the Animal Turn

The *animal turn* has led to a broadening of perspective, which now considers the animal in art and in visual testimonies as an autonomous image object. Before that, animals were merely symbols in a Christian or mythological context, or accessories to depictions by humans. The animal was analyzed in order to make statements about the humans who were represented. Animal-human relations accompanying the representation of an animal are also now of interest. If we apply two classical questions of art history regarding the production of an image, as well as the modes of depiction, to the animal, we can analyze historical concepts of treatment, nature-historical knowledge and the value of the animal. The *animal turn* also regards the animal in its role as supplier of materials for paints, brushes and parchment, and therefore for the production of art. This raises awareness of the role of the animal as material for human art production. Art history after the *animal turn* shows that depictions of animals always serve to also impart knowledge about animals. Depictions of animals in the early modern period can be seen as forerunners of representations in the modern era, both in artistic images and in images in the service of science.

In the early modern period, the ideal of mimetic representation dominated the emergence of a visual language which allowed the animal to appear alive in paintings, prints and drawings as well as in nature-historical illustrations. This ideal developed its greatest evidential value in bringing materials and animals equally to life in images. These forms of reflection for an animal aesthetic, according to the ideals of *naer het leaven, after the life, aprés nature* or *abcontertfect* had their starting point in both dead and living animals. On the pictorial surfaces of paintings and drawings, the means of convincing lay in a spectrum of mimetic references. This spectrum showed which modes of representation were of a nature-historical interest on the one hand, and could produce a nature-historical aesthetic on the other.[38] This interplay between the generation of knowledge and the fixation with results depended on the selection of animal species; their aesthetics determined pictorial arrangements.

In the arrangement of motifs on the parchment, the canvas and the printing plate, there is always a guide to dealing with different animals. Animals and parts of animals are placed on the surface of the image just like in a guidebook: criteria for their arrangement are provided through the reading of images. In this way, instructions for the arrangement of collections and also of nature-historical evidence

---

38 Kusukawa, *Picturing the Book of Nature*.

emerge.[39] In the early modern period, images are part of the dispositive of natural history. Images clarify arrangements of *naturalia* and parts of dead animals in collections as well as living animals and their observation in menageries. The nature-historical aesthetic is equally entwined with the rules of artistic representation and with the science of natural history. Images provide their own "aesthetic meaning"[40] of the animal. The better form and content are related to one another, the more concentrated the substance of the artwork.[41] To constitute this relatedness, a special meaning is attached to "comparative vision." The practice of comparative vision is not only significant for the interpretation of images in the "cultural studies laboratories of pictorial history,"[42] but also for the classificatory approach to animals in natural history. Investigating images of animals does not just enable the reading of animal-human relations. For the animal is also a *tertium comparationes* ['the third of the comparison'] that enables images of art and of science, image content and its bearers, the represented and its nature, to be analyzed together.

Selected Bibliography

Aloi, Giovanni. *Speculative Taxidermy. Natural History, Animal Surfaces and Art in the Anthropocene*. New York, NY: Columbia University Press, 2018.

Grén, Roni. *The Concept of the Animal and Modern Theories of Art*. New York, NY and London: Routledge, 2018.

Kusukawa, Sachiko. Conrad Gessner on an Ad Vivum Image. In *Ways of Making and Knowing: The Material Culture of Empirical Knowledge*, Pamela H. Smith, Herold J. Cook, Amy R. W. Meyers (eds.), 330–356. Ann Arbor, MI: The University of Michigan Press, 2014.

Leonhard, Karin. *Bildfelder. Stillleben und Naturstücke des 17. Jahrhunderts*. Berlin: Akademie Verlag, 2013.

Neri, Janice. *The Insect and the Image. Visualizing Nature in Early Modern Europe, 1500–1700*. Minnesota, MN: University of Minnesota Press, 2011.

Pollock, Mary S. and Catherine Rainwater (eds.). *Figuring Animals. Essays on Animals Images in Art, Literature, Philosophy and Popular Culture*. New York, NY: Palgrave Macmillan, 2005.

Quinsey, Katherine M. (ed.). *Animals and Humans: Sensibility and Representation, 1650–1820, Oxford University Studies in the Enlightenment*. Oxford: Voltaire Foundation, 2017.

Schleif, Corine. Who Are the Animals in the Geese Book? In *Animals and Early Modern Identity*, Pia F. Cuneo (ed.), 209–242. Farnham and Burlington: Routledge, 2014.

---

39 Kusukawa, *Picturing the Book of Nature*, 118.

40 Panofsky, Erwin. Kunstgeschichte als geisteswissenschaftliche Disziplin. Translated by Wilhelm Höck. In *Sinn und Deutung in der bildenden Kunst*, Erwin Panofsky, 7–36. Cologne: DuMont, 2002, 16 [first published in 1957].

41 Panofsky, Erwin. Zum Problem der Beschreibung und Inhaltsdeutung von Werken der bildenden Kunst (1931). In *Aufsätze zu Grundfragen der Kunstwissenschaft*, Hariolf Oberer, Egon Verheyen (eds.), 85–87. Berlin: Wissenschaftsverlag Spiess 1992 [first published in 1964].

42 Cf. Bredekamp, Horst and Franzsika Brons. Fotografie als Medium der Wissenschaft. Kunstgeschichte, Biologie und das Elend der Illustration. In *Iconic Turn. Die neue Macht der Bilder*, Hubert Burda, Christa Maar (eds.), 365–381. Cologne: DuMont, 2005, 366 [translation mine].

Spickernagel, Ellen. *Der Fortgang der Tiere. Darstellungen in Menagerien und in der Kunst des 17.–19. Jahrhunderts.* Cologne: Böhlau, 2010.
Stiftsbibliothek St. Gallen (eds.). *Schafe für die Ewigkeit. Handschriften und ihre Herstellung. Katalog zur Jahresausstellung in der Stiftsbibliothek St. Gallen.* St. Gallen: Verlag am Klosterhof, 2013.
Swan, Claudia. From Blowfish to Flower Still Life Paintings. In: *Merchants & Marvels. Commerce, Science, and Art in Early Modern Europe.* Pamela H. Smith, Paula Findlen (eds.), 109–136. New York, NY and London: Routledge 2002.

Laura McLauchlan
Multispecies Ethnography

1 Introduction and Overview

From reflections on interspecies love and distance, to considerations of the sociality of compost and artistic collaborations with more-than-human animals, it is attention to lived experiences of and with other species that coheres the wide range of research that comes under the rubric of multispecies ethnography. In approaching more-than-human[1] animals as subjects, multispecies ethnography (MSE) brings the *animal turn* into the practice of ethnographic fieldwork. The multiple species upon whom such ethnography focuses, however, include more than just animals. Such ethnographies also attend to plants, bacteria, larger ecosystems and the interplay of cultural and economic forces with the living world.[2] In seeing the interconnection and co-shaping of life, much of MSE is marked by an attention to the fundamental relationality of existence. Increasingly – and particularly through the influence of indigenous scholars – MSE is also attending to questions of the borders of life/non-life and to the more-than-secular.[3]

Both the methodologies and themes of MSE may be of use to scholars in historical animal studies. As Erica Fudge has noted, "[t]he limits of understanding animals of the past lie in both the fragmented nature of historical documentation and the reality that such records are human-created."[4] While rigorous historiographical analysis thus requires careful attention to the limitations of representation, engagement with methodologies and content of MSE can potentially supplement such work. In particular, MSE offers novel methodologies for attending to the lives of more-than-human animals, expanding modes of possible interspecies connection well beyond classical ethology. At the same time, through emphasizing the biological and sociocultural positionalities of researchers, MSE also tends to pay careful attention to the necessary limitations of such knowledge. Finally, through its relational focus, MSE prompts questioning of just what an animal (whether human or more-

---

[1] Abram, David. *The Spell of the Sensuous: Perception and Language in a More-Than-Human World.* New York, NY: Vintage Books, 1996.
[2] Kirksey, Eben and Stefan Helmreich. The Emergence of Multispecies Ethnography. *Cultural Anthropology* 25, no. 4 (November 2010): 545–576, 545.
[3] TallBear, Kim. Beyond the Life/Not Life Binary: A Feminist-Indigenous Reading of Cryopreservation, Interspecies Thinking, and the New Materialisms. In *Cryopolitics: Frozen Life in a Melting World*, Emma Kowal, Joanna Radin (eds.), 179–202. Cambridge, MA: MIT Press, 2017.
[4] Fudge, Erica. What Was it Like to be a Cow? History and Animal Studies. In *The Oxford Handbook of Animal Studies*, Linda Kalof (ed.), 258–282. Oxford: Oxford University Press, 2017.

than-human) *is*, encouraging focus on the interplay of forces and connections of which we are all comprised.

The term multispecies ethnography (MSE) was coined in 2010 by Eben Kirksey and Stefan Helmreich to refer to an emerging collection of ethnographic practices attending to other-than-human lives.[5] While many scholars have since adopted this term for their work, a range of scholars are, and have been, practicing ethnographic study with more-than-human beings under labels other than MSE. Such rubrics include the 'anthropology of life',[6] 'ecologies of selves',[7] 'anthropology beyond the human',[8] 'zooethnography',[9] 'etho-ethnology and ethno-ethology'.[10] Furthermore, outside of anthropology, 'more-than-human geography' often shares both methods and key theoretical concerns with MSE [→Historical Geography].[11] For the purposes of this chapter, ethnographic works attending to a 'more-than-human'[12] subject will be considered as part of the category of MSE. Although such ethnographies potentially attend to a vast array of non-animal species and forces, this chapter will focus on ethnographies attending to animal others. However, as will be outlined, due to the tendency of MSE to focus on the importance of relationality, even those ethnographies ostensibly focusing on animal species typically also take into account a vast array of influences and interactions which make and shape animal lives.

2 Topics and Themes

More-Than-Human Representation

To understand the themes and theoretical interests of MSE, it is important to be aware both of its roots in humanistic anthropology as well as of the influence of the 'animal' and 'posthuman' turns as they have played out in anthropology and cog-

---

[5] Kirksey and Helmreich, Multispecies Ethnography.
[6] Kohn, Eduardo. How Dogs Dream: Amazonian Natures and the Politics of Transspecies Engagement. *American Ethnologist* 34, no. 1 (February 2007): 3–24.
[7] Kohn, Eduardo. *How Forests Think: Toward an Anthropology Beyond the Human*. Berkeley, CA: University of California Press, 2013.
[8] Ingold, Tim. Anthropology Beyond Humanity. *Suomen Anthropologi* 38, no. 3 (2013): 5–23; Kohn, *How Forests Think*.
[9] Pedersen, Helena. Follow the Judas Sheep: Materializing Post-Qualitative Methodology in Zooethnographic Space. *International Journal of Qualitative Studies in Education* 26, no. 6 (June 2013): 717–731.
[10] Lestel, Dominique, Florence Brunois and Florence Gaunet. Etho-Ethnology and Ethno-Ethology. *Social Science Information* 45, no. 2 (June 2006): 155–177.
[11] Panelli, Ruth. More-Than-Human Social Geographies: Posthuman and Other Possibilities. *Progress in Human Geography* 34, no. 1 (June 2010): 79–87.
[12] Abram, *Spell of the Sensuous*.

nate disciplines. With some exceptions,[13] earlier anthropological work that attended to more-than-human lives typically did not approach them as subjects to be focused on in their own right.[14] Instead, more-than-human lives were typically studied for the ways in which they served humans. In particular, this involved ways in which they were materially important for human subjects – such as through the provision of protein or clothing – or to the extent to which animals were, as Lévi-Strauss famously said, 'good to think with' – holding symbolic importance for a group of people.[15]

In line with the *animal turn*, MSE can be seen to both decenter the human as the obvious subject of ethnographic focus as well as to challenge the sorts of human exceptionalism which has enabled animals to be considered in terms of their service to humans.[16] MSE, however, is also influenced by posthumanism, a mode of thinking in which not only is anthropocentrism challenged but so, too, are humanist ontologies resting on dualisms such as those separating (and hierarchizing) nature/culture, human/nonhuman and subject/object.[17] MSE generally takes the stance that, regardless of imagined divisions between humans and animals – particularly in Euro-American imaginaries – that, in reality, humans have 'shared lives' with other animals.[18] Approaching life as something 'in common' challenges the sorts of human/animal oppositions common in the West since antiquity.[19] As Piers Locke and Ursula Münster have argued, MSE is "part of a larger quest in the social sciences and humanities to replace dualist ontologies by relational perspectives."[20] This emphasis on relationality has implications for the subject matter of MSE: within such ethnographies, biologies are considered in relation to – and part of – broader ecologies, political economic concerns, cultures and meanings. As a way of creating meaning, the production of MSE is itself a mode of participation in an ever-emergent world.

---

13 For example, Ingold, Tim. On Reindeer and Men. *Man* 9, no. 4 (1974): 523–538; Morgan, Lewis Henry. *The American Beaver and His Works*. Philadelphia, PA: J. B. Lippincott, 1868.
14 Mullin, Molly. Animals in Anthropology. *Society & Animals* 10, no. 4 (2002): 378–393.
15 Tsing, Anna. More-Than-Human Sociality: A Call for Critical Description. In *Anthropology and Nature*, Kirsten Hastrup (ed.), 27–43. London: Routledge, 2013; Lévi-Strauss, Claude. *The Savage Mind*. Chicago, IL: University of Chicago Press, 1970, 204–208.
16 See especially Feinberg, Rebecca, Patrick Nason and Hamsini Sridharan. Introduction: Human–Animal Relations. *Environment and Society: Advances in Research* 4, no. 1 (September 2013): 1–4.
17 Locke, Piers and Ursula Münster. Multispecies Ethnography. In *Oxford Bibliographies*. Oxford: Oxford University Press, 2015; Cary Wolfe, in particular, has argued that "at full force", animal studies are not just those studies that have "converged on an object of study called 'the animal'" but, rather, is a mode of study that questions humanism, fundamentally unsettling and reconfiguring "the question of the knowing subject and the disciplinary paradigms and procedures that take for granted its form and reproduce it". Wolfe, Cary. *What Is Posthumanism*. Minneapolis, MN: University of Minnesota Press, 2009, xxix.
18 Lestel, Brunois and Gaunet, Etho-Ethnology and Ethno-Ethology, 156.
19 Lestel, Dominique and Hollis Taylor. Shared Life: An Introduction. *Social Science Information* 52, no. 2 (2013): 183–186.
20 Locke and Münster, Multispecies Ethnography.

As part of both offering appreciation for more-than-human lives and challenging human exceptionalism, multispecies ethnographies typically include recognition that the fullness of more-than-human species' lives and experiences cannot be fully comprehended by humans. Vinciane Despret – an ethnographer of ethologists – has been particularly influential in helping multispecies ethnographers to both take seriously the ways in which we are able to know one another while avoiding the assuming that we might be able to directly experience the world of another. In this regard, Despret writes of partial affinities, in which particular stories, biologies, abilities and disabilities and practices of attending to others all play roles in allowing for vital and never complete experiences of being-with other species.[21] The question of what it is to represent the other is thus a key aspect of many multispecies ethnographies. Multispecies studies thus shares with traditionally human-focused disciplines such as anthropology, literary studies, sociology and history, an interest in questions of representation and of claims to speak for others.[22] While matters of translation, representation, and avoiding ventriloquism are vital questions when working with human subjects, there are additional challenges when one's subject is of a different species.[23]

Many ethnographic projects have experimented with novel and artistic modes of representation in order to attend to more-than-human lifeways without assuming to fully know such beings.[24] Examples include the work of Karin Bolender, making soap with small quantities of the milk of the American Spotted Asses she lives with as a way to give space to the donkeys' ways of knowing and processing environments without claiming to fully 'know' their ways.[25] Zachary Caple has approached the relational worlds of limpkins in ethnographic poetry[26] and attending to the latrines of

---

[21] Despret, Vinciane. Responding Bodies and Partial Affinities in Human–Animal Worlds. *Theory, Culture & Society* 30, no. 7–8 (August 2013): 51–76.
[22] Roscher, Mieke. New Political History and the Writing of Animal Lives. In *The Routledge Companion to Animal–Human History*, Hilda Kean, Philip Howell (eds.), 53–75. London and New York, NY: Routledge, 2018; Fudge, Erica. A Left-Handed Blow: Writing the History of Animals. In *Representing Animals*, Nigel Rothfels (ed.), 3–18. Bloomington, IN: Indiana University Press, 2002; Swart, Sandra. "But Where's the Bloody Horse?": Textuality and Corporeality in the "Animal Turn." *Journal of Literary Studies* 23, no. 3 (September 2007): 271–292.
[23] Appadurai, Arjun. Introduction: Place and Voice in Anthropological Theory. *Cultural Anthropology* 3, no. 1 (February 1988): 16–20, 20.
[24] Kirksey, Eben. *The Multispecies Salon*. Durham, NC: Duke University Press, 2014.
[25] Bolender, Karin. R.A.W. Assmilk Soap. In *The Multispecies Salon*, Eben Kirksey (ed.), 64–86. Durham, NC: Duke University Press, 2014.
[26] Caple, Zachary. The Limpkin: A Poem and Short Essay. *Engagement: A blog published by the Anthropology and Environment Society*, October 13, 2015. URL: aesengagement.wordpress.com/2015/10/13/the-limpkin-a-poem-and-short-essay/ (August 18, 2020).

water voles has been used as a way to give presence to traces of these critters without reductive interpretation of their lifeways.[27]

In careful tension with not assuming full comprehension of the experiences of members of other species, many multispecies ethnographies also seek to consider – at least speculatively – what might be "meaningful to the animals themselves."[28] Eduardo Kohn, for example, has attended to questions of meaning-making from a more-than-human perspective, considering what it is that dogs dream[29] and arguing more generally for the 'provincializing' of language, taking seriously the many symbolic systems that matter for other species.[30] Thom van Dooren and Deborah Bird Rose look at the attachment of little penguins and flying foxes to their homes in an urban Australian context. Rather than assuming it is only humans who find meaning in particular places, the authors look at the "non-human storying of places" evidenced by the site fidelity of these animals, despite the obstructions and harm introduced by some of the area's human residents.[31] In such considerations, animals shift from being objects of study to subjects and actors in their own meaningful (if never fully comprehensible) worlds.

Challenging Categories

Despite the term 'species' being in its title, many multispecies ethnographies challenge the idea of species as a 'natural' category.[32] As Tim Ingold (working under the rubric of 'anthropology of life') has argued: "[o]nly in the purview of a universal humanity – that is, from the perspective of species-being – does the world of living things appear as a catalogue of biodiversity, as a plurality of species".[33] While MSE often does use commonly-accepted species categorizations, so that one might refer to an ethnography on meerkats or house cats,[34] other multispecies ethnographies also attend to the ways in which species concepts are *made,* following such processes in

---

27 Hinchliffe, Steve, Matthew B. Kearnes, Monica Degen and Sarah Whatmore. Urban Wild Things: A Cosmopolitical Experiment. *Environment and Planning D: Society and Space* 23, no. 5 (October 2005): 643–658.
28 Buchanan, Brett. *Onto-Ethologies: The Animal Environments of Uexküll, Heidegger, Merleau-Ponty, and Deleuze.* Albany, NY: State University of New York Press, 2008, 2.
29 Kohn, How Dogs Dream.
30 Kohn, *How Forests Think.*
31 Van Dooren, Thom and Deborah Bird Rose. Storied-Places in a Multispecies City. *Humanimalia* 3, no. 2 (Spring 2012): 1–27, 1.
32 Kirksey and Helmreich, Multispecies Ethnography, 563.
33 Ingold, Anthropology Beyond Humanity, 19.
34 Candea, Matei. I Fell in Love with Carlos the Meerkat: Engagement and Detachment in Human-Animal Relations. *American Ethnologist* 37, no. 2 (May 2010): 241–258; Bussolini, Jeffrey. Toward Cat Phenomenology: A Search for Animal Being. *Found Object* 8 (2000): 155–185.

labs and broader social settings.³⁵ In accord with philosophers of biology such as John Dupré, many multispecies ethnographies acknowledge that while 'species' may be a useful tool for biological categorization, assuming a species to be, as per Ernst Mayr's definition, an "interbreeding population", does not reflect the reproductive diversity of a great many organisms.³⁶ Species are also frequently recognized as a problematic mode of categorization because of the extent of genetic material shared between apparently different species. When it comes to the human animal, for example, many multispecies ethnographies are greatly influenced by genomic analysis revealing that only around 10% of genetic material comprising human cells is of human origin.³⁷

The work of feminist science studies scholar, Donna Haraway, has also been deeply influential on the development of multispecies ethnography, particularly in terms of exploding mainstream species notions. In thinking with the figure of the cyborg, Haraway argues that, rather than pure biologies ever being possible, that the "[t]he machinic and the textual" are internal to such apparently 'biological' kinds, and vice-versa. In later considerations, Haraway came to include cyborg relations within a broader consideration of co-shaping in which she considers the ways in which a multitude of 'companion species' come together to shape our world. Rather than being 'companion animals' or 'pets', these companion 'species' include the vast array of "technologies, commerce, organisms, landscapes, peoples, practices" of which life is comprised. Here, 'species' as a concept breaks down, not only in that such types are in no way recognizable as biological 'species', but also in that they are also ultimately not separable. Such realities trouble both organism/environment boundaries and human/animal divisions or, as Haraway writes, they "make a mess out of categories in the making of kin and kind."³⁸

Alongside challenging species categorizations, multispecies ethnographies also tend to participate in questioning what an animal *is*, particularly questioning the extent to which any animal is ever a singular, coherent entity. A theorist who has greatly impacted MSE is anthropologist Gregory Bateson. Emphasizing the fundamental relationality of being, Bateson argued that the organism, rather than just being the unit of survival, is instead the "organism-in-its-environment", with environment and organism shaping and co-creating each other.³⁹ As Anna Tsing notes, "no organ-

---

35 Kirksey, Eben. Species: A Praxiographic Study. *Journal of the Royal Anthropological Institute* 21, no. 4 (December 2015): 758–780.
36 Dupré, John. In Defence of Classification. *Studies in History and Philosophy of Biological and Biomedical Sciences* 32, no. 2 (June 2001): 203–219.
37 Haraway, Donna. *When Species Meet*. Minneapolis, MN: University of Minnesota Press, 2008, 3–4.
38 Haraway, *When Species Meet*, 19.
39 Bateson, Gregory. *Steps to an Ecology of Mind: Collected Essays in Anthropology, Psychiatry, Evolution, and Epistemology*. San Francisco, CA: Chandler, 1972, 456.

ism can become itself without the assistance of other species".[40] In this light, attending to any particular animal in isolation is to overlook the relationships of which it is comprised.

Natural-Cultural Ontologies

While challenging the "constructed discontinuity between humans and non-humans",[41] MSE also looks at the very real effects of such notions of discontinuity. Ideas about 'nature' have immense impacts on what Joanna Latimer and Mara Miele refer to as the "choreographies of everyday life".[42] Such ideas become part of the broader, entangled, landscape of materials, meanings and forces that MSE attends to. This also includes attending to the ways in which human categorizations affect more-than-human lives.[43] One particular way in which the interrelation of nature and culture has been approached in MSE is through an emphasis on naturecultures. The term refers to the ultimate inseparability of the elements of life which are, nonetheless, imagined to be separable within Western thought as evidenced in the categories of 'culture' and 'nature'.[44] Much MSE attempts to avoid re-inscription of such dichotomies by using concepts such as naturecultures or similar dualism-challenging frameworks, including Karen Barad's notions of "intra-actions" and the "material-discursive" or Haraway's attention to "material-semiotic" actors.[45]

While such thinking may be a relatively recent shift within Western scholarship, much indigenous scholarship has never relied on dualistic framings.[46] Scholars such as Juanita Sundberg and Zoe Todd have pointed out that Western post-humanist arguments often fail to acknowledge that the division of nature and culture is not a

---

[40] Tsing, Anna. Catachresis for the Anthropocene: Three Papers on Productive Misplacements. *AURA's Openings. AURA's Working Papers* 1 (2013): 1–10.
[41] Latimer, Joanna and Mara Miele. Naturecultures? Science, Affect and the Non-Human. *Theory, Culture & Society* 30, no. 7 (October 2013): 5–31.
[42] Latimer and Miele, Naturecultures, 5.
[43] Lowe, Celia. *Wild Profusion: Biodiversity Conservation in an Indonesian Archipelago*. Princeton, NJ: Princeton University Press, 2006.
[44] Strathern, Marilyn. No Nature, No Culture: the Hagen Case. In *Nature, Culture and Gender*, Carol MacCormack, Marilyn Strathern (eds.), 195–203. Cambridge: Cambridge University Press, 1980; Haraway, Donna. *The Companion Species Manifesto: Dogs, People, and Significant Otherness*. Chicago, IL: Prickly Paradigm, 2003; Latour, Bruno. *We Have Never Been Modern*. Cambridge, MA: Harvard University Press, 1993.
[45] Barad, Karen. Meeting the Universe Halfway: Realism and Social Constructivism without Contradiction. In *Feminism, Science, and the Philosophy of Science*, Lynn Hankinson Nelson, Jack Nelson (eds.), 161–194. Dordrecht: Kluwer Academic Publishers, 1996; Haraway, Donna. Otherwordly Conversations, Terran Topics, Local Terms. In *Material Feminisms*, Stacy Alaimo, Susan Hekman (eds.), 157–187. Bloomington, IN and Indianapolis, IN: Indiana University Press, 2008.
[46] Sundberg, Juanita. Decolonizing Posthumanist Geographies. *Cultural Geographies* 21, no. 1 (January 2014): 33–47, 35.

universal.[47] Kim TallBear has noted that non-indigenous scholarship tends to reinforce life/non-life dualisms, restricting consideration of 'life' to organisms.[48] In contrast, indigenous metaphysics understands animacy in non-hierarchical ways, recognizing the liveliness of rocks, landforms and the cosmos and actively attending to the "co-constitutive entanglements between the material and immaterial".[49] TallBear has thus noted that multispecies scholarship "will benefit from indigenous standpoints that never forgot the interrelatedness of all things."[50]

3 Methods and Approaches

Multisensory Practice

MSE both employs and challenges key orientations of the human-focused ethnography it evolved from. The term 'ethnography' derives from the Greek *ethnos*, referring to a people, tribe, or nation and *graphie* meaning "to write". Somewhat confusingly, within anthropology, 'ethnography' refers both to the fundamental anthropological method – a mode of study that emphasizes direct engagement or "being there" as the main mode of learning – as well as to the written account of a group of people produced by such methods. Ethnography that is 'multispecies' retains this focus on both dwelling with others and on writing from the basis of such observations. Like ethnographers focusing on human subjects, *multispecies* ethnographers typically spend a period of time with their subjects using the core ethnographic method of participant observation, a method at times colloquially (and descriptively) referred to as 'deep hanging out' in which the ethnographer develops a sense of the lifeways of others through participating in aspects of their lives and reflecting upon the experience.[51] Like ethnographers generally, multispecies ethnographers typically rely not only on direct experience as a form of knowledge but also embed their studies within analysis of broader political/economic and social structures.[52]

There are also ways in which the practice of multispecies ethnography necessarily departs from humanist ethnographic practices. In attending to more-than-human lives, ethnographic techniques such as interviewing and entering into the social

---

47 Todd, Zoe. An Indigenous Feminist's Take on the Ontological Turn: Ontology is Just Another Word for Colonialism. *Journal of Historical Sociology* 29, no. 1 (March 2016): 4–22, 19.
48 TallBear, Beyond the Life/Not Life Boundary, 188.
49 TallBear, Beyond the Life/Not Life Boundary, 191–2.
50 TallBear, Beyond the Life/Not Life Boundary, 180.
51 Geertz, Clifford. Deep Hanging Out. *The New York Review of Books* 45, no. 16 (October 1998): 69–72.
52 Pedersen, Helen. Unstable Mixtures: Zooethnographic Educational Relations as Difference, Contagion, Critique, and Potential. *Other Education: The Journal of Educational Alternatives* 1 (2012): 152–165.

worlds of others may require more creative approaches. Multispecies ethnographies thus frequently employ creative ways of getting to know aspects of one's more-than-human participants and have used methods as diverse as playing music with bird subjects,[53] to composing and living-with dung earthworms[54] and analyzing the ear-bones (otoliths) of salmon.[55]

So, how does one *do* a multispecies ethnography? Alongside vital recognition of the limits and partialities of human knowing of other species, multispecies ethnography is marked by an openness to the ways in which researchers might come to know members of other species. Many MSE-approaches challenge the sorts of 'mechanomorphic' modes of framing the behavior of animals that Carol Crist has identified as being fundamental to much of classical ethology. Such modes, influenced by Cartesian thought, tend to interpret more-than-human animals in mechanical, behaviorist ways.[56] In contrast, rather than focusing solely on the observable actions of such creatures, MSE tends to be interested in careful speculation as to the lived experience of more than humans.

The work of German biologist Jakob von Uexküll has been important for many multispecies ethnographers. Famous for studying the ways in which ticks sense and act upon the world, von Uexküll noted in his 1934 *A Stroll through the Worlds of Animals and Men* that the bodies of different beings sense, act and are acted upon by their surroundings in particular ways.[57] Von Uexküll referred to this realm of meaningful action and impact as a species' environment or '*Umwelt*'.[58] Although von Uexküll's tick science is overly-simplistic, his vivid descriptions of tick *Umwelten* as being comprised largely of three main components – namely, warmth, sunlight and the butyric acid emanating from mammalian sweat – remains a source of helpful inspiration for many multispecies ethnographers attempting to "step out of ourselves and into the strange environments of bees, sea anemones, dogs, ticks, bears, and many others".[59]

Attending to how the world might be for other species requires that one make use of potentially "unfamiliar sensoriums, with different kinds of touch, smell,

---

[53] Taylor, Hollis. Blowin in Birdland: Improvisation and the Australian Pied Butcherbird. *Leonardo Music Journal* 20 (2010): 79–83.
[54] Abrahamsson, Sebastian and Bertoni Filippo. Compost Politics: Experimenting with Togetherness in Vermicomposting. *Environmental Humanities* 4, no. 1 (2014): 125–148.
[55] Swanson, Heather. Methods for Multispecies Anthropology: Thinking with Salmon Otoliths and Scales. *Social Analysis* 61, no. 2 (June 2017): 81–99.
[56] Crist, Eileen. *Images of Animals: Anthropomorphism and Animal Mind*. Philadelphia, PA: Temple University Press, 1999.
[57] Von Uexküll, Jakob. A Stroll Through the Worlds of Animals and Men: A Picture Book of Invisible Worlds. Translated by Claire H. Schiller. In *Instinctive Behavior: The Development of a Modern Concept*, Claire H. Schiller (ed.), 5–80. New York, NY: International Universities Press, 1957, 6–9.
[58] Uexküll, A Stroll Through the Worlds.
[59] Buchanan, *Onto-Ethologies*, 1–2.

taste, and vision".[60] At times this has required that multispecies ethnographers make use of sensory engagements seldom used in academia, or make novel use of sensory prostheses. In particular, the work of Eva Hayward has made use of thought about what it is to know the world through the sense of touch.[61] The sorts of expertise required to attend to the lives of others means that many multispecies ethnographers also prioritize collaboration in their work. Multispecies ethnographic approaches to a diverse range of subjects, including ocean systems, soil science and extra-terrestrial life have all required that multispecies ethnographers interact with scholars from other fields.[62]

Making use of these different disciplinary knowledges and perspectives also introduces challenges. In MSE, it is vital that researchers are able to, as Heather Swanson has argued, "engage natural science tools while remaining alert to the politics of knowing".[63] In this, science studies, and especially feminist science studies, is vital to MSE. The scholarship of Haraway, Barad and Isabelle Stengers, in particular, have been used by multispecies ethnographers to attend to the effect of the positionality of researchers (and the institutions to which they belong) on the ways in which research subjects are constituted.[64]

Bodily Methods: Partial Affinities and the Miracles of Attunement

In working to attend to the worlds of others, multispecies ethnographies generally hold a tension between experimenting with ways of extending one's ability to enter into the worlds of others while also attending to the limits of such connection. As Ingold has noted, one "cannot enter directly into the *umwelten* of other creatures".[65] While empathy as direct, unmediated, insight is generally questioned within

---

60 Kirksey and Helmreich, Multispecies Ethnography, 565.
61 Hayward, Eva. Fingereyes: Impressions of Cup Corals. *Cultural Anthropology* 25, no. 4 (November 2010): 577–599.
62 Helmreich, Stefan. *Alien Ocean: Anthropological Voyages in Microbial Seas*. Berkeley, CA: University of California Press, 2009; Helmreich, Stefan. An Anthropologist Underwater: Immersive SounScapes, Submarine Cyborgs, and Transductive Ethnography. *American Ethnologist* 34, no. 4 (November 2007): 621–641.
63 Swanson, Methods for Multispecies Anthropology, 81.
64 Haraway, Donna. Situated Knowledges: The Science Question in Feminism and the Privilege of Partial Perspective. *Feminist Studies* 14, no. 3 (Autumn 1988): 575–599; Barad, Meeting the Universe Halfway; Stengers, Isabelle. *Cosmopolitics 1*. Minneapolis, MN and London: University of Minnesota Press, 2010.
65 Ingold, Tim. *The Perception of the Environment: Essays on Livelihood, Dwelling and Skill*. London and New York, NY: Routledge, 2000, 176; cf. Haraway, Donna. Staying with the Trouble: Xenoecologies of Home for Companions in the Contact Zones. David Schneider Memorial Lecture at Meetings of the Society for Cultural Anthropology, Santa Fe, New Mexico, May 7, 2010.

MSE, researchers tend to take the stance that bodily attunement with other beings is a vital avenue for gaining insight into our relationships with other animals. Despret argues that in contrast to attempts of empathy, in processes of attunement one attends to what it is to be with the other with each body developing "a responsiveness to the other," rather than talking about what it is like to *be* the other.[66] One is changed through one's interactions with other species, coming to know oneself and the world in different ways. As Despret has noted, in such connection, both parties are affected by one another, with attunement being the process 'of being transformed and gaining understanding as bodies come into rhythm with one another'.[67]

A vital aspect of attending to attunement, however, is to note that one's connection with others is always mediated by certain values and assumptions. In recognizing that we all exist in naturalcultural worlds – worlds of materiality and meaning – it becomes apparent that the relationship that emerges between field scientists and meerkats, for example, is not a-cultural or somehow value-neutral. Instead, such connections are particular relationships, shaped by cultural sentiments including ideals of scientific distance.[68] The love of hedgehogs demonstrated by volunteer urban conservationists in the United Kingdom, for example, is mediated by particular notions of wildness that emphasize the need for space.[69] The questions our worlds ask of other species greatly affects the answers they can give as to who they are.[70] In another example, the assumption that cats might be able to achieve anything seems to have an impact on the highly skilled and seemingly eccentric cats who are able to unlock fridges and have a liking for hot chilis in the New York apartment and feline study center of Jeffery Bussolini. Here, "the question becomes one of fashioning the space, manner and language to experiment with ways of being-together where the human is not the center of activity or meaning-making."[71] An important aspect of MSE methodologies is thus to attend to the stories and expectations mediating the relationships between researchers and the lives one is attending to.

---

[66] Despret, Vinciane. The Body We Care For: Figures of Anthropo-Zoo-Genesis. *Body & Society* 10, no. 2–3 (June 2004): 111–134.
[67] Despret, The Body We Care For, 128.
[68] Candea, I Fell in Love with Carlos the Meerkat.
[69] McLauchlan, Laura. Wild Disciplines and Multispecies Erotics: On the Power of Wanting Like a Hedgehog Champion. *Australian Feminist Studies* 34, no. 102 (November 2019): 509–523.
[70] Despret, Vinciane. *What Would Animals Say If We Asked the Right Questions?* Minneapolis, MN: University of Minnesota Press, 2015.
[71] Lestel, Dominique, Jeffrey Bussolini and Matthew Chrulew. The Phenomenology of Animal Life. *Environmental Humanities* 5, no. 1 (May 2014): 125–148.

4 Implication(s) of the Animal Turn

The impact of the *animal turn* on anthropology and other ethnography-employing disciplines offers more than just ethnography with animals in it. Instead, MSE can be seen to be part of a broader shift in scholarship towards better recognizing and working with the multiple agencies at play in worlds we care for and are comprised of. Influenced by both the 'animal' and the 'posthuman' 'turns', MSE typically not only attends to the lives of more-than-human subjects but also frequently questions the divisions between nature and culture by paying ethnographic attention to the fundamental entanglement of modes of representation and 'the animals themselves' [→Post-Domestication: The Posthuman]. MSE shares such theoretical commitments with a range of disciplines, including historical animal studies. Through paying ethnographic attention to relationality, MSE offers situated, sustained attention to what researchers and other animals might become – together – in particular contexts.

Situating Ethologies

A vital methodological question of historical animal studies centers around how to take the limits of representation seriously while not unnecessarily limiting the possibilities of scholarly insight into the lives of other species.[72] As Mieke Roscher has argued, "what is required is a fusion of historiographical approaches that take representation seriously, but which go further by also including the material life of the animal, namely the life of specific animals in historical contexts".[73] Multispecies ethnographies tend to be mindful that animal behavior does not necessarily follow simple species lines but, rather, is shaped by a myriad of social and cultural interactions (and thus may be considerably different in the past). Such awareness is also already present in historical animal studies scholarship. As Susan Nance has noted in her historical attention to elephants, "different environments or communities of captivity will produce different kinds of animals and people, and humans use those processes to produce human cultures and identities."[74] MSE is potentially a helpful source for theoretical reorientation to such relationality of life. One major impact of the *animal* (and posthuman) *turn* entering into ethnography is the consideration of the ways in which just such social and ecological conditions affect and shape lives. As Locke and Münster have argued, multispecies ethnography asserts that any "theoretically integrated account of existence" must consider the entanglements of multiple lifeforms

---

[72] Fudge, What Was it Like to be a Cow?
[73] Roscher, New Political History and the Writing of Animal Lives, 53.
[74] Nance, Susan. *Entertaining Elephants: Animal Agency and the Business of the American Circus.* Baltimore, MD: Johns Hopkins University Press, 2013, 10–13.

with technologies and landscapes.⁷⁵ Such work acknowledges that life is emergent, arising from inter/intra-connected forces of signification, biochemical, spoken and gestural language.⁷⁶ While some MSEs do focus on specific macro-level "animals", many also question the idea that critters are ever singular, attending to both the social and cultural 'companion species' of which we are all comprised. Paying ethnographic attention to the intra-actions of which life is comprised may thus fundamentally unsettle many common categories of animal studies scholarship, including that of 'individual', 'organism', as well as 'species'.⁷⁷

In acknowledging the naturalcultural relationality of life, MSE also takes into account the mediating distances, attitudes and technologies at play in any interspecies relationship. This has particular implications for how one reads apparently 'neutral' studies of animals, including biological and ethological studies. Through recognizing the behavior of animal scientists as a culturally-specific way of relating to other species, attempts to create 'objective' distances between those studying animals and the animals themselves is no longer a 'neutral' behavior, but, rather, a *particular* one, encouraging particular kinds of interspecies relationships and interactions.⁷⁸ Thus, for animal historians interested in extending the "glimpses of animals in the past"⁷⁹ through engagement with such studies, MSE may be helpful for challenging the apparent objectivity of scientific studies of animal lives.

Expanding Methods and Tending to Limits

Both the attention paid to the limits of knowing other species in MSE as well as the willingness to engage in speculative ethological practices may be useful to animal historians who are interested in attending not only to the behaviors and historical 'agency' of more-than-human animals, but also to their lived experience. As Jennifer Adams Martin has noted, although not attempting to enter the experiential worlds of other species can seem to be a gesture of respect, in practice, avoiding such knowing not only makes it difficult to attend to the agency of such beings. It also potentially reinforces species boundaries, overlooking "the historicity, diversity, and agency of wild animals".⁸⁰

One example of multispecies ethnographic approaches informing historical animal scholarship can be seen in the work of animal historian, Chris Pearson. Pearson

---

75 Locke and Münster, Multispecies Ethnography.
76 Lestel, Brunois and Gaunet, Etho-Ethnology and Ethno-Ethology, 155; Hoffmeyer, Jesper and Barbara J. Haveland. *Signs of Meaning in the Universe*. Bloomington, IN: Indiana University Press, 1996.
77 Barad, Meeting the Universe Halfway.
78 Candea, I Fell in Love with Carlos the Meerkat; Despret, Responding Bodies and Partial Affinities.
79 Fudge, What Was It Like to be a Cow?, 262.
80 Martin, Jennifer Adams. When Sharks (Don't) Attack: Wild Animal Agency in Historical Narratives. *Environmental History* 16, no. 3 (July 2011): 451–455.

asks how living and walking with dogs (or, in this case, Pearson's close canine companion, Timmy) through urban environments might inform historical research into human-animal histories of cities. Pearson asks, "what are the possibilities for 'real/actual' dogs to enliven and become partners in historical research?".[81] While, as Pearson notes, there are limits to one's ability to know the world from the perspective of one's more-than-human collaborators, it is still possible to write histories that give space to and are informed by more-than-human modes of engagement in the present.

An important aspect of rigor in MSE is acknowledging the limits to what we can know from our particular primate bodies.[82] As a researcher it is vital to situate oneself both in terms of one's social positioning but also the bodily limits of one's knowing of other species, even when one's ability to register the sounds and smells of others is technologically enhanced. As part of acknowledging aspects of disconnection and misunderstanding and difference, it is important to attend to the ways in which not all knowledge can necessarily be easily translated into prose. Thus, the poetics and art practices used in some multispecies ethnographies may be of use to animal historians considering not only speculative ways of connecting with and representing aspects of the lives of others, but who are also interested in forms of expression that do not reduce the presence of the more-than-human to what can be explained in academic prose.

This dual approach of both rigorously attending to the partialities of interspecies connections while simultaneously being open to novel modes of connecting suggests additional possibilities for historical animal studies. While historical animal studies focuses on documentary evidence from largely human sources, what sorts of more-than-human records might also be available from the past? If the latrines and stamping grounds of water voles might give 'voice' to aspects of more-than-human presence and experiences in the present, what traces might be available from past more-than-human lives? The combination of multispecies ethnographic approaches with multispecies archaeological methods of reading material histories for more-than-human traces may be a particularly fruitful approach for animal historians interested in making space for interruption of their scholarship and of human expertise by allowing for the (often mysterious) presence of more-than-human traces within their scholarship [→Material Cultural Studies;→Pre-Domestication: Zooarchaeology].[83]

---

81 Pearson, Chris. Walk in the Park with Timmy: History and the Possibilities of Companion Species Research. *The Wild* 1 (2009): 87–96.
82 Haraway, Situated Knowledges.
83 Pilaar Birch, Suzanne E. *Multispecies Archaeology*. Oxon: Routledge, 2018.

Attending to Relationality: Questioning 'Animals' and Scholarship as Emergence

Conducting one's own ethnographic studies in the present may assist historical scholars to develop a bodily sense of and attunement with species being studied. This, in turn, subtly, but potentially powerfully, reconfigures that researcher. As Despret has noted, this relationality raises key questions: "What affects you, and whom does your way of living, your manner of being, affect in turn?"[84] There is a strong politics to this, namely, as animal historian, Erica Fudge has argued, that "if we change our position in relation to them perhaps we change our understanding of our shared realities."[85] An important humility to such thinking, however, is to acknowledge that we never entirely know what it is that might matter and what might emerge from such connections.[86] Considering the ways in which coming to know a member of another species is a case of mutual shaping has implications for understanding what a researcher is more generally. Particularly in doing the on-the-ground work of attending to the attachments that make the world, the world transforms as one's own attachments to others in the world shift.

The *animal turn,* as it has affected anthropology, has led to a mode of studying that does more than study animals using ethnographic method. Alongside attending to larger ecosystems and the interplay of economics, cultures, bacteria, spirits, stories and technologies, MSE tends to also see the world of the researcher as part of the work of emergent practice. Alongside such work of situating and paying careful attention to the limits and possibilities of knowing other species, MSE also offers modes of reconsidering both species and animality, as well as for thinking about what it is that scholarship *does.* Through its relational focus – including to the naturalcultural aspects of life, in which culture and story are seen as intra-acting with – co-shaping the living world – representation literally *matters.* The work of storytelling – academic or otherwise – thus takes on a clear political and ethical valence as it calls the audience to connect with the world and with particular others. Attending to the importance of perspectives in such a way requires one to think and feel more carefully about which lives 'matter' and to attend to which lives count as lives. The work of challenging representations and of examining the contingencies of historical relationships between species thus becomes the vital politics of attending to and participating in the always more-than-human lives of which we, as scholars, are a part.

---

84 Despret, Vinciane and Michel Meuret. Cosmological Sheep and the Arts of Living on a Damaged Planet. *Environmental Humanities* 8, no. 1 (May 2016): 24–36, 35.
85 Fudge, What Was it Like to be a Cow?, 270.
86 Despret and Meuret, Cosmological Sheep, 35.

Selected Bibliography

Despret, Vinciane. Responding Bodies and Partial Affinities in Human–Animal Worlds. *Theory, Culture & Society* 30, no. 7–8 (August 2013): 51–76.

Haraway, Donna. *When Species Meet*. Minneapolis, MN: University of Minnesota Press, 2008.

Ingold, Tim. Epilogue: Towards a Politics of Dwelling. *Conservation and Society* 3, no. 2 (2005): 501–508.

Kirksey, Eben and Stefan Helmreich. The Emergence of Multispecies Ethnography. *Cultural Anthropology* 25, no. 4 (October 2010): 545–576.

Kohn, Eduardo. *How Forests Think: Toward an Anthropology Beyond the Human*. Berkeley, CA: University of California Press, 2013.

Ogden, Laura, Billy Hall and Kimiko Tanita. Animals, Plants, People, and Things: A Review of Multispecies Ethnography. *Environment and Society: Advances in Research* 4, no. 1 (2013): 5–24.

Swanson, Heather. Methods for Multispecies Anthropology: Thinking with Salmon Otoliths and Scales. *Social Analysis* 61, no. 2 (2017): 81–99.

TallBear, Kim. Beyond the Life/Not Life Binary: A Feminist-Indigenous Reading of Cryopreservation, Interspecies Thinking, and the New Materialisms. In *Cryopolitics: Frozen Life in a Melting World*, Emma Kowal, Joanna Radin (eds.), 179–202. Cambridge, MA: MIT Press, 2017.

Tsing, Anna Lowenhaupt. Arts of Inclusion, or, How to Love a Mushroom. *Australian Humanities Review* 50 (2011): 5–22.

Van Dooren, Thom and Deborah Bird Rose. Storied-Places in a Multispecies City. *Humanimalia* 3, no. 2 (Spring 2012): 1–27.

Sarah D. P. Cockram
History of Emotions

1 Introduction and Overview

According to the cognitive ethologist Marc Bekoff, it is through a common world of emotions that we can bridge the species gap and gain connection with animals:

> It's because animals have emotions that we're so drawn to them; lacking a shared language, emotions are perhaps our most effective means of cross-species communication. We can share our emotions, we can understand the language of feelings [...] Emotions are the glue that binds.[1]

There is much for a historian of human-animal interactions to unpick in Bekoff's statement. Are human beings, present and past, drawn to animals because they display emotion? Are emotions the most effective way, then as now, for humans and animals to communicate and reach interspecific understandings? Do we have emotions that we share, what are these, and can we straightforwardly read these in others, including in other animals? How universal are emotions across variables such as individual difference, and difference in species, and in cultures across place and time? For historical animal studies, work on the emotions can provide a key analytical lens and a set of extremely valuable tools, suggesting answers we might propose to the questions above; putting forward means to understand human and animal emotions and communication through "the language of feelings"; and finding new ways to think about multispecies lived experience in the past. This chapter takes the opportunity to assess the potential to the historian of the cross-pollination of the *emotional turn* and the *animal turn*, discussing theoretical underpinnings of work in the history of emotions and how these might interplay with historical animal studies.

The historian can approach the history of animals and emotion in a number of, potentially-interlocking, manners. The scholar may be interested predominantly in human emotions about animals, studying ways that people have felt about animals in historical context (attachment to pets, awe of captive lions, fear of sharks, annoyance at vermin).[2] Research may investigate how humans in the past have conceived

---

[1] Bekoff, Marc. *The Emotional Lives of Animals*. Novato, CA: New World Library, 2007, 15.
[2] For example, Tague, Ingrid H. The History of Emotional Attachment to Animals. In *The Routledge Handbook of Animal-Human History*, Hilda Kean, Philip Howell (eds.), 345–366. London: Routledge, 2019; Howell, Philip. When Did Pets Become Animals? In *Historical Animal Geographies*, Sharon Wilcox, Stephanie Rutherford (eds.), 11–22. London: Routledge, 2018; Ben-Ami, Ido. Emotions and the Sixteenth-Century Ottoman Carnival of Animals. In *Interspecies Interactions Animals and Humans between the Middle Ages and Modernity*, Sarah Cockram, Andrew Wells (eds.), 17–33. London: Routledge, 2018; Maglen, Krista. 'The Monster's Mouth ...': Dangerous Animals and the European Settle-

of emotions in animals, including classic focus on the ideas of Michel de Montaigne, René Descartes, Jeremy Bentham, and Charles Darwin; with studies on beliefs about animal pain and emotions within the history of science; and work on animal training [→History of Science].³ The historian might interrogate interactive ways that humans and animals have elicited emotions in each other (for example, the affective benefits or drawbacks to both owners and pets of the companion animal relationship; interspecies bonding in military environments; or levels of trust/mistrust between communities of people and wild animals).⁴ Critical attention can be paid to how we might try to access the emotional experience of animal subjects of our historical research (such as the life of the early modern cow).⁵ These projects develop from a range of objectives and ways of coming at historical animal-human interaction – with different weight accorded to the animal side of the equation; they also necessitate a range of approaches to meet such varied aims. This chapter will question available research methods and theories that can be applied in work on emotion in historical animal studies and discuss advantages and risks of these. Turning to Topics and Themes, it may be useful to begin with an example.

---

ment of Australia. In *Interspecies Interactions, Animals and Humans between the Middle Ages and Modernity*, Sarah Cockram, Andrew Wells (eds.), 214–229. London: Routledge, 2018; Fissell, Mary. Imagining Vermin in Early Modern England. *History Workshop Journal* 47, no. 1 (Spring 1999): 1–29.
3 Tague, The History of Emotional Attachment to Animals, provides an overview of ideas about animal emotion in early modern Britain and of thinkers including Montaigne, Descartes and David Hume. See also Richardson, Angelique. *After Darwin: Animals, Emotions, and the Mind*. Amsterdam: Rodopi, 2013. Further publications on animals in the history of science and emotions include Dror. Otniel E. The Affect of Experiment: The Turn to Emotions in Anglo-American Physiology, 1900–1940. *Isis* 90, no. 2 (June 1999): 205–237; Gray, Liz. Body, Mind and Madness: Pain in Animals in Nineteenth-Century Comparative Psychology. In *Pain and Emotion in Modern History*, Rob Boddice (ed.), 148–163. Basingstoke: Palgrave Macmillan, 2014; Mayer, Jed. The Expression of the Emotions in Man and Laboratory Animals. *Victorian Studies* 50, no. 3 (February 2008): 399–417; Sandra Swart points out efforts given to understanding the horse mind in training. Swart, Sandra. *Riding High: Horses, Humans and History in South Africa*. Johannesburg: Wits University Press, 2010, 217; Chris Pearson analyses dog training and pet-keeping as "emotional practices" (following Monique Scheer). Pearson, Chris. "Four-Legged Poilus": French Army Dogs, Emotional Practices and the Creation of Militarised Human-Dog Bonds, 1871–1918. *Journal of Social History* 52, no. 3 (Spring 2019): 731–760.
4 Cockram, Sarah. Sleeve Cat and Lap Dog. Affection, Aesthetics and Proximity to Companion Animals in Renaissance Mantua. In *Interspecies Interactions Animals and Humans between the Middle Ages and Modernity,* Sarah Cockram, Andrew Wells (eds.), 34–65. London: Routledge, 2018; McEwen, Andrew. He Took Care of Me: The Human-Animal Bond in Canada's Great War. In *The Historical Animal*, Susan Nance (ed.), 272–287. Syracuse, NY: Syracuse University Press, 2015; Pearson, Four-Legged Poilus; Hediger, Ryan. Dogs of War: The Biopolitics of Loving and Leaving the US Canine Forces in Vietnam. *Animal Studies Journal* 2, no. 1 (2013): 55–73; Rangarajan, Mahesh. Animals with Rich Histories: The Case of the Lions of Gir Forest, Gujarat, India. *History and Theory* 52, no. 4 (December 2013): 109–127.
5 Fudge, Erica. Milking Other Men's Beasts. *History and Theory* 52, no. 4 (December 2013): 13–28, 23.

2 Topics and Themes

On the morning of 28 October 1462, a large, often fierce, rusty-colored dog called Rubino ran away. Rubino had been accidentally left behind in his bed while his owner Lodovico II Gonzaga (marquis of Mantua in Northern Italy) readied to board a boat for the marquis' palace of Revere on the river Po. While Lodovico and his entourage embarked, his men brought Rubino along to catch up. But they let the dog off his leash and Rubino bolted in search of Lodovico, the men chasing after. Rubino could not be found. The marquis wrote to his wife, Barbara of Brandenburg, the following day from Revere, requesting that she put out a search party for the lost dog. The marquis guessed that Rubino, not seeing his master and wishing only to be with him, would have headed back home towards Mantua. He gave instructions that Rubino should be well chained when found, a difficult task as the dog would usually allow no one but Lodovico to touch him. Lodovico wrote, "under no circumstances would we wish to lose [Rubino], and we know that if he should see us then he would come to us and no other chain would be needed". Barbara was able to write and put her husband's mind at rest. She informed Lodovico that Rubino "arrived here [at the palace in Mantua] yesterday evening, all wet" after his long day out, and she straightaway dispatched a groom to escort Rubino safely back to the marquis. Lodovico was greatly relieved to be reunited with his dog and told Barbara that Rubino "gave us many caresses and begged forgiveness, and so we have forgiven him". Barbara was pleased Rubino and Lodovico were now back together, Rubino when in Mantua having been "all miserable [... pacing] from room to room looking for Your Excellency".[6] The relationship between Lodovico and Rubino is one that in many respects

---

6 Rodolfo Signorini brought Rubino's story to light and proposed that a portrait of Rubino is found in Mantegna's *Camera degli Sposi* (1465–1474). See Signorini, Rodolfo. A Dog Named Rubino. *Journal of the Warburg and Courtauld Institutes* 41 (1978): 317–320. Lodovico Gonzaga to Barbara of Brandenburg, 29 October 1462, Archivio di Stato, Mantua, Archivio Gonzaga, Busta 2097: "Ulterius quando hery matina montassemo in nave el cane nostro Rubino se smarite ché non vene cum nui. Li nostri lo conducevano per terra ligato e quando foreno a Pontemolino lo desligoreno. E perché el non vole stare se non cum nui, come sapeti, el se partite da loro e vene traversando verso Mantua. Ge venero drieto e non lo sepeno atrovare. Vogliamo che faciati vedere si l'è lì e ch'el ne sia mandato in catena. S'el non fusse venuto a Mantua, mandare in quelle ville verso san Zorzo a cercare s'el se ritrova e vedere de haverlo, e ligarlo, ché credemo serà difficile perché el non se lassa tochare ad altri che nui. Ma faciano ciò che pono per pigliarlo, e s'el non fusse pur longi che X miglia ge veneressemo nui, perché a modo alcuno non lo voressemo perdere, e sapiamo che come el ne vedesse veneria da nui e non ge bisogneria altra catena". Barbara to Lodovico, 29 October: "Per Luca da Mariana mando a la Illustre Signoria Vostra el suo cane chiamato Rubino, el qual, secondo me fu dicto, gionse heri sera qui tuto bagnato circa le 24 hore. Non so s'el se partesse cum bona licentia de la Excellentia Vostra e par ch'el se ne sia fugito per qualche suo mancamento". Lodovico to Barbara, 30 October: "Rubino è gionto questa matina. El povereto se domentigò nel lecto quando se partessemo per venire a la nave. Conducendolo poi li nostri e non vedendone nui, se parti da loro. El ne ha facto careze assai e ne ha domandato perdonanza, e così gli habiamo perdonato". Barbara to Lodovico: "Me

might seem familiar to us. A strong emotional bond between this dog and his human is suggested to us by these documents as well as by the depiction of Rubino contentedly sitting under the marquis' chair in Mantegna's *Camera degli Sposi*; by letters detailing Lodovico's response to Rubino's illness later in life; and by the memorial tomb Lodovico erected after Rubino's death in 1467.[7]

But how should historians interpret the emotions presented to us in Barbara and Lodovico's 1462 epistolary exchange over the lost Rubino? Are the worry, affection, relief, and tactile joviality at reunion to be equated with how we would understand these feelings? And what of Rubino's affective experience? We access Rubino's emotions through the ruling couple's evaluation and reporting of these. Do the sources indeed reflect a dog who really seeks his owner, misses him, despairs and pines on not finding him at home, and responds to seeing Lodovico with joy, and perhaps the guilt of a runaway dog done-wrong? How do we read the emotions of someone else, human or animal, past or present?

If the historian believes that their field of enquiry is based on revealing lived experience of the past, then whether the research comes from a perspective of social, political, intellectual or cultural history, or any other, the feelings of the subjects of study are likely to be germane to analysis. However, emotions as identified in historical source material must be handled with care. Scholars in the dynamic area of the history of emotions set out foundations of how to approach these, starting with core decisions about the nature of historical emotions.[8] The history of emotions has seen debate between opposing camps. On one side are universalists, influenced by the life sciences, for whom emotions are essentially innate, biologically driven and unaltered across cultures and time; on the other side are social constructivists, influenced by anthropological models, who see emotions as learnt, situated in their context, and determined by cultural forces which are of no lesser importance than biology. There are also attempts to reconcile these polarities.[9] It may be easier to understand the emotion we read in our historical sources from a universalist standpoint: "what Lodovico expresses for Rubino is the same as I might feel for my favorite dog". If we accept that Lodovico's love is culturally-bound and applicable only in his time and place (and in its essence really only to that one man), then we have a far greater leap to make to try to understand it. And that is before we attempt to ask about the ani-

---

piace che Rubino sia ritornato e certo el stasevo ben qui tuto gramezoso, et andava de camera in camera cercando la Celentia Vostra".

[7] See Signorini, A Dog Named Rubino; also Jonietz, Fabian. Animal Deaths, Commemoration, and Afterlives at the Gonzaga Court and Beyond. In *Animals and Courts: Europe, c. 1200–1800*, Mark Hengerer, Nadir Weber (eds.), 361–396. Berlin: De Gruyter Oldenbourg, 2020.

[8] For recent debate and new work across historical periods see Broomhall, Susan, Jane W. Davidson and Andrew Lynch. *A Cultural History of the Emotions*. London: Bloomsbury, 2019.

[9] Plamper, Jan. *The History of Emotions: An Introduction*. Translated by Keith Tribe. Oxford: Oxford University Press, 2015.

mal: what of Rubino's misery when parted from Lodovico or his physical affection when reunited?

Historical animal studies is already well acquainted with the problem of accessing another's *Umwelt:* that unique sensory environment of an individual, the way that any given organism experiences – and feels – the world. We reference, for instance (maybe in the first seminar of a university course on historical animals), the philosopher Thomas Nagel's reflection on the impossibility of knowing how it feels to be an animal, more specifically in his example a bat. Even if we can imagine ourselves as bats, we cannot ever know "what it is like for a bat to be a bat".[10] Perhaps we shrug and move on, or seek support from the theologian Andrew Linzey, who responds by levelling the playing field, albeit by possibly destabilizing us further. For Linzey, we can understand something of the bat's world, and "know these things at least as reasonably as we know them in the case of most humans".[11] After all, we might argue that we cannot know what it is like for Lodovico to be Lodovico really much better than we know what it is like for Rubino to be Rubino.

At this point, the historian's science and art is under intense, uncomfortable scrutiny. As Hilda Kean points out, we recognize the impossibility of understanding historical human lives while trying to do so anyway, and historical animal studies faces "the ongoing concern historians have about the nature of experience [...] with the additional layer of another species".[12] Jan Plamper asks how, in a social constructivist model in which we acknowledge the pivotal significance of cultural context, we can imagine to understand another culture's emotions at all.[13] This situation presents a serious hindrance to comprehension of human cultures different to ours in time and/or space, let alone to any endeavor to think across species. And if following a universalist model, the historian nevertheless remains at a loss to effectively understand a bat's inner life. An intertwined history of the emotions with the senses cannot fully help us here either, fruitful though such an approach can so very often be. We hit a stumbling block with those animal senses beyond human perception in type and capability, such as echolocation or the role of Rubino's sensescape in finding home,[14] and this is before we even allow for ways in which senses as we find them represented in our historical sources should be viewed as acculturated too.[15] In sum, there is a danger that, with Nagel on one shoulder and

---

10 Nagel, Thomas. What is it Like to be a Bat? *Philosophical Review* 83, no. 4 (October 1974): 435–450.
11 Linzey, Andrew. *Why Animal Suffering Matters: Philosophy, Theology and Popular Ethics.* Oxford: Oxford University Press, 2009, 50.
12 Kean, Hilda. Animal–Human Histories. In *New Directions in Social and Cultural History,* Sasha Handley, Rohan McWilliam, Lucy Noakes (eds.), 173–189. London: Bloomsbury, 2018.
13 Plamper, *The History of Emotions,* 116.
14 On dog sensescapes see Pearson, Chris J. A Walk in the Park with Timmy: History and the Possibilities of Companion Species Research. *The Wild* 1 (2009): 87–96.
15 For an introduction to the history of the senses, see Classen, Constance. *A Cultural History of the Senses.* London: Bloomsbury, 2014; Howes, David and Constance Classen. *Ways of Sensing: Understanding the Senses in Society.* New York, NY: Routledge, 2013; Jütte, Robert. *A History of the Senses:*

Plamper on the other, we are in a double whammy if we want to understand what an animal in the past might ever have felt.

For Ingrid Tague,

> we can only explain animal behavior with reference to our own, human thoughts and feelings. [...] As a historian, then, [...] I focus unapologetically on humans [...] not because I think animals are unimportant, but because ultimately the study of history must be a study of humans [...] the history we can know and write requires access – no matter how indirect – to the motives, beliefs, and feelings of individuals in the past, which we cannot achieve with animals.[16]

Chris Pearson shares this restraint in his recent study of French army dogs: "I consider canine emotional experiences as beyond my grasp. I instead examine how human actors experienced and made sense of canine emotions and engaged them in their emotional practices".[17] The perils of anthropomorphism are ones that each historian of animals has to face in their own manner.[18] For some, the threat of misconception and unsound projection is too great; access to animal inner lives in our sources is not possible; and the responsible historian's role has limits at the species boundary. For others, as Bekoff would have it for contemporary understanding of animal emotion, "we must be imaginative in our interactions with other animals".[19] What part should imagination, empathy and our own feelings play in any historian's toolbox? While our intuition may prove valuable, if the historian is prepared to abandon the caution held by Tague, any reading that might lead, even indirectly, to the animal must be carefully conducted. In analysis of a historical source which contains human-authored perception of animal emotion in the past, the source is made up of layers that are dense and several and require nuanced, contextualized interpretation. While historians may not agree on the limits of their role, each must commit to their own uncompromising adherence to disciplinary principles, as they see them. From this basis, further tools might be presented by the interdisciplinary collaborations and methods discussed below.

From Antiquity to Cyberspace. Cambridge: Polity Press, 2005; Smith, Mark M. *Sensing the Past: Seeing, Hearing, Smelling, Tasting, and Touching in History*. Los Angeles, CA and Berkeley, CA: University of California Press, 2008.

16 Tague, Ingrid H. *Animal Companions: Pets and Social Change in Eighteenth-Century Britain*. University Park, PA: Pennsylvania State University Press, 2015, 9.

17 Pearson, Four-Legged Poilus, 734.

18 On anthropomorphism see Daston, Lorraine and Gregg Mitman (eds.). *Thinking with Animals: New Perspectives on Anthropomorphism*. New York, NY: Columbia University Press, 2005; Smith, Julie A. and Robert W. Mitchell. *Experiencing Animal Minds: An Anthology of Animal–Human Encounters*. New York, NY: Columbia University Press, 2012; Also Fudge, Erica. *Quick Cattle and Dying Wishes: People and Their Animals in Early Modern England*. Ithaca, NY: Cornell University Press, 2018, 217.

19 Bekoff, *The Emotional Lives of Animals*, xxi.

3 Methods and Approaches

The sources available to the historian of animals and emotions determine methodologies that can be employed upon these. Difficulties that can attend the finding and analysis of animal traces in historical documents have long been pored over in historical animal studies, and suggestions for reading "against the grain" have been proposed.[20] Archival sources typically deployed by historians such as diaries, government documents, and wills have been used to examine human-animal emotional relationships.[21] The historian's research questions about animals and emotion might also be illuminated through analysis which includes rich cultural representations of animals in literary sources, and which goes beyond the textual to make use of visual and material evidence, including taxidermy and archaeological specimens [→Material Culture Studies;→Visual Culture Studies and Art History].[22] Depending on the research objectives, questions can then be asked in relation to the reading of human emotion in our sources, or of how to interpret animal emotions, behavior, or body language. One difficulty is that our sources are likely to provide a record of emotions caught at one moment in time. How to account for emotions that are in a state of transformation? As Lawrence Stone put it: "human feelings are so changeable and

---

[20] In his essay On the Concept of History, Walter Benjamin urged the historian to "brush history against the grain" to allow counter readings subvert dominant narratives. Benjamin, Walter. On the Concept of History. Translated by Edmund Jephcott. In *Walter Benjamin: Selected Writings, 4: 1938–1940*, Howard Eiland, Michael W. Jennings (eds.), 389–400. Cambridge, MA: Harvard University Press, 2003, 392. Scholarship reflecting on the animal in historical sources includes: Fudge, Erica. A Left-Handed Blow: Writing the History of Animals. In *Representing Animals*, Nigel Rothfels (ed.), 3–18. Bloomington, IN: Indiana University Press, 2002; Fudge, Erica. What Was it Like to be a Cow? History and Animal Studies. In *The Oxford Handbook of Animal Studies*, Linda Kalof (ed.), 258–278. Oxford: Oxford University Press, 2017; Kean, Hilda. Challenges for Historians Writing Animal-Human History: What is Really Enough? *Anthrozoös* 25, Supplement (April 2012): 57–72; Swart, Sandra. "But Where's the Bloody Horse?": Textuality and Corporeality in the "Animal Turn". *Journal of Literary Studies* 23, no. 3 (September 2007): 271–292; Swart, Sandra. The World the Horses Made: A South African Case Study of Writing Animals into Social History. *International Review of Social History* 55, no. 2 (August 2010): 241–263.
[21] Recent examples include White, Benjamin Thomas. Humans and Animals in a Refugee Camp: Baquba, Iraq, 1918–20. *Journal of Refugee Studies* 32, no. 2 (May 2018): 216–236, 224–225; Fudge, *Quick Cattle*.
[22] Poliquin, Rachel. *The Breathless Zoo: Taxidermy and the Cultures of Longing*. University Park, PA: Pennsylvania State University Press, 2012; DeMello, Margo. *Mourning Animals: Rituals and Practices Surrounding Animal Death*. East Lansing, MI: Michigan State University Press, 2016. As an example of the use of literature and art see Cuneo, Pia. Equine Empathies: Giving Voice to Horses in Early Modern Germany. In *Interspecies Interactions Animals and Humans between the Middle Ages and Modernity*, Sarah Cockram, Andrew Wells (eds.), 66–86. London: Routledge, 2018.

evanescent that interpretation of them is a most hazardous exercise",[23] quite, and for animal feelings a wealth of other methodological problems are thrown in.

A starting point is the use of methodologies from the history of emotions. Foundational attention in the history of emotions is paid to the interactive, mutually constitutive, socially-overseen and performative nature of human emotional life. Carol and Peter Stearns' early work on "emotionology" focused on the prescription of norms and expectations of the expression of emotion, and how these might differ from lived emotional behavior.[24] William Reddy has emphasized the performance of emotional speech acts ("emotives") and given the name "emotional regimes" to the ideals and modes of emotional expression that prevail in given periods and places, such as in eighteenth and nineteenth-century France.[25] Barbara Rosenwein's seminal work established co-existing medieval "emotional communities": with members expected to keep to shared beliefs about emotions and how these should be expressed.[26]

In an interview conducted by Jan Plamper for *History and Theory*, William Reddy, Barbara Rosenwein, and Peter Stearns suggested that, rather than entrenching itself in a specialized field, work in the history of emotions is instead to be integrated with other fields in historical study (such as social, cultural and political history), with emotion as an enriching category of analysis.[27] In such an approach, species can also be added as a key category – working alongside emotion, and the core intersectional categories of gender, race and class – to give a richer understanding of the past [→Feminist Intersectionality Studies]. If the history of emotions can reveal ideas about the political and non-political; the rational and irrational; the natural and the civilized; the feeling and the un-feeling, such thinking is a clear place for work on animals, whose position within such binaries is culturally defined, subject to power dynamics and impactful on lived experience. Susan Matt has flagged up ways in which ongoing research in the history of emotions might concentrate on sub-

---

[23] Stone, Lawrence. *The Family, Sex and Marriage in England 1500–1800.* Harmondsworth: Penguin, 1979, 233–234.

[24] Stearns, Peter N. and Carol Z. Stearns. Emotionology: Clarifying the History of Emotions and Emotional Standards. *The American Historical Review* 90, no. 4 (October 1985): 813–836.

[25] Reddy, William M. Against Constructionism: The Historical Ethnography of Emotion. *Current Anthropology* 38, no. 3 (June 1997): 327–351; Reddy, William M. *The Invisible Code: Honor and Sentiment in Postrevolutionary France, 1815–1848.* Berkeley, CA: University of California Press, 1997; Reddy, William M. *The Navigation of Feeling: A Framework for the History of Emotions.* Cambridge: Cambridge University Press, 2001.

[26] Rosenwein, Barbara H. *Emotional Communities in the Early Middle Ages.* Ithaca, NY: Cornell University Press, 2006. See also Boddice, Rob. *The History of Emotions.* Manchester: Manchester University Press, 2018. The latter aims to "take the temperature of the field as a whole as it now stands" (2), and chapter 3 discusses methodologies of the history of emotions, evaluating the influence of Reddy, Rosenwein and Stearns.

[27] Plamper, Jan. The History of Emotions: An Interview with William Reddy, Barbara Rosenwein, and Peter Stearns. *History and Theory* 49, no. 2 (March 2010): 237–265.

altern and marginalized lives, with a view not only from the bottom up but also from the inside out,[28] the animal is surely also to be an important focus for such work.

Rob Boddice has recently summed up forceful ideas of historicized human emotions: "emotions are at the center of the history of the human being, considered as a biocultural entity that is characterized as a worlded body, in the worlds of other worlded bodies".[29] Within such worlds exist nonhuman "biocultural entities", animal bodies and minds "worlded" in their own ways including in being influenced by human culture around them – such as by practices of farming, breeding, transportation, training and so on [→Cultural History; →Social History]. Might we incorporate animals within multispecies emotional communities?[30] Such work could aim not only to explore emotional communities of humans which are built around aspects of human-animal relations, but could also think through multispecies emotional communities in which animal emotional lives are accorded equally meaningful status in analysis. An approach which reaches out across disciplines, influential on the field of animal studies more broadly, may be stimulating for the historian here. In dialogue with Donna Haraway, the anthropologist Anna Tsing suggests the richness of thinking in terms of multispecies ethnography [→Multispecies Ethnography], and reiterates that 'Human nature is an interspecies relationship'.[31] Ideas of entanglement and mutable webs of relationality and dependence, that give equal space to wide varieties of lifeforms, including plants, fungi and bacteria, are useful in flattening anthropocentric power hierarchies, although there are limits to how far into an eco-system the researcher of multispecies emotional entanglement may tread. Development of histories of more-than-human emotional communities will need careful theorizing and standards of responsible interdisciplinarity, with the maintenance of rock-hard historical rigor in tandem with creative use of theory and method.

Beyond collaboration with sister disciplines such as anthropology and philosophy, within historical animal studies as within the history of emotions, the life sciences suggest tantalizing possibilities. Recent studies claim increasing knowledge of the animal mind, including in the ability of animals to express emotion and to

---

28 Matt, Susan J. Current Emotion Research in History: Or, Doing History from the Inside Out. *Emotion Review* 3, no. 1 (January 2011): 117–124, 120–123.
29 Boddice, *The History of Emotions*, 1.
30 On this question see Tague, The History of Emotional Attachment to Animals, 359–360; Webb, Thomas, Chris Pearson, Penny Summerfield and Mark Riley. More-Than-Human Emotional Communities: British Soldiers and Mules in Second World War Burma. *Cultural and Social History* 17, no. 2 (April 2020): 245–262.
31 Tsing, Anna. Unruly Edges: Mushrooms as Companion Species. *Environmental Humanities* 1, no. 1 (November 2012): 141–154, 144; Haraway, Donna. *When Species Meet*. Minneapolis, MN: University of Minnesota Press, 2008, 19. See also Haraway, Donna. *The Companion Species Manifesto: Dogs, People and Significant Otherness*. Chicago, IL: Prickly Paradigm Press, 2003. For multispecies ethnography see Kirksey, S. Eben and Stefan Helmreich. The Emergence of Multispecies Ethnography. *Cultural Anthropology* 25, no. 4 (October 2010): 545–576, especially 553.

read human emotion.[32] But how to use these findings, if at all? Should we take up methods and data offered by ethology, animal welfare science, or neuroscience for interpretation of animal emotions and behavior? While bringing to the table the historical contextualization of scientific claims, can historians take away from engagement with the sciences a greater nuance in our readings of historical material? In Brett L. Walker's words: "the expression of animal emotions [as recorded in our sources] can be read by the historian as a kind of 'text'".[33] For Michael Glover, in drawing on scientific disciplines "the historian is guided to ask better and more just questions about historical documents and to interpret more richly and accurately those documents about animals".[34] Scientific work may point to ways of approaching human-animal emotional communities, for instance in Erica Fudge's use of the research of the animal welfare scientist Françoise Wemelsfelder in relation to the culture of the early modern farmyard.[35] What if we wish to think of animals in emotional communities beyond the human? If we propose that the emotional worlds of animals are both innate and cultural – in that they are shaped by their environment and social interactions – we may draw on scientific studies to conclude that wild elephants or chimpanzees, for example, have a way they are supposed to act and feel in given social situations.[36] We may argue that there is certainly a history of animal emotion, even a multispecies one that can include or not include human beings, we just might not understand it with ease. Arguably, this has strayed beyond the historian's place, unless historical evidence can be found to address precisely such concerns, and this would even then likely be mediated through human perception.

---

[32] Recent popular publications include de Waal, Frans. *Mama's Last Hug: Animal Emotions and What They Teach Us about Ourselves*. London: Granta, 2019; Wohlleben, Peter. *The Inner Life of Animals: Surprising Observations of a Hidden World*. London: Bodley Head, 2017; and, with a focus beyond the mammal, Ackerman, Jennifer. *The Genius of Birds: The Intelligent Life of Birds*. London: Corsair, 2016; and Balcombe, Jonathan. *What a Fish Knows: The Inner Lives of Our Underwater Cousins*. London: Oneworld, 2016; see also Masson, Jeffrey Moussaieff and Susan McCarthy. *When Elephants Weep: The Emotional Lives of Animals*. New York, NY: Delta, 1996; on horses and dogs reading human facial expression see Smith, Amy Victoria, Leanne Proops, Kate Grounds, Jennifer Wathan and Karen McComb. Functionally Relevant Responses to Human Facial Expressions of Emotion in the Domestic Horse (Equus caballus). *Biology Letters* 12, no. 2 (February 2016): 20150907; Siniscalchi, Marcello, Serenella d'Ingeo and Angelo Quaranta. Orienting Asymmetries and Physiological Reactivity in Dogs' Response to Human Emotional Faces. *Learning & Behavior* 46, no. 4 (June 2018): 574–585; on mice see Dolensek, Nejc, Daniel A. Gehrlach, Alexandra S. Klein and Nadine Gogolla. Facial Expressions of Emotion States and their Neuronal Correlates in Mice. *Science* 368, no. 6486 (April 2020): 89–94.
[33] Walker, Brett L. *The Lost Wolves of Japan*. Seattle, WA: University of Washington Press, 2005, 13.
[34] Glover, Michael. A Cattle-Centred History of Southern Africa. In *Nature Conservation in Southern Africa: Morality and Marginality: Towards Sentient Conservation*, Jan-Bart Gewald, Marja Spierenburg, Harry Wels (eds.), 25–47. Leiden: Brill, 2018, 27.
[35] Fudge, *Quick Cattle*, 215–218.
[36] Moss, Cynthia. *Elephant Memories: Thirteen Years in the Life of an Elephant Family*. Chicago, IL: University of Chicago Press, 2001; Whiten, Andrew. Culture Extends the Scope of Evolutionary Biology in the Great Apes. *PNAS* 114, no. 30 (July 2017): 7790–7797.

There are pitfalls to the use of scientific data to the historian. Susan Nance – writing on the history of American circus elephants – advises that her decision to draw from research in ethology and animal welfare science "does require some caution", as environmental factors and differences across time and place will influence the animal and human behaviors under investigation.[37] Furthermore, Erica Fudge warns that: "Science, we must remember, is itself historically and culturally constructed for a particular reason; it is not a source of objective 'truth'".[38] As Saurabh Mishra puts it:

> as someone with a keen interest in the History of Science and Medicine, I cannot help being sceptical of this tendency to take scientific claims about nonhuman behavior at face value. Scientists might claim an ability to reveal the internal lives of nonhumans, but scholars working within the richly interdisciplinary field of 'animal studies' must question these claims.

Our training as historians should lead us to carefully query assertions of scientific fact. Mishra continues: "there is no doubt that [scientific] hypotheses allow historians to tease meanings out of the faint traces of nonhuman lives in the archives [but] an uncritical acceptance of such claims often leads to aberrations", and Mishra then points the finger at Fudge's use of the work of Temple Grandin.[39] Fudge defends her innovative methods in What Was it Like to be a Cow?: "One way to address the issue of the anachronism of applying findings from animal welfare science to past animal behavior is, obviously, to ensure that the animal welfare science fits the historical evidence: that it may, in fact, not be wholly anachronistic".[40]

In this way a critical approach must be the starting point to interdisciplinary methodologies. From there scientific studies might alert us, for instance, to wrong interpretation of animal behavior that comes from accessing nonhuman emotions through our own human point of view. A study in animal behavior science found, for example, that owners often misread guilty behavior in dogs, and were more lenient in punishment of contrite-looking dogs.[41] Should we re-visit Lodovico's fond embrace of the apparently guilty Rubino in light of such conclusions? To what extent should we trust or make use of such scientific research? Might it help us make critical sense of human-canine relations and emotions, and gain new understandings of the past? For Bekoff "the truth is simply that a dog has a rich emotional and cognitive experience of the *dog kind*. Ethological studies and research in social neuroscience

---

[37] Nance, Susan. *Entertaining Elephants: Animal Agency and the Business of the American Circus.* Baltimore, MD: Johns Hopkins University Press, 2013, 13.
[38] Fudge, What Was it Like to be a Cow?, 262.
[39] Mishra, Saurabh. History Writing, Anthropomorphism, and Birdwatching in Colonial India. *History Compass* 15, no. 8 (August 2017): 1–8, 2; referring to Fudge, Milking Other Men's Beasts.
[40] Fudge, What Was it Like to be a Cow?, 275.
[41] Hecht, Julie, Ádám Miklósi and Márta Gácsi. Behavioral Assessment and Owner Perceptions of Behaviors Associated with Guilt in Dogs. *Applied Animal Behaviour Science* 139, no. 1–2 (June 2012): 134–142. Study cited by Tague, The History of Emotional Attachment to Animals, 347.

show that humans aren't the sole occupants of the emotional arena [...] animals talk to us using a myriad of behavior patterns – postures, gestures, and gaits – along with their mouths, tails, eyes, ears, and noses".[42] Scientific data is one thing; theories another; lived, worlded, experience another again. While recognizing the species gap, should we trust the cues that animals give us and our instinctive response to these? This is easier in domestic animals with whom we have co-evolved [→Domestication: Coevolution;→History of Agriculture;→History of Pets], and the emotional lives of the wild, the hostile, the disgusting, the tiny, the strange, and the otherwise inscrutable are yet another matter. In questions of emotion, the mammalian bias that dogs animal studies is at least understandable, and can be explained in part by the apparent relative ease of identifying the emotions of some companion animals. The further from the mammalian and domestic we seek to venture – into territory of birds, reptiles, fish, insects and beyond – the ever greater the limitations to fathoming nonhuman emotion. Dangers of reductive anthropomorphism or misinterpretation abound, and the more we may decide that collaboration with scientists offers the greatest chance of insight into animal emotional worlds. The challenge remains of how to valuably relate to multispecies bodies and minds in the emotional cultures of the present and the past.

4 Implication(s) of the Animal Turn

In this volume's chapter on Social History, Mieke Roscher asks if there can be a social history that goes beyond the human [→Social History]; here we have asked the same of emotional history. Research going forward will surely continue to develop understanding of how people in the past have felt about animals; how they have thought that animals feel; how humans and animals have made each other feel; and how we might know. To quote Boddice, emotions "are not merely the effect of historical circumstances [...] but are active causes of events and richly enhance historiographical theories of causation",[43] and further research can illuminate ways in which: things happening have caused emotions in or about animals; things have taken place because of emotions about animals or as a result of animal emotion. Histories of human-animal emotions can focus on contextually-grounded individuals and groups, taking into account the big picture as well as human and animal individuality and personality, encompassing a range of emotions, everyday and extraordinary, and examining what emotion reflects and makes happen in lived relationships.

In the field of emotions, as elsewhere, thinking with animals exposes fundamental questions, requiring the historian to thoughtfully set out their stall, weighing up research theories and methods, from the humanities and beyond, and probing to the

---

42 Bekoff, *The Emotional Lives of Animals*, 15.
43 Boddice, *The History of Emotions*, 1.

heart of what historians might be for. Kari Weil describes animals as "a limit case for theories of difference, otherness, and power";[44] this applies clearly for work on human and animal emotions and on the history of emotions in particular here. The very thorniness of the questions at stake over human and animal emotions are where productive and exciting future research may lie. What may happen if we, like Haraway, "stay with the trouble",[45] and in this case, for instance, test the limits of conceiving of multispecies social emotional communities in the past?

For Harriet Ritvo, the place of animal studies at the margins (still, though ever less?) and at borders between disciplines – not just within the humanities and social sciences but also with the sciences – is "the source of much of its appeal and power. Its very marginality allows the study of animals to challenge settled assumptions and relationships – to re-raise the largest issues – both within the community of scholars and in the larger society to which they and their subjects belong".[46] There are highly political ethical implications and responsibilities in scrutinizing human emotions towards animals, in following Bekoff to consider emotions as "the glue that binds",[47] and in acceptance of animal emotion and its value.

Selected Bibliography

Bekoff, Marc. *The Emotional Lives of Animals*. Novato, CA: New World Library, 2007.
Boddice, Rob. *The History of Emotions*. Manchester: Manchester University Press, 2018.
Cockram, Sarah. Sleeve Cat and Lap Dog. Affection, Aesthetics and Proximity to Companion Animals in Renaissance Mantua. In *Interspecies Interactions Animals and Humans between the Middle Ages and Modernity*, Sarah Cockram, Andrew Wells (eds.), 34–65. London: Routledge, 2018.
Dror, Otniel E. The Affect of Experiment: The Turn to Emotions in Anglo-American Physiology, 1900–1940. *Isis* 90, no. 2 (June 1999): 205–237.
Fudge, Erica. What Was it Like to be a Cow? History and Animal Studies. In *The Oxford Handbook of Animal Studies*, Linda Kalof (ed.), 258–278. Oxford: Oxford University Press, 2017.
Pearson, Chris. "Four-Legged Poilus": French Army Dogs, Emotional Practices and the Creation of Militarised Human-Dog Bonds, 1871–1918. *Journal of Social History* 52, no. 3 (Spring 2019): 731–760.
Plamper, Jan. *The History of Emotions: An Introduction*. Translated by Keith Tribe. Oxford: Oxford University Press, 2015.
Richardson, Angelique. *After Darwin: Animals, Emotions, and the Mind*. Amsterdam: Rodopi, 2013.
Tague, Ingrid H. The History of Emotional Attachment to Animals. In *The Routledge Handbook of Animal–Human History*, Hilda Kean, Philip Howell (eds.), 345–366. London: Routledge, 2019.

---

44 Weil, Kari. *Thinking Animals: Why Animal Studies Now?* New York, NY: Columbia University Press, 2012, 5.
45 Haraway, Donna. *Staying with the Trouble: Making Kin in the Chthulucene*. Durham, NC: Duke University Press, 2016.
46 Ritvo, Harriet. On the Animal Turn. *Daedalus* 136, no. 4 (Fall 2007): 118–122, 122.
47 Bekoff, *The Emotional Lives of Animals*, 15.

Webb, Thomas, Chris Pearson, Penny Summerfield and Mark Riley. More-Than-Human Emotional Communities: British Soldiers and Mules in Second World War Burma. *Cultural and Social History* 17, no. 2 (April 2020): 245–262.

Part V: **History of Human-Animal Interactions**

Amir Zelinger
History of Pets

1 Introduction and Overview

The first systematic discussion of pets as a subject of historical scholarship appeared more than forty years ago – in the form of a parody. In an article published in 1974 in the *Journal of Social History*, "Charles Phineas" (pseudonym) stated that historians should start investigating how pets became "a vital focus of modern society." For readers then, this claim was clearly not meant to be taken seriously: "Dr. Phineas," introduced as a researcher at "Boxer College," wrote that his article "is based on no factual material whatsoever," and that, should records on the history of people's relations with pets become available, they "would lend themselves more to anal than to oral history." Pets, he predicted, would be the next fad in social history [→Social History] once other marginalized groups such as left-handlers or homosexuals lost their allure, namely due to "personal tensions among historians" (at least regarding the latter group). Despite the parodic nature of this initial pseudo-discussion of the history of pet-keeping, the question at the root of its inquiry – how pets became "a vital focus of modern society" – points at the key feature of this history as actually written in the decades following Phineas' joke: its almost exclusive association with a specific historical period and culture, specifically that of Western modernity. Among possible reasons for the emergence of pet-keeping, Phineas cited urbanization, the rise of the middle class and the nuclear family, and the expansion of capitalist consumerism.[1] Most historians of pet-keeping followed his approach inasmuch as they connected their subject matter to core developments in modern Western society.

In the following analysis of the history of pet-keeping, I will reveal how its association with Western modernity has proven enormously influential in terms of how this history has been written thus far. Most fundamentally, it has informed the definition most pet historians have chosen for their subject matter. Animal historians are inclined to blur the boundaries between different categories (e.g. "human" and "animal" or "nature" and "culture"), yet works on the history of pets nearly always include in their introductory sections a rigorous definition of the term that is the focus of their attention. Scholars have mostly defined "pets" as animals that live at home and are kept for pleasure and recreation rather than practical purposes, are close companions of their human owners, and are treated by owners with friendliness and care that privileges them over all other types of animals. According to the

---

[1] Phineas, Charles. Household Pets and Urban Alienation. *Journal of Social History* 7, no. 3 (Spring 1974): 338–343.

prominent narrative about the rise of pet-keeping in modern Western society, certain conditions in this society (i.e. a more rigid separation between the public and private spheres and between work and leisure) rendered it possible for certain animals to escape the exploitation that characterizes people's treatment of most other animals to become classified as pets per the aforementioned definition.

In this chapter, I contend that the association between pet-keeping and Western modernity, along with the rigid definition of "pets" this association favors, has influenced most pet historians' approaches and hence the kind of animal history they decided to write. If a key objective of animal history is to show that human societies were never purely human and that they evolved and transformed through highly intimate relations with nonhuman creatures,[2] then pet-keeping should have presumably become one of its most prominent topics. According to the narrative around the rise of pet-keeping in modern times, the last few centuries have witnessed no other animals integrate more closely into humans' lives than those they opted to keep as pets. While livestock animals vanished from the immediate surroundings of most people in the modern Western world, other animals were invited to enter their homes and bedrooms and become humans' closest companions.[3] In essence, the history of pet-keeping is a history of human–animal entanglement. Yet so far, this history has been written in a human-centered way. Rather than emphasizing the obvious association between humans and animals that owner-pet relations embodied, it has focused on how pet-keeping merely reflected changes within human society itself – those already captured in Phineas' parody. Despite its potential, the history of pet-keeping has not been told as a history of intimate human–animal relations; instead, it looked at animals to explain what has happened among humans.

2 Topics and Themes

Pets in Pre-Modern Times

The link between pets and Western modernity is so prominent that even works on pet-keeping in pre-modern times were largely written against the backdrop of this association. These works sought to show that pets – corresponding to the above definition of animals kept purely for pleasure and as companions of their owners who treat them with great care – already existed in historical periods that had long preceded modern society. For example, Belgian historian Liliane Bodson opens her article about pets in Ancient Greece and Rome with the statement that "pet-keeping was a widespread and well-accepted phenomenon in classical antiquity." She goes

---

2 Walker, Brett L. Animals and the Intimacy of History. In *The Oxford Handbook of Environmental History*, Andrew C. Isenberg (ed.), 52–75. Oxford: Oxford University Press, 2014.
3 Fudge, Erica. *Animal*. London: Reaktion Books, 2002, 28.

on to show that already in that period, people from all social backgrounds had developed attitudes towards certain animals, mostly dogs and cats, that transformed those creatures into full-fledged pets. As Bodson notes, ancient pet-keepers defined their relationships with such animals "in the same way" as "modern dog owners." They were highly sentimental when writing about these relationships (e. g., in gravestone epitaphs), referring to their animal companions as sources of great joy and speaking of the reciprocal affection between themselves and their animals and the devotion they felt for each other. Similar to modern pet-keepers, these people frequently anthropomorphized their pets, most notably calling them "children" or "foster children." Hence, Bodson concludes, the "strong and selfless affection" at the heart of pet-keeping "should no longer be considered a uniquely modern phenomenon."[4]

This statement was reinforced by the most comprehensive study on medieval pets. In her book of the same name, Kathleen Walker-Meikle wrote about animals that lived in the Middle Ages but were equally suited to the modern definition of pets. These animals, which could be dogs, cats, monkeys, or caged birds, had "no function other than to offer companionship," were "pampered and treated like members of the household," and received "as much care as that accorded to humans." Their owners belonged to two groups – upper-class women and clerics – whose unique living circumstances made it possible for them to cultivate relationships with animals that were analogous to those with pets. Because people from these groups spent most of their time indoors and were secluded from the outside world where humans' relations with animals were based on utility and exploitation (i.e., farming and hunting), they were able to keep animals who were their friends and provided nothing but delight. Living in a clearly distinct private sphere, detached from lifestyles that necessitated abusing and killing animals for material reasons, these people already fulfilled in the Middle Ages crucial conditions that came to be associated with the rise of pet-keeping in the modern age.[5]

---

[4] Bodson, Liliane. Motivations for Pet-Keeping in Ancient Greece and Rome: A Preliminary Survey. In *Companion Animals and Us: Exploring the Relationships between People and Pets*, James A. Serpell, Elizabeth S. Paul, Anthony Podberscek (eds.), 27–41. Cambridge: Cambridge University Press, 2000. Another animal historian of ancient Greece, Louise Calder, wrote that Greeks occasionally described a "two-way relationship" with pet dogs "in terms that could be those of [...] any modern dog-owner." Calder, Louise. Pet and Image in the Greek World: The Use of Domesticated Animals in Human Interaction. In *Interactions between Animals and Humans in Graeco-Roman Antiquity*, Thorsten Fögen, Edmund Thomas (eds.), 61–88. Berlin and Boston, MA: De Gruyter Oldenbourg, 2017, 64.
[5] Walker-Meikle, Kathleen. *Medieval Pets*. Woodbridge: Boydell Press, 2012. Esther Pascua similarly claimed that in the Middle Ages "certain animals emerged as something similar to the modern pet, lacking functional use and forming part of the inner circle of the family." Pascua, Esther. From Forest to Farm and Town: Domestic Animals from ca. 1000 to ca. 1450. In *A Cultural History of Animals in the Medieval Age*, vol. 2, Brigitte Resl (ed.), 81–102. Oxford: Berg, 2007, 100. Sophia Menache argued that during the late Middle Ages, only aristocrats kept pets because only they were engaged in an activity – hunting – that used certain animals (dogs) for recreational, not economical, reasons. Medieval ar-

Not all historians who wrote explicitly about "pets" in antiquity or the Middle Ages concurred that such animals corresponded to the modern definition of the term though. Michael MacKinnon, in his article "Pets" for the *Oxford Handbook of Animals in Classical Thought and Life*, points out that "While certainly cases of pet-keeping in antiquity exist, caution should be exercised in linking such bonds too closely with modern culture." He cites the modern definition of pets in asserting that the pets about which he wrote corresponded only partially to this definition.[6] Ingvild Saelid Gilhus avers that "Graeco-Roman society was not a pet-keeping society in a similar way to [...] advanced capitalist societies today" and thus recommends avoiding the term "pets" altogether in discussions about human–animal relations in antiquity. She acknowledges that ancient Greeks and Romans sometimes developed intimate relationships with individual animals, but to avoid confusion with the "modern pet culture," these animals should not be called pets but rather, more matter-of-factly, "personal animals."[7] However, while these scholars allege that no real pets existed in the periods they studied but only animals either resembling pets in some respects or otherwise quasi-pets, the association of pets with modern society still constitutes the background against which their narratives are told. If the earlier cited historians opened their books and articles by stating that ancient or medieval pets were akin to modern pets, then these more skeptical historians asserted that ancient and medieval pets were not like modern pets. In the scholarship about them, pre-modern pets have been consistently compared with the animals who succeeded them in later periods in history.

The Rise of Modern Pets: Alienation and Domination

Almost all pet historians of the modern period agree with the skeptical pet historians of antiquity and the Middle Ages that prior to modernity, there were no real pets. In fact, this assumed absence of pets in the pre-modern past marks a point of departure in most research on modern pets. Such scholarship has generally been dedicated to

---

istocratic hunting thus paved the way for non-instrumental human–animal relations based on affection and friendship. Menache, Sophia. Hunting and Attachment to Dogs in the Pre-Modern Period. In *Companion Animals and Us*, Anthony L. Podberscek, Elizabeth S. Paul, James A. Serpell (eds.), 42–60. Cambridge: Cambridge University Press, 2000.

6 MacKinnon, Michael. Pets. In *The Oxford Handbook of Animals in Classical Thought and Life*, Gordon Lindsay Campbell (ed.), 269–281. Oxford: Oxford University Press, 2014.

7 Gilhus, Ingvild Saelid. *Animals, Gods, and Humans: Changing Attitudes to Animals in Greek, Roman, and Early Christian Thought*. London and New York, NY: Routledge, 2006, 29. For a similarly skeptical position regarding the Middle Ages see O'Connor, Terry. Animals in Urban Life in Medieval to Early Modern England. In *The Oxford Handbook of Zooarchaeology*, Umberto Albarella, Mauro Rizzetto, Hannah Russ, Kim Vickers, Sarah Viner-Daniels (eds.), 214–229. Oxford: Oxford University Press, 2017, 220.

explaining "the emergence of modern pet-keeping";[8] its purpose is to demonstrate how certain developments in the modern world evoked "significant changes in the status of particular domestic animals in Western cultures":[9] to relations that no longer were characterized by human exploitation but by affection and companionship, hence aligning with our present-day definition of pets.

The explanations offered for the rise of modern pets comprise a broad narrative that places this development in the context of more general transformations underpinning the modern period in terms of relations between human society and the natural world. According to this narrative, pet-keeping is a consequence of two fundamental features of people's relations with nature in Western modernity: alienation and domination. Art critic John Berger once wrote that the keeping of pets is part of the "withdrawal into the private small family unit, decorated or furnished with mementoes from the outside world, which is such a distinguishing feature of consumer societies."[10] Many pet historians have adopted this argument to develop a kind of "compensation theory" for the rise of modern pet-keeping. They contend that pets became widespread because these animals compensated for the alienation of more and more people in modern society, especially those living in urban environments, from the natural world and the greater part of the animal kingdom. In this case, pets were like surrogates – they replaced for modern, urbanized people the intimate contact with nature and the "lost world of human/animal communion" allegedly experienced by their predecessors in the pre-modern past.[11]

At the same time, the domination that people in Western society gradually acquired over the natural world thanks to scientific and technological progress made it possible to view nature not as a constant threat but as an object of affection. When nature was tamed, it could be loved; pets were the embodiment of this tamed nature. As representatives of the animal world that were subjected to complete

---

8 Ritvo, Harriet. The Emergence of Modern Pet-Keeping. In *Social Creatures: A Human and Animal Studies Reader*, Clifton P. Flynn (ed.), 96–106. New York, NY: Lantern Books, 2008.
9 Kean, Hilda. The Moment of Greyfriars Bobby: The Changing Cultural Position of Animals, 1800–1920. In *A Cultural History of Animals in the Age of Empire*, vol. 5, Kathleen Kete (ed.), 25–46. Oxford: Berg, 2007, 35.
10 Berger, John. Why Look at Animals? In *About Looking*, John Berger, 1–28. New York, NY: Vintage, 1980, 14.
11 Raber, Karen. From Sheep to Meat, From Pets to People: Animal Domestication, 1600–1800. In *A Cultural History of Animals in the Age of Enlightenment*, vol. 4, Matthew Senior (ed.), 73–99. Oxford: Berg, 2007; Steinbrecher, Aline. Die gezähmte Natur in der Wohnstube: Zur Kulturpraktik der Hundehaltung in frühneuzeitlichen Städten. In *Die Natur ist überall bey uns: Mensch und Natur in der Frühen Neuzeit*, Aline Steinbrecher, Sophie Ruppel (eds.), 125–142. Zürich: Chronos, 2009; DeMello, Margo. *Animals and Society: An Introduction to Human-Animal Studies.* New York, NY: Columbia University Press, 2012, 153; Brown, Frederick L. *The City Is More than Human: An Animal History of Seattle.* Seattle, WA and London: University of Washington Press, 2016, 150.

domination by humans, they managed to become friends for humans.[12] It is in this context of an increasingly dominated nature that new sensibilities towards certain animals emerged, reversing centuries of despising them as brute creations and placing them on a path to be transformed into people's most cherished companions. Already in the early modern period people began to ascribe to some domesticated animals mental and emotional characteristics that rendered them capable of becoming partners to humans. These animals were now suddenly appreciated for their intelligence and trustworthiness, and they were said to have "character and individual personality."[13] This development seems to have intensified towards the end of the eighteenth-century with the "emergence of a radically new vision of animal emotions and thus human–animal relationships." Now, a new culture of sensibility was on the rise that stressed the superiority of feelings and even passions over reason. This culture prompted a new discourse on the emotional capacity of animals. Certain domesticated animals, most notably dogs, were said to have in their nature the capacity to experience such complex social emotions as affection and attachment to individuals of other species and could thus form genuine friendships with humans. As tamed creatures that expressed unconditional loyalty to humans, they were now on their way to become truly their best friends [→History of Emotions].[14]

This process through which humans' love of animals followed on their domination peaked in the nineteenth century, when a rigid hierarchy between "good" tame animals and "bad" unruly animals was constructed. As Harriet Ritvo revealed with regard to Victorian England, people in this period sought the companionship of animals who, in their eyes, had a subordinate character and subjugated themselves to human needs. By contrast, animals that resisted domination by humans or continued to pose a threat to humans (e.g., wild predatory animals) were considered the most despicable representatives of the animal kingdom. These were the animals that, for nineteenth-century civilized societies, were the anti-pets – the ones humans rejected most vehemently as unworthy of becoming their companions.[15] This connection between pet love and pet domination was even more pronounced in the United States during the same period. There, a "domestic ethic of kindness to animals" developed, that, according to Katherine Grier, conflated caring for pets with a perception of them as defenseless creatures that required human stewardship and protec-

---

[12] Ritvo, The Emergence of Modern Pet-Keeping, 101. For an anthropological summary of this hypothesis see Tuan, Yi-Fu. *Dominance & Affection: The Making of Pets.* New Haven, CT: Yale University Press, 1984.
[13] Thomas, Keith. *Man and the Natural World: A History of the Modern Sensibility.* New York, NY: Pantheon Books, 1983, 100–120.
[14] Tague, Ingrid H. The History of Emotional Attachment to Animals. In *The Routledge Companion to Animal–Human History*, Hilda Kean, Philip Howell (eds.), 345–366. London and New York, NY: Routledge, 2018.
[15] Ritvo, Harriet. *The Animal Estate: The English and Other Creatures in the Victorian Age.* Cambridge, MA: Harvard University Press, 1987, 15–30. See also Amato, Sarah. *Beastly Possessions: Animals in Victorian Consumer Culture.* Toronto: University of Toronto Press, 2015, 39–49.

tion. In this new moral ideology, human–pet relations came to resemble ideal family relations as envisaged by middle-class Americans, ones that were based on the dependence of the weak upon the strong and in which the benevolence of parents was reciprocated by obedient and disciplined children. Similar to love for children, love for pets was grounded in the obligations of the powerful towards the subordinate.[16]

3 Methods and Approaches

The inclination in scholarship about modern pet-keeping to explain its rise by referring to another, much larger phenomenon such as the new relationship of domination between people and nature is indicative of the fact that, surprisingly, only a small portion of this work has focused on actual human–pet relations as its main topic. Because many historians associate the emergence of modern pet-keeping with other, more prominent developments that lie at the heart of modern history, they tend to build their narrative around these other events instead of pet-keeping itself. For instance, in *Pets and Social Change in Eighteenth-Century Britain*, Ingrid Tague explores the spread of pet-keeping by connecting it to such major transformations as the rise of Britain as an empire, the emergence of a consumer culture in this country, and the intellectual revolution generated by the Enlightenment. She asks how the new culture of pet-keeping was reflective of these transformations and in turn helped people ponder them and the "major problems of their day" such as racial, gender, and class boundaries. In this narrative, the development of human–pet relations is not an interesting subject in its own right but only inasmuch as it contributes to debates about changing relations within human society itself.[17]

Similarly, Kathleen Kete interprets the proliferation of pet-keeping in her book about *Petkeeping in Nineteenth-Century Paris* as an expression of a much more pivotal transformation in contemporary Western society – the rise of bourgeois culture. Pet-keeping is an interesting historical theme because it encapsulates many of the core features associated with this emergent culture such as withdrawal into the private sphere, isolation of the family home against the outside world, and sentimentalization of the relations within this small family universe. Caring for pets in the home was, according to Kete, a statement made by the bourgeoisie about its unique way of life – designed as a response to "the problems of modernity" and the "heart-

---

[16] Grier, Katherine C. *Pets in America: A History*. Chapel Hill, NC: University of North Carolina Press, 2006, 127–181. See also Pearson, Susan J. *The Rights of the Defenseless: Protecting Animals and Children in Gilded Age America*. Chicago, IL and London: University of Chicago Press, 2011, 32–43.
[17] Tague, Ingrid H. *Animal Companions: Pets and Social Change in Eighteenth-Century Britain*. University Park, PA: Pennsylvania State University Press, 2015.

less world" they created. This is what makes it such an important topic of modern history.[18]

The emphasis in the historiography of modern pet-keeping on how people used pet-keeping to reflect upon the radical transformations in their societies is also connected to a certain perception in human–animal studies that regards pets as mere symbols rather than "real" animals. As implied by the abovementioned "compensation theory," modern pets are mere surrogates for the "real" animals with whom humans had intimate contact in the past when they were immersed in nature and the animal world.[19] According to Susan Pearson and Mary Weismantel, it is impossible to have with pets the "lived experience of quotidian and extraordinary interactions" that characterize relations between humans and animals in traditional societies; instead, pets serve an "almost exclusively symbolic function."[20] This is one of the reasons why pet historians usually focus on representations of pets in such sources as paintings, children's books, lexicons, and natural history treatises, whereas comparatively few of them refer to ego-documents that record actual interactions between owners and their animal companions. When it comes to pets, historians are not concerned with "validating past lives" by analyzing the actions and experiences of the animals they study – a key goal of animal history today as defined by Hilda Kean. In contrast to other branches of animal history, most historiography on pet-keeping does not "acknowledge interest in an animal existence,"[21] and it fails to highlight the importance of actual human–animal relations in the history of human societies; it is content with revealing how representations of animals were used by these societies to contemplate their own, human-centered problems. This is a key consequence of the association created in this historiography between pets and Western modernity.

Although this association has become quite powerful in historical scholarship on pets, it has been questioned by scholars from other disciplines, who, unlike most historians, focused on actual human–animal relations in their studies. Anthropologists especially challenged the idea that pet-keeping is unique to modern Western societies. They found that even people in hunting and gathering societies have often treated animals in ways fundamentally similar to the standard treatment of pets in the modern West. In many Amerindian tribes, for instance, people regularly capture young wild animals that they raise as their tamed companions. They do so with as

---

18 Kete, Kathleen. *The Beast in the Boudoir: Petkeeping in Nineteenth-Century Paris*. Berkeley, CA: University of California Press, 1994. For the nexus between pet-keeping and nineteenth-century bourgeois culture, see also Frykman, Jonas and Orvar Löfgren. *Culture Builders: A Historical Anthropology of Middle-Class Life*. New Brunswick, NJ: Rutgers University Press, 1987, 78–79.
19 See Fudge, Erica. *Pets*. Stocksfield: Acumen, 2008, 23–25.
20 Pearson, Susan J. and Mary Weismantel. Does the Animal Exist? Toward a Theory of Social Life with Animals. In *Beastly Natures: Animals, Humans, and the Study of History*, Dorothee Brantz (ed.), 17–37. Charlottesville, VA: University of Virginia Press, 2010, 30–31.
21 Kean, Hilda. Finding a Man and His Horse in the Archive. In *Animal Biography: Re-framing Animal Lives*, André Krebber, Mieke Roscher (eds.), 41–55. Cham: Palgrave Macmillan, 2018.

much affection and devotion as expressed by modern pet owners towards their favorites. They care for them in ways that ensure the animals' greatest well-being and happiness, and they caress, kiss, and coddle them as if they were their own children. Sometimes they even dress them and give them human names.[22]

The anthropological challenge to the idea that pet-keeping is a unique phenomenon of Western modernity presents an opportunity to write a more multidimensional history of pets. In this history, pets are not the byproduct of structural developments associated with the emergence of modern society but instead characterize a wide variety of creatures, who were transformed into human companions for myriad reasons and motivations. As indicated in the introduction, the association of pets with Western modernity went hand-in-hand with the rigid definition of the term "pets" that most historians of pet-keeping readily adopted. Some pet historians, however, embraced a more open definition of pets,[23] referring to certain animals as pets on the ground that these animals were treated with great kindness and care, even if they did not fully meet all criteria of the standard definition of the term. These historians' readiness to include unusual pets and, most importantly, unusual human–pet relations in their narratives, underscores the potential to write an alternative history of pet-keeping – one in which its emergence is not interpreted as a mere reflection of the transformations that made (Western) human society modern.

Unusual pets that have garnered a certain amount of attention in histories of pet-keeping are so-called *useful pets* – animals that were treated and viewed by their owners as companions but at the same time served material interests for humans and were in one way or the other seen productive. Historians of antiquity and the Middle Ages have often referred to the porous boundary between companion animals and livestock in these periods when seeking to justify their hesitation to speak of genuine "pets" in their works.[24] Narratives about the rise of modern pet-keeping, by contrast, identify the demarcation between companionship and usefulness in people's relationships with animals as a critical step in this process. Only when, in the last few centuries, certain animals were no longer kept for utilitarian reasons but only for pleasure could they become real, "pure pets."[25] Other historians, how-

---

22 Serpell, James A. Pet-Keeping in Non-Western Societies: Some Popular Misconceptions. *Anthrozoös* 1, no. 3 (January 1987): 166–174; Erikson, Philippe. On Native American Conservation and the Status of Amazonian Pets. *Current Anthropology* 38, no. 3 (June 1997): 445–446; Erikson, Philippe. The Social Significance of Pet-Keeping among Amazonian Indians. In *Companion Animals and Us: Exploring the Relationships between People and Pets*, James A. Serpell, Elizabeth S. Paul and Anthony Podberscek (eds.), 7–26. Cambridge: Cambridge University Press, 2000.
23 For an argument about the need to embrace a more open definition of "pets," see Eddy, Timothy J. What Is a Pet? *Anthrozoös* 16, no. 2 (April 2003): 98–105.
24 See MacKinnon, Pets, 279; O'Connor, Animals in Urban Life, 214–216.
25 Ritvo, The Emergence of Modern Pet-Keeping, 99; Grier, Pets in America, 201–207; Möhring, Maren. Das Haustier: Vom Nutztier zum Familientier. In *Das Haus in der Geschichte Europas: Ein Handbuch*, Joachim Eibach, Inken Schmidt-Voges (eds.), 389–405. Berlin and Boston, MA: De Gruyter Oldenbourg, 2015.

ever, tried to complicate this dichotomous narrative by pointing out that also in modern human–animal relations, usefulness and companionship were not always mutually exclusive but sometimes even complementary.

Several studies have shown, for example, that the keeping of small or medium-sized farm animals like chickens and goats in the home or its immediate surroundings was a common phenomenon in urban households until at least the first decades of the twentieth-century – especially among poor and working-class families.[26] As Andrea Gaynor points out in her works about the history of urban farming in Australia, these animals were perceived by their owners as "legitimate occupants" of the "domestic sphere." As such they upset middle class attempts to construct the home as a space free of work and production – and as the dominion of the useless, purely pleasure-affording pet. But, as Gaynor demonstrated, these domestic productive animals were also not purely productive. While as providers of milk and eggs they were crucial for the economic survival of their low-income owners, human relationships with these animals were also characterized by emotional attachment. In accounts of owners about their daily work with these animals, they often refer to them as "friend[s] and family member[s]," emphasizing how much they loved them. Gaynor thus concludes that these workers–livestock relations were complex social relations in which material, recreational, and sentimental elements co-existed.[27] Such perceptions call into question the more simplified narrative about the duality of useful animals and pets in modernity.

This duality is challenged even more strongly by historical accounts indicating that usefulness was occasionally the reason for pet-keeping. Even in modern urban settings, the appreciation of people for the service they received from animals was often what motivated them to keep animals in their homes and treat them as companions [→Urban (and Rural) History]. Frederick Brown argued that in early twentieth-century Seattle, a prime motive for the keeping of dogs and cats was the work they did for their owners in the home: dogs guarded them and cats protected their food storages against mice. As Brown claimed, dogs and cats were not pure pets – they "were often beloved, yet they typically worked." Even more importantly, he notes that these two aspects, love and work, were not diametrically opposed in contemporary people's minds but rather complementary. In newspaper ads, for instance, dogs were presented in the same sentence as being ideal companions for chil-

---

[26] Zelinger, Amir. *Menschen und Haustiere im Deutschen Kaiserreich: Eine Beziehungsgeschichte*. Bielefeld: transcript, 2018, 50–58; Dyl, Joanna. The War on Rats versus the Right to Keep Chickens: Plague and the Paving of San Francisco, 1907–1908. In *The Nature of Cities: Culture, Landscape, and Urban Space*, Andrew C. Isenberg (ed.), 38–61. Rochester, NY: University of Rochester Press, 2006, 48–49; Gaynor, Andrea. Fowls and Contested Productive Spaces of Australian Suburbia, 1890–1990. In *Animal Cities: Beastly Urban Histories*, Peter Atkins (ed.), 205–219. Farnham and Burlington, VT: Ashgate, 2012.
[27] Gaynor, Andrea. Animal Agendas: Conflict over Productive Animals in Twentieth-Century Australian Cities. *Society & Animals* 15, no. 1 (2007): 29–42.

dren as well as excellent watchdogs. The affection these animals received was not the antithesis of the work they performed but rather its consequence.[28] This combination was even more meaningful in relations between African-American slaves and their dogs. As John Campbell contended, slaves loved dogs because they received from their animal friends "concrete benefits" in a life marred by hardship. Dogs helped slaves hunt opossums and raccoons and steal from their masters foods that improved their meager diets; they protected their guardians against dangerous animals that invaded their quarters; and they warned them against the approach of slave catchers when they sought to escape. Due to such "vital assistance," these dogs were involved in their guardians' life to a degree perhaps unimaginable by their masters or, for that matter, white middle-class families. Dogs were, as Campbell puts it, slaves' "constant companions." They were the recipients of an especially "deep affection" from slaves because they were critically useful to them.[29]

Other unusual pets that transgress the narrow boundaries set by the standard definition of "pets" are animals that are treated by humans as their companions despite spending much of their lives in public spaces instead of in people's homes. Works that include discussions of the relations between people and such public pets also help outline a more complex historical picture of pet-keeping by loosening its association with the emergence of the nuclear family home. Several historians of the Ottoman Empire, for example, wrote about the intimate interactions and mutual attachment between people and the street dogs that were ubiquitous in cities like Istanbul and Cairo until at least the beginning of the twentieth century. Although the cities' inhabitants were not these dogs' owners, they were their caregivers: like bona fide pet-keepers they provided them regularly with food and water and were especially benevolent towards them and attentive to their needs.[30]

However, in the early twentieth century, free-ranging dogs were omnipresent not only in the Middle East but also in the streets of Western cities. Brown stated that

---

[28] Brown, *The City Is More than Human*, 151–155. For a discussion of the combination of work and companionship as part of the complexity of pet-keeping, see Hart, Lynette A. Pets along a Continuum: Response to What Is a Pet. *Anthrozoös* 16, no. 2 (June 2003): 118–122.

[29] Campbell, John. My Constant Companion: Slaves and their Dogs in the Antebellum South. In *Working Toward Freedom: Slave Society and Domestic Economy in the American South*, Larry E. Hudson (ed.), 53–76. Rochester, NY: University of Rochester Press, 1994. See also Rice, Kym S. Dogs. In *World of a Slave: Encyclopedia of the Material Life of Slaves in the United States*, vol. 1, Kym Rice, Martha B. Katz-Hyman (eds.), 181–183. Santa Barbara, CA: Greenwood, 2011; Meacham, Sarah Hand. Pets, Status, and Slavery in the Late-Eighteenth-Century Chesapeake. *Journal of Southern History* 77, no. 3 (August 2011): 521–554.

[30] Gündoğdu, Cihangir. The State and the Stray Dogs in Late Ottoman Istanbul: From Unruly Subjects to Servile Friends. *Middle Eastern Studies* 54, no. 4 (February 2018): 555–574, 557; Mikhail, Alan. A Dog-Eat-Dog Empire: Violence and Affection on the Streets of Ottoman Cairo. *Comparative Studies of South Asia, Africa, and the Middle East* 35, no. 1 (May 2015): 76–95, 81–83; Pinguet, Catherine. Istanbul's Street Dogs at the End of the Ottoman Empire: Protection or Extermination. In *Animals and People in the Ottoman Empire*, Suraiya Faroqhi (ed.), 353–371. Istanbul: Eren, 2010.

until the post-World War II period, the streets of Seattle were "dog commons": dogs roaming freely in public areas were an integral part of urban life – and, for many citizens, a welcome one. As Brown argued, this freedom of movement granted to dogs helped them "establish a type of companionship" with humans that was not exclusively connected to the home but also "to the neighborhood." In the community, the dogs had friendly interactions with many people who were not their owners. Some dogs were playmates of the children in the neighborhood, not the exclusive partners of just one person or family. These public human–canine companionships started to disappear only after World War II; at this point, the rise of a more home-centered middle-class family culture in the United States heralded the transformation of dogs, at least in hegemonic perceptions of pet-keeping, into genuine household pets.[31] Interestingly, however, even the "domestic ethic of kindness to animals" that originated in the middle-class family home sometimes crossed the boundaries of this home to extend to wild animals living outdoors. Such was the case, for instance, with the gray squirrel, an animal that became the public pet par excellence in the United States. According to Etienne Benson, a drastic increase in the number of squirrels in American cities in the late nineteenth-century resulted from people's desire to integrate these animals into the urban community by nurturing their populations in public parks and other recreational green spaces. There, squirrels were supposed to become a source of entertainment and pleasure for urbanites – just like the pets in their homes. In return, they were also cared for like pets. Benson explained that a main driver behind the popularity of squirrel-feeding in parks was that it gave devoted feeders the opportunity to "demonstrate their generosity towards the needy" and their "compassion for the weak," in the same manner that pet-keepers expressed their paternalistic benevolence towards the defenseless creatures that lived with them at home.[32] As this story shows, sometimes the same values that dictated people's relations with pets in their homes also informed their treatment of animals they encountered in the outside world.

4 Implication(s) of the Animal Turn

In *When Species Meet*, Donna Haraway wrote that "'companion species' does not mean smallish animals treated like indulged children in fur-coats (or fins or feathers) in late imperial societies. Companion species is a permanently undecidable category, a category-in-question that insists on the relation as the smallest unit of being and of

---

[31] Brown, *The City Is More than Human*, 165–180. See also Sanders, Jeffrey C. Animal Trouble and Urban Anxiety: Human-Animal Interaction in Post-Earth Day Seattle. *Environmental History* 16, no. 2 (April 2011): 226–261, 238–250.
[32] Benson, Etienne. The Urbanization of the Eastern Gray Squirrel in the United States. *Journal of American History* 100, no. 3 (December 2013): 691–710.

analysis."[33] By connecting its subject matter to structural developments in modern Western society, the historiography of pet-keeping has largely focused on the stereotyped image of pets as "smallish animals" and surrogate children, rather than contributing to an animal history that is first and foremost a history of actual human–animal relations. In the historical scholarship about them, pets are still presented as a byproduct of Western affluence and domination of nature; accounts have exhibited far less interest in the actual everyday interactions these animals had with their human owners and how the patterns of such interactions morphed throughout history. This diminished focus on relations has made pet-keeping historiography more anthropocentric in its approach than most other branches of animal history. The emphasis on how pet-keeping reflected major transformations within human society means that this historiography treats animals in a fashion typical of academic scholarship in the days before the *animal turn* when, as Cary Wolfe described, animals were regarded merely "as a vehicle or symptom for some other, deeper problematic: often race or gender, sometimes class or nationality."[34] It is hence unsurprising to read the following statement from one of the most prominent pet-keeping historians today: "As a historian, then, I focus unapologetically on humans. I am interested in animals primarily because of their impact on human life, rather than the other way around – not because I think animals are unimportant, but because ultimately the study of history must be a study of humans."[35]

In this respect, historians who embrace a more open definition of pets and include in their works unusual pets pose a welcome challenge to the standard composition of the history of pet-keeping – not only because they add complexity to this history, but also because they help to reduce its anthropocentrism. Their works regard pets not as symptoms of some "deeper problematic" that merely touches "human life" but as real animals whose intricate and often paradoxical relations with humans had far-reaching outcomes both for them and for the people who were their companions and caregivers. These narratives hint at the possibility of finally writing the history of pet-keeping as one of human–animal intimacy.

Selected Bibliography

Bodson, Liliane. Motivations for Pet-Keeping in Ancient Greece and Rome: A Preliminary Survey. In *Companion Animals and Us: Exploring the Relationships between People and Pets*, James A. Serpell, Elizabeth S. Paul, Anthony Podberscek (eds.), 27–41. Cambridge: Cambridge University Press, 2000.

---

[33] Haraway, Donna J. *When Species Meet*. Minneapolis, MN: University of Minnesota Press, 2008, 165.
[34] Wolfe, Cary. Moving Forward, Kicking Back: The Animal Turn. *Postmedieval: A Journal of Medieval Cultural Studies* 2, no. 1 (March 2011): 1–12, 2.
[35] Tague, *Animal Companions*, 9.

Brown, Frederick L. *The City Is More than Human: An Animal History of Seattle*. Seattle, WA and London: University of Washington Press, 2016.
Campbell, John. My Constant Companion: Slaves and their Dogs in the Antebellum South. In *Working Toward Freedom: Slave Society and Domestic Economy in the American South*, Larry E. Hudson (ed.), 53–76. Rochester, NY: University of Rochester Press, 1994.
Grier, Katherine C. *Pets in America: A History*. Chapel Hill, NC: University of North Carolina Press, 2006.
Kete, Kathleen. *The Beast in the Boudoir: Petkeeping in Nineteenth-Century Paris*. Berkeley, CA: University of California Press, 1994.
MacKinnon, Michael. Pets. In *The Oxford Handbook of Animals in Classical Thought and Life*, Gordon Lindsay Campbell (ed.), 269–281. Oxford: Oxford University Press, 2014.
Ritvo, Harriet. *The Animal Estate: The English and Other Creatures in the Victorian Age*. Cambridge, MA: Harvard University Press, 1987.
Ritvo, Harriet. The Emergence of Modern Pet Keeping. In *Social Creatures: A Human and Animal Studies Reader*, Clifton P. Flynn (ed.), 96–106. New York, NY: Lantern Books, 2008.
Tague, Ingrid H. *Animal Companions: Pets and Social Change in Eighteenth-Century Britain*. University Park, PA: Pennsylvania State University Press, 2015.
Walker-Meikle, Kathleen. *Medieval Pets*. Woodbridge: Boydell Press, 2012.

Takashi Ito
History of the Zoo

1 Introduction and Overview

In early modern Europe, a number of monarchs and nobles indulged in the collection of natural and artificial objects. Their collections are called "Wunderkammer" or "cabinet of curiosities," including not only gemstones, religious artifacts and mechanical inventions, but also animal specimens [→History of Taxonomy/Animal Collections].[1] In parallel with the culture of collecting grew the desire to collect living animals, rare and exotic ones in particular. As the difficulty of obtaining and keeping animals such as lions and tigers alive was large, they were highly valued and treated as a symbol of wealth and power.[2] In England, the Tower of London had already begun to house animals under King John's reign, and by the late eighteenth century, the royal collection had opened to the public as the "Tower Menagerie."[3] In late-seventeenth-century France, the Versailles menagerie was founded under the direction of Louis XIV, who aspired to the universal monarchy at a time when France held a political and cultural hegemony.[4] Although it was demolished at the outbreak of the French Revolution, the Tiergarten Schönbrunn constructed in Vienna under the Habsburg monarchy in the mid-eighteenth century has remained open at the same site until today, and claims to be the world's oldest zoo.[5]

This is what can be summarized as the origin of zoos, based on relatively recent publications on the subject.[6] But it is also a construct of historical narrative that reflects how zoos have recounted their histories to the public, which itself should be put into historical perspective. Bearing this in mind, this chapter begins with an out-

---

[1] Pomian, Krzystof. *Collection and Curiosities: Paris and Venice, 1500–1800*. London: Polity Press, 1990; Findlen, Paula. *Possessing Nature: Museums, Collecting and Scientific Culture in Early Modern Italy*. Berkley, CA: University of California Press, 1994.
[2] See, for example, Belozerskaya, Marina. *The Medici Giraffe: And Other Tales of Exotic Animals and Power*. New York, NY: Little, Brown and Company, 2006, 87–129.
[3] Grigson, Caroline. *Menagerie: The History of Exotic Animals in England, 1100–1837*. Oxford: Oxford University Press. 2016; Plumb, Christopher. *The Georgian Menagerie: Exotic Animals in Eighteenth-Century London*. London: I. B. Tauris, 2015.
[4] Robbins, Louise E. *Elephant Slaves and Pampered Parrots: Exotic Animals in Eighteenth-Century Paris: Animals, History, Culture*. Baltimore, MD: Johns Hopkins University Press, 2002, 37–67.
[5] Ash, Mitchell and Lothar Dittrich. *Menagerie des Kaisers: Zoo der Wiener*. Vienna: Pichler Verlag, 2002.
[6] Hoage, Robert J. and William A. Deiss. *New Worlds, New Animals: From Menagerie to Zoological Park in the Nineteenth-Century*. Baltimore, MD: Johns Hopkins University Press, 1996; Kisling, Vernon N. *Zoo and Aquarium History: Ancient Animal Collections to Zoological Gardens*. Boca Raton, FL: CRC Press, 2000; Baratay, Eric and Elisabeth Hardouin-Fugier. *Zoo: A History of Zoological Gardens in the West*. London: Reaktion Books, 2002.

line of the development of zoological gardens from the late eighteenth to the twentieth century. It then discusses approaches to understanding the ideas and meanings that traveled with animals from fields to zoos. The last section considers the implication(s) of the *animal turn* for zoo historiography. How and to what extent historians can understand and reconstruct the agency of animals is a key issue. As a whole, this chapter invites a rethinking of zoos as a product of modernity as well as a window through which we can glimpse its essence with regard to human-animal relationships. Now in the age of the Anthropocene, the euphoric vision of nineteenth-century zoological gardens has long faded into history, but many zoos are still facing problems inherited from the past.

As it is undeniable that many elements of present-day zoos originate in the European Enlightenment, and as it is in western European and North American countries where zoos today most noticeably face criticism and inspire scholarly inquiry, the content of this chapter tends to be more or less confined in geographical terms. But to offer a balance, zoos and their precursors in other regions are also discussed in this chapter.

2 Topics and Themes

The enthusiasm for obtaining exotic animals was not exclusively expressed by European monarchs and nobles. In early-fifteenth-century China, Zheng He, a court eunuch and mariner during the Ming dynasty crossed the Indian Ocean to reach the African east coast and returned with lions, leopards, zebras and giraffes. These animals were presented to the emperor, and giraffes became his favorite.[7] In eighteenth-century Japan, a pair of Indian elephants were imported from Vietnam by a Chinese merchant to fulfill the *shōgun*'s desire to see the animals with his own eyes. Crowds of spectators had a chance to watch one of them travelling overland from Nagasaki, where they disembarked, to Edo, the capital city.[8] This "elephant sensation" can be compared to the "giraffe sensation," which occurred a century later in Paris. When the Pasha of Egypt sent a giraffe to Charles X of France as a diplomatic gift, over 100,000 people gathered at the menagerie of the Jardin des Plantes to see the animal in the summer of 1827 [→Diplomatic History].[9] In both Asia and Europe exotic animals were valued and sometimes shown for popular entertainment, but it was the latter where an institutional basis for collecting, keeping and displaying animals

---

[7] Ringmar, Erik. Audience for a Giraffe: European Expansionism and the Quest for the Exotic. *Journal of World History* 17, no. 4 (December 2006): 389–393.
[8] The other one of the pair died soon after arrival in Nagasaki. Minoru, Wada. *Kyōhō 14nen Zō Edo he Yuku* [an elephant going to tokyo in the 14th year of the reign of *Kyōhō*]. Tokyo: Iwata Shoin, 2015.
[9] Allin, Michael. *Zarafa: A Giraffe's True Story, from Deep in Africa to the Heart of Paris*. London: Delta, 1998.

was laid out. What caused "the great divergence" in the history of animal collection and led to the birth of modern zoos?[10]

Enlightenment Origins

In the Enlightenment era, natural history developed as a study of plants, animals and minerals. Naturalists kept records of a variety of specimens collected from various parts of the world, as for them it was key to understanding the system of nature. While the knowledge of natural history was represented by the collecting of stuffed animals, the collecting of living animals could be equally associated with the pursuit of zoological science. In this regard, the menagerie of the Jardin des Plantes in Paris is most telling. In 1794, it opened to the public as a branch of the Muséum national d'Histoire naturelle, which had been founded in the previous year by the revolutionary government.[11] Public education aside, the Muséum offered chairs of professorship to prominent zoologists including Georges Cuvier, Jean-Baptiste Lamarck and Étienne Geoffroy Saint-Hilaire, and set the stage for them to debate evolutionary theories based on observations of living animals in the collection.[12]

The association with science was also embedded in the origin of the London Zoo. In 1826, the Zoological Society of London was founded by Thomas Stamford Raffles, former governor of Singapore, and Humphry Davy, president of the Royal Society. The society's prospectus remarked that its objective was to advance the study of zoological science, and for this and other purposes to establish a zoo and a zoological museum. When the zoo opened at a corner of Regent's Park, London, in 1828, it was soon referred to as the "Zoological Gardens."[13] Following in its footsteps, an increasing number of zoos appeared in major cities in Europe from the 1830s onwards: including Dublin (1831), Bristol (1836), Amsterdam (1838), Antwerp (1843), Berlin (1844) and Frankfurt (1858).[14] Although science was not necessarily the most visible

---

10 The expression is borrowed from Pomeranz, Kenneth. *The Great Divergence: China, Europe and the Making of the Modern World Economy*. Princeton, NJ: Princeton University Press, 2000.
11 Spary, Emma C. *Utopia's Garden: French Natural History from Old Regime to Revolution*. Chicago, IL: University of Chicago Press, 2000, 230–231.
12 Osborne, Michael A. *Nature, the Exotic and the Science of French Colonialism*. Bloomington, IN: Indiana University Press, 1994, 62–97; Outram, Dorinda. New Spaces in Natural History. In *Cultures of Natural History*, Nicholas Jardine, James A. Secord, Emma C. Spary (eds.), 249–265. Chicago, IL: University of Chicago Press, 1996.
13 Bastin, John. The First Prospectus of the Zoological Society of London: New Light on the Society's Origins. *Journal of the Society for the Bibliography of Natural History* 5, no. 5 (October 1970), 369–388; Desmond, Adrian. The Making of Institutional Zoology in London 1822–36. *History of Science* 23 (1985): 133–185; Ito, Takashi. *London Zoo and the Victorians, 1828–1859*. Woodbridge: Boydell, 2014.
14 Flack, Andrew. *The Wild Within: Histories of a Landmark British Zoo*. Charlottesville, VA: University of Virginia Press, 2018; Cowie, Helen. *Exhibiting Animals in Nineteenth-Century Britain: Empathy, Education, Entertainment*. London: Palgrave Macmillan, 2014; Colley, Ann C. *Wild Animal Skins in Vic-*

element in their administration, the language of science became a cliché that articulated their raison d'être in society [→History of Science; →History of Animal Collections/Animal Taxonomy].[15]

In the late nineteenth century, with the progress of the industrial revolution, skepticism about science and civilization increased. A highlight reflecting this social climate was the anti-vivisection movement, which became prominent initially in Britain and spread out internationally if not globally [→History of Experimental Animals and the History of Animal Experiments].[16] Although the vivisection debate rarely concerned zoo animals, it helped to establish an animal-ethical perspective from which zoos were to be evaluated. By the turn of the century, popular attitudes towards zoos and their animals had become more clearly ambivalent. A fin-de-siècle critique likening the London Zoo to "the Bastille of the beasts" was a notable example that signaled a shift towards negative perceptions of zoos and animal captives.[17] In the early twentieth century, many zoos and their associations have come to proclaim that their primary goal was nature conservation and the advancement of related sciences.

A claim to science characterized the institutional origin and development of zoological gardens in Europe. As such, many, if not all of the zoos emerging in the nineteenth century were able to secure their own space and status in the changing urban environments. Indeed, another important theme inseparable from a zoo's historiography is urbanization [→Urban (and Rural) History].

---

torian Britain: Zoos, Collections, Portraits and Maps. Burlington, VT: Ashgate, 2014; Mehos, Donna C. *Science and Culture for Members Only: The Amsterdam Zoo Artis in the Nineteenth Century*. Amsterdam: Amsterdam University Press, 2006; Bruce, Gary. *Through the Lion Gate: A History of the Berlin Zoo*. Oxford: Oxford University Press, 2017; Sakurai, Ayako. *Science and Societies in Frankfurt am Main*. London: Pickering and Chatto, 2013; Hochadel, Oliver and Agustí Nieto-Galan. *Barcelona: An Urban History of Science and Modernity, 1888–1929*. London: Routledge, 2016.

15 It was, of course, more than mere rhetoric. For the scientific study of acclimatization, see Osborne, Michael A. Acclimatizing the World: A History of the Paradigmatic Colonial Science. *Osiris* 15 (2000): 135–151. For medical research, see Cassidy, Angela, Rachel Mason Dentinger, Kathryn Schoefert and Abigail Woods. Animal Roles and Traces in the History of Medicine, c. 1880–1980. *British Journal for the History of Science Themes* 2 (2017): 11–33.

16 French, Richard D. *Antivivisection and Medical Science in Victorian Society*. Princeton, NJ: Princeton University Press, 1975; Rupke, Nicholaas A. *Vivisection in Historical Perspective*. London: Croom Helm, 1987.

17 Burkhardt, Richard W. Constructing the Zoo: Science, Society, and Animal Nature at the Paris Menagerie, 1794–1838. In *Animals in Human Histories*, Mary Henninger-Voss (ed.), 231–257. Rochester, NY: University of Rochester Press, 2002, 249.

Urban Life

Urban life was much more visibly connected with different types of animals in the nineteenth century than it is today.[18] While a scene of cattle being driven to and assembled at a livestock market was routinely observed, travelling menageries continued to dominate fairs and festivals in European cities.[19] Moreover, the expansion of consumer culture prompted the commercialization of pet-culture by making dogs and cats, rabbits and pigeons objects of consumption and affection.[20] The zoos' development during this period can be related to both of these public and private spheres, where animals became important resources for the cultural and emotional life of urban populations.

On the other hand, there were types of animals unwelcome to urban life. While major European cities suffered from several epidemics, newly implemented sanitary reforms affected popular perceptions of animals.[21] In London, the Smithfield livestock market was criticized for producing "miasma" – believed to be a cause of diseases until then denied by germ theory. Sanitary reform contributed to the relocation of the meat market from inside to outside the residential areas. With the construction of the abattoir complex, the processing of livestock to meat and other products was made invisible to consumers, whereas the London Zoo became the privileged site of animal spectacle.[22] The majority of urban citizens came to perceive zoos as a popular destination for family excursions, a refuge from everyday social realities, and an ideal contact zone with the natural world.

Across the Atlantic, the United States showed a distinct pattern of zoos' development. Only in the last quarter of the nineteenth century did American zoos begin to be constructed with the support of local communities and authorities. The public opening of the Philadelphia Zoo, America's first zoo, was delayed until 1874, due to the outbreak of the Civil War of 1861 to 1865.[23] Other northeastern cities such as Cincinnati, Providence and New York, soon established their own zoos. The openings

---

[18] Atkins, Peter J. *Animal Cities: Beastly Urban Histories*. Farnham: Ashgate, 2012; Almeroth-Williams, Thomas. *City of Beasts: How Animals Shaped Georgian London*. Manchester: Manchester University Press, 2019.

[19] Philo, Chris. Animals, Geography and the City: Notes on Inclusions and Exclusions. In *Animal Geographies: Place, Politics, and Identity in the Nature-Culture Boundaries*, Jennifer Wolch, Jody Emel (eds.), 51–71. London: Verso, 1998; Altick, Richard D. *The Shows of London*. Cambridge, MA: Harvard University Press, 1978.

[20] Amato, Sarah. *Beastly Possessions: Animals in Victorian Consumer Culture*. Toronto: University of Toronto Press, 2015, 26–55.

[21] Porter, Dorothy. *Health, Civilization and the State: A History of Public Health from Ancient to Modern Times*. London: Routledge, 1999, 77–127.

[22] Ito, Takashi. Locating the Transformation of Sensibilities in Nineteenth-Century London. In *Animal Cities*, Peter J. Atkins (ed.), 189–204. Farnham: Ashgate, 2012.

[23] Hanson, Elizabeth. *Animal Attractions: Nature on Display in American Zoos*. Princeton, NJ: Princeton University Press, 2004, 83.

in 1891 of the Bronx Zoo in New York City, as well as of the National Zoological Park as part of the Smithsonian Institution in Washington, DC heralded the spread of the zoo movement to the west and the south. In principle, American zoos emphasized urban park planning, as middle-class Americans longed for a "middle landscape" -- "gardens, parks, or other natural landscapes situated outside the overstimulating city but short of the primitive wilderness."[24]

As the United States narrowed the gap from Western European nations in terms of economic and military power in the last quarter of the nineteenth century, zoos were described as an emblem of the rise of the nation [→American Studies]. The Smithsonian Institution collected comments of support for the establishment of the National Zoological Park, which were then submitted to the House of Representatives. Among them the *Pittsburgh Dispatch* remarked: "That a nation so far in advance in the march of progress as the United States should be entirely without some such institution under Government protection seems almost incredible."[25] American zoos weathered the years of the Great Depression since public work projects under Roosevelt's New Deal policy involved the construction and renovations of zoo facilities.[26] This investment paved the way for them to consolidate resources and expertise for further reforms to come in the 1960s and the 1970s. The zoo reform movements during these decades had a global influence and they will be discussed further with reference to the *animal turn* in the section below.

3 Methods and Approaches

For the history of modern zoos outlined above, the other important theme is "empire." The period called "the age of empire(s)," roughly from the late-eighteenth to the nineteenth centuries and beyond, was the period of zoos' expansion. The military and mercantile networks that supported the process of colonization excited Europeans' interest in exotic animals in Asia and Africa and helped them collect and transport living animals. Quite a few zoos in former colonial countries developed from their networks during the nineteenth century. While Alipore in present Kolkata started as a supply station of exotic species, the Jardin d'esseai (present Botanical Garden Hamma in Algiers) served as a test site of economic zoology.[27] The Adelaide and Mel-

---

24 Hanson, *Animal Attractions*, 28.
25 Annual Report of the Board of Regents of the Smithsonian Institution to July 1888. Washington, DC: Government Printing Office, 1990, 44. Collected in *The Miscellaneous Documents of the House of Representatives for the Second Session of the Fiftieth Congress, 1888–1889*, 14 (1890); see also, Kisling, *Zoological Gardens of the United States*, 159.
26 Bender, Daniel E. *The Animal Game: Searching for Wildness at the American Zoo*. Cambridge, MA: Harvard University Press, 2016, 119–128.
27 Mittra, D. K. Ram Brahma Sanyal and the Establishment of the Calcutta Zoological Gardens. In *New Worlds, New Animals*, Robert J. Hoage, William A. Deiss (eds.), 86–93. Baltimore, MD: Johns Hop-

bourne Zoos envisioned transforming the colonial landscape into what white settlers had been familiar with in their former home country.[28] In Meiji Japan and Qing China, the Tokyo Ueno Zoo and the precursor of the Beijing Zoo, Wansheng Yuan, were founded in 1882 and 1906 respectively as part of the state-run Westernization program, whose prime objective was to resist and rival the European and American empires.[29] While the Enlightenment helps to explain the European origin of zoos, "empire" helps to explain their spread across the world, while also raising methodological issues.

Empire

First of all, disambiguation is necessary because the use of the term "empire" and "imperialism" often varies among scholars and is therefore applicable to different aspects of modern zoos.[30] In principle, an empire postulates to be a political entity, whether or not it officially bears the title of empire. As a historiographical term, it refers to the ideology of imperialism or the geopolitical ambition to establish dominance over foreign lands and peoples, often with the justification of a "civilizing mission."[31] Furthermore, either a tool or a product of colonization, modern science facilitated the surveying of colonial land and prospecting for mineral and biological resources.[32] The unsymmetrical power relation between the colonizers and the colonized led to the other imperialistic relation between humans (mainly on the side of the colonizers) perceiving nature as a resource for exploitation, which increasingly subjected nature to their rapacious enterprise [→(Post)Colonial History].[33] Indeed, the

---

kins University Press, 1996; Osborne, *Nature, the Exotic and the Science of French Colonialism*, 163–171.

28 Anderson, Kay. Culture and Nature at the Adelaide Zoo: At the Frontiers of Human Geography. *Transactions of the Institute of British Geographers* 20, no. 3 (1995): 275–294, 275–280; Ritvo, Harriet. *The Animal Estate: The English and their Creatures in the Victorian Age*. Cambridge, MA: Harvard University Press, 1987, 240–242.

29 Miller, Ian Jared. *The Nature of the Beasts: Empire and Exhibition at the Tokyo Imperial Zoo*. Cambridge, MA: Harvard University Press, 2013; Xianhua, Zhu. Qingdai Dongwuyuan [forbidden city]. *Zinjin Cheng* 23 (1984): 34–39.

30 Reynolds, Susan. Empire: A Problem of Comparative History. *Historical Research* 79, no. 204 (May 2006): 151–165.

31 Conklin. Alice L. *A Mission to Civilize: The Republican Idea of Empire in France and West Africa 1895–1930*. Stanford, CA: Stanford University Press, 1998; Hall, Catherine. *Civilizing Subjects: Metropole and Colony in the English Imagination 1830–1867*. Cambridge: Polity Press, 2002.

32 Macleod, Roy (ed.). Nature and Empire: Science and the Colonial Enterprise. Special issue of *Osiris* 15 (2000); Raj, Kapil. *Relocating Modern Science: Circulation and the Construction of Knowledge in South Asia and Europe, 1650–1900*. Basingstoke: Palgrave, 2007.

33 Grove, Richard H. *Green Imperialism: Colonial Expansion, Tropical Island Edens and the Origins of Environmentalism, 1600–1860*. Cambridge: Cambridge University Press, 1995; McKenzie, John M. *The*

synthesis of political and environmental imperialism has directed historians' attentions to botanic gardens, archeological galleries and, not least, zoological gardens. The convergence of the two forms of imperialism was illustrated by "human zoos," which claimed to pursue ethnographical inquiries into "uncivilized peoples" and their "primitive societies" and were typically staged in expositions and zoos from the late nineteenth to the early twentieth century. This was at a time when social Darwinism powerfully influenced scientific discourse, social policy and public education.[34]

For further consideration, it is useful to separate two distinct activities that enabled zoos to operate. One is the collecting of animals. Until relatively recently, most zoo animals have been collected from the wild. This entailed a series of processes from capturing animals alive in savannah to shipping them to a zoo in a home country. The other core activity is the displaying of animals. The exhibition of collected animals could be affected by a variety of factors ranging from rarity and value to media coverage. Since the two dimensions – collecting and displaying – were not seamlessly connected, what follows initially discusses them separately. Based on the analysis of each aspect it then becomes possible to consider their connection.

Collecting

The making and maintaining of an empire required establishing a sustainable political organization responsible for the governance of colonial land. An instance in which such a colonial institution was deployed to collect animals was the British East India Company. Although the Company was generally unwilling to provide strategic backup to scientific enterprises, many of its officers and officials were individually interested in natural history and collecting activities.[35] Around 1801, a menagerie was constructed at Barrakpore, West Bengal, by a medical officer.[36] In 1849, another Company officer wrote to the Zoological Society of London: "England has a right to expect from her sons in the colonies contributions to our National Zoological Society in London."[37] The phrase was adapted from Nelson's message at the Bat-

---

Empire of Nature: Hunting, Conservation and British Imperialism. Manchester: Manchester University Press, 1988.
34 Blanchard, Pascal, Nicolas Bancel, Charles Forsdick, Gilles Boetsch, Sandrine Lemaire and Eric Deroo. *Human Zoos: Science and Spectacle in the Age of Colonial Empires.* Liverpool: Liverpool University Press, 2008.
35 Arnold, David. *Science, Technology and Medicine in Colonial India.* Cambridge: Cambridge University Press, 2000.
36 Kisling, Vernon N. Colonial Menageries and the Exchange of Exotic Faunas. *Archives of Natural History* 25, no. 3 (October 1998): 303–320, 312–314.
37 *Proceedings of the Zoological Society of London* 17 (1849): 106.

tle of Trafalgar, suggesting that the military ethos was essential to the collecting campaign.

An exclusive focus on a single empire and its colonies, however, may risk overshadowing an even wider geopolitical setting in which collecting took place. The giraffe sensation discussed above was an outcome of international rivalries between European empires and the Ottoman empire. In one sense, Egypt, from which three giraffes were sent, one apiece, to the monarchs of Britain, France and Austria, officially belonged to the Ottoman empire, but in fact Muhammad (Mehmed) Ali, Pasha of Egypt under the Sultan's authority, sought to build his own dynastic empire. He advanced the Western-style modernization, took an expansionist foreign policy and invaded and ruled present Sudan.[38] The act of giraffe diplomacy can best be explained with reference to his negotiations with European powers, as well as to his territorial ambitions [→Diplomatic History; →Global History].

Moreover, the fact that Carl Hagenbeck expanded his family business as animal broker and dominated a global network of animal trade from the late nineteenth to the early twentieth century warns against looking exclusively at one empire and its overseas colonies. Despite his eponymous zoo founded at the outskirt of Hamburg, Hagenbeck's geographical reach extended far beyond the territory of the German empire. The Tokyo Ueno Zoo's first giraffe was purchased from Hagenbeck in 1907 at its quarter-century anniversary.[39] At the outbreak of the First World War, Hagenbeck's animal trade empire declined, and instead American dealers intervened to dominate the business. Among them, Frank Buck became an iconic figure by publishing many books on his "adventures" as well as running his own animal exhibits at world's fairs such as Frank Buck's Jungle Camp in Chicago in 1933 to 1934. American zoological institutions also conducted large-scale collecting campaigns; for example, in 1937, William Mann, director of the National Zoological Park, organized the National Geographic Society-Smithsonian Institution East Indies Expedition and returned with nearly 900 live specimens.[40]

In the aftermath of the Second World War, the collecting of zoo animals reached a turning point. While decolonization proceeded in many parts of the world in the 1960s, the status of wild animals shifted from the object of collection to the object of protection.[41] The changing intellectual and social milieu was enhanced by the

---

38 Fahmy, Khalid. *All the Pasha's Men: Mehmed Ali, his Army and the Making of Modern Egypt.* Cambridge: Cambridge University Press, 1998, 310; Mikhail, Alan. *The Animal in Ottoman Egypt.* Oxford: Oxford University Press, 2014, 137–139; Ito, *London Zoo and the Victorians*, 63–71.
39 Miller, *The Nature of the Beasts*, 76–78.
40 Bender, *The Animal Game*, 51–114.
41 The growth of ecological concerns and conservation movements had already partially led to legislation by the beginning of the twentieth century. See, for example, Ritvo, Harriet. Going Forth and Multiplying: Animal Acclimatization and Invasion. *Environmental History* 17, no. 2 (April 2012): 404–414; Doughty, Robin W. *Feather Fashions and Bird Preservation: A Study in Nature Protection.* Berkeley, CA: University of California Press, 1975.

publication of *The IUCN Red List of Threatened Species* starting in 1964. In 1975, the Convention on the International Trade in Endangered Species of Wild Fauna and Flora (CITES) entered into force with 10 countries having ratified the treaty.[42] How the change in the status of wildlife affected zoos will be further discussed in the section on the *animal turn*.

Displaying

It might be tempting to think that, because zoos were designed to represent a particular perception of nature, such as a vison of human conquest of wildness or a sense of responsibility for nature conservation, we can achieve a better understanding by analyzing their layout plan, architectural styles and exhibition philosophies. The process of constructing and operating zoos was, however, constricted by many physical and pragmatic circumstances.[43] These restraints were wide-ranging, from the shapes of the allotted land, the quality of soil and the topographical properties of the construction site to the timing of animal acquisition. Accordingly, the effect of any design principle and exhibition philosophy was often sporadic and elusive.

Still a chronological pattern of the design and animal exhibits can be briefly outlined. At an earlier stage, English landscape gardening characterized zoos' spatial fabrication. The key elements of picturesque gardening, such as winding paths and ornamental plants helped to create an aesthetically pleasing natural setting and included fenced-in animals as an integral component.[44] Architectural style varied from Alpine rustic to neoclassicism and Gothic, and later Orientalism added to the diversity. Meanwhile, cages and railings gradually came to be seen as a symbol of suffering in captivity. In the 1890s, Carl Hagenbeck developed a new style of exhibition called "panorama" to hide moats from visitors' view so that no sign of captivity could be observed.[45]

In the United States, Hagenbeck's panoramic exhibition was referred to as a "habitat" zoo, but in fact Hagenbeck himself had little intention of creating a truly faithful natural environment.[46] In the 1920s, the design orientation towards functionalism accorded with the rise of ethology. The synthesis of modernist architecture and experimental psychology was highlighted by the Penguin Pool at the London Zoo. It tried to encourage penguins to exhibit their "natural behaviors" by creating a stimulating

---

42 Bender, *The Animal Game*, 267.
43 On the other hand, theoretical approaches inspire and deepen source-based analysis. See, for example, McDonald, Tracy and Daniel Vandersommers. *Zoo Studies: A New Humanities*. Montreal: McGill-Queen's University Press, 2019.
44 Ito, *London Zoo and the Victorians*, 29–30.
45 Rothfels, Nigel. *Savages and Beasts: The Birth of the Modern Zoo*. Baltimore, MD: Johns Hopkins University Press, 2002, 161–188.
46 Hanson, *Animal Attractions*, 144–147.

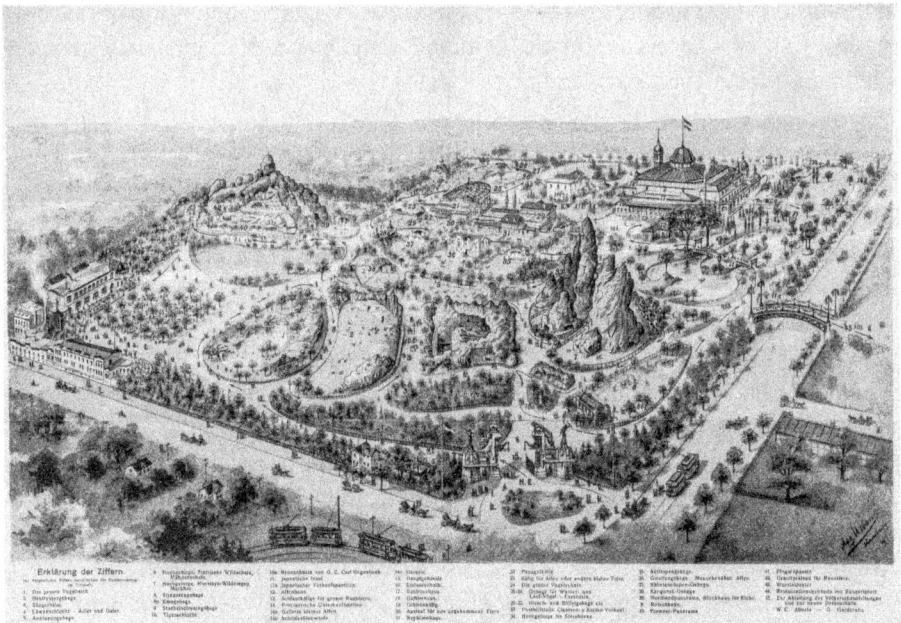

Figure 1: The bird's-eye view of the Tierpark Hagenbeck in the early twentieth century confirms that the specified perspective (from the middle left edge of the print) encompassed different animals unhindered by screening, so that animals of different species and the background landscape would be integrated into the zoo's total scenery. Source: author's collection.

environment: two narrow curved ramps interlocking each other over an oval swimming pool.[47] The allure of minimalism and functionalism finally waned by the 1960s. Desmond Morris, a zoologist and popular writer on sociobiology, criticized the zoo's adaption of modernist architecture and its philosophy. His disdain for the "naked cage" – where animals were enclosed by iron bars, constantly exposed to the human gaze, and enforced to pace lifelessly – emphasized a movement of zoo reform.[48]

In the mid-1970s, the idea of landscape immersion was inspired by the renovation program of the Woodland Park Zoo in Seattle. As its Gorilla Exhibit shows, the commissioned designers not only re-created the animal's natural habitat but allowed visitors to become immerged in the landscape, so that they could feel as if they would be touring the African savanna. The immersion exhibit combined the re-

---

[47] Guillery, Peter. *The Buildings of London Zoo*. London: Royal Commission on the Historical Monuments of England, 1993, 82–85; Shapland, Andrew and David Van Reybrouck. Competing Natural and Historical Heritage: The Penguin Pool at London Zoo. *International Journal of Heritage Studies* 14, no. 1 (November 2008): 10–29.

[48] Uddin, Lisa. *Zoo Renewal: White Flight and the American Ghetto*. Minneapolis, MN: University of Minnesota Press, 2015, 46–52.

alities of on-site experience and the power of the imagination in a way that would impress visitors with the importance of animal welfare as well as humanity's responsibility for ecological conservation.[49]

Reception Analysis

How were the ideologies underlying the process of collecting animals translated into the space used for display? And how was it perceived by the zoo's public? To follow this line of inquiry, we cannot avoid the question of reception analysis. In general, though not always, three different types of sources are available. First, when a zoo in question had a governing body, its official documents such as unpublished minutes of meetings and published reports provide a starting point for research. Although the quantity and quality of information obtained from them greatly vary, ideally, they recorded the purpose and intent of the animal exhibit.

Second, the media coverage of the exhibit provides an idea of how the exhibit was generally viewed by the public, although it should not be interpreted as reflecting the zoo-goers' experience per se. Editorial policies mattered; for example, zoo proprietors might invite journalists to write a favorable article. A repertoire of images of zoo animals such as smart elephants and ferocious tigers proliferated with a variety of stories offered by magazine articles, commercial advertisements and children's literature. One of its cultural effects was the proliferation of anthropomorphic art, as shown in Figure 2.

And third, a question then arises as to how individuals were affected by such media representations when reflecting on their personal experience of an animal exhibit, as well as of the complexities of the world that enabled that exhibit. Certainly, if these individuals left any related record in their letters and diaries, they become firm evidence in which reception analysis can be grounded. Although the insight gained from a few individuals' records should not be overstated, it helps to gauge the gaps between the intentions attached by zoo proprietors to animal exhibits, and the meanings extracted by zoo visitors from them.[50]

In the case of the London Zoo's giraffe exhibit in 1836, the archival documents of the Zoological Society testify to its intention to commemorate the giraffe exhibit as the landmark of scientific achievement. Although this message was reflected in some journal articles, especially in one written anonymously by the society's vice-president, other reviewers freely elaborated a repertoire of descriptions that praised the beauty and singularity of the animals. Above all, a letter from a country gentleman records that he had visited the zoo after church and pondered on the giraffe's

---

49 Hancocks, David. *A Different Nature: The Paradoxical World of Zoos and Their Uncertain Future.* Berkeley, CA: University of California Press, 2001, 117–127; Hanson, *Animal Attractions*, 159–161.
50 Mizelle, Brett. Contested Exhibitions: The Debate Over Proper Animal Sights in Post-Revolutionary America. *Worldviews: Environment, Culture, Religion* 9, no. 2 (January 2005): 219–235.

Figure 2: The collection of the French caricaturist J. J. Grandville titled *Scenes from the Public and Private Life of Animals* (1840–1842) testified how the menagerie of the Jardin des Plantes, which he frequented, acted as a catalyst for bringing animal characters into the center of the national political theater. Kashiwagi, Takao and Nobuyuki Kobayashi. *A Supplementary Volume to the Reprint Edition of J. J.* Grandville, *Scènes de la vie privée et publique des animaux.* Tokyo: Athena Press, 2009, p. facing 23.

creation within the frame of natural theology: "It is difficult to imagine their destined use unless for food either for man or some other animal."[51] This example suggests that the meanings that zoo animals could convey were so variable and fluid that historians should not consult just a few of them and argue as if they would preside over the possibilities of other interpretations and presume to be exhaustive. What is important is regarding this as a space for multiple, if not unlimited, interpretations. Precisely this hermeneutical diversity has encouraged people through generations to think and rethink how they look through zoos as a window to understand the world beyond its confines, what animals are vis-à-vis humans both in reality and the imagination, and whether and where the exact boundary can be drawn between these two forms of life.[52]

---

51 Ito, *London Zoo and the Victorians*, 71–72.
52 See, for example, Lee, Keekok. *Zoos: A Philosophical Tour.* New York, NY: Palgrave Macmillan, 2006; Malamud, Randy. *Reading Zoos: Representations of Animals in Captivity.* New York, NY: New York University Press, 1998.

4 Implication(s) of the Animal Turn

The zoo's hermeneutical space has given a momentum to the *animal turn* within the fields of humanities and the social sciences. Controversies concerning the life and post-life of Knut, the hand-reared polar bear in Berlin Zoo, offer an ideal source for the reappraisal of its impact. Nurtured in a man's embrace, Knut acquired a tamed and pampered expression of agency through his interactions not only with his caretaker, but also with the artificial environment set up for his care and a number of zoo-goers gathering around the exhibit. Being a polar bear turned Knut into an icon of the "fight against global warming."[53]

Long before Knut was iconized, a series of events and efforts had transformed zoos' exhibition philosophies and ethical standards. Bramwell's principle of "five freedoms" for protecting the welfare of livestock animals was adopted as the basis for the animal welfare of zoo animals.[54] Under the initiative of WAZA and other regional zoo associations, steps have been taken towards the standardization of conditions and measuring criterion of zoo animal welfare across the world.[55] Meanwhile, the theory of animal rights questioned the entire value system that had continued to create and operate the institution of animal captivity.[56] It empowered zoo opponents to argue for the abolition of zoos. In response, zoos have increasingly drawn on their missions to serve conservation and promote environmental education.[57]

The ethical change in zoo exhibits has entailed an equally significant change in the understanding of the agency of zoo animals. Accredited zoos no longer obtain animals from the wild, except in special cases, for example, where a conservation program for an endangered species cannot be carried out otherwise. As a result, zoo animals have moved far from being "type specimens" of wild species. Zoo-goers are encouraged to focus on the individual characters of each animal resident. A notable example is Binti-jua, the western lowland gorilla, who was born and bred in America. In August 1996, when a three-year-old boy fell into the gorilla enclosure in Brookfield Zoo in Chicago's suburb, she rescued him from her inquisitive mates and returned him to a zookeeper. This incident signifies an end point of the social evolution of the mod-

---

[53] Flinterud, Guro. Child Stars at the Zoo: The Rise and Fall of Polar Bear Knut. In *Zoo Studies*, Tracy McDonald, Daniel Vandersommers (eds.), 191–210. Montreal: McGill-Queen's University Press, 2019; Flinterud, Guro. Polar Bear Knut and His Blog. In *Animals on Display: The Creaturely in Museums, Zoos and Natural History*, Liv Emma Torsen, Karen A. Rader, Adam Dodd (eds.), 192–213. University Park, PA: Pennsylvania State University Press, 2013.
[54] Hosey, Geoff, Vicky Melfi and Sheila Pankhurst. *Zoo Animals: Behavior, Management and Welfare*. Oxford: Oxford University Press, 2009, 244–245.
[55] Maple, Terry and Bonnie M. Perdue. *Zoo Animal Welfare*. Heidelberg: Springer, 2013, 122–123.
[56] Singer, Peter. *Animal Liberation: A New Ethics for Our Treatment of Animals*. New York, NY: HarperCollins, 1975.
[57] Donahue, Jesse and Erik Trump. *The Politics of Zoos: Exotic Animals and Their Protectors*. Dekalb, IL: Northern Illinois University Press, 2006.

ern zoo starting in the age of the Enlightenment. Zoo animals begin and end their lives in the artificially controlled environment of human civilization and have therefore been deprived of their supposedly original agency: wildness.[58]

The *animal turn* has brought methodological changes too. Harriet Ritvo's pioneering work, *The Animal Estate* (1987), showed that animals are not negligible as actors that have constituted the history of human civilization.[59] In the meantime, like "estate" or "class", "animal" in animal history is more than a descriptive social category. Thanks to the linguistic turn, historians have realized that the conceptual categorization of "animal" itself – together with that of "human-animal" as opposed to "nonhuman-animal", and sub-categorizations such as "sentient animal" and "insentient animal" – has to be historically situated, which in its own right deserves scholarly inquiry. After all, we need to decide how to deal with the essential question: can animal history be written "from below," like Thompsonian social history [→Social History], with the premise that animals have their own voice and volition in the historical narrative?[60]

On the theoretical side of the issue, answering the question is far from easy as it leads to the even larger question of what agency is.[61] On the practical side, however, there are some experimental case studies to learn from. One approach is to assume that agency takes form by its response or resistance to the roles that zoo animals were expected to play. Stepping back into late Victorian Britain, Sarah Amato's analysis of the white elephant Toung Taloung reveals how his not-quite-white skin contradicted spectators' expectations of confirming the whiteness of the white elephant and became associated with the glowing concerns of racial hygiene. Susan Nance's reappraisal of the celebrity elephant Jumbo of the London Zoo discusses his agency as a construct of modern consumerism. Lisa Uddins' study of a herd of white rhinoceros released in the San Diego Wild Animal Park shortly prior to its opening in 1972 casts light on the elusive agency of the animals as a reflection of the anxieties of white middle-class suburbanites about their racial status and security. The zoo historian Nigel Rothfels turns to elephants in the present Oregon Zoo, suggesting the possibility of imagination to find out a new form of animal-human relationships other than that of animal curiosities and human spectators. These studies demonstrate that the zoo-animal's agency should be explored in light of its association with gender, race and other themes of social and cultural history [→Feminist Intersectionality Studies;→African Studies].[62]

---

[58] Bender, *The Animal Game*, 320–321.
[59] Ritvo, *The Animal Estate*.
[60] Hribal, Jason C. Animals, Agency and Class: Writing the History of Animals from Below. *Human Ecology Review* 14 (2007): 101–112.
[61] Howell, Philip. Animals, Agency and History. In *The Routledge Companion to Animal–Human History*, Hilda Kean, Philip Howell (eds), 197–221. London and New York, NY: Routledge, 2018.
[62] Amato, *Beastly Possessions*, 139–181; Nance, Susan. *Animal Modernity: Jumbo the Elephant and the Human Dilemma*. New York, NY: Palgrave, 2015; Uddin, *Zoo Renewal*; Blau, Dick and Nigel Roth-

Zoos have staged the contradicted agency of the animals. While they have been physically confined in cages and enclosures, we are encouraged to perceive them as individuals whose interests should be taken into consideration. While we sympathize with them through anthropomorphism, we need to set it aside to understand them on their own terms. The recent controversy regarding Marius the giraffe in the Copenhagen Zoo epitomized a further paradox. Marius' agency became most visible in the public demonstration in which his body was dissected and fed to lions.[63] The distant cause of his death arguably lies in the fact that zoos ceased to collect animals from the wild and therefore reduced the genetic diversity of their populations (this is not to say that zoos should again obtain wild animals). The decision to euthanize the surplus giraffe consciously problematized the biopolitics behind the operation of zoos. Looking back to their origin in the Enlightenment era, biopower was already operative as zoos were essentially about selecting and reproducing animals worthy of life as specimens.[64] In the age of the Anthropocene, the modern zoo confronts the challenge of its history and struggles to change to ensure a better future for both animals and humans.

Selected Bibliography

Bender, Daniel E. *The Animal Game: Searching for Wildness at the American Zoo*. Cambridge, MA: Harvard University Press, 2016.
Bruce, Gary. *Through the Lion Gate: A History of the Berlin Zoo*. Oxford: Oxford University Press, 2017.
Cowie, Helen. *Exhibiting Animals in Nineteenth-Century Britain: Empathy, Education, Entertainment.* London: Palgrave, 2014.
Flack, Andrew. *The Wild Within: Histories of a Landmark British Zoo*. Charlottesville, VA: University of Virginia Press, 2018.
Hanson, Elizabeth. *Animal Attractions: Nature on Display in American Zoos*. Princeton, NJ: Princeton University Press, 2004.
Ito, Takashi. *London Zoo and the Victorians, 1828–1859*. Woodbridge: Boydell, 2014.
McDonald, Tracy and Daniel Vandersommers (eds.). *Zoo Studies: A New Humanities.* Montreal: McGill-Queen's University Press. 2019.
Methos, Donna C. *Science and Culture for Members Only: The Amsterdam Zoo Artis in the Nineteenth Century.* Amsterdam: Amsterdam University Press, 2006.
Miller, Ian Jared. *The Nature of the Beasts: Empire and Exhibition at the Tokyo Imperial Zoo*. Berkeley, CA: University of California Press, 2013.

---

fels. *Elephant House*. Pennsylvania, PA: Pennsylvania State University Press, 2015; see also Cowie, Helen. Exhibiting Animals: Zoos, Menageries and Circuses. In *The Routledge Companion to Animal–Human History*, Hilda Kean, Philip Howell (eds.), 298–321. London and New York, NY: Routledge, 2018, 309–314.

63 For the history of the Copenhagen Zoo, see Gjerløff, Anne Katrine. When Zoo became Nature: Copenhagen Zoo and Perceptions of Animals and Nature around 1900. *Tidsskrift for kulturforskning* 9, no. 1 (2010): 22–37.

64 Braverman, Irus. *Zooland: The Institution of Captivity*. Palo Alto, CA: Stanford University Press, 2012, 159–185.

Rothfels, Nigel. *Savages and Beasts: The Birth of the Modern Zoo*. Baltimore, MD: Johns Hopkins University Press, 2002.

Uddin, Lisa. *Zoo Renewal: White Flight and the American Ghetto*. Minneapolis, MN: University of Minnesota Press, 2015.

Peta Tait
History of Circus Animals

1 Introduction and Overview

This chapter presents a historical overview of nonhuman animals (animals) in the nineteenth-century circus and travelling menagerie, and trained acts in the twentieth-century circus and cinema.[1] The early modern circus began in the late eighteenth century around the time public menageries and zoos were emerging to influence scientific and social attitudes. Circus benefited from the way exotic animal display stimulated a growing nineteenth-century colonial business of hunting, capturing, transporting and trading wild animals to colonial centers; this ran parallel with campaigns to improve the welfare of domesticated animals.[2] Exhibiting practices reached a monumental scale by the turn-of-the-twentieth-century with hundreds of thousands of animals as dead specimens in museums and live specimens in menageries and circus [→History of Animal Collections/Animal Taxonomy]. By then, prominent hunters in parts of Africa and Asia were publicly warning of a worrying decline in species numbers in the wild where circus animals were obtained.

This brief history of animal acts is based on English language archival research that addresses a question asked within the field of performance studies: what did spectators see the animals doing in the performance and how was this achieved? It seeks to avoid collapsing together widely divergent historical and national practices into an amorphous general idea of the circus. Even where it is associated with the travelling menagerie, circus can be distinguished from other types of captivity by the ring performance [→History of the Zoo]. This history draws conclusions from the material practices of historically specific performances and developments and what prominent tamers and trainers did to create performance illusions. The chapter emphasizes big cats and elephants because these became the lead trained acts in the twentieth-century circus and the focus of efforts to ban animal acts.

Tamer performances with lions and tigers (big cats) were first presented in small cages in travelling menageries and these cages were sometimes wheeled into the circus ring from the accompanying menagerie. In arguing for animal rights in England in 1892, Henry Salt specifies menagerie entertainment rather than circus to rid-

---
[1] For an extended history of animals in nineteenth-century travelling menageries and in the circus and related practices within British colonies, see Tait, Peta. *Fighting Nature: Travelling Menageries, Animal Acts and War Shows*. Sydney: Sydney University Press, 2016.
[2] See the seminal work: Ritvo, Harriet. *The Animal Estate: The English and Other Creatures in the Victorian Age*. Cambridge, MA: Harvard University Press, 1987.

icule claims that animals enjoyed captivity.³ Circus performance was actually dominated by equestrian acts throughout the nineteenth century and while there were historical variations between countries, the travelling menagerie was largely a separate enterprise from the ring performance even where it travelled with a circus. While big cats, elephants and other large animals made periodic walk-through appearances in the circus ring and on theatre stages, sometimes over several weeks, they were unpredictable and constituted costly purchases for most travelling circuses. Thus, these animals were not routinely included in the actual circus performance until the advent of trained animal acts from the 1880s. As a consequence, nineteenth-century anxiety about exotic animal performance was largely focused on the travelling menagerie.

The advent of fully trained acts meant that large animals including elephants could be routinely included in the circus performance and with big cats behind the newly invented arena barrier surrounding the ring. Animals associated with exotic geographical regions and considered dangerous were reliably trained for performance, and training influenced the zoological study of animal behavior well into the twentieth century. Popular trained big cat and elephant acts and trainer lineages originating with the Hagenbeck and Bostock family businesses shared top billing in the early twentieth-century circus with flying trapeze acts invented in 1859. Although criticism of animals in performance was longstanding, organized opposition to the circus was a twentieth-century phenomenon with animal rights protests to ban animal acts gaining momentum during the 1980s. This coincided with the invention of the animal-free new circus from the 1970s, now known as contemporary circus and exemplified by, for example, Canada's Cirque du Soleil, France's Cirque Baroque and Cirque Plume, Australia's Circus Oz, and Britain's No Fit State Circus. Contemporary circus continues to develop as a human art form alongside traditional circus.⁴ In 2018, the Feld family-managed, iconic Ringling Bros. Barnum and Bailey Circus The Greatest Show on Earth® closed, marking the end of an era in the traditional American circus presenting large animal acts, and an outcome that was partly the result of targeted public protests. At this time, however, circuses with trained animals continue to tour in a number of countries.

---

3 Salt, Henry. *Animals' Rights: Considered in Relation to Social Progress*. Clarks Summit, Pennsylvania, PA: Society for Animal Rights Inc., 1980, 50 [first published in 1892].

4 For an overview of the traditional circus and the contemporary circus, see Tait, Peta and Katie Lavers (eds.). *The Routledge Circus Studies Reader*. London: Routledge, 2016. See also Weber, Susan, Kenneth Ames and Matthew Wittman (eds.). *The American Circus*. New York, NY: Bard Graduate Centre and Yale University Press, 2012; Leroux, Louis Patrick and Charles Batson. *Cirque Global: Quebec's Expanding Circus Boundaries*. Montreal: McGill–Queen's University Press, 2016.

2 Topics and Themes

Topics associated with circus animal history include: (a) the development of circuses as an entertainment attraction alongside the menagerie/zoo; (b) the connection to the growing animal trade and colonialism; (c) opposition to travelling menageries and then circuses because of perceived cruelty; and (d) performance as the symbolic display of human superiority and power over nature. Human dominance was demonstrated in the control of large bodied lions, tigers and elephants in performance and implicitly conveyed ideas of social Darwinism.

Circus animal acts were developed initially as displays of human acrobatic skill and prowess with horses. The early modern circus emerged out of the riding school of ex-soldier, Philip Astley in 1768 in London, and subsequently in Paris. Astley's Circus and its competitor, the Royal Circus, soon added rope and clown acts, and enlarged the program with theatrical pantomimes and acts with other domesticated species such as dogs and clever pigs.[5] The capacity of some animals to seem human-like might have appealed but this also unsettled ideas of human superiority. Monkey jockeys did remain popular nineteenth-century acts because they were framed as comic. The appearances of large imported animals, however, were intermittent until late in the nineteenth century. On occasion, exotic animals from menageries in walk-through action in a pantomime would shift the narrative to a distant geographical location; they embodied an exotic fantasy.

Marius Kwint explores how accomplished horsemanship in the early circus demonstrated the triumph of human civilization and science.[6] The treatment of circus horses was considered to be in advance of other areas of society completely reliant on horses. Astley, for example, advocated sympathetic and kind attitudes to horses and even positive reinforcement, and denounced poor horsemanship and cruelty in his widely read manuals on horse training that targeted the military.[7] As the main resource of the circus business, horses held economic significance that can be grouped under Nicole Shukin's concept of "Animal Capital" and her explanation of "biopower" within societies reliant on horses for agricultural production and all modes of transport.[8] The ownership of horses and their labor created economic value, and

---

[5] For reliable histories, see, for example, Speaight, George. *A History of the Circus*. London: The Tantivy Press, 1980; Thétard, Henry. *La Merveilleuse Histoire du Cirque*, 2 vols. France: Prisma, 1947; Hoh, LaVahn G. and William H. Rough. *Step Right Up. The Adventures of Circus in America*. White Hall, VA: Betterway Publications, 1990. Also: Jando, Dominique. *Circopedia: The Free Encyclopedia of the International Circus*. URL: www.circopedia.org/Main_Page (August 21, 2020).
[6] Kwint, Marius. The Circus and Nature in Late Georgian England. In *The Routledge Circus Studies Reader*, Peta Tait, Katie Lavers (eds.), 331–348. London: Routledge, 2016.
[7] For example, see Astley, Philip. *Astley's System of Equestrian Education, Exhibiting the Beauties and Defects of the Horse*. Dublin: Thomas Burnside, 1802, 9, 18, 43. URL: www.biodiversitylibrary.org/bibliography/37724#/summary (December 6, 2020).
[8] Shukin, Nicole. *Animal Capital*. Minneapolis, MN: University of Minnesota Press, 2009, 16.

Shukin elaborates on the literal and metaphoric exploitation of the horse body, and within modernity the transition to mechanized power. Circus horses worked to create the skilled performance and transported the circus prior to transport mechanization.

The pastime of viewing animals coincided with increased leisure time, and travelling circuses and menageries reached small towns. As Brett Mizelle explains about the parallel development of menagerie and circus in the USA, moral judgments about good and bad depended on the social context so that initially menageries were "good", viewed as socially educational and for children.[9] Menagerie practices, however, did become controversial because, despite promotional rhetoric about education and kindness, the harsh treatment was often apparent to visitors. The animals were caged and tethered and, with only knowledge by their keepers of specific animal species and their needs, many died quickly. Moreover, cages and restraints often encouraged spectators to poke and prod animals, so the animals had to be protected from humans.[10]

Menageries were initially reliant on a haphazard trade but increased in size and number as the trading businesses in major cities grew from the 1860s.[11] An increased supply made animals more affordable and larger circuses began to maintain a menagerie in a separate tent; circuses continued to perform in tents outside of Europe. The demand for animals converged with idealized masculine hunting practices in colonized regions. The expansion of British and European imperial rule over parts of Africa and Asia during the nineteenth century created the conditions for the expansion of zoos and menageries [→(Post)Colonial History]. As colonial shipping routes facilitated circus from England, Europe and then the USA to tour globally, adventurers and military men in the colonies shipped back exotic animals. Colonial hunter identity became a dominant motif in both tamer and trainer acts through costuming and props such as guns that fired blanks.

Performances with lions and tigers in small cages developed in the menagerie from 1825, and well-known tamers such as Isaac van Amburgh wielding a crowbar appeared in theatres and circus pantomimes demonstrating power over large animals considered dangerous. The tamer act overtly manifest aggression and attracted large audiences even though some spectators declared it cruel. Big cats in small cages with male presenters bodily handling them were popular menagerie attractions from the 1830s to the 1870s and included loud noise and rapid movement to heighten the impression of danger. There were female tamers such as Ellen Chapman in the late 1840s who also proved to be popular attractions and therefore lucrative despite the moral condemnation of women performing in tamer acts.

---

9 Mizelle, Brett. Horses and Cat Acts in the Early American Circus. In *The American Circus*, Susan Weber, Kenneth Ames, Mathew Wittmann (eds.), 250–275. New York, NY: Bard Graduate Centre and Yale University Press, 2012.
10 On unruly spectators and hooligans see Tait, *Fighting Nature*, 103–119.
11 Simons, John. *The Tiger that Swallowed the Boy: Animals in Victorian England*. Farringdon: Libri Press, 2012.

While animals in tamer acts were habituated, they were not systematically conditioned until techniques of training developed after the 1880s. Circus performance was less obviously associated with the inhumane practices of menageries until these animals became routinely included in the performance, the forceful methods of elephant submission had happened elsewhere. The systematic training of large exotic animals to mirror that of trained horses was gradually realized. For example, John Cooper had trained a small group of elephants by 1876 to reliably undertake feats (tricks).[12] Training was based on patient careful observation of the bodily reactions of a species and involved time spent in close proximity in order to accustom animals to the trainer's presence and to closely observe each individual's personality and unique capacity. Histories of training are species-specific and involve the intensive efforts of determined trainers. Leading early trainers and employers of trainers included Carl and Wilhelm Hagenbeck in Germany from a family of animal traders and zoo owners and, for a brief time, circus owners, and Frank and Edward Bostock from a family who ran menagerie businesses in England and also became prominent in the USA.[13] The procurement of exotic animals became regularized by the Hagenbeck business appointing agents and putting in orders for the species in demand in an economic chain. When entrepreneur P. T. Barnum expanded his business from a museum menagerie into the travelling circus in 1872, he began buying animals including elephants from the Hagenbecks.[14] As Susan Nance explains, it was trained elephants in particular – and some became famous – who became synonymous with circus as it became a central part of American mass culture [→American Studies].[15]

Trained acts in the circus heightened and refined impressions of human dominance over nature. The Hagenbecks created the dominant business by training a whole act in mixed-species configurations for high-profile appearances. By the 1890s a circus could purchase the complete act from Hagenbecks, with or without a human presenter since the routine was established. The animal performers were accustomed to the cues for the action in the act and often took cues from each other. This made Hagenbecks the dominant circus animal business internationally for over fifty years. Drawing on Paul Bouissac, Gillian Arrighi points out that while

---

[12] For example, see Hippisley Coxe, Antony. *A Seat at the Circus*. Hamden, CT: Archon Books, 1980, ch. 9; Tait, Peta. *Wild and Dangerous Performances: Animals, Emotions, Circus*. Basingstoke: Palgrave Macmillan, 2012.
[13] For a history of Hagenbecks business, see Rothfels, Nigel. *Savages and Beasts: The Birth of the Zoo*. Baltimore, MD: The Johns Hopkins University, 2002. For an account of the Hagenbecks' contribution to circus training and the Bostocks' importance, see Tait, *Wild and Dangerous Performances*, 15–27.
[14] Saxon, A. H. *P.T. Barnum: The Legend and the Man*. New York, NY: Columbia University Press, 1989.
[15] Nance, Susan. Elephants and the American Circus. In *The American Circus*, Susan Weber, Kenneth Ames, Mathew Wittmann (eds.), 232–249. New York, NY: Bard Graduate Centre and Yale University Press, 2012.

such acts purchased from Hagenbecks symbolized nature and wildness, paradoxically, they were highly structured and managed, and they embodied further contradictory tensions within a national political context.[16] For example, Arrighi finds these acts enhanced how the circus business in Australia underlined its capitalist imperative and global reach.

As Janet Davis outlines in her analysis of the cultural significance of the modernist American circus, animal acts were criticized from the late nineteenth century because of animal welfare concerns and notions of human benevolence that were developing alongside the social acceptance of evolutionary connections with animals.[17] Criticism in newspapers meant that owners developed strategies for publicly defending their treatment of big cats and elephants. Yet concern about animals in theatrical performance by the 1890s and into the early twentieth century did continue to emphasize the treatment of dogs.[18] During the early twentieth century, however, distinct organizations focused on the welfare of large circus animals; the timing of changes in the welfare law and governmental policies varied between countries. David Wilson's history of the movement against animals in performance in Britain and the campaigns of Edmund MacMichael reveals the cogent resistance of the twentieth-century animal entertainment industry, and how campaigns received pro-animal support from organizations fighting for social justice more broadly, including, for example, women's rights.[19] British newspapers were presenting debates about circus animals by the 1960s, with coverage not evident in other countries until animal rights campaigns of the 1980s. The shift away from public acceptance of live animal acts in circus does coincide with increasing numbers of film and television documentaries about animals in the wild.

3 Methods and Approaches

Performance Analysis: Taming to Training

Animal acts in circuses followed distinct performative formulas. The composition of the performance and how this was achieved is illustrated here with representative examples of influential American acts. The working life of American performer George Conklin exemplifies the major nineteenth-century transition from taming to training performance, his memoir usefully naming individual animals and explain-

---

[16] Arrighi, Gillian. Political Animals: Engagements with Imperial and Gender Discourses in Late-Colonial Australian Circuses. *Theatre Journal* 60, no. 4 (December 2008): 609–629.
[17] Davis, Janet. *The Circus Age*. Chapel Hill, NC: The University of North Carolina Press, 2002, 153.
[18] For example, see Bensuan, Samuel L. The Torture of Trained Animals. *English Illustrated Magazine* 15 (1896): 25–30.
[19] Wilson, David. *The Welfare of Performing Animals: A Historical Perspective*. Heidelberg: Springer, 2015, 68.

ing practices in an unguarded way. After working as a cage attendant, he became a tamer in a small cage act in 1867, and progressed to training animals for the circus by the 1880s. After 1886 he was head trainer of elephants with the major American circus, the Barnum and Bailey Circus The Greatest Show on Earth® (BB), and on its international tour.[20] Over four decades, Conklin habituated twenty-five cages of lions; worked with elephants often acquired from Hagenbecks; conditioned a number of other species including bears; and made spotted hyenas jump and run, and zebras pull a cart.

By the 1870s, as Conklin recounts, the menagerie's lion cage could be pulled into the circus ring on a wagon – and sometimes by an elephant. Tamer Conklin wore a Roman-style shift, a costume made famous by Van Amburgh in the 1830s, and Conklin admitted that his own motivation was the high pay. The act consisted of jumping and chasing and the tamer handling an animal. For example, by applying pressure to the lion's back, the tamer would put his head in the lion's mouth. This feat attracted large crowds and Conklin did this feat regularly with the lion, Pomp. Conklin held Pomp's nose with one hand and the lower jaw with the other and he would "open his mouth as wide as possible and put my head in it as far as it would go which was about halfway".[21] The way that he held Pomp's jaw meant that he could detect muscle movement to evade being bitten. The ending of the act involved a wrestle over a piece of meat and Pomp growling before Conklin fired a pistol and left the cage. There was ongoing criticism of the feat of the head in the lion's jaw. For example, "This is a fool-hardy feat [...] without exhibiting any intelligence, grace or docility on the part of the lion".[22] Attendants were placed around the cage during the performance ready with long rods to prod the animals. While the practice of feeding bloodied meat to the animals before or during an act was also part of the tamer era, and was gradually discontinued, the theatrical embellishment of safari hunting continued until the 1970s [→History of Hunting].

Conklin became aware that lions followed their own leader within the group.[23] His movement would frame the animal movement; for example, Conklin danced so the lion appeared to waltz. Conklin also noted that humans mistook a roar for an indication of aggression when it was a sign of presence. Conklin probably exaggerated his claim that he discovered what became a standard defensive ploy in trained big cat acts of holding up a stool or a chair because the big cats could not focus on multiple legs quickly.

During the 1870s, Conklin worked for a circus with a menagerie business with seven tents and fifty cages of animals on the pathway toward the main circus tent,

---

20 Conklin, George. *The Ways of the Circus*. New York, NY: Harper & Brothers Publishers, 1921, 11–12, 251.
21 Conklin, *The Ways of the Circus*, 37.
22 Lions and Lion Tamers. *The New York Clipper*, April 13, 1872, 12.
23 Conklin, *The Ways of the Circus*, 38, 44–45, 52–53, 73, 152–153. Includes information on other training.

and Conklin learnt how to work with elephants from Stuart Craven around 1868. In the 1870s Conklin taught four elephants to move and stop on cue, a feat that was put to military marching music. He worked with thirty-five-year-old Queen Annie who appeared to dance when the band accompanied her.[24] He used a whip and an elephant hook (goad) to get Queen Annie to lie down, which meant her movements were intended to avoid the hook in her skin. The brutal and physically torturous early treatment that forced an elephant into submission happened away from public view. Over months Conklin coached the elephant, Tom Thumb, to walk a low timber plank carrying a pole with his trunk and gradually raised the height, and Tom Thumb learnt to turn a key and to throw objects.

An elephant taught to sit at a table and ring a bell created a clown act and with Conklin sitting opposite and leaving behind a hat, which the elephant then sat on.[25] It might have made the elephant endearing, but this was achieved by fastening the elephant's back legs and then forcing him or her into the chair and using an elephant hook. A head stand was achieved with a rope sling through a pulley that allowed the back of the elephant to be lifted against the wall until the elephant was used to doing this feat on command. By the time BB toured to London's Olympia Hall in 1896–1897, Conklin was responsible for three groups of elephants described as "wonderfully educated" to do feats.[26] There was also an act with "Comicalities and Humorous Feats" with Juno, the baby elephant – baby elephants proved popular but were rarely born or survived in captivity. While positive reinforcement was being advocated and used for the training of big cats and other animals by the 1890s, the long-living circus elephants had all undergone a process of forced submission. This type of performance was possible because individual animals complied with such regimes.

Confronting Circus in Cinema and in Gendered Demarcation

Circus trainers recognized that animals could learn complex action. Training relied on the capacity of animals to understand and interpret human gesture and spoken language and it appreciated an animal's sensory attunement to detail and slight changes in the environment. Trainers valued the intelligence of the animals and studied their behavior closely, and they used psychological approaches to the interaction between humans and animals, some trainers having read Darwin on animal emotions.[27] They recognized animal personalities. This appreciation of animal individuality and emotional responsiveness and admiration for their intelligence happened

---

24 Conklin, *The Ways of the Circus*, 115, 117, 119. Also on Tom Thumb.
25 Conklin, *The Ways of the Circus*, 122, 124, 132. And about the use of pulleys and harnesses.
26 Barnum and Bailey Greatest Show on Earth® (BB) Official Programme 1896–1897, Joe E. Ward Collection, Harry Ransom Library Special Collections, University of Texas at Austin, Box 55, items 63–68.
27 Tait, *Wild and Dangerous Performances*.

well in advance of twentieth-century science. Such positive values, however, did not offset the way trainers required animals to lead restricted lives in captivity and remain submissive to commands, and constantly obedient. In assuming the right to control animal lives and emotions, training promoted a more complicated version of humanity conquering nonhuman nature in which the animals appeared to comply willingly.

The popularity of trained circus acts grew in the first half of the twentieth century, as engagement with live circus performance expanded through cinema, and cinema relied on circus trainers and animals for film stunts. The most well-known twentieth-century big cat trainer was American Clyde Beatty, in part because his circus acts featured in cinema under his name, and he worked with the Ringling Bros. Barnum and Bailey Circus The Greatest Show on Earth® (Ringling Bros.) as it was named between 1931 and 1934.[28] Beatty's acts replicated the pistol-firing, safari-hunting confrontation with his quasi military costume, a whip and a gun firing blanks – he actually carried a chair for defense. The filmed versions of Beatty's acts remain confronting to watch because Beatty presented 'fighting acts', in which the trainer and trained animals appear to be in conflict in a performance of ferocity. Conversely, European acts were known for performances of docility.

Since the animals were trained, they knew the routine and trainers did not generally touch the animals to move them – in contrast to tamer acts and nineteenth-century lion-wrestling stunts. There were trained carrying feats, however, such as shoulder carrying that did involve bodily contact. Beatty used his body positions and gestures to instruct the animals when and where to move. This was a major difference with training even in the presentation of noisy confrontation.

Beatty first presented animals trained by other trainers, initially a bear act trained by Louis Roth who worked with Hagenbeck-trained animals, and Roth had been also part of the Bostock lineage of trainers. Beatty was offered an opportunity to take over an existing act with big cats, and there were approximately fifty exotic animal acts in 1925 in the USA.[29] In 1934 he started training animals himself. Between the 1920s and the 1960s, Beatty worked with over two thousand lions and tigers and several bears and reached an audience estimated at forty million.[30] He performed with up to thirty – and for a short time forty – big cats in a fifteen-minute act. Beatty's act involved creating a standard pyramid of lions and tigers standing on graduated pedestals and standard feats such as leaping, two-legged poses, and balancing. For a time, he presented a tiger on the back of an elephant, originally a Hagenbeck act.

It was an animal attack that first gave Beatty media attention in 1930. A female lion, Trudy, attacked Beatty, but another lion, Prince, defended him and attacked

---

28 Beatty, Clyde and Earl Wilson. *Jungle Performers*. London: Robert Hale, 1946, 144–146.
29 Joys, Joanne Carol. *The Wild Animal Trainer in America*. Boulder, CO: Pruett Publishing Co., 1983, 108.
30 Beatty, Clyde and Edward Anthony. *Facing the Big Cats*. London: Heinemann, 1965, 2.

Trudy.³¹ Beatty was injured and underwent surgery before returning to the evening performance. Or so the story goes; it was repeated in publicity over decades, and Prince became Nero in some versions, which undermines its veracity. This attack was restaged for the film, *The Big Cage* (1934).

Beatty, like other trainers, claimed that he worked with the animals through kindness and patience and big cats were trained with positive reinforcement. The act, however, staged confrontation. His was a masculine persona and Beatty's contemporary, Mabel Stark, working with tigers, adopted an attitude more like that of a teacher giving verbal instruction, although also costumed in an equestrian military uniform.³² Women presenters and trainers had to seem more caring.

Claire Heliot was probably the most well-known female trainer of big cats prior to Stark.³³ The German-born Heliot (Klara Haumann) rose to prominence at the end of the nineteenth century and made triumphant appearances touring European cities, reaching New York by 1905. Her act included up to 14 lions and was a mixture of tamer actions such as hand feeding and trainer feats such as balancing on a raised platform and a rolling barrel. The climax involved Heliot's feat of carrying 10-year-old, 159 kilograms lion Sicchi across her shoulders. Heliot performed in a full-length fashionable dress like most female presenters to convey an impression of femininity that offset the social irregularity of the performance, and she sat down at a dinner table with the animal performers, which created an impression of domesticity [→Feminist Intersectionality Studies]. Heliot was actually the trainer of the animals in her act and while she was described as working with a psychology of kindness, she carried a steel rod and a whip for protection and worked for the income. The female trainer was perceived in gendered ways, and Heliot's casual gesture of putting her arm around a lion's neck required long hours of careful training.

Billed as the only woman who 'breaks' tigers, Stark became the most well-known twentieth-century female trainer and she mainly worked with tigers over a long 57-year career that included international touring and performing in Japan for several years.³⁴ She was working with Roth-trained tigers by 1912 and soon after began training tigers herself. Stark instructed the tigers verbally and with her tone of voice, and distinctively moved constantly around the space in an act that presented up to 16 tigers on pedestals in a pyramid. Stark did present a unique feat with Rajah who leapt down from his pedestal in what seemed to be an attack, and they rolled together on the ground before he returned to his pedestal. Other feats included balancing on balls and moving on instruction. Some of the tigers from Stark's act can be viewed in the film, *Demetrius and the Gladiators* (1954). Tigers lead solitary lives in comparison with lions, which makes them more difficult to train for group work, although

---

31 Joys, *The Wild Animal Trainer in America*, 112.
32 Tait, *Wild and Dangerous Performances*, 116. Image.
33 Tait, *Fighting Nature*, 229–223.
34 Stark, Mabel and Gertrude Orr. *Hold That Tiger*. Caldwell, ID: The Caxton Printers Ltd, 1940, 18; Tait, *Wild and Dangerous Performances*, 115–122.

Stark claimed she found them easier to work with and that individuals were loyal to her. Stark appeared at Ringling Bros., in 1922 with five tigers and a leopard in a program that included a Hagenbeck trained act, and Stark was soon one of the leading circus stars. She trained a tiger to stand up on his back legs and put his front paws on her shoulders, and by the late 1920s, Stark's teacher persona had become implicitly sexualized as that of a femme fatale and she appeared in publicity with the film star, Mae West. An erotic identity for a female presenter in an act with animals had been covertly implied in a leopard act in which the performer wore a burlesque costume consisting of a jacket reaching only to the top of the thigh, tights, and boots up to the mid-thigh in the first decade of the twentieth century.[35] But Stark was fully clothed, and the sexualized implication of this type of act arose from a general masculinized concept in which the danger of feminine seduction aligned with ideas of dangerous animals.

Stark's choice of an ankle-to-neck trouser and jacket quasi military costume, however, had a practical purpose as the thick material protected her skin and hid the numerous mostly accidently scratches and slashes that she had received. Stark's scarred skin carried the ongoing risks of performing in close proximity with tigers.

While the trainer proclaimed persistence, patience and even kindness, there were less publicized factors. Only a small number of individual personalities proved suitable for complex tricks. Starting out, Carl Hagenbeck found that only four out of twenty-one lions could be trained.[36] Training remained species specific even when animals were put into multiple-species acts, and it took advantage of individual bodily movement. Animals such as big cats needed to co-operate for the act, and if an animal showed resistance or defied the trainer, he or she was simply removed. The information about what happened to these animals is vague, as only some were sent to a farm where animals lived outside the performance season. Older individuals accustomed to life in the circus did not like to be left out, and there are anecdotal examples of deliberate activity in front of an audience. These animals must be admired for their high level of accomplishment as performers. The cultural significance of the trained act with a female presenter, however, was attributed a vague feminine capacity to beguile dangerous animals rather than skill in training.

4 Implication(s) of the Animal Turn

Philosophical perspectives recognizing animal subjectivity, and the need for physical environments suited to a species body confronted the limitations of a captive life in

---

[35] Tait, Peta. Burlesque Costuming and Sensationalist Circus Animal Acts. *Australasian Drama Studies* 63 (December 2013): 84–95.
[36] Hagenbeck, Carl. *Beasts and Men, being Carl Hagenbeck's Experiences for Half a Century Among Wild Animals.* Abridged and translated by Hugh Samual Roger Elliot and Arthur Gordon Thacker. New York, NY: Longman Green and Company, 1909, 32.

the circus. Circus animals became a focus of the animal rights movement influenced by Peter Singer's *Animal Liberation* and its philosophical expansion of Richard Ryder's idea of speciesism.[37] The development of the animal-free circus coincides with a widespread social shift to these pro-animal values. Animal performance in the twentieth-century circus was indicative of "human-made replicas of nature" and a manifestation of the "vast and self-serving misrecognition of animals by humans".[38] The *animal turn* and its political stance was matched by a growing public discomfort with trained anthropomorphic acts. In arguing for a pro-animal zooësis, Una Chaudhuri seeks to overturn the facelessness of animals in performance and their objectification. Individual animals can be recognized as contributing to material practices of performance while the purpose of such entertainment is interrogated and rejected. As Derrida (2004) argues, humans should see animals as they present themselves to be seen, rather than framed as cultural metaphor.[39]

Animal performance historically involved the sight, sound and smell of bodies, which was part of the experience; it created a sensorium. The highly artificial, confined environments of menageries and the trained self-regulation in circus performance went against how free-roaming species live and congregate at a safe distance. There was an underlying physical violence to exhibition and performance. Although Western culture and thought pays only limited attention to the significance of sensory body engagement and affect and emotional feelings, this is the substance of animal experience and intelligence.[40] In his phenomenological philosophy about embodied experience, Maurice Merleau-Ponty contends that humans are bodily attuned to seeing the action of other bodies and following movement in habitual patterns.[41] Animal movement in the circus was conditioned for human sensory interest and species-determined action removed. As Merleau-Ponty explains, the animal body exists within an *Umwelt* through bodily movement and differentiated sensory responses.[42] The animal's loss of a species sensory world inflicts cruelty, a loss that is magnified in circus performance with its subservience to human sensory experience.

An appreciation of human phenomenological experience, however, can benefit animals when it recognizes embodied entities within shared environments. In his phenomenological exploration of a shared world, and Nietzsche's rejection of the

---

37 Singer, Peter. *Animal Liberation*. London: Pimlico, 1995 [first published in 1975].
38 Chaudhuri, Una. *The Stage Lives of Animals: Zooësis and Performance*. London: Routledge, 2017, 27.
39 Derrida, Jacques. The Animal that Therefore I Am. Translated by David Wills. In *Animal Philosophy*, Matthew Calarco, Peter Atterton (eds.), 113–128. London: Continuum, 2004.
40 See, for example, Bekoff, Marc. *The Emotional Lives of Animals*. Novato, CA: New World Library, 2007.
41 Merleau-Ponty, Maurice. *Phenomenology of Perception*. Translated by Donald A. Landes. London: Routledge, 1996.
42 Merleau-Ponty, Maurice. *Nature*. Translated by Dominique Séglard. Robert Vallier (ed.). Evanston, IL: Northwestern University Press, 2003, 175.

Christian taming of human animality, Ralph Acampora builds the argument for compassion in human–animal encounters.[43]

As circus animal performance was completely humanized in feats that went against the habitual movement, it was distorted further with an overlay of human-centric emotions. This helped make animal acts watchable.[44] Circus added a theatrical layer through the presenter's costume, props, music and action to create short emotive narratives that further masked animal subjectivity. Even given emotional bonds with humans, training circumvented species bonding, and performance camouflaged the absence of a self-determining life, one reduced to what Giorgio Agamben calls "a bare life"[45].

While training recognized agency, as philosopher and horse trainer Vicki Hearne explains, the subjective experience of the other is unknowable whether animal or human.[46] Wild animals stood for freedom within modernity, and, Luc Ferry suggests, within a continuum with pre-modernist precepts that relayed "a phenomenology of the enigmatic nature of animals and of the contradictory sentiments".[47] The circus performance, however, presented wild animals theatrically framed by familiar narratives of human emotional experience. Emotively evocative performance erased an idea of enigmatic subjectivity and freedom. At least in the circus, however, the individual personality and emotional responses of an animal were long recognized, if not an innate right to a free-roaming life.

Selected Bibliography

Chaudhuri, Una (ed.). Animals and Performance. Special issue of *The Drama Review* 51, no. 1 (193) (Spring 2007).
Chaudhuri, Una and Holly Hughes (eds). *Animal Acts: Performing Species Today*. Ann Arbor, MI: The University of Michigan Press, 2014.
Goodall, Jane R. *Performance and Evolution in the Age of Darwin*. London: Routledge, 2002.
Grant, Teresa, Ignacio Ramos Gay and Claudia Alonso Recarte (eds.). Real Animals on the Stage. Special issue of *Studies in Theatre and Performance* 38, no. 2 (June 2018).
Ham, Jennifer. Taming the Beast: Animality in Wedekind and Nietzsche. In *Animal Acts: Configuring the Human in Western History*, Jennifer Ham, Matthew Senior (eds), 145–163. New York, NY: Routledge, 1997.
Knowles, Ric (ed.). Interspecies Performance. Special issue of *Theatre Journal* 65, no. 3 (October 2013).
Read, Alan (ed.). On Animals. Special issue of *Performance Research* 5, no. 2 (August 2000).

---

43 Acampora, Ralph R. *Corporal Compassion: Animal Ethics and Philosophy of Body*. Pittsburgh, PA: University of Pittsburgh Press, 2006.
44 Desmond, Jane. *Staging Tourism*. Chicago, IL: University of Chicago Press, 1999, 174.
45 Agamben, Giorgio. *Homo Sacer*. Stanford, CA: Stanford University Press, 1998.
46 Hearne, Vicki. *Adam's Task*. New York, NY: Alfred A. Knopf, 1986.
47 Ferry, Luc. *The New Ecological Order*. Translated by Carol Volk. Chicago, IL: University of Chicago Press, 1995, 44.

Orozco, Lourdes and Jennifer Parker-Starbuck (eds). *Performing Animality: Animals in Performance Practices*. London: Palgrave, 2015.
Orozco, Lourdes. *Theatre and Animals*. London: Palgrave Macmillan, 2013.
Parker-Starbuck, Jen (ed.). Theatre and the Nonhuman. Special issue of *Theatre Journal* 71, no. 3 (September 2019).

Linda Kalof
History of Animal Iconography

1 Introduction and Overview

Iconography is a misunderstood, abused and flexible term with undefinable boundaries, but as an intellectual activity it has a clear definition: the description or reading of images and the subjects depicted in art, with the goal of documenting and trying to understand what is seen using verbal means from elaborate descriptions to short succinct words or codes.[1] Visual arts are critical to the understanding of culture and knowledge of cultural context is essential in interpreting imagery. Thus, the practice of iconography has moved away from traditional art history toward cultural studies, such as Michael Camille's proposed "anti-iconography" that pursued the meaning of animal forms for medieval secular folk that pushed against the official interpretation.[2] In this chapter I describe representative animal images produced in the western world in an attempt to understand the relationships we humans have had with animals throughout history. I follow Charles Sanders Peirce's definition of an icon (from the Greek word *eikon*): a representation that resembles or has a likeness to its object, such as a photograph or a portrait; "a pure icon represents whatever it may represent, and, whatever it is like."[3] The term animal icon also captures the reverence humans have always had for animals. Because visual images and archeological artifacts are the only evidence we have of preliterate human-animal relationships, I begin with a detailed account of prehistoric animal iconography. After this review of the evidence from prehistory, I then cover representative examples of animal iconography from Antiquity to the present day.[4] I conclude with a discussion of issues in the contemporary approach to animal iconography, with a focus on implications for the *animal turn*.

---

[1] Hourihane, Colum. Iconography in the Western World. In *Oxford Bibliographies Online* in Art History, Thomas DaCosta Kaufmann (ed.), published 30 January 2014, last modified 24 June 2020. doi.org/10.1093/OBO/9780199920105–0044.
[2] Roberts, Ann. Review of Iconography at the Crossroads. *Studies in Iconography* 16 (1994): 228–231.
[3] Peirce, Charles Sanders. *The Essential Peirce, vol. 2: Selected Philosophical Writings (1893–1913)*. Nathan Houser (ed.). Bloomington, IN: Indiana University Press, 2003, 163.
[4] Portions of this chapter are revised and updated from my earlier works: Kalof, Linda. *Looking at Animals in Human History*. London: Reaktion Books, 2007; Kalof, Linda. Animal Images in Paleolithic Cave Art. *Actual Archaeology* Spring 13 (2015): 26–35.

2 Topics and Themes

Animal Iconography in Cave Art

Nowhere is the importance of animals to humans more compelling than in Paleolithic cave art. Hundreds of animal figurative wall paintings and examples of portable art have been discovered, and the art is very ancient. New cave art is found regularly[5] (and old finds are re-dated) in many regions of the world, including Indonesia, South Africa, China, Australia, Argentina, India, and Russia. I focus on European cave art from southwestern France and northern Spain, often referred to as Franco-Cantabrian cave art.

Most prehistoric animal art discovered thus far was created by the Paleolithic Europeans who painted some 350 caves in Europe, with each cave telling a different story of the cultural and physical conditions of the time the art was created. Over the many thousands of years of the production of parietal (wall) and portable animal art, the figures did not develop from simple to complex or from abstract to naturalistic. The depicted animal behaviors vary widely from cave to cave, even in the same geographical area. In France's Chauvet cave the animals were drawn in ways suggesting movement, speed and strength. But in the Cosquer cave, the positions of the animals are fixed and immobile, never leaping, running or charging. A few animals are depicted motionless, often with turned-up feet and protruding tongues, leading some to the conclusion that these images are representations of dead or dying animals. For example, it has been argued that the curled-up bison on the Altamira ceiling is dead. On the other hand, as is often the case, scholars see different things in the same imagery. The Altamira curled bison has also been interpreted as not dead at all, but rather wallowing or rolling on the ground to mark his scent. Indeed, Randall White argues that animals are rarely depicted in postures of suffering and pain in Paleolithic cave paintings, and there is a marked absence of violence in the pictures.[6] The one famous depiction of violence is the Shaft Scene at Lascaux that includes the only human figure drawn on the Lascaux cave walls – a human lying on the ground, likely dead from an attack by a wounded bison. The bison is pictured hovering over the human with a shaft piercing his side and entrails pouring from his wound – clearly a hunt gone awry. The species most often depicted in cave paintings were ungulates (horse, bison, aurochs, deer, ibex and mammoth). Bears, felines and rhinos constitute about 10 percent of the images. Horses and bison were drawn larger and in

---

[5] In 2019 scientists reported a 44,000-year-old scene of a pig-and-buffalo hunt on a cave wall in Sulawesi, Indonesia, that is considered to be currently the oldest pictorial record of storytelling and the earliest figurative artwork in the world (Aubert, Maxime, Rustan Lebe, Adhi Agus Oktaviana, Muhammad Tang, Basran Burhan, Hamrullah, Andi Jusdi et al. Earliest Hunting Scene in Prehistoric Art. *Nature* 576, (2019): 442–445. doi.org/10.1038/s41586–019–1806-y).

[6] White, Randall. *Prehistoric Art: The Symbolic Journey of Humankind*. New York, NY: Harry N. Abrams, 2003.

more detail than other species, and there are some images of smaller animals, such as a weasel in the Niaux cave (c. 14,000 BP),[7] a penguin in the Cosquer cave (c. 27,000 BP), and owls in the Chauvet (c. 32,000 BP) and Le Portel (c. 11,600 BP) caves.

Chauvet is the oldest cave art found in Europe thus far at approximately 32,000 BP. The Chauvet cave is filled with complex scenes – confronting rhinoceroses, snarling lions, groups of animals drawn as if rapidly moving through the cave – 420 animal figures in all (and only 6 human images). The Chauvet cave artists used sophisticated artistic techniques to render the carnivores in their environment as spectacularly lifelike. They depicted animal figures with motion, speed, strength and power by using shading, perspective and rock surfaces for three-dimensionality and volume. The remarkable anatomical detail of some species is evidence that prehistoric humans had substantial knowledge of certain animals in their environment. An example of careful observation of animals by our Paleolithic ancestors is the "dappled horse" image (c. 25,000 BP) in the Pech-Merle cave (Fig. 1). The image was clearly drawn from real-life experience of the animal, since the depicted color schemes match the genotype of ancient horse populations.[8] Terry O'Connor of the University of York noted that the findings indicate that we can have greater confidence in understanding Paleolithic depictions of animals as naturalistic illustrations because people drew what they observed.[9]

Much of the familiarity humans had with some animal species was likely the result of the short flight distance between human hunters and their prey. Thus, carnivores are consistently inaccurately represented in Paleolithic art, probably because they were not as easy to observe from short distances as were herbivores. Paul Bahn and Jean Vertut write that the most common mistake in the representation of carnivores is the incorrect position of the canine teeth (the only teeth ever depicted in carnivore images and usually exaggerated in size) in both bears and felines, with the lower canines depicted in such a way that they would come in contact with the upper canines when the animal's mouth is closed.[10]

Bears had a unique relationship with the Paleolithic Europeans. In the Chauvet cave, drawings of bears never included their eyes, and while we do not know the significance of that absence, the bear seems to have had an important role in the cave. There are traces of bear everywhere: there are bear prints and claw marks on the

---

[7] BP is calculated by adding 1950 to the BC year.
[8] Pruvost, Melanie, Rebecca Bellone, Norbert Benecke, Edson Sandoval-Castellanos, Michael Cieslak, Tatyana Kuznetsova, Arturo Morales-Muñiz, Terry O'Connor, Monika Reissmann, Michael Hofreiter and Arne Ludwig. Genotypes of Predomestic Horses Match Phenotypes Painted in Paleolithic Works of Cave Art. *PNAS* 108, no. 46 (Nov 2011): 18626–18630. doi.org/10.1073/pnas.1108982108.
[9] Cohen, Jennie. Cave Painters Didn't Dream Up Spotted Horses, Study Shows. *History*, November 8, 2011, updated August 22, 2018. URL: www.history.com/news/cave-painters-didnt-dream-up-spotted-horses-study-shows (December 6, 2020).
[10] Bahn, Paul G. and Jean Vertut. *Journey Through the Ice Age*. Berkeley, CA: University of California Press, 1997.

Figure 1: Dappled Horses, Pech-Merle Cave, France, c. 25,000 BP. Replica in the Brno museum Anthropos. Own Work.

walls and hollows of the cave, hundreds of bear bones and bear paw prints are preserved in the cave floor. Analysis of the mitochondrial DNA of the bear remains in the cave indicates that the population was small and perhaps as old as 37,000 BP. Humans likely venerated the bear because of the animal's similarity to them in form (in gait, in the ability to stand on the hind feet, and in possessing a palm with five digits) and in omnivorous diet.

The emphasis on large, powerful animals declined around 25,000 years ago, when parietal art represented primarily herbivores with few detailed drawings of carnivores. While it has been argued that this stylistic difference could be a reflection of the abundance of animal species in the older period which later became rare in the environment and thus also in art, it is largely unknown why the motifs changed from carnivores to herbivores. Herbivores of course accounted for most of the diet of Paleolithic humans. And of the numerous theories proposed over the past 125 years to account for the cave art, the most common is that the paintings and carvings of animals represent hunter art – part of a ritual in which visual representations of hunted prey provided a means of communicating hunting information to other humans. But we simply do not know why they went to such great lengths to paint, draw and carve animal images. In addition to the theory that the animal art represented a kind of hunter art, some argue that perhaps animals were in short supply and the purpose of the cave art was to *make* animals, with the art encouraging their reproduction and thus ensuring a source of food.[11] It is logical that they would celebrate animals whom they held in high esteem. For most of early history, humans were more often prey than they were predators and occasionally scavenged off the carcasses killed and abandoned by other, more powerful carnivores. And it is hardly surprising that humans would admire the carnivores who took precedence in the feeding hierarchy. At most cave sites there is no direct relationship between the species depicted on the walls and ceilings and the bones found scattered about, indicating that the artists did not usually draw the species eaten by the human group. For example, at Altamira they ate red deer and drew bison. The Lascaux humans consumed primarily reindeer, but only one of the pictures they drew is clearly a reindeer, and yet Lascaux has a stunning mural of prehistoric bulls and horses.

Animal Iconography in Portable Art

Animal iconography in portable art also shows how Paleolithic artists crafted representations of animal behaviors. Three small figures, none longer than 2 cm and each carved from mammoth ivory, were found in 1999 in Hohle Fels, Germany: a water bird, the head of a horse and a part-human and part-animal therianthrope with feline characteristics. Some of the oldest representatives of figurative art in the world, the three-dimensional carvings were sculpted with remarkable attention to details of animal physical characteristics and polished from constant handling. The Hohle Fels bird (c. 31–33,000 BP) has an extended neck suggestive of flying or diving, with wings close to the body, feathers along the back, clearly recognizable eyes and a sharp conical beak. The Hohle Fels horse, 30,000 BP, also has an extended neck suggestive of running or galloping, fine cross hatching and parallel lines along the sides of the face and jaw, with a recognizable mouth, nostrils and eyes.

The ibex spearthrower of Le Mas d'Azil is a particularly interesting piece of portable art. The ibex is looking back at his hindquarters at what appears to be emerging feces with two birds perched on top. It has been argued that defecation imagery was depicted in prehistoric art because the presence of prey feces leads to vital information about securing game food.[12] The ibex image dates to 16,000 BP and, like the hybrid animal-human statuette mentioned above, it was probably mass produced because fragments from ten figures of the same image have been found, suggesting that hundreds of the design were made. Most figurines were carved from stone or ivory, since carving in hard bone is very difficult, but there are some splendid examples of bone portable art, particularly horse heads. Scholars suggest that early hu-

---

[11] Janson, Dora, Horst Woldemar Janson and Joseph Kerman. *A History of Art and Music.* Englewood Cliffs, NJ: Prentice Hall, 1968, 4.
[12] Mithen, Steven J. *Thoughtful Foragers: A Study of Prehistoric Decision Making.* Cambridge: Cambridge University Press, 1990, 245.

mans may have attached a mystical role to carvings or engravings on material that was once alive, particularly antler because of its continual growth and annual shedding,[13] such as the Montastruc mammoth spearthrower that was made from reindeer antler c. 13,000 BP, and a reindeer antler tool decorated with figurative art. As in the cave paintings, there is no relationship between the animals sculpted and those in the diet of the Paleolithic artists, further evidence that prehistoric animal representations are not a roster of hunted prey.

The last prehistoric paintings ever made were primarily of horses and bison that were drawn on the walls of the Le Portel cave in France, which dates to c. 11,600 BP. Steven Mithen argues that as a result of global warming the tradition of painting and carving animals disappeared.[14] During the cold climate of the Ice Age, cave art, along with mythology and religious rituals, sustained hunting information and how to survive in a harsh environment. But around 11,500 BP the climate grew warmer, life was not as demanding, hunting could be practiced anywhere, at any time and by anyone, and there was no longer a need to convey hunting information.[15]

The end of cave art marked a major change in the relationship between humans and other animals. Excavations at Çatalhöyük, one of the first true cities inhabited 9000 years ago in the area of present-day Turkey, document the transition from hunting and gathering subsistence to the domestication of animals [→Domestication: Co-evolution]. Art is everywhere among the remains of Çatalhöyük, and most of the images are of animals – leopards, foxes, cranes, a mural of a wild bull being baited by hunters – and stunning installations of animal remains such as bull bucrania that decorated the main room of the houses.[16]

In addition to the change in art that came about with domestication, around 5450 BP Sumerian stone cylinder seals depict a motif of animals and humans engaged in fighting. The only visual record we have of life in ancient Mesopotamia, the cylinder seals illustrate the perception of human-animal relationships as linked to struggle and violence. As more and more humans came to live in cities, wealth, trading and fighting increased, and humans began to use wild, ferocious animals, particularly bulls, as symbolic of struggle, violence, warring kingdoms, and the ongoing cycles of battle between the uncivilized and civilized and nature and culture.

One thousand years later, c. 4550 BP, Mesopotamian artists created lifelike, naturalistic sculptures of animals, giving form and substance to individual species such as lapis lazuli and shell for eye inlays that showed animal vitality [→Material Culture Studies]. As in prehistoric iconography, animal behaviors are also represented, such as the illustration of free movement in a series of animal images that progress from

---

13 Bahn and Vertut, *Journey*, 103.
14 Mithen, Steven. *After the Ice: A Global Human History, 20,000 – 5000 BC*. Cambridge, MA: Harvard University Press, 2004.
15 Mithen, *After the Ice*.
16 German, Senta. Çatalhöyük. *Khan Academy*. URL: www.khanacademy.org/humanities/prehistoric-art/neolithic-art/a/atalhyk (December 6, 2020).

walking to cantering to full gallop. Most of the animal artefacts from this period come from elaborately decorated musical instruments and sculptures found in royal tombs and burial sites, such as the well-known sculpture of a male goat rearing up to reach the leaves on a budding plant, an image believed to have been a common sight along the banks of the Tigris and Euphrates rivers, in addition to possibly representing the Sumerian concern with the fertility of plants and animals.[17]

No animal species has been so critical to human civilization as cattle. Indeed, it is argued that historically, with the possible exception of gold, the most enduring object of desire has been the cow.[18] The domestication of cattle approximately 11,000 years ago was a crucial step in human history, leading to modifications in diet, behavior and socioeconomic structure of human populations.[19] The importance of cattle is conveyed in the cultural landscape of art, from prehistory to the nineteenth century. While much of the ancient art were depictions of human-animal confrontation, there are many peaceful scenes of domestication and friendship between humans and cattle in ancient iconography (Fig. 2). Ancient Egyptian limestone carvings show images of nurturing and bonding among cattle, such as mother cows who gaze at or lick their calves and calves glancing back at mothers for reassurance while being carried over a river on a cowherd's back (Tomb of Ti, Saqqara, c. 4350 BP). Domestic animals are often represented in situations that document the daily routine of country life in ancient Egypt. One excellent example is the Voyage to Punt (c. 3425 BP), a beautiful limestone carving of a herd of cattle grazing under trees that gives the illusion of the movement of foraging animals and of the leaves overhead rustling in the wind.

Animal Iconography and the Medieval Bestiary

Naturalistic representations of animals waned as Antiquity came to a close. The detailed descriptions of animal physiology and behavior produced by the classics such as Pliny the Elder and other observations of real animals were rare until the thirteenth century when firsthand observations of animals finally began to be used again to generate scientific knowledge and works of art.

---

17 Hafford, William B. Object Lesson: Ram in the Thicket. *Current World Archaeology* 91 (September 20, 2018). URL: www.world-archaeology.com/issues/object-lesson-ram-in-the-thicket/ (December 6, 2020); Collins, Paul. A Goat Fit for a King. *ART News* 102 (2003): 106–108; Hansen, Donald P. Art of the Royal Tombs of Ur: A Brief Interpretation. In *Treasures from the Royal Tombs of Ur*, Richard L. Zettler, Lee Horne (eds.), 43–59. Philadelphia, PA: University of Pennsylvania Museum, 1998.
18 Sharpes, Donald K. *Sacred Bull, Holy Cow: A Cultural Study of Civilization's Most Important Animal.* New York, NY: Peter Lang, 2006.
19 Beja-Pereira, Albano, David Caramelli, Carles Lalueza-Fox, Cristiano Vernesi, Nuno Ferrand, Antonella Casoli, Felix Goyache et al. The Origin of European Cattle: Evidence from Modern and Ancient DNA. *PNAS* 103, no. 21 (May 2006): 8113–8118. doi.org/10.1073/pnas.0509210103.

Figure 2: Old Egyptian hieroglyphic painting showing an early instance of a domesticated animal (cow being milked), date unknown. From *1000 Fragen an die Natur*, via The Metropolitan Museum of Art, Rogers Fund, 1948.

Exceptions to the penchant for animal iconography to be based at the time on secondhand observations are found in the marginalia of medieval devotional books. For example, the Luttrell Psalter of the mid-1300s illustrates rural life in England and the daily activities of animals and humans. The drawings are remarkable lifelike animal representations. One can easily make out the whiskers and stubby snout of bulls who pull a plough, the heavy working shoes on the hooves of draft animals, the crowded conditions of sheep penned up in a sheepfold, and even emotional states as shown in a tired, exhausted "grumpy" bull.

Animal imagery was often used to convey moral messages and religious principles to those who could neither read nor write. Excellent examples of how animals were used to teach folklore and Christian dogma are in the bestiaries that were popular during the ninth to the fourteenth centuries. These "books of beasts" were visual illustrations of the Greek *Physiologus*, a text on the supposed natural characteristics of animals (but not based on direct observation, instead drawing on the classic writings of Aristotle and Pliny the Elder) and allegorized those characteristics according to Christian principles. For example, in a drawing from the Aberdeen Bestiary, a wolf sneaks into a sheepfold while the shepherd sleeps, teaching the principle of the necessity of tending to one's task (Fig. 3). One can see the exaggerated physical characteristics of the wolf that emphasize the predator's yellow eyes, sharp teeth and claws and unusually large size compared to the sheep.

Medieval anti-Jewish attitudes and misogyny were often narrated alongside pictures of animals, with the texts "de facto expressions of hostility toward women and

Figure 3: Aberdeen Bestiary MS 24, Folio 16v: Wolf sneaking up to the sheepfold, Twelfth Century. Special Collections, University of Aberdeen.

Jews, albeit marketed as moral guidance."[20] Artistic representations of the "backwards ride" ritual are good examples of the use of animals to humiliate those devalued in medieval society. Nude women (symbolic of lewd behavior) and apes (symbolic of Jews) are pictured in devotional books riding backwards on animals considered filthy, evil or undesirable, such as rams, pigs or goats, and the popular appeal of the ritual was linked to the principle of amusing by abusing and the ambivalence of a world turned upside down.[21] Representations of the backwards ride have been found in historical documents and folklore in a wide range of countries across the globe and in contemporary popular culture in US political cartoons.[22]

In the thirteenth century, the Benedictine monk, cartographer and artist Matthew Paris produced one of the first naturalistic representations of an animal since Antiquity, with the exception of the few medieval pictures of rural life mentioned earlier. Paris' sketches of an elephant in the Tower of London are remarkably accurate in proportional relationships, illustrations of motion, and in details of the animal's jointed legs and flexible trunk. Even when animals were "drawn from life" most me-

---

[20] Hassig, Debra. Sex in the Bestiaries. In *The Mark of the Beast: The Medieval Bestiary in Art, Life and Literature*, Debra Hassig (ed.), 71–93. New York, NY: Garland, 1999, 82.
[21] Mellinkoff, Ruth. Riding Backwards: Theme of Humiliation and Symbol of Evil. *Viator* 4 (1973): 153–176.
[22] Mellinkoff, Riding Backwards.

dieval artists had difficulty with realistic anatomical detail, such as the depiction of elephants with trumpet-like trunks, sharply curved boar-like horns and cloven hooves. However, Matthew Paris' realistic representations of animals were harbingers of a transformation in attitudes from a symbolic to a naturalistic view of the natural world that created a science devoted to the exact imitation of nature that dominated western art for the next seven hundred years.[23]

Animal Iconography and Renaissance Realism

Naturalistic animal iconography was at its peak in the Renaissance, and some of the most realistic animal images from this period are of domestic animals. Antonio Pisanello, the first great Renaissance artist of animals, sketched stunningly life-like animal portraits of horses, cows and dogs, many of which are collected in the Codex Vallardi, a portfolio of drawings housed in the Louvre. Pisanello's sketch of a horse head shows his concentration on the shape of the horse's head, his body hair, the flowing hair of the mane, and the shorter hair of the muzzle whiskers (Fig. 4).

Figure 4: Horse Head, Pisanello (1395–1455), c. 1440, Catalogue de l'exposition à Paris: Musée du Louvre, 1996.

---

23 White, Lynn, Jr. *Medieval Religion and Technology and Social Change.* London: Oxford University Press, 1962.

Leonardo da Vinci's sketches of animals were unique in illustrating their individuality, behaviors and especially their liveliness. Leonardo's ability to convey visually the moods and emotions of animals was unique among the artists of his time.[24] His drawings of horses in battle scenes are particularly vivid in the depiction of emotions including rage and hatred, such as his 1505 Battle of Anghiari (a lost painting at times referred to as "The Lost Leonardo") that shows horses and riders in such close contact that even the horses appear to bite one another (Fig. 5), with man and animal becoming "a single creature, whose uncontrolled rage finds appropriate expression in an unnaturally twisted body."[25]

In the sixteenth century Albrecht Dürer also sketched animals as mirrors of nature, such as his stunning drawing of Two Seated Lions (Fig. 6). Dürer's realistic representations were based on his sketches of live animals held in menageries and dead animals from curiosity cabinets, such as the Dead Blue Roller and the Wing of a Eu-

Figure 5: Reproduction of Leonardo da Vinci's The Battle of Anghiari, 1505, by Gérard Edelinck, c. 1666, Museum of Fine Arts, Houston, Google Cultural Institute.

---

24 Clark, Kenneth. *Leonardo da Vinci: An Account of his Development as an Artist.* London: Penguin, 1959.
25 Zöllner, Frank. *Leonardo da Vinci 1452–1519: The Complete Paintings and Drawings.* Cologne: Taschen, 2003, 252.

Figure 6: Two Seated Lions, Albrecht Dürer, 1521, Kupferstichkabinett Berlin. Google Cultural Institute.

ropean Roller [→History of Animal Collections/Animal Taxonomy]. While direct observations were the mark of scholarly credibility and essential to the advancement of scientific knowledge,[26] Dürer's woodcut of a rhinoceros (1515) is famous for becoming the standard visual representation for two hundred years, albeit an incorrect one – there is no second horn on the animal's shoulder as is depicted in the woodcut. Dürer did not draw the animal from his own observations but based his work on a printed description of a rhino in a zoo. Using a printed source rather than his own observation was highly unusual for him, and the non-existent shoulder horn was copied in natural history representations of the rhino until the nineteenth century.

Animal Iconography of Food, Hunting and Breeding

The epidemics of plague in the fourteenth and fifteenth centuries had a substantial impact on animal iconography. In the sixteenth century, animals were often depicted as dead and waiting for human consumption in the aesthetic illustration of food

---

[26] Baratay, Eric and Elisabeth Hardouin-Fugier. *Zoo: A History of Zoological Gardens in the West.* London: Reaktion Books, 2002.

abundance. Even though meat was largely unavailable to a starving public, artists such as Pieter Aertsen and Joachim Beuckelaer painted lavish market and kitchen scenes that illustrated the objectification of agricultural produce and food animals, although, as Bendiner argues, the market scenes are "fantasies" because such abundance of food did not exist before the nineteenth century, and none of the scenes show rotten, discolored or insect-spoiled food.[27]

Well known in the iconography of food abundance is Aertsen's 1551 "Butcher's Stall," an elaborate display of butchered animals and animal flesh – cows, chickens, fish, pigs, animal entrails and slabs of meat and fat – with the religious narrative of the Flight into Egypt buried in the background. In addition to the caricature of peasants who engage in morally and socially inferior activities such as trading in meat and vegetables, Norman Bryson argues that the discourse of the painting associates lower social status with the bodies of animals and with human body functions such as consumption and ingestion.[28] Bendiner notes that Aertsen "uses food to give a smell of the world [...] (t)his food may be dead matter, but it seems to throb with the pulse of bodily life."[29]

Dead animal portraiture continued the tradition of painting animals in elaborate displays of food abundance in the form of gamepiece art that had been popular as still life trophy pictures since the early 1500s, as in Jacopo de' Barbari's Still-life with Partridge and Gauntlets (1504) [→Visual Culture Studies and Art History]. The hunting still life was a safe artistic endeavor. With the onset of the Protestant Reformation, artists, particularly in the Netherlands, specialized in themes that were not religious and thus not considered objectionable, such as scenes of nature, daily life and other genre pictures illustrating the surface of things. Dead game art was popular among the rising middle class. While they were not allowed to hunt game animals, they could own a gamepiece that represented the sport of hunting and the owner's social status.[30] The gamepiece displayed animals only the nobility were allowed to hunt – stag, boar, roe deer, pheasant and swan – prized aristocratic animals who replaced the mundane domestic cows, pigs and chickens of the kitchen and market scenes of the time.[31] Thus, paintings of dead game conveyed a discourse of privilege in the masculine sphere of hunting, an iconography of masculinity removed from the trivial culture of the domestic sphere, the kitchen and the market [→Feminist Intersectionality Studies].[32] While market scenes and the gamepiece co-existed

---

27 Bendiner, Kenneth. *Food in Painting: From the Renaissance to the Present*. London: Reaktion Books, 2004, 48.
28 Bryson, Norman. *Looking at the Overlooked: Four Essays on Still Life Painting*. London: Reaktion Books, 1990.
29 Bendiner, *Food in Painting*, 40.
30 Sullivan, Scott A. *The Dutch Gamepiece*. Totowa, NJ: Rowman & Allanheld, 1984.
31 Sullivan, *Dutch Gamepiece*, 17.
32 Bryson, *Looking at the Overlooked*, 160–161.

Figure 7: Still Life with a Dead Stag, Frans Snyders, 1650. Mauritshuis, Den Haag, 794.

for centuries, the distinction between the two is one of social class.[33] Dead game animals belonged to the upper class who acquired the meat using strength, skill and the social right to hunt, while the animals sold in the market belonged to those common people who could afford to pay for meat.[34]

The gamepiece genre reached its peak in the mid-1600s with the work of Frans Snyders and Peter Paul Rubens. Frans Snyders was a prolific painter of dead game and his work often included a large animal who centered the painting, such as the gutted stag in "Still Life with a Dead Stag" (Fig. 7) who is posed in human positions (lying on his back, legs crossed, head canted to the side, neck stretched out from his body). Typical of Snyder's gamepiece art, a snarling boar's head is positioned to the right of the center figure, along with a lobster and on the left a confrontation between a live cat and bird.

Gamepiece iconography transitioned after 1650 as artists focused on dead animals pictured in decorative outside settings, such as hunting pavilions or landscapes, and included the display of hunting weapons and sporting dogs.[35] Popular in the eighteenth century among the French aristocracy, Jean-Baptiste Oudry's work stands as an exemplar of the decorative trophy gamepiece, such as his "Nature

---

33 Bendiner, *Food in Painting*, 44.
34 Bendiner, *Food in Painting*, 44.
35 Sullivan, *Dutch Gamepiece*, 21, 78.

Figure 8: The Meadow, Paulus Potter, 1652, Musée du Louvre, Atlas database: entry 25801.

morte au buste de l'Afrique" [Still Life with the Bust of Africa]. In a picturesque garden scene, dead game animals are piled against a hunting rifle while hooded hawks perch overhead and a hunting dog gazes upon the trophies. A tribute to the art of hunting, the image depicts the three types of hunts practiced by the royalty: the shooting hunt symbolized by the rifle, the high-flying hunt represented by the hawks, and hunting with hounds symbolized by the presence of the dog.[36]

The early Enlightenment brought increased attention to the lived experience of animals, and much of the iconography of the time pictured animals in the fields and meadows where they lived and labored and not foregrounded or backgrounded in a human context. Paulus Potter's work in the mid-seventeenth century is an exemplar of the depiction of animals with the landscape only a backdrop, such as "The Meadow" (Fig. 8) and his numerous "Two Cows" paintings including "Landscape with Two Cows," "Two Cows Seen from Behind," "Two Cows on a Hill," "Two Cows beside a Stream," and so on. The milk cow was a symbol of pride in the Dutch milk industry, reflecting prosperity and stable social and economic conditions.[37] Potter's work and the paintings of Aelbert Cuyp (that often included shep-

---

36 Le Chateau de Versailles en 100 Chefs-d'oeuvre, dossier de presse. URL: www.aphg.fr/IMG/pdf/160320-versailles-arras.pdf (December 6, 2020).
37 Turnbull, Alexandra. The Horse in Landscape: Animals, Grooming, Labour and the City in the Seventeenth-Century Netherlands. *Shift: Queen's Journal of Visual & Material Culture* 3 (2010): 1–24.

herds tending cows against a panoramic Dutch landscape) convey a message of "reciprocal partnerships" between humans and work animals.[38]

Potter's horse and cattle paintings displaced the human as owner, groomer or viewer in the scene, reflecting seventeenth-century Dutch concerns about changes in the land, commerce, and labor.[39] For example, Turnbull argues that Potter's visual language in his horse images represent the horse as disengaged from his traditional role as a laborer – there are no pictures of horses at work or wearing harnesses and their bodies are framed in the landscape "in a way that mediates between the viewer and the pristine urban settlements situated along the horizon [...] (and thus are) a crucial site for reflecting on urban concerns."[40]

In the early 1800s, there was a return to the objectification and corporealization of animals with the popularity of livestock portraiture. Wealthy breeders commissioned portraits of their prized animals, with size and pedigree serving as proof of their value, thus emphasizing the owner's social status [→Social History].[41] The celebrated animal was often pictured against a distant landscape and from the side with small heads and broomstick legs (Fig. 9), and if humans appeared in the picture, they were tiny in comparison to the animal.[42]

Animal iconography was a masculine endeavor[43] until the appearance of Rosa Bonheur on the French art scene in 1841. Bonheur was the first woman to win the Grand Cross by the French Legion of Honor, and her paintings of animals in natural habitats are known for their lack of sentimentalism and anthropomorphism, such as her "Stag Listening to the Wind" (1867) that contrasts with Landseer's popular "Monarch of the Glen" (1851).[44] While Landseer's stag is portrayed against distant mountains as a proud animal with an outward, almost confrontational gaze, Bonheur's stag represents a cautious deer in a wooded background poised listening for danger. Her paintings in the 1860s were of domestic animals, deer, horses and dogs from the collection of animals on her estate. Her most famous masterpiece, "The Horse Fair," included a self-portrait in the form of a central male figure in the painting that provided visual testimony of a public lesbian identity.[45] Two American women artists,

---

38 Turnbull, The Horse, 9.
39 Turnbull, The Horse, 17.
40 Turnbull, The Horse, 12.
41 Berger, John. *Ways of Seeing*. London: Penguin Books, 1972.
42 Ritvo, Harriet. *The Animal Estate: The English and Other Creatures in the Victorian Age*. Cambridge, MA: Harvard University Press, 1987.
43 We do not know if the animal art of Prehistory and Antiquity was created primarily by men (although some scholars suggest it to be so in the Paleolithic, see Guthrie, Russel Dale. *The Nature of Paleolithic Art*. Chicago, IL: University of Chicago Press, 2005). However, since the Middle Ages animal iconography has been attributed to men until the nineteenth century.
44 Gaze, Delia (ed.). *Concise Dictionary of Women Artists*. New York, NY: Routledge, 2013.
45 Saslow, James M. Disagreeably Hidden: Construction and Constriction of the Lesbian Body in Rosa Bonheur's Horse Fair. In *The Expanding Discourse: Feminism and Art History*, Normal Broude, Mary D. Garrard (eds.), 187–206. New York, NY: Routledge, 1992.

Figure 9: Gloucestershire Old Spot, James Ward, 1805, Yale Center for British Art.

Lily Irene Jackson (1848–1928) and Matilda Lotz (1858–1923), were also celebrated for their realistic paintings of domestic animals, primarily dog portraits (Fig. 10).

Animal Iconography of Animal Abuse

Concern for the inward animal fueled iconographic representations of animal welfare, particularly cruelty and injustice at the hands of humans in a variety of venues from hunting to urban living. For example, in the mid-1800s the hunting scenes painted by Gustave Courbet (himself an avid hunter) reconceptualized the traditional genre of hunting as a noble pursuit of the wealthy. At a time when hunting was no longer reserved for the nobility, Courbet's work illustrated the torture and suffering of the hunted animal (Fig. 11), contradicting the view of the glorified hunter and reflecting on the morality of the hunt.[46] In Tseng's words, Courbet combined "naturalistic, psychological, and ethical implications [...] (forging) a new kind of history painting

---

[46] Tseng, Shao-Chien. Contested Terrain: Gustave Courbet's Hunting Scenes. *Art Bulletin* XC, no. 2 (2008): 218–243, 218.

as an interchange between artistic subjectivity, lived experience, and social attitudes toward animals."[47]

Visual messages against animal cruelty existed well before the nineteenth century. Paulus Potter's "Punishment of a Hunter" (c. 1650), a 14-panel painting that depicts a hunter captured by animals, condemned to death and roasted alive, is considered to be "a moment of transition in cultural attitudes towards human-animal relationships."[48] The Dutch art world was substantially influenced by Michel de Montaigne's sixteenth-century treatise against human superiority over other animals.[49]

Most of the animal art of the Renaissance and Enlightenment was commissioned by and for the aristocracy. It was rare for iconography to depict animals in the lives of common people. A notable exception was William Hogarth's eighteenth-century

Figure 10: Study of a Dog, Rosa Bonheur, 1860s. Princeton University Art Museum.

---

47 Tseng, Contested Terrain, 231.
48 Beirne, Piers and Janine Janssen. Hunting Worlds Turned Upside Down: Paulus Potter's Life of a Hunter. *Tijdschrift over Cultuur & Criminaliteit* [magazine about culture and crime] 4, no. 2 (2014): 15–28.
49 Beirne and Janssen, Hunting Worlds, 26.

Figure 11: Fox Caught in a Trap, Gustave Courbet, 1860, National Museum of Western Art, Tokyo.

prints of animal abuse in "The Four Stages of Cruelty" (1751) that illustrated not only the graphic torture and suffering of animals but also the progression from animal abuse to cruelty to other humans. Hogarth's intention was to reach the largest possible audience at an affordable price – he wrote that "neither correctness of drawing or fine engraving were at all necessary, but on the contrary would set the price of them out of the reach of those for whom they were chiefly intended."[50] Thus, there was a wide distribution of Hogarth's prints; they were nailed to the walls of inns, taverns and other public places and conveyed the horror of animal abuse throughout Britain and beyond.

Issues in Contemporary Approaches to Animal Iconography

Contemporary animal iconography is fraught with contradictory representations of both domestic and exotic animals. According to Steve Baker, contemporary postmodern art is about creating distance from animals to produce an unsettled view of their bodies that is fractured, awkward and wrong.[51] For example, Damien Hirst's work of "botched taxidermy" (skinned dogs and cats with modeled heads, pigs and sharks preserved in formaldehyde, a sculpture of dead animals tarred, feathered and dis-

---

50 William Hogarth, quoted in Gordon, Ian R.F. The Four Stages of Cruelty (1750). *The Literary Encyclopedia*, first published November 5, 2003. URL: www.litencyc.com/php/sworks.php?rec=true&UID=807 (December 6, 2020).
51 Baker, Steve. *The Postmodern Animal*. London: Reaktion Books, 2000.

played hanging from a tree) are among the most popular of art exhibitions. As Joe Zammit-Lucia notes, sensationalism has launched the career of many artists; misuse of the animal is an easy way for artists to achieve notoriety, with negative reactions guaranteed to become part of the artwork.[52] Live animals are also part of the genre of contemporary animal art that some artists have used paradoxically to critique animal abuse. For example, Guillermo "Habacuc" Vargas stated that his 2007 exhibition in a Nicaraguan art gallery that displayed an emaciated street dog whom he tied to a wall by a short length of wire, was intended to highlight the hypocrisy in ignoring the suffering of both animals and humans.[53]

In a very different and yet compelling critique of animal abuse, Britta Jaschinski's photojournalism documents the abuse of wildlife for entertainment, status and profit. For example, Jaschinski has captured the essence of captivity and imprisonment in her photographs of animals exploited in China under horrific conditions of confinement and cruelty. A poignant example is her photograph of Oliver, a brown bear (Fig. 12) who, before he was rescued by Animals Asia in 2010, was kept in a metal corset in a tiny cage for 30 years on a Chinese bear bile farm. Many of her photographs are troubling as she highlights the global problem of wildlife contraband in photographs of taxidermied zebra heads and tiger fetuses, elephant foot ashtrays, and bear paw slippers.

Approaching animal conservation from a different perspective are two artists who emphasize human kinship with animal others rather than animals as victims of human exploitation, Joe Zammit-Lucia and Tim Flach. Both artists use a representational approach that frames the animal in a way that mimics the human studio portrait. Zammit-Lucia's photographs are taken in a studio-like setting rather than in the wild or in a captive setting, thus extracting the subject from any habitat or landscape. He is interested in capturing the animal as an individual rather than a mere specimen of species and, most importantly, conveying something of the experience of being the animal.

The iconography of animal conservation is the life work of the animal portrait photographer, Tim Flach, who bridges animal "otherness" with "sameness," emphasizing their personality, behavior and kinship with humans (Fig. 13). However, Flach's work also incorporates the material aspects of the ecosystems of endangered animals, and his mission is to engage our moral compass in promoting animal conservation. When we view his striking portraits that give human characteristics to endangered animals, we develop an emotional connection to their vulnerability. Flach argues that portraiture can help tell stories about the natural world, uncover the human elements in animals and make the issue of wildlife conservation salient.

---

52 Zammit-Lucia, Joe. Practice and Ethics of the Use of Animals in Contemporary Art. In *The Oxford Handbook of Animal Studies*, Linda Kalof (ed.), 433–455. New York, NY: Oxford University Press, 2017.
53 Yanez, David. You Are What You Read. *Art 21 Magazine*, 4 March, 2010. URL: magazine.art21.org/2010/03/04/you-are-what-you-read/#.XC4wcFxKiUk (December 6, 2020).

History of Animal Iconography — 491

Figure 12: Oliver, a bear exploited for his bile for three decades on a Chinese bear bile farm, by Britta Jaschinski, reproduced with artist's permission.

Figure 13: Crested Macaque Celebes Monkey Eyes by Tim Flach, reproduced with artist's permission.

In a very real sense, Flach's art shows the potential role of animal iconography in the historical process of environmental conservation.

3 Implication(s) of the Animal Turn

In prehistory and Antiquity, animal iconography was devoted to realistic representations of animals based on observations of fauna in the natural environment, such as images of predator and prey animals and daily scenes of domestication. With the exception of a resurgence of naturalistic images in the Renaissance to meet a scientific ideal, animal iconography since the Middle Ages has focused primarily on the representation of animals as symbolic of hostility toward devalued others, objects of elaborate consumption, illustrative of national superiority, and more recently, victims of human exploitation. The iconography of animal exploitation and abuse that arose in the seventeenth century was concerned primarily with the cruelty of hunting and was a major change in the cultural attitudes toward human-animal relationships that previously glorified hunting as an aristocratic pursuit. In the next century, William Hogarth's mid-1700 illustrations of the graphic torture and suffering of animals were notable in their inclusion of animals in the lives of common people with the warning that the abuse of animals progressed to the abuse of other humans. Contemporary animal iconography has produced contradictory messages. Some present-day animal iconography focuses on animal abuse as a subject of sensationalism, others attempt to address animal abuse and the need for conservation with an emphasis on animal similarity with humans, bridging animal otherness with sameness. The contemporary turn toward a kinship/empathetic approach to animal iconography brings sensitivity to the way that nonhumans are represented, enhancing feelings of empathy, sympathy, and connectedness with the natural world, thus encouraging altruistic coexistence for all beings in earth's ecosystem.

Selected Bibliography

Bahn, Paul G. and Jean Vertut. *Journey Through the Ice Age*. Berkeley, CA: University of California Press, 1997.

Baker, Steve. You Kill Things to Look at Them: Animal Death in Contemporary Art. In *Killing Animals*, The Animal Studies Group, 69–95. Urbana, IL and Chicago, IL: University of Illinois Press, 2006.

Berger, John. *Ways of Seeing*. London: Penguin Books, 1972.

Bryson, Norman. *Looking at the Overlooked: Four Essays on Still Life Painting*. London: Reaktion Books, 1990.

Flach, Tim. *Endangered*. New York, NY: Abrams, 2017.

Hourihane, Colum. Iconography in the Western World. In *Oxford Bibliographies Online* in Art History, Thomas DaCosta Kaufmann (ed.), published 30 January 2014, last modified 24 June 2020. doi.org/10.1093/OBO/9780199920105–0044.

Jaschinski, Britta. *Photographers Against Wildlife Crime*. [Great Britain]: Photographers Against Wildlife Crime, 2018.

Morphy, Howard (ed.). *Animals into Art*. London: Unwin Hyman, 1989.

Sullivan, Scott A. *The Dutch Gamepiece*. Totowa, NJ: Rowman & Allanheld, 1984.

Zammit-Lucia, Joe. *First Steps: Conserving Our Environment*. New York, NY: Matte Press, 2008.

Andrew Gardiner
History of Veterinary Medicine

1 Introduction and Overview

The history of veterinary medicine has recently entered the academy. A similar change took place in the histories of science and of medicine in the 1970s when universities opened departments of Science Studies, Social Studies of Science, and History and Philosophy of Science, and academics began to investigate the methods and social structures of science and medicine.[1] Prior to this, most histories had been internal and took the form of biographies or institutional histories.[2] Academic veterinary history is developing on similar lines; in many ways veterinary history can be considered part of the histories of science and of medicine, as shown by where historians choose to publish veterinary papers. Most historians of veterinary medicine are veterinarians with an interest in the development of their profession. They are likely to belong to a local (national) veterinary history society, which may be affiliated to their national veterinary association, and/or the World Association for the History of Veterinary Medicine (WAHVM). WAHVM was founded in 1969 following a series of national and then international symposia organized by the German Veterinary Medical Association. WAHVM's main activity is the organization of its biennial congress. WAHVM acts as a federation of national veterinary history societies, but individual and institutional membership is also possible. The congresses attract both academic[3] and veterinarian-historians[4] – a collaboration also evident in WAHVM's Board structure. WAHVM organizes and funds an 'early career scholar award' to encourage new researchers of any age.[5] Most veterinarian-historians work in veterinary medicine; there are few posts in history within veterinary schools and none in the UK or North America. Veterinarian-historians usually publish in the journals of their national veterinary history societies, in international veterinary history journals and occasionally in general veterinary publications. Academic historians are unlikely to re-

---

1 The University of Edinburgh was active in this area and an *Edinburgh School* within Science Studies emerged. In the 1970s/80s the University offered interdisciplinary undergraduate degrees in Science Studies (Physics) and Science Studies (Zoology).
2 The same applies in veterinary history, for example Pattison, Iain. *John McFadyean. A Great British Veterinarian*. London: J.A. Allen, 1981; Ware, Jean and Hugh Hunt. *The Several Lives of a Victorian Vet*. London: Bachman & Turner, 1979; Cotchin, Ernest. *The Royal Veterinary College London: A Bicentenary History*. London: Barracuda Books, 1990.
3 By 'academic historian' I mean an individual trained in history, usually to doctorate level, undertaking historical research associated with a university department.
4 By 'veterinarian-historian' I mean an individual primarily trained in veterinary medicine.
5 Refer to the Association's webpage for further information: *World Association for the History of Veterinary Medicine*. URL: www.wahvm.co.uk/ (August 11, 2020).

gard veterinary medicine as their sole area of research; they may or may not be veterinarians themselves, although the number of veterinarians formally trained in history is growing. Related areas of research might include medical history, history of animals, history of science and technology, cultural history, environmental history, etc. [→History of Science;→Cultural History;→Environmental History]. Veterinary history may also be published in animal studies journals such as *Anthrozoös* and *Society & Animals*, and in journals of agricultural history, amongst others [→History of Agriculture].

An important question, and one that is pertinent to an encyclopedia of historical animal studies, is how much veterinary history is concerned with animals rather than with veterinarians and their institutions. Most historiography has concerned the latter.

The Western European veterinary profession arose in the eighteenth century in response to a need to understand devastating livestock pandemics like cattle plague (rinderpest). The profession also served to maintain the health of the horses keeping commerce, agriculture, industry, and armies moving. Involvement in public health, including the inspection of meat and milk, was another important part of the veterinary role, and often one which brought veterinarians into dispute with medical doctors.[6] The profession was either state-supported and controlled, as in Germany, or a private profession, as in Britain. This determined how veterinarians were educated, where they worked and how they went about it. Given this variation, and with the object of reviewing veterinary history in the context of animal studies, it is useful to examine what veterinary historians have studied, and then to consider what the animal studies perspective may offer history and animals.

2 Topics and Themes

This section provides an overview of work in the field of veterinary history by describing and commenting on research presented at the four most recent biannual conferences of WAHVM as well as a content analysis of the British journal *Veterinary History*.

A typical WAHVM conference receives most of its papers from individuals working within veterinary or allied science departments, followed by academic historians or inter-disciplinary scholars (Table 1). The conference in London in 2014 was an exception, with academic historians making up the majority of the speakers (50.6%). This reflects a developing academic focus on veterinary history within active departments in London (History, King's College), Manchester (Centre for the History

---

[6] Laxton, Paul. This Nefarious Traffic: Livestock and Public Health in Mid-Victorian Edinburgh. In *Animal Cities: Beastly Urban Histories*, Peter Atkins (ed.), 154–171. Abingdon: Ashgate Publishing, 2012.

of Science, Technology and Medicine, University of Manchester) and a small but growing network of academic veterinary historians elsewhere in Britain. At the four recent WAHVM conferences, around 30% of speakers have been academic historians (nearer 20% if the unusual London conference is ignored) and around 45% veterinarian-historians. Representatives of national veterinary history societies (often practicing or retired veterinarians), those working in state veterinary services, archivists/librarians and independent scholars make up the remainder, at around 4–8% each.

Table 1: Primary author affiliation of papers and posters at WAHVM congresses, 2012–2018

Affiliation	Bergen, 2018	Vienna, 2016	London, 2014	Utrecht, 2012	Total
Academic veterinary or allied science department*	20 (46.5%)	27 (64.3%)	26 (31.3%)	32 (51.6%)	105 (45.7%)
Academic history or interdisciplinary department	8 (18.6%)	7 (16.7%)	42 (50.6%)	14 (22.6%)	71 (30.9%)
National veterinary history society	3 (6.9%)	3 (7.1%)	5 (6.0%)	8 (12.9%)	19 (8.3%)
State veterinary service or institute	8 (18.6%)	3 (7.1%)	0	2 (3.2%)	13 (5.7%)
University archivist or librarian	2 (4.7%)	2 (4.8%)	7 (8.4%)	1 (1.2%)	12 (5.2%)
Independent scholar	2 (4.7%)	0	3 (3.6%)	5 (8.1%)	10 (4.3%)
Total	43 (18.7%)	42 (18.3%)	83 (36.1%)	62 (27.0%)	230

* Includes departments/divisions of veterinary history or allied subjects in veterinary schools. There are few of these.

Topics

Most WAHVM conferences have a theme as well as an open call, so an analysis of topics must be read in this light (Table 2). However, across all conferences and themes, the commonest topics for papers were 'Disease and health' (26.1%) and 'Institutions' (23.9%). Significant coverage was given to 'Public/One Health' (13.0%) and 'Biographies' (12.6%). Papers on archives and libraries made up about 10% of papers, papers on animal welfare and ethics about 5%, and papers on practices about 4%. Studies of how ordinary veterinarians made their living are lacking in veterinary history. This hole in the research might comprise a kind of veterinary history 'from below' [→Social History]. These findings suggest that veterinary history has some of the characteristics of the histories of science and of medicine before the 1970s, with a tendency to focus on disease and elites. This type of focus is consistent with history written from within a discipline, often presenting a progressivist narrative of discovery and emphasizing revolutions, breakthroughs, great white men and ever-advancing knowledge, and a concomitant tendency to minimize 'wrong tracks', peripheral yet interesting personalities, women, minorities, and the social and cultural structures within which scientists operate. From the 1970s, ideas from the Sociology of Scientific Knowledge (SSK) influenced ways of writing about science and medicine. Rather than treating scientists as unbiased investigators uncovering concrete truths about the world, SSK portrayed scientific knowledge as socially constructed and susceptible to cultural influences, including those of science itself.[7] As a branch of science, veterinary medicine can be studied in this way, but its historiography lags behind that of science and medicine. A key concern within SSK is the concept of agency, a subject that is of both historical and contemporary interest in veterinary medicine with regard to the nonhuman patient.

Table 2: Topic themes at WAHVM congresses, 2012–2018

Theme	Bergen, 2018	Vienna, 2016	London, 2014	Utrecht, 2012	Total
Disease & health	18 (41.9%)	7 (16.7%)	21 (25.3%)	14 (22.6%)	60 (26.1%)
Institutions	3 (7.0%)	16 (38.1%)	5 (6.0%)	31 (50.0%)	55 (23.9%)
Public & 'One Health'	5 (11.6%)	4 (9.5%)	19 (22.9%)	2 (3.2%)	30 (13.0%)
Biographies	2 (4.7%)	8 (19.0%)	10 (12.0%)	9 (14.5%)	29 (12.6%)
Archives, libraries & sources	8 (18.6%)	4 (9.5%)	7 (8.4%)	3 (4.8%)	22 (9.6%)
Other	0	0	13 (war) (15.7%)	0	13 (5.7%)
Welfare & ethics (animal)	5 (11.6%)	2 (4.8%)	5 (6.0%)	0	12 (5.2%)
Practices	2 (4.7%)	1 (2.4%)	3 (3.6%)	3 (4.8%)	9 (3.9%)
Total	43	42	83	62	230

---

7 See, for example, Latour, Bruno and Steve Woolgar. *Laboratory Life: The Construction of Scientific Facts*. Princeton, NJ: Princeton University Press, 1986; Pickstone, John V. *Ways of Knowing: A New History of Science, Technology and Medicine*. Manchester: Manchester University Press, 2000; Chalmers, Alan Francis. *What Is This Thing Called Science?* Maidenhead: Open University Press, 2003.

Species

If we examine the species emphasis across all conference paper subjects, some interesting features arise (Table 3). Overall, the commonest species was the human (44.4%), followed by farmed animals (28.1%, principally cattle), horses (14.4%) and companion animals (8.5%). Fish and wildlife/exotic animals were present in much lower numbers. Species emphasis is affected by conference themes. For example, the Utrecht 2012 conference theme was 'veterinary societies', so the 80.4% figure for humans is not surprising, but the high value for humans in all conferences reflects the anthropocentric nature of veterinary history. Of nonhumans, farmed animals and equids are the commonest species. Fish rarely appear. Their presence in 2018 was because the host society stipulated a fish farming theme to reflect regional interests. This type of confounding factor will affect any analysis of this kind, being an example of how any empirical or 'scientific' approach – in this case the simple enumeration of papers at veterinary conferences – is susceptible to social, cultural, economic and political influences.

Veterinary history might be said to be "specied"[8] and noting species highlights how much veterinary historiography privileges the human. We also see that, within the category of nonhumans, certain animals (e.g. cattle, horses) take precedence over others (e.g. cats, poultry, fish). It might be argued that veterinary history can and should only be about humans. However, this produces a rather narrow view of a profession which is, after all, dedicated to animals. Not only that, but the lens of species reflects how veterinarians see themselves, as I explain below, so that figuring in the animal makes for stronger 'human' history.

Table 3: Species represented in papers at WAHVM congresses, 2012–2018

Species	Bergen, 2018	Vienna, 2016	London, 2014	Utrecht, 2012	Total
Human	3 (12%)	9 (36%)	15 (28.8%)	41 (80.4%)	68 (44.4%)
Farmed	13 (52%)	6 (24%)	17 (32.7%)	7 (13.7%)	43 (28.1%)
Equid	2 (8%)	8 (32%)	11 (21.2%)	1 (2%)	22 (14.4%)
Companion	1 (4%)	2 (8%)	9 (17.3%)	1 (2%)	13 (8.5%)
Fish	5 (20%)	0	0	1 (2%)	6 (3.9%)
Wildlife & Exotic	1 (4%)	0	0	0	1 (0.7%)
Total	25	25	52	51	153

---

8 The term is used in Kirk, Robert and Michael Worboys. Medicine and Species: One Medicine, One History. In *The Oxford Handbook of the History of Medicine*, Mark Jackson (ed.), 561–577. Oxford: Oxford University Press, 2011.

How do these findings resonate in the journal *Veterinary History?* The content of papers in *Veterinary History* was analyzed and coded in three ways to afford cross-checking and triangulation:
1. The 26-year period 1992–2018 using journal Contents pages
2. The 10-year period, 2007–2017 using journal Contents pages
3. The 34-year period 1973–2007 using the published journal Index.

The main headings used for coding were: 'Humans', 'Companion animals', 'Horses', 'Farmed animals', 'Institutions', 'Diseases', 'Practice' and 'Other'. Problems inevitably arose, for example how to code a paper on the history of rabies in British dogs? Generally, species was given precedence over other criteria in order to locate the animal in veterinary history. The results are summarized in Table 4. Again, we see the dominance of humans, institutions, and diseases from the analysis.

Table 4: Results of content analysis of *Veterinary History* counted in three ways. The 'Other' category has been omitted.

Rank order	1992–2018	2007–2017	Index 1973–2007
1	Humans (34%)	Humans (48%)	Humans (35%)
2	Institutions (12%)	Institutions (20%)	Institutions (19%)
3	Diseases (10%)	Diseases (12%)	Diseases (14%)
4	Companion animals = horses (8%)	Horses (10%)	Farmed animals (12%)
5	Farmed animals (7%)	Companion animals = farmed animals (8%)	Horses (10%)
6			Companion animals (4%)

The results of the analysis of conferences and one long-running journal clearly suggest that current veterinary history is anthropocentric, and that institutions and diseases are consistently favored as research topics. Among nonhuman animals, farmed animals and horses are favored at conferences; differences between nonhumans are less evident within the journal.

These results are partly explained by the nature of the sources that determine what historians of veterinary medicine may study. However, it is also true that sources may be interrogated in different ways. Two examples illustrate this. A paper entitled, "The History of the Norwegian Veterinary Association"[9] outlines the origins of this national society, provides quantitative details such as regions and membership fluctuations, describes the evolution of veterinary specialisms including one partic-

---

[9] Folkestad, Per. The History of the Norwegian Veterinary Association. Paper presented at the 40th International Congress of the World Association for the History of Veterinary Medicine. Utrecht University, Netherlands, August 22–25, 2012.

ular to Norway (aquaculture), discusses Norway's acceptance of European Union legislation on the free movement of meat and animals, and explores links between practicing veterinarians, the national society, veterinary education and certain livestock and breeding organizations. So, this is a paper about veterinarians, how they organized themselves, how they divided up veterinary labor and how they interacted with the state and with other institutions with which they collaborated and/or competed. Animals here feature as objects within veterinary medicine, the substrate upon which veterinary medicine as a human activity is acted out. The paper is essentially about people; animals are present only insofar as they are part of the landscape which veterinarians inhabit. By contrast, another paper on the same subject adopts a different approach. In "Enacting biopower with care: the history of the Norwegian veterinary profession and the construction of human-animal welfare"[10] the authors undertake a Foucauldian analysis of jurisdiction and control to dissect the role of veterinarians as mediators between animals and human society. The paper applies a framework drawn from science and technology studies, using concepts of 'care' to investigate aspects of professional practice.

Veterinary historiography can, therefore, fill gaps in the record or take a more exploratory and theoretical approach. Either way, histories of disease, of institutions, and of prominent individuals have been predominant.

3 Methods and Approaches

Veterinary history has focused on the history of the profession as a (human) social history. Given that the patient is always an animal, a key question must be, 'Where is the animal?' One problem, mentioned above, is the lack of studies on veterinary practice, where animals naturally appear. The constant presence of animals in the everyday lives of working veterinarians merits attention in veterinary history, yet in institutional histories and human biographies, and even in disease histories, animals are conspicuously absent. From an animal studies perspective, figuring-in the animal becomes important. How best to do this?

Animals may be made visible in several ways. They are likely to appear in histories of veterinary practices, and interesting data can emerge when sources are interrogated. For example, an examination of 'day books' recording the work of a country practice in Dalkeith, Scotland, during the 1930s found that more cats and dogs were seen than any other animal, but that their contribution to practice income was

---

10 Druglitrø, Tone and Kristin Asdal. Enacting Biopower with Care: The History of the Norwegian Veterinary Profession and the Construction of Human–Animal Welfare. Paper presented at the 41st International Congress of the World Association for the History of Veterinary Medicine. Imperial College London, UK, September 10–13, 2014.

disproportionately low.[11] The dogs and cats were not valued highly in terms of fees – sometimes no fees were charged – yet they were clearly valued in other ways since professional time and care was devoted to them, quite aside from the significance they apparently had for their owners who took them to the vet in such numbers in the first place. This pattern, seen elsewhere in practices of the same type and period, says something about how dogs and cats were regarded in mixed, general practice before the main 'turn' to companion animal practice in the mid to late twentieth century [→History of Pets]. It is usually stated that companion animals were marginal before 1950; practice data suggest that they were marginal only in certain ways: numerically, they were very much 'in the room'. As a historiographical approach, this type of empirical analysis aligns with Kean's statement:

> Given the dearth of history writing on animals in the past, it is neither surprising nor, I believe, an issue in itself that books and articles continue to be written within a framework that is intended to 'reveal' the past role of animals within established historical methodologies. [...] We are still at very early stages of 'retrieving' the hidden pasts of animal existences and their impact upon humans. Work that is intended to bring to light previously unknown subject matter is being produced alongside research that attempts more historiographically imaginative approaches.[12]

Veterinarians have long identified themselves primarily by the species of animal they mainly treat: species is still the first-level category definition in clinical specialization. In this sense, animals occupy a prominent, even defining, position within veterinary cosmology. In the twentieth century, the 'type species', and hence the 'type veterinarian', shifted from the horse (at the beginning of the century), through the dairy-cow (in the middle) to the companion animal (in the last half to quarter). This is evidenced not just by the numbers of veterinarians specializing in them, but also by their coverage in veterinary curricula.[13] Qualities associated with the different species defined professional identity, for example noble and useful (the horse), important to the economy and public health (the dairy cow), of sentimental value and as a means of practicing advanced medicine and surgery (the companion animal). These definitions were not static across time and always depended on context. For example, an early twentieth-century male veterinarian who *only* treated dogs would have been considered unusual, if not dubious on account of the gendered nature of different species: horses and cattle were 'manly', cats and lapdogs were 'effeminate' [→Feminist Intersectionality Studies]. As late as the 1980s, it was

---

11 Gardiner, Andrew. Small Animal Practice in British Veterinary Medicine 1920–56. PhD diss.: University of Manchester (2010), ch 3.
12 Kean, Hilda. Challenges for Historians Writing Animal–Human History: What is Really Enough? Anthrozoös 25, Supplement (April 2012): 57–72.
13 Interestingly, this does not relate to the actual numbers of animals present in veterinary schools themselves. Even in the horse and cow eras, there were large numbers of companion animals in college clinics and hospitals. This does not seem to be only a British phenomenon. See Gardiner, Small Animal Practice, ch 2.

likely that a male veterinarian working only with cats, for example, would be assumed to be homosexual; this was a time when societal and professional stigma surrounding homosexuality was rife.[14] While certain species were favored in particular eras, others were not. For example, the author remembers a farm animal practitioner referring to the then emerging trend for the treatment of rabbits, mice, rats and other small mammals as the veterinary medicine of 'roadkill and vermin' when addressing undergraduates at the Royal (Dick) School of Veterinary Studies in Edinburgh in the 1990s. Surfacing this aspect of veterinary history reveals a *form* of animal agency: different animals determine distinct modes of veterinary work and practice, generating different 'tribes' of veterinarian who identify with their chosen species in interesting ways yet to be fully investigated. In that sense, animals might be said to co-construct veterinary medicine by virtue of their species. If this is agency, it is an unconscious form that partly results from the ways that humans relate to species categories and use them for their own ends.

Ideas of agency more intuitively relate to allowing animals' preferences to influence their treatment. Treatment is only possible when animals 'consent' or yield to it. In everyday practice, such consent becomes a practical, everyday matter, governing what veterinarians can and cannot do, based on what individual animals will or will not tolerate. Equipment and instruments get developed, animal-friendly handling techniques evolve, animal-palatable medication gets manufactured, all of which allow the easier practice of veterinary medicine in the different species, and each species has its own requirements. Evidence for this may be found in diverse sources: student notes, images, drug and equipment catalogues, case records, textbooks. Interpretation of the evidence helps 'animate' the patient in historical or contemporary practice, although veterinary knowledge might be needed to decode the sources.

For example, a popular textbook on companion animal surgery, first published in 1900, describes innovative techniques and, through informal case studies, contains some clues as to how some treatments were received by animals. A Schipperke dog, on being fitted with a set of dentures, resented the false teeth and "on numerous occasions successfully removed them with his paws".[15] Various changes to the procedures for inserting and retaining the teeth were tried. The patient eventually became used to the teeth and wore them regularly for eighteen months. Canine dentures, however, did not catch on, because overall patient compliance was low; the dogs 'rejected' dentures (spat them out). However, another innovation – the fitting of limb prostheses – was eventually accepted and found its way into canine and feline orthopedic practice, albeit not until the end of the twentieth century (Fig. 1).

---

14 Gardiner, Andrew. Prescription for a Broad-Spectrum Veterinary Profession. Paper presented at the *50 Years After Stonewall* conference. University of Edinburgh, September 10 – 13, 2019 [forthcoming in print].
15 Hobday, Frederick. *Hobday's Surgical Diseases of the Dog and Cat.* London: Baillière, Tindall & Cox, 1947, 150 – 151.

THE LOCOMOTOR SYSTEM 391

is the well-known Whitehead's varnish. The stitches should be removed about the fourth or fifth day, or sooner if it is suspected that infection has occurred and pus is present.

FIG. 316.—BLENHEIM SPANIEL WITH FALSE LEG. (MR. FRANK LEIGH.)

FIG. 317.—A CHEAP PATTERN OF PLAIN LEATHER FALSE LEG.

False legs, consisting merely of a plain leather socket, or a more elaborate arrangement of silver, rubber, chamois leather and vulcanite, can be fitted afterwards; but it is astonishing to see how soon an animal

Figure 1: Historical dog prostheses from *Hobday's Surgical Diseases of the Dog and Cat*, p. 391. The page also shows another interesting feature: the difference between public and private veterinary medicine. The 'middle class' Blenheim spaniel in the top picture, one of Hobday's private clients from his practice in Bloomsbury, London, sports a cosmetic and life-like prosthesis. The mixed breed dog below, probably treated at the Royal Veterinary College's Poor People's Clinic, where Hobday also practiced, has a simple wooden prop (image from Hobday 1947, reprinted with permission).

In both these examples – dentures and false legs – the animal treatment was analogical and arose from treating the patient as one would a human with similar problems. With limb protheses, it was soon discovered that most dogs coped perfectly well with three, and occasionally even two, legs, so early limb prosthetics fell away (as they often did literally). However, this move away from prostheses could only happen once general anesthetics became safe enough to use for limb amputations. It was only by removing limbs that veterinarians, owners, and indeed dogs themselves, realized that they could cope without them. Later in the twentieth century, when dogs started living much longer and began suffering from age-related degenerative joint disease and cancers, and when surgical techniques and materials advanced first in human and then in veterinary medicine (the human advances depending, of course, on animal experimentation)[16], limb protheses became a viable option for companion animals when three legs were no longer sufficient because the remaining legs were themselves compromised by arthritis or damaged by major trauma. The same process may ultimately happen with false teeth; dogs and cats now routinely receive root canal treatments and orthodontics. Techno-medicine may yet solve the problem of the comfortable retention of animal dentures. Interestingly, in the 1980s there was a brief fashion for fitting dentures in sheep, since premature loss of teeth has significant welfare and economic implications in the sheep industry. However, the sheep denture experiment was abandoned for animal welfare reasons: the prosthetic teeth caused more problems than they solved in a species in which the prosthesis was never going to be sufficiently individualized to provide a unique, 'personal' fit. This does not apply to companion animals whose owners are willing to pay more.

4 Implication(s) of the Animal Turn

Considering the animal as an active participant in treatment in terms of toleration of treatment (compliance) – even when that compliance is less about conscious choice and more about what animals will put up with – prompts ethical consideration and aligns with the animal studies approach of linking scholarship to advocacy.[17] However, just because an animal tolerates or allows a certain treatment does not make it right. The agency and autonomy of domestic animals is controlled by humans and animals are severely limited in their ability to express 'compliance' in the sense in which we usually think of it. That does not mean, however, that it is never

---

16 Schlich, Thomas. Surgery, Science and Industry: A Revolution in Fracture Care 1950s–1990s. London: Palgrave Macmillan, 2002.
17 Gardiner, Andrew. The Animal as Surgical Patient: A Historical Perspective in the Twentieth Century. *History and Philosophy of the Life Sciences* 31, no. 3–4 (January 2009): 355–376; Jones, Susan. Framing Animal Disease: Housecats with Feline Urological Syndrome, Their Owners and Doctors. *Journal of the History of Medicine* 52, no. 2 (April 1997): 202–235.

shown. Companion animal owners will frequently complain that their animal will not let them apply treatment in the way prescribed. In that situation, the veterinarian must work with owner and animal to find a mutually agreeable solution. To some respect, this type of 'negotiation' occurs throughout history and across all species.

Animal-attentive approaches may yield data and explanations not available through traditional biographical, institutional, and disease-based approaches to veterinary history, albeit with important complications and qualifications. Veterinarian-historians are ideally placed to read veterinary sources using accumulated clinical experience. Non-veterinary researchers adopting more or less anthropological approaches to the veterinary encounter can also reveal aspects unrecorded so far. However, new approaches are needed to further explore animal subjectivities. Scholars' theoretical approaches, even those directed at nonhumans, are always grounded in *human* subjectivity. In his thesis, Animal-attentive Queer Theories, Tyler T. Wuthmann states, "[Human subjectivity] has prevented a deep engagement with the lives, experiences and suffering of animal others due to their fundamental lack of human speech and discursive cultural practices".[18] In a response to J.M. Coetzee's *The Lives of Animals*, the primatologist Barbara Smuts discusses this problem of language. She notes, "My own life has convinced me that the limitations most of us encounter in our dealings with other animals reflect not their shortcomings, as we so often assume, but our own narrow views about who they are" – hence the longstanding fascination with trying to get animals to understand human language.[19]

Current veterinary history, and much animal history, tells us how things were. In reading it, we trust an author who has done their research and presented their conclusions; in some ways it becomes a closing down process. Animal-attentive veterinary history has the potential to be an opening up process, allowing for the possibility of histories in which animals' subject status is approached in new ways. This type of history has relevance beyond disciplinary concerns and narrative reconstructions. Examples are beginning to emerge in broader animal history.[20] An important methodological and ethical issue for veterinary history is the necessity for symmetrical approaches across species of equivalent sentience, that is, writing the history of a pig should not differ markedly from that of a dog, as the subjectivities of those two animals are comparable. Speciesism is thereby revealed in terms of how animals of equivalent sentience are constructed as veterinary patients. Another issue to consider is whether veterinarians are uniquely able to interpret certain source materi-

---

18 Wuthmann, Tyler. Animal-attentive Queer Theories. PhD diss.: Wesleyan University (2011). doi.org/10.14418/wes01.1.658.
19 Smuts, Barbara. Barbara Smuts. In *The Lives of Animals*, John Maxwell Coetzee, 107–120. Princeton, NJ: Princeton University Press, 1999; Krebber, André. Washoe: Das Subjekt in der Tierforschung. In *Philosophie der Tierforschung 3: Milieus und Akteure*, Martin Böhnert, Kristian Köchy, Matthias Wunsch (eds.), 187–220. Freiburg and Munich: Verlag Karl Alber, 2018.
20 Kean, Hilda. *The Great Cat and Dog Massacre: The Real Story of World War Two's Unknown Tragedy*. Chicago, IL: University of Chicago Press, 2017.

als, for example practice records, and whether such insider knowledge permits a nuanced appreciation of animal agency. Given the importance of agency for historical animal studies, one solution could be a co-disciplinary approach, with the historian or animal studies scholar working closely alongside a veterinarian to identify instances in sources that require specialist knowledge to interpret.

Furthermore, animal studies scholars have begun to consider the subject of animal biography.[21] This approach lends itself to veterinary history since any treatment of the individual animal starts with biography: what the veterinarian already knows about the patient (in terms of disease history, temperament, preferences); what the owner or carer says about the animal; and what the animal says about herself/himself as revealed by attitude, response and behavior. The challenge lies in locating the source materials. Practice records are increasingly being collected and archived and much is likely to be found in forgotten, dusty boxes.[22] Rich history can come from items like appointment diaries, day books and correspondences. The text in Figure 2, a card sent from a veterinary practice to a bereaved client in the late 1800s or early 1900s, reads as follows:

> We have just sent the 'Chow' to sleep, he was quite quiet & just slept away, you could have watched it done it was so painless. The gardener is coming tomorrow morning for him & if at any time you wish to see where he is buried I will lend you our key to get into the garden.

Of course, the Chow is now dead, and his agency is no longer apparent. Nevertheless, the act of sending the card, the tone of the message and the sense of managing the 'afterlife' of a companion dog tells us something about this particular human-animal relationship and the role of the veterinarian. Further research could indicate the likely method used to carry out the euthanasia; the name and age of the person who sent the card might be traced via the Veterinary Register. The fact that the patient is not named but referred to by the breed name in quotation marks (the 'Chow') is interesting in itself. If practice day or accounting books were available, it may be possible to reconstruct the life of this dog in a meaningful way in the known context of veterinary practice of the period.

Oral history can also be a valuable means of surfacing animals. Veterinarians habitually discuss cases among themselves and are articulate in recalling them in a conversational setting that values subjective impressions and reflections as well as practical details. Human attitudes and feelings are revealed in contexts that can be historicized, and animals can materialize as individuals.[23]

---

21 Krebber, André and Mieke Roscher (eds.). *Animal Biography: Re-framing Animal Lives*. London: Palgrave Macmillan, 2018.
22 Hunter, Pamela. *Veterinary Medicine. A Guide to Historical Sources*. Aldershot: Ashgate Publishing Ltd, 2004.
23 Brancker, Mary. Mary Brancker interviewed by Sue Bradley. An Oral History of Veterinary Practice. 2009. C1519/01. British Library, London, UK.

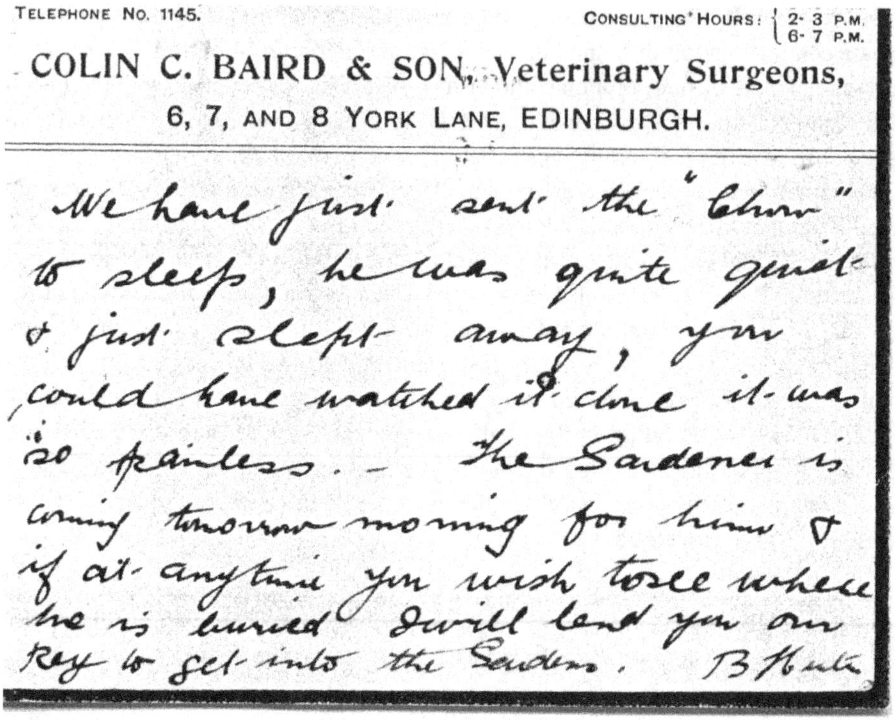

Figure 2: Condolence card sent to the owner of a Chow after the dog was euthanized at a practice in Edinburgh in the early 20th century.

In conclusion, veterinary history has been written as social history of the veterinary profession, with most work focusing on diseases, institutions, and human biography. The history of veterinary practice – the ordinary lives of working veterinarians – is under-researched and poses challenges in terms of accessing and interpreting source materials. However, the history of practice provides unique opportunities to locate the animal in veterinary history and to explore new ways of writing veterinary history that align with approaches in animal history and animal studies. Such approaches have the potential to reveal not only what practicing veterinarians did, but also how animals responded to and influenced practice and, in the process, to produce insights about the nature of veterinary patient-hood that can inform current and future veterinary medicine.

With thanks to Sue Bradley.

Selected Bibliography

Bradley, Sue. Hobday's Hands: Recollections of Touch in Veterinary Practice. *Oral History* 49, no. 1 (Spring 2021): 35–48.
Elvbakken, Kari Tove. Veterinarians and Public Health: Food Control in the Professionalization of Veterinarians. *Professions & Professionalism* 7, no. 2 (June 2017): 1–15.
Gardiner, Andrew. The Dangerous Women of Animal Welfare: How British Veterinary Medicine Went to the Dogs. *Social History of Medicine* 27, no. 1 (January 2014): 466–487. doi.org/10.1093/shm/hkt101.
Hipperson, Julie. Professional Entrepreneurs: Women Veterinary Surgeons as Small Business Owners in Interwar Britain. *Social History of Medicine* 31, no. 1 (February 2018): 122–139. doi.org/10.1093/shm/hkx058.
Jones, Susan. *Valuing Animals. Veterinarians and their Patients in Modern America*. Baltimore, MD: Johns Hopkins University Press, 2003.
Mitsuda, Tatsuya. Entangled Histories: German Veterinary Medicine, c. 1770–1900. *Medical History* 61, no. 1 (December 2017): 25–47. doi.org/10.1017/mdh.2016.99.
Skipper, Alison. The 'Dog Doctors' of Edwardian London: Elite Canine Veterinary Care in the Early Twentieth Century. *Social History of Medicine* 33, no. 4 (November 2020): 1233–1258. doi.org/10.1093/shm/hkz049.
Woods, Abigail, Michael Bresalier, Angela Cassidy and Rachel Mason Dentinger. *One Health and its Histories: Animals and the Shaping of Modern Medicine*. Cham: Palgrave Macmillan, 2018. URL: link.springer.com/content/pdf/10.1007%2F978-3-319-64337-3.pdf (December 6, 2020).
Woods, Abigail. Animals and Disease. In *Routledge History of Disease,* Mark Jackson (ed.), 147–164. London: Routledge, 2016.
Woods, Abigail. Animals in the History of Human and Veterinary Medicine. In *Routledge Companion to Animal–Human History,* Hilda Keane, Philip Howell (eds.), 147–170. London: Routledge, 2018.

Axel C. Hüntelmann
History of Experimental Animals and the History of Animal Experiments

1 Introduction and Overview

This chapter provides an overview of the history of experimental animals. The subject needs clarification. What seems to be a small linguistic distinction has an important effect on the subject: A history of animal experiments would have to focus on the experiment and the role animals played within the experimental setting, whereas the history of experimental animals makes the animal the subject of its narration. It raises the questions: What is an experimental animal? And is an experimental animal the same as a, often used synonymously, laboratory animal?

Animal experiments are part of studies on historical epistemology from a history of science perspective, especially on the history of medicine and the life sciences, and the human-animal relationship is one of many aspects in the history of knowledge production [→History of Science]. In contrast, the history of experimental animals deals explicitly with human-animal relationships in experimental settings – which again raises questions about the experiment. Do natural observations of animals in antiquity or demonstrations, for instance by Galenus of Pergamon, qualify as experiments? If so, the history of experimental animals would cover a period of two thousand years – or even more! In contrast, the history of the laboratory animal is bound to its human-made artificial habitat, the scientific laboratory where experiments are carried out under (supposedly) controlled conditions, and which according to Andrew Cunningham and Perry Williams emerged with the laboratory revolution at the beginning of the nineteenth century.

While a history of animal experiments differs from one of experimental animals due to its perspective, the latter differs from a history of laboratory animals due to space and time. Whereas a history of experimental animals would cover a larger period of time, a history of laboratory animals might include all living creatures housed – voluntarily or forced, intended or unintended by their human companions – in a laboratory. Many of these creatures are often overlooked by historians or cultural scientists working on laboratory animals: regular inhabitants of agar plates under investigation, like bacteria, monads and parasites, as well as such common cohabitants of human habitats as flies or other insects and pets accompanying their human companions working as scientists in the laboratory. In addition, laboratory animals might also include those animals who are not experimented on themselves but are used in a secondary capacity as bioindicators in testing procedures, for example, to test a chemical compound's toxicity or to evaluate the efficacy of pharmaceuticals.

In general, any animal species and in principle any animal, depending on the experiment, can be an experimental animal. In a narrower sense, however, Gerda Opitz has defined experimental animals as "animals bred specifically for scientific experiments."[1] The experimental animal is therefore not a specific species, but an unspecific animal that is attributed certain functional purposes and characteristics, such as being easy to handle, uncomplicated to keep and cheap to feed, susceptible to human diseases or marked by high reproductive rates. But these characteristics and requirements are defined by humans and depend on social and cultural factors – and as such they are subject to historical transformations. Thus, the idea of what, or who, an experimental animal is has changed fundamentally over time: the frog, the "old martyr of science,"[2] and as "ranae exploratae" the best studied animal from the seventeenth to the nineteenth centuries,[3] was displaced by the mouse at the end of the nineteenth century. But since the first decades of the twentieth century, the mouse has been essentially outnumbered by *Drosophila melanogaster* (fruit flies) and *caenorhabditis elegans* (nematode worms), which became the leading and best studied model organisms in developmental genetics and biology.[4]

2 Topics and Themes

In the history of experimental animals, topics, themes, approaches and methods are deeply entangled: The history of animal experiments is, for instance, related to a history of discourses about the justification of experiments versus their cruelty to animals or the history of the anti-vivisection movement and of animal welfare. Thus, cultural studies of certain animals such as the frog or the dog and their use in animal experiments have to be related and contextualized in the history of the scientific experiment. This subchapter provides an overview of the main topics and themes in the history of experimental animals by summarizing the relevant historical literature. Although sketching animal experiments from antiquity to modern times, this overview focuses on the nineteenth and twentieth centuries and the entangled relationships and interactions between human and nonhuman animals in the laboratory.

---

[1] Opitz, Gerda. Tierversuche und Versuchstiere in der Geschichte der Biologie und Medizin. PhD diss.: Friedrich Schiller University (1968), 2–3.
[2] Holmes, Frederic L. The Old Martyr of Science. The Frog in Experimental Physiology. *Journal of the History of Biology* 26, no. 2 (Summer 1993): 311–328.
[3] Rothschuh, Karl E. Laudatio ranae exploratae. *Sudhoffs Archiv* 57, no. 3 (1973): 231–244.
[4] See the contributions in Creager, Angela, Elizabeth Lunbeck and M. Norton Wise (eds.). *Science without Laws. Model Systems, Cases, Exemplary Narratives.* Durham, NC: Duke University Press, 2007.

From Demonstrations to Experiments

Since antiquity, humans studied animals and investigated the functioning of their bodies. Aristotle observed the natural life of animals and later the Alexandrians Herophilus and Erasistros dissected human bodies and performed vivisections of living animals – but only rarely and primarily to discuss matters of natural philosophy. A few centuries later, animals played a more prominent role in the work of the Roman Galen: Besides natural observations, he performed a large number of dissections and vivisections for teaching purposes and for public demonstrations.[5] Although Galen's work was more practical and systematic than what came before him, it has to be distinguished from a modern understanding of "experiment." In contrast to Galen's observations and demonstrations, experiments in a "modern" sense are planned in advance, embedded in a broader scientific program, guided by a hypothesis and research questions on a defined problem, and take place in controlled settings under monitored physical conditions, as Anita Guerrini shows in her study of the history of the animal in human research. The empirical research results are recorded and discussed with and verified or falsified by other scientists. As a turning point in scientific medicine, Guerrini considered William Harvey and his experiments on living animals to describe the function of the heart and blood circulation in the first decades of the seventeenth century, because his experiments included all of the characteristics mentioned above, whereas scientists in the middle ages mainly reproduced Galen's findings, and performed dissections and animal vivisections only for demonstrational purposes.[6]

Harvey stimulated discussions about method and the validity of experiments as well as the functioning of the body. In the seventeenth and eighteenth centuries natural scientists performed experiments in this "modern" sense all over Europe, and animals played a vital role in their empirical research: Robert Boyle conducted research on respiration and the nervous system of living animals and their metabolism; Robert Hooke injected dyes into the blood vessels of living animals to demonstrate their branching and to differentiate between venous and arterial blood by pumping air in and out of a living dog's lungs with an opened thorax, and observed the animals' reactions to various chemical substances; Marcello Malpighi utilized similar methods to investigate the blood stream in the lungs and to explain the respiratory system; Stephen Hales investigated the circulatory system of the blood by attaching cannulas to arteries of dogs measuring the differential pressure of the blood when filling and emptying the ventricules and studied the relationship of blood pressure and heart rate and the effects of stress and pain on the blood pressure; and Albrecht von Haller experimented on the irritability and sensitivity of muscles, skin and body

---

[5] Opitz, Tierversuche, 11–32; Guerrini, Anita. *Experimenting with Humans and Animals. From Galen to Animal Rights.* Baltimore, MD: Johns Hopkins University Press, 2003, 23–33.
[6] Guerrini, *Experimenting*, 23–33.

organs of animals.[7] Furthermore, animals have also played a significant role in pharmacology and toxicological tests. In the seventeenth century, Johann Jakob Wepfner, for example, tested the effects of drugs and poisons on animals and subsequently dissected them.[8]

Early Debates About Animal Suffering and Human-Animal Analogies

The aforementioned experiments of Boyle, Hooke and Hales caused horrifying pain in animals. Aristotle had already recognized their agony but defended the vivisection of animals as a necessary evil to gain knowledge about the functioning of the (human) body, assuming an analogy between the human and the mammalian body structure. The infliction of pain happened for a higher purpose and Aristotle justified the killing of animals with their supposed lack of reasoning and a rational soul. They were considered inferior to humans in the hierarchy of organisms and the "great chain of being."[9]

Despite discussions and criticism of the assumed analogy between the human and the animal body,[10] comparative anatomists in the seventeenth and eighteenth centuries continued to point to the similarity of human and animal bodies, and because vivisections on humans were not allowed, animals were considered important research objects to investigate vital functions of the (human) body. Differences in the anatomical structure diminished in the description of physiological functions in mechanistic ways: Harvey, for instance, compared the workings of the heart to that of a mechanical pump. Boyle, Hooke, Hales and other natural scientists in the seventeenth and the eighteenth centuries performed their experiments on the premise that the body functioned according to mechanical laws. A merely mechanistic view of the living organism was developed by the French philosopher René Descartes. He created a dualism between mind and body that reduced the body to mechanical processes and modes of functioning – according to him, the body functioned like a machine [→History of Ideas]. But while the "human machine"

---

[7] Guerrini, *Experimenting*, 57–65. On Boyle see Shapin, Steven and Simon Schaffer. *Leviathan and the Air-Pump. Hobbes, Boyle and the Experimental Life*. Princeton, NJ: Princeton University Press, 1985.
[8] Opitz, Tierversuche, ch. 3. On Wepfner see Maehle, Andreas-Holger. Zur wissenschaftlichen und moralischen Rechtfertigung toxikologischer Tierversuche im 17. Jahrhundert. Johann Jakob Wepfer und Johann Jakob Harder. *Gesnerus* 43 (1985): 213–221.
[9] Guerrini, *Experimenting*, 7–11. On the argument about the superiority of the human being within the animal kingdom see Lovejoy, Arthur O. *The Great Chain of Being. A Study of the History of an Idea*. Cambridge, MA: Harvard University Press, 1936.
[10] In the mid-sixteenth century, Andreas Vesalius, after many dissections, emphasized the structural anatomical differences between the human and the animal body and criticized Galen's descriptions of the human body that were based on the dissection of animals only, see Guerrini, *Experimenting*, 24–28.

was animated by a mind and soul, the animal machine was not and instead followed purely mechanical laws. Without a reasoning mind and soul, Descartes concluded, animals were not able to experience feelings such as pain, and his mechanistic philosophy helped to legitimize animal experiments. Although Descartes himself experimented only rarely on (living) animals, he had an enormous impact on the life sciences and animal welfare. He paved the way for a mathematical approach to solve scientific problems that led to the quantification of evidence and the proliferation of animal studies in the life sciences.[11]

Beginning in the late seventeenth and eighteenth centuries the number of animal experiments rose rapidly, and with increasing coverage of these experiments, published in newspapers and discussed in coffeehouses and salons, vivisection became the focus of public debates and voices were raised that sparked a discussion whether animals were able to feel pain or not.[12] With experiments like those of Albrecht von Haller on nerves and their sensitivity, doubts rose in the second half of the eighteenth century about a dualistic separation between mind and soul; and with observations of polyps that were able to regenerate body parts, doubts were also expressed about a purely mechanistic model to explain organic processes.[13] Arguments about the capacity of animals to suffer and moral concerns about the cruelty of animal experiments, brought forward by Jeremy Bentham and others, became major topics in the history of animal experiments and were thus at the heart of human-animal relationships during this time.

Experimenting (With) the Animal

With the scientification of medicine and its establishment as an experimental life science at the turn of the nineteenth century, experiments became a key issue in the process of knowledge production. Referring to the natural sciences, experiments in medicine had to meet certain requirements: they had to be performed in a stable setting in order to exclude coincidences and to ensure the condition of *ceteris paribus* ['all things being equal']. Moreover, for the sake of objectivity, an experiment had to be repeated several times to achieve valid research results. During the nineteenth century, this experimental environment became more technical and standardized, and experiments were no longer performed in ordinary study rooms, but in specially

---

11 Maehle, Andreas-Holger and Ulrich Tröhler. Animal Experimentation from Antiquity to the End of the Eighteenth Century. Attitudes and Arguments. In *Vivisection in Historical Perspective*, Nicolaas A. Rupke (ed.), 14–47. London: Routledge, 1987; Shanks, Niall. *Animals and Science. A Guide to the Debates*. Santa Barbara, CA: ABC-Clio, 2002, 56–62; Guerrini, *Experimenting*, 23–24, 28–37.
12 Maehle and Tröhler, Animal Experimentation, 28–36; Guerrini, *Experimenting*, 46–47, 59–60.
13 Guerrini, *Experimenting*, 63–67. On regeneration and reproduction and changing views on the principle of the living see Jacob, François. *Die Logik des Lebenden. Von der Urzeugung zum genetischen Code*. Frankfurt a.M.: Fischer, 1972.

equipped laboratories. This had consequences for the animals used in experiments: they had to be healthy, of the same kind and in similar condition.[14] With the laboratory revolution of the nineteenth century, ordinary animals used for experimental purposes transformed into laboratory animals.

Thus, in contrast to previous research on individual functions of the living body, François Magendie claimed to have established a systematic research program investigating the functioning of the body *in toto*. He himself performed numerous dissections, vivisections and animal experiments on the nervous system, the functioning of the spinal cord and the motoric and sensory functions of the spinal nerve roots. For example, he investigated the deficits dogs manifested after he removed or disconnected parts of the brain. Claude Bernard continued Magendie's scientific program and propagated quantifiable methods in scientific medicine. He also gave practical and methodological advice on the performance of animal experiments. Bernard himself had carried out vivisections associated with feeding experiments studying metabolic processes and the connection between digestion and the vegetative nervous system.[15] In Germany, Emil Du Bois-Reymond did research on "animal electricity" and electrophysiology. He electrically stimulated the muscles and nerves of frogs to observe the neuromuscular transmission and electrical conductivity of the frogs' body. Also in the second half of the nineteenth century, Rudolf Heidenhain investigated the influence of a frog's vagus nerve on the heart rate and the heart muscle. While he, as well as later on Ivan Pavlov in Russia, carried out experiments on the functioning of numerous glands in the liver, stomach and pancreas and on salivary secretion in dogs.[16]

In their publications, Bernard, Heidenhain and others emphasized their long-lasting practice of animal experiments and their experience in the technical arrangements of these experiments, their knowledge in the use of instruments and their special surgical skills. In surgical operations, tubes and instruments were implanted in animals, for example to harvest and measure inner secretion of organs. The experiments and the often horrendous toll they took on the animals' well-being required significant effort to keep their bodies functioning, because the experiment did not end with one operation or vivisection but could drag on for days, weeks or even months. At the end of the century, the laboratory animals had emerged as part of

---

[14] Cunningham, Andrew and Perry Williams. *The Laboratory Revolution in Medicine*. Cambridge: Cambridge University Press, 1992.

[15] On France see Elliott, Paul. Vivisection and the Emergence of Experimental Physiology in Nineteenth-Century France. In *Vivisection in Historical Perspective*, Nicolaas A. Rupke (ed.), 48–77. London: Routledge, 1987; Guerrini, *Experimenting*, 81–86; Shanks, *Animals and Science*, 95–115; Holmes, Frederic L. *Claude Bernard and Animal Chemistry. The Emergence of a Scientist*. Cambridge, MA: Harvard University Press, 1974.

[16] For Du Bois-Reymond's experiments see Hüppauf, Bernd. *Vom Frosch. Eine Kulturgeschichte zwischen Tierphilosophie und Ökologie*. Bielefeld: transcript, 2011. On Pavlov see Todes, Daniel P. *Pavlov's Physiological Factory. Experiment, Interpretation, Laboratory Enterprise*. Baltimore, MD: Johns Hopkins University Press, 2002.

a comprehensive "scientific enterprise," in which numerous scientists and technical assistants were involved and a minute organization of processes, such as maintenance of technical devices or care-taking for experimental animals, became necessary.[17]

Microorganisms in the Laboratory

The status of the experimental animal changed again at the end of the nineteenth century with a paradigm shift in medicine and the rise of microbiology and bacteriology. Microbiologists investigated the life and living conditions of microorganisms and especially pathogen microorganisms. Since the 1870s, various microorganisms have been identified as causal agents for infectious diseases, for example black spots in silkworms that were observed by Louis Pasteur to cause an epizootic disease called pébrine, or little rods in sputum that caused, according to Robert Koch, tuberculosis. In bacteriology, experimental animals were constitutive for the scientific subject: the mode of action and infection as well as the life cycle of pathogenic microorganisms could only be observed *in vivo* ['in a living organism']. Especially after the establishment of the so-called *Koch's postulates*, animal experiments became indispensable: Once a specific microorganism had been identified in the tissue, blood or body fluids of a (non-)human animal showing specific clinical symptoms, Koch and his followers suspected this specific microorganism as the causal agent of a disease. In a next step, the identified microorganism was isolated and bred in a pure culture and, in a third step, the cultured microspecies was injected into a healthy animal. If the animal then showed the same clinical symptoms as the primary patient, it was concluded that the microorganism caused the disease. But until such causal relationships could be clearly proven, thousands of laboratory animals had been infected.[18]

Bacteriological research changed animal experiments in three ways: Firstly, the focus of research shifted from the observation of physiological processes in the "normal" (meaning healthy) body to the investigation of pathological alterations – from the normal to the pathological. Secondly, the subject of investigation shifted to the micro level. Although the disease and its symptoms manifested in the laboratory animal, the focus of research was on the microscopic pathogen, since only the exami-

---

17 Such a scientific enterprise is described by Todes, *Pavlov's Physiological Factory.*
18 Gradmann, Christoph. Das Maß der Krankheit. Das pathologische Tierexperiment in der medizinischen Bakteriologie Robert Kochs. In *Maß und Eigensinn. Studien im Anschluß an Georges Canguilhem*, Cornelius Borck, Henning Schmidgen, Volker Hess (eds.), 71–90. Munich: Wilhelm Fink, 2005; Hüntelmann, Axel C. Mäuse, Menschen, Menagerien. Laborchimären und ihre wechselvolle Beziehung im Königlich Preußischen Institut für Experimentelle Therapie nach 1900. In *Philosophie der Tierforschung 3: Milieus und Akteure*, Martin Böhnert, Kristian Köchy, Matthias Wunsch (eds.), 221–262. Freiburg and Munich: Verlag Karl Alber, 2018.

nation of the pathogenic microorganism provided certainty about the nature of the disease. The laboratory animal only functioned as a living environment for the pathogen, a storage room and passageway necessary to observe alterations in virulence over time or to investigate measures to attenuate or kill the pathogen and control the disease. Thirdly, animal experiments in bacteriological research were very demanding and the laboratory animal had to meet certain conditions. For instance, research on tuberculosis required an animal that was susceptible to tuberculosis, and as a large part of bacteriological research was concerned with (human) epidemics, animals were infected and sickened with pathogens of human diseases (becoming some sort of chimera). Thus, mammals like rats and mice were preferred as laboratory animals for bacteriological research, superseding the frog as the "martyr of science."

But laboratory animals also played an important role in research on disease control and in the search for cures. In pharmacological research and the pharmaceutical industry that emerged in the decades around 1900, laboratory animals were used to test the sterilizing effects of chemical compounds and synthesize preparations of pathogens, as well as to evaluate the curative effects on the course of a disease; and they also served as bio-indicators for toxicity tests to figure out a compound's harmful or unintended effects. In the following decades and through large screenings, the efficacy of thousands of compounds on various specific pathogens – involving an uncountable number of laboratory animals – were tested.[19]

Genetics and the Life Sciences

During the twentieth century, the life sciences diversified into various sub-disciplines. There was a multiplication of scientific actors, and the research fields in which laboratory animals and animal experiments played a crucial role expanded: experimental cancer research and oncology, molecular biology and biochemistry, population and molecular genetics, hormone research and reproductive medicine, neurology as well as cognitive and behavioral sciences. In oncology, for instance, artificial tumors were transferred and bred in animal bodies necessary for systematic research on the tumors' morphology, structure, growth and virulence. Laboratory animals were also used in long-term experiments that studied the causes of cancer or influential factors on tumor growth, such as toxic substances or radioactivity in the environment.

A new emerging discipline – genetics – changed the need and requirements for laboratory animals again. Various species of animals were bred for generations to observe the inheritance of traits, or they were genetically 'optimized' and designed to meet special purposes. From the end of the twentieth century onwards, they were

---

19 Hüntelmann, Mäuse.

also cloned.[20] This step required experimental animals with a short life-cycle sequence and a high production rate, for which the fruit fly *Drosophila melangastor* was considered particularly suitable. Hermann J. Muller observed not only a high rate of natural mutations in fruit flies but also that mutations could be caused artificially by X-rays. Since the 1920s, *Drosophila melangastor*, cultivated as laboratory animal in (population) genetics, became the most commonly used experimental subject worldwide.[21] Later in the twentieth century laboratory animals also played an eminent role in neurology and brain research of cognitive or behavioral sciences, where they were involved, for instance, in investigations about the effects of stress and fear (on the body), or in research on learning processes.[22]

Emergence of a New (Laboratory) Animal

In the twentieth century, the need for laboratory animals in the life sciences rose rapidly, partly because newly founded non-university research institutions were explicitly dedicated to experimental research. Thus, the costly maintenance, housing, care and feeding of laboratory animals gained increasing economic importance.[23] During the nineteenth century, the purchase of experimental animals was difficult, because they were mainly exchanged between scientists, bred in the research institutions or purchased from regular pet shops as well as, in the case of exotic animals, from zoos.[24] But aside from quantitative aspects and the rising number of animals populating laboratories, they also changed in a qualitative way. According to the serialized character of the experiments, animals were required to be as similar as possible to prevent individual discrepancies and achieve comparable, reproducible, "objective" research results, independent from the performing person, research object, place and time. In order to meet the increasing demand for laboratory animals, research institutes specialized in breeding animals that were considered standardized,

---

[20] Franklin, Sarah. *Dolly Mixtures. The Remaking of Genealogy*. Durham, NC: Duke University Press, 2007.
[21] Kohler, Robert E. *Lords of the Flies. Drosophila Genetics and the Experimental Life*. Chicago, IL: University of Chicago Press, 1994; Brookes, Martin. *Fly. An Experimental Life*. London: Weidenfeld & Nicolson, 2001.
[22] See, for example, Ramsden, Edmund and Jon Adams. Escaping the Laboratory. The Rodent Experiments of John B. Calhooun & Their Cultural Influence. *Journal of Social History* 42, no. 3 (March 2009): 761–792.
[23] On laboratory animals' maintenance see Schlünder, Martina, Christian Reiß, Axel C. Hütelmann and Susanne Bauer. Cakes and Candies – zur Geschichte der Ernährung von Versuchstieren. *Berichte zur Wissenschaftsgeschichte* 35, no. 4 (January 2012): 275–285.
[24] On the cultivation of the Mexican Axolotl in Europe see Reiß, Christian. *Der Axolotl. Ein Labortier im Heimaquarium, 1864–1914*. Göttingen: Wallstein, 2020.

which became valuable commodities.[25] The Jackson Laboratory in the United States, for instance, developed specific lines of inbred mice that were related to corresponding research topics, among them mice disposed for cancer that were susceptible to the cultivation of tumor tissue, marketed as JAX-mouse and established as a standard laboratory animal.[26] These changing requirements were related to large-scale research and long-lasting series of experiments, the emergence of big science and the industrialization of scientific enterprises.[27] Life scientists had to deal with the paradox that, although the laboratory environment was artificial, its animals were not calculable technical devices but living organisms with their own agency. As such, on the one hand they were an essential precondition to model and reproduce vital processes, but on the other hand they were expected to work like animal machines – calculable and adjustable.

Other prerequisites for laboratory animals in the experimental life sciences were their small size, a high reproductive rate and short rhythms of generation sequences. Moreover, from a human researcher's perspective, they had to be tame, easy to maintain and cheap to feed. Hence, mice and fruit flies were the ideal targets. With the differentiation of the life sciences throughout the twentieth century and highly specialized fields of research, laboratory animals had to comply with various other scientific demands. For tumor research, for example, animals had to have a high disposition for certain tumor types. As a consequence, specific animals became the dominant model organism depending on the research area or question, such as the fruit fly for genetics, certain strains of mice, rats and zebrafish for tumor research,[28] and guinea pigs susceptible to various infectious diseases for early bacteriology. Nevertheless, science was always searching for "the 'right' organism for the job."[29] Animals with a disposition for specific diseases were especially bred or,

---

[25] Clause, Bonnie T. The Wistar Rat as a Right Choice. Establishing Mammalian Standards and the Ideal of a Standardized Mammal. *Journal of the History of Biology* 26, no. 2 (June 1993): 329–349; Logan, Cheryl A. Before There Were Standards. The Role of Test Animals in the Production of Empirical Generality in Physiology. *Journal of the History of Biology* 35, no. 2 (June 2002): 329–363; Kirk, Robert G.W. Wanted – Standard Guinea Pigs. Standardization and the Experimental Animal Market in Britain, ca. 1919–1947. *Studies for the History and Philosophy of Biological and the Biomedical Sciences* 39, no. 3 (September 2008): 280–291; Kirk, Robert G.W. Standardization Through Mechanization. Germ-Free Life and the Engineering of the Ideal Laboratory Animal. *Technology and Culture* 53, no. 1 (January 2012): 61–93.
[26] Rader, Karen. *Making Mice. Standardizing Animals for American Biomedical Research, 1900–1955.* Princeton, NJ: Princeton University Press, 2004.
[27] Cf. Todes, *Pavlov's Physiological Factory.*
[28] Rader, *Making Mice*; Endersby, Jim. *A Guinea Pig's History of Biology.* Cambridge, MA: Harvard University Press, 2007; Meunier, Robert. Stages in the Development of a Model Organism as a Platform for Mechanistic Models in Developmental Biology: Zebrafish, 1970–2000. *Studies for the History and Philosophy of Biological and the Biomedical Sciences* 43, no. 2 (June 2012): 522–531.
[29] Clause, The Wistar Rat, 330.

like the so-called *OncoMouse* for tumor research, genetically designed and established as model organisms with distinct features.[30]

As the supply of specific and standardized laboratory animals in sufficient numbers became more complicated and their care more demanding, animal care and preservation was institutionalized and professionalized from the middle of the twentieth century onwards, and the welfare of animals turned into a scientific discipline in itself.

3 Methods and Approaches

Until the 1980s, animal experiments and laboratory animals usually served as anecdotal prodigy in biographies of "famous" scientists and histories of scientific institutions or "famous inventions" and "discoveries." With the adoption of approaches from the social and cultural sciences and the emergence of new concepts in the field of history like the practical, cultural and material turn as well as interdisciplinary cooperations among philosophers, historians, cultural, social and natural scientists – especially in the history of science – the focus shifted from mere events, individual persons and their discoveries to ideas and epistemologies, practices and processes, objects and actors – human and nonhuman [→Social History;→Cultural History;→History of Ideas].

In the vein of the emerging cultural studies and ever since the publication of Harriet Ritvo's *Animal Estate* in 1987, research on (the history of) animals has intensified. But it took some time before the history of experimental animals came into focus. This is especially true for German historiography, where a strong focus on social history delayed the cultural and *animal turn* by about a decade. Neither the first historical anthology on the history of human-animal relations, published in 1998,[31] nor a similar volume published a decade later[32] contains any chapters on experimental animals – despite controversial public debates on animal experiments. In the Anglo-American scientific discourse, in contrast, we already find studies on the history of laboratory animals in the 1990s.[33] Since the new millennium especially, various publications have appeared,[34] and animal studies anthologies[35] or monographs

---

30 Endersby, *Guinea Pig's History*; Meunier, Stages.
31 Münch, Paul and Rainer Walz (eds.). *Tiere und Menschen. Geschichte und Aktualität eines prekären Verhältnisses*. Paderborn: Ferdinand Schöningh, 1998.
32 Brantz, Dorothee and Christof Mauch (eds.). *Tierische Geschichten. Die Beziehung von Mensch und Tier in der Kultur der Moderne*. Paderborn: Ferdinand Schöningh, 2009.
33 See, for example, Clause, The Wistar Rat; or Holmes, The Old Martyr.
34 See, for example, Brookes, *Fly*; Logan, Before There Were Standards; Shanks, *Animals and Science*; Guerrini, *Experimenting*; Rader, *Making Mice*.
35 See, for example, the related contributions in Kalof, Linda (ed.). *A Cultural History of Animals*, 6 vols. Oxford: Berg, 2007; Kalof, Linda (ed.). *The Oxford Handbook of Animal Studies*. Oxford: Oxford University Press, 2017; Krüger, Gesine, Aline Steinbrecher and Clemens Wischermann (eds.). *Tiere und*

on certain animal species usually contain chapters on laboratory and experimental animals.³⁶

Today, human-animal relationships are debated in various historical fields. Aside from a primarily historical or chronological narrative, as presented above, or a cultural studies perspective with a focus on specific animals, the history of animal experiments and laboratory animals could equally be written from the perspective of the history of emotions, as experimenters were trained to ignore any emotions and become "objective" [→History of Emotions]. In general, however, investigations on laboratory animals and animal experiments are conducted from a history of science perspective with a focus on the establishment of institutions and scientific disciplines or on epistemological questions that interrogate the role of animals within the process of knowledge production [→History of Science].

Historical Epistemology

Historical epistemology reflects on knowledge production and the historical conditions under which an object came into the focus of scientific interest and became, according to Hans-Jörg Rheinberger, an "epistemic thing." In this process, "epistemic things" were distinguished from "technical things" that are, for instance, part of technical arrangements in experiments. Within an experimental framework, all parameters and "technical things" remain steady while one variable changes, thus producing new information and new knowledge.³⁷ While the life scientists Paul Ehrlich and Sahachiro Hata, for example, were searching for a cure for syphilis, they studied the effects of arsenic compounds to find the maximum quantity that was still tolerated by a laboratory animal without showing lethal toxic reactions. Within the experimental setting, all parameters and the laboratory environment, including the laboratory animal (same species, health constitution, weight, and age), the way of applying the compound (intramuscular or subcutaneous), and its composition remained constant – only the amounts administered to an animal were regularly changing and the effects recorded.³⁸ Within such an experimental setting, it is inevitable that beyond the variable (the dosage), all parameters and "technical things"

Geschichte. Konturen einer Animate History. Stuttgart: Franz Steiner, 2014; or Böhnert, Martin, Kristian Köchy and Matthias Wunsch (eds.). *Philosophie der Tierforschung*, 3 vols. Freiburg: Verlag Karl Alber, 2016–2018.
36 For the frog see Hüppauf, *Frosch;* for the rabbit Davis, Susan E. and Margo Demello (eds.). *Stories Rabbits Tell. A Natural and Cultural History of a Misunderstood Creature.* New York, NY: Lantern Books, 2003.
37 Cf. Rheinberger, Hans-Jörg. *Experiment – Differenz – Schrift. Zur Geschichte epistemischer Dinge.* Marburg: Basilisken-Presse, 1992; Rheinberger, Hans-Jörg. *Historische Epistemologie zur Einführung.* Hamburg: Junius, 2007.
38 Hüntelmann, Mäuse.

involved in the experiment are stable to exclude coincidences influencing the outcome. The need for a stable laboratory environment facilitates (if not presupposes) the standardization of all apparatuses involved, including the animals that appear in these experiments, as "technical things." Studies in historical epistemology challenge the conventional dichotomy between a human subject performing an experiment on an animal object, emphasizing instead the historical conditions of and the uncertainty and indeterminacy produced in such experimental settings. However, in interrogating the broader process of knowledge production, historical epistemology focuses on the interrelation between a diversity of subjects and not primarily on human-animal relationships.

Related to historical epistemology are studies on models and model organisms like *drosophila melangastor*, nematodes, or mice. For example, in bacteriology the infection of a laboratory animal with pathogens represented the disease of a sick human, and animal experiments in immunology modelled the process of recovery, imitating immunological processes of the (human) body fighting the disease. Studies on model organisms investigate the design, establishment, role and function of animal models within the experimental framework as well as difficulties that arose when animal organisms were compared to or used to model those of humans.[39]

Science and Technology Studies

The works of Bruno Latour alongside actor-network theory elaborated by him, John Law, and Michel Callon exerted a huge impact on the practical, material and ontological turn in the history of science, the social sciences and the philosophy of science. Latour developed his approach from empirically observing "laboratory life" and the interactions of various human and nonhuman actors within the process of knowledge production, something that Latour describes as "Science in Action." Latour not only refers to animals, however, but to all kinds of beings and "things." Animals are considered to be actors, who can be human and nonhuman alike while all have "agency" and the capacity to influence each other.[40]

Experimental animals have agency here in two ways: Firstly, although there is a strong power-asymmetry in the relation between human scientists and laboratory animals, they often oppose or thwart human actions and goals – they resist, bite

---

[39] Chardarevian, Soraya de and Nick Hopwood. *Models. The Third Dimension of Science.* Stanford, CA: Stanford University Press, 2004; Creager, *Science without Laws*; Gradmann, Maß; Roelcke, Volker. Repräsentation – Reduktion – Standardisierung: Zur Formierung des Tiermodells menschlicher Krankheit in der experimentellen Medizin des 19. Jahrhunderts. In *Tier – Experiment – Literatur, 1880–2010*, Roland Borgards, Nicolas Pethes (eds.), 15–36. Würzburg: Königshausen & Neumann, 2013.
[40] Latour, Bruno. *Reassembling the Social. An Introduction to Actor-Network-Theory.* Oxford: Oxford University Press, 2005.

and try to escape, for example. Their human handlers react insofar as they construct facilities like cages, operating tables or other devices to fixate animals. Furthermore, they learn special techniques to deal with and treat animals, for example the docking of a mouse's tail to take blood samples. Secondly, the animal's agency refers to the *re*-action within the experimental setting, which might potentially differ from human intentions. For instance, in experiments on the development of chemotherapeutics after 1900, laboratory animals reacted differently after an infection with pathogens, depending on their individual health status, strain of spirochaetes, or other reasons: some died, some got seriously ill and some only showed signs of discomfort. As a consequence, Sahachiro Hata started a long-term experimental series to standardize both the infection and laboratory animal as such by using a specific strain of pathogen, a standardized way of application and blood samples that only exhibited a specific amount of pathogens and, after numerous animal deaths, a specific virulence.[41] Although animals in the Latourian laboratory no longer appear as passive, instrumentalized objects but as actors involved in research that relate to and influence other (human) animals, the attribution of agency to a laboratory animal should not conceal the profound power asymmetry between humans and animals within their relationship.

4 Implication(s) of the Animal Turn

In contrast to Latour, Donna Haraway looks specifically at animals and the interactions between humans and animals. These relations are not constant but change over time. In such a "co-constitutive relationship," laboratory animals are as involved in the performances of the laboratory as their human companions, described by Haraway as "work companions."[42] Haraway's analysis of historical and contemporary human-animal relations, including her personal example of human-dog partnerships in agility sports, highlights the importance of both species and contextual circumstances. This in turn has various implications for an animal history, and especially the history of laboratory animals. Firstly, animal studies are generally biased towards the human engaged in the study and how they look at animals. Secondly, historical accounts are passed down through time and thus the studies described cannot be reinspected anymore, whereas in contemporary inquiries an investigator could, in principle, screen the accounts through observation. Furthermore, studies in the history of human-animal relationships are predetermined by their sources – and this applies especially to laboratory animals, who are absent and at the same time omnipresent in the historical sources. Animal-sensitive historical accounts of

---

41 Hüntelmann, Mäuse.
42 Haraway, Donna. *When Species Meet.* Minneapolis, MN: University of Minnesota Press, 2008, 69–93.

pets, livestock or exotic animals at the zoo can rely on sources that actually feature the investigated animals. The same cannot be said for laboratory animals. Although scientists and technicians in laboratories work and interact with animals all day, they are not the center of their attention. The scientists' focus is, depending on the specific purpose of their experiments, on the research subject such as pathogens, tumor cells, chemical compounds or genetic variations.

Thus, although laboratory animals are ubiquitous and constantly observed, historical accounts of the lives of laboratory animals (and especially of a single life) are rare and difficult to trace. They appear as short notices in laboratory notebooks, as marks or numbers in tables or graphs, or as photographs in scientific publications. Only if unintended events occur, such as when an animal develops severe symptoms where it should not, and corrections within the experimental setting become necessary, are notes about an animal's behavior or health condition further elaborated. This means that the animal's existence and agency is primarily filtered through human intentions, limited in its effects and impact on the experiment, and appears only as a re-action, highlighting the severe power asymmetry within this special human-animal relationship.[43] The animal's own agency (in the sense of a capacity to act), their own intentions, collaborations as well as resistances, have to be extracted from notebooks or indirectly from the need to "fix" them during an experimental intervention.

An experimental animal is, however, often much more than just an experimental animal. One individual animal, living in a laboratory environment, might be evaluated from different human perspectives and, depending on the viewpoint, varying ontological statuses. Following the ontological turn, proponents of historical ontology like Ian Hacking, John Law, Annemarie Mol and Bruno Latour have questioned the pre-givenness of things and ontological entities. They suggest instead considering entities as materializations of practices and processes with mutable ontologies that may change over time. This perspective allows us to investigate the socio-cultural conditions of knowledge production as well as the generation of scientific objects (and subjects) and how they produce each other. For instance, in Sarah Franklin's study of the cloned sheep Dolly, sheep were much more than mere *Ovis aries*, emerging as precious breeders, scientific objects, or wool suppliers depending upon whether they were seen through the lenses of biology, farming, economics, or culture. And a simple grey mouse, commonly living in the house or garden and considered vermin, could become – through her inclusion in bacteriological experiments – a laboratory animal. This animal might be infected with pathogens, injected with serum or

---

[43] The asymmetric human–animal power relationship could be incorporated in a "body history" of laboratory animals with reference to Michel Foucault, as suggested by Guerrini, *Experimenting*; Eitler, Pascal. Animal History as Body History: Four Suggestions from a Genealogical Perspective. *Body Politics* 2, no. 4 (2014): 259–274; Merdes, Dominik. Co-constitutive Relationships in Modern Medicine. Körper-Werden um die Geburtsstunde der modernen Chemotherapie. *Body Politics* 2, no. 4 (2014): 329–364.

have tumor substances of another animal planted into her, and thus be transformed into a hybrid, and, together with the inscribed information of the experiment, into a valuable epistemic thing. Such a specific mouse could equally serve as an animal passage, through which a strain of bacteria or a tumor cell is passed on to observe their alterations, that increases the virulence of tumor cells, become a storage vessel for bacteria, or be sent to other laboratories and become a valuable commodity in a gift economy among scientific colleagues.

Selected Bibliography

Endersby, Jim. *A Guinea Pig's History of Biology*. Cambridge, MA: Cambridge Harvard University Press, 2007.
Franklin, Sarah. *Dolly Mixtures. The Remaking of Genealogy*. Durham, NC: Duke University Press, 2007.
Guerrini, Anita. *Experimenting with Humans and Animals. From Galen to Animal Rights*. Baltimore, MD: Johns Hopkins University Press, 2003.
Haraway, Donna. *When Species Meet*. Minneapolis, MN: University of Minnesota Press, 2008.
Holmes, Frederic L. The old Martyr of Science. The Frog in Experimental Physiology. *Journal of the History of Biology* 26, no. 2 (Summer 1993): 311–328.
Rader, Karen. *Making Mice. Standardizing Animals for American Biomedical Research, 1900–1955*. Princeton, NJ: Princeton University Press, 2004.
Reiß, Christian. *Der Axolotl. Ein Labortier im Heimaquarium, 1864–1914*. Göttingen: Wallstein, 2020.
Rupke, Nicolaas A. (ed.). *Vivisection in Historical Perspective*. London: Routledge, 1987.
Schlünder, Martina, Christian Reiß, Axel C. Hütelmann and Susanne Bauer. Cakes and Candies – zur Geschichte der Ernährung von Versuchstieren. *Berichte zur Wissenschaftsgeschichte* 35, no. 4 (January 2012): 275–285.
Shanks, Niall. *Animals and Science. A Guide to the Debates*. Santa Barbara, CA: ABC-Clio, 2002.

Veronika Settele
History of Agriculture

1 Introduction and Overview

Farm animals play a major role in everyone's daily lives. Even as "meat" production, particularly in Western Europe and North America, begins to move away from actual animals due to in vitro meat or plant-based meat alternatives, agricultural animals will continue to play that major role. Since the emergence of an animal-centered diet in the second half of the nineteenth century, the agricultural system of industrialized countries has become oriented towards livestock. Globally, the consumption of animal products increased rapidly. Expanding foreign trade and a dependency on fodder imports accompanied the focus on keeping animals (instead of growing crops). Techniques of livestock handling in modern times gave rise to formerly unknown rates of productivity, but also produced new threats, such as the spreading of animal epidemics or damage to soil and water supply caused by massive amounts of liquid manure. Finally, since the 1980s, these modern farming techniques have reinforced ethical concerns about the animals kept for food production.

Against this background, one would expect farm animals to occupy a prominent place within the historiography of agriculture. However, this is not the case. Neither have animals advanced to a sub-section within agricultural history nor does animal history seem to show a particular interest in agricultural animals. This chapter presents approaches to overcoming this double blind-spot. Agricultural animals are seldomly looked at in and for themselves. They are instead lumped in with other animals, even though their living conditions differ decisively from animals kept for other reasons than food production.[1] I follow Susan Merrill Squier's "Manifesto for Agricultural Studies," in which she asks not only for a better communication of "high-density, large-scale industrial animal farming" in positivist modes of knowing but also for a broader understanding of current practices of animal farming. The historical explanation of modern animal agriculture with its benefits as well as limitations might help to "find ways of growing food that are both sustainable and equitable."[2]

Note: For their critical perspectives and helpful feedback, I would like to thank the participants of the Princeton-Harvard-MIT Workshop on the History of the Physical Sciences 2017, where I presented an earlier version of this essay, especially Mickey McGovern, and William San Martin, as well as Norman Aselmeyer, Erika L. Milam, and Harriet Ritvo.

1 Squier, Susan Merrill. *Poultry Science, Chicken Culture: A Partial Alphabet*. New Brunswick, NJ: Rutgers University Press, 2011, 7.
2 Squier, *Poultry Science*, 6. On p. 11 she states: "We need the explicit, critical, and cultural analysis of agriculture".

Historiographically, the *animal turn* suggests conceptualizing animals as "agentive subjects and autonomous individuals."[3] Accordingly, scholars are trying to grasp animal behavior as a "crucial part in the development of human cultures" and "seek out the activities of nonhumans as factors of historical causation."[4] In short, they look for powerful animals in history. In agricultural settings, such creatures do not stare us in the face. However, as I will argue, this does not mean that agricultural animals are not important for explaining the history of agriculture. Humans keep agricultural animals for economic reasons. In this setting, the animal's raison d'être is to serve human needs. They are controlled, dominated, subjugated, and used by humans throughout their lives and beyond. Stories of coevolution, togetherness, mutual influence and cooperation are not obvious at first glance [→Domestication: Coevolution]. Livestock agriculture, cast in the mold of industrial modernity, has understandably been an unappealing topic for animal history, considering its political links to "the modern animal-rights-movement and the controversies surrounding factory farming."[5] Agricultural development in modern times has systematically decreased the economic and cultural value of the single animal and its room for manoeuver. Simultaneously, human power and control over animals has increased.[6] This devaluation of the (individual) animal might be the reason why agricultural animals have not entered the mind of animal historians looking for substantial human-animal-interactions in the first place and why modern agriculture, despite its continued political, economic, and ecological relevance, does not play a dominant role within animal history.[7] Farmed animals, however, have continuously influenced the history of agriculture. This chapter will show how to grasp historically significant human-animal-relationships in the setting of modern livestock production.

---

[3] Ománska, Ewa. Animal History. *History and Theory* 56, no. 2 (June 2017): 267–287, 281; see also Cox, Lisa. Finding Animals in History: Veterinary Artefacts and the Use of Material History. In *The Historical Animal*, Susan Nance (ed.), 99–117. Syracuse, NY: Syracuse University Press, 2015, 101; Zehnle, Stephanie. Of Leopards and Lesser Animals: Trials and Tribulations of the Human-Leopard-Murders in Colonial Africa. In *The Historical Animal*, Susan Nance (ed.), 221–239. Syracuse, NY: Syracuse University Press, 2015, 223; Nance, Susan. Introduction. In *The Historical Animal*, Susan Nance (ed.), 1–16. Syracuse, NY: Syracuse University Press, 2015, 8.
[4] Ománska, Animal History, 269, 280.
[5] Tortorici, Zeb and Martha Few. Writing Animal Histories. In *Centering Animals in Latin American History*, Martha Few, Zeb Tortorici (ed.), 1–30. Durham, NC and London: Duke University Press 2013, 9.
[6] Pearson, Chris. History and Animal Agencies. In *The Oxford Handbook of Animal Studies*, Linda Kalof (ed.), 240–257. Oxford: Oxford University Press 2017, 252.
[7] Krüger, Gesine, Aline Steinbrecher and Clemens Wischermann (eds.). *Tiere und Geschichte: Konturen einer Animate History*. Stuttgart: Franz Steiner Verlag, 2014. This has resulted in significantly more literature on the ethics of food animal production than on their history, for example Friend, Catherine. *The Compassionate Carnivore: Or, How to Keep Animals Happy, Save Old MacDonald's Farm, Reduce your Hoofprint and Still Eat Meat*. Cambridge, MA: Da Capo Press, 2009; Thomson, Paul B. The Ethics of Food Animal Production. In *The Oxford Handbook of Animal Studies*, Linda Kalof (ed.), 364–379. Oxford: Oxford University Press, 2017.

There is a second reason that has prevented animal-sensitive approaches from gaining ground in the realm of the history of agriculture: farm animals have disappeared. During the twentieth century, they were not only kept more and more outside of human settlements. In industrialized countries they also started to spend their life entirely inside stables and barns.[8] Under industrial management, mechanically enhanced all-year indoor farming became more cost-effective. Among other things, this appeared necessary, because available human labor was in decline – and would continue to be so. Fewer people worked on farms, potentially because they hoped for better jobs in the industrial and service sectors. Moreover, farm work and especially livestock production, which was traditionally strenuous and dirty, has become less lucrative and lost its esteem within society.

This chapter concentrates on conceptual as well as empirical ideas for a history of animal agriculture in modern times, mainly in Europe and the United States. To justify this focus, in temporal terms, pre-modern and early modern farm animals have so far already garnered significantly more attention.[9] Erica Fudge and Harriet Ritvo, two pioneering animal historians, have extensively dealt with early modern human-animal relationships and inspired numerous scholars to follow suit.[10] In spatial terms, the development of modern techniques of livestock production in industrialized countries began to spread rapidly all over the world as the most cost-effective way to produce animal products.

I will argue for looking at the micro-level of interaction between humans and farm animals. By doing so, the history of farm animals offers a chance for a new agri-*cultural* history to reclaim attention within historiography. With the decreasing economic relevance of agriculture in industrialized countries, agricultural history lost territory in historiography. As there were fewer people working in agriculture, looking at farming practices no longer provided valuable insights into human soci-

---

8 Mizelle, Brett. Unthinkable Visibility. Pigs, Pork, and the Spectacle of Killing and Meat. In *Rendering Nature. Animals, Bodies, Places, Politics*, Marguerite S. Shaffer, Phoebe S. K. Young (eds.), 263–286. Pennsylvania, PA: University of Pennsylvania Press, 2015; Otto, John Solomon. Cattel-Grazing in the Southeastern United States, 1670–1949. In *Animals in Human Histories: The Mirror of Nature and Culture*, Mary J. Henninger-Voss (ed.), 56–82. Rochester, NY: University of Rochester Press, 2002, 76–77.
9 Beginning with zooarchaeology of ancient times, see Clutton-Brock, Juliet. *Animals as Domesticates: A World View through History*. East Lansing, MI: Michigan State University Press, 2012; Macgregor, Arthur. *Animal Encounters: Human and Animal Interaction in Britain from the Norman Conquest to World War One*. London: Reaktion Books, 2012, 415–495; Stone, David. Medieval Farm Management and Technological Mentalities: Hinderclay Before the Black Death. *Economic History Review* 54, no. 4 (November 2001): 612–638.
10 Fudge, Erica. What Was it Like to be a Cow? History and Animal Studies. In *Oxford Handbook of Animal Studies*, 258–278; Ritvo, Harriet. *Noble Cows and Hybrid Zebras: Essays on Animals and History*. Charlottesville, VA and London: University of Virginia Press, 2010; Anderson, Virginia DeJohn. *Creatures of Empires: How Domestic Animals Transformed Early America*. Oxford: Oxford University Press, 2004; Edwards, Peter. *Horse and Man in Early Modern England*. London: Hambledon Continuum, 2007.

ety. Today, however, expanding parts of Western societies are concerned about the practices of modern animal farming. Their worries render animal agriculture once again a historically promising topic – regardless of its relative loss of economic significance.

2 Topics and Themes

Breeding

Agricultural animals were supposed to function for specific production targets and those targets narrowed in modern animal agriculture. Animals whose bodies were specialized in one production target were most profitable. For this reason, breeding, which preceded actual animal farming, gained attention among farmers, agricultural politicians and economists, and became a constitutive part of modern animal agriculture. In 1948, Vantress Hatchery from California was the lucky winner of the nationwide "Chicken of Tomorrow" contest. Vantress was able to grow a heavier, meatier chicken faster than 39 competitors. Not only did the company receive $5,000 in prize money, but after this win the orders for their breed skyrocketed. Ten years later, Vantress-produced birds had become the standard used by the nation's poultry farmers.[11] The "Chicken of Tomorrow" contests ran in three-year cycles from 1946 to 1961, starting at the regional level every first and second year and culminating in a big national contest in the third. Howard C. Pierce, the marketing specialist of A&P Food Stores and the man behind the idea, successfully convinced the American poultry industry and the US Department of Agriculture to launch this long-term contest for small-scale farmers as well as large commercial breeders to foster the development of bigger, faster-growing chickens.[12] In comparison to yesterday's chicken, the new bird would be defined by the meatiest animal that producers could raise for the lowest feed cost. Egg yield no longer mattered because tomorrow's chickens were only bred for meat production. Heightened demand for chicken meat on the consumers' side was the impetus for this shift: Around 1950 more people wanted to consume chicken regularly. They regarded lean meat as a sign of social advance-

---

[11] Horowitz, Roger. Making the Chicken of Tomorrow: Reworking Poultry as Commodities and as Creatures, 1945–1990. In *Industrializing Organisms: Introducing Evolutionary History*, Susan R. Schrepfer, Philip Scranton (ed.), 215–235. New York, NY: Routledge, 2004, 215–216.

[12] Squier, *Poultry Science*, 44–45; Orland, Barbara. Turbo-Cows: Producing a Competitive Animal in the Nineteenth and Early Twentieth Century. In *Industrializing Organisms: Introducing Evolutionary History*, Susan R. Schrepfer, Phillip Scranton Schrepfer (ed.), 167–190. New York, NY: Routledge, 2004; Nisly, Jadon. Er komme von seinem Viehe nicht hinweg: Mensch-Nutztier-Beziehung in einem volksaufklärerischen Mustergut (1782–1795). In *Tiere nutzen: Ökonomien tierischer Produktion / Jahrbuch für die Geschichte des ländlichen Raumes* 12, Lukasz Nieradzik, Brigitta Schmidt-Lauber (eds.), 88–104. Innsbruck: Studien Verlag, 2016.

ment and wealth.¹³ Clever tradesmen such as Howard C. Pierce immediately recognized the enormous business opportunity. They just had to get the breeders and farmers first to create, and then produce the new birds.

Putting the producers in competition was the first step toward the altered animals, the systematic application of science and technology the second. Within three decades after World War II, the breeding of agricultural animals was radically transformed. Humans started to define and shape animals to become the agricultural animals they wanted to use in a more strategic way. Applications of science and technology, such as progeny-testing, artificial insemination, quantitative genetics and synthetic breeding lines increased the influence of biomedical experts. These biotechnological shifts in animal breeding have stimulated cultural studies scholars to look at animal farming.¹⁴ Inspired by Michel Foucault's biopolitics, their work posits that the animals remained central protagonists – no matter how technologically refined their breeding became.

Just as during World War I, the question of agricultural supplies was high on the political and administrative agenda during the Second World War. The animals' productivity became a matter of victory or defeat. Later it was perpetuated as a condition of economic growth and social progress in the post-war era.¹⁵ Unlike thirty years earlier, however, in the middle of the twentieth-century new scientific knowledge on genetics, heredity, and scientific management had become available.¹⁶ Bert Theunissen has shown for the Netherlands¹⁷ how the "art of breeding" became a "sci-

---

13 Horowitz, Making the Chicken of Tomorrow, 215.
14 Chrulew, Matthew and Dinesh Wadiwel. *Foucault and Animals*. Leiden: Brill, 2016; Holloway, Lewis. Subjecting Cows to Robots: Farming Technologies and the Making of Animal Subjects. *Environment and Planning D: Society and Space* 25, no. 4 (August 2007): 1041–1060; Shukin, Nicole. *Animal Capital: Rendering Life in Biopolitical Times*. Minneapolis, MN: University of Minnesota Press, 2009; Taylor, Chloë. Foucault and Critical Animal Studies: Genealogies of Agricultural Power. *Philosophy Compass* 8, no. 6 (June 2013): 539–551; Asdal, Kristin, Tone Druglitrø and Steve Hinchliffe. *Humans, Animals and Biopolitics: The More-Than-Human Condition*. Abingdon and New York, NY: Routledge 2017.
15 Gaudillière, Jean-Paul. The Farm and the Clinic: An Inquiry into the Making of our Biotechnological Modernity. *Studies in History and Philosophy of Biological and Biomedical Sciences* 38, no. 2 (June 2007): 521–528, 525; Saraiva, Tiago. *Fascist Pigs: Technoscientific Organisms and the History of Fascism*. Cambridge, MA: MIT Press, 2016, 13, 101–135; Roscher, Mieke. Das nationalsozialistische Tier: Projektionen von Rasse und Reinheit im Dritten Reich. *TIERethik. Zeitschrift zur Mensch-Tier-Beziehung* 8 (2016): 30–47.
16 For a longer perspective on the history of animal breeding, see Derry, Margaret E. *Masterminding Nature: The Breeding of Animals*. Toronto: University of Toronto Press 2015; as well as her work on cattle and chicken breeding, see Derry, Margaret E. *Art and Science in Breeding: Creating Better Chickens*. Toronto: University of Toronto Press 2012; Derry, Margaret E. *Ontario's Cattle Kingdom: Purebred Breeders and Their World, 1870–1920*. Toronto: University of Toronto Press 2001.
17 Together with Denmark and Ireland, the Netherlands were among the first European countries to restructure their agriculture towards export-oriented livestock production at the end of the nineteenth century, and since then became leaders in so-called high-performance cows, see Mai, Gunther. Die Agrarische Transition: Agrarische Gesellschaften in Europa und die Herausforderungen der industri-

ence," and how "craftsmanship was displaced by scientific expertise."[18] However, the effects of animal breeding were no predictable outcome of a modernization project. Sources suggest that the feeding situation in the 1950s and 1960s had been too austere to allow the cows to exploit their genetic potential.[19] The animals' bodies showed unintended reactions to new forms of breeding. Pigs bred in the same period for putting on more muscle to provide more pork in a cheaper way, for example, first complicated this process. Their legs were not yet strong enough to carry the added weight of muscle tissue growing on their back.[20] Eventually, the breeding aims set by humans were reached, but unintended bodily reactions of the animals structured the means to get there.

In addition to the biomedical dimension of breeding, there are two more domains in which animal breeding has already entered the historical picture: firstly, colonial expansion fostered ideas of breeding for the colonies [→(Post)Colonial History]. Scholars such as Virginia DeJohn Anderson, John R. Fischer, Rebecca Woods and Sarah Franklin have shed light on the relationship between transformations of livestock and expansions of empire.[21] The animals' blood lines tell us about global entanglement, show how imperialism was instantiated in the bodies of sheep and cattle and explain how colonial suppression of natives was tied to settler colonialists' modes of animal farming.[22] Secondly, the connection between animal breeding and human eugenics highlights the transfer and transformation of interspecies politics:

---

ellen Moderne im 19. und 20. Jahrhundert. *Geschichte und Gesellschaft* 33, no. 4 (January 2007): 471–514, 477–478.

18 Theunissen, Bert. Breeding for Nobility or for Production: Cultures of Dairy Cattle Breeding in the Netherlands, 1945–1990. *Isis* 103, no. 2 (June 2012): 278–309, 278, 307.

19 Theunissen, Breeding for Nobility, 308.

20 Schilling, Erich. Biologische Probleme in der Tierzuchtforschung. In *Zwanzig Jahre Tierzuchtforschung in Mariensee: Bericht über wissenschaftliche Arbeiten aus dem Max-Planck-Institut für Tierzucht und Tierernährung in den Jahren 1948 bis 1968*, Max-Planck-Institut für Tierzucht und Tierernährung (ed.), 22–27. Mariensee: Max-Planck-Institut, 1970.

21 Woods, Rebecca. The Herds Shot Round the World: Native Breeds and the British Empire, 1800–1900. PhD diss.: MIT, Boston, MA (2013); Franklin, Sarah. Crook Pipettes: Embryonic Emigrations from Agriculture to Reproductive Biomedicine. *Studies in History and Philosophy of Biological and Biomedical Sciences* 38, no. 2 (May 2007): 358–373; Fischer, John R. *Cattle Colonialism: An Environmental History of the Conquest of California and Hawai'i*. Chapel Hill, NC: The University of North Carolina Press, 2015; Anderson, *Creatures of Empire*; see also Saha, Jonathan. Milk to Mandalay: Dairy Consumption, Animal History and the Political Geography of Colonial Burma. *Journal of Historical Geography* 54 (October 2016): 1–12.

22 Garagarza, León Garcia. The Year the People Turned into Cattle: The End of the World in New Spain, 1558. In *Centering Animals in Latin American History*, Martha Few, Zeb Tortorici (eds.), 31–61. Durham, NC and London: Duke University Press 2013; McGrea, Heather. Pest to Vector: Disease, Public Health, and the Challenges of State-Building in Yucatán, Mexico, 1833–1922. In *Centering Animals in Latin American History*, Martha Few, Zeb Tortorici (eds.), 123–148. Durham, NC and London: Duke University Press 2013.

Gabriel Rosenberg shows for hog breeding between 1865 and 1930 how "human and swine racial ascents and declines converged and conversed."[23]

Together, this research forms a starting point for a history of farm animal breeding alongside the key questions of animal history: Where were the animals? What did they do? What happens if we consider animals as playing a role in history? How did they affect, control and influence human actions?[24]

Keeping and Handling

To farm as profitably as possible has long been the logic of animal agriculture. It was with the animal itself that money was made. Its bodily processes were the object of desire for an ideology of increasing efficiency and return. These bodies, however, did not always do what breeders, farmers, and policymakers sought to make them do.[25] Moments where farmers faced unintended challenges are often the easiest way to grasp the animal and its stakes in the history of animal farming.

The new German milking machines of the 1950s presented one such challenge. They threatened the health of the animal and thus the farmer's profits, as they did not stop after emptying the udder. If the human supervising the machine was late or distracted, the continued suckling of the machine damaged the udder's tissue and resulted in inflammation of the animal's body – and losses to the farmer's income.[26] The machine was supposed to increase productivity. However, if it harmed the body of the cow, the intended profit increase suddenly became a liability. Reduced milking performance, expenses for veterinarian treatment and medication

---

[23] Rosenberg, Gabriel N. A Race Suicide Among the Hogs: The Biopolitics of Pork in the United States, 1865–1930. *American Quarterly* 68, no. 1 (March 2016): 49–73, 51.

[24] Krüger, Gesine, Aline Steinbrecher and Clemens Wischermann. Animate History: Zugänge und Konzepte einer Geschichte zwischen Menschen und Tieren. In *Tiere und Geschichte: Konturen einer Animate History*, Gesine Krüger, Aline Steinbrecher, Clemens Wischermann (eds.), 9–34. Stuttgart: Steiner, 2014.

[25] See for a great synopsis, Wilmot, Sarah (ed.). Between the Farm and the Clinic: Agriculture and Reproductive Technology in the Twentieth Century. Special issue of *Studies in History and Philosophy of Science Part C: Studies in History and Philosophy of Biological and Biomedical Sciences* 38, no. 2 (June 2007), particularly: Wilmot, Sarah. From Public Service to Artificial Insemination: Animal Breeding Science and Reproductive Research in Early Twentieth-Century Britain: 411–441; Grasseni, Cristina. Managing Cows: An Ethnography of Breeding Practices and Uses of Reproductive Technology in Contemporary Dairy Farming in Lombardy (Italy): 488–510; Brassley, Paul. Cutting Across Nature: The History of Artificial Insemination in Pigs in the United Kingdom: 442–461. See also Van der Laan, Steven. *Een varken voor iedereen: De modernisering van de Nederlandse varkensfokkerij in de twintigste eeuw* [a pig for everyone: the modernization of Dutch pig breeding in the twentieth century]. Utrecht: Xerox, 2017, 209–226.

[26] For empirical details of this paragraph see Settele, Veronika. Cows and Capitalism: Humans, Animals and Machines in West German Barns, 1950–80. *European Review of History* 25, no. 6 (September 2018): 849–867.

or, at worst, the cow's death were reflected in the budget. Its relation to the farmer's money made the animal a historical factor impossible to ignore within the stables. Starting with the problem of sustaining animal health in the setting of industrialized animal production, soon the full potential of animal compliance was channeled into increasing efficiency.[27] This process, however, was marked by trial and error. In the cowshed, the animal's performance was the knife's edge of the farmer's benefit.

The link between animal health and profit challenges the narrative that capitalism made its unhampered way into animal farming as one of its last stages. Instead, the history of animal agriculture reveals a unique version of capitalist expansion in which the bodily processes of the animals became sites of resistance and affinity. To rediscover the animal in the history of agriculture, I suggest looking closely at what happened in the stables. Focusing on everyday micro-practices of both the animals and humans permits the telling of a history in which the animals' role cannot be overlooked. The animals' influence was not limited to unintended bodily reactions towards new techniques of keeping and working animals. As long as the economic value of single animals was high enough, they were able to influence farming practices massively. In Germany, cows, to stick with this example, who refused to yield milk with the new machines in the 1950s were milked by human hands.[28] Twenty years later, however, the very same animal behavior was no longer able to alter the behavior of the farmers. Instead, cows who refused the milking machine were slaughtered.[29] Adapting the milking method to the animals' needs had become too costly for the farmers and thus economically irrational. This episode illustrates how closely the animals' room for historical manoeuver was tied to its cash-value within the setting of animal agriculture. The cheaper the single animal became, the smaller its possibilities to influence the way it was kept. In Europe and North America, farm animals were continuously made cheaper throughout the nineteenth and twentieth century as inexpensive milk, meat and eggs were simultaneously consumer wishes and at the center of nutrition politics.

Compared to the history of breeding, there are fewer studies on the history of livestock handling during its industrialization.[30] There is, however, brilliant scholarship on the industrialization of arable farming. Deborah Fitzgerald showed in *Every Farm a Factory: The Industrial Ideal in American Agriculture* how new knowledge on

---

27 See, for example, Grandin, Temple and Mark Deesing. *Humane Livestock Handling*. North Adams, MA: Storey Publishing, 2008.
28 G. D. aus T. Das frage ich das Wochenblatt. Kuh und Melkmaschine. *Bayerisches Landwirtschaftliches Wochenblatt* 146, no. 40 (1956): 8.
29 Weber, Willi. Mehr Technik um die Kuh – weniger Stunden im Stall. *Bayerisches Landwirtschaftliches Wochenblatt* 166, no. 5 (1976): 12.
30 Farm animals are missing in historical surveys which cover food systems, hunting, the zoo, petkeeping and laboratories, but not agriculture, see Henninger-Voss, Mary J. (ed.). *Animals in Human Histories: The Mirror of Nature and Culture*. Rochester, NY: University of Rochester Press, 2002; Kalof, Linda. *Looking at Animals in Human History*. London: Reaktion Books, 2007.

quantification and mechanization became promoted by authorities and reshaped US agriculture between 1920 and 1940.³¹ Frank Uekötter took a longer perspective and showed in *Die Wahrheit ist auf dem Feld: Eine Wissensgeschichte der deutschen Landwirtschaft* how chemistry and engineering changed German soil cultivation starting in the nineteenth century.³² Empirically, both scholars focus on crops, suggesting, however, parallel developments in the history of livestock farming.³³ In doing so, they provide valuable inspiration for reconstructing the industrialization of animal farming, which faced the same paradigms of quantification, mechanization and professionalization.

In the narrower realm of the history of animal agriculture, we have the work of Abigail Woods, who highlights the role of animals in the multidirectional paths of agricultural modernity. Woods shows for British pig production between 1910 and 1965 that producers aimed at matching "the pig's natural needs and desires" regardless of whether their production system was small-scale mixed outdoor farming or large, specialized, and intensive.³⁴ This approach contradicts prevailing views that modern farming was only concerned with dominating and manipulating nature, while moral views of pigs as active, sentient individuals pre-dated industrialized livestock production.³⁵ That the ambivalence of modernization ideals and their problems of implementation deserves scholarly attention was recently demonstrated by Anett Laue and Thomas Fleischman through their studies of socialist industrialization of livestock in East Germany.³⁶

To sum up: focusing on the bodies and practices of the animals kept for business in agricultural settings seems the most promising way to grasp those animals within history. The literature on animal diseases and their ties to human health offers another foothold for this undertaking.³⁷ The animals' bodies were the most necessary and

---

31 Fitzgerald, Deborah. *Every Farm a Factory: The Industrial Ideal in American Agriculture*. New Haven, CT: Yale University Press, 2003.
32 Uekötter, Frank. *Die Wahrheit ist auf dem Feld: Eine Wissensgeschichte der deutschen Landwirtschaft*. Göttingen: Vandenhoeck & Ruprecht, 2010.
33 Fitzgerald, *Every Farm a Factory*, 116; Uekötter, *Die Wahrheit ist auf dem Feld*, 16, 19.
34 Woods, Abigail. Rethinking the History of Modern Agriculture: British Pig Production, c. 1910–1965. *Twentieth Century British History* 23, no. 2 (June 2012): 165–191, 165.
35 DeMello, Margo. The Present and Future of Animal Domestication. In *A Cultural History of Animals in the Modern Age*, vol. 6, Randy Malamud (ed.), 67–94. Oxford: Berg, 2007.
36 Laue, Anett. *Das sozialistische Tier: Auswirkungen der SED-Politik auf gesellschaftliche Mensch-Tier-Verhältnisse in der DDR (1949–1989)*. Cologne: Böhlau, 2017; Fleischman, Thomas J. Three Little Pigs: Development, Pollution, and the Greening of East Germany, 1970–1989. PhD diss.: New York University (2013).
37 Woods, Abigail. Is Prevention Better than Cure? The Rise and Fall of Veterinary Preventive Medicine, c. 1950–1980. *Social History of Medicine* 26, no. 1 (February 2013): 113–131; Asebe, Getahun, Bizelew Gelayenew and Ashwani Kumar. The General Status of Animal Welfare in Developing Countries: The Case of Ethiopia. *Veterinary Science & Technology* 7, no. 3 (April 2016): 1–6; Finlay, Mark R. Hogs, Antibiotics, and the Industrial Environments of Postwar Agriculture. In *Industrializing Organ-*

precious component of animal agriculture and thus able to threaten the operability of the whole system. Historians should look out for animals resisting the farming practices imposed upon them but also take the numerous situations of compliance into account.[38]

Slaughter and Eating

A recent boom in food history[39] and substantial historical research on the industrialization of slaughter offers additional links for renewing agricultural history through an animal lens [→History of Animal Slaughter]. In modern times, agricultural animals are kept for one single reason: to be consumed. Tractors, machines, and artificial fertilizers narrowed the animals' duties within agriculture. Cattle were no longer expected to do fieldwork, provide manure and additionally produce meat and milk, but to do only the latter. The narrowing of the animal's tasks was a precondition for the industrialization of animal farming. The products they were to become increasingly determined their living conditions.[40] The industrialized slaughtering processes preceded and affected livestock handling. Large slaughterhouses with industrialized workflows had opened from the 1860s onwards in the US and Europe. Since then, abattoirs asked for a steady supply of animals resembling each other.[41] Despite profound works on the industrialization of slaughtering, what also remains to be done in

---

isms: *Introducing Evolutionary History*, Susan Schrepfer, Philip Scranton (eds.), 237–260. New York, NY: Routledge, 2004.
38 I tried to do so in Settele, Veronika. *Revolution im Stall: Landwirtschaftliche Tierhaltung in Deutschland 1945–1990*. Göttingen: Vandenhoeck & Ruprecht, 2020.
39 See on the one hand empirical studies such as Vester, Katharina. *A Taste of Power: Food and American Identities*. Oakland, CA: University of California Press, 2015; and on the other hand, a variety of handbooks organizing the emerging field, such as Helstosky, Carol (ed.). *The Routledge History of Food*. London and New York, NY: Routledge, 2015; Pilcher, Jeffrey M. (ed.). *The Oxford Handbook of Food History*. Oxford: Oxford University Press, 2012.
40 Smith-Howard, Kendra. *Pure and Modern Mild: An Environmental History since 1900*. Oxford: Oxford University Press, 2014, especially 107–115, 123–128; or for a critical-animal-studies perspective Fitzgerald, Amy J. *Animals as Food: (Re)connecting Production, Processing, Consumption, and Impacts*. East Lansing, MI: Michigan State University Press, 2015.
41 On the modernization and industrialization of slaughtering see, for example, Pacyga, Dominic A. *Slaughterhouse: Chicago's Union Stock Yard and the World it Made*. Chicago, IL: University of Chicago Press, 2015; Brantz, Dorothee. Animal Bodies, Human Health, and the Reform of Slaughterhouses in Nineteenth Century Berlin. In *Meat, Modernity, and the Rise of the Slaughterhouse*, Paula Young Lee (ed.), 71–88. Durham, NH: University of New Hampshire Press, 2008; Nieradzik, Lukasz. *Der Wiener Schlachthof St. Marx. Transformation einer Arbeitswelt zwischen 1851 und 1914*. Vienna: Böhlau, 2017; Erickson, Ken C. Beef in a Box: Killing Cattle on the High Plains. In *Animals in Human Histories: The Mirror of Nature and Culture*, Mary J. Henninger-Voss (ed.), 83–111. Rochester, NY: University of Rochester Press, 2002; Specht, Joshua. *Red Meat Republic: A Hoof-to-Table History of How Beef Changed America*. Princeton, NJ: Princeton University Press, 2019.

this realm is to take a close view of the animal's body and practices. What was going on between humans and animals through the arrival of living beings and their departure as meat as a fait accompli?

The rules of the game within agriculture then were as evident as they were vicious: The requirement for an agricultural animal's right to stay alive was subordination and fulfilment of performance expectations – as it always has been within animal farming, although the performance expectations have changed. A closer look at agricultural human-animal interactions, however, also reveals the animal's relevance for understanding the past in this hierarchical setting.

3 Methods, Approaches, and Implication(s) of the Animal Turn

The central methodological question of animal history is: What place do we, as historians, want to attribute to the animal?[42] This question touches upon historiography's foundations, and my final thoughts are on how to address agricultural animals in methodological terms to best explain the occurrence and cultivation of animal farming in its historical change. I suggest that the historical approach can be fruitfully combined with "[seeking] out the activities of nonhumans as factors of historical causation in a necessarily interspecific past."[43] Demonstrating the animal's agency is a central purpose of animal history. Agency ranges between intentional acting, the capability to act, and measurable effects of an animal's behavior.[44] Talking about animal agency was crucial for opening up history's doors for nonhumans. Today, it has become widely accepted that animals shaped the world around them. Moreover, not only animals but a wide range of nonhuman actors, for example market forces, the church or the climate, have been warmly welcomed as historical factors effecting past change. I join Joshua Specht and suggest continuing to move

---

[42] Pooley-Ebert, Andria. Species Agency: A Comparative Study of Horse-Human Relationships in Chicago and Rural Illinois. In *The Historical Animal*, Susan Nance (ed.), 148–165, 149. Syracuse, NY: Syracuse University Press, 2015.
[43] Nance, Introduction, 7.
[44] Kurth, Markus, Katharina Dornenzweig and Sven Wirth. Handeln nichtmenschliche Tiere? Eine Einführung in die Forschung zu tierlicher Agency. In *Das Handeln der Tiere. Tierliche Agency im Fokus der Human-Animal Studies*, Sven Wirth, Anett Laue, Markus Kurth, Katharina Dornenzweig, Leonie Bossert, Karsten Balgar (eds.), 7–42. Bielefeld: transcript, 2015, 16; on agency within animal history see further Pearson, History and Animal Agencies; Shaw, David Gary. A Way with Animals: Preparing History for Animals. *History and Theory* 52, no. 4 (December 2013): 1–12; McFarland, Sarah E. and Ryan Hediger. Approaching the Agency of Other Animals: An Introduction. In *Animals and Agency: An Interdisciplinary Exploration*, Sarah E. McFarland, Ryan Hediger (ed.), 1–20. Leiden: Brill, 2009.

further beyond the idea of the agency of autonomous individual actors, human or animal.⁴⁵ The history of agricultural animals shows why.

Farm animals were, like all animals, potential agents. The range of agricultural settings influenced the impact of animal agency, which varied widely. Taking a closer look, the history of agriculture reveals that even if dominated, animals influenced the ways in which they were kept. The degree to which their individual characteristics formed farming methods depended on whether and how those characteristics interrelated with profits. Hence, the historian's task is to sort out how autonomous as well as imposed actions operated within and were constrained by surrounding structures. Following this approach, agency is useful as a starting point for "mapping the varied economic, political, social, and cultural contexts in which animals are embedded," but not as the examination's outcome.⁴⁶

In this regard, Pascal Eitler's suggestion of a body history approach to past animals seems particularly useful for agricultural animals. "[C]ases where humans and animals have formed [...] some sort of collective" are of particular interest.⁴⁷ We as historians "have to clearly determine the time period and social conditions in which certain animals became actors" – and how.⁴⁸ The way we can do this, Eitler suggests, is by analyzing bodies: Animal and human bodies likewise were never just there, they were steadily about to become. During this process, "bodies have been differently shaped and changed in their very concrete modes of existence."⁴⁹ Worries about human obesity and health, for example, led to leaner hogs and a rise in chicken meat in the second half of the twentieth century. Hence, reconstructing how and why various bodies within the sphere of human-animal relations were transformed reveals a closely tied interspecies history, despite the enormous power imbalance.

Finally, the way farm animals influenced the history of agriculture contains a double bottom. Cattle, chicken, and hogs continued to be efficacious as animals when their practical significance within the agricultural production process had significantly decreased. From the 1960s onwards, however, intensive keeping methods troubled some consumers. In countries of the Global North consumers started to worry about the consequences of industrial production techniques on the living beings therein. In the subsequent decades, such concerns gained political momentum. Agricultural animals played their game also indirectly, one could say. Their behavior not only had a direct impact on their treatment in the stalls. It also caused political trouble which in some places changed practices of animal farming. In this way since

---

45 Specht, Joshua. Animal History after Its Triumph: Unexpected Animals, Evolutionary Approaches, and the Animal Lens. *History Compass* 14, no. 7 (July 2016): 326–336.
46 Specht, Animal History after Its Triumph, 332.
47 Eitler, Pascal. Animal History as Body History. Four Suggestions from a Genealogical Perspective. *Body Politics* 2, no. 4 (2014): 259–274, 261.
48 Eitler, Animal History as Body History.
49 Eitler, Animal History as Body History, 263.

2012 some forms of caged egg production came to an end in the member states of the European Union. Traditional agricultural history did not take the animal as a living being into account and thus did not explain why growing parts of Western societies were concerned about animal farming. An animal sensitive history of agriculture that focuses on material and discursive practices at the same time illuminates the lasting significance of animals within the history of agriculture – including for times where scarcely anybody still has personal ties to agricultural animals.

Selected Bibliography

Fitzgerald, Deborah. *Every Farm a Factory: The Industrial Ideal in American Agriculture*. New Haven, CT: Yale University Press, 2003.
Fleischman, Thomas. *Communist Pigs: An Animal History of East Germany's Rise and Fall*. Seattle, WA: University of Washington Press, 2020.
Harriet, Ritvo (ed.). *Noble Cows and Hybrid Zebras: Essays on Animals and History*. Charlottesville, VA: University of Virginia Press, 2010.
Laue, Anett. *Das sozialistische Tier: Auswirkungen der SED-Politik auf gesellschaftliche Mensch-Tier-Verhältnisse in der DDR (1949–1989)*. Cologne: Böhlau, 2017.
Nieradzik, Lukasz and Brigitta Schmidt-Lauber (eds.). *Tiere nutzen: Ökonomien tierischer Produktion / Jahrbuch für die Geschichte des ländlichen Raumes* 12. Innsbruck: Studien Verlag, 2016.
Schrepfer, Susan R. and Philip Scranton (eds.). *Industrializing Organisms: Introducing Evolutionary History*. New York, NY: Routledge, 2004.
Settele, Veronika. *Revolution im Stall: Landwirtschaftliche Tierhaltung in Deutschland 1945–1990*. Göttingen: Vandenhoeck & Ruprecht, 2020.
Smith-Howard, Kendra. *Pure and Modern Milk: An Environmental History since 1900*. Oxford: Oxford University Press 2014.
Specht, Joshua. *Red Meat Republic: A Hoof-to-Table History of How Beef Changed America*. Princeton, NJ: Princeton University Press 2019.
Squier, Susan Merrill. *Poultry Science, Chicken Culture: A Partial Alphabet*. New Brunswick, NJ: Rutgers University Press, 2011.
Woods, Abigail. Rethinking the History of Modern Agriculture: British Pig Production, c. 1910–1965. *Twentieth Century British History* 23, no. 2 (June 2012): 165–191.

Annette Leiderer
History of Animal Slaughter

1 Introduction and Overview

To slaughter an animal is neither to kill nor to murder an animal. To slaughter an animal is to take the life of an animal and to transform a dead nonhuman body into meat or, as Noélie Vialles has put it, to turn an *Animal to Edible*.[1] In historical sources we usually find restrictions in relation to the animals and humans involved as well as in relation to the acts of slaughter: Only a select group of animals is considered appropriate, only a designated group of persons has the right to slaughter, and there are restrictions considering the place, space and time of slaughter. Thus, it is fair to say that no historical society is indifferent towards animal slaughter. People are interested in their own culture's mode of slaughter and they draw ethical and political conclusions from how other cultures execute it. Hence, it is no coincidence that there are emotionally charged processes of othering happening through discourses about animal slaughter. There is misogynist contempt towards females slaughtering chicken, there is anti-Semitic discrimination against Jewish butchers, and there is orientalist reporting about the dog meat festival in the city of Yulin. Thus, animal slaughter defines a culture and is of great interest to researchers in the humanities.

The history of animal slaughter is part of social, cultural and political history because it reflects dynamics and imbalances of power [→Social History; →Cultural History; →Political History]. The persons who are allowed to slaughter embody social norms and hierarchies. A butcher, a priest or a hunter demonstrate the human power over the lives and deaths of animals. At the same time, all these acts of slaughter are intertwined with a certain degree of social or political power: a master butcher has authority over his apprentices and the access to the butcher craft.[2] Ancient priests supervised their religious community and its order.[3] And hunters not only control other animals but also land. Still, in the cases of butchers an advanced social status was in general only true for male ox-butchers before the industrialization of animal slaughter.

The history of animal slaughter is an economic history [→Economic History]. Animals are part of all human economies and they were always a measure for wealth; the Latin word "pecunia" means wealth or money and derives from the word for cat-

---
[1] Vialles, Noélie. *Animal to Edible*. Translated by J. A. Underwood. Cambridge: Cambridge University Press, 1994.
[2] Lerner, Franz. *Lebendiges Fleischerhandwerk. Ein Blick in Vergangenheit und Gegenwart*. Frankfurt a. M.: Deutscher Fleischerverband, 1975.
[3] Burkert, Walter. *Homo Necans*. Berlin: De Gruyter, 1972.

tle. Animal slaughter is closely connected to agriculture, animal trade and all businesses that use or *harvest* dead animal bodies. Animal husbandry and trade with cattle and pigs changed globally in the 1840s and industrialized slaughter began in the 1860s.[4] This change came during a time of increasing isolationism, in which state officials and journalists reported with patriotic or even nationalist bias about animal health standards and the meat production of other countries. The industry of animal slaughter sparked conflicts such as the pork war between Germany and the USA from 1871 to 1891.[5]

The history of animal slaughter is a history of ideas [→History of Ideas]. Every society has a different set of rules to declare an animal a wild beast, a scientific case or an animal fit for slaughter, and these norms depend on various, interdependent belief systems. Therefore, even in a specific historical context, there is more than just one answer to the question: is there a difference between humans and animals and if so, where and how do we draw the line? The answers are not only given in theory, but also and maybe even more so in practice by the way humans make use of animals. Every human-animal-interaction reproduces the human-made interspecies hierarchy. Thus, animal slaughter for producing meat stabilizes the gap that divides humans from cows and calves, pigs and piglets, sheep and lambs. It also separates animals marked for slaughter from zoo animals that are caught in order to be exhibited and looked at or from a boar that is shot by hunters in order to conserve a forest's tree population.[6]

The history of animal slaughter is a history of emotions [→History of Emotions]. Sources show that especially those humans whose job it is to kill animals declare interest in the emotional state of the living beings whose lives they will end. There are historical accounts of butchers that state that there is no other human being more interested in the psychological state of animals for slaughter and their wellbeing than butchers. In contrast to the first point, European societies have a tradition of scrutinizing the psyche of persons who show violence against animals. Kant famously wrote in his *Metaphysics of Morals* that humans had to abstain from violent and cruel treatment of animals, because it would weaken compassion and even eradicate morality in relations with other people. The underlying prejudice or assumption is that butchers must lack certain emotions or empathy to perform their job, that they are brutal by nature or that the craft necessarily makes them brutal. Furthermore, the declaration of certain emotions (among them compassion, anger, grief)

---

[4] Perren, Richard. *Taste, Trade and Technology. The Development of the International Meat Industry since 1840.* Aldershot: Ashgate, 2006.
[5] Spiekermann, Uwe. Dangerous Meat. German-American Quarrels over Pork and Beef 1870–1900. *Bulletin of the German Historical Institute* 46 (Spring 2010): 93–110.
[6] Baker, Steve. You Kill Things to Look at Them: Animal Death in Contemporary Art. In *Killing Animals*, The Animal Studies Group (eds.), 69–95. Urbana, IL and Chicago, IL: University of Illinois Press, 2006.

is a vital part of the construction of group identity inside animal protection societies which emerged internationally in the nineteenth century.[7]

The history of animal slaughter is a history of human and nonhuman bodies. Breeding societies, nationally funded scientists, and corporations have created today's few omnipresent pig and cattle races.[8] This standardization of animal bodies has a predecessor in aristocratic horse breeding developed long before the nineteenth century. During the late 1800s, scientists also defined the protein-doctrine, that is, the conviction that animal protein is the most valuable substance for human bodies. The doctrine was integrated in food policies of industrialized countries: only meat would make industrial workers and soldiers strong enough for economic and military challenges. Meat for the masses thus became a political goal that was an important stimulus for the industrialization of animal slaughter.

While this chapter is called "History of Animal Slaughter" it is important to state that its objective is not to tell the history of all forms of animal slaughter. This would be an impossible undertaking. This entry will instead highlight developments in Germany in the late nineteenth century and early twentieth century which are to a certain degree representative of all industrialized countries: Already in the 1830s there was a disassembly line for piglets in Cincinnati and in 1865 the Chicago Union Stockyards opened their gates. In Europe the completion of Paris' La Villette in 1867 was especially impactful: This "abattoir" would become a model for city planners in Berlin and other European capitals.[9] In Germany the rise of modern slaughterhouses also began in the late 1860s and took off in the 1880s. When Munich's slaughterhouse-delegation in 1873 looked for examples of how to build a rationally organized large-scale facility, they could already visit establishments in northern Italy (Milan, Geneva and Turin), Switzerland (Genève, Zurich and Basel) and in Austria-Hungary (Pest and Vienna).[10]

By looking at the German case of introducing the industrialized slaughter of animals one can understand the mindset that was the foundation of today's animal farming and mass meat consumption. What unifies the countries that took part in

---

7 Eitler, Pascal. Übertragungsgefahr: Zur Emotionalisierung und Verwissenschaftlichung des Mensch-Tier-Verhältnisses im Deutschen Kaiserreich. In *Rationalisierungen des Gefühls. Zum Verhältnis von Wissenschaft und Emotionen 1880–1930*, Uffa Jensen, Daniel Morat (eds.), 171–187. Paderborn: Fink, 2008; Roscher, Mieke. *Ein Königreich für Tiere: die Geschichte der Britischen Tierrechtsbewegung*. Marburg: Tectum, 2009.
8 Blanchette, Alex. *Porkopolis. American Animality, Standardized Life and the Factory Farm*. Durham, NC: Duke University Press, 2020; for breeding in 18th and 19th century Britain see Ritvo, Harriet. *The Animal Estate. The English and other Creatures in the Victorian Age*. Cambridge, MA: Harvard University Press, 1987, 45–81.
9 Brantz, Dorothee. Risky Business. Disease, Disaster and the Unintended Consequences of Epizootics in Eighteenth- and Nineteenth-Century France and Germany. *Environment and History* 17, no. 1 (February 2011): 35–51.
10 Aybar, Canan-Aybüken. Geschichte des Schlacht- und Viehhofs München. PhD diss.: University of Munich (2005), 55.

the industrialization of animal slaughter is the reciprocal cycle of declaring meat to be of the greatest nutritious value, building a meat supply infrastructure and therefore creating a growing demand for meat especially in urban settings.[11]

This book is published at a moment when laboratory meat or meat production *without* slaughter is no longer just an idea but has been implemented. To understand this current development, it is all the more relevant to look at the decades when the now-questioned meat production *with* animal slaughter was invented and institutionalized.

2 Topics and Themes

History of Industrialized Animal Slaughter

The history of animal slaughter from the middle of the nineteenth century until World War I is characterized by the rationalization, scientification and nationalization of breeding, trade, butchering, selling meat and utilization of dead animal bodies. In Germany, its most obvious manifestation was the rise of modern slaughterhouses after 1881: 700 of these mostly public buildings were erected in cities larger than 10,000 inhabitants.[12] The second consequence was the rise of meat consumption: If we include all meat products, the overall consumption of meat between 1859 and 1909 grew from 24 kg to 54 kg per head per year. In the case of Germany, this meant the advent of mass pork consumption. Germans consumed around 7 kg per year in the 1840s, in 1906 this number had quadrupled. In comparison: during the same timespan the annual consumption of beef on average rose from around 10 kg to around 15 kg per person.[13]

The industrialization of animals would not have been possible without the rationalization and scientification of farming. In the late eighteenth century, England was the pioneering country with regard to farming and animal husbandry and inspired researchers abroad.[14] In Germany, the physician Albrecht Thaer became the founder of scientific agriculture. Regarding animal husbandry, Thaer was a supporter of indoor housing of animals during summer months and also developed an assessment system for the nutritional value of animal fodder ["Heuwerttheorie"]. Thaer's

---

11 Perren, Richard. The Meat Trade in Britain 1840–1914. London: Routledge & Paul, 1978; Perren, Taste, Trade and Technology, 8.
12 Tholl, Stefan. *Preußens blutige Mauern. Der Schlachthof als öffentliche Bauaufgabe im 19. Jahrhundert*. Wiesbaden: Europäische Food Edition, 1993.
13 In comparison, today's meat consumption in Germany has reached around 59 kg meat per head per year: 37 kg pork, 12 kg chicken, 9 kg beef and 2 kg other kind of meat. Referat Marktinformation, Kritische Infrastruktur Landwirtschaft. *Bericht zur Markt- und Versorgungslage Fleisch*. Bonn: Bundesanstalt für Landwirtschaft und Ernährung, 2018, 14.
14 For the transformation of animal bodies in Britain, see Ritvo, *The Animal Estate*, 46–81.

work gained additional importance from the 1820s, when agricultural academies in rural areas began to teach his principles. In the 1860s, the chemist Justus von Liebig affiliated these institutions to universities. The scientific approach to farming led to a growth of crops, enriched animal feed and changes in agriculture and animal farming.

The industrialization of animal slaughter was also a process of scientification and rationalization of animal bodies, above all that of cows: Around 1850, the rise of the milk cow began.[15] Until then cows worked in the fields or were used for transport. But after the 1850s cows became specialized milk producers. This change in the utilization of bovine species marks one of the most significant transformations of nineteenth-century farming. Cow breeds from the coastal and alpine regions, where they had already been used for milk production, became popular all over the German provinces. These breeding efforts led to a change in cattle bodies: The average weight of a milk cow grew from 150 kg in the 1830s to 500 kg at the turn of the nineteenth to the twentieth century. The effects of rationalized feed and breeding could also be observed in the increase in the amount of milk a cow could produce per year: around 1800, a cow produced from 350 to 750 liters of milk per year. For 1892, there are reports of cows that produced on average 2,727 liters milk per year. While this amount is large, today's turbo-cows provide up to 6,000 liters per year [→History of Agriculture].

The systematization of breeding caused a nationalized perception of livestock. As a result, the German regulation of live animal or meat imports grew more and more isolationist and protectionist.[16] Sanitary protocols for the transporting of foreign animals became stricter. This protectionism of national livestock was fed by Germany's goal of reaching autarky from animal imports. In contrast, other countries such as Great Britain increased the number of imported animals or meat products. Furthermore, the industrialization of animal bodies meant that the differences in cultural appreciation of cows and pigs increased. Agrarian scientists such as the Hallensian professor Karl Steinbrück considered the landed gentry to be the only class intelligent enough to be breeders of cows, bull or oxen, while smaller farmers and peasants were in his opinion not up to the task. This attitude reflects the model character that aristocratic horse breeding had for the breeders of domestic animals [→History of Pets], all the while pigs and pig breeders were painted as a lower social group. This hierarchical perception and the corresponding difference in status of the social groups associated with cows or pigs was even reinforced by the promotion of pig farms with several hundred animals and the encouragement of urban dwellers to raise individual pigs in their townhouses. Pigs became mass products.

---

15 Orland, Barbara. Turbo-Cows: Producing a Competitive Animal in the Nineteenth and Early Twentieth Century. In *Industrializing Organisms. Introducing Evolutionary History*, Philip Scranton, Susan R. Schrepfer (eds.), 167–190. New York, NY: Routledge, 2004.
16 Spiekermann, Dangerous Meat.

In parallel to the transformation of animal bodies, food standards also became subject to rationalization, scientification and nationalization. Justus von Liebig, Carl von Voit and their student Max Rubner propagated a mechanist and industrialized conception of the human body: food became fuel.[17] In this era, animal-based protein was declared the most valuable source of protein, a conviction that influenced the goals of national food industries providing access to meat to as many people as possible. While the value of meat was scientifically shored up, irrational beliefs concerning meat continued to exist that equally drove meat production: In Britain the consumption of "red meat" was seen as essential for the military in order to keep up the strength of soldiers. In contrast, the lower meat consumption of the French was seen as inferior and France's soldiers therefore were considered "puny" and "sniveling."[18] With the founding of the German nation state and its administrative structure in 1871 the protein-doctrine became a patriotic endeavor: German agriculture, the affiliated food trades, livestock breeding and the infrastructure of animal slaughter (and milk producers) was to provide every industrial worker with 118 g of protein a day, or the equivalent of 3,000 calories. These numbers were so high that even a proponent of the protein doctrine like Rubner accepted – even though only implicitly – the criticism of large protein consumption by Russell Chittenden and other scientists. In 1908, Rubner debated the common interpretation of Voit's standards and suggested that it was better not only to include the profession of a person for nutrition recommendations but also their individual metabolism and cultural upbringings. He even criticized Germans who were too fond of meat and called their behavior a meat-cult ["Fleischkultus"].[19] But Rubner never questioned the value of protein, and in 1914, when Germany faced the question of food supply during war time, he recommended lower but the still high number of 80 g of protein for men and 68 g of protein for women.[20]

The changes in agriculture, animal farming, breeding of animals and food science were fundamental to the rationalization, scientification and nationalization of animal slaughter. In Berlin, the transformation of animal slaughter started as a legislative project. Early on the local medical society established a delegation that advised the city council. The most prominent figure of this board was the internationally acclaimed physician Rudolf Virchow, who was also member of the Prussian parliament and of Berlin's city council. The main concern of the delegation was to fight possible health hazards caused by animal slaughter and affiliated trades. Three arguments were brought forward in favor of public slaughterhouses: to remove

---

[17] Tanner, Jakob. *Fabrikmahlzeit. Ernährungswissenschaft, Industriearbeit und Volksernährung in der Schweiz 1890–1950*. Zürich: Chronos Verlag, 1999.
[18] Ritvo, *The Animal Estate*, 47.
[19] Rubner, Max. *Volksernährungsfragen*. Leipzig: Akademische Verlagsanstalt, 1908.
[20] Today, the WHO still recommends that a male person of 70 kg eats 58 g of protein per day. World Health Organization (WHO). *Protein and Amino Acid Requirements in Human Nutrition. Report of a JointWHO/FAO/UNU Expert Consultation. WHO Technical Report Series* 935 (2007): 242.

evaporations and inconveniences caused by private slaughterhouses, to remove cattle from the city and to control the health of animals destined for slaughter.

Four years later, in 1868, Prussia introduced the "slaughterhouse law" that after 1871 would become adapted by every German state.[21] It urged communities to ban slaughter from private facilities, hence the full name: law for the erection of public, exclusively to be used slaughterhouses. But it was a law open to interpretation. Only a few communities in the very south of Germany, Mannheim and the Ruhr area started to build slaughter facilities. But political pressure to improve conditions for animal slaughter grew. Important lobbies were societies for the improvement of public health who propagated sanitation and called for stricter laws. In 1881, the decisive amendment to the "slaughterhouse law" was enacted: German communities then had the right to close or ban private slaughterhouses inside their municipal district and to introduce meat inspection of any imported meat. Between 1881 and 1918, at least 700 such institutions were built across all German states, but above all in industrialized areas with a large working class. There were even 16 cities that built two new facilities during the period of the German Empire, among them smaller places like Bamberg or Ratibor, but also political centers like Dresden and Düsseldorf. That the German numbers are high is best understood when compared with Britain, where the erection of public slaughterhouses had started in Edinburgh in 1851. The number of public abattoirs, as they were called, because they focused on killing and not the whole process of slaughtering, had grown to 54 in 1895 and to 135 in 1908.[22] In the eyes of their supporters, the slaughterhouse boom was a success: no other measure pushed the quantity of consumed meat as a new slaughterhouse did.

Aside from food supply and health concerns as motivations for building a public and new slaughterhouse in Berlin, the magistrate was also motivated by contributing to the international reputation of the German capital.[23] The politicians looked with great interest to the meatpacking industry in the USA and the slaughterhouses in Vienna and Paris. Officials and urban planners were well informed about international developments in the slaughter business and undertook expeditions to gather information about the status quo of animal slaughter inside and outside of Europe. The goal was clear. Berlin should not become the taillight of European animal slaughter [→Urban (and Rural) History].

The Berlin magistrate was also driven by a paternalist attitude: A public slaughterhouse was an instrument to increase control over animals, butchers and the slaughter process. The politicians were convinced that only a public institution would meet scientific, medical and economic standards and could guarantee the capital's meat supply. As a consequence, Berlin built a public cattle market and

---

21 Brantz, Risky Business, 35–51.
22 Otter, Chris. Cleansing and Clarifying: Technology and Perception in Nineteenth-Century London. *Journal of British Studies* 43, no. 1 (January 2004): 40–64, 50.
23 Schindler-Reinisch, Susanne. *Berlin-Central-Viehhof. Eine Stadt in der Stadt*. Berlin: Berlin Verlag, 1996.

slaughterhouse [Berliner Central Vieh- und Schlachthof] in the eastern part of the city and starved private competitors.

Corresponding to the rationalization and scientification of animal slaughter, a public German slaughterhouse was usually led by veterinarians. Once a year they published administrative reports ["Verwaltungsbericht"] that shows the industrial standards applied to animal slaughter.[24] They inventoried the animals traded through the cattle market, the amount of meat that was slaughtered and the long list of people of more than 20 professions that worked there. They also provide insights into the large variety of businesses that used the facilities of a public slaughterhouse.

The *Verwaltungsberichte* are an essential source for studying the political and cultural meaning of industrialized slaughter before World War I. Historically the reports assured the government that the German system of public slaughterhouses was capable of providing meat for the masses and was not inferior to privately run businesses. Their content also became public knowledge: Journalists regularly summed up the reports for daily newspapers.

Expanding Markets, New Industries

At the end of the nineteenth century, the growing number of cows and pigs not only produced more milk, meat and manure, but were the basis for other industries. The slaughterhouses had become the hinge of a growing animal-based economy: they provided felt and bristles for textiles and brushes; animal bones were cooked to produce gelatin, glue or bone oil; animal fats went to producers of soap, ink and sebum; horns went to manufacturers of combs and buttons; rennet was used for making cheeses; lymphatic fluids used for vaccines; and intestinal skin to cover sausages [→Economic History]. Only the eyes and uteruses of cows were not re-used and were thrown out. Finally, unhealthy animals had also been slaughtered – once inside a slaughterhouse they could not leave the premises alive. Instead, they were led to the emergency slaughterhouse ["Sanitätsschlachthaus," "Seuchenhof" or "Kontumazhof"]. The non-edible body parts went to the knackery ["Abdeckerei" or "Kadavererwertungsanlagen"] where they were transformed into tar, manure, animal meal or bone meal. The meat of the carcasses that stemmed from sick or unhealthy animals still could be deemed fit for consumption. If this was the case it was sold at the *Freibank*, a butcher shop owned by the municipality. Only a designated butcher was allowed to sell this kind of meat and it had to be sold at a lower price than standard meat. Although the concept of the Freibank seems to fit perfectly into the economic rationale of the era, it was not an international phenomenon. Quite the con-

---

24 Kaiserliches Gesundheitsamt. *Schlachthäuser im Deutschen Reiche:* 1876–1932, R 86/3360–3386 and R 86/3500–3503, Federal Archives Berlin Lichterfelde.

trary: In Britain a Freibank was unthinkable. To sell or consume meat that stemmed from an animal that had previously been diagnosed with a disease was considered "revolting."[25]

The political interest in the efficiency of slaughterhouses explains why their heads, such as Berlin's slaughterhouse-director Otto Hausburg, dutifully recorded and reported the number of slaughtered animals. The importance of growth becomes highly evident in the documentation of the day with the largest number of animals killed in any year ["Hauptschlachttag"]. In 1886, that day saw 5,831 animals slaughtered. In 1898, the number had almost doubled and amounted to 1,403 cows, 3,984 calves and 5,297 pigs. While the numbers of animals slaughtered rose, Berlin's meat consumption at one point began to stagnate: Berlin's public slaughterhouse had become a meat exporting business. In 1884, around 200,000 pigs had been slaughtered in Berlin. The institution surpassed the number of one million slaughtered pigs per year in 1904, and in 1913 Berlin butchers even slaughtered 1.5 million pigs. Between 1905 and 1913, around 10% of all pigs that were slaughtered in the German Empire were killed in Berlin alone.

Compared to the meatpacking industry in the United States these numbers are relatively low: Cincinnati had already surpassed the slaughtering of one million pigs in 1874. In 1900, 25,000 workers slaughtered 75,000 cows, 80,000 sheep and 300,000 pigs per day in the Chicago stockyards.[26] European slaughterhouse experts knew these numbers, but still were proud of their own "modernized" slaughter systems. In fact, they intentionally chose to maintain their own slaughter culture. Nationalism and imperialist competition between European countries and the rising power of the USA favored different industrialized slaughter cultures: In Germany, slaughterhouses had become sources of local pride. They were listed in guidebooks for tourists, and openings of a new cattle market and slaughterhouse were publicly celebrated.[27] They were seen as an urban achievement alongside new train stations or electric streetlamps or zoological gardens [→History of the Zoo]. Their administration buildings were fashionable and inviting.[28]

In this sense, they differed significantly from what German readers learned about Chicago's grim meatpacking district in Upton Sinclair's *The Jungle*.[29] This 1906 novel highlighted the exploitation of eastern European immigrants in the United States. There it inspired stricter meat control standards, while German officials used it to

---

25 Rossie, C. Smith. Cheap Meat: The German Freibank. *The Contemporary Review* 98 (1910): 661–670.
26 Gräser, Marcus. Chicagos Eingeweide. Schlachthöfe als Image. In *Versorgung und Entsorgung der Moderne. Logistiken und Infrastrukturen der 1920er und 1930er Jahre*, Wiebke Porombka, Heinz Reif, Erhard Schütz (eds.), 105–122. Frankfurt a.M. and Bern: Peter Lang, 2011.
27 Buchner-Fuhs, Jutta. Hinter den Mauern des Schlachthauses: Anmerkungen zur historischen Konstruktion von Wirklichkeit. *Hessische Blätter für Volks- und Kulturforschung* 32 (1997): 87–104.
28 Tholl, *Preußens blutige Mauern*.
29 Sinclair, Upton. *The Jungle*. New York, NY: Double Day, 1906.

decry American meat production: The industrialization of animal slaughter incorporated not only an economic, but also a cultural rivalry.

Transforming the Butcher's Craft

The industrialization of animal slaughter in the nineteenth century was not necessarily a process in which butchers were replaced by meatpackers. This is true for Germany, Austria and even more so in the case of France, where Napoleon had centralized the meat production while not questioning the artisanal character of butchering – and this stayed true while new reforms were introduced in the late 1800s.[30]

German butchers had to adapt to new actors in their workplace. Since the early nineteenth century, freedom of trade and occupation ["Gewerbefreiheit"] had started to change the character and organization of the butcher craft. This economic liberalization was not welcomed by all butchers. Hence, in the following decades and especially during the German Empire butchers tried – successfully – to regain power over access to the craft, training and exams.[31] Simultaneously the butchers adapted to working in public slaughterhouses. There, veterinarians and public inspectors observed their work. In larger facilities, animal welfare groups and societies for public health visited slaughterhouses from time to time and observed the butchers' work.

Already in 1902, the change of the butcher craft in urban areas was described as a "vertical split."[32] This means that butchers were no longer involved in every step of the meat production process from buying, killing to slaughtering and selling the meat. Butchers of the industrialized era in general either focused on the killing or on the meat selling. The public interest in animal slaughter shown by politicians, scientists, health organizations, animal welfare groups and architects changed the places, spaces and processes of their craft. But it did not replace them. Butchers were included, though not as equal contributors, in the planning of slaughterhouses in Germany.[33] Inventors and engineers who tried to create the fastest killing techniques depended on the judgement of butchers: inventors and businessmen sent flyers of

---

30 Claflin, Kyri. La Villette: City of Blood (1867–1914). In *Meat, Modernity and the Rise of the Slaughterhouse*, Paula Young Lee (ed.), 27–45. New Hampshire, NH: University of New Hampshire Press, 2008; Nieradzik, Lukasz. *Der Wiener Schlachthof St. Marx. Transformation einer Arbeitswelt zwischen 1851 und 1914.* Cologne: Böhlau, 2017.
31 Haase, Henner. Von der Fleischwirtschaft zur Marktwirtschaft. In *Lebendiges Fleischerhandwerk. Ein Blick in Vergangenheit und Gegenwart*, Franz Lerner (ed.), 81–110. Frankfurt a. M.: Deutscher Fleischerverband, 1975.
32 Rothe, Arthur. *Das Deutsche Fleischergewerbe.* Jena: Gustav Fischer, 1902.
33 See for the case of Berlin: Magistrat zu Berlin. *Beratungen mit Sachverständigen aus dem Schlächtergewerk über die innere Einrichtung der Schlachthäuser usw.:* 1879–1880, A Rep 005–03–02 nr. 9, State Archive Berlin; see for the case of Munich: Stadtmagistrat München. *Der Schlacht- und Viehhof in München.* Munich: Franz'sche Buchdruckerei, 1878, 2.

their apparatuses for stunning animals to the directors of the public slaughterhouses and asked that their inventions be introduced at the facilities. But if the butchers did not find the techniques useful, convenient or fast, the head of the slaughterhouse rejected the offer.[34]

Slaughter techniques for pigs and cattle were not advanced in the same way. The difference is telling for the cultural hierarchy of pigs and cattle. The most common techniques to stun a pig before cutting the animal's throat to let it bleed to death were using an axe, club or an ironclad hammer. This technique was popular throughout the German Empire, because it was efficient: In 1911, Hamburg's head veterinarian was proud to announce that one butcher alone could blow and bleed dry 75 pigs per hour with this method. The most notable invention of the Empire was the introduction of the bolt shot pistol and the less successful bullet shot pistol. Both devices induce immediate brain trauma by either penetrating or percussing the pig's head. Other inventions such as the Swedish pig trap or the slaughter-machine were only debated, but never installed. While stunning pigs became obligatory, this did not mean that all painful procedures were forbidden. Many slaughterhouses continued to trap pigs with a hook that pierced through their ears or feet, in order to fixate them for stunning. During testing periods of new slaughter methods on pigs, neither their pain nor the point of losing consciousness was monitored: only velocity of stunning and cutting was tested. In contrast, when slaughter techniques for cattle were tested, veterinarians monitored the minutes that passed between cutting, losing consciousness and dying.

Various techniques to slaughter cattle were standardized before World War I. The mechanical method of stunning cattle with axes, clubs and hammers did not disappear, but was professionalized: Berlin's slaughterhouse installed a school for instruction on how to properly stun larger animals by blowing ["Schlagschule"].

Another method was the use of the "Hakenbouterole" – a hammer whose metallic ends had a hook on one side and a bolt on the other side for catching and stunning a cow. It lost popularity but did not go out of use. Starting in the 1890s, engineers propelled inventions in the field of slaughter. Drowning, electrical execution or choking was debated, but only three new methods for cattle slaughter were admitted: The slaughter-mask, the shooting-mask and the bullet shot apparatus.

The industrialization of animal slaughter normalized one form of slaughter, while stigmatizing others. Cooperation between engineers, animal welfare groups, slaughterhouse administrations and butchers in order to find the most advanced slaughter techniques created an image of "normal" slaughtering. More specifically, it led to the inaccurate idea of a homogeneous German tradition of slaughter that was most fervently instrumentalized by animal protection societies and anti-Semites.

---

34 Notiz des Direktors Hausburgs, 4. April 1894, Magistrat der Stadt Berlin, *Deputation für den städ-tischen Vieh-und Schlachthof, Schlachtmethoden:* 1882–1906, A Rep. 013–02–02 nr. 53, State Archive Berlin.

They falsely stated that cattle slaughter without stunning was only typical among Jewish butchers and even declared it a danger to German "civilization." In 1892, Saxony banned slaughter without stunning and nearly all other German states applied double standards to Jewish butchers.[35]

Since animal slaughter was charged with nationalist sentiment, there was also a tendency to degrade the butchering culture of other countries. This becomes very clear after the publication of Sinclair's *The Jungle* in 1906. While decades earlier American efficiency had occupied the minds of German reformers, now the German system of publicly monitored meat production was considered superior.

3 Methods and Approaches

No other history of animal slaughter has attracted more researchers from all branches of the humanities than that of the United States. Chicago and Cincinnati still are synonymous with the meatpacking industry. *The Jungle* made American beef culture not only famous, but also a notorious symbol of exploitation of humans and animals by capitalism. This critical line of thought has never left the discourse about animal slaughter.[36]

With *Mechanization takes Command*, Siegfried Giedion introduced a second narrative about the modern slaughterhouses. Three years after the end of World War II, Giedion raised the question about the connection between industrialized killing of animals and humans. A historian of architecture, he guides his readers through the history of mechanization as a history of changing ideas and mentalities in science, agriculture, labor and art between 1770 and 1939. The chapter titled "Mechanization of Death" describes slaughterhouses in Cincinnati, Chicago and La Villette as culmination points of a mechanized culture that created and established a problematic and, in his view, dangerous neutrality and indifference towards death and killing.[37]

With a new wave of environmentalism and animal welfare activism in the late 1960s, critical investigations of the current state of animal farming and slaughter emerged. Ruth Harrison's account of Britain's chicken farming and slaughter industry even led to new animal welfare laws in 1968.[38]

---

[35] Judd, Robin. *Contested Rituals. Circumcision, Kosher Butchering and Jewish Political Life in Germany, 1843–1933*. Ithaca, NY: Cornell University Press, 2007.
[36] Rifkin, Jeremy. *Beyond Beef. The Rise and Fall of the Cattle Culture*. New York, NY: Dutton, 1992; Sebastian, Marcel. Deadly Efficiency – The Impact of Capitalist Production on the Meat Industry, Slaughterhouse Workers and Nonhuman Animals. In *Animal Oppression and Capitalism*, David Nibert (ed.), 167–183. Santa Barbara, CA: Praeger Press, 2017.
[37] Giedion, Siegfried. *Mechanization Takes Command: A Contribution to Anonymous History*. New York, NY: Oxford University Press, 1948, 246.
[38] Harrison, Ruth. *Animal Machines. The New Factory Farming Industry*. London: Stuart, 1964.

Still, the history of animal slaughter continued to a large part to be told as an economic history. Richard Perren first analyzed the history of Britain's meat trade between 1840 and 1914. In this book, Perren focuses on a history of supply, government regulations and technological innovations with great insights into the history of refrigeration and transportation. Nearly 30 years later in a book called *Taste, Trade and Technology* he kept his economic focus on the global meat trade with lengthy debates of numbers. But Perren also explains colonialist expansion of the American meat industry to South America, Australia, and New Zealand, and explicates the suppression of native cultures alongside modern criticism of meat production and consumption by the vegetarian movement.[39]

In the 1990s, the history of animal slaughter on the one hand continued to be told as an economic and socio-political history.[40] On the other hand, food history and the question of meat consumption attracted more interest and led to new critical narratives. In 1986, anthropologist Noélie Vialles published *From Animal to Edible*, a close study of animal slaughter in the Ardour region that provides a critical introduction to the cultural meaning of meat and the ambivalent relationship to animal slaughter in industrial societies. Carol J. Adams' *The Sexual Politics of Meat* (1990) analyzed US-American "beef culture."[41] She describes how the process of meat-eating not only let the body of animals disappear, but also extinguished their social existence and occluded the system of oppression they lived in. That she sees a connection between patriarchal values, meat eating, and the oppression of women and animals has led equally to praise and polemics from the public and academia.

Until the early 2000s the history of animal slaughter was, thus, mainly told as a history of scientific, political, technical and economic progress in Europe and the USA. Yet this historical narrative was continuously challenged by anti-capitalist, environmentalist and feminist perspectives that tended to focus on the current state of animal slaughter and not on historical variations or predecessors.

4 Implication(s) of the Animal Turn

Since the early 2000s, animal studies, human-animal-studies and critical animal studies acknowledge the historical impact, social roles and agency of animals.

---

39 Perren, *Meat Trade*; Perren, *Taste, Trade and Technology*.
40 Mohrmann, Ruth-E. Blutig wol ist Dein Amt, o Schlachter. Zur Errichtung öffentlicher Schlachthäuser im 19. Jahrhundert. In *Mensch und Tier. Kulturwissenschaftliche Aspekte ihrer Sozialbeziehung*, Siegfried Becker, Andreas C. Bimmer (eds.), 101–118. Marburg: Jonas Verlag, 1991; Reulecke, Jürgen. Die Politik der Hygienisierung: Wandlungen im Bereich der kommunalen Daseinsvorsorge als Elemente fortschreitender Urbanisierung. In *Stadtgesellschaft und Kindheit im Prozeß der Zivilisation*, Imbke Behnke (eds.), 13–25. Opladen: Leske und Budrich, 1990.
41 Adams, Carol J. *The Sexual Politics of Meat. A Feminist-Vegetarian Critical Theory*. London and New York, NY: Continuum, 1990.

They challenge the definitions of *the animal*, debate new methodological approaches and have rethought the relationships between human and nonhuman animals.

Regarding the history of animal slaughter, we continue to see that American and global beef cultures dominate the research, but the *animal turn* has diversified the research interests on animal slaughter: There is more anthropological, ethnographic and sociological interest in twentieth and twenty-first century-animal slaughter. These studies take a closer look at animal farms, slaughterhouses and meatpacking industries by analyzing human-animal-interactions or the effects of the conditions of slaughter on humans, animals and the environment.[42] In the last one and a half decades, research has been conducted on private and small-scale animal slaughter that continued and continues to exist next to industrialized forms of slaughter.[43] Additionally, not only mammals attract interest but other species, too: there is growing scholarly research on the global chicken industry and on the killing of maritime animals, farming of mussels and salmon, and hunting of whales or tuna.[44] Furthermore, we see that the history of religious, gender or racial conflicts around slaughter are researched on a global scale.[45] But what has grown exponentially is the number of books on the ethics of animal slaughter, meat-eating and human dominance over animals. In a similar manner as the first animal rights movements of the nineteenth century, many of these scholars combine anti-capitalist criticism with their study of animal slaughter and animal use.[46] Many of these writers are committed to transforming the production and consumption of meat, but not all researchers are dedicated to the full abolition of animal-exploiting industries.[47]

The *animal turn* and the rethinking of animals as historical actors is especially interesting for the history of animal slaughter, because researchers deal with two subaltern histories: both slaughter-animals and most humans that slaughter animals

---

[42] Pachirat, Timothy. *Every Twelve Seconds. Industrialized Slaughter and the Politics of Sight.* New Hampshire, NH: Yale University Press, 2013.

[43] MacLachlan, Ian. A Bloody Offal Nuisance: The Persistence of Private Slaughterhouses in Nineteenth Century London. *Urban History* 34, no. 2 (August 2007): 227–254.

[44] Kalof, Linda, Seven Mattes and Amy Fitzgerald (eds.). *Animal Studies Bibliography.* Michigan, MI: Michigan State University, 2020. URL: www.animalstudies.msu.edu/bibliography.php (December 7, 2020).

[45] Ballard, Richard. Slaughter in the Suburbs: Livestock Slaughter and Race in Post-Apartheid Cities. *Ethnic & Racial Studies* 33, no. 6 (June 2010): 1069–1087; Gómez, Mauricio Alejandro. Pigs and Social Control of the Poor: 18th Century Province of Antioquia. *Anuario Colombiano de Historia Social y de la Cultura* 43, no. 1 (January 2016): 31–59; Mirza, Shireen. Cow Politics: Spatial Shifts in the Location of Slaughterhouses in Mumbai City. *Journal of South Asian Studies* 42, no. 5 (August 2019): 861–879; Vinicius Erichsen da Rocha, Lucas and Alessandra Izabel de Carvalho. Mapeando Cerceamentos e o Lugar da Matança Animal: o Caso do Matadouro Municipal de Ponta Grossa. *Antíteses* 10 (2017): 397–424.

[46] Potts, Annie (ed.). *Meat Culture.* Leiden: Brill, 2016; Wadiwel, Dinesh. *The War Against Animals.* Leiden: Brill, 2015.

[47] Watt, Yvette M., Siobhan O'Sullivan and Fiona Probyn-Rapsey. Should We Eat Our Research Subjects? Advocacy and Animal Studies. *Animal Studies Journal* 7, no. 1 (2018): 180–205.

have become marginalized in industrialized societies. From both, there are too few primary documents available. Historiography for a long time neglected the history of the killing and slaughter process, and the impact it had on butchers, their craft and their relations to the animals. In the case of Germany, butcher guilds started to write their own history in the 1970s, while academia was researching the social history of industrial labor, where slaughterhouse workers were a marginalized group in comparison to millions of coal and steel workers.[48]

An integration of butchers and animals into the history of animal slaughter becomes tangible in Lukasz Nieradzik's cultural analysis of the transformation of the butcher craft in the slaughterhouse of St. Marx/Vienna.[49] He makes butchers part of history – which is still uncommon – and by doing this, animals for slaughter, too, become part of this history. He explains transformations of spaces, technologies and bodies. Through this approach, a comprehensive social history of butchers unfolds, whose relationships to animals are interdependent with their social relationships to humans. Though Nieradzik's focus is on the butchers and their history, his study is a good example for the scientific impact of the *animal turn*, because it points out how the agency of humans and animals is limited by the practice of slaughter.

Selected Bibliography

Adams, Carol J. *The Sexual Politics of Meat. A Feminist-Vegetarian Critical Theory*. London and New York, NY: Continuum, 1990.
Cronon, William. *Nature's Metropolis. Chicago and the Great West*. New York, NY and London: Norton, 1992.
Giedion, Siegfried. *Mechanization Takes Command: A Contribution to Anonymous History*. New York, NY: Oxford University Press, 1948.
Judd, Robin. *Contested Rituals. Circumcision, Kosher Butchering, and Jewish Political Life in Germany, 1843–1933*. Ithaca, NY: Cornell University Press, 2007.
Luke, Brian. *Brutal: Manhood and the Exploitation of Animals*. Chicago, IL: University of Illinois Press, 2007.
Nieradzik, Lukasz. *Der Wiener Schlachthof St. Marx. Transformation einer Arbeitswelt zwischen 1851 und 1914*. Cologne: Böhlau, 2017.
Perren, Richard. *Taste, Trade and Technology. The Development of the International Meat Industry since 1840*. Aldershot: Ashgate, 2006.
Ritvo, Harriet. *The Animal Estate. The English and other Creatures in the Victorian Age*. Cambridge, MA: Harvard University Press, 1987.
Vialles, Noélie. *Animal to Edible*. Translated by J. A. Underwood. Cambridge: Cambridge University Press, 1994.
Young, Paula. *Meat, Modernity and the Rise of the Slaughterhouse*. New Hampshire, NH: University of New Hampshire Press, 2008.

---

48 Lerner, *Lebendiges Fleischerhandwerk*.
49 Nieradzik, *Der Wiener Schlachthof St. Marx*.

Gesine Krüger
History of Hunting

1 Introduction and Overview

More than four million animals fall prey to various forms of hunting each year. The usually highly emotional discussions about the pros and cons of hunting mostly concern trophy hunting, animal suffering due to unprofessional hunters, and the justification of hunting as a leisure activity.[1] Moreover, the killing of living beings was, and will always be, tied to taboos and rituals as well as to rules of access or exclusion.[2] This is also true for hunting and it is therefore useful to bear in mind the difference between the killing of non-domesticated animals – as in an act of defense or with a poisoned decoy – and hunting as a comprehensive practice. Hence, not every act of killing an animal would be a hunt, whereas a hunt always has the objective of killing an animal. And moreover, with a praxeological approach, we can define hunting as "an appropriation of uncultivated space by means of capturing animals."[3] If hunting is conceptualized as a performative act, we should look at its beginning and its end, because the ritual and ritualized preparations which belong to the daily hunt for subsistence,[4] as well as the elaborate courtly hunt with its manifold symbolic forms and meanings, begin long before the actual hunt itself. And, after the death of the hunted animal, trophies and forms of ritual care for the quarry's mortal remains demonstrate an ongoing relation to either the specific animal or to the category of animals it represents.[5]

Concerning the differentiation between hunting and killing, it is also important to distinguish whether individual animals are hunted (even if this results in a large number of shot animals), or whether the aim is to exterminate entire herds most effectively, such as the buffaloes on the American plains.[6] This large-scale butchery had the objective to make space for the cattle of the settlers as well as to destroy

---

1 Bode, Wilhelm and Elisabeth Emmert. *Jagdwende. Vom Edelhobby zum ökologischen Handwerk*. Munich: C.H. Beck, 2000.
2 Marvin, Garry. Wild Killing. Contesting the Animal in Hunting. In *Killing Animals*, The Animal Studies Group (ed.), 10–29. Urbana, IL and Chicago, IL: University of Illinois Press, 2006.
3 Morsel, Joseph. Jagd und Raum. Überlegungen über den sozialen Sinn der Jagdpraxis am Beispiel des spätmittelalterlichen Franken. In *Jagd und höfische Kultur im Mittelalter*, Werner Rösener (ed.), 255–288. Göttingen: Vandenhoeck & Ruprecht, 1997, 286.
4 Nadasdy, Paul. The Gift of the Animal. The Ontology of Hunting and Human–Animal Sociality. *American Ethnologist* 34, no. 1 (January 2007): 25–42.
5 Hill, Erica. Archaeology and Animal Persons. Toward a Prehistory of Human–Animal Relations. *Environment and Society. Advances in Research* 4, no. 1 (September 2013): 117–136.
6 Isenberg, Andrew C. *The Destruction of the Bison. An Environmental History 1750–1920*. Cambridge: Cambridge University Press, 2001.

the livelihood of the indigenous population, who had become a modern hunting society themselves thanks to the appropriation of horses and firearms [→American Studies]. Forms of hunting with the sole purpose of killing most effectively – either for economic reasons or in order to cull animal populations such as wolves[7] – can already be found in the preindustrial societies of Europe.

A rough periodization of hunting in Europe could be sketched as follows: During the Roman era, hunting could presumably be performed by anyone and was not prohibited, "because game was considered as *res nullius* (nobody's thing) and the *saltus* (forest) was therefore not a closed hunting ground."[8] Hunting as a representation of lordship – which could already be found in African, Asian and Latin American civilizations as well as in ancient Greece and Rome – intensified in Europe during the Carolingian period as their realms were characterized by monarchy and aristocracy.[9] Even though hunting might have been a general right until the early Middle Ages, the introduction of "banned forests" (comparable to royal forests in Britain at the time) reserved hunting and forest rights for the Merovingian kings and their vassals. Henceforth, hunting became the independent prerogative of lordship [the "Jagdregal"] which was only abolished in Germany during the revolution of 1848/49. The (symbolic) significance of hunting is well displayed by the fact that 13 percent of the petitions to the Frankfurt Parliament (Paulskirche) considered noble hunting rights. Along with the abolishment of the noble hunting prerogative, paid or unpaid hunting services were also terminated and after a short transition period the hunting law ["Jagdrecht"] and the right to perform hunting ["Jagdausübungsrecht"] were split. The right to perform hunting was granted to municipalities or communities of landowners. This legal arrangement of hunting is being debated and criticized even today and it is the main component of the reform of the hunting law in Germany. This legal history of hunting, however, merely forms the framework for the complex relationships between humans and nonhuman animals performed in the hunt.

Popular and academic discourses deploy a number of historical and biological arguments to proof the central role of hunting in the history of humankind that need to be carefully deciphered, especially regarding the underlying assumptions about the "nature" of man. The multi-relational approach taken in this chapter tries to include various perspectives, which means different humans, different animals and different social and spatial settings. Since nowadays for most people in industrialized or post-industrialized societies hunting is a hobby and a sport, not a necessity, there seems to be a powerful mystification of the performance of hunting. No one *has* to hunt, neither to provide food nor to defend themselves against animals,

---

[7] Dasler, Clemens. Funktionen der Jagd im Hochmittelalter, unter besonderer Berücksichtigung der Pelztierjagd. *Vierteljahrschrift für Sozial- und Wirtschaftsgeschichte* 91, no. 1 (March 2004): 1–19, 2.
[8] Zotz, Thomas. Beobachtungen zu Königtum und Forst im früheren Mittelalter. In *Jagd und höfische Kultur im Mittelalter*, Werner Rösener (ed.), 95–122. Göttingen: Vandenhoeck & Ruprecht, 1997, 98.
[9] Fenske, Lutz. Jagd und Jäger im frühen Mittelalter. Aspekte ihres Verhältnisses. In *Jagd und höfische Kultur im Mittelalter*, Werner Rösener (ed.), 29–94. Göttingen: Vandenhoeck & Ruprecht, 1997, 36.

apart from very rare exceptions. Yet, the perception of an evolutionary legacy, a genetic disposition of humans – or rather of men and boys – towards hunting has nevertheless shaped the academic as well as popular discourses for a long time and is currently experiencing a revival.[10] Taking this perception seriously, the chapter aims first at providing insights into central themes such as the "man the hunter" theory. Equally important are power structures inherent in hunting in Europe and the colonies alike. This is why in "Methods and Approaches" insights from social, cultural and historical anthropology are presented. Finally, it will be argued that a spatial approach in animal history might help to shape our understanding of hunting.

2 Topics and Themes

As the subject of hunting includes so many aspects of human life – from the history of human evolution to post-industrial leisure activities – the publications from different disciplines have become multitudinous. At the same time, there are only a few general historiographical overviews focusing on the social and cultural aspect of hunting in different historical periods.[11] Hence, the perspective on the history of hunting is often narrowed down to either extremely specialized forms of hunting, or picked up by proponents of a distinct school of theories of evolution trying to explain modern human behavior by looking into the "deep history" of evolution, and conceptualizing man as hunters to explain everything from gender relations to war.

While general historians have paid more attention to the subject of hunting in recent years, research still predominantly focuses on the Middle Ages and the early modern period.[12] This can be explained by the great significance of hunting during medieval times and its close linkages with noble representation [→Diplomatic History]. After the bourgeois revolution of 1848, a form of hunting was established that included ideals such as "fair chase" as well as care for and protection of wild animals. This form of hunting challenged the courtly hunting style implicitly and explicitly, but was, however, itself ideologically dressed up.[13] Not only have social historians increasingly been engaging with the subject of hunting but also

---

10 Riechelmann, Cord. Von Jägern und Stammlern. *Jungle World* 30 (2006). URL: jungle.world/artikel/2006/30/von-jaegern-und-stammlern (December 7, 2020).
11 Rösener, Werner. *Die Geschichte der Jagd. Kultur, Gesellschaft und Jagdwesen im Wandel der Zeit*. Düsseldorf and Zürich: Artemis und Winkler, 2004; Radkau, Joachim. *Natur und Macht. Eine Weltgeschichte der Umwelt*. Munich: C.H. Beck, 2000, 68; Knoll, Martin. Hunting in the Eighteenth Century – An Environmental History Perspective. *Historical Social Research* 29, no. 3 (November. 2004): 9–36, 10.
12 This is also true for the overview of Rösener, *Die Geschichte*; for an ethnological perspective see Hiller, Hubertus. *Jäger und Jagd. Zur Entwicklung des Jagdwesens in Deutschland zwischen 1848 und 1914*. Münster: Waxmann, 2003.
13 For a perspective of legal history see Kohl, Gerald. *Jagdrecht und Revolution. Das Jagdrecht in den Jahren 1848 und 1849*. Frankfurt a.M.: Peter Lang, 1993.

environmental historians.¹⁴ Colonial history also has engaged with the idea of a political ecology of hunting to expose the underlying power structures hunting entails [→Social History;→Environmental History;→(Post)Colonial History].¹⁵ While hunting as a means of colonial rule and exercise of power has increasingly become a field of research, the subject of hunting and gender can as yet hardly be found in publications. Hunting still carries a male connotation, despite Artemis and Diana being its goddesses and the existence of many famous huntresses in history.¹⁶ There are, however, specific topics and themes that seem to have resonated more widely with academics in the field.

Man, the Hunter

As mentioned above, the "man the hunter" theory is one of these topics. The theory claims that hunting had been the evolutionary driving force behind hominid development and that even "family organization may be attributed to the hunting way of life."¹⁷ This has been the starting point for many extensive hypotheses on human evolution. Generally, they portray hunting as a source for innovations, such as collaborative planning and the ability to cooperate, the development of tools and social techniques as well as the development of gender-related division of labor.¹⁸ Such theories combine perspectives of cultural history – for example that hunting has shaped human behavior until modernity – with ideas of biology and evolutionary history of a still existing "hunting instinct." This "hunting legacy" mainly concerns men, while women apparently solely contributed to the feeding of the family as "gatherers" and otherwise mainly concentrated on rearing the children.

However, these theories have not remained uncontested and their conclusions were criticized by various academic disciplines, including from feminist perspec-

---

14 Knoll, Hunting.
15 Gissibl, Bernhard. Jagd und Herrschaft. Zur politischen Ökologie des deutschen Kolonialismus in Ostafrika. *Zeitschrift für Geschichtswissenschaft* 56, no. 6 (June 2008): 502–520; Gissibl, Bernhard. Das kolonisierte Tier. Zur Ökologie der Kontaktzone des deutschen Kolonialismus. *WerkstattGeschichte* 56, no. 2 (July 2011): 7–28.
16 See Giese, Martina. Rezension zu: Fietze, Katharina: *Im Gefolge Dianas. Frauen und höfische Jagd im Mittelalter (1200–1500)*. Köln 2005. H-Soz-Kult, August 16, 2005. URL: www.hsozkult.de/publicationreview/id/reb-7539 (December 7, 2020); about the figure of the huntress see Schmitz, Sigrid. Man the Hunter / Woman the Gatherer – Dimensionen der Gender-Forschung am Beispiel biologischer Theoriebildung. *Freiburger FrauenStudien* 13 (2003): 151–174.
17 Washburne, Sherwood L. and C.S. Lancaster. The Evolution of Hunting. In *Man the Hunter*, Richard B. Lee, Irvan DeVore (ed.), 293–303. New York, NY: De Gruyter, 1968, 295.
18 Dart, Raymond A. and Dennis Craig. *Adventures with the Missing Link*. London: Hamish Hamilton, 1959.

tives.[19] Robert Sussman and Donna Hart, for example, think that early humans must have been prey rather than hunter for millions of years,[20] and Pat Shipman developed the theory of man the scavenger in the 1980s.[21] According to her theory, early humans were not predominantly hunters, but searched for remains of prey left by predators. Shortly after Shipman's publication, Robert Blumenschine was able to experimentally ascertain during an 11-month field trip to the Serengeti that it is possible to obtain a substantial amount of meat, that is entrails and bone marrow, by searching for remains of lion hunts.[22] The feminist critique of the hunting hypothesis also questions the overall importance of hunting and the consumption of meat in general for human development. A significant connection between the development of hunting and the evolution of the brain, in particular, is yet to be proven. It seems instead that the "close relationship between mother and child and the closely related gathering and sharing activities were the starting point for human inventions, orientation and communication, for the development of language and complex social structures."[23] However, the trope of humankind as naturally prone to hunting has become a sort of sediment for all kinds of narrations that also claim to be historical.

Distress and Passion – Poaching

Another topic historians have dealt with, albeit from a societal rather than cultural perspective, is the impact of "game" laws. Infringements of hunting regulations and access to animals and territories have been committed during all ages. Nevertheless, for certain periods upturns can be detected in poaching and similar offences, for example when agrarian capitalism became prevalent in England in the eighteenth and nineteenth centuries.[24] And as hunting rights were also part of the seigneurial, the noble culture, poaching not only formed an illegal food supply, but also a critique of dominion and authority in different forms and manners. However, in periods of crisis and war, animals are killed without permission out of sheer necessity. After the Thirty Years' War, for instance, there was an increase in poaching and a concomitant tightening of penalties against illegal hunting. Also, after the two World Wars,

---

19 See Pickering, Travis Rayne. *Rough and Tumble. Aggression, Hunting and Human Evolution*. Berkeley, CA: University of California Press, 2013; Sussman, Robert. The Myth of Man the Hunter, Man the Killer and the Evolution of Human Morality. *Zygon – Journal of Religion and Science* 34, no. 3 (September 1999): 453–471.
20 Hart, Donna and Robert Sussman. *Man the Hunted. Primates, Predators, and the Human Evolution*. New York, NY and London: Routledge, 2009.
21 Shipman, Pat. Scavenger Hunt. *Natural History* 4, no. 93 (1984): 20–28.
22 See Blumenschine, Robert J. Man the Scavenger. *Archaeology* 42, no. 4 (July/August 1989): 26–33.
23 Schmitz, Man, 164.
24 See Shakesheff, Timothy. *Rural Conflict, Crime, and Protest. Herefordshire, 1800 to 1860*. Rochester, NY: Boydell and Brewer, 2003.

poaching increased, especially as there were more weapons circulating, and encounters between hunters and poachers became more brutal.

The poacher is a central figure in peasant folklore. Biographical narratives about poaching not only concern revolts against authorities or purely economic reasons for illegal hunting, but they always speak of passion, if not addiction, too.[25] In Europe as well as in the colonial territories, the figure of the poacher is closely entangled with the history of hunting laws and hunting as a practice of power. Thereby, conflicts about the ownership of animals are closely intertwined with conflicts about the use of land and the increasing juridification of access to natural resources. Hunting and nature conservation acts abridged the customary use of fields and forests, which affected the entire peasant population far beyond the issue of hunting.[26] Hence, a scholarly debate emerged regarding the thesis of poachers or illegal hunters as being social rebels, who sought appreciation and protection through their rebellious acts and the distribution of meat and hides.[27] Yet, poaching as part of alpine culture was so deeply embedded in the social structure of village communities that Norbert Schindler speaks of a "theatre of masculinity" in which almost all young men engaged in Alpine peasant societies,[28] especially younger sons without inheritance claims or farmhands without property.

The sentences for poaching, which was regarded a capital crime and an attack on the sovereignty of the territorial prince, were still severely draconian in the early modern period.[29] But there are indications that such threats of punishment were often imposed with mitigated force, as it could have been problematic to impose such harsh sentences when poaching was such a common phenomenon and therefore the punishment could have challenged the peasant community and rural culture in general. Criminal investigation moreover failed, because it foundered on a wall of silence or on the tricks used by poachers and their supporters.[30] If women were involved in crimes related to poaching, they could usually talk their way out of trouble or hope for mitigating circumstances. As poaching was "such male preserve or a conflict of power staged in a masculine category of honor,"[31] women were not even considered criminally liable for such offences.

Poaching in wildlife reserves in Africa or Asia, by contrast, is often reduced either to a conflict between agriculture and nature conservation that can be solved

---

25 See Girtler, *Wilderer: Rebellen in den Bergen.* Vienna: Böhlau, 2004; Odermatt, Taïs, director. *Nid hei cho.* Lucern University of Applied Sciences and Arts, 2009.
26 For North America see Jacoby, Carl. *Crimes against Nature. Squatters, Poachers, Thieves, and the Hidden History of American Conservation.* Berkeley, CA: University of California Press, 2001.
27 Girtler, *Wilderer*, 41.
28 Schindler, Norbert. *Wilderer im Zeitalter der Französischen Revolution. Ein Kapitel alpiner Sozialgeschichte.* Munich: C.H. Beck, 2001, 177.
29 Schindler, *Wilderer im Zeitalter der Französischen Revolution*, 14.
30 Schindler, *Wilderer im Zeitalter der Französischen Revolution*, 13.
31 Schindler, *Wilderer im Zeitalter der Französischen Revolution*, 217.

by education and inclusion of the local communities – or the exploitation of (endangered) animals is considered to be a global business against which only international sentences or agreements could be a measure [→(Post)Colonial History]. Although the protection of rhinoceros in South Africa, for example, is of course an agreeable goal, the enforcement of legal actions "often runs up against the economic frustrations and temptations of a large, predominantly black, under-class, which for generations has been excluded from wildlife management and conservation."[32] The competition for game and access to resources is therefore rooted in the colonial period, during which professional "white hunters" realized that the game population had declined drastically in certain areas, due to commercial exploitation of ivory, fur, and feathers. Thus, making conservation measures a requirement, the law limited commercial hunting but at the same time turned almost all African hunters into poachers, because it declared the exercise of their ancestral right a criminal practice.[33]

Not only in a colonial or postcolonial context are the lines between hunters and poachers fluid as well as dependent on legal frameworks, negotiation processes, and the "moral economy."[34] In Europe, roles were changing as well: Poachers could become hunters and especially the professional hunters could use their skills and knowledge to hunt sometimes illegally for the sake of excitement or out of necessity.[35] In the archdiocese of Salzburg for instance, the double identity of hunters who increased their meagre wages with poaching as a side-line was a common phenomenon at the end of the eighteenth century.[36] A hundred years earlier in Bavaria, we can detect veritable poaching networks in which not only poor people with rural backgrounds took part, but also officials and members of the landed gentry. This speaks more of a dissent against norms between the princes of Bavaria and their subjects, than of the peasant counterculture mentioned above.[37] These different forms and modes of poaching not only represent different hunting cultures. They also shape and influence the ways in which animals are considered within different societies and times as well as the relationship between people and animals. The regulations and socio-political context in which they materialize determine the form, extent and qualities of interactions with animals.

---

32 Humphreys, Jasper and M.L.R. Smith. The Rhinofication of South African Security. *International Affairs* 90, no. 4 (July 2014): 795–818, 818.
33 Steinhart, Edward I. *Black Poachers, White Hunters. A Social History of Hunting in Colonial Kenya.* Athens, OH: Ohio University Press, 2006.
34 Thompson, Edward P. The Moral Economy of the English Crowd in the Eighteenth Century. *Past & Present* 50 (February 1971): 76–136.
35 See Girtler, *Wilderer*, 242.
36 Schindler, *Wilderer im Zeitalter der Französischen Revolution*, 240.
37 Freitag, Winfried. Das Netzwerk der Wilderei. Wildbretschützen, ihre Helfer und Abnehmer in den Landgerichten um München im späten 17. Jahrhundert. In *Kriminalitätsgeschichte. Beiträge zur Sozial- und Kulturgeschichte der Vormoderne*, Andreas Blauert, Gerd Schwerhoff (eds.), 707–757. Konstanz: Universitätsverlag Konstanz, 2000.

Hunting, Globalization and the Sea

Animals should become the focal point when historians look at the practices of hunting that have been exercised. While fishing with nets and lines is usually not regarded as hunting, this is different concerning whales, seals, and walruses. This is probably because they are mammals who are more readily perceived as individuals, although industrialized whaling degraded these highly admired animals nonetheless to a commodity. Whaling and industrialization are interconnected in three ways: First, whale oil served as lubricant for machines, fuel for lighting and basic material for margarine, secondly, whaling was the starting point for industrial development, for example on the east coast of the United States, and thirdly whaling itself was industrialized by factory ships.

From an economic and geographical perspective, whaling is a globalized form of hunting. In the past, processing and consumption, especially of whale oil, was controlled by global cartels. Today, hunting regulations are bound by international agreements. This opens up new fields of research that shift the focus from land to seascape, as "no other form of hunting took humans over such immense distances than whaling in the 18th and 19th centuries."[38] Modern commercial whaling spans over 100 years – from the invention of the harpoon cannon around 1860, which enabled the hunting of the quicker blue and fin whales, until 1960, when populations decreased dramatically. The harpoon cannon also ended the "Moby-Dick age of the duel between man and whale, in which the animal opponent still stood at least a minimal chance."[39] Mark Cioc compares the commercial deep-sea fishery with an extractive industry and draws a comparison between whaling and the exploitation of ivory, hides and feathers in the colonies.[40] Thereby another field of research, the globalized hunt, is introduced. The basis for the simultaneous "whaling Olympics"[41] and the extermination of vast wildlife populations in Africa in the second half of the nineteenth century was the conjunction of a scientific and technical revolution, colonialism, and an expanding capitalist global economy.

---

[38] Schürmann, Felix. Die Wale, ihre Jäger und der Strand von Annobón. In *Afrikanische Tierräume. Historische Verortungen*, Winfried Speitkamp, Stephanie Zehnle (eds.), 43–75. Cologne: Rüdiger Köppe, 2014, 43.
[39] Osterhammel, Jürgen. *The Transformations of the World. A Global History of the Nineteenth Century*. Princeton, NJ: Princeton University Press, 2014, 387.
[40] See Cioc, Mark. *The Game of Conservation. International Treaties to Protect the World's Migratory Animals*. Athens, OH: Ohio University Press, 2009.
[41] Cioc, *The Game of Conservation*, 6.

Representation and Power – Dominion Over People, Animals, and Spaces

Hunting aims at the appropriation of an animal, of its body, of its attributes and probably of its essence or soul. The economic significance of hunting is dependent on the definition of hunting (do we include trapping and fishing?), and also highly disputed in different periods. Thus, no straightforward distinction can be drawn between profitable hunting (meat, fur, and other usable parts) and "leisure hunting," which demands substantial expenditure. Neither can we consider hunting as developing from necessity to luxury, as the representational courtly hunts for instance did not primarily aim at the yield of meat. At the same time, hunting is a spatial practice,[42] where appropriation of an animal is interlinked with (re-)ordering and dominating territories. The persistence hunt, for example, created an entirely new architecture of the countryside: "a scenery of princely presence."[43] One of the sensuous elements of such a presence was the "acoustic permeation" of the hunting noise caused by hounds, horses, and humans.[44]

According to medieval hunting treatises, diverse forms of hunting existed, carried out either individually or in groups, "among which the falconry was primarily considered as the most befitting craft."[45] During the fourteenth century we can detect an increasing preference towards hunting with hounds, especially at the princely courts where deer-stalking enjoyed the highest esteem. Hunting with hounds allowed lavish hunting parties with vast entourages, and simultaneously enabled the control of "harmful" predators, regarded as "pest." However, an argument against hunting with hounds was that it would be asking too much of noble women.[46] Hound and bird, man and woman, forest and pasture were facing each other in a "discursive polarization."[47] In analyzing the debates about the pros and cons of different forms of hunting – which also included arguments about visual and acoustic relishes – we see that not only the hunt itself, with its preparations and festive framing, but also the acts of talking or writing about hunting, hounds, hawks, deer, rules and prerogatives were used as a platform to stage a competition for "prestige and the favor of the prince."[48]

Additionally, deer-stalking could be seen as a ritualized appropriation of land with the aid of the aristocracy – as a symbol of warfare, subject to an extensive co-

---

42 See Morsel, Jagd.
43 Knoll, Hunting, 28.
44 Teuscher, Simon. Hunde am Fürstenhof. Köter und "edle wind" als Medien sozialer Beziehungen vom 14. bis 16. Jahrhundert. *Historische Anthropologie* 6, no. 3 (1998): 347–369, 367.
45 Teuscher, Hunde, 364.
46 Teuscher, Hunde.
47 Rösener, *Jagd*, 277.
48 Rösener, *Jagd*, 365.

dification, akin to tournaments and dances.[49] Treatises from the early modern period in particular speak of an analogy between hunter and sovereign: "The hunter decides on life and death of the quarry as does the sovereign on life and death of his subjects. This is why each and every hunt can be used as a representation of sovereign power; and that is why the big feudal hunting parties were an essential component of absolutistic demonstration of power."[50] This demand for representation meant oppression and exploitation for the peasants, not just the animals. Arguments about entitlements to hunt and obligations the peasants had to meet were often cause for protests and a symbolic reason for revolts. We can detect evidence for hunting bans as early as Thomas Müntzer's manifesto during the German Peasants' War or in the demands of the peasants from Upper Swabia from 1525. Yet, it seems that grazing rights in the forest as well as access to wood and other produce of the forest were evidently more important for the peasant economy than the right to hunt, especially as the latter required weapons.

After 1848, hunting for trophies became popular as a new bourgeois form of hunting emerged with a stronger link to the city than the countryside, even though aristocrats and royals also displayed their trophies in hunting chateaus. Although the presentation of antler heads and other trophies also denoted an appropriation of forest and field, the hunting ground was no longer a political territory as such (even if it served as a venue for politics). Instead, it was a stage on which power, obtained elsewhere, could be showcased in earlier forms of representation.

A different form of bourgeois hunting, which imitated the noble hunting parties, were the colonial safaris that formed part of the hunting cult in imperial Africa and Asia.[51] While in the nineteenth and twentieth centuries safaris in Africa served mainly as colonial demonstrations of power – although Steinhart stresses that the expertise of the indigenous population was of great importance[52] – in Asia, especially in India, the existing forms of aristocratic hunting were entwined with imperial forms of demonstrations of power. At the same time, hunting in colonial territories was a means to exploit ivory, ostrich plume, fur, pelts, hides, and other animal products. Beyond the representational, colonial-aristocratic and economic aspects, science played an important role because numerous hunts and expeditions were carried out for the sole purpose of acquiring exhibits and research objects for the newly established museums of natural history of the nineteenth century.[53] Furthermore, the

---

49 Nelson, Janet L. The Lord's Anointed and the People's Choice. Carolingian Royal Ritual. In *Rituals of Royalty. Power and Ceremonial in Traditional Societies*, David Cannadine, Simon Price (eds.), 137–180. Cambridge: University of Cambridge Press Syndicate, 1987.
50 Borgards, Roland. Tiere jagen. *TIERethik* 5, no. 7 (February 2013): 7–11, 9.
51 See Mackenzie, John M. *The Empire of Nature. Hunting, Conservation and British Imperialism*. Manchester: Manchester University Press, 1988; Sussman, The Myth of Man.
52 Steinhart, *Black*, 111.
53 For Theodore Roosevelt's hunting parties for the Smithsonian Institute see Kalof, Linda. *Looking at Animals in Human History*. London: Reaktion Books 2007, 148.

zoological gardens mainly fulfilled their demand for exotic wildlife through organized hunts [→History of the Zoo].

In the colonies, hunting was part of a political ecology that linked the representation of dominion with the exercise of power by means of acts and laws. Hence trophies were of particular importance, because they were – as colonial objects – "manifestations of a desire to possess and control nature."[54] Colonial hunting should not only be considered a symbolic staging of dominion, but as a form and practice of colonial rule in its own right.[55] The appropriation and control of nature was, however, concomitant with its destruction. It is therefore no coincidence that big-game hunters were involved from the very beginning in nature conservation movements, as they were well aware of the fact that unmonitored hunting and the colonial extermination of animals, for example in the context of malaria control, would soon result in a lower number of trophies for them to shoot. However, more research in all eras needs to be carried out on regional and local levels in order to show who was allowed to hunt which game and how this translated into practice, that is, which offences were punished, and which actions were tolerated.

3 Methods and Approaches

When looking at historical developments, historians of hunting always have to distinguish between normative sources and actual practices: Although every hunt aims at catching and killing an animal, the phenomenon of the hunt is nonetheless characterized by considerably divergent practices. A hunt can be performed alone or in a group, with or without weapons, with or without animal company, in different surroundings and on different prey. Whereas individual hunting includes stalking and hiding, social hunting, for instance, includes beating and battue. In many forms of hunting a network of relationships is created between and by animals and humans, all of which participate in the hunt through different functions and distinct roles. By taking a praxeological approach that has been adapted from both anthropology and social theory, historians can gain a broader perspective on what was happening during the hunt. This includes the participating animals, for instance auxiliary animals like falcons and other birds of prey for falconry, cheetahs for hunting hoofed mammals, cormorants and Eurasian otters for fishing, horses for the chase and transport, ferrets for rabbit hunting, and elephants for riding during a tiger hunt. On all continents and through all periods, however, the most important animal as companion of the hunter is the dog, domesticated ten thousand years before other animals or plants. This means that the coevolution of human and dog as well as the

---

[54] Ryan, James R. Hunting with the Camera. In *Animal Spaces, Beastly Places: New Geographies of Human–Animal Relations*, Chris Philo, Chris Wilbert (ed.), 202–220. New York, NY and London: Routledge, 2000, 209.
[55] Gissibl, Das kolonisierte Tier, 19.

general engagement with animals in order to gain control over them, appears to have been a more important key factor for human evolution than the hunt alone.[56]

Simon Teuscher analyses why hunting with as many dogs as possible became so popular at the transition from medieval times to the early modern period. He understands dogs as a medium of social relations and as carrier of meaning – an approach that can also be applied to other eras.[57] These social relations again come to the forefront when looking at concrete practices. Dogs had various roles and tasks, serving, for example, as aggressive but regulating mediators between the hunter and the game.[58] At the same time, the categorization and description of the different types of dogs mirrored the perception of hierarchy, model of order, as well as dynastic concepts.[59] If the purpose of uniting packs of dogs at courtly hunting parties was to make hierarchy "perceptible by the senses" and "re-negotiable,"[60] this could only be achieved with cooperation between humans and animals and among the animals themselves – among dogs as well as among dogs and horses.

Such human-inflicted collaborations and relations among animals also occurred during cheetah hunts, which had already been performed for a long time by Sumerian and Egyptian rulers as well as in India and Persia before they were introduced in Europe by the Hohenstaufen emperor Frederik II.[61] Hunting with cheetahs was, like falconry and the persistence hunting of the seventeenth and eighteenth centuries, a court ceremonial, and its main purpose therefore was not the provision of meat but representation. Catching and breaking in the cheetahs was very laborious as they were always caught as cubs in the wild. They had already learned how to hunt from their mothers and had to become accustomed to humans. In Europe, the hunting technique was to set the cheetah from the croup of a galloping horse on the game. In India, by contrast, the cheetahs were driven to the hunt by an ox-cart, which was followed or led by hunters on horseback. Here, dogs were used to track down the game. In this example, humans and animals interacted with each other as well as various animals with one another – horses, oxen, dogs, and cheetahs.

Moreover, the various human-animal relationships during hunting have much more than just an auxiliary function. Such relationships are cooperative, sometimes they are forced, sometimes voluntary, they can be trained or instinctive, or culturally adapted. And although induced by humans, these relationships also depend on specific characteristics of the animal/s and their agency. This includes the recognition of

---

56 Shipman, Pat. The Animal Connection and Human Evolution. *Current Anthropology* 51, no. 4 (August 2010): 519–538.
57 For ancient history see Hoppe, Anja. Zur Rolle des Hundes bei der Liebeswerbung auf attischen Vasenbildern zwischen dem 6. und dem 4. Jahrhundert v. Chr. PhD diss.: University of Stuttgart (2010).
58 Teuscher, Hunde, 352.
59 Teuscher, Hunde, 358.
60 Teuscher, Hunde, 364.
61 Giese, Martina. Tierische Jagdhelfer – einst und jetzt. In *Bericht über die 15. Österreichische Jägertagung*, Brunhilde Egger (ed.), 1–6. Irdning: Raumberg-Gumpenstein, 2009, 4.

the prey as other subjects, even if this subject is killed in the hunt. Arctic archaeological sites show carefully arranged skulls of belugas, which apparently are places "of reciprocal exchange between human hunters and beluga prey."[62] As the belugas are perceived to have sacrificed themselves to the hunters, they could expect an appropriate treatment of their bones. This in turn was the precondition for other belugas, seals, whales, and walruses to offer themselves in the future to the hunters, "with each species having its own set of taboos and preferred treatment."[63]

There are various archaeological and ethnological examples for a special, reciprocal relationship towards a prey animal and its (rather his or her) dead body. Such findings as well as discoveries from animal burials, for example of dogs or wolves,[64] contribute substantially to a history of hunting, as they broaden the Western academic horizon geographically, and culturally. However, we have to be careful not to juxtapose "Western ontologies" with a supposed universal "indigenous ontology" that would always care for nature and for curating the mortal remains of their animal brothers and sisters. Ayo Adeduntan, Nigerian scholar in cultural studies, for example, does not place emphasis on the harmony but on the antagonism between human civilization and the wild, as a space of the unknown, in his studies of the Yoruba society: "The hunter is in the vanguard of his society's eternal quest to domesticate the unknown."[65]

Whether early humans were hunters or prey, whether they were eating the remains of quarry from predators or lived on a mainly vegetarian diet continues to remain under debate. Yet, efforts to anchor contemporary hunting practices to early periods of human evolution and to a genetic legacy ignore the cultural transformations which have shaped hunting since millennia. This in turn becomes evident when we consider both the practices and relationships they foster.

4 Implication(s) of the Animal Turn

One result of looking at hunting in a multi-relational way is to recognize its profound spatiality. The importance of space and place has been already acknowledged by various animal historians and could possibly even be marked as one of the main characteristics of the *animal turn* [→Historical Animal Geographies]. The relationship between animal history and spatial history is particularly evident when considering whaling, because "along the imaginary trails – guided by planetary currents – of their quarry, mid-eighteenth-century whalers developed a new network of hunting grounds, traffic routes, ports of transshipment, and transport chains across all

---

62 Hill, Archaeology, 125.
63 Hill, Archaeology.
64 Hill, Archaeology, 123.
65 Adeduntan, Ayo. *What the Forest Told Me. Yoruba Hunter, Culture and Narrative Performance.* Braamfontein: Unisa Press, 2014, 3.

oceans." As the persistence hunt created new landscapes, "whaling generated a specific topography spanning across the oceans that did not mirror the already established structures of cargo and passenger shipping. It were the routes of the whales that set those of their hunters."[66] The actions of the travelling whales – their speed, their direction, in short, their behavior – determined the "complex geography and chronology"[67] of this global business. Therefore, different spheres of time and space that were defined by the whales should be examined. And following the practice of whaling, another element that defines human-animal relationships in the context of the hunt after the *animal turn* can be outlined. The question of the *agency* of whales – grappled with in Western literature, for example, in Herman Melville's *Moby Dick* – can also be found in oratures of the Arctic people,[68] for example in stories of collaborative hunts of orcas and humans on belugas and bowhead whales, all of which are animal persons with an existing reciprocal relationship.

Killing and loving, fear and awe, are thus all closely connected in the hunt. There is no straightforward development from the cruel fox-tossing (a hunting game where small, previously captured, animals were tossed into the air with the aid of a panel of fabric, sustaining fatal wounds when hitting the ground) to the expert craft of hunting; or from the perilous hunt on walruses to the convenient shooting of a lion, bred solely for this purpose on touristic hunting farms in Namibia and South Africa. The *animal turn* has drawn attention to the fact, however, that the hunt, no matter where or in which form it is performed, and no matter for what purpose or in connection with what cultural or ritual background, always aims for the death of an animal as its ultimate conclusion. Yet it equally makes clear how much the hunt as a practice and operation has deeply implicated human history, and how substantially the animal – even as prey – has shaped human cultures.

Selected Bibliography

Gissibl, Bernhard. Hunting and Empire. In *The Encyclopedia of Empire*, Nathan Dalziel, John M. MacKenzie, (eds.), January 11, 2016. URL: doi.org/10.1002/9781118455074.wbeoe403.

Jacoby, Carl. *Crimes against Nature. Squatters, Poachers, Thieves, and the Hidden History of American Conservation*. Berkeley, CA: University of California Press, 2001.

Knight, John. The Anonymity of the Hunt. A Critique of Hunting as Sharing. *Current Anthropology* 53, no. 3 (June 2012): 334–355.

Knoll, Martin. Hunting in the Eighteenth Century – An Environmental History Perspective. *Historical Social Research* 29, no. 3 (November 2004): 9–36.

Mackenzie, John M. *The Empire of Nature. Hunting, Conservation and British Imperialism*. Manchester: Manchester University Press, 1988.

---

66 Schürmann, Die Wale, 43.
67 Osterhammel, *The Transformation*, 386.
68 See Hill, Archaeology, 126.

Marvin, Garry. Wild Killing. Contesting the Animal in Hunting. In *Killing Animals*, The Animal Studies Group (eds.), 10–29. Urbana, IL and Chicago, IL: University of Illinois Press, 2006.

Pickering, Travis Rayne. *Rough and Tumble. Aggression, Hunting and Human Evolution*. Berkeley, CA: University of California Press, 2013.

Rösener, Werner. *Die Geschichte der Jagd. Kultur, Gesellschaft und Jagdwesen im Wandel der Zeit*. Düsseldorf and Zürich: Artemis und Winkler, 2004.

Steinhart, Edward I. *Black Poachers, White Hunters. A Social History of Hunting in Colonial Kenya*. Athens, OH: Ohio University Press, 2006.

Washburne, Sherwood L. and C.S. Lancaster. The Evolution of Hunting. In *Man the Hunter*, Richard B. Lee, Irvan DeVore (eds.), 293–303. New York, NY: De Gruyter, 1968.

Ryan Hediger
History of War

1 Introduction and Overview

In a climactic scene of Steven Spielberg's film *War Horse* (2011), the equine protagonist Joey narrowly escapes death by a tank in the decimated terrain of World War I's no-man's land, only to become tangled in barbed wire.[1] A British and a German soldier see the unfortunate animal and, indicating peaceable intentions, meet at the horse, working together across national enmity to cut him loose from the wire. After a coin flip, Joey rejoins the British, eventually to be reunited with the young man who first trained him to work an agricultural field before the war. The barbed wire scene epitomizes, perhaps too neatly, many of the traumas that characterize that conflict: the interminable horrors of the trenches, the inhuman power of machinery like tanks and technologies like barbed wire, the seeming madness of mutual hostility organized around national identity when shared international projects (evoked in cutting Joey free) always remain possible. And of course, the scene centers not only on the surprising and persisting importance of horses and other animals in the ostensibly technological Great War, but also on the fundamental tragedy of animal suffering. Joey's entanglement in the barbed wire is legible as an icon of the experiences of horses – and by extension of other animals – in human wars, especially modern wars.

But how do we understand this icon? What competing meanings does it create or reinforce? For many viewers of the film, this scene is part of a larger sentimental treatment of the horrors of the Great War.[2] Viewers sympathize with the horse(s), but the suffering of animals is often taken to be of marginal importance next to the other tragedies of the war. *War Horse* thus appears for some as a sincere but naïve effort to memorialize human and animal suffering in a historical context when war seems increasingly inevitable. In *Multitude: War and Democracy in the Age of Empire*, Michael Hardt and Antonio Negri echo the view of a number of scholars – that we live in a "seemingly permanent state of conflict across the world."[3] For them, this problem is grave not only because of war's specific and concrete harms,

---

[1] Spielberg, Steven, director. *War Horse*. Dreamworks, 2011.
[2] Burr, Ty. Classic Hollywood Rides Again: Spielberg Packs War Horse Full of Nostalgia. *The Boston Globe*, December 23, 2011. URL: www.bostonglobe.com/arts/2011/12/23/war-horse-review-classic-hollywood-rides-again-steven-spielberg-new-movie/OubBhIWPBZI35i9QF5z0gL/story.html (August 25, 2020).
[3] Hardt, Michael and Antonio Negri. *Multitude: War and Democracy in the Age of Empire*. New York, NY: Penguin, 2004, xi.

but more generally because "war has always been incompatible with democracy."⁴ They argue that the hopes for a genuinely global democracy rest on coming to terms with this interminable status of war, no easy project. As they confess concluding their preface, "the current state of war and global conflict [...] can easily seem to be an insurmountable obstacle to democracy and liberation."⁵

Indeed, the sense of war's inevitability is often reinforced by attending to animals' experiences in it. War becomes a monstrous and unstoppable force, swallowing everything in its path, including animals. From this perspective, the tragedy of animals in war, in the words of the Animals in War memorial installed near Hyde Park in London and unveiled in 2004, derives from the idea that they "had no choice" about it. That approach to animals in war also underlies the 2014 essay collection *Animals and War: Confronting the Military-Animal Industrial Complex*.⁶ However, Hilda Kean, in her 2017 book *The Great Cat and Dog Massacre: The Real Story of World War II's Unknown Tragedy*, takes some issue with this memorial language of "no choice," arguing that it "implies that animals have and had no agency, no ability to think or to choose how to relate to the wars of which they were an integral part."⁷ I agree with Kean that nonhuman animals can think and demonstrate agency, and I further agree with her claim that in war, humans often lack agency too. Perhaps "no choice" overstates matters. However, in another sense, the wording on the memorial hinges on a crucial point. Animals cannot be understood to have chosen to engage in war in the same sense that humans have chosen. Even when humans are conscripted into war, our ability to access more of the communications and information about war positions us differently in conflict than other animals. It seems difficult to claim that any nonhuman animal chooses to be in war, per se. I hope this point chastens a usage I often fall into below, calling human reliance on animals in war a form of "partnership," when it is often conscription at best.

Furthermore, war, defined as the systematic organization of a larger group of people and animals for the purpose of violent conflict with another group or groups, usually as a method of settling some dispute, seems distinctively human. Again, recognizing this idea can seem to deepen the sense that human war is inevitable. In her history *The Animals' War: Animals in Wartime from the First World War to the Present Day*, for example, Juliet Gardiner states this widely held assumption: "war seems to be a perennial and almost universal part of the human condition."⁸ Yet, another important possibility exists: considering animals in war can produce a powerful es-

---

4 Hardt, Negri, *Multitude*, xi.
5 Hardt, Negri, *Multitude*, xviii.
6 Nocella, Anthony J. II, Colin Salter and Judy K.C. Bentley (eds.). *Animals and War: Confronting the Military-Animal Industrial Complex*. Lanham, MD: Lexington Books, 2014.
7 Kean, Hilda. *The Great Cat and Dog Massacre: The Real Story of World War II's Unknown Tragedy*. Chicago, IL: University of Chicago Press, 2017, 165.
8 Gardiner, Juliet. *The Animals' War: Animals in Wartime from the First World War to the Present Day*. London: Portrait, 2006, 10.

trangement effect, as I have already implied above. War can be exposed in its unnecessary, constructed, historical character; it can suddenly appear as other than inevitable. Joey's movement between "sides" in the Great War, and his inspiration of sympathetic, humane cooperation between the German and British soldiers who free him, makes more possible a view of war as unnecessary.

Indeed, the icon of Joey caught in the barbed wire of no-man's land can also be read to evoke the distinctly *historical* containments and entrapments of life more generally in what can be called the age of enclosure. As Robert P. Marzec shows in *Militarizing the Environment: Climate Change and the Security State*, the logic and practices of enclosure in Britain helped to inaugurate the radical changes in the management of land and property that facilitated the triumph of capitalism, imperialism, modern war, and the security state of neoliberalism.[9] These changes have impacted animals – both wild and domesticated – hugely, including but also reaching well beyond war. So, Joey, caught in the barbed wire in no-man's land, epitomizes the technologies and practices of enclosure played out on the global stage, incorporating not only repurposed animal-agricultural tools like barbed wire but the whole ensemble of war machinery. Joey demonstrates how the logic of global biopolitics, the management and production of all life, sweeps up not only humanity, but the nonhuman, the blasted trees of the trenches, the decimated grasses, shrubs, and their associated fauna, the animals imported into the war effort, including Joey himself.

Such considerations are too large for a single essay to treat fully, so this chapter uses a *longue durée* approach to evoke broad patterns and focuses largely on horses, whose involvement in war has been central across time. While even the importance of horses has been neglected in histories of war, less obviously charismatic animals have received even less attention: As I briefly show, insects, diseases, and small forms of life have hugely influenced history and the history of war, undermining the strong tendency to imagine war solely as a confrontation between closely-matched equals on the anthropocentric world stage.

2 Topics and Themes

A central challenge of the crucial topic of animals and war involves defining and dating war. If war is understood in a kind of minimalist way as violence orchestrated by a few individuals, then it likely reaches far back into human prehistory and probably before, and can be understood to be practiced beyond the human, including, for instance, by chimpanzees, meerkats, and social insects like ants and bees. But commonly, the behaviors of such animals are understood to be distinct from human

---

9 Marzec, Robert P. *Militarizing the Environment: Climate Change and the Security State*. Minneapolis, MN: University of Minnesota Press, 2015, 11.

war, more in line with predation and the like, as Dan Dembiec notes.[10] More commonly, war itself – not merely violence, but organized and systematic conflict – is understood to have arisen with agriculture some 12,500 years ago.[11] Thorough scrutiny of such considerations is too involved for our purposes here, but it is worth underscoring how these competing ideas can radically shift broad-brush understandings of life. War can seem inevitable and natural if we define it broadly enough to include the actions of many nonhuman animals. On the other hand, if war began with agriculture, some 12,500 years ago, then war seems much more confined to certain regimes of life. The latter case is made by Frans de Waal, who argues that humans are not *necessarily* or *naturally* war-makers; rather, war is merely one possible human undertaking, one that seems to result from the power of human technologies, beginning, de Waal suggests, with agriculture and its creation of surplus, of accumulated wealth. In *Dark Ecology*, Timothy Morton calls this suite of agricultural tools and behaviors "agrilogistics," and ascribes much of what we take to be fundamentally human to this historical – not inevitable – form of human life.[12] While we tend to associate humanity with agriculture, work, tools, writing, and war, in fact, most of those attributes seem to have arisen in the past 12,500 or so years; the human species, by contrast, is roughly 200,000 years old.

It is not large animals alone who have seriously impacted war. Entomologist Jeffrey A. Lockwood reveals in his book *Six-Legged Soldiers* the huge importance of insects in the history of conflict, discussing "an unholy trinity of strategies – transmission of pathogenic microbes, destruction of livestock and crops, and direct attacks on humans – through which six-legged soldiers have wreaked havoc on human society."[13] He suggests that the "earliest hypothesized use of insects as weapons of war was around 100,000 years ago during the Upper Paleolithic era."[14] But that dating – reaching back 100,000 years – probably precedes the advent of war as such, according to common definitions. In any event, Lockwood notes that such possibilities are difficult to ascertain given the scarcity of early, specific archeological evidence, but the advent of writing "about 5,000 years ago" clarifies the use and importance of insects in human war.[15] He notes, for one thing, that insects often functioned as models for war-making mentalities and strategies in their loyalty, courage, and

---

10 Dan Dembiec, quoted in Langley, Liz. Do Animals Go to War? *National Geographic*, January 30, 2016. URL: www.nationalgeographic.com/news/2016/01/160130-animals-insects-ants-war-chimpanzees-science/ (December 7, 2020).
11 De Waal, Frans. *The Age of Empathy: Nature's Lessons for a Kinder Society*. New York, NY: Harmony, 2009, 22–23.
12 Morton, Timothy. *Dark Ecology: For a Logic of Future Coexistence*. New York, NY: Columbia University Press, 2016.
13 Lockwood, Jeffrey A. *Six-Legged Soldiers: Using Insects as Weapons of War*. Oxford: Oxford University Press, 2009, 1.
14 Lockwood, *Six-Legged Soldiers*, 10.
15 Lockwood, *Six-Legged Soldiers*, 11.

more.¹⁶ He also discusses several examples of weaponized insects within the temporality of writing, mentioning that the Tiv people of Nigeria had developed a "bee cannon" that directed the insects towards enemies, perhaps doing so as early as 2500 BCE. Similarly, he notes, "By 2600 BCE, the Mayans had weaponized bees or wasps,"¹⁷ and he details much subsequent history.

Despite this long history, Lockwood underscores that military leaders have often failed to recognize insects' impact on the course of war,¹⁸ and that "few history books" give sufficient attention to this larger history.¹⁹ This neglect may derive from a certain version of human agency supported by ideas of war that revolve around human power, often distinctly masculine power. However, the use and importance of insects and other smaller forms of life in war run contrary to those norms. In that way, the case of insects and microorganisms only clarifies as a more extreme case the larger reality discussed in this chapter: that the importance of animals in war has been under-recognized. When acknowledged at all, animals in war tend to be folded into anthropocentric – androcentric – histories, with men as omnipotent heroes or villains.

Consider the example of the influenza of 1918–1920. It is often described as being more deadly even than the astonishingly gruesome Great War. Geoffrey W. Rice notes that while World War I brought some "10 million military deaths," the number of deaths from influenza between 1918 and 1920 "has been estimated at between 50 and 100 million, or 3 to 5% of the world's population at that time."²⁰ These figures reiterate the tremendous power of the nonhuman. Yet this framing, calling influenza more deadly than the war, relies upon a false distinction between them. In fact, the conditions of that war were central to the grim mortality of the influenza, facilitating its rapid global spread. The disease tended to be dispersed by military movements, then spread into the civilian population via contact between soldiers and loved ones.²¹ This very mobility – of weapons, but also of people and thereby of disease – is central to what the Great War was. This connection between war and disease is true more generally. As Rice notes in opening his short piece, "It is now a truism among historians that before the 20ᵗʰ Century far more people died in wartime from disease than from combat."²²

Whereas small forms of nonhuman life have had vast, often under-recognized impacts on war then, the animal that has been inescapably significant in the long

---

16 Lockwood, *Six-Legged Soldiers*, 9.
17 Lockwood, *Six-Legged Soldiers*, 17.
18 Lockwood, *Six-Legged Soldiers*, 1.
19 Lockwood, *Six-Legged Soldiers*, 5.
20 Rice, Geoffrey W. A Disease Deadlier than War. *The New Zealand Medical Journal* 126, no. 1378 (July 2013): 12–14, 12.
21 Radusin, Milorad. The Spanish Flu – Part II: The Second and Third Wave. *Vojnosanitetski Pregled. Military Medical & Pharmaceutical Journal of Serbia* 69, no. 10 (October 2012): 917–927, 919.
22 Rice, A Disease Deadlier than War, 12.

history of war, coming forward all the way to the present, is the horse. Indeed, narrating the history of the horse's role in war is a serviceable way to tell the history of war more generally. Again, information about the early millennia is patchy, relying upon anthropological findings as well as, later, somewhat limited written and pictorial records.[23] One key early change that set the stage for subsequent events is humans beginning to *ride* horses, often to herd other animals used for food. David W. Anthony surmises that horse riding began around 5,000 BCE. This form of action with horses likely led to a need for more land, Anthony argues, since working with horses permitted people to manage more animals for food. That success, in turn, may have rippled through the human social structures as new territorial boundaries were worked out along with new grievances and alliances.[24]

Add to the horse a wheeled cart, and this assemblage of horse, human, tools, and weapons takes a key further step. In *A History of Warfare*, John Keegan writes about this change, underscoring the importance of the ongoing revisions to the war carts as they went from four wheels to two, becoming the chariot.[25] He writes, "Charioteers were the first great aggressors in human history. Aggression, by an opposite if not always equal reaction, stimulates defence [...]".[26] This means, Keegan notes, that the inhabitants of the agriculturally settled valleys "in Mesopotamia, Egypt, the Indus Valley," and elsewhere were deeply affected by the raiding horse peoples of the Eurasian steppe, the area of unforested grassland that reaches across the center of Eurasia.[27] Those steppe environments facilitated mobility among raiding groups (and in the Middle Ages made possible the Silk Road). Chariot warfare dominated for some 1,000 years, from 1,700 to 700 BCE, reports Anthony.[28] Louis DiMarco emphasizes that this technological ensemble of the horse and chariot was distributed widely,[29] but these war systems would be abandoned in the face of a new set of tactics: mounted archers. Anthony dates this change beginning at around 800 BCE, noting that the success of archery from horseback unworked the war chariot.[30]

During this period, many other animals were recruited into human wars. Elephants have been used in wars at least since 1500 BCE, when they were engaged

---

[23] Clutton-Brock, Juliet. *Horse Power: A History of the Horse and the Donkey in Human Societies*. Cambridge, MA: Harvard University Press, 1992, 68; Keegan, John. *A History of Warfare*. New York, NY: Vintage, 1993, 157.
[24] Anthony, David W. *The Horse, the Wheel and Language: How Bronze-Age Riders from the Eurasian Steppes Shaped the Modern World*. Princeton, NJ: Princeton University Press, 2007, 222.
[25] Keegan, *History of Warfare*, 136.
[26] Keegan, *History of Warfare*, 139.
[27] Keegan, *History of Warfare*, 140.
[28] Anthony, *The Horse, the Wheel, and Language*, 18.
[29] DiMarco, Louis A. *War Horse. A History of the Military Horse and Rider*. Yardley, PA: Westholme Publishing, 2008, 4.
[30] Anthony, *The Horse, the Wheel, and Language*, 18. I make similar points introducing Animals and War: Studies of Europe and North America, Ryan Hediger (ed.), 1–25. Leiden: Brill, 2013.

in Syria,[31] and they have continued to be employed in many wars, including the World Wars. Birds have been used as messengers in war since around 1150 BCE,[32] an effort that persisted again through the World Wars. Gardiner notes that while most of the pigeons involved were asked to carry messages during the World Wars, others were "fitted with miniature cameras to take reconnaissance photographs over enemy or occupied territory."[33] "By the end of the Great War," Gardiner writes, "there were some 22,000 pigeons in service with British forces, looked after by 400 pigeoneers." The birds were not replaced by technology as one might assume, but were often used in cooperation with it, in one of the characteristic ironies of the technological World Wars. Thus, in World War I, Britain dropped birds *by aircraft* on the European continent as a method of learning about events there. Similarly, World War II did not obviate the need for birds: "something like a quarter of a million messenger pigeons were active in the Second World War."[34] Other animals saw activity in war from early on. Donkeys, mules, oxen, camels, buffalo, and many more often performed as draft animals, along with other roles.[35]

So, many animals have been active human wars, for a long time. But in the archer on horseback, a particularly resilient and potent combination had been attained. DiMarco argues the "comprehensive military force" that the mounted archers of the East represented, dominating for some 2,500 years, would only be displaced by the arrival of another revolutionary technology: gunpowder.[36] Gunpowder, discovered by the Chinese in the 9th century CE as a medicine or elixir, ironically,[37] took several centuries to become an obviously powerful entity, as the problems of harnessing its explosive force were worked out. Alfred Crosby writes of this effort, "The Chinese tried all manner of missile delivery systems, catapults and trebuchets and others, even live birds and oxen." He speculates, following the work of historian Joseph Needham, that the notion of the gun became possible in part due to the ubiquity in Chinese daily experience of bamboo, strong hollow tubes which clearly resemble the guns that would finally appear.[38]

But gunpowder and explosives did not end the importance of animals, as already implied above by the consideration to use birds and oxen to deliver explosives; instead, the animals often simply performed different roles – or the same roles in different ways. The Ottoman Turks of 1300 CE and following offer a good example as a culture that advanced its power "by exercise of sword, composite bow, horseman-

---

31 Kistler, John M. *War Elephants*. Westport, CT: Praeger, 2005, 8.
32 Cooper, Jilly. *Animals in War*. London: Corgi Books, 1983, 2000, 97.
33 Gardiner, *Animals' War*, 99.
34 Gardiner, *Animals' War*, 102–103.
35 Gardiner, *Animals' War*; Cooper, *Animals in War*.
36 DiMarco, *War Horse*, 116.
37 Crosby, Alfred W. *Throwing Fire: Projectile Technology Through History*. Cambridge: Cambridge University Press, 2002, 96.
38 Crosby, *Throwing Fire*, 98–99.

ship, and statecraft."³⁹ These ingredients, again including horses but more especially a culture of "horsemanship" alongside the other elements, created a context in which the advent of artillery and firearms could become successful, permitting the Turks to win "battle after battle," notes Crosby, and aiding in their campaign to conquer Constantinople. Relying on huge cannons – bombards – as well as smaller ones to conquer the city famous for its sturdy walls, the Turks defeated Constantinople in about two months during the spring of 1453 CE.⁴⁰ Thus, to some extent the effectiveness of war animals, and the cultures and activities built around them, facilitated the rise and success of gunpowder.

Crosby notes that even so powerful a technology as guns and gunpowder did not immediately impact all cultures equally for a range of reasons; it took time to percolate through the world's different peoples.⁴¹ Thus, even as Chinese alchemists were developing gunpowder in the 9th century CE, Western Europe was following another important historical trajectory: that of feudalism, held together to a significant extent by the role of horses. Conventionally, feudalism has been understood to involve a structure of power that steadily coalesced around the 9th century CE or so, or perhaps a little later, into the familiar, rigid hierarchy centered upon nobility and the mounted, armored knight who served the crown.⁴² Though scholars have debated just how consistently this social order operated,⁴³ it remains safe to say that the military unit of horse and rider was deeply influential. The social organization of feudalism, for instance, involved military forces that, over an also-debated period of time, shifted "from armies of foot soldiers to mounted warriors." In the German tradition, the foot soldiers were lower-class but free farmers "who lacked the wealth necessary to participate in mounted combat."⁴⁴ Due to this deficiency, they saw their access to power erode, until "farmers no longer had a right to participate in offensive military campaigns," according to one dominant account summarized by Bachrach. The farmers suffered a decline in prestige thereby, and, by contrast, the importance and power of the mounted warrior was cemented into this historical milieu [→Social History].⁴⁵

Another crucial episode of the second millennium CE is the European raiding and conquest of the Americas, an undertaking in which the horse again played a primary role. Indeed, at first consideration, the conquest seems implausible in some

---

39 Crosby, *Throwing Fire*, 114.
40 Crosby, *Throwing Fire*, 115–116.
41 Crosby, *Throwing Fire*, 107–129.
42 Bachrach, David Stewart. *Milites* and Warfare in Pre-Crusade Germany. *War in History* 22, no. 3 (July 2015): 298–343, 300; Keen, Maurice. Introduction: Warfare and the Middle Ages. In *Medieval Warfare. A History*, Maurice Keen (ed.), 1–13. Oxford: Oxford University Press, 1999, 7.
43 Reynolds, Susan. *Fiefs and Vassals: The Medieval Evidence Reinterpreted*. Oxford: Oxford University Press, 1994.
44 Bachrach, *Milites* and Warfare, 302.
45 Bachrach, *Milites* and Warfare, 302–303.

sense, as Inga Clendinnen emphasizes opening her article on the archetypical events of Hernán Cortes' attack in Mexico. Clendinnen charts several of that period's own explanations of European success that hinge on essentially racist or ethnocentric accounts of cultural difference between the Europeans and the Native Americans, accounts which *obscure* how cultural differences likely functioned in the encounters.[46] By contrast, clearly the ensembles of military prowess that had been gestating in Eurasia – tools, techniques, and animal "partnerships" – played a massive role, as did their particularly brutal use. Crosby also emphasizes the theatrics made possible by gunpowder.[47] Clendinnen emphasizes the same theme: Cortés' displays of power, "plays concocted to terrify Moctezoma's [sic] envoys – a stallion, snorting and plunging as he scented a mare in estrus; a cannon fired to blast a tree."[48]

But displays were just the beginning, Crosby underscores. Once the "propaganda effect" of such shows wore off, "gunpowder's effectiveness as a killer came into play."[49] Writing in *Military History*, Justin D. Lyons similarly underscores the importance of "gunpowder weapons" in Spanish conquest, along with many other Spanish military implements, swords, pikes, armor, crossbows, generally of iron and steel against the Indians' weapons of wood and bone. These European ensembles of weaponry, I add, must be understood as originally motivated in part by the powerful, horse-centered histories of Eurasian warfare. Lyons goes on to note, "The Spaniards also benefitted from their use of the horse, which was unknown to Mesoamericans. Though the conquistadors had few mounts at their disposal, tribal foot soldiers simply could not match the speed, mobility or shock effect of the Spanish cavalry, nor were their weapons suited to repelling horsemen."[50] These dynamics were important to Christopher Columbus' incursions further north, as Crosby underscores,[51] and to the many other encounters by a range of peoples across the Americas north, south, and central.

It is nonetheless further testimony to the power of the horse that it was quickly and widely adopted by many Amerindians after European arrival. This history is complex, however. Pekka Hämäläinen discusses the common historiographic view of the horse as reverse ecological imperialism, the rare counterexample of "the destructiveness of the Columbian Exchange" [→Environmental History].[52] That is, while many of the results of European arrival in the Americas beginning in the

---

46 Clendinnen, Inga. Fierce and Unnatural Cruelty: Cortés and the Conquest of Mexico. *Representations* 33 (1991): 65–100, 65.
47 Crosby, *Throwing Fire*, 128.
48 Clendinnen, Fierce and Unnatural Cruelty, 72.
49 Crosby, *Throwing Fire*, 128.
50 Lyons, Justin D. Master of the Conquest: Hernan Cortes Himself – not Spanish Arms, Smallpox or Mesoamerican Allies – Was the Catalyst behind the Stunning Defeat of the Aztec Empire. *Military History* 33, no. 6 (March 2017): 30–38, 32.
51 Crosby, *Throwing Fire*, 127.
52 Hämäläinen, Pekka. The Rise and Fall of Plains Indian Horse Cultures. *Journal of American History* 90, no. 3 (December 2003): 833–862, 833.

late 1400s resulted in harm not only to Amerindian peoples, but also to American ecologies, the horse is often seen as "the ultimate anomaly," "a straightforward success story," "an equestrian experiment that lifted the Indians, both materially and figuratively, to a new level of existence, while uniquely equipping them to resist future Euro-American invasions."[53] Hämäläinen's essay disrupts that narrative, arguing that the "transformational power of horses was simply too vast," that the horse's presence had not only positive effects on Indian peoples, but also brought "destabilization, dispossession, and destruction." He explains that horses not only helped Indians do many things; they also "disrupted subsistence economies, wrecked grassland and bison ecologies, created new social inequalities, unhinged gender relations, undermined traditional political hierarchies, and intensified resource competition and warfare" [→American Studies].[54]

Hämäläinen's account recalls the story of the horse's first impacts on humans offered by Anthony, cited above. It also recalls Keegan's argument that "the charioteers and the horse-riding peoples who succeeded them altered the world in which civilized arts of peace had begun to flourish" in Mesopotamia, Egypt, the Indus Valley, and elsewhere.[55] We can date those events some 3,000 years after the first use of the horse and roughly 3,000 years before the European invasion of the Americas. Despite the significant differences of time and context, then, this consistent theme returns: that horses especially but also other animals have had massive impacts on human life and human conflict across time.

As these events roiled the Americas, Europeans were learning to leverage the power of new war weapons and strategies on the continent, plying them again and again on horseback. Napoleon Bonaparte is the quintessential figure later in this history, playing a major role in some twenty-three years of nearly perpetual war from 1792, with the French attack of Austrian-ruled Belgium, to France's more general capitulation in 1815. Historian David A. Bell calls this period the true first era of "total war," contrasting his view with the common notion that total war began with the World Wars. He underscores the change that war underwent, spurred in large part by the French Revolution. War had been a relatively controlled and restrained phenomenon that "seemed entirely natural and proper to the noblemen who led Europe's armies under the Old Regime," a kind of "theater of the aristocracy" of the sort discussed above, in which horses played an essential role [→Political History;→Diplomatic History].[56] But the advent of the Napoleonic Wars wrought a change in the scale of conflict, one which Bell calls, pointedly and carefully, "apocalyptic." He notes that the change was less about war technology than about the

---

53 Hämäläinen, Plains Indian Horse Cultures, 833.
54 Hämäläinen, Plains Indian Horse Cultures, 834.
55 Keegan, *History of Warfare*, 136–139.
56 Bell, David A. *The First Total War: Napoleon's Europe and the Birth of Warfare as We Know It*. Boston, MA: Houghton Mifflin, 2007, 5.

"scope and intensity of warfare," citing the much larger numbers of combatants, leading to massive casualties:

> During the Napoleonic period, France alone counted close to a million war deaths, possibly including a higher proportion of its young men than died in World War I. The toll across Europe may have reached as high as 5 million. In a development without precedent, the wars brought about significant alterations in the territory or the political system of every single European state. [...] This, then, was the first total war.[57]

One important example from this period is Napoleon's 1812 campaign against Russia, which clarified the decline of the French empire that was already underway and saw massive losses of life, both human and nonhuman. The French began their offensive with a huge force, perhaps as many as 600,000 men, reportedly the biggest army in history up to that point.[58] The French army included many oxen and some 250,000 horses used to pull carts of weaponry and supplies[59] and included a train of "possibly 50,000 cattle" for food.[60] The army quickly encountered trouble in feeding these crucial animals, in part because "it was a late, hot, dry spring; streams were low, and forage was scarce."[61] Then, "a severe storm at the very beginning of the campaign killed approximately 10,000 horses."[62] The campaign involved a long Russian retreat through the summer, eventually permitting the French to take Moscow and burn much of it. But the Russians would not formally relent, and the French began their own retreat as the fall arrived and the weather grew colder. The problems of transportation for the French had been significant from the start, Elting notes, with Russia's "narrow back roads" providing challenges for the animal-drawn carts full of supplies. Additionally, in a theme common to many big armies, assembling the sheer numbers of horses and oxen necessary meant for a "hasty scramble" in which "horses and oxen were bought up at random and put to work before they were properly broken in," pulling wagons "built in a slapdash rush" that "began coming apart early in the campaign."[63]

When the weather turned snowy, notes military historian Saul David, the horses were not properly equipped, having only summer horseshoes in anticipation of a shorter conflict. Without the "little spikes" for traction in the snow, David explains,

---

57 Bell, *The First Total War*, 7.
58 Clodfelter, Michael. *Warfare and Armed Conflicts: A Statistical Encyclopedia of Casualty and Other Figures, 1492–2015*. Jefferson, NC: McFarland, 2017, 161; Joes, Anthony James. Continuity and Change in Guerilla War: The Spanish and Afghan Cases. *The Journal of Conflict Studies* 16, no. 2 (1996): 64–74, nt. 27. URL: www.erudit.org/en/journals/jcs/1996-v16-n2-jcs_16_2/jcs16_02art04/ (December 6, 2020).
59 David, Saul. Napoleon's Failure. For the Want of a Winter Horseshoe. *BBC News Magazine*, February 9, 2012. URL: www.bbc.co.uk/news/magazine-16929522 (August 25, 2020).
60 Elting, John R. *Swords Around a Throne: Napoleon's Grande Armée*. New York, NY: Da Capo Press 1988, 566.
61 Elting, *Swords Around a Throne*, 567.
62 Elting, *Swords Around a Throne*, 566; DiMarco, *War Horse*, 208.
63 Elting, *Swords Around a Throne*, 569–570.

the horses faced a "disaster" of falls and injuries.⁶⁴ These problems were possibly compounded by "some train troops [...] deliberately neglecting their animals; the sooner their horses and oxen died, the sooner their responsibilities were over."⁶⁵ These failures of *esprit de corps* were surely intensified by another huge nonhuman impact on Napoleon's army. As Lockwood puts it, "Napoleon's worst defeat by insects came in 1812. Rather than taking Russia, his Grande Armée lost 200,000 men to louse-borne typhus."⁶⁶ While many factors were involved in this defeat, Lockwood is clearly correct that the power of insects over human life in that scenario is significant. These factors culminated in a stunning decline of Napoleon's army: Famously, that huge invasive French force shrank dramatically, reducing the 600,000 or so invading men to around 70,000 survivors who marched back out of Russia.⁶⁷ Napoleon's 1812 Russian campaign is an especially clear example of the biopolitics of war, with entire regimes of life organized around its execution, and massive impacts arriving from human and nonhuman activities alike.

As the nineteenth century advanced, ever-improving war technologies continued to be remixed with new techniques, including horses and other animals of war. Two more brief examples will have to suffice to evoke the case. First, the United States Civil War (1861–1865) saw improving weapon technologies combine with large-scale conflict of the Napoleonic scale, joined with the persistent need for horses.⁶⁸ This scenario contributed to the astonishing numbers of horse deaths in that war, some fifty percent of all the animals in service.⁶⁹ In this form of large-scale, mass war, horses became expendable in a way that would continue to haunt subsequent conflicts. A particularly galling example is the Boer War, or South African War of 1899–1902 [→African Studies]. This war, Sandra Swart writes, is "widely regarded as proportionally the most devastating waste of horseflesh in military history up until that time." The British side lost 66.88% of its horses (326,073 fatalities) and 35.37% of its mules (51,399 fatalities).⁷⁰

Of course, in terms of sheer numbers, far more horses were lost in the World Wars. Although use of these animals proved entirely ineffective in familiar cavalry techniques against the power of machine guns, horses were utterly crucial as motive forces for transporting guns and ammunition, moving provisions, pulling carts of water, moving injured people (and other animals) in ambulances, and much

---

64 David, Napoleon's Failure.
65 Elting, Swords Around a Throne, 570.
66 Lockwood, Six-Legged Soldiers, 2.
67 Joes, Continuity and Change in Guerilla War, nt. 27.
68 Kistler, John M. *Animals in the Military: From Hannibal's Elephants to the Dolphins of the U.S. Navy.* Santa Barbara, CA: ABC-Clio, 2011, 174.
69 DiMarco, War Horse, 302.
70 Swart, Sandra. *Riding High: Horses, Humans, and History in South Africa.* Johannesburg: Witwatersand University Press, 2011, 104.

more.⁷¹ Horses also pulled cable carts to place telegraph line, a task for which dogs were likewise recruited on a smaller scale.⁷² Elephants, camels, mules, donkeys, reindeer, bullocks, oxen and other animals were similarly used to move a range of supplies.⁷³

In an arresting passage, Keegan contrasts the hopes of a technological war with the reality of the Great War: when the huge armies and the "several million horses" arrived on the battlefield, "they found that the almost miraculous mobility conferred by rail movement evaporated. Face to face with each other, they were no better able to move or transport their supplies than Roman legions had been; forward of railhead, soldiers had to march, and the only means of provisioning them was by horse-drawn vehicles. Indeed, their lot was worse than that of the well-organized armies of former times, since contemporary artillery created a fire-zone several miles deep with which re-supply by horse was impossible and re-provisioning of the infantry – with ammunition as well as food – could be done only by man-packing."⁷⁴

The same problem appeared in the Second World War, Keegan notes, leading the Germans, for example to bring into the effort some 2.75 million horses, almost double the number they used in the Great War. Discussing these events, Keegan writes, "most [horses] died in service, as did the majority of the 3.5 million horses mobilized by the Red Army between 1941 and 1945."⁷⁵ Keegan generalizes to emphasize the challenges of war logistics and supply, attributing many of the difficulties to the industrial capacities of the nations involved. This situation involves a significant irony, then. Industrialization changed the nature of war in significant ways, yet the horse in such cases can also be viewed as analogous to guns and gunpowder in a specific sense: much as cannon technology persisted mostly unchanged from around 1400 to 1870 or so,⁷⁶ horses were an enduringly potent force, so potent that their importance and even many of their specific forms of use changed little across time.⁷⁷ Guns and horses functioned as paired hubs in the wheels of historical change for long periods as cultures, strategies, modes of organization, deployment, and so on all turned around them.

Many other organisms were brought into the effort of the World Wars. Lockwood notes that in World War II, the Japanese had "weaponized" insects, releasing across China "hundreds of millions" of fleas and flies infected with typhus, and also attempting "to infiltrate the United States" with insects. "By the end of the war," he concludes, these insects "were responsible for more deaths than the atomic bombs dropped on Japan." Similarly, Lockwood writes that by 1944, "Germany had stock-

---

71 Gardiner, *Animals' War*, 42.
72 Gardiner, *Animals' War*, 94–95.
73 Gardiner, *Animals' War*, 38–85.
74 Keegan, *History of Warfare*, 307–308.
75 Keegan, *History of Warfare*, 308.
76 Crosby, *Throwing Fire*, 131.
77 DiMarco, *War Horse*, ix; Gardiner, *Animals' War*, 12.

piled 30 million secret weapons: the Colorado potato beetle." [78] It is not clear, he notes, whether those beetles were actually used against British agriculture, but it is entirely clear that the power of insects in modern war was recognized and mobilized.

Marzec's study of the militarization of the environment moreover demonstrates the many ways in which all manner of forms of life have been folded into the military machine, famously called "the military industrial complex" by Dwight Eisenhower, World War II army general and, later, U.S. president.[79] For example, "In 1949 Secretary of Defense James Forrestal hired entomologist Caryl Haskins to chair a committee devoted to exploring the potential of biological warfare." This inquiry and others like it explored ways to "destroy enemy crops," manipulate "the world's climate," and even make "use of nuclear bombs to reconfigure the sea floor and change the course of ocean currents."[80] In Marzec's account, such astonishing plans are of a piece with the logic of security and militarization that has become ever-more dominant since the British enclosure acts of the seventeenth and eighteenth centuries. Those acts, he writes, "introduced modern conceptions of privatization, surveillance, and environmental manipulation." They are a key step in the larger reality Marzec summarizes: "the relationship between the military and the environment spans the entire length of the modern era and extends even several centuries before Columbus' arrival in America."[81] It has had truly massive impacts on animals and the ecologies that make us all possible.

3 Methods and Approaches

Considering the roles of animals in war involves a number of new approaches. At bottom, perhaps the central recalibration involves revising the usual sense of human dominance. While humans do in fact have significant control over wars, it is also the case that the whole history of war has always involved human partnerships, with animals and with other nonhuman objects, tools, landscapes, and so on. I noted above, for instance, that insect behavior was often understood as a model for both theory and practice of early war, and war language remains rife with animalistic and naturalistic phrases and terms, including such (unjust) usages as "dogs of war," "dog-eat-dog," "rules of the jungle"; imagery on airplanes and the like has often involved animals; and many animal-centered expressions, such as "pigs," "dogs," "vermin," and so on have been used to denigrate opponents and thereby facilitate and excuse violence against them.

---

[78] Lockwood, *Six-Legged Soldiers*, 2–3.
[79] Marzec, *Militarizing the Environment*, 32.
[80] Marzec, *Militarizing the Environment*, 34–35.
[81] Marzec, *Militarizing the Environment*, 11.

Rethinking this history is not just theoretical or philosophical, though. A more traditional key method has involved scrutinizing extant sources with new questions. Hilda Kean, for instance, relies significantly on diaries, institutional records, advertisements, and other traditional sources to produce a new history of animals in war. She makes a compelling case that much of the history of animals in war is actually available but has not been sought by historians.[82] This method of revisiting extant sources is visible in other historians' work, as when John M. Kinder revisits the events of 2006 during wartime, or when Brian Lindseth rethinks the meaning of nuclear testing with a focus on animals.[83]

Clearly one difficulty of such work is our inability to receive direct reports from animals themselves, as distinct from the opportunity to discuss war experience with human veterans and civilians. The problem of interpreting the experiences and feelings of animals, one that afflicts practically all work in animal studies, is especially severe in the case of animals and war because of war's dynamics: its tendency to unfold rapidly and surprisingly, with little time or place to pause and take account of events. This problem is compounded by the traditional hierarchies of species, in which human experiences are frequently prioritized over nonhuman ones, and hierarchies even subdivide animals, with more attention given to horses, dogs, and other beings whose experiences are more directly related to human activities, while other creatures often receive no account at all. The unusually severe traumas that attend war, many of which have been evoked above, present obstacles to interpretation for all who experience them, human and nonhuman alike.

4 Implication(s) of the Animal Turn

I have gestured above at many of the implications of the *animal turn* in war studies, so here, I will return to key themes from the introduction. Attention to the presence of an ever-growing range of animals in war facilitates a surprising rethinking of the function of sympathy, which opens up a way toward a more radical rethinking of war and violence. While rehabilitating sympathy and affect has a long and varied genealogy reaching through feminism, gender studies, and elsewhere, in animal studies, sympathy for animals caught up in the traumas of war has been key [→Feminist Intersectionality Studies].[84] The blooming of sympathy in the midst of war's horrible violence exemplifies – and concretely advances – the denaturalization of the very

---

82 Kean, *Great Dog and Cat Massacre*, 14–15.
83 Kinder, John M. Zoo Animals and Modern War: Captive Casualties, Patriotic Citizens, and Good Soldiers. In *Animals and War: Studies of Europe and North America*, Ryan Hediger (ed.), 45–75. Leiden: Brill, 2013; Lindseth, Brian. Nuclear War, Radioactive Rats and the Ecology of Exterminism. In *Animals and War: Studies of Europe and North America*, Ryan Hediger (ed.), 151–174. Leiden: Brill, 2013.
84 See, for example, Kean, *Great Dog and Cat Massacre*, 99–114.

logic of war, exposing its constructed, contingent, and, in a sense, unnatural character. This recognition offers resources for resisting war in general, and for rethinking the regimes of life associated with war and the ecological destruction it wreaks.

As Marzec argues, in a point resonant also with Hardt and Negri's work in *Multitude*, our notions of humanity in contemporary times are "summoned into existence by the militarized milieu of scientific exploration in the age of environmentality," growing out of the age of enclosure.[85] Confronting Joey, our equine protagonist in the film *War Horse*, as he is captured in the barbed wires above the trenches of the Great War, then, provokes not only human compassion for animals, but also a recognition of ourselves, caught in these same logics and practices. The scene makes possible a serious rethinking of the ideologies we are immersed in. Freeing Joey from those barbed wires includes the also-important possibility of freeing ourselves.

Selected Bibliography

Cooper, Jilly. *Animals in War*. London: Corgi Books, 1983.
DiMarco, Louis A. *War Horse: A History of the Military Horse and Rider*. Yardley, PA: Westholme Publishing, 2008.
Gardiner, Juliet. *The Animals' War: Animals in Wartime from the First World War to the Present Day*. London: Portrait, 2006.
Hediger, Ryan (ed.). *Animals and War: Studies of Europe and North America*. Leiden: Brill, 2013.
Kean, Hilda. *The Great Cat and Dog Massacre: The Real Story of World War II's Unknown Tragedy*. Chicago, IL: University of Chicago Press, 2017.
Kistler, John M. *Animals in the Military: From Hannibal's Elephants to the Dolphins of the U.S. Navy*. Santa Barbara, CA: ABC-Clio, 2011.
Kistler, John M. *War Elephants*. Westport, CT: Praeger, 2006.
Lockwood, Jeffrey A. *Six-Legged Soldiers: Using Insects as Weapons of War*. Oxford: Oxford University Press, 2009.
Nocella, Anthony J. II, Colin Salter and Judy K.C. Bentley (ed.). *Animals and War: Confronting the Military-Animal Industrial Complex*. Lanham, MD: Lexington Books, 2014.
Sax, Boria. *Animals in the Third Reich: Pets, Scapegoats, and the Holocaust*. New York, NY: Continuum, 2002.

---

85 Marzec, *Militarizing the Environment*, 232.

Janet M. Davis
History of Animal Fights and Blood Sports

1 Introduction and Overview

The history of animal fights and blood sports is ancient and transnational. In diverse, far-flung cultural settings, people have groomed, trained, and staged animals to fight each other, often resulting in the death of one or both combatants. In some cases, such as the cockfight and the beetle fight, these contests have been inspired by an animal's natural behavior. The male Rhinoceros beetle (*Xylotrupes socrates*), for one, fights other males during mating season in Thailand.[1] Staged animal fights can involve the same species, such as dogfighting and cockfighting, while others have pitted members of different species against each other. Human beings have also fought matches against other animals, such as lions, crocodiles, elephants, and bears. In Ancient Rome, *bestiarii* (beast fighters) were often Christians, prisoners of war, or convicted criminals, who faced certain death through forced interspecies combat, while *venatores* (skilled hunters) participated voluntarily to bolster their socioeconomic status.[2] In other historical settings, such spectacles have been (arguably) more benign. For example, on May 22, 1894, celebrity strongman Eugen Sandow wrestled a lethargic, muzzled lion wearing gloves on his paws in a padded cage at Central Park in New York City.[3]

By definition, animal fights are "blood sports," a term – often used pejoratively – that describes any contest of athletic skill, physical dexterity, or brute force resulting in physical injury or death to an animal. An audience is usually present, ranging in size from 250,000 people at the Circus Maximus racetrack in Ancient Rome, to countless makeshift cockpits worldwide forged in a circle of dirt with a handful of spectators.[4] Yet animal fights and blood sports are not automatically synonymous because many blood sports contain no fighting, but still result in injury or death, such as

---

[1] See Snell-Rood, Emilie C. and Armin P. Moczek. Horns and the Role of Development in the Evolution of Beetle Contests. In *Animal Contests*, Ian C. W. Hardy, Mark Briffa, (eds.), 178–198. Cambridge: Cambridge University Press, 2013.
[2] Coley, Jacob. Roman Games: Playing with Animals. In *Heilbrunn Timeline of Art History*. New York, NY: The Metropolitan Museum of Art (September 2010). URL: www.metmuseum.org/toah/hd/play/hd_htm (December 7, 2020); Grout, James. Bestiarius. *SPQR Encyclopaedia Romana*, last updated September 12, 2020. URL: penelope.uchicago.edu/~grout/encyclopaedia_romana/gladiators/bestiarii.html (December 7, 2020).
[3] Man Against Beast. Sandow and the Lion to Try Conclusions. *San Francisco Chronicle*, May 22, 1894. Microfilm Collections, Washington State University Pullman Interlibrary Loan.
[4] Beard, Mary. *SPQR: A History of Ancient Rome*. New York, NY: Liveright Publishing, 2015, 446, 462; Davis, Janet M. Cockfight Nationalism: Blood Sport and the Moral Politics of American Empire and Nation Building. *American Quarterly* 65, no. 3 (September 2013): 549–574, 552–553.

https://doi.org/10.1515/9783110536553-043

goose-pulling, fox-tossing, or bear-whipping.[5] Other blood sports, such as foxhunting and coursing (hunting a hare with gazehounds), involve pursuing and eventually overwhelming an unconfined animal, usually with dogs.

Baiting, confining, and beating live animals has been banned in most countries since the nineteenth century. But the status of other animal fights and blood sports has remained wildly uneven. Some animal fights, such as dogfighting and cockfighting, are banned in some countries and celebrated in others. Some blood sports have faced escalating opposition over time. Foxhunting, for one, historically has been the province of powerful elites around the world, but in England and Wales, the Hunting Act 2004 bans the use of dogs to hunt foxes, even though the law is riddled with loopholes.[6] Still other blood sports, such as hunting and fishing, which are discussed in greater detail in other chapters of this book [→History of Hunting], are legal, regulated, and widely accepted worldwide because they allow an animal "fair chase," or the possibility of escape in its own habitat. And lastly, some animal sports, such as horseracing and rodeo, occasionally result in injury or death, but do not meet the definition of a blood sport because any harm is accidental, rather than intentional. Overall, this chapter encourages animal historians to connect these gruesome cultural forms to broader intersectional questions concerning identity formation, anthropomorphism, animal agency, and ideologies of civilizational progress.

This chapter will suggest that the history of animal fights and blood sports is inseparable from the history of social and moral reform. This history is dialectical-shaped as much by opposition movements as it is by blood sports enthusiasts. The rise of organized animal protection movements during the nineteenth century played a critical role in outlawing specific blood sports in distant locations around the world. Additionally, other social movements, including temperance, anti-vice, and anti-gambling activism joined forces with animal advocates to shape this history. Social control was at the heart of these oppositional movements because blood sports were popular sites of unruly behavior, which occasionally erupted into drunken brawling. As a result, moral reformers often lost sight of the suffering animal body in favor of targeting human depravity. This disjuncture reflects wider methodological tensions in the diverse body of historiography comprising the *animal turn*: some scholars approach historical animals primarily as sentient, embodied beings possessing some degree of historical agency, while others treat animals in representational terms, as conduits to human ideologies of self, other, culture, and nation.

---

5 In goose-pulling, a goose is tethered upside down and contestants try to pull its neck while galloping on horseback; in fox-tossing, contestants trap a fox, force it into a cloth sling, and then bounce it high into the air; in bear-whipping, a bear is chained and whipped to death. See Thomas, Keith. *Man and the Natural World: A History of the Modern Sensibility*. New York, NY: Pantheon Books, 1983.
6 UK foxhunters are still allowed to follow a scent trail with dogs in order to flush a fox. Once the fox emerges, a falconer is permitted to set a bird of prey upon the exposed animal. Baker, Neal. For Fox Sake. *The Sun*, July 4, 2019. URL: www.thesun.co.uk/news/3533139/fox-hunting-uk-law-ban-theresa-may-bird-of-prey-latest/ (August 20, 2019).

2 Topics and Themes

Although the history of animal fights and blood sports is global, this chapter will focus on America and its transnational past: Colonial America was an English colony; indigenous groups and settler colonialists alike shaped the history of blood sports on American soil; moreover, American empire building played a significant role in making this transnational history [→American Studies;→Global History]. To complement the *animal turn* in the humanities, this chapter uses the interdisciplinary methodological insights of intersectional social and cultural history, foreign relations history, religious studies, literature, legal studies, and the biological sciences to reach a broader historical and anthrozoological understanding of these violent cultural practices.

This chapter contains several interconnected topics that structure its analysis and organization: Welfare, entertainment, training, and empire building. The history of animal welfare appears throughout this chapter in the form of legislation, organized animal protection societies, and grassroots mobilization. The persistent popularity of blood sports and animal fights over time must be understood through their status as forms of entertainment, which often involved large crowds whose composition and behavior reflected broader social hierarchies. An analysis of training practices gives voice to the complex relationships that blood sports participants forged with animals, thus illuminating the paradoxical nature of cultural practices involving daily regimens of exercise, care, and attention in the service of inevitable pain and suffering. As the United States became an overseas empire at the turn of the twentieth century, blood sports and animal fights played a surprising role in shaping systems of formal governance abroad, culturally specific forms of entertainment, and ideologies of belonging and exclusion.

Blood Sports and Animal Welfare

Puritan settlers in Massachusetts Bay Colony tacitly addressed blood sports as part of their broader concern for animal protection in their legal code, the Body of Liberties, authorized by the General Court in 1641. Specifically, Section 92 decreed: "No man shall exercise any Tirranny or Crueltie towards any bruite Creature which are usuallie kept for man's use."[7] While Section 92 did not mention English blood sports by name, this measure served as a tacit rebuke of animal sports across the Atlantic that were popular among royals and commoners alike.[8]

---

7 The Liberties of the Massachusetts Collonie in New England, 1641. *Hanover Historical Texts Project*, Hanover College History Department, last modified March 8, 2012. URL: history.hanover.edu/texts/masslib.html (December 7, 2020).
8 Thomas, *Man and the Natural World*, 144–147.

Historian Kathleen Kete observes that English Puritans objected to blood sports for theological, social, and political reasons. Puritans argued that Adam and Eve's removal from the Garden of Eden in the Book of Genesis demonstrated a biblical commitment to gentle stewardship because Adam's sin unleashed an age of tremendous cruelty to animals. In response to rising Puritan criticism, King James I issued the King's Declaration of Sports in 1618, which defended royal blood sports – an edict that his son and successor, King Charles I, reissued in 1633 amid escalating Puritan opposition to his reign. At the height of Puritan rule, Oliver Cromwell issued the Protectorate's order of 1654, which banned animal fights and blood sports. However, the order was quickly rescinded during the reign of Charles II (1660–1685), as part of the broader reimplementation of royal power during the Restoration.[9] Fundamentally, the Body of Liberties in Massachusetts Bay affirmed colonial New England solidarity with their English Puritan allies across the Atlantic.

Yet blood sports remained popular in Colonial and Revolutionary America – so popular that the Continental Congress moved to ban them in 1774 as proof of American republican virtue vis-à-vis English decadence. The First Continental Congress adopted the ban on October 20, 1774, as part of its sweeping Articles of Association in response to a punitive set of new British laws – dubbed the "Coercive Acts," or the "Intolerable Acts" – designed to punish the colonists economically and politically for the Boston Tea Party (1773). Among other measures to promote American agriculture and manufacturing, the Articles of Association stressed the patriotic significance of frugality. As a result, "every species of extravagance and dissipation" was condemned, including cockfighting, horseracing, "and all kinds of games" inseparably linked to gambling.[10]

The Article of Association's ban on cockfighting condemned "extravagance and dissipation," rather than animal suffering. Yet during the evangelical Second Great Awakening (1790–1840), revivalist theologians and lay reformers alike increasingly admonished blood sports for causing pain. Rejecting predestination in favor of free moral agency, they interpreted biblical dominion as God's call to serve as gentle stewards to animals, rather than agents of domination. Here, they built upon earlier generations of social thought, including philosopher John Locke, who exhorted parents to teach their children creaturely kindness in *Some Thoughts on Education* (1693).[11] Timothy Dwight, a Congregational theologian, minister, and president of Yale College, condemned blood sports in his *Theology: Explained and Defended in*

---

9 Kete, Kathleen. Animals and Ideology: The Politics of Animal Protection in Europe. In *Representing Animals*, Nigel Rothfels (ed.), 19–34. Bloomington, IN: Indiana University Press, 2000, 20–25.
10 Journals of the Continental Congress – The Articles of Association; October 20, 1774. *The Avalon Project: Documents in Law, History and Diplomacy*, Yale Law School. URL: avalon.law.yale.edu/18th_century/contcong_10-20-74.asp (December 7, 2020).
11 Smith, Bill Leon. Animals Made Americans Human: Sentient Creatures and the Creation of Early America's Moral Sensibility. *Journal of Animal Ethics* 2, no. 2 (Fall 2012): 126–140, 129.

a Series of Sermons, published posthumously in 1823.[12] Echoing Locke, Dwight urged parents to teach their children kindness, "to shun cruelty even to an insect." He decried the popular practice of stoning poultry as "gradual torture," and vigorously condemned cockfighting, "so widely and shamefully extended in some parts of this country [...] abominable for its cruelty, and detestable for its fraud."[13] Dwight's condemnation of animal cruelty and suffering resonated with the era's growing constellation of social reform movements that targeted bodily violence, including abolitionism, temperance, asylum reform, and the movement against corporal punishment.[14]

The growing moral rejection of animal pain coincided with the passage of the first anticruelty laws in Great Britain and the United States. Great Britain's first animal welfare law (1822) was amended in 1835 to give the (Royal) Society for the Prevention of Cruelty to Animals the powers of prosecution against "wanton cruelty" in animal sports.[15] That same year, Massachusetts became the first U.S. state to ban dogfighting.[16] Other states were slow to prohibit blood sports and animal fights, but they joined the Bay State in passing anticruelty laws that punished people for publicly harming livestock. Legal scholars Vivian Tseng and David Favre observe that antebellum animal welfare laws in Maine (1821), New York (1829), and elsewhere, treated acts of cruelty as property crimes. They note that a person could escape prosecution in Vermont (1846) if beating his or her own livestock but could face prosecution if the animals belonged to someone else.[17]

Blood Sports, Popular Entertainment and Animal Training

Despite the creeping growth of animal protection laws, animal fights and blood sports remained widely popular forms of entertainment across antebellum America. In 1850, San Francisco possessed two bullfighting arenas, as well as facilities for bear baiting and cockfighting. In San José, California's first capital, the state legislature recessed during a ring fight between a bull and a grizzly bear.[18] On April 26, 1840,

---

[12] Dwight, Timothy. *Theology: Explained and Defended in a Series of Sermons*. New Haven, CT: S. Converse, 1823, 273–274.
[13] Dwight, *Theology*, 303–304.
[14] Davis, Janet M. *The Gospel of Kindness: Animal Welfare and the Making of Modern America*. New York, NY: Oxford University Press, 2016, 26–49.
[15] Ritvo, Harriet. *The Animal Estate: The English and Other Creatures during the Victorian Age*. Cambridge, MA: Harvard University Press, 1987, 150–151.
[16] Kim, Claire Jean. *Dangerous Crossings: Race, Species, and Nature in a Multicultural Age*. Cambridge: Cambridge University Press, 2015, 256.
[17] Favre, David and Vivien Tsang. The Development of Anti-Cruelty Laws during the 1800's. *Detroit College of Law Review* 1993, no. 1 (Spring 1993): 1–36, 6–8.
[18] Betts, John Richard. *American Sporting Heritage*. Reading, MA: Addison-Wesley, 1974, 25; Davis, *The Gospel of Kindness*, 191.

the New Orleans *Daily Picayune* beckoned readers to attend a "Great Fight between Some French Dogs, a Bear, an Ass, and a Bull. Admittance $1. Children, half-price."[19] Animal fights and blood sports were just as common on the East Coast, even though New York State outlawed them in 1856. Kit Burns (1830–1870), a notorious Irish American dogfighter and saloon keeper in New York City, recalled a barefoot antebellum childhood immersed in dogs and the excitement of catching a glimpse of them in training at local fighting venues.[20]

Burns proved to be an able student of blood sports. During his brief life, he owned a liquor store and saloon, which served as fronts for Sportsman's Hall, his profitable dogfighting and rat-baiting emporium. Burns and other fighters kept their dogs on exacting – and often excruciating – training and dietary regimens. Fighting dogs were often fed meat and were subjected to long, punishing hours on a treadmill to build endurance and muscle mass. Each dog was cajoled to keep pace, either by the presence of other dogs, or by threat of force, "compelling him to keep in motion a wheel which was made to revolve by the action of the dog's feet."[21] Dogs in training were also set upon stray curs to hone their fighting instincts to make them more "gamey." Prior to entering the ring, dogs were routinely washed; a neutral party "tasted" their fur to make sure an unethical owner hadn't laced the animal's coat with poison to endanger an opponent.[22] Cockfighters engaged in similarly rigorous training programs. A fighting cock typically spent the first two years of his life in training before entering the cockpit. He ate a special diet, exercised vigorously, and bonded with his handler, who often cradled and stroked him daily to promote close communication. Writer Tim Pridgen unwittingly observed a striking degree of homoerotic interspecies intimacy between cockfighters and their roosters when he interviewed American cockers during the 1930s: "A cocker always is stroking a rooster, firm, easy strokes from the hackles to the rump, stroking and feeling, sizing up the quiver of him, knowing his hard, corky spring, with the cock sitting steady in his arms, loving it, looking out with wild brown and yellow gleams of arrogance, demanding trouble."[23] When a rooster was injured during a match, his handler often licked his wounds, or submerged his head in his own mouth in the belief that human saliva sterilized and healed the laceration.[24]

---

19 Kaplan-Levenson, Laine. In the Mid-19th Century, Vicious Animal Combat Drew Thousands to Algiers. *New Orleans Public Radio*, February 11, 2016. URL: www.wwno.org/post/mid-19th-century-vicious-animal-combat-drew-thousands-algiers (September 25, 2019).
20 Kaufman, Martin and Herbert J. Kaufman. Henry Bergh, Kit Burns, and the Sportsmen of New York. *New York Folklore Quarterly* 28, no. 1 (March 1972): 15–29, 15–16.
21 Alleged Cruelty to the Animal. *New York Herald*, February 17, 1867, 7.
22 Kaufman and Kaufman, The Sportsmen of New York, 20–21.
23 Pridgen, Tim. *Courage: The Story of Modern Cockfighting*. Boston, MA: Little, Brown and Company, 1938, 4–5.
24 Davis, Cockfight Nationalism, 556.

The proprietors of blood sports and animal fights faced mounting challenges after the Civil War, when a nationwide social movement in defense of animals flowered amid the Reconstruction-era human rights revolution. In April 1866, Henry Bergh, an influential New York shipping heir, successfully lobbied his state legislature to incorporate a new animal protection organization, the American Society for the Prevention of Cruelty to Animals (ASPCA). Significantly, the ASPCA's state charter granted it policing powers to "bring the cruelest to justice." As an instantly recognizable symbol of their powers of arrest, Bergh and fellow ASPCA officers wore police-like uniforms and badges as they patrolled the streets.[25] Weeks later, Bergh and his allies in the New York State Legislature amended the state's antebellum anticruelty law to recognize an animal's right to avoid pain and suffering.[26] In 1867, Bergh's supporters amended the new law to include tougher prosecution of blood sports and animal fights. Nonetheless, the law was often ineffective because blood sport proprietors maintained cozy relationships with law enforcement. Political scientist Claire Jean Kim observes that the *Police Gazette* (1845–1977) routinely published dogfighting rules and news of notable matches, along with its coverage of crime and police news.[27]

Bergh immediately set his sights on Kit Burns. For the next three years, Bergh staged a series of highly publicized raids on Sportsman's Hall, which were met with fierce protest whenever Burns was taken into police custody.[28] Burns was a cagey opponent: his popular fighting pit was concealed in a back basement connected to his saloon by a circuitous hallway and stairs. Guards were stationed throughout and warned Burns and his crew to "Douse the glim!" as Bergh and ASPCA officials approached the pit. This advance warning system meant that little physical evidence of the fights remained after Bergh arrived on the scene, which stymied prosecution. Burns was equally clever in the courtroom: he avoided conviction at one trial by claiming that no animal cruelty laws had been violated because rats, in his words, were "vermin," not animals. He later moved his operations to a Christian revival building as a cover to avoid prosecution. Yet his luck finally ran out. Shortly after his last arrest, Burns abruptly fell ill with pneumonia before his court date and died at the age of thirty-nine on December 19, 1870. As a testament to his fame, Burns' death was widely reported in newspapers around the nation.[29]

---

25 Davis, *The Gospel of Kindness*, 8–9.
26 Pearson, Susan. *The Rights of the Defenseless: Protecting Animals and Children in Gilded Age America*. Chicago, IL: University of Chicago Press, 2011.
27 Kim, *Dangerous Crossings*, 256.
28 See Kaufman and Kaufman, The Sportsmen of New York.
29 See General News. *Buchanan County Bulletin* (Independence, Iowa), December 23, 1870. URL: chroniclingamerica.loc.gov/lccn/sn84027186/1870-12-23/ed-1/seq-2 (November 2, 2019.); City of New York. *Wheeling Daily Intelligencer* (West Virginia), December 21, 1870. URL: chroniclingamerica.loc.gov/lccn/sn84026844/1870-12-21/ed-1/seq-1/ (November 2, 2019). Both from Chronicling America: Historic American Newspapers. Library of Congress.

Henry Bergh's crusade against blood sports and animal fights in New York City was hardly an isolated effort. New state and local SPCAs across the country lobbied to ban cockfighting, dogfighting, bullfighting, and gander pulling. By 1920, forty-one states (out of forty-eight) had passed legislation that criminalized dogfighting and cockfighting. The American Humane Association successfully campaigned for new bans on gander pulling and thwarted local attempts to legalize bullfighting.[30] But animal protectionists faced fierce challenges when they tried to enforce new laws – especially laws concerning elite blood sports. Wealthy fox hunters, pigeon shooters, and coursing enthusiasts could afford expensive legal representation and thus, often escaped prosecution. Animal protection leaders engaged in costly multiyear court battles with affluent pigeon shooters. State legislatures in Rhode Island (1874), Massachusetts (1879), and New York (1902) eventually passed laws banning pigeon shoots, but in Pennsylvania, gun and hunt clubs thwarted such legislation. Similarly, the Women's Pennsylvania SPCA (WPSPCA) fought tirelessly to prosecute the Philadelphia Hunt Club on charges of cruelty to foxes in the 1880s, but their successful conviction was overturned on appeal in 1889.[31]

The Imperial Dimension of Blood Sports

Blood sports became a subject of intense global interest for American enthusiasts and opponents alike when the United States became an overseas empire after the Spanish-American War (1898). Cockfighting, in particular, was ubiquitous in territories under formal American colonial control, as well as in quasi-independent Cuba, where the U.S. periodically redoubled its economic and political influence with military force. Global cockfighting traditions predated colonization, but the expansion of bullfighting was tied to Spanish empire building. Bullfighting required expansive pasturing and elaborate *Plaza de Toros* (stadium-like bullrings), which made it a relatively stationary blood sport, in contrast to the highly portable cockfight.[32]

Across the new empire, American military governments banned cockfighting for ideological and practical reasons. Evoking the rhetoric of the Revolutionary-era First Continental Congress, they argued that cockfighting undermined the expansion of sober American values. Animal protectionists in the United States and in colonial territories joined forces with missionaries, temperance advocates, and local Protestant leaders to condemn animal suffering and to castigate cockfighting as a form of moral depravity. As a practical concern, American officials deemed the cockpit to be a wanton site for dangerous fraternization, where explosive political movements, such as

---

30 Beers, Diane L. *For the Prevention of Cruelty: The History and Legacy of Animal Rights Activism in the United States.* Athens, OH: Ohio University Press, 2006, 76–77.
31 Beers, *For the Prevention of Cruelty*, 77–78.
32 See Mitchell, Timothy. *Blood Sport: A Social History of Spanish Bullfighting.* Philadelphia, PA: University of Pennsylvania Press, 1991.

the Philippine insurrection against American rule (1899–1902), could be funded and fueled. Yet they also made strange bedfellows with Philippine nationalists, who were otherwise their fiercest enemies, such as Emilio Aguinaldo, who rejected cockfighting as a form of vice.[33]

Still, cockfighting continued to flourish among local residents in the American empire – often in collaboration with American soldiers and businessmen, who were equally enthusiastic about animal fights. Consequently, the cockpit was a site of colonial cultural exchange: American soldiers, for example, introduced a fierce line of popular fighting roosters known as the Texas to Philippine cockfighting.[34] Cockfighters deployed nationalistic language to oppose American colonial bans as a form of oppressive imperial overreach. Across the empire, indigenous cockfighters argued that rooster fights were part of their cultural heritage to be vehemently protected, rather than denied. In 1907 and 1908, Cuban cockfighters staged huge rallies across the Island to protest American cockfight bans, and by association, American military control.[35] Eventually American authorities capitulated across the empire – lifting the bans in favor of regulation and taxation, a status quo that remained until 2018, when Senator Susan Collins of Maine successfully inserted an amendment into the Farm Bill, which bans cockfighting in American territories overseas, effective December 2019. Like past prohibitions, this ban has been met with jubilation and fierce condemnations of imperial overreach.[36]

Twentieth Century Legal Battles

Blood sports and animal fights slowly moved out of public view in the United States during the twentieth century. The gradual passage of new state bans reflected a growing public intolerance for such activities. Cockfighting, for one, finally became illegal in all fifty U.S. states in 2008, when Louisiana passed anti-cockfighting legislation. Yet these laws remain uneven: thirteen states levy misdemeanor charges for a first offense, while the remaining thirty-seven impose felony charges.[37] During the twentieth century, dogfighting became a particularly volatile social minefield because the dog's cultural status as "man's best friend" became evermore powerful in a modern

---

[33] Davis, Cockfight Nationalism.
[34] Lansang, Angel J. *Cockfighting in the Philippines (Our Genuine National Sport)*. Baguio City: Catholic School Press, 1966, 54.
[35] Davis, *The Gospel of Kindness*, 134–138.
[36] Davis, Cockfight Nationalism; Davis, *The Gospel of Kindness*, chs. 4 and 6; Mazzei, Patricia. Cockfighting Ban by Congress Breeds Anxiety in Puerto Rico. *New York Times*, November 10, 2019. URL: www.nytimes.com/2019/11/09/us/puerto-rico-cockfighting-ban.html (August 24, 2020).
[37] Griffin, Jonathan. Cockfighting Laws. *National Conference of State Legislatures* 22, no. 1 (January 2014). URL: www.ncsl.org/research/agriculture-and-rural-development/cockfighting-laws.aspx (December 7, 2020).

consumer culture of pet-keeping [→History of Pets].³⁸ The United Kennel Club recognized specific fighting breeds at its inception in 1898, but – as a reflection of broader cultural shifts – the organization stopped endorsing dogfighting in the 1930s. In 1976, dogfighting had become illegal in all fifty states, although Wyoming and Idaho only levied misdemeanor charges. Both state legislatures made dogfighting a felony offense in 2007, only after Michael Vick, the star quarterback for the Atlanta Falcons, was convicted (and subsequently imprisoned) for operating an extensive dogfighting operation in Virginia.³⁹

The Vick case highlights a paradoxical trend. Claire Jean Kim observes that in the face of heightened state criminalization, the internet has spawned robust underground dogfighting communities among working-class men, especially southern whites and urban African Americans, since the 1990s. Dedicated websites sell dogs, equipment, training services, betting opportunities, and thinly veiled "fictional" recaps of actual dogfights.⁴⁰ To this day, blood sports and animal fights remain a pervasive facet of human and animal entanglements. Furthermore, these durable, if disturbing, practices illuminate key methodological, epistemological, and ontological challenges in the historiography of animal studies.

3 Methods and Approaches

Blood sports and animal fights are complex – if gruesome – forms of interspecies interaction and conflict, which require interdisciplinary fields of historical analysis to reveal their significance across time. Scholars of intellectual and political history illuminate the relationship between blood sports and the history of political thought and state formation [→Political History]. Susan Pearson argues that post-bellum animal and child protection societies expanded the responsibilities of the state through an ideological project she calls "sentimental liberalism," which coupled an older, classical liberal language of universal rights and limited government with a sentimental belief that "beasts and babes" had a right to protection because they could feel and suffer. She contends that this fusion of universal rights and the right to pro-

---

38 See Grier, Katherine. *Pets in America: A Cultural History*. Chapel Hill, NC: University of North Carolina Press, 2006. Grier observes that medical innovations have accelerated the practice of modern pet-keeping. Sulfonamides, penicillin, and anti-parasitic flea collars and baths promote canine health, longevity, comfort, and have enabled dogs to live in close proximity to their owners inside the house.
39 Kates, Brian. Quarterback Michael Vick Released from Prison After Serving 18 Months for Dogfighting Ring. *New York Daily News*, May 20, 2009. URL: www.nydailynews.com/sports/football/quarterback-michael-vick-released-prison-serving-18-months-dogfighting-ring-article-1.395626 (November 16, 2019).
40 Kim, *Dangerous Crossings*, 257.

tection helped propel a growing public expectation of the state's obligation to prevent suffering.⁴¹

The fields of foreign relations history and transnational American Studies demonstrate the significance of blood sports and animal fights to intersectional transnational histories of nation building, military strategy, and moral reform. While the field of diplomatic history traditionally analyzed formal mediation between nation states, a growing body of scholarship after World War II, most notably the Wisconsin School, expanded the field to include nonstate actors, social movements at home and abroad, business, culture, and the social history of everyday life [→Diplomatic History].⁴² Emily Rosenberg's pathbreaking study, *Spreading the American Dream: American Economic and Cultural Expansion, 1890–1945* (1982), expanded these insights to include popular entertainments, such as Wild West Shows, as ideological formations of empire building and statecraft.⁴³ In his examination of Cuban resistance to Spanish colonialism in the late nineteenth century, Louis A. Pérez analyzes baseball and blood sports as catalysts for anticolonialism: he observes that Cuban students who had studied abroad in the United States brought baseball to Cuba in the 1860s, where it flourished as a popular pastime and as a critique of Spanish rule. Nationalist leaders, such as José Martí, demanded that Spanish bullrings – "a futile bloody spectacle" – should be replaced by baseball bullpens as Cubans fought for self-determination.⁴⁴

Blood sports and animal fights are complex transnational entertainments that exist in multiple cultural registers – live performance, fiction, film, painting, song, and poetry, among other locations. Moreover, they are intersectional and geographically situated. Consequently, they require interdisciplinary historical analysis. In the 1990s, the interdisciplinary field of American Studies introduced new methodological interventions in the historiography of empire building, which are useful to the study of blood sports and animal fights [→American Studies]. Building upon the insights of the "cultural turn" in the history of American foreign relations, transnational American Studies draws upon intersectional considerations of race, gender, sexuality, and class, as well as close textual readings using the literary analysis of the New Historicism, ethnic studies, and borderlands history to explore the history of American empire building [→Feminist Intersectionality Studies]. Amy Kaplan, for example, locates Perry Miller's most famous work, *Errand into the Wilderness*

---

41 Pearson, *The Rights of the Defenseless*, 16.
42 See, for example, Williams, William Appleman. *The Tragedy of American Diplomacy*. New York, NY: Norton, 1959; LaFeber, Walter. *The New Empire: An Interpretation of American Expansion, 1860–1898*. Ithaca, NY: Cornell University Press, 1998 [first published in 1963]; McCormick, Thomas. *China Market: America's Quest for Informal Empire*. Chicago, IL: Quadrangle Books, 1967.
43 Rosenberg, Emily. *Spreading the American Dream: American Economic and Cultural Expansion, 1890–1945*. New York, NY: Hill and Wang, 1982.
44 Pérez, Louis A. *On Becoming Cuban: Identity, Nationality, & Culture*. Chapel Hill, NC: University of North Carolina Press, 2001, 75–83.

(1956), on the banks of the Congo, where Perry unloaded oil drums as a college dropout. Kaplan contends that Perry's imaginary of the Congolese jungle shaped his conception of the New England wilderness as an incubator for Puritan thought. Tellingly, Miller's jungle imaginary was devoid of human inhabitants and human agency; moreover, his conception of American exceptionalism ignored the looming presence of enslavement and the slave trade in virtually every aspect of the American experience.[45]

This welter of interdisciplinary approaches to the study of empire – formal diplomacy, ideology, culture, everyday social life, racial formations, gender, sexuality, class, and economic life – offer fresh insights for the historical study of animals. For example, Aaron Skabelund demonstrates that the history of Japanese nationalism and ideologies of racial purity can be traced through the bodies of Japanese dogs. When Japanese leaders engaged in rapid technical and industrial modernization during the Meiji Restoration, they collaborated with American and European business leaders, who used racist civilizational discourses to characterize indigenous Japanese dogs as "savage." Yet with the rise of nationalism and military expansionism during the early twentieth century, Japanese leaders began to prize native dog breeds as corporeal embodiments of the nation's strength, loyalty, beauty, and racial "purity."[46] Harriet Ritvo's pioneering text, *The Animal Estate* (1987), likewise demonstrates how a range of animal entanglements – from urban rabies control in London to colonial hunts in India and Rhodesia, among other colonies – were manifestations of social and political control rooted in class conflict and civilizational ideologies of racial supremacy.[47] Moira Ferguson similarly explores women's animal activism in Great Britain from 1780 to 1900 in relation to moral reform and imperial ideologies of uplift and the "white man's burden."[48] Collectively, this scholarship helps illuminate the ways in which blood sports in the American empire became a flashpoint for ideologies of cultural nationalism, claims of sovereignty, moral progress, and racial "fitness" for future citizenship.

The most productive methods and approaches to histories of blood sports and animal fights are intersectional – following Kimberlé Crenshaw's lead in making race, gender, and class symbiotic, mutually constitutive categories of analysis.[49] Moreover, these works engage in symbiotic interspecies analysis. Claire Jean Kim

---

[45] Kaplan, Amy. Left Alone with America: The Absence of Empire in the Study of American Culture. In *Cultures of United States Imperialism*, Amy Kaplan, Donald E. Pease (eds.), 3–21. Durham, NC: Duke University Press, 1993.

[46] Skabelund, Aaron. *Empire of Dogs: Canines, Japan, and the Making of the Modern Imperial World*. Ithaca, NY: Cornell University Press, 2011.

[47] Ritvo, *The Animal Estate*, 243–288.

[48] Ferguson, Moira. *Animal Advocacy and Englishwomen, 1780–1900: Patriots, Nation, and Empire*. Ann Arbor, MI: University of Michigan Press, 1998.

[49] Crenshaw, Kimberlé. Mapping the Margins: Intersectionality, Identity Politics, and Violence against Women of Color. *Stanford Law Review* 43, no. 6 (July 1991): 1241–1299.

forcefully shows how Michael Vick's overlapping identities as a working-class African American man engaged in two deeply violent, masculinist enterprises – football and dogfighting – shaped public responses to his arrest on multiple charges of animal cruelty. Simultaneously, she demonstrates how "pit bull" terrier breeds have been similarly racialized and demonized through breed ban legislation and targeted euthanasia at animal shelters. Kim observes that animal advocates and other critics used an "optic of cruelty" to underscore Vick's guilt, while Vick's defenders deployed an "optic of racism" to frame his imprisonment as evidence of a racist white society all too eager to incarcerate black men.[50] She contends that all too often, these optics argue past each other: the former engaging in racial innocence to deny racism in the Vick case, and the latter deemphasizing canine trauma to amplify the primacy of human suffering.[51] Fundamentally, Kim calls for an "ethics of avowal" to recognize the connections between institutionalized violence, racism, structural inequality, and cruelty to animals.[52]

4 Implication(s) of the Animal Turn

In November 2016, Dan Vandersommers published *The 'Animal Turn' in History* in *Perspectives on History*, the newsmagazine of the American Historical Association. Vandersommers observes that this "turn" has occurred gradually: "Over the past thirty years, nonhuman animals have crept slowly, but persistently, from the margins of history to its center."[53] Vandersommers contends that seismic changes, such as the earth's intensifying climate crisis, have prompted many historians to decenter humanity as the exclusive agent of change over time in favor of symbiotic interspecies and environmental analyses. Put another way, many historians have begun to interrogate bedrock anthropomorphic epistemologies – thereby treating human agency more relationally in the web of life.

And yet Erica Fudge cautions that historians can unwittingly exclude the animal from the *animal turn* when exploring human attitudes toward animals, or by examining animals only in relation to human beings [→Post-Domestication: The Posthuman]. Both approaches are fundamentally anthropocentric. Fudge encourages historians to decenter the human past through "holistic" historical analysis that treats animals not just representationally, but as historical agents.[54] Virginia DeJohn Ander-

---

50 Kim, *Dangerous Crossings*, 266–271.
51 Kim, *Dangerous Crossings*, 276–277.
52 Kim, *Dangerous Crossings*, 278–279.
53 Vandersommers, Dan. The Animal Turn in History. *Perspectives on History*, November 3, 2016. URL: www.historians.org/publications-and-directories/perspectives-on-history/november-2016/the-animal-turn-in-history (December 7, 2020).
54 Fudge, Erica. A Left-Handed Blow: Writing the History of Animals. In *Representing Animals*, Nigel Rothfels (ed.), 3–18. Bloomington, IN: Indiana University Press, 2000.

son, for one, forcefully examines how invasive cattle and pigs were vanguards of colonial expansion in North America. Because colonists allowed livestock to wander, these animals frequently rooted, trampled, and ate Native American crops, which invariably triggered political conflict.[55] Other writers, including Robert Sullivan, Nate Blakeslee, and Marc Bekoff, have likewise preserved the centrality of the animal by making ethology a methodological cornerstone of their studies of rats, wolves, and dogs, respectively.[56]

Of course, fighting bulls, cocks, and dogs do not generate their own historical paper trail of memoirs and other traditional documents. Their histories variously are written by breeders, promoters, novelists, and animal protectionists. Despite this filter of human representation, the physical presence of these animals remains: their bodies, behavior, affect, and sensory worlds are the beating heart of partisan blood sport writing. Cockers, bullfighters, and dogmen argue that animal combatants "love" to fight; they are simply doing what comes naturally.[57] Cocking enthusiast Tim Pridgen claims, "A cock asks for no better death than with steels [gaffs] on."[58] Such claims of animal agency, however, ignore the active hand of human dominion in breeding, socializing, and training animals to fight.[59] Moreover, anthropocentric interpretations of animal affect can be wildly misleading. For example, Gregg Mitman notes that the "smile" of a dolphin at cetacean shows masks serious stress – captive "smiling" dolphins engage in aggressive sexual behavior with other animals as a way to reduce anxiety.[60]

Beyond the ontological difficulties of interpreting animal affect and agency, historians face other challenges in writing blood sport histories. As this chapter has suggested, the history of animal sports is inexorably tied to the history of their prohibition. Therefore, some writers might assume that this history is one of progressive human betterment because virtually all blood sports in the United States (and other countries) are now banned. Former CEO of the Humane Society of the United States Wayne Pacelle includes blood sport bans in his analysis of the twenty-first century

---

[55] Anderson, Virginia DeJohn. *Creatures of Empire: How Domestic Animals Transformed Early America*. New York, NY: Oxford University Press, 2006.

[56] See Sullivan, Robert. *Rats: Observations on the History and Habitat of the City's Most Unwanted Inhabitants*. New York, NY: Bloomsbury, 2004; Blakeslee, Nate. *American Wolf: A True Story of Survival and Obsession in the West*. New York, NY: Broadway Books, 2017; Bekoff, Marc. *Canine Confidential: Why Dogs Do What They Do*. Chicago, IL: University of Chicago Press, 2018.

[57] See Pridgen, Tim. *Courage: The Story of Modern Cockfighting*. Boston, MA: Little, Brown and Company, 1938; Lansang, *Cockfighting in the Philippines*.

[58] Pridgen, *Courage*, 174.

[59] Kim, *Dangerous Crossings*, 260; Haraway, Donna. *When Species Meet*. Minneapolis, MN: University of Minnesota Press, 2007.

[60] Mitman, Gregg. *Reel Nature: America's Romance with Wildlife on Film*. Cambridge, MA: Harvard University Press, 1999, 169–171.

"humane economy."[61] Yet scholars should avoid Whig history when interpreting the temporal arc of these violent cultural practices. Blood sports render animal suffering concretely intelligible in ways that can inadvertently pull scholars away from analyzing distant, structurally embedded forms of animal cruelty. Philosopher Gary Francione pointedly claims that "We Are All Michael Vick" to implicate a society that condemns blood sports, but supports a factory farming system that kills 25 million animals per day in the United States alone.[62] Indeed, the same 2018 Farm Bill that has abolished cockfighting in Puerto Rico, Guam, and other U.S. overseas territories also generously subsidizes industrial agriculture, which sustains systemic confinement, debeaking, forced molting, tail docking, beatings, and other forms of unmitigated suffering for billions of domestic animals every year. Anthrozoologist Hal Herzog provocatively asks, "What would you rather be, a gamecock or a broiler?" He flatly rejects cockfighting, but notes that gamecocks are allowed to forage outside for insects; they exercise every day; and they bond with their trainers in a life typically lasting two years. By contrast, a broiler hen spends her brief forty-two days lifespan crammed with thousands of other birds inside a dark, stifling "grow-out house" caked with ammonium feces; they are tended by poorly paid human workers whose health suffers in this toxic environment.[63] Herzog's troubling question begs for an ethically satisfying answer. Nonetheless, his call for philosophical contemplation adds an important method to the historian's toolkit in gaining a richer understanding of blood sports in relation to other human and animal entanglements in the Anthropocene.

Selected Bibliography

Beers, Diane L. *For the Prevention of Cruelty: The History and Legacy of Animal Rights Activism in the United States*. Athens, OH: Ohio University Press, 2006.
Davis, Janet M. *The Gospel of Kindness: Animal Welfare and the Making of Modern America*. New York, NY: Oxford University Press, 2016.
Herzog, Hal. *Some We Love, Some We Hate, Some We Eat: Why It's So Hard to Think Straight about Animals*. New York, NY: Harper, 2010.
Kim, Claire Jean. *Dangerous Crossings: Race, Species, and Nature in a Multicultural Age*. New York, NY: Cambridge University Press, 2015.
Lansang, Angel J. *Cockfighting in the Philippines (Our Genuine National Sport)*. Baguio City: Catholic School Press, 1966.
Mitchell, Timothy. *Blood Sport: A Social History of Spanish Bullfighting*. Philadelphia, PA: University of Pennsylvania Press, 1991.

---

[61] Pacelle, Wayne. *The Humane Economy: How Innovators and Enlightened Consumers are Transforming the Lives of Animals*. New York, NY: William Morrow, 2016.
[62] Facts – Farm Animals. Animal Matters. URL: www.animalmatters.org/facts/farm/ (December 7, 2020); Kim, *Dangerous Crossings*, 262.
[63] Herzog, Hal. *Some We Love, Some We Hate, Some We Eat: Why It's So Hard to Think Straight About Animals*. New York, NY: Harper Collins, 2010, 155–170.

Pearson, Susan J. *The Rights of the Defenseless: Protecting Animals and Children in Gilded Age America*. Chicago, IL: University of Chicago Press, 2011.

Pridgen, Tim. *Courage: The Story of Modern Cockfighting*. Boston, MA: Little, Brown and Company, 1938.

Ritvo, Harriet. *The English and Other Creatures in the Victorian Age*. Cambridge, MA: Harvard University Press, 1987.

Rothfels, Nigel (ed.). *Representing Animals*. Bloomington, IN: Indiana University Press, 2002.

Andrew Wells
History of Animal Collections/Animal Taxonomy

1 Introduction and Overview

Collections of animals range from individual pets and isolated fossils to the vast repositories of natural history museums and the living inhabitants of menageries, nature reserves, national parks, and zoos around the world. These examples immediately suggest a number of questions. Most fundamental is what counts as an animal: do the taxidermically preserved remains of an extinct creature, for example? What about a fossil, the geological shadow of an animal that decomposed millennia ago? Upon the answers to such questions depends our understanding of an animal collection, which is also related to its size (is a single pet a "collection"?), status, location, ownership, accessibility, and so on. The principles by which collections were organized and their objects classified also had repercussions far beyond the individual cabinet, museum, menagerie, or zoo. Ambitious Renaissance collectors seeking to assemble a (limited) microcosm of the natural world deployed taxonomic categories that either influenced those used to describe the wider world or were directly copied from them. Paradoxically, this continued even when collections became more specialized from the late seventeenth century: cabinets and museums limited to fish, insects or shells might have little to say on taxonomic concepts beyond their immediate purview but helped to elaborate and more closely define those that pertained to their objects.

The following discussion addresses these issues by concentrating on the four historically most significant forms of animal collection – cabinets of curiosities, (natural history) museums, menageries, and zoos – and on the history of taxonomy since the early modern era. With a geographical focus on Western Europe, the chapter will outline the key themes in their history before dealing with the principal methodological approaches and challenges faced by scholars interested in these topics. A final section looks at the role of the *animal turn* in this scholarship, which as elsewhere has been profound. The discussion highlights a number of the key insights to have emerged from recent scholarship, from foregrounding the agency of nonhuman animals (and its limitations), to demonstrating the co-existence of contradictory attitudes towards them, and underscoring the intersection of sensory, emotional, and other histories with those of taxonomy and animal collections.

2 Topics and Themes

Animal Collections

Neatly distinguishing between collections of living and dead animals is easier said than done. Both shared common purposes, proprietors, suppliers, and underwent a similar change as they shifted over the course of the long eighteenth century (c.1660 – c.1840) from a baroque set of priorities and institutions (cabinets of curiosities, menageries) to "modern" counterparts (museums, zoos). This transition was no linear march of progress, nor did it occur at the same pace for both sorts of collection – museums emerged in the mid-eighteenth century, zoos outside Paris only after 1825 – and their common ground is most evident from the range of purposes they served.

The oldest of these was to display wealth and power. The expense of purchasing, transporting, housing, and maintaining living or dead specimens made animal collections an ideal showcase for the riches and influence of the collector. The ability to procure rare animals signified broad, deep, and global economic and diplomatic ties. Rare and exotic animals were prized as diplomatic gifts that were more distinctive than the usual presents of hunting animals (especially horses and falcons) [→Diplomatic History].[1] Other sources of animals and animal objects included mariners (whose opportunistic purchases stocked pet shops in port towns), private individuals, government officials, and the owners and operators of private menageries and zoos such as Edward Cross or Carl Hagenbeck, both of whom described themselves as purveyors of wild animals.[2]

Just as important as the number and rarity of objects was the manner in which they were displayed. The *Kunstkammer* of Duke Albrecht V of Bavaria (1528 – 79), for example, was housed in large, bright, specially decorated galleries that contained a range of purpose-built cabinets for displaying artefacts. Visitors walked through the rooms as on a "triumphal progress", in which the wealth, self-confidence, and stability of Bavaria was conveyed via the rich collection. Cabinets owned by private citizens also emphasized their prosperity and social standing, often by aping noble fashions in the display of objects, such as suspending exotic animals (e. g. crocodiles) from the ceiling.[3]

---

[1] Heal, Felicity. *The Power of Gifts: Gift-Exchange in Early Modern England.* Oxford: Oxford University Press, 2014, 154 – 160.

[2] Robbins, Louise E. *Elephant Slaves and Pampered Parrots: Exotic Animals in Eighteenth-Century Paris.* Baltimore, MD: Johns Hopkins University Press, 2002, chs. 1 – 2; Hoage, R. J., Anne Roskell and Jane Mansour. Menageries and Zoos to 1900. In *New Worlds, New Animals: From Menagerie to Zoological Park in the Nineteenth Century*, Robert J. Hoage, William A. Deiss (eds.), 8 – 17. Baltimore, MD: Johns Hopkins University Press, 1996.

[3] MacGregor, Arthur. *Curiosity and Enlightenment: Collectors and Collections from the Sixteenth to the Nineteenth Century.* New Haven, CT: Yale University Press, 2007, chs. 1 – 2.

On an altogether grander scale was the Versailles menagerie, built on the grounds of Louis XIV's absolutist showcase in 1664. This came to feature a panopticon-like central building whose balconies looked out onto an octagonal courtyard divided into seven enclosures and a high-walled path leading to the building; once inside, the natural world was suddenly revealed beneath the spectator, evoking a fantasy of mastery and omniscience.[4] Notwithstanding the menagerie's popularity, it fell into decline under Louis XV (1715–74), only to be briefly revived before final closure in 1793.[5]

Its animals were then transported to the new, national menagerie at the Jardin des Plantes in Paris. Widely described as the first modern zoo, two of its characteristics were widely copied in the wave of zoo foundations after 1825: the good of the nation was its central purpose, and animals were shown in landscaped grounds, in which human intervention was designed to remain hidden. Unlike pre-modern menageries, which reflected the power and wealth of individual patrons, zoological gardens – all founded in the era of the nation state – broadcast national and imperial prowess that derived from the size, diversity, and scientific credentials of their animal collections [→History of the Zoo; →Political History].[6]

Closely related was the second principal reason for collecting and exhibiting animals: entertainment. Only the richest could afford to spend vast sums on mere leisure, and the reputation of cabinets assembled for the private amusement of their collectors, such as Francesco I de' Medici (1541–87) in Florence, soared because of their exclusivity. But entertainment was the central motive for menageries that lacked institutional support and were aimed at the paying public. The largest such menagerie in Britain opened in 1788 on the first floor of a building at Exeter 'Change on the Strand, and was extremely popular, boasting a variety of exotic species, including an elephant.[7]

Mobile animal collections reached even more people, varying enormously in size from a solitary beast or handful of creatures exhibited in taverns or coffee houses, to the vast travelling menageries of entrepreneurs such as George Wombwell. These travelled the length and breadth of Britain, often setting up at major fairs, especially London's Bartholomew Fair. Showmen competing for public attention used an arsen-

---

4 On the Versailles menagerie and Bentham's Panopticon see, above all, Foucault, Michel. *Discipline and Punish: The Birth of the Prison.* Translated by Alan Sheridan. New York, NY: Vintage, 1995, 203.
5 Robbins, *Elephant Slaves*, ch. 2; Senior, Matthew. The Ménagerie and the Labyrinthe: Animals at Versailles, 1662–1792. In *Renaissance Beasts: Of Animals, Humans, and Other Wonderful Creatures,* Erica Fudge (ed.), 208–232. Urbana, IL: University of Illinois Press, 2004; Baratay, Éric and Elisabeth Harouin-Fugier. *Zoo: A History of Zoological Gardens in the West.* London: Reaktion Books, 2004, 48–52.
6 Ritvo, Harriet. *The Animal Estate: The English and Other Creatures in the Victorian Age.* London: Penguin, 1990, ch. 5; Donald, Diana. *Picturing Animals in Britain, c.1750–1850.* New Haven, CT: Yale University Press, 2007, ch. 5; Baratay and Hardouin-Fugier, *Zoo,* ch. 5.
7 Grigson, Caroline. *Menagerie: The History of Exotic Animals in England, 1100–1837.* Oxford: Oxford University Press, 2016, 98–99.

al of tactics to encourage visitors to spend money on what were often the most expensive fairground attractions. Advertisements and handbills peppered with superlatives were widely distributed, and one late-Victorian commentator recalled the vibrant appeal to the fairgoer's senses, particularly

> the gorgeously-uniformed bandsmen, whose brazen instruments brayed and blared from noon till night on the exterior platform, and the immense pictures, suspended from lofty poles, of elephants and giraffes, lions and tigers, zebras, boa constrictors, and whatever else was most wonderful in the brute creation, or most susceptible of brilliant coloring.[8]

As these comments indicate, the supplementary use of visual art was an important aspect of animal exhibitions, serving to attract visitors, to enhance displays, or to provide a souvenir or ersatz experience of visiting an exhibition.[9]

The final major purpose behind animal collections was "improvement", which comprised education; utilitarian efforts at acclimatization, breeding, and domestication; and science, especially natural history and zoology. Educating the public was not a monopoly of modern zoos. For all the razzmatazz of the travelling menagerie, most paid lip service to educational goals in their publicity and several produced souvenirs that doubled as informative literature on their exhibits. Yet education was a priority for modern zoological gardens in a way it never was for menageries. The Zoological Society of London (ZSL) ran a library and museum alongside its menagerie in Regent's Park, and its publicity material was saturated with educational themes. This hinted at the tension between educational goals and social exclusivity (London Zoo was open for all fee-paying visitors only from 1846) by being unapologetically bourgeois in orientation.[10]

In support of utilitarian goals, the ZSL opened a stud farm in 1833 at Kingston Hill in suburban London, but low demand and the ascendancy of the scientific wing of the Society forced its closure within a few years. Similarly, domestication and acclimatization were abandoned as goals by most zoos after 1870 because of the difficulty of measuring, let alone achieving these aims.[11]

---

[8] Frost, Thomas. *Old Showmen and the Old London Fairs*. London: Tinsley Brothers, 1875, 259.
[9] Altick, Richard D. *The Shows of London*. Cambridge, MA: Harvard University Press, 1978, ch. 3; Cowie, Helen. *Exhibiting Animals in Nineteenth-Century Britain: Empathy, Education, Entertainment*. Basingstoke: Palgrave, 2014; Sahlins, Peter. The Royal Menageries of Louis XIV and the Civilizing Process Revisited. *French Historical Studies* 35, no. 2 (April 2012): 237–267; Senior, Ménagerie and the Labyrinthe.
[10] Ritvo, *Animal Estate*, 213–214.
[11] Brantz, Dorothee. The Domestication of Empire: Human-Animal Relations at the Intersection of Civilization, Evolution and Acclimatization in the Nineteenth Century. In *A Cultural History of Animals in the Age of Empire*, vol. 5, Kathleen Kete (ed.), 73–93. Oxford: Berg, 2010, 90–92; Ritvo, Harriet. The Order of Nature: Constructing the Collections of Victorian Zoos. In *New Worlds, New Animals*, Robert J. Hoage, William A. Deiss (eds.), 43–50. Baltimore, MD: Johns Hopkins University Press, 1996; Baratay and Hardouin-Fugier, *Zoo*, 141–146.

This is not to say that there were not individual successes in breeding, even in the unpropitious setting of a commercial menagerie: lion cubs were born at Exeter 'Change in 1817, a feat Cross was not slow to exploit. But "domestication", when it occurred, was often more metaphoric than literal. For example, celebrity animals often owed their fame to being "domesticated" as British. The immense popularity of Jumbo, the elephant sold by the ZSL to P. T. Barnum's circus in 1882 [→History of Circus Animals], was stoked by the flurry of patriotic outrage provoked by his sale.[12] Jumbo's Britishness and his celebrity were mutually reinforcing in this sort of "domestication", which also suggests how zoos – and especially travelling menageries, which had their heyday during the high-water mark of British imperialism (1870–1900) and reached a wider public than zoos – helped to "domesticate" and "bring home" empire [→(Post)Colonial History].[13]

Britain's global power was also projected by its scientific prowess, for which cabinets and museums were the principal vehicle, despite two key limitations. First, they remained dominated by *artificialia*, "whatsoever the hand of man by exquisite art or engine hath made rare in stuff, form, or motion"; *naturalia*, while always present, only became broadly popular from the mid-eighteenth century.[14] And second, early modern collections were overwhelmingly focused on rarities.

By the late seventeenth century, there were exceptions to this exceptionalism. The Royal Society's Repository in the early 1680s contained "not only Things strange and rare, but the most known and common amongst us."[15] But the Repository was unusual, underfunded, and attacked for its miscellaneous character and intellectual superficiality.[16] The same criticisms dogged Sir Hans Sloane, whose collection formed the kernel of the British Museum after his death in 1753. The creation of a national museum open to its citizens as a matter of principle was an important new development that helped to proclaim and develop a sense of national identity.[17] This was imperfectly realized in the British Museum before 1840, where natural history collections were so disorganized that visitors drew unfavorable comparisons with foreign institutions, leading the ZSL to open its own short-lived museum. After 1837, the Natural History Department of the British Museum was reorganized into separate departments of botany, zoology, and geology, resulting in major improvements and expanded collections. Specialization proceeded apace and a cam-

---

12 Ritvo, *Animal Estate*, 232; Baratay and Hardouin-Fugier, *Zoo*, 125; Kotar, S. L. and J. E. Gessler. *The Rise of the American Circus, 1716–1899*. Jefferson, NC: McFarland & Co., 2011, 245.
13 Brantz, Domestication of Empire, 89–90; Ritvo, Order of Nature; Hall, Catherine and Sonya O. Rose (eds.). *At Home with the Empire: Metropolitan Culture and the Imperial World*. Cambridge: Cambridge University Press, 2006.
14 Bacon, Francis. A Device for the Gray's Inn Revels. In *The Major Works*, Brian Vickers (ed.), 55. Oxford: Oxford University Press, 2008.
15 Grew, Nehemiah. *Musæum Regalis Societatis*. London: 1685, sig. [A4]ᵛ.
16 Arnold, Ken. *Cabinets for the Curious: Looking Back at Early English Museums*. Aldershot: Ashgate, 2006, 217.
17 MacGregor, *Curiosity and Enlightenment*, 237.

paign to establish a dedicated Natural History Museum was ultimately successful in 1880, when the British Museum's collections were calved off to be placed in a new institution based in South Kensington.[18]

Notwithstanding their popularity and range of uses, animal collections were a constant target of criticism. They were expensive: as Diderot complained in the *Encyclopédie* (1765), menageries "must be destroyed when the people lack bread; it would be disgraceful to feed beasts at great expense while people all around are dying of starvation."[19] Ethical objections to methods of collection were also common, particularly in the nineteenth century when the hunting cult was in full swing and the acquisition of trophies and (living) animals involved killing on an appalling scale [→History of Hunting].[20] Complaints were levelled at menageries as a foreign intrusion, and class-based objections were frequently heard, although grumbling about the admittance of the lower orders to zoos and museums died down after the mid-nineteenth century as these institutions became loci of patriotic sentiment.

The questionable benefits of collections were a regular topic for discussion, especially if they fueled "idle" curiosity or popular credulity. Hoaxes often involved mythical or artificial creatures that were fashioned from parts of individuals from different species. Financially motivated, they sought to capitalize on the enduring fascination with both fabulous animals and exotic new lands, and they persisted well into the nineteenth century. Fakes were often taken seriously by naturalists, such as the mermaid exhibited in 1822 that consisted of the remains of "an orangutan, a baboon and a salmon."[21] Despite being unmasked as a fake, it remained a popular attraction, which shows that counterfeits are not necessarily evidence of a credulous audience, but could be "designed to intrigue and vex rather than to deceive" the spectator.[22] But suspicion and folklore could endanger genuine zoology, as when the platypus was thought a fake or when the narwhal's tusk was taken to be a unicorn horn.[23]

---

18 Arnold, *Cabinets*, 211–234; Delbourgo, James. *Collecting the World: The Life and Curiosity of Hans Sloane*. London: Penguin, 2017, 290–292, 303–342; MacGregor, *Curiosity and Enlightenment*, 139–149, 258–268.
19 *Encyclopédie, ou Dictionnaire raisonné des sciences, des arts et des métiers*. Neufchâtel: 1765, s.v. "ménagerie."
20 Baratay and Hardouin-Fugier, *Zoo*, 117–122; MacKenzie, John M. *The Empire of Nature: Hunting, Conservation and British Imperialism*. Manchester: Manchester University Press, 1988.
21 Ritvo, Harriet. *The Platypus and the Mermaid and Other Figments of the Classifying Imagination*. Cambridge, MA: Harvard University Press, 1997, 179.
22 Ritvo, *Platypus*, 178–182; MacGregor, *Curiosity and Enlightenment*, 47.
23 Benedict, Barbara M. *Curiosity: A Cultural History of Early Modern Inquiry*. London: University of Chicago Press, 2001, ch. 3; Ritvo, *Platypus*, 3–15.

Taxonomy

The Bible itself records that taxonomy was the very first human activity. By bestowing on "every living creature" names that "were very proper, and significant of their Natures", Adam evoked a number of themes that recur throughout the history of taxonomy.[24] The first of these is the importance of terminology. The sheer diversity of names for animals and plants, even within a single linguistic domain, was prohibitive: different terms might exist for the same creature and the same name might apply to a number of animals or plants. This situation was not helped by the influx of natural specimens from the New World, and the number of recognized species rapidly grew to make this terminological confusion unsustainable. The number of known quadrupeds doubled in the forty years from the mid-1690s and by the late nineteenth century more than 1,000 new genera of animals *per year* were being described.[25]

The most important advance in the history of taxonomy before 1800 was therefore the Linnaean system of binomial nomenclature, introduced in the 1730s and dominant by the end of the century. Linnaeus' system was not the first to use (binomial) Latin terms, nor was his taxonomic method anywhere near as popular as its form. He was frequently attacked on personal, chauvinistic, and intellectual grounds, against which his tendency to regularly amend his system and break his own rules were a poor defense. Nevertheless, his achievement in developing a regular, ordered form of botanical and zoological nomenclature that won widespread acceptance was a major step forward.[26]

Agreeing on an overall system of nomenclature was, however, quite different from agreeing on the names of individual animals. Naturalists competed for priority in the naming of species and genera, increasingly along national lines as the nineteenth century progressed. Some created the space for new names by splitting species into sub-species, while others lumped these categories together to obliterate their competitors' terms. In addition, categories like "species" and "genus" themselves came under sustained pressure as others – especially but not exclusively "race" – were added to the nomenclatural mix, and the meaning of these categories was debated. The ancient definition of biological species as a reproductive community of interfertile creatures was challenged by the apparent existence of transspecific hybrids capable of reproduction. Consequently, some naturalists, such as Johann

---

[24] Thomas, Keith. *Man and the Natural World: Changing Attitudes in England, 1500–1800*. London: Penguin, 1984, 71; Genesis 2:19–20, English Standard Version; Stackhouse, Thomas. *A New History of the Holy Bible*. London: 1742–1744, 1:10.
[25] Ritvo, *Platypus*, ch. 1; Sloan, Phillip R. The Gaze of Natural History. In *Inventing Human Science: Eighteenth-Century Domains*, Christopher Fox, Roy Porter, Robert Wokler (eds.), 112–151. Berkeley, CA: University of California Press, 1995.
[26] Ritvo, *Platypus*, 15–26, 51–59; Koerner, Lisbet. *Linnaeus: Nature and Nation*. Cambridge, MA: Harvard University Press, 1999, chs. 1–2.

Friedrich Blumenbach (1752–1840), moved away from reproductive community to morphological affinity as the key criterion of species membership.[27]

Additional challenges to terminological clarity came from differences over the content of individual categories and, indeed, over the perception of nature itself. One of the longest lasting and most influential metaphors used to explain the (super)natural world was the "great chain of being" [→History of Ideas], in which everything from angels to dirt was linked hierarchically. This image was popular and authoritative, but it depended largely on a static idea of creation in which no new species were created, and it was possible to assert some alarming linkages depending on the key characteristic used to differentiate the individual elements. One critic, as Harriet Ritvo has shown, demonstrated that by concentrating on nails and external covering (scales and shells), it was possible to move from a human to a tortoise in five steps.[28] As a result, an arboreal metaphor became more popular, particularly as Darwin's theories highlighted reproductive success and genealogical descent, although the tree of nature never entirely replaced the chain of being.[29]

The taxonomic principles used in classifying animals was the second key theme raised by Adam's naming of the animals. Scholars have tended to divide such principles into "folk" and "scientific" taxonomies, which were respectively characterized by morally and culturally subjective criteria often involving vernacular labels, and less anthropocentric, "objective" concepts, usually expressed in Latinate terms. This distinction is, however, less neat than it appears. For one thing, "folk" taxonomic categories, such as "vermin" or the (at times paramount) "wild"/"tame" distinction, persisted into the high summer of nineteenth-century zoology. "Useful" and "useless" were also widely used, even for undomesticated animals such as bees. On the other hand, latinate terminology was mobilized by non-scientists – such as breeders of livestock, dogs, or racehorses – because of its scientific prestige [→History of Pets;→History of Agriculture]. The religious and cultural contexts of taxonomy also placed tight constraints on the "objectivity" and non-anthropocentrism of "scientific" taxonomy.[30]

As with nomenclature, it was easier to agree on the abstract need for a classification system than on any particular candidate. Some focused on the blood (warm-blooded or cold-blooded) as the decisive characteristic by which to differentiate animals; others concentrated on the skeleton (vertebrate or invertebrate) or reproductive mode (viviparous or oviparous). There were more radical possibilities, such as the "quinary" system that enjoyed a brief window of popularity in the 1830s. Here, animals were arranged in a set of embedded circles, each of which consisted of five subsidiary circles, which were each subdivided into five smaller ones (and so on), with the aim of identifying continuities through the animal kingdom that

---

27 Sloan, Philip R. The Buffon-Linnaeus Controversy. *Isis* 67, no. 238 (October 1976): 356–375.
28 Ritvo, *Platypus*, 29.
29 Lovejoy, Arthur O. *The Great Chain of Being*. Cambridge, MA: Harvard University Press, 1936.
30 Thomas, *Man and the Natural World*, 51–70; Ritvo, *Platypus*, 38–46, 68–70.

were not reflected in linear models of a tree or chain. Still more unorthodox were the synecdochic systems of the likes of John Hunter (1728–1793), which sought to base taxonomies not on the entire animal but only constituent organs or tissues. Given all of these possibilities, Ritvo is surely right to describe "the internal history of zoological classification [as ...] as much a constantly shifting kaleidoscope of competing systems and principles as a steady evolution and elaboration of a dominant paradigm."[31]

Far more unified were the cultural and religious contexts of taxonomizing (the third theme highlighted by Adam's onomastics) in early modern Europe. Natural theology, the idea that God's wisdom, benevolence, and existence were all manifested in His creation, was the dominant paradigm. The ideal fit between animals and their food and habitats, and by mysterious creatures or processes such as reproduction, all displayed God's perfection and mystery. As the creation of a loving and benevolent God, nature was a source of moral guidance; aspects of the natural world that could not be readily aligned with Christian morality were explained away. Wildness resulted from the Fall and the consequent estrangement of man from wild beasts, and the discomfiture arising from obvious and widespread destruction and predation in the natural world was soothed by the pious recognition that we do not understand God's benign purposes.[32]

Alternatives to natural theology took these difficulties more seriously. Orthodox natural theology taught that humanity was but one part of creation, ultimately expressed in Linnaeus' taxonomy that classed humans as mammals alongside apes. Defenders of human dignity, above all Linnaeus' nemesis Buffon, consequently adopted an unapologetically anthropocentric perspective. Buffon's *Histoire naturelle* (1749–67) ordered the animal creation according to human priorities of use, beauty, and tameness. But we should not overstate the difference between these two perspectives, which both emphasized man's limited understanding when compared to God's.[33]

More self-confident yet disconcerting was the view based on predation, that the struggle for existence was a central facet of nature. Increasingly widespread after 1750, this darker vision emphasized competition and conflict in a natural world not characterized by harmony and plenitude. It raised the possibility of species extinction, something believed to derogate from the perfection of God's creation yet seemingly confirmed by the disappearance of the dodo and the discovery of fossils. Meanwhile, comparative anatomy identified a growing number of affinities between species while geological studies, which established the great antiquity of the earth, rendered transformations between them more plausible. Furthermore, these affinities between man and animals were not merely anatomical as "the dark view of nature

---

[31] Ritvo, *Platypus*, 38.
[32] Thomas, *Man and the Natural World*, ch. 1; Donald, *Picturing Animals*, ch. 1.
[33] Donald, *Picturing Animals*, 33.

that gained predominance at this time owed much to changing notions of *human* capacity for cruelty and bloodshed."[34] Where natural theology viewed nature as a repository of God's moral teaching, this new, bleaker, picture of the natural world identified human corruption with natural cruelty, leading thinkers like John Stuart Mill to reject (and the Marquis de Sade to champion) the idea that nature could teach moral lessons.[35]

The greatest challenge to imposing taxonomic order on nature was posed by creatures that straddled established boundaries or were *sui generis*. "Monsters" might seem a case in point, although these paradoxically became less threatening at the same time as their study was ennobled with its own discipline – teratology. By the end of the seventeenth century, there was a consensus among the learned that monsters and marvels should evoke neither horror nor wonder, only distaste; indeed, the study of monsters was increasingly justified in the eighteenth century by the insights it offered on healthy organisms.[36] As we have seen, exotic animals like the platypus engendered confusion that could persist for decades: confirmation that they laid eggs came only in 1884 and was even then not universally believed.[37] Some otherwise unexceptional animals became problematic because of taxonomic innovation. The coining of *mammalia* by Linnaeus, for example, begged the question why aquatic creatures and bats should be included alongside the older "quadrupeds".

In devising the "mammal", Linnaeus also, of course, cemented his earlier erasure of the boundary between humans and the animal kingdom, which remained controversial and demonstrates that the most serious taxonomic problems were caused at the borderlines between major categories. That between the animal and vegetable kingdoms was no less problematic. Not only was it difficult to decide whether the freshwater polyp (*hydra vulgaris*) was animal or vegetable, but the discovery in 1739 of its ability to regenerate when divided into two (or more) parts also jeopardized received wisdom concerning animal reproduction. This was no small matter given the importance of reproduction for the concept of "species" and for Linnaeus' own method of botanic classification, in which the sexual organs of plants were crucial.[38]

Reproduction may be the single most important strand of continuity in the early modern history of taxonomy. It was present in ancient definitions of "species"; it was inherent to the genealogical connections implied by viewing nature as a tree; people

---

[34] Donald, *Picturing Animals*, 74.
[35] Donald, *Picturing Animals*, ch. 2; Warman, Caroline. *Sade: From Materialism to Pornography*. Oxford: Voltaire Foundation, 2002, ch. 4.
[36] Daston, Lorraine and Katherine Park. *Wonders and the Order of Nature*. New York, NY: Zone, 1998, ch. 5.
[37] Ritvo, *Platypus*, 14–15.
[38] Dawson, Virginia P. *Nature's Enigma: The Problem of the Polyp in the Letters of Bonnet, Trembley and Réaumur*. Philadelphia, PA: American Philosophical Society, 1987.

whose business it was (i.e. breeders and fanciers) underlined the privileged status of Latinate terminology through imitation; and reproductive success was the central feature of Darwinian natural selection (Darwin himself stated that "all true classification is genealogical").[39] As such, reproduction was also a crucial marker of change, because it underscored the transition of perceptions of nature from fixed and unchanging to a living and dynamic natural world.

3 Methods and Approaches

Aside from works on menageries and zoos, there are few dedicated histories of animal collections. Much of the older scholarship on *Wunder-* and *Kunstkammern*, cabinets of curiosities, and museums is in the shape of broader surveys or studies of specific institutions or key figures.[40] More recent historical work has formed two overlapping strands: one deals with the formation of collections and the sociology of knowledge represented by collectors. Here the focus has been on the scholarly and commercial networks, behavioral norms, and social organization that allowed individuals to collect objects, either by interpersonal exchange or on the market.[41] The shape of those collections and the uses to which they were put is a second field of enquiry, in which the layout, use, and representation of museums and collections is investigated. These studies point to the development of a 'museum science' and the complex relationship between science and art, much of which was produced to enable the (temporally and/or spatially) remote experience of a collection.[42]

The layout of collections and use of space – including by nonhumans – are vitally important topics. One of the most important developments in historical animal studies on collecting concentrates on the experience of collections from the perspective of its objects, what can be termed an "exhibit's eye-view". More relevant source material exists for collections of living animals, and recent studies have engaged

---

39 Darwin, Charles. *On the Origin of Species by Means of Natural Selection.* London: John Murray, 1859, 420.
40 Examples from Daston and Park, *Wonders,* 426–427 nt. 17.
41 Ogilvie, Brian W. *The Science of Describing: Natural History in Renaissance Europe.* Chicago, IL: University of Chicago Press, 2006; Findlen, Paula. *Possessing Nature: Museums, Collecting, and Scientific Culture in Early Modern Italy.* Berkeley, CA: University of California Press, 1994; Egmond, Florike. *The World of Carolus Clusius: Natural History in the Making, 1550–1610.* London: Pickering & Chatto, 2010.
42 MacGregor, *Curiosity and Enlightenment*; Arnold, *Cabinets*; Felfe, Robert. Spatial Arrangement and Systematic Order. In *Worlds of Natural History,* Hellen Anne Curry, Nicholas Jardin, James Andrew Secord, Emma C. Spary (eds.), 185–204. Cambridge: Cambridge University Press, 2018; Neri, Janice. *The Insect and the Image: Visualizing Nature in Early Modern Europe, 1500–1700.* Minneapolis, MN: University of Minnesota Press, 2011.

with the animal experience of collections [→History of the Zoo].⁴³ But a growing body of scholarship also deals with the "afterlife" of animals that have been preserved (whole or in part) post-mortem [→Material Culture Studies]. Approaches such as object biography or analyses of the "social life of things" demonstrate just how varied and full the life of animal artefacts can be.⁴⁴

This has yet to earn animals a more prominent part in the history of taxonomy, which remains dominated by studies of botany for periods before the later eighteenth century. Studies of zoology in the era before Linnaeus tend to concentrate more on prominent figures (such as Aristotle, Gesner, Topsell, Edward Wotton, and Ulisse Aldrovandi) than on zoology's contribution to taxonomic ideas.⁴⁵ This picture is different after 1800 thanks to the towering figure of Darwin, but three aspects of early modern taxonomy have received the lion's share of attention. Historians of the life sciences have tended to focus on the development of taxonomic principles and particularly on the development and use of particular categories (for example, species) by early modern thinkers.⁴⁶ A broader current of interest, especially in recent cultural history, has been on borderline or marginal groups such as monsters and hybrids.⁴⁷ Finally, taxonomy has been of profound interest for scholars who have explored its influence and interplay with categories of identity such as gender and race.⁴⁸

From this background emerge two methodological challenges, both relating to sources, for histories of taxonomy and animal collections that seek to center the animal. In general, the survival of sources on animal collections is somewhat patchy. Major institutional menageries, particularly Versailles, enjoy a substantial documentary footprint, but historians studying their public – especially mobile – counterparts are largely reliant on ephemera, newspapers, and the occasional letter or diary entry. Likewise, some major collections such as the cabinet at Schloss Ambras, Innsbruck,

---

43 Rothfels, Nigel. *Savages and Beasts: The Birth of the Modern Zoo*. Baltimore, MD: Johns Hopkins University Press, 2002; Baratay and Hardouin-Fugier, *Zoo*; Robbins, *Elephant Slaves*; Grigson, *Menageries*; Plumb, Christopher. *The Georgian Menagerie: Exotic Animals in Eighteenth-Century London*. London: I. B. Tauris, 2015.
44 Alberti, Samuel J. M. M. (ed.). *The Afterlives of Animals: A Museum Menagerie*. Charlottesville, VA: University of Virginia Press, 2011; Ross, Alan S. Recycling Embryos: Old Specimens in New Museums, 1660–1840. *Journal of Social History* 52, no. 4 (Summer 2019): 1087–1109; Poliquin, Rachel. *The Breathless Zoo: Taxidermy and the Cultures of Longing*. University Park, PA: Pennsylvania State University Press, 2012.
45 Enenkel, Karl A. E. and Paul J. Smith (eds.). *Zoology in Early Modern Culture*. Leiden: Brill, 2014.
46 Wilkins, John S. *Species: A History of the Idea*. Berkeley, CA: University of California Press, 2009.
47 Douthwaite, Julia V. *The Wild Girl, Natural Man, and the Monster: Dangerous Experiments in the Age of Enlightenment*. Chicago, IL: University of Chicago Press, 2002; Huet, Marie-Hélène. *Monstrous Imagination*. London: Harvard University Press, 1993; Enenkel, Karl A. E. The Species and Beyond: Classification and the Place of Hybrids in Early Modern Zoology. In *Zoology*, Karl A. E. Enenkel, Paul J. Smith (eds.), 57–148. Leiden: Brill, 2014.
48 Schiebinger, Londa. *Nature's Body: Gender in the Making of Modern Science*. New Brunswick, NJ: Rutgers University Press, 1993, especially ch. 2; Bernasconi, Robert (ed.). *Concepts of Race in the Eighteenth Century*. Bristol: Thoemmes, 2001.

have left wonderfully detailed sources that allow us to reconstruct them in detail, including their contents, layout, and even reception. For many other collections, we have (at most) a list of contents, and for not a few only scraps of evidence that they existed at all. The survival of animal specimens and objects is still more subject to the vagaries of fate, even where collections remain extant. Not only have once-existing sources disappeared (such as Sloane's taxidermic specimens, almost entirely destroyed by zealous nineteenth-century curators), some never existed: early modern taxonomy was a largely armchair pursuit that relied on a substantial body of textual data and therefore left little evidence of sustained human-animal interactions.

Against this highly variable picture must be set evidence that has survived beyond the ken of humans, such as zooarchaeological remains that attest to other forms of animal "collection" (particularly livestock for farming, slaughter or breeding) and are relatively uncontaminated by human purposes, not having been preserved, displayed, stored, or destroyed in accordance with human designs. Another way of mitigating the patchy survival of sources is the second methodological challenge: to use evidence innovatively by reading it against the grain. This has been most effectively realized in intersectional work that combines the history of animal collections and taxonomy with social, sensory, emotional, and other histories [→Social History;→History of Emotions]. Examples include studies of working-class interactions with horses, of the sensory experience of early modern animal exhibitions, and of emotional interaction with carnivals, high art, live animals, and monsters.[49]

4 Implication(s) of the Animal Turn

Using sources innovatively is doubly necessary not only because almost all of our sources derive from humans, but also historical animal studies have tended to focus on questions that prioritize elucidation of the human – rather than animal – experience, due in part to ongoing skepticism towards the field.[50] Several themes

---

[49] Edwards, Peter. The Tale of a Horse: The Levinz Colt, 1721–29. In *Interspecies Interactions: Animals and Humans Between the Middle Ages and Modernity*, Sarah Cockram, Andrew Wells (eds.), 89–106. Abingdon: Routledge, 2018; Plumb, Christopher. Reading Menageries: Using Eighteenth-Century Print Sources to Historicise the Sensorium of Menagerie Spectators and their Encounters with Exotic Animals. *European Review of History–Revue européenne d'histoire* 17, no. 2 (April 2010): 265–286; Ben-Ami, Ido. Emotions and the Sixteenth-Century Ottoman Carnival of Animals. In *Interspecies Interactions*, Sarah Cockram, Andrew Wells (eds.), 17–33; Cockram, Sarah. Sleeve Cat and Lap Dog: Affection, Aesthetics and Proximity to Companion Animals in Renaissance Mantua. In *Interspecies Interactions*, Sarah Cockram, Andrew Wells (eds.), 34–65; Donald, *Picturing Animals*; Daston and Park, *Wonders*.
[50] Baratay, Éric. Building an Animal History. In *French Thinking About Animals*, Louisa Mackenzie, Stephanie Posthumus (eds.), 3–14. East Lansing, MI: Michigan State University Press, 2015. See also Fudge, Erica. A Left-Handed Blow: Writing the History of Animals. In *Representing Animals*, Nigel Rothfels (ed.), 3–18. Bloomington, IN: Indiana University Press, 2002, 11.

have emerged from the battles fought against this background. The first and most serious of these concerns anthropocentrism, which is not automatically to be rejected. Indeed, there exists at present reasonable justification on pragmatic grounds for limited anthropocentrism within historical animal studies. As long as our language of ethics continues to associate the highest ethical standards with terms like "humane" and "humanitarian", it would be unwise to jettison in their entirety ethical concepts that are human-focused. This has taxonomic implications, as the philosopher of science John Dupré made clear when discussing why whales are (not) fish. He argued that one good reason for not classing aquatic mammals as fish is that we have become accustomed to viewing all mammals as fundamentally similar to ourselves and consequently treat them better. Furthermore, he warns against the juggernaut of scientific taxonomy, arguing that placing limits on the authority of science is beneficial, a point made – using the same creatures – in the nineteenth century by Mill.[51]

A second theme is the thorny question of "agency". Recent calls to scrap this concept seem somewhat premature, especially given that many of its shortcomings are a result of poor formulation or misuse. One critique of the idea is that it depends too heavily on an absolute distinction between a rational, autonomous agent and the structure within and by which their choices and actions are framed and governed. In response, environmental historians have called for models of agency in which the isolation of the individual agent is broken down such that "so-called human agency cannot be separated from the environments in which that agency emerges."[52] These enable interpretations of ethology in terms of agency, and show how animals whose behavior was interpreted as "wild" or "tame", for example, might exercise control over their own taxonomic fate. What is true for a species (the level at which ethology usually operates) is doubly so for an individual, hence Éric Baratay's recent call for an animal history that pays serious attention to the behavior of individual animals.[53]

Highlighting individuality also helps to overcome a major problem in the use of "agency", whereby mere activity is uncritically interpreted as an act of resistance yet "historically and culturally situated acts of resistance [are reduced] to manifestations of a larger, abstract human capacity – agency".[54] Such sleight of hand relies on viewing agents as an undifferentiated collective, such as "slaves" or "animals", whose actions are abstracted from particular contexts. Focusing on individuals means concentrating on individual behavior, which provides a far sounder basis for examining "agency" and highlights the ethical implications of group categories and taxonomies. As Thierry Hoquet has pointed out, speaking systematically about animals denies their individuality, and the breadth and abstractness of the concept "animal"

---

51 Ritvo, *Platypus*, 49.
52 Nash, Linda. The Agency of Nature or the Nature of Agency? *Environmental History* 10, no. 1 (January 2005): 67–69, 69.
53 Baratay, Building an Animal History.
54 Johnson, Walter. On Agency. *Journal of Social History* 37, no. 1 (Autumn 2003): 113–124, 117.

has facilitated the emergence of a moral universe in which pet cemeteries can exist alongside the industrialized slaughter of livestock.[55]

The "illogic" of this relationship, in which some animals are members of our families while others are straightforward chattel or even disease vectors to be eradicated, is one of the principal insights of the *animal turn*. Both animal collections and taxonomy pose in stark form the question that gives rise to such radically different, co-existing attitudes to our fellow creatures and with which this chapter began: what counts as an animal? If everything from humans to bacteria, from fossils to stuffed fauna, can be included in this category, then it is far too vast for the fine distinctions necessary to order our subjects of study. To this extent, historical animal studies is itself an echo of the task that collectors and taxonomists themselves undertook.

Selected Bibliography

Alberti, Samuel J. M. M. (ed.). *The Afterlives of Animals: A Museum Menagerie*. Charlottesville, VA: University of Virginia Press, 2011.

Baratay, Éric and Elisabeth Hardouin-Fugier. *Zoo: A History of Zoological Gardens in the West*. London: Reaktion Books, 2004.

Cowie, Helen. *Exhibiting Animals in Nineteenth-Century Britain: Empathy, Education, Entertainment*. Basingstoke: Palgrave, 2014.

Curry, Hellen Anne, Nicholas Jardine, James Andrew Secord and Emma C. Spary (eds.). *Worlds of Natural History*. Cambridge: Cambridge University Press, 2018.

Delbourgo, James. *Collecting the World: The Life and Curiosity of Hans Sloane*. London: Penguin, 2017.

Findlen, Paula. *Possessing Nature: Museums, Collecting, and Scientific Culture in Early Modern Italy*. Berkeley, CA: University of California Press, 1994.

Grigson, Caroline. *Menagerie: The History of Exotic Animals in England, 1100–1837*. Oxford: Oxford University Press, 2016.

MacGregor, Arthur. *Curiosity and Enlightenment: Collectors and Collections from the Sixteenth to the Nineteenth Century*. New Haven, CT: Yale University Press, 2007.

Ritvo, Harriet. *The Platypus and the Mermaid and Other Figments of the Classifying Imagination*. Cambridge, MA: Harvard University Press, 1997.

Rothfels, Nigel. *Savages and Beasts: The Birth of the Modern Zoo*. Baltimore, MD: Johns Hopkins University Press, 2002.

---

[55] Hoquet, Thierry. Animal Individuals: A Plea for a Nominalistic Turn in Animal Studies? *History and Theory* 52, no. 4 (December 2013): 68–90.

List of Contributors

Abel Alves is Chair of the Ball State University History Department. He is an early modernist and author of *The Animals of Spain* (Brill, 2011), *Brutality and Benevolence: Human Ethology, Culture, and the Birth of Mexico* (UVP, 1996), and Oxford Bibliographies' "Pets and Domesticated Animals in the Atlantic World" (2017). Other works include articles in history, anthropology, literary studies and theology journals. He is currently writing an essay on animals, the Anthropocene and imperialism since 1492.

Mitchell G. Ash is Professor Emeritus of Modern History at the University of Vienna, Austria, and a member of the Berlin-Brandenburg Academy of Sciences. He has published on the political, social and cultural relations of the sciences in the nineteenth and twentieth centuries, including the history of animal-human relationships, most recently the article "Zoological Gardens" in *Worlds of Natural History* (eds. H. Curry et al.) (Cambridge University Press, 2018). He is working on a study of scientific and political changes in German-speaking Europe in the twentieth century.

Barbara Rossetti Ambros is a Professor of East Asian religions at the University of North Carolina at Chapel Hill. She has been serving as co-chair of the Animals and Religion Group of the American Academy of Religion since 2014. Her publications include *Bones of Contention: Animals and Religion in Contemporary Japan* (University of Hawai'i Press, 2012) as well as a special issue of *Religions* titled *Buddhist Beasts: Reflections on Animals in Asian Religions and Culture* (co-edited with Reiko Ohnuma, 2019). She is currently working on a book project on animal releases in early modern Japan.

Etienne S. Benson is Associate Professor of the History and Sociology of Science at the University of Pennsylvania. He is the author of *Wired Wilderness: Technologies of Tracking and the Making of Modern Wildlife* (Johns Hopkins University Press, 2010) and *Surroundings: A History of Environments and Environmentalisms* (University of Chicago Press, 2020) as well as a number of articles on the history of ecology, environmentalism, and human-animal relations. His current book project is a history of the science of rivers and landforms.

Anna Boswell is a Lecturer in English and Drama in the School of Humanities (Te Puna Aronui) at the University of Auckland (Te Whare Wānanga o Tāmaki Makaurau). Her research deals with the fraught environmental legacies of settler colonialism in Australia and Aotearoa/New Zealand, with particular focus on animal worlds and indigenous histories. Anna's work in this area has been supported by a Marsden Fast-Start grant awarded by the Royal Society of New Zealand (2016–2019). Her most recent research is published in *Humanities*, *Pacific Dynamics* and *Animal Studies Journal*.

Dorothee Brantz is Professor of Urban History and Director of the Center for Metropolitan Studies at the Technische Universität Berlin, Germany. She has published widely on human-animal relations, including the history of slaughterhouses in 19th-century Berlin, Paris and Chicago; the history of acclimatization and urban environmental history. Currently she is writing a book on seasons in the city and she also just started a new research project on multispecies urbanism and planetary health.

Aritri Chakrabarti is a doctoral candidate at the Department of History, Ashoka University, India, and has also served briefly as a research assistant for the Mailman School of Public Health, Columbia University, USA. Her research interests can be located at the intersection of animal histories, colonial science and medicine, and environmental histories of South Asia. Some of her forthcoming publications explore the role of laboring animals in the making of British Empire in India.

Sarah D.P. Cockram is a historian at the University of Edinburgh. Sarah's publications in historical animal studies include *Interspecies Interactions: Animal and Human between the Middle Ages and Modernity*, with Andrew Wells (Routledge, 2018) and a Special Issue of *Renaissance Studies* (31:2), co-edited with Stephen Bowd. These volumes include Sarah's work on companion animals and on exotic megafauna. Sarah's current research continues her interest in courtly creatures, and she is writing on transnational animals and on animals of war.

Helen Cowie is Professor of History at the University of York. Her research focuses on the history of animals and the history of natural history. She is author or *Conquering Nature in Spain and its Empire, 1750–1850* (MUP, 2011), *Exhibiting Animals in Nineteenth-Century Britain: Empathy, Education, Entertainment* (Palgrave, 2014) and *Llama* (Reaktion, 2017). Her current project, 'Victims of Fashion: Animal Commodities in Victorian Britain', is forthcoming with Cambridge University Press.

Janet M. Davis is Professor of American Studies and History at the University of Texas at Austin. Her publications include: *The Gospel of Kindness: Animal Welfare and the Making of Modern America* (OUP, 2016); *The Circus Age: Culture and Society under the American Big Top* (U North Carolina P, 2002), and *Circus Queen and Tinker Bell: The Life of Tiny Kline* [editor] (UIP, 2008). Her current book project is a transnational American cultural history of human and shark entanglements from the Age of Sail to the *Jaws* Age.

Joanna Dean is Associate Professor of History at Carleton University in Ottawa, Canada, where she teaches environmental history and animal history. She is co-editor of *Animal Metropolis: Histories of Human-Animal Relations in Urban Canada* (University of Calgary Press, 2017) and ran a lecture series, *Beastly Histories*, in 2014. She has published numerous articles on the history of urban trees and curated an exhibit, *Six Moments in the History of an Urban Forest* at Ottawa's Bytown Museum in 2012.

Silke Förschler has just finished her second book entitled *Tierdarstellungen in der Frühen Neuzeit. Ästhetiken und Praktiken der Naturgeschichte*. Her publiactions include the edited volumes *Heim/Tier. Tier-Mensch Relationen im Wohnen* (together with Christiane Keim and Astrid Silvia Schönhagen, transcript, 2019) and *Akteure, Tiere, Dinge. Verfahrensweisen der Naturgeschichte* (together with Anne Mariss, Böhlau, 2017). She currently works at Bauhaus Archiv Berlin on Otti Berger, weaving artist and her cooperation with the Roßhaar factory Schriever.

Andrew Gardiner is Senior Lecturer at the Royal (Dick) School of Veterinary Studies, University of Edinburgh. He is a practising veterinarian, having worked in veterinary practice for 20 years, and has additional clinical qualifications in surgery and a PhD in veterinary history. He has wide interdisciplinary interests in animal welfare, ethics and veterinary education and has published in these fields.

Julia Hauser is Associate Professor of Global History and the History of Globalization Processes at the University of Kassel. She has published on the history of missions, gender, and imperialism in the "long" nineteenth century, most notably a monograph on *German Religious Women in Late Ottoman Beirut. Competing Missions* (Brill, 2018). In her current research project, she is looking at the entangled history of vegetarianism between Europe and South Asia in the nineteenth and twentieth century.

Ryan Hediger is Associate Professor of English at Kent State University in Ohio. His research focuses primarily on the environmental humanities and animal studies. Publications include *Homesickness: Of Trauma and the Longing for Place in a Changing Environment* (University of Minnesota Press, 2019) as well as essays on military dogs, Hemingway, and the film *Grizzly Man*. He co-edited *Animals and Agency* (Brill, 2009) and edited *Animals and War* (Brill, 2013). He is currently editing an essay collection on labor norms in the Anthropocene.

Kit Heintzman is a Society of Fellows postdoctoral scholar at the University of Southern California. Kit's PhD in the History of Science at Harvard University demonstrated the link between livestock and nationalism in Enlightenment France through the emergence of standardized veterinary medicine. Kit's research, financially supported by SSHRC, the Smithsonian, the Institute for Cultural Inquiry, Chateaubriand, and others, has received awards from the Society for the Social History of Medicine and the British Society for the History of Science.

Erica Hill is Professor of Anthropology at the University of Alaska Southeast. She is the editor of *Iñupiaq Ethnohistory* (University of Alaska Press, 2013) and *The Archaeology of Ancestors* (with Jon Hageman, University Press of Florida, 2016). Her research focuses on the archaeology of Arctic human–animal relations, animal geographies, and zooarchaeology. She is working on an edited volume focused on animals in the circumpolar North.

Philip Howell is Reader in Historical Geography at the University of Cambridge, UK. He has specialised in the historical geography of nineteenth-century Britain, writing a monograph on the regulation of prostitution in Britain and the British empire, as well as a book on the historical and cultural geography of the dog in Victorian Britain. His work in historical animal geography has led to two co-edited volumes on animal history. His more recent work on animal-human relations has moved to twentieth-century history and also to contemporary concerns.

Andreas Hübner is Lecturer of North American Studies and Didactics at Leuphana University Lüneburg. His research focuses on Cultural and Global History as well as History Didactics and Human-Animal Studies. In 2015, he received his Ph.D. from Justus Liebig University Giessen. His publications include numerous monographs and articles, among them an essay on "panda diplomacy" (together with Mieke Roscher, *Zeitschrift für Geschichtsdidaktik*, 2019). His latest project explores the role of Human Animal Studies and the Anthropocene in secondary education.

Axel C. Hüntelmann is Postdoctoral Research Fellow at the Institute for the History of Medicine at the Charité Medical School in Berlin. He has worked and published on the German Imperial Health Office (PhD 2007) and other European public health institutions between 1850 and 1950, the history of laboratory animals, the production, marketing and regulation of pharmaceuticals in Germany and France and on the immunologist Paul Ehrlich. As a trained historian and economist, he is currently finishing a book on accounting and bookkeeping in medicine (1730–1930).

Takashi Ito is Associate Professor at the Institute of Global Studies, Tokyo University of Foreign Studies, Japan. He is concerned with the history of human-animal relationships in Britain and Japan. His publications include *London Zoo and the Victorians, 1828–1859* (Boydell, 2014) and 'Flying Penguins in Japan's Northernmost Zoo', in Tracy McDonald and Daniel Vandersommers (eds), *Zoo Studies: A New Humanities* (McGill-Queen's University Press, 2019).

Linda Kalof is Professor of Sociology at Michigan State University. Her books in animal studies include *Looking at Animals in Human History* (Reaktion, 2007), *The Oxford Handbook of Animal Studies* (Oxford University Press, 2017), the six-volume *Cultural History of Animals* (Bloomsbury, 2011), and the second edition of *The Animals Reader* is forthcoming with Routledge. She edits *The Animal Turn* book series for Michigan State University Press, and she is currently working on a monograph, Animal Blood Sport.

André Krebber is Lecturer for Social and Cultural History and Human-Animal Studies at the University of Kassel, Germany. He has published widely in the areas of history of science and ideas, environmental humanities and human-animal studies. Together with Mieke Roscher he has published the edited volume *Animal Biography* (Palgrave, 2018). He currently works on a monograph on ecological agency and the nonhuman in the Anthropocene and on a collaborative project on the aesthetics of octopuses.

Gesine Krüger is Professor of Modern and African History at the University of Zurich, Switzerland. She has published on human-animal relations, colonial animals and hunting. Her publications include the edited volumes *Tiere und Geschichte. Konturen einer Animate History* (together with Aline Steinbrecher and Clemens Wischermann, Steiner, 2014), and *"Ich Tarzan". Affenmenschen und Menschenaffen zwischen Science und Fiction* (together with Marianne Sommer and Ruth Mayer, transcript, 2008). She is founding member of the online magazine *Geschichte der Gegenwart*.

Heinrich Lang holds a German Research Foundation Grant for Economic History and is teaching at the University of Leipzig. His publications are on Late Medieval and Early Modern economic history and history of accounting like his *Wirtschaften als kulturelle Praxis* (Franz Steiner, 2020), and political history of the Italian Renaissance. He is currently working on a monograph on "Investment and Practices of Refinance: Accounting of Capitalist Rentiers and Enterprises from Florence and Augsburg in the Sixteenth Century".

Laura McLauchlan is a multispecies ethnographer and lecturer at NYU (Sydney) and at UNSW. She has a particular interest in emergent and marginal ontologies and practices which might allow for greater responsiveness to life. Recent articles include 'A multispecies collective planting trees' (Cultural Studies Review) and 'Wild disciplines and multispecies erotics' (Australian Feminist Studies). She is currently working on a monograph exploring the ways in which humans learn to love and hate hedgehogs in the UK and Aoteaora/New Zealand.

Annette Leiderer is a PhD Student at the University of Freiburg, Germany. Her dissertation *Die Industrialisierung des Tieres* focusses on the beginnings of industrialized slaughter in 19th century Germany and analyzes the changing roles and perceptions of cows, pigs and their slaughter in economic, political, scientific and artistic discourses. She currently is teaching history and politics at a private high school and is collaborating with a state research commissioner in order to facilitate the education of human-animal-history in secondary education.

Ian Jared Miller is Professor of History and Dean of Cabot House at Harvard University. He is the author of *The Nature of the Beasts: Empire and Exhibition at the Tokyo Imperial Zoo* (University of California Press, 2013), co-editor of *Japan at Nature's Edge: The Environmental Context of a Global Power* (University of Hawaii Press, 2013), and co-editor of *Oceanic Japan: The Archipelago in Pacific and Global History* (University of Hawaii Press, forthcoming). His current work recasts the history of Tokyo as a history of energy.

List of Contributors — 623

Brett Mizelle is Professor of History and Director of the American Studies Program at California State University Long Beach. His publications include the book *Pig* (Reaktion, 2011) and numerous articles, book chapters, and reviews in the fields of nineteenth-century American history and the history of human-animal relationships. He is currently working on a critical animal studies project on the discursive and material making and taking of animal life in nineteenth-century America and on the Reaktion "Animal Series" book *Squirrel*.

Dominik Ohrem is a Research Associate in the Department of English II at the University of Cologne, Germany. He is editor of *American Beasts: Perspectives on Animals, Animality and U.S. Culture, 1776–1920* (Neofelis, 2017) and co-editor of two volumes published in the Palgrave Studies in Animals and Literature series (together with Roman Bartosch and Matthew Calarco, respectively). He is currently finishing his dissertation and working on a multivolume collection of historical sources on human-animal relations in the long-nineteenth-century United States.

Mieke Roscher is Associate Professor for Cultural and Social History and the History of Human-Animal-Relations at the University of Kassel, Germany. Her publications include *Ein Königreich für Tiere* (Tectum, 2009) on the British animal rights movement, the edited volume *Animal Biography* (together with André Krebber, Palgrave, 2018) and articles on animal historiography and agency, the twentieth-century European zoo and gender in animal welfare. She is currently writing a monograph on the political history of animals in the Third Reich.

Veronika Settele is Assistant Professor for Modern History at Bremen University, Germany. Her publications include *Revolution im Stall* (Vandenhoeck & Ruprecht, 2020) on the history of animal farming in both German states in the second half of the twentieth century and the edited volume *Geschichte des Nicht-Essens* (together with Norman Aselmeyer, De Gruyter, 2018) on the history of food renunciation, avoidance, and refusal in the Modern Era. She is currently working on international trade-conflicts and their impact on democratic institutions.

Aaron Skabelund is an Associate Professor in the History Department at Brigham Young University, located in Provo, Utah. He is the author of *Empire of Dogs: Canines, Japan, and the Making of the Modern Imperial World* (Cornell University Press, 2011), and more recently, "'Bite, Bite against the Iron Cage': The Ambivalent Dreamscape of Zoos in Colonial Seoul and Taipei," *Journal of Asian Studies* (May 2020), co-authored with Joseph Seeley of the University of Virginia.

Sandra Swart is Professor in the Department of History at Stellenbosch University. She received her DPhil in Modern History from Oxford University in 2001, while simultaneously obtaining an MSc in Environmental Change and Management, also from Oxford. She focuses on the socio-environmental history of southern Africa, especially the shifting relationship between humans and animals. She authored *Riding High – Horses, Humans and History in South Africa* (Witwatersrand UP, 2010). She is currently working on the long, entangled histories of humans and baboons.

Peta Tait, Professor at La Trobe University, is an academic scholar, playwright and a Fellow of the Australian Academy of the Humanities. She has written 60 scholarly articles and chapters and recent books include: the authored *Theory for Theatre Studies: Emotion* (Bloomsbury, 2021); the co-edited *Feminist Ecologies: Changing Environments in the Anthropocene* (Palgrave, 2018); the authored *Fighting Nature: Travelling Menageries, Animal Acts and War Shows*; the co-edited *The Routledge Circus Studies Reader*; the authored *Wild and Dangerous Performances*.

Nadir Weber leads the SNSF Ambizione project "Falcons in Court Society" at the University of Bern, Switzerland, and is currently a visiting scholar at the Humboldt University of Berlin. He has published widely on diplomatic history, intellectual history, and the history of human-animal relations in the early modern period. He has co-edited the volume *Animals and Courts* (De Gruyter, 2020) and is currently writing a monograph on the life courses and symbolic roles of raptors in European princely courts.

Andrew Wells is an independent scholar based in Leipzig, Germany. A cultural and intellectual historian of the British Atlantic world, his publications include the edited collection *Interspecies Interactions: Animals and Humans between the Middle Ages and Modernity* (with Sarah Cockram, Routledge, 2018), as well as essays on animals, nuisance, and globalization in the early modern city. He is currently working on a monograph on urban cultures of freedom in Britain and America between the English and American Revolutions.

Anna-Katharina Wöbse is an environmental historian and curator. She works and lectures at the University of Gießen. Her book "Weltnaturschutz" (Campus, 2012) explored the environmental diplomacy of the League of Nations and early UN. She has extensively published on the history of international environmental movements and biographies, national parks, animal-human-relations, and visual history. Currently, her research focuses on wetlands, and she is editing a handbook on European environmental history (with Patrick Kupper).

Amir Zelinger is a historian of pet-keeping and animal breeding. His book, *Menschen und Haustiere im Kaiserreich: Eine Beziehungsgeschichte*, was published in 2018 (transcript). His most recent article, "Race and Animal Breeding: A Hybridized Historiography", was published in *History and Theory* in 2019.

Index

A
abattoir   118, 443, 534, 541, 545
aboriginal   22, 26, 107, 328
Abul Abbas (Elephant)   198, 221
acclimatization / acclimatization societies   104, 106, 112, 264, 606
actor   7, 9–14, 25, 31, 63, 78, 126–127, 133–135, 140, 142–145, 147, 195, 206, 208–209, 214, 216–217, 219–220, 224–225, 238, 254, 293, 301, 304, 318–319, 326, 338, 354–355, 367, 386, 397, 399, 414, 453, 516, 519, 522, 535–536, 548, 552, 597
Actor-Network Theory   140, 209, 244, 272, 338, 521
Ad vivum   376, 388
Adams, Carol   344, 551
Adorno, Theodor W.   275, 287
affection   40–42, 153–155, 350, 412–413, 427, 429, 435
agency   7, 14, 24–27, 34, 38, 42, 44, 50, 65, 78, 80–82, 89, 94, 117, 126–127, 129, 133–134, 138, 140, 144, 178, 200, 207–208, 220–221, 224, 236, 254, 269, 272–273, 282, 288, 300–301, 305, 315, 319, 405, 440, 452–454, 469, 496, 501, 503, 505, 518, 521, 523, 535–536, 553, 566, 568, 572, 575, 590, 599–600, 616
agriculture   21, 54, 77, 111, 120, 166, 184, 206, 235, 253, 334, 364, 525–528, 531–536, 542, 544, 560, 574, 584, 590, 601
Akeley, Carl (1864–1926)   231, 239
Aloi, Giovanni   386
American Society for the Prevention of Cruelty to Animals (ASPCA)   75, 593
Amerindian, →*see also* Native American   40, 46, 157, 432, 579–580
ancient   23, 29, 33, 98, 104, 115, 119–120, 122–123, 128, 149, 170, 247, 359–360, 426, 428, 472–473, 476–477, 539, 556, 587, 609, 612
Anderson, Virginia DeJohn   43, 49, 70, 78, 157–158, 530, 600
animal collection   441, 603–606, 608, 613–615, 617
animal colonialism   347

animal experiments, →*see also* vivisection   260–261, 264, 503, 509–510, 513–516, 519, 521
animal minds   281–282, 345, 417
animal protection movement   588
animal rights   206, 266, 268, 452, 457–458, 462
– movement   70, 76, 78, 88, 215, 468, 552
animal spaces   74, 255, 309, 312, 314, 317, 323
animal subjectivity   75, 349, 467, 469, 504
animal trade   118, 124, 181, 187–188, 217, 219, 254, 298, 330, 332, 365, 367, 369, 447, 457, 459–460, 540, 542
animal training, →*see* training / animal
animal welfare   74, 78, 218, 418–419, 450, 452, 462, 487, 503, 513, 548–550, 589, 591
Annales school   133, 141, 296
anteater   150, 158–159
– depicted   152
Anthropocene   34, 48–49, 145, 241, 306, 440, 454, 601
anthropocentric   24–25, 33, 48, 64, 79, 233, 288, 306, 309–310, 320, 322, 343–345, 354, 359, 417, 437, 498, 573, 575, 599–600, 610–611
– post-   341, 345–346, 348–349, 352
anthropocentrism   25, 73, 79, 81, 119, 143, 215, 302, 319, 387, 395, 437, 616
– non-   610
anthropomorphism   81, 143, 152, 162, 267, 281, 319, 414, 420, 450, 454, 468, 599
anti-cruelty   74–75, 337
antiquity   30, 144, 197, 311, 395, 426, 428, 433, 477, 492, 509, 511
aquatic life   101, 105, 296, 612, 616
archaeology   22–24, 27, 29, 34–35, 89
archive   89–90, 113, 126, 214, 224, 227, 232, 241, 304, 327, 343, 359, 372, 419, 495
aristocracy   136, 484, 488, 556, 563, 580
Aristotle (384–322 BCE)   478, 511–512, 614
Armstrong, Philip   34, 111, 115, 220, 338, 353
Armstrong Oma, Kristin   34
ass   592
Atlantic / trans   69, 77–78, 139, 158, 160, 297, 589–590

aurochs   30, 270
– depiction   472

B
baboon   85–86, 93–98, 145, 608
Bacon, Francis (1561–1626)   259, 261, 271
Baker, Steve   166, 489
Balto (Dog)   237
Barnum, P.T. (1810–1891)   160, 239, 458, 461, 463, 465, 607
barter economy   186
Bastet (Cat)   149
bat   2, 31, 128, 413, 612
bear   125, 150, 249, 452, 463, 465, 473, 500, 587–588
– depiction   472–473, 490, 501
– grizzly   233–234, 591
– polar   208, 229, 452
beaver   25, 61, 205
becoming / becoming with   94, 109, 141, 387
bed bugs   244
bee   243, 262, 266–267, 573, 575, 610
beetle   584, 587
– depiction   377
Bekoff, Marc   409, 414, 419, 421, 600
beluga   27, 567–568
Benjamin, Walter   14, 141, 287–288, 376
Benson, Etienne   303, 436
Bentham, Jeremy (1748–1832)   410, 513
Berger, John   3, 69, 82, 166, 236, 278, 287, 429
Bernard, Claude (1813–1878)   263, 514
big cats   174, 457–458, 460, 462–467
biogeography   101, 229–230
biography   81, 110, 137, 177, 220–221, 233, 305, 358, 493, 505, 519
biopolitics / biopolitical   299, 312, 454, 529, 573, 582
biopower   122, 454, 459, 499
biotechnology   54, 56, 228
bird   21, 29, 37, 40, 72, 105, 112, 115, 119, 124, 243, 249, 260, 265–266, 268, 273, 298, 306, 327, 401, 427, 528, 563, 565, 577
– depiction   384, 475, 484
– wings   24
Bird Rose, Deborah   109, 114, 397
bison   6, 30, 70, 72, 74, 147, 231, 294–295, 472, 580
– depiction   472, 475–476
Black Beauty (Horse)   138, 150

Black Panther Party   173
Blondie (Dog)   170
boar   28, 38, 191, 202, 540
– depiction   483–484
body history   536
Boer War   582
Bonheur, Rosa (1822–1899)   486
bourgeoisie / bourgeois   40, 118, 136, 148, 160, 251–252, 254, 275, 431, 557, 564, 606
bovine   45, 106, 173, 184, 187–189, 191, 194, 286, 334, 351, 543
Brantz, Dorothee   3, 144
Braudel, Fernand   139, 216
breed   9, 45, 122, 124, 136–137, 170, 176, 232, 236, 238, 265, 330–331, 528, 543, 596, 598–599
breeding   9, 49, 104, 136, 138, 162, 176, 183–184, 187, 191, 200, 203, 225, 262, 264, 270, 272, 286, 330–331, 363, 369, 417, 517, 528–531, 541–544, 600, 606–607, 615
Buddhism   43, 126–129
buffalo   189–191, 231, 331, 555, 577
Buffon, Comte de, Georges-Louis Leclerc (1707–1788)   262, 611
bulldog   136, 173
bullfighting   591, 594, 600
Bulliet, Richard   17, 38, 44, 47–48, 167, 216

C
cabinet of curiosities   9, 261, 389, 439, 481, 603–605, 613
camel   49, 106, 110, 112, 216, 244–245, 331, 577, 583
canaries
– monuments   238
canine   28, 32, 126, 136, 160, 167–168, 174, 176, 178, 331, 406, 414, 419, 436, 599
– dentures   473, 501
– nuptials   154
capybara   150
– depiction   152
carnivore
– depiction   473–474
Carson, Rachel (1907–1964)   76–77, 119, 298
cat   41–42, 45, 124, 148–149, 153–154, 162, 168, 248, 250, 280, 347, 397, 403, 427, 434, 443, 489, 497, 499–500, 503
– cafés   124

– depiction 484
– museum 232
caterpillars 262
– depiction 377
cattle 21, 44–45, 71, 74, 88, 122, 185, 187–188, 220, 231–232, 294, 329, 331, 334, 336, 477, 497, 500, 530, 534, 536, 543, 545, 549–550, 555, 581
– colonial 600
– depiction 486
– feral 157
– hides 43
– markets 187–188, 545–547
– pastoralism 332
– races 541
– sacrificial 128
– trade 540
cavalry 122, 166, 579, 582
cave art 236, 472–474, 476
Chakrabarty, Dipesh 325
cheetah 202, 565–566
chicken 46, 48, 182, 185, 229, 248–249, 434, 528, 536, 539, 550, 552, 601
– depiction 483
chimpanzee 42, 50, 62, 163, 281, 418, 573
Chow (Dog) 505
circus 81, 162, 239, 253, 319, 457–464, 467–469, 607
– act 459, 462
– acts 465
– Maximus, ancient Rome 587
– ring 457, 463
– trainers 464
CITES (Convention on International Trade in Endangered Species of Wild Fauna and Flora) 206, 448
Civil War 75, 443, 582, 593
Clara (Rhinoceros) 383–384
class 69, 94, 98, 107, 118, 135–137, 142, 144–145, 149, 246, 250, 254, 312, 425, 427, 431, 435–437, 444, 453, 483, 543, 561, 578, 598, 608
classification 112, 262, 611–612
Clutton-Brock, Juliet 38, 48
co-constitution 8, 329, 343, 400, 522
co-dependencies 6, 21, 28, 34
co-domestication 21
co-habitation 31, 194, 256
cockfighting 148, 587–588, 590, 591, 594–595, 600–601

cod 296
coevolution 37, 39, 41, 44, 137, 526, 565
Cold War 75, 220
collaboration
– across sciences 298, 402, 414, 417, 420, 493
– with other animals 273, 393, 523, 566
colonial 22, 43, 46, 59, 70–72, 74, 78, 88, 92, 94, 101, 110, 112, 126, 129, 137, 142, 150, 157, 171, 175, 214, 217–218, 223–224, 321, 325–335, 337–338, 347–348, 361, 366–367, 383, 444–446, 457, 530, 558, 560–561, 564–565, 590, 594–595, 598, 600
colonialism / colonialist 77, 91, 206, 216, 222, 234, 317, 325–327, 331–332, 335–337, 348, 367, 459, 551, 562, 589, 597
colonization / colonizer 21, 28, 58, 72, 77, 79, 92, 101, 137, 158, 216, 223, 321, 326, 331, 334, 337, 348, 366, 444–445, 460, 594
Columbian exchange 49, 77, 579
companion animal 39–40, 124, 182, 185, 331, 346, 398, 406, 410, 420, 433, 497, 500–501, 503, 505, 565
– depiction 384
companion species 398, 405, 436
companionship 41, 49, 117, 168, 266, 298, 427, 429–430, 433, 436
comparative psychology 61, 63
consciousness 61, 234, 237, 266, 281–282, 288, 350, 387, 549
conservation 93, 108–109, 135, 208, 216, 218, 224, 240, 252, 296, 314, 328, 403, 442, 452, 490, 540, 560
– movement 73, 565
consumer culture 252, 431, 443, 596
cormorant 202, 565
COVID-19 1–2, 15, 128–129
cow 44–45, 48, 157, 182, 184, 187, 189–190, 201, 228, 245, 248, 253, 410, 477, 500, 530–532, 543, 546–547, 549
– cultural appreciation 543
– depiction 480, 483
– protection 337
Cowie, Helen 49
coyote 243, 321
critical animal studies 78, 232, 343, 551
crocodile 106–107, 587, 604
Cronon, William 78, 294

Crosby, Alfred 49, 77–78, 113, 216, 299, 577–579
cultural turn 597

D
dairy 45, 106, 122, 331, 500
Darnton, Robert 153, 160, 285
Darwin, Charles (1809–1882) 37, 42, 51, 54–55, 86, 92–96, 113, 126, 225, 260, 262, 265–266, 376, 410, 464, 613–614
Daston, Lorraine 150, 277
Davis, Janet 462
DDT 76, 298
de-centering 293
deep history 34, 125, 367, 557
deer 28, 30, 43, 49, 202, 475, 563
– depiction 472, 483, 486
dehumanization 91, 98, 353
DeMello, Margo 4, 80
Derrida, Jacques 3, 285, 468
Descartes, René (1596–1650) 43, 45, 271, 279, 281–282, 410, 512
Despret, Vinciane 396, 403, 407
Dickens, Charles (1812–1870) 158
dingo 103, 111
diorama 230–231
disease 1, 22, 25, 47, 161, 219, 263–264, 331, 333–334, 364, 498–499, 503, 506, 510, 515–516, 518, 521, 533, 547, 575, 617
Disney 147
DNA 33, 55, 57, 89, 240, 474
dodo 111, 611
dog, →see also canine and first dogs 24, 27–28, 31–32, 34, 39–42, 45–47, 97, 103–104, 121–122, 135–138, 148, 153–154, 160, 167–168, 170, 172, 176, 185, 198, 202, 245, 248, 250–251, 280, 286, 317, 320, 331, 397, 406, 411–412, 419, 427, 430, 434, 443, 459, 462, 489, 499–500, 503, 505, 522, 565–566, 583, 588, 592, 596, 598, 600
– colonial 142
– dentures 501
– depiction 384, 480, 484, 486
– experimental 264, 273, 511, 514
– fighting 587–588, 591–596, 599–600
– insult 584
– lapdogs 124, 154–155, 500
– meat 337, 539
– military 177–178, 414
– monuments 237–238
– museum 232
– police 80, 167, 174–175, 178
– show 136
– street 435
Dolly (Sheep) 259, 523
dolphin 75, 268, 600
domesticated 22, 40, 42, 45–46, 48–50, 77, 118, 122, 157–158, 247–248, 300, 330, 345, 347, 349–351, 367, 430, 457, 459, 466, 565, 607
domestication 22, 25, 37–38, 41, 43–45, 48–50, 121, 216, 331, 476–477, 567, 606–607
domination 45, 58, 90, 311, 313–314, 429–431, 437, 590
donkey 47, 110, 111, 139, 150, 244, 248, 331, 396, 577, 583
Douglass, Frederick (1818–1895) 350–352
dove 162, 261
Dreamings 109, 114
Drosophila melanogaster (fruit fly) 268, 271, 510, 517–518
Dürer, Albrecht (1471–1528) 200, 383, 388, 481

E
eagle 171, 203
early modern 8, 16, 45, 49, 122, 138, 153, 191, 198, 202–205, 207, 249–250, 278–279, 282, 288, 360–361, 367, 369, 376–378, 381, 389–390, 430, 439, 457, 459, 557, 560, 564, 566, 607, 611–612, 614–615
– imagery 375, 390–391
earthworm 401
ecology 24, 34, 57–61, 72, 101, 109–110, 112–113, 118, 209, 255, 293–294, 297, 302, 326, 336, 394–395, 574, 580, 584
ecosystem 24, 37, 77, 101–102, 105–106, 109, 206, 211, 295, 297, 300, 370, 393, 490
– ecology 59–60
ecotourism 111
eel 105–106, 111
egg 21, 181, 183, 185, 192, 249, 267, 359, 361, 434, 528, 532, 537, 612
elephant 26, 31, 81, 118, 125, 160, 162, 164, 173, 198, 200–201, 203, 207, 221, 234, 239–241, 244–245, 281, 329, 332, 361, 368–369, 404, 418, 440, 450, 453, 457–

459, 461–465, 479, 490, 565, 576, 583, 587, 605–607
embodiment   63, 326, 351, 468
emotion / emotional   32, 34, 40, 43, 47, 51, 63, 91–92, 139, 154, 162–163, 200–201, 236–237, 240–241, 251, 281, 302, 316, 319, 329, 335, 409, 412–417, 419–421, 430, 434, 443, 464, 468–469, 478, 481, 490, 520, 540, 555
empire   47, 58, 171, 177, 222, 325–327, 329–332, 334, 364, 367, 431, 435, 444–447, 530, 548–549, 581, 589, 594–595, 597–598, 607
Enlightenment   42, 66, 86, 213, 250–251, 285, 361, 440–441, 445, 453–454, 485, 488
entangled history   79, 87, 222, 231, 337, 510
entanglement   3, 6, 24–25, 37, 45, 56, 60, 86, 139, 182, 221, 223, 231, 322–323, 327, 331, 333, 368, 386, 400, 404, 417, 426, 530, 560, 596, 598
environmental humanities   305
epistemology   71, 364, 509, 520–521, 599
epizootic   219, 249, 333–334, 515
ethnography   89, 285, 327, 393–398, 400–402, 404, 406, 417
ethology   15, 29, 31, 35, 61–62, 162, 220–221, 225, 260, 266, 268, 273, 282, 302, 318–319, 393, 401, 405, 418–419, 448, 600, 616
– cognitive   162
eugenics   266, 530
Eurocentrism   11, 214–215, 224
evolution   8, 37, 55–57, 113, 211, 302
– brain   559
– cultural / human   265, 558, 566–567
– theory of   54–55, 57, 59, 85, 92–93, 263, 265–266, 557
exotic animals   102, 118, 158, 200, 207, 217, 261, 263–264, 378, 384, 440, 444, 457–461, 465, 497, 517, 604, 612
– depiction   376, 378, 381, 384–385, 489
extermination   72, 74, 103, 108, 167, 295, 562, 565
extinction   34, 72, 77, 106, 108–109, 111, 113, 172, 203, 243, 253, 295–296, 368, 611
– de-   106

F
Fala (Dog)   169
falcon   194, 198, 200, 202, 565, 604
falconry   200, 563, 565–566
farming   21–22, 94, 118, 120–122, 184, 206, 334, 417, 427, 497, 523, 525, 527–536, 541–544, 550, 615
– factory   48, 526
– silk, →see also sericulture   366
– urban   434
fascism   125, 170–171, 237, 270
feathers   30, 46, 265, 298, 361, 387, 561–562
– depiction   475
feline   149, 162, 403, 501
– depiction   472–473, 475
feminism   270, 287, 341–342, 585
ferret   565
first dogs   167–170, 177
fish   27, 107–108, 119, 122, 124, 206, 245, 260, 268, 297, 302, 497, 603, 616
– depiction   483
fishing   21, 108, 111, 119–121, 205, 296, 562–563, 565
– over-   206
fleas   2, 244, 583
flying foxes   397
Fossey, Dian (1932–1985)   93
fossils   85–86, 93, 111, 121, 294–295, 603, 611, 617
Foucault, Michel   286, 321, 529
fox   39, 49, 106, 108, 397, 568, 588, 594
– depiction   476
fox-tossing   568, 588
foxhunting   588, 594
Franklin, Adrian   107
Franklin, Sarah   523, 530
Freckles (Transgenic Spider Goat)   227–229
frontier   73, 115
Fudge, Erica   3, 41, 177, 250, 277–278, 393, 407, 418–419, 527, 599
fur   29, 72, 104, 205, 265, 361, 365, 369, 561, 564
– depiction   385

G
gamepiece   483–484
Geertz, Clifford   148
gender   26, 72, 90, 98, 123, 140, 142, 145, 154, 207, 215, 246, 342, 344, 347–348, 350, 353, 437, 453, 557–558, 580, 598, 614
genetics   55–57, 93, 164, 510, 516, 518, 529

Ginzburg, Carlo   148, 386–387
giraffe   148, 155, 158–159, 198, 440, 447, 450, 454, 606
globalization   206, 217–219, 223–225, 253
GloFish   229
goat   21, 122, 228–229, 237, 241, 248, 434, 479
– depiction   477
goldfish   124
Goodall, Jane   50, 62, 93, 163
gorilla   94, 231, 449, 452
Grandin, Temple   419
Great War, →see also World War I   571, 573, 575, 577, 583, 586
Greene, Ann Norton   70, 138
Greyfriars Bobby (Dog)   237
guinea pig   44, 46, 244, 518

H

Hachikō (Dog)   176–178, 237–238
Hagenbeck, Carl (1844–1913)   81, 217, 447–448, 458, 461, 465, 467
Hanno (Elephant)   198, 200
Haraway, Donna   3, 33, 80, 82, 111, 209, 231, 270, 273, 277, 287, 332, 398, 402, 417, 421, 436, 522
Harvey, William (1578–1657)   260, 511–512
hawk   202, 563
– depiction   485
Heidegger, Martin   310
Hobsbawm, Eric   133
Hoefnagel, Joris (1542–1601)   376–380, 386
hog   43, 185, 294, 531, 536
Hogarth, William (1697–1764)   488, 492
Horkheimer, Max   275, 287
horse   32, 44–45, 49, 72, 87, 120, 122, 137–138, 143, 150, 166, 183–184, 188, 191, 198, 200–201, 228, 244, 247–248, 251, 317, 330–331, 333, 459, 461, 494, 497–498, 500, 541, 543, 556, 565–566, 571, 573, 576–583, 604, 615
– burial   32
– depiction   384, 472–473, 475–476, 480–481, 486
– diplomacy   204
– monuments   238
– museum   232
– police   174
– Przewalski's   236
– sacrificial   128

Howell, Philip   7, 160, 245
human dominance   38, 43, 45, 59, 245, 459, 461, 552, 584
human exceptionalism   53, 56–60, 64, 66, 74, 276, 282, 304, 395–396
human zoos   446
Hüntelmann, Axel   272
hunter   27–28, 233, 329, 387–388, 457, 460, 473, 487, 539–540, 555, 557, 559–561, 564–568, 587
– depiction   474, 488
– man-the-hunter theory   557–558
hunter-gatherer   21–23, 26, 29–31, 120
hunting   21–22, 24, 27–29, 32, 40, 72, 108, 119, 121, 135, 150, 157, 166–167, 194, 202, 206, 248, 263, 271, 295, 297–298, 327–329, 369, 387, 427, 432, 457, 460, 463, 465, 474, 476, 552, 555–568, 588, 608
– depiction   375, 384, 483–484, 487, 492
– dogs   28, 39, 41, 566
– trophies   197, 564
– truffle   38, 194
husbandry   71, 118, 121, 183, 191, 334, 336, 540, 542
Huxley, Julian (1887–1975)   55
hybrid   10, 225, 293, 318–319, 321, 524, 609, 614
– hybridization   30, 40, 334
hyena   161, 463

I

ibex
– depiction   472, 475
icon   108, 125, 159, 173, 175, 231, 306, 312, 447, 452, 458, 571, 573
– theory   471
iconography   150, 171, 375, 377, 386, 471, 475–478, 480, 482–485, 488–490, 492
imperialism   54, 58, 72, 77, 124, 126, 213, 216, 326, 331, 445, 530, 579, 607
indigenous   71, 87, 95, 101–103, 105, 107–108, 110, 115, 150, 171, 176, 233, 280, 335, 347, 364, 366–367, 556, 564, 567, 589, 595, 598
– scholarship   399
industrialization   48, 73, 119, 121, 138, 182, 192, 247, 251, 253, 295, 364, 518, 525, 527, 532, 534, 539–550, 552–553, 556, 583, 617
– industrial revolution   137–139, 442

– whaling 562
Ingold, Tim 310, 397, 402
insect 21, 37, 82, 88, 105, 124, 243–244, 262, 270, 298, 359, 370–372, 483, 509, 573–575, 582–584, 591, 601, 603
International Whaling Commission (IWC) 208
interspecies 10, 32, 34, 39, 43, 134, 143, 193, 203, 287, 310, 345, 393, 405–406, 409, 417, 530, 536, 540, 587, 592, 596, 598–599
intimacy / intimate 30, 107, 110, 124, 162, 182, 341, 426, 428–429, 432, 435, 437, 592
invasive species 25, 72, 104, 205, 346
Isenberg, Andrew 70
Ito, Takashi 217
ivory 31, 197, 227, 367–370, 475, 561–562, 564

J
Joey (Horse) 571, 573, 586
Jørgensen, Dolly 230, 234, 236
Jumbo (Elephant) 160, 227, 239–241, 453, 607

K
Kalof, Linda 3, 40, 82
kangaroo 108, 203
Kean, Hilda 3, 7, 13, 142, 232, 234, 238, 354, 413, 432, 500, 572, 585
Kennel Club 136, 596
Kete, Kathleen 40, 148, 253, 431, 590
kinship 85, 94, 123, 257, 490, 492
Kirksey, Eben 394
kiwi 108
Knut (Polar Bear) 452
koala 203
Koch, Robert (1843–1910) 264, 515
Krebber, André 81

L
labor 34, 44, 123, 192–193, 330, 351, 486, 499, 550, 553
– animal 166, 181–182, 184, 187, 191–192, 194, 244, 247, 251, 332, 459
– division of 123, 138, 558
– human 47, 74, 122, 184, 189, 361, 527
laboratory animal 238, 273, 509, 514–523
lac insect 361, 370, 372
Laddie Boy (Dog) 168, 178

ladybird 377
– depiction 379
Laika (Dog) 177
Landseer, Edward (1802–1873) 232, 486
language 55, 63–65, 73, 148, 210, 246, 265, 269, 284, 318, 349, 355, 361, 390, 397, 403, 405, 409, 415, 442, 464, 486, 504, 559, 572, 584, 616
Latour, Bruno 3, 80, 140, 209, 272, 306, 521–523
Lefebvre, Henry 316
leopard 171, 174, 201, 440, 467
– depiction 476
life sciences 389, 412, 417, 509, 513, 516–518, 614
linguistic turn 246, 253–254, 353, 453
Linnaeus, Carl (1707–1778) 92, 110, 261, 270, 376, 609, 611–612, 614
lion 160, 171, 197, 203, 234, 243, 253, 281, 328, 439–440, 454, 457, 459–460, 463, 465–467, 559, 568, 587, 606–607
– depiction 161, 473
literary theory 148, 284
llama 49, 149–150, 244–245
longue durée 16, 89, 216, 360, 573
Lorenz, Konrad (1903–1989) 61, 266–267, 273

M
macrohistory 214
Maharaja (Elephant) 160
mammal 27–28, 30, 40, 45, 74–75, 105, 136, 163, 198, 200, 206, 240, 243–244, 249, 260, 270, 295, 361–362, 401, 420, 501, 512, 516, 552, 562, 565, 611–612, 616
mammoth 31, 475
– depiction 472, 476
mapping 104–105, 303, 311–312, 315, 325, 536
Marius (Giraffe) 454, 459
Martina (Goose) 267, 273
Marx, Karl (1818–1883) 192
masculinity 148, 231, 328, 344, 350, 352, 460, 466–467, 483, 486, 560, 575, 599
material iconography 386
material-semiotic 110, 399
materialism / materialist 34, 51, 75, 126, 140, 286
Mbembe, Achille 349, 353

meat   21, 32, 43, 46, 118, 122, 128, 149, 182–188, 192–194, 219, 265, 294, 312, 314, 337, 361, 369, 483–484, 494, 499, 525, 528, 532, 534, 536, 539–548, 550–551, 559–560, 566
– depiction   483
– market   443
– packing   74, 545, 548, 550, 552
mediality   207, 375
meerkat   397, 403, 573
megafauna   218, 328–329
Melville, Herman (1819–1891)   295, 568
menagerie   9, 155, 158, 201, 261, 383–384, 389, 391, 439–441, 446, 457, 459–461, 463, 468, 603–608, 614
– depiction   384
– queer   114
– travelling   443, 457, 459–460, 605–607
Merian, Maria Sibylla (1647–1717)   262
Merleau-Ponty, Maurice   468
mermaid   112, 608
microbe   77, 264, 299, 305, 574
microhistory   141, 148, 214, 219, 224
microorganism   264, 515–516, 575
Middle Ages   122, 128, 150, 183–184, 191, 201–202, 204, 248–249, 280, 416, 427–428, 433, 478–479, 492, 511, 556–557, 563, 566, 576
migration / migrating   29, 108, 296, 303, 314, 318, 365, 367
– human   46, 104, 359
milk   21, 45, 181–183, 185, 187, 189–191, 228, 331, 361, 396, 434, 485, 494, 532, 534, 543–544, 546
milking   21, 531–532
Mill, John Stuart (1806–1873)   612, 616
Mizelle, Brett   80, 460
Möbius, Karl (1825–1908)   58, 300
Moby Dick (Whale)   295, 568
modernity / modern   16, 26, 31, 47, 75, 110, 117, 119, 125, 138, 171, 175, 205, 236–237, 250–252, 279–280, 285, 288, 303, 316, 431, 437, 453, 469, 557, 595
– activist movements   76, 78
– agriculture   235, 525–526, 528, 533–534
– animal collections   604
– animal husbandry   118
– animal labor   460
– animal space   69, 311–312
– borders   204
– circus   319
– environmental consciousness   106, 306
– environmental diplomacy   208, 218
– fishing   121
– genetics   55–56
– humans   85
– hunting   119
– imagery   184, 390
– imperialism   326
– indigenous   556
– Japan   122–123, 128
– liminality   245
– metropolis   252
– military complex   584
– pets   124, 425–429, 431–434
– preservation movements   298
– science   57, 162, 260, 262, 264, 302, 389, 511
– slaughter   541–542, 550–551
– urban/rural divide   247
– war   571, 573, 584
– whaling   295, 562
– zoological garden   127, 440–441, 444, 453–454, 606
mollusk   300, 359
monkey   40, 85, 90, 92, 96, 98, 157, 265, 268, 427, 459
more-than-human   255, 306, 309, 314, 318–320, 322–323, 325–326, 335, 337, 341, 344, 393–397, 399–401, 404–407, 417
Morgan, C. Lloyd (1852–1936)   61, 112, 266
Morgan, Louis H. (1818–1881)   61
mosquito   115, 219, 229
mouse   124, 229, 250, 264, 268, 272, 280, 434, 501, 510, 516, 518, 521, 523
– experimental   264
– monuments   238
Muir, John (1838–1914)   73
mule   49, 139, 144, 182, 245, 331, 577, 582–583
– monuments   238
multispecies ethnology   105, 359
multispecies studies   305, 396
museum   82, 103, 105, 111, 227, 236, 241, 358, 461, 604, 613
– *American Museum of Natural History* 231, 240
– *Barker Museum of Veterinary History* 238
– *British Museum*   607

– Canada Agriculture and Food Museum 228–229, 237
– DDR Museum 362
– display 230–232, 238, 286
– empire 607
– ethnographic 361
– exclusion 608
– Hunterian Zoology Museum 230
– indigenous 233
– multispecies 367
– Museum of English Rural Life 235
– Museum of the History of Others 231–232
– natural history 160, 229, 252, 261–263, 564, 603
– Pittsburgh Center for PostNatural History 228
– specimens 230, 239, 264, 273, 358, 457
mussel 552
– depiction 377
mustang 49
myth 22, 73, 149, 171, 235–236, 245, 305, 476, 608
– mythmaking 238

N

Nagel, Thomas 3, 413
Nance, Susan 3, 133, 160, 232–233, 404, 419, 453, 461
national park 73, 164, 203, 218, 233, 314, 603
Native American 40, 49, 71–72, 157, 205, 351, 579, 600
natural history 50, 142–143, 213
– collections 229, 389, 607
– colonial 327, 446
– imagery 391, 482
– science 260, 262, 265, 273, 375–376, 388, 441, 606
– specimens 239, 358
Neolithic 40, 120–121
New Criticism 284
New Historicism 597

O

object biography 614
objectification 267, 271, 273, 344, 359, 468, 483, 486
OncoMouse 259, 519
oral history 89, 113, 303, 364, 425, 505
orangutan 608

orca 233, 568
ornithology 263
osteobiography 29, 33
othering 90, 97–98, 344, 539
Oudry, Jean-Baptiste (1636–1755) 383–384, 484
ox 77, 117, 120, 122, 139, 166, 181–184, 187, 248, 351–352, 577, 581, 583
oyster 58, 300

P

Pacific 106, 112, 115, 300
– southern 101–110, 112–114
Paleolithic 30, 40, 275, 360, 472–476, 574
panda 125, 172, 200, 203
– diplomacy 125
pangolin 1–3
parasite 219, 244, 253, 509
Pavlov, Ivan (1849–1936) 264, 514
Pearson, Chris 42, 175, 178, 234, 405, 414
Pearson, Susan 75, 142–144, 432, 596
penguin 111, 397, 448
– depiction 473
performance 24, 81–82, 141, 154, 319, 416, 457–466, 468–469, 597
– agricultural 531, 535
– Earth's 59
– experiments 514, 522
– hunting 556
– studies 457
– subversive 154
person / personhood 25–27, 30, 32–33, 93, 230, 232, 282, 568
– Black people 350
pest 1, 106, 119–120, 127, 249, 253, 256, 270, 563
pesticide 76, 298
pet-keeping 118, 148, 244, 249, 425–427, 429, 431–437, 596
– alternative history of 433, 437
– antiquity 428
– in East Asia 124, 126
– pre-modern 426
pets 9, 40, 43, 48–49, 135, 244–245, 248, 254, 256, 280, 425–426, 428–429, 432, 437, 517, 603–604
– agency 50
– Amerindian 157
– anti- 430
– cemeteries 237, 372, 617

- chickens 46
- culture 443
- dog 47, 167–169, 177, 337
- exotic 155
- food 111
- medieval 427
- modern 429
- prehistoric 24
- protection 206
- Renaissance 162
- unusual 433, 435
- useful 433
- useless 434
pig 38, 43, 45, 48–49, 78, 111, 121, 157, 159, 183, 185, 191, 237, 245, 248–249, 253, 479, 489, 530, 533, 546–547
- circus 459
- colonial 600
- cultural appreciation 543
- depiction 479, 483
- heritage 237
- insult 584
- races 541
- slaughter 549
- trade 540
pigeon 50, 72, 75, 443, 577, 594
- monuments 238
place-making 309, 313, 315–317, 323
plants 21–22, 24, 26, 43, 77–78, 112, 139, 216, 265, 299, 305, 359, 393, 417, 441, 448
- taxonomy 609, 612
- use by animals 95
platypus 110, 198, 279, 608, 612
plumage, →see also feathers
- depiction 385
poaching 49, 327, 559–561
political ecology 107, 209, 558, 565
pollution 48, 76, 118, 246, 306
Pomp (Lion) 463
porcupine 361, 364, 366
possum
- brushtail 6, 104, 106, 108, 110
- opossum 435
poststructuralism 78, 270
Potter, Paulus (1625–1654) 485–486, 488
practical turn 269
praxeology 182, 189, 192–193, 555, 565
praxis 141–142, 145, 342–343
pre-modern 201, 245, 280, 429, 469
- economy 186

- Europe 183, 204
- Japan 120, 122, 128
- menageries 605
- pets 426, 428
- sericulture 123
primate 86, 92–93, 95, 98, 145, 265, 273, 406
- experiment 270
primatology 98, 266, 268, 273, 287

Q
Queen Annie (Elephant) 464

R
rabbit 103, 106, 108, 248, 261, 263, 443, 501, 565
raccoon 49, 435
race 54, 79, 90, 92–95, 98, 129, 136, 142, 145, 215, 281, 317, 327, 343, 345–350, 353–354, 437, 531, 598, 609, 614
- critical race studies 341
- racial hygiene 453
racism 86, 90, 92, 95, 129, 335, 343, 347–348, 579, 598–599
- anti- 344
Rangarajan, Mahesh 234, 328
rare breed 235
rat 1–2, 50, 103–104, 106, 108, 229, 249–250, 501, 516, 518, 592–593, 600
reindeer 27, 475–476, 583
- monuments 238
relationality / relation 7, 17, 26, 66, 78, 87, 140, 182, 267, 298, 301, 341, 385, 388, 393–396, 398, 404–405, 407, 415, 417, 436, 521, 556, 567, 599
- aesthetic 378, 381
Renaissance 158, 162, 250, 480, 488, 492, 603
representation 8–9, 13, 23, 63, 66, 75, 82, 150, 159, 177–178, 238, 247, 256, 278, 316, 319, 322, 326, 338, 363, 375, 377–378, 385–386, 390–391, 393, 396, 404, 407, 415, 432, 450, 474, 600
- diplomatic 201, 210, 221
- mimetic 390
- mode of 376–378, 388–389
- of collections 613
- of power 556–557, 564–566
- scientific 381
reptile 243, 352, 420

rhinoceros 200, 221, 361, 381, 383, 453, 561
– depiction 384, 388, 473, 482
rhinoceros beetle 587
Rin Tin Tin (Dog) 168
rinderpest 205, 219, 334, 494
ritual 23–24, 27, 32, 119, 123–124, 126–127, 147, 150, 154, 474, 476, 555, 568
– depiction 479
Ritvo, Harriet 3, 5, 15, 79, 110, 112, 135 136, 235, 253, 277, 279, 301, 304, 326, 421, 430, 453, 519, 527, 598, 610–611
rodent 40, 105
Roscher, Mieke 81, 404, 420
Rothfels, Nigel 3, 217, 236, 277, 453
royal hunt 119, 202, 248–249, 328, 485
Royal Society for the Prevention of Cruelty to Animals (RSPCA) 162, 591
Royal Society for the Protection of Birds 298
Royal Society of London for Improving Natural Knowledge 378, 381, 441, 607

S

Saha, Jonathan 234, 325, 331–332
salmon 208, 401, 552, 608
science and technology studies 114, 499, 521
sea lion 162
seal 27, 562, 567
sentience / sentient 26, 39, 41–44, 50, 103, 237, 453, 504, 588
– pig 533
sericulture 123, 366
Serpell, James 40, 43, 46, 48, 167
settler 71, 78, 87, 96, 103–106, 109–110, 142, 157, 216, 336, 347, 366–367, 445, 530, 555, 589
– culture 107, 113
– governments 101, 104
– nation 110
Sewell, Anna (1878–1968) 137, 150
Shapiro, Kenneth 80
shark 30, 409, 489
sheep 21, 34, 103, 111, 122, 186, 220, 235, 248, 362–364, 366–367, 376–377, 386, 503, 523, 530, 547
shell 32, 198, 300, 359, 361, 376, 476, 603, 610
shellfish 120
shellac 370–372
silk 123, 128, 228, 360, 363–364, 366–367
Silk Road 576

silkworm 6, 123, 128, 362, 366–367, 515
simian 90–93, 97, 105
– almost-simian 91
Sinclair, Upton (1878–1968) 74, 547, 550
Skabelund, Aaron 125, 129, 238, 331, 598
Skinner, B.F. (1904–1990) 62
slaughter 43, 182, 185, 230, 254, 539–542, 544–548, 550–553, 615, 617
– animal protection 206
– colonial 327
– industrial techniques 549
– industrialization 534, 543, 549
– Japan 118
– nationalism 550
– United States 550
– walrus 369
– zoo 125
slaughterhouse 185–186, 193, 311–312, 314, 534, 546–547, 552
– butchers 548, 553
– Chicago 74
– colonial 337
– de-privatization 544–545
– European 547
– methods of killing 549
– modern 541–542, 550
– public 545–546
– urban 252, 254, 295
slavery 42, 46, 91, 138, 272, 317, 348–353, 367–368, 435, 598, 616
– enslavement 79, 349–350, 352, 598
Sloane, Hans (1660–1753) 607, 615
social media 50, 86, 97, 234–235, 241
sparrow 243
spirit world 127
squirrel 303, 436
subaltern 133, 214, 331, 338, 417, 552
– studies 338
symbol / symbolism 8, 30, 48, 119, 126–127, 138, 141, 143, 149–150, 156, 166–168, 170–171, 173–174, 176, 178, 197, 201, 203, 249, 275, 370, 395, 397, 432, 439, 459, 462, 476, 479–480, 485, 492, 550, 555–556, 563

T

Tague, Ingrid 154, 414, 431
Tait, Peta 319
taming / tamer 171, 457, 460 463, 465–466
– human animality 469

tapir   150, 157, 162
taxidermy   82, 178, 228–230, 239, 261, 286, 331, 386, 415, 489, 603, 615
– depiction   490
taxonomy   105, 111, 229, 603, 609–610, 612, 614, 616
theology   47, 213, 336, 451, 590, 611–612
thick description   89, 148, 153
Thomas, Keith   8, 250, 253, 278
Thompson, Edward P.   133
Thorndike, Edward (1874–1949)   266
thylacine   106
Tinbergen, Niklaas (1907–1988)   61, 266–267
toad   108, 112, 114
– depiction   377
Tom Thumb (Elephant)   464
Topsy (Elephant)   81
tortoise   162, 198, 610
totem / totemism   94, 108, 280
tracks   104, 233, 268, 297
training / animal   362, 417, 457–458, 460–462, 464–465, 467–469
– analogy to slavery   350
– animal fighting   587, 592, 596, 600
– critique   468
– dog   287
– female trainers   466
– hunting   202, 566
– techniques   459, 461, 463–464, 466, 589
travelling menagerie, →see menagerie
trophy
– hunting   328, 555
– painting   483–484
tuna (fish)   105–106, 111, 118, 552
tuna (New Zealand longfin eel)   105–106, 111

U
Uexküll, Jakob von (1864–1944)   401
Ullrich, Jessica   386
Umwelt   401–402, 413, 468
unicorn   158, 608
urbanization   137, 166, 219, 244–245, 425, 442

V
vermin   88, 106, 110, 409, 501, 523, 593, 610
– insult   584
veterinary medicine   252, 334, 358, 493, 496, 498, 501, 503
– colonial   333–335

vivisection, →see also animal experiments   260, 266, 279, 511–512, 514
– anti-vivisection movement   135, 238, 268, 270, 442, 510
– critique   252, 261, 513
– defence   263, 512

W
Walker, Brett L.   126, 145, 418
wallaby   108
Wallace, Alfred Russel (1823–1913)   113
Wallace (Lion)   160
walrus   29, 361, 369, 562, 567–568
war horse   138, 238
– film   238, 571, 586
warthog   361
wasp   108, 359, 575
Weismantel, Mary   142–144, 432
wet market   1, 129
whale   26, 29, 72, 110, 121, 163, 208–209, 268, 295–296, 552, 562, 567–568, 616
whaling   121, 126, 208–209, 218–219, 295–296, 315, 562, 567
wild animal   21, 44, 48–49, 157, 233, 236, 294, 328, 405, 410, 432, 436, 469
– categorization   245, 248, 447
– diplomatic history   208
– fascism   171
– hunting   119, 557
– in circus   457
– in zoological gardens   217, 454
– protection   298
– sacrificial   128
wilderness   73, 248, 327–328, 444
– urban   255, 304
wildlife   88, 295, 298, 329, 448, 490, 497
– African   92, 562
– colonial   327
– depiction   490
– hunting   565
– management   203, 321, 326
– poaching   560
– sanctuaries   109
– tourism   106
– urban   255
Wöbse, Anna-Katharina   218
wolf   32, 39–40, 126, 159, 170–171, 176, 233, 321, 556, 600
– coy-wolf   321
– depiction   384, 478

– she-wolf of Rome 171
Wolf (Dog) 170
Wolfe, Cary 437
wool 21, 149, 186, 220, 248, 360, 362–364, 366–367, 369, 523
working class 133, 136, 138, 346, 434, 545, 596, 615
World War I / First World War 59, 202, 218–219, 238, 253, 363, 371, 436, 447, 529, 542, 546, 549–550, 571–572, 575, 577, 581, 583–584, 597
World War II / Second World War 59, 122, 173, 198, 219, 237, 253, 363, 371, 436, 447, 529, 550, 577, 583–584, 597
World Wildlife Fund (WWF) 125, 203

X
xenotransplantation 111

Y
Yellowstone National Park 218, 233

Z
zebra 440, 463, 490, 606
zebrafish 518

zoological garden 9, 69, 79, 94, 103, 105, 107, 125, 127, 162, 200, 217–218, 236, 252–253, 263, 265, 312, 314, 439–441, 443, 446, 450, 457, 565, 603–606
– acclimatization 606
– animal agency 453–454
– animal welfare 452
– collection of animals 447
– design 448
– empire 444–445, 460, 607
– exclusion 608
– Great Zoo Massacre (Japan) 125
– Hagenbeck 217, 447–448
– immersion 449
– North American 443–444
– opposition 442
– panoramic exhibition 448
– science 252, 266, 268, 273, 441–442, 517
zoological science 441
zoology 162, 285, 606, 608, 610, 614
– economic 444
– societies 263

www.ingramcontent.com/pod-product-compliance
Lightning Source LLC
Chambersburg PA
CBHW081821230426
43668CB00017B/2343